AF333439

OBSERVATIONAL COSMOLOGY

INTERNATIONAL ASTRONOMICAL UNION
UNION ASTRONOMIQUE INTERNATIONALE

OBSERVATIONAL COSMOLOGY

PROCEEDINGS OF THE 124TH SYMPOSIUM OF THE
INTERNATIONAL ASTRONOMICAL UNION,
HELD IN BEIJING, CHINA, AUGUST 25-30, 1986

EDITED BY

ADELAIDE HEWITT

*University of California, San Diego,
La Jolla, California, U.S.A.*

GEOFFREY BURBIDGE

*University of California, San Diego,
La Jolla, California, U.S.A.*

and

LI ZHI FANG

*University of Science and Technology,
Hefei, Anhui, China*

D. REIDEL PUBLISHING COMPANY

A MEMBER OF THE KLUWER  ACADEMIC PUBLISHERS GROUP

DORDRECHT / BOSTON / LANCASTER / TOKYO

Library of Congress Cataloging in Publication Data

International Astronomical Union. Symposium (124th: 1986: Beijing, China)
    Observational cosmology.

    Includes indexes.
    1. Cosmology—Observations—Congresses.   I.   Hewitt, Adelaide.
II. Burbidge, Geoffrey R. III. Fang, Li Zhi. IV. Title.
QB980.I57   1986   523.1   87–9561
ISBN 90–277–2475–X
ISBN 90–277–2476–8 (pbk.)

Published on behalf of
the International Astronomical Union
by
D. Reidel Publishing Company, P.O. Box 17, 3300 AA Dordrecht, Holland

Sold and distributed in the U.S.A. and Canada
by Kluwer Academic Publishers,
101 Philip Drive, Assinippi Park, Norwell, MA 02061, U.S.A.

In all other countries, sold and distributed
by Kluwer Academic Publishers Group,
P.O. Box 322, 3300 AH Dordrecht, Holland

Printed in The Netherlands

TABLE OF CONTENTS

(Titles of invited papers are given in capital letters)

PREFACE.. xv

ORGANIZING COMMITTEES.. xvii

PARTICIPANTS.. xix

OBSERVATIONAL COSMOLOGY 1920-1985: AN INTRODUCTION TO THE
CONFERENCE /
    A. Sandage.. 1

I. THE MICROWAVE BACKGROUND RADIATION

THE COSMIC MICROWAVE BACKGROUND /
    R.B. Partridge.. 31

High Sensitivity Observations of the Microwave Background
Radiation /
    R. D. Davies & A. N. Lasenby............................ 55

Measurement of the Microwave Background Temperature from CN
Absorption /
    N. Mandolesi, P. Crane, & D. J. Hegyi.................. 59

Measurement of the Dipole Moment of the Cosmic Background
Radiation at mm and Sub-mm Wavelengths /
    M. Halpern.. 63

Rocket Experiment to Search for the Near-Infrared Extragalactic
Background Light /
    T. Matsumoto, M. Akiba, & H. Murakami.................. 69

Large Distances, Hidden Mass, and Fluctuations of the RELIC
Radiation /
    V. N. Lukash & I. D. Novikov........................... 73

The Sunyaev-Zel'Dovich Effect and H_0 /
    M. Birkinshaw.. 83

II. THE ORIGIN AND ABUNDANCES OF THE LIGHT ELEMENTS

ON THE ORIGIN OF THE LIGHT ELEMENTS (D, ^{3}He, ^{4}He, and ^{7}Li) /
J. Audouze.. 89

Primordial Nucleosynthesis of ^{7}Li /
D. K. Duncan & L. M. Hobbs.. 119

III. THE CLASSICAL QUANTITIES OF COSMOLOGY

THE STATUS OF THE HUBBLE DIAGRAM IN 1986 /
H. Spinrad & S. Djorgovski....................................... 129

The Radio and Infrared Luminosities of 3CR Radio Galaxies –
Are They Correlated? /
M. G. Yates, L. Miller, & J. A. Peacock.................. 143

A Study of the Hubble Flow /
E. J. Wampler.. 147

THE COSMIC DISTANCE SCALE / G. A. Tammann...................... 151

A Distance Scale from the IR Magnitude/HI Velocity Width
Relation /
M. Aaronson... 187

Local Calibrators and H_0: The Distance to M31 Using RR Lyrae
Stars /
C. J. Pritchet.. 197

Distance Moduli from the Tully-Fisher Relation /
E. Giraud... 199

ESTIMATES OF Ω BASED ON MOTIONS WITHIN THE LOCAL SUPERCLUSTER /
R. B. Tully... 207

A New Measurement of the Geometry of Space /
E. D. Loh... 217

Non-Uniformities in the Hubble Flow: Results from a Survey of
Elliptical Galaxies /
R.L. Davies, D. Burstein, A. Dressler, S.M. Faber, D.
Lynden-Bell, R. Terlevich, & G. Wegner.................. 223

DISTRIBUTION OF IRAS GALAXIES /
M. Rowan-Robinson... 229

Contributions to the Local Gravitational Field from Beyond the
Virgo Supercluster /
A. Yahil.. 247

THE ANGULAR SIZE – REDSHIFT RELATION AS A COSMOLOGICAL TOOL /
    V. K. Kapahi.. 251

Radio Bending at High Redshifts – A New Probe of
Protogalaxies? /
    G. K. Miley.. 267

IV. THE LARGE SCALE DISTRIBUTION OF GALAXIES

THE DISTRIBUTION OF BRIGHT GALAXIES /
    G. Chincarini & G. Vettolani.............................. 275

LARGE-SCALE STRUCTURE: THE CENTER FOR ASTROPHYSICS REDSHIFT
SURVEY /
    M. J. Geller, J. P. Huchra, & V. de Lapparent............ 301

The Galaxian Surface Density in the Nearby Universe /
    C. Balkowski, P. Chamaraux, & P. Fontanelli.............. 315

Tracing Superclusters and Voids with Abell Clusters /
    J. O. Burns & D.J. Batuski............................... 319

Simulations of Large-Scale Structure Compared to Abell Cluster
Distribution /
    D.J. Batuski, A. Melott, & J.O. Burns.................... 323

Shape of Superclusters /
    G. Vettolani, G. Chincarini, R.E. de Souza.............. 327

An Analysis of Fifty-Five Bright Southern Clusters of Galaxies /
    R.P. Olowin.. 331

Large-Scale Structure in the Universe: Spatial Distribution and
Peculiar Velocities /
    N. A. Bahcall.. 335

SPATIAL DISTRIBUTION OF GALAXIES: BIASED GALAXY FORMATION,
SUPERCLUSTER-VOID TOPOLOGY, AND ISOLATED GALAXIES /
    J. Einasto and E. Saar...................................... 349

The Void Probability Function as a Statistical Indicator /
    M. Lachièze-Rey & S. Maurogordato.................... 359

Large-Scale Distribution of Galaxies with Different
Luminosities /
    X.Y. Xia, Z.G. Deng, and Y.Y. Zhou.................... 363

THE DISTRIBUTION OF FAINT GALAXIES / R. Ellis.................. 367

Evolution of Very Faint Field Galaxies and Quasars /
    D.C. Koo & R.G. Kron.. 383

V.   THE THEORY OF GALAXY FORMATION AND LARGE-SCALE STRUCTURE

GALAXY FORMATION: CONFRONTATION WITH OBSERVATIONS /
    J. Silk... 391

TOWARDS UNDERSTANDING THE LARGE-SCALE STRUCTURE? /
    A. Dekel.. 415

The Sponge-Like Topology of Large Scale Structure in the
Universe /
    J. R. Gott, III... 433

Systematic Properties of Galaxies: Implications for Galaxy
Formation /
    S. Okamura, K. Kodaira, & M. Watanabe...................... 437

Search for Proto-Clusters at Meter Wavelengths /
    G. Swarup & R. Subrahmanyan................................ 441

VI.   THEORIES OF COSMOLOGY

ALTERNATIVE COSMOLOGIES /
    J.V. Narlikar.. 447

TOPOLOGY OF THE UNIVERSE /
    L.Z. Fang & H.J. Mo.. 461

VII.   THE NON-STANDARD APPROACH

OBSERVATIONS REQUIRING A NON-STANDARD APPROACH /
    H. Arp... 479

Is There an Alignment of Quasars Near NGC 520?
    E. Gosset, J. Surdej, & J.P. Swings....................... 499

Redshift Asymmetries and the Missing Mass /
    G.G. Byrd & M.J. Valtonen.................................. 503

A Possible Tired-Light Mechanism /
    J.-C. Pecker & J.-P. Vigier................................ 507

VIII. GALAXIES AND CLUSTERS

The Luminosities and Sizes of Disk Galaxies in Clusters /
  G. Giuricin, F. Mardirossian, & M. Mezzetti.............. 515

Luminosity Segregation in Distant Galaxy Clusters /
  G. Mathez, O. Le fèvre, Y. Mellier, G. Soucail,
  & A. Mazure... 519

Elemental Abundances and Temperature Distributions in the
Intra-cluster Medium of the Perseus Cluster /
  M.P. Ulmer, R.G. Cruddace, E. Fenimore, W.A. Snyder &
  G. Fritz.. 523

Tidal Triggering of Seyfert Galaxies and Quasars: Physical
Sufficiency /
  G.G. Byrd, M.J. Valtonen, B. Sundelius,
  & L. Valtaoja.. 527

An Interesting Pair of Active Galaxies /
  X.-T. He, X.H. Xiao, S. Okamura, T. Noguchi, H. Maehara, &
  R.D. Cannon.. 531

A Comment on the Seyfert Environment /
  Z. -L. Zou & J. -S. Chen.................................. 535

Search for Structural Changes in the Cores of "Nearby"
Radio Galaxies /
  E. Preuss & W. Alef....................................... 537

IX. THE DISTRIBUTION OF RADIO SOURCES

RADIO SOURCE COUNTS AND THEIR INTERPRETATION /
  K.I. Kellermann & J.V. Wall............................... 545

The 10 GHz Log N - Log S Curve Obtained at NRO /
  K. Aizu, M. Inoue, H. Tabara, & T. Kato.................. 565

Molonglo Deep Survey at 843 MHz /
  C.R. Subrahmanya & B.Y. Mills............................ 569

Ultradeep Optical Identifications and Spectroscopy of Faint
Radio Sources /
  R.A. Windhorst, A. Dressler, & D.C. Koo................. 573

X.   THE DISTRIBUTION OF X-RAY SOURCES AND THE X-RAY BACKGROUND

THE DISTRIBUTION OF X-RAY EMITTING QUASARS IN SPACE /
  G. Setti.. 579

X-Ray Surveys as Tools to Investigate the Cosmological Evolution
of Quasars, BL Lac Objects and Clusters of Galaxies /
  I.M. Gioia, T. Maccacaro, & A. Wolter..................... 593

Log N – Log S Slope Determination in Imaging X-Ray Astronomy /
  T. Maccacaro, S. Romaine, & J.H.M.M. Schmitt........... 597

The EXOSAT High Galactic Latitude Survey /
  P. Giommi & G. Tagliaferri............................... 601

Correlation of X-Ray and Infrared Emission of Radio-Quiet
QSOs /
  D.M. Worrall.. 607

The Cosmic X-Ray Background /
  E. Boldt.. 611

XI.   QUASI-STELLAR OBJECTS: DISTRIBUTION AND GENERAL PROPERTIES

COSMIC DISTRIBUTION OF OPTICALLY SELECTED QUASARS /
  M. Schmidt... 619

SPATIAL DISTRIBUTION OF QUASARS /
  Y. Chu & L.Z. Fang... 627

The Cosmic Evolution of Quasars at High Redshifts /
  P. Véron... 639

The Clustering and Evolution of Optically-Selected QSOs /
  B.J. Boyle, T. Shanks, R. Fong, & B.A. Peterson........ 643

On Cosmological Evolution of Quasars /
  J. Machalski... 649

The Space Distribution of Faint CFHT Quasars /
  D. Crampton, A.P Cowley, & F.D.A. Hartwick............. 655

Counts of Optically Selected Quasars in the Magnitude Range
19 < J < 22 /
  V. Zitelli, B. Marano, & G. Zamorani.................... 657

A Multi-Colour Survey for High-Redshift Quasars /
  S.J. Warren, P.C. Hewett & M.J. Irwin.................. 661

The Infrared Spectra of Quasars – A Luminosity Dependence /
   B.J. Wills.. 665

Spectroscopy of QSOs from the Texas Radio Survey /
   D. Wills & B.J. Wills.. 669

A Complete Sample of Flat–Spectrum Radio Sources from the
Parkes 2.7 GHz Survey /
   A. Savage, C.J. Chandler, D.L. Jauncey, M.J. Batty,
   G.L. White, B.A. Peterson, W.L. Peters, S. Gulkis,
   & J.J. Condon.. 673

The Evolution of Quasars Selected by Slitless Technique /
   Z.G. Deng, Y.Y. Zhou, & Y.Z. Liu........................... 677

An Exploration of Volume Test V/Vm Under Various Values of
Deceleration Parameter q_0 /
   S.M. Gong, H.J. Li, & C.L. Xia............................. 681

The Application of the Study of Galaxies Associated With
Quasars to Observational Cosmology /
   H.K.C. Yee... 685

Complete Samples of Variable Quasars /
   M.R.S. Hawkins... 691

Quasars, Disks and Cosmology /
   H. Netzer... 695

XII.   DARK MATTER

DARK MATTER / R. Sancisi & T.S. van Albada................. 699

Axions in the Universe /
   S. Tsuruta & K. Nomoto..................................... 713

The Neutrino Mass and the Cellular Large Scale Structure of
the Universe /
   R. Ruffini, D.J. Song, & S. Taraglio....................... 719

Determination of "inos" Masses Composing Galactic Halos /
   D.J. Song & R. Ruffini..................................... 723

XIII.   GRAVITATIONAL LENSES

GRAVITATIONAL LENSES AS TOOLS IN OBSERVATONAL COSMOLOGY /
   C.R. Canizares... 729

A VLA Gravitational Lens Survey /
    J.N. Hewitt, E.L. Turner, B.F. Burke, C.R. Lawrence,
    C.L. Bennett, G.I. Langston & J.E. Gunn.................... 747

3C324 : A Probable New Gravitational Lens /
    F. Hammer, O. Le fèvre, & L. Nottale.............. 751

1146+111B,C: A Giant Gravitational Lens? /
    E.L. Turner...................................... 755

The "Gravitational Lens" 3C 321: A Remarkable Impostor /
    A.V. Filippenko................................. 761

Astrophysical Applications of Gravitational Micro-Lensing /
    R. Kayser, S. Refsdal, R. Stabell & B. Grieger........ 767

Gravitational Lens Effects of the Dark Matter of the
Milky Way /
    C.M. Xu & X.J. Wu............................... 771

XIV. ABSORPTION IN QUASI-STELLAR OBJECTS

QSO ABSORPTION LINES AND COSMOLOGY /
    W.L.W. Sargent................................. 777

ABSORPTION LINES IN QSOs /
    J.-S. Chen..................................... 793

Absorption Line Systems at z > 3 in the QSO 2000-330 /
    R.W. Hunstead, M. Pettini, J.C. Blades & H.S. Murdoch.. 799

Clustering of Lyα Absorbing Clouds at High Redshift /
    J.K. Webb...................................... 803

XV. CLUSTERING OF QUASI-STELLAR OBJECTS

Clustering of Quasars from the ROE/ESO Large-Scale AQD Survey
for Quasars /
    R.G. Clowes, A. Iovino, & P. Shaver.................... 809

Quasar Clustering / P.A. Shaver............................. 815

SUMMARY

OBSERVATIONAL COSMOLOGY 1986 /
    M.S. Longair.. 823

POSTER PAPERS... 841

AUTHOR INDEX.. 845

SUBJECT INDEX... 849

PREFACE

The Symposium was held at the Great Wall Sheraton Hotel in
Beijing, China in the period August 25-30, 1986. The decision to
concentrate on the observational aspects of modern cosmology was
taken in part because this conference has come in a period when there
have been several international meetings on one aspect of modern
cosmology, namely the early universe and its possible relationship to
particle physics. While that approach is extremely exciting, it has
the disadvantage that its connection with much of observational
cosmology is very indirect. Thus there has been little opportunity
to discuss critically the wealth of new data that are now becoming
available which bear on the structure and evolution of the Universe
but not always on its early history.

This Symposium was planned to cover all aspects of observational
cosmology, with only comparatively minor excursions into theory.

Nearly 200 participants attended from 21 countries. A total of
26 invited papers and 73 contributed papers were given. This meant
that everyone worked hard and long from 9 A.M. to about 5:30 P.M.
for five of the six days of the conference. In addition to oral
contributions, space was made available for poster papers and 56 of
these were available for study for the duration of the conference.

Thursday, August 28 was a free day, and the delegates and those
who had accompanied them visited the Beijing Ancient Observatory and
the Forbidden City. On Sunday, August 31, following the end of the
meeting, there was a visit to the Great Wall and the Ming Tombs.

The host organizations for the meeting were the China
Association for Science and Technology (CAST) and the China
International Conference Center for Science and Technology (CICCST).
Our very hardworking colleagues on the Local Organizing Committee, in
particular Yao Quan Chu and Fu Zheng Cheng from the University of
Science and Technology of China in Hefei, and Ming Hu from CICCST
were especially helpful.

The majority of the participants provided us with manuscripts
before the prescribed deadline. Of the 100 papers given 99 are
reproduced here. A few of the papers, and the majority of the
questions and answers were typed at the University of California, San
Diego. In some cases papers were edited to make them more
comprehensible. We would like to thank Debra Bomar for her help in
the preparation of the final manuscript.

A. Hewitt
G. Burbidge
L.-Z. Fang

SCIENTIFIC ORGANIZING COMMITTEE

V.A. Ambartsumian
J. Audouze
G. Burbidge (Chairman)
J. Einasto
L.Z. Fang
J. Gunn
A. Hewitt
M. Kodaira
M. Longair
G. Marx
J.V. Narlikar
V.C. Rubin
A.R. Sandage
G. Setti
J.V. Wall
S.G. Wang

LOCAL ORGANIZING COMMITTEE

Fu Zheng Cheng
Yaoquan Chu
Li Zhi Fang (Chairman)
Ming Hu (CICCST)
Shu-Mu Gong
Xiang-Tao He
Tan Lu
Guo-Xuan Song
Guang-Zhong Xie
Yun-Qiang Yu
Zhen-Long Zhou

LIST OF PARTICIPANTS

AARONSON M.	University of Arizona, Tucson, Arizona, U.S.A.
AIZU K.	Rikkyo University, Tokyo, Japan
ARP H.C.	Max-Planck-Institut für Astrophysik, Munich, F.R.G.
AUDOUZE J.	Institut d'Astrophysique, Paris, France
BAHCALL N.A.	Space Telescope Science Institute, Baltimore, Maryland, U.S.A.
BALKOWSKI C.	Observatoire de Meudon, Meudon, France
BATUSKI D.J.	University of New Mexico, Albuquerque, New Mexico, U.S.A.
BERGERON J.	Institut d'Astrophysique, Paris, France
BIAN Y.-L.	Beijing Astronomical Observatory, Beijing, China
BIRKINSHAW M.	Harvard University, Cambridge, Massachusetts, U.S.A.
BOKSENBERG A.	Royal Greenwich Observatory, E. Sussex, England, U.K.
BOLDT E.	NASA/Goddard Space Flight Center, Greenbelt, Maryland, U.S.A.
BONOLI F.	Università di Bologna, Bologna, Italy
BOYLE B.J.	University of Edinburgh, Edinburgh, Scotland, U.K.
BURBIDGE G.	University of California, San Diego, La Jolla, California, U.S.A.
BURKE B.F.	Massachusetts Institute of Technology, Cambridge, Massachusetts, U.S.A.
BURNS J.O.	University of New Mexico, Albuquerque, New Mexico, U.S.A.
BURSTEIN D.	Arizona State University, Tempe, Arizona, U.S.A.
BYRD G.G.	University of Alabama, University, Alabama, U.S.A.
CANIZARES C.R.	Massachusetts Institute of Technology, Cambridge, Massachusetts, U.S.A.
CAO L.	Purple Mountain Observatory, Nanjing, China
CAVALIERE A.G.	Università degli Studi di Roma, Rome, Italy
CHEN L.-F.	Shanghai Normal University, Shanghai, China
CHEN J.S.	Beijing Astronomical Observatory, Beijing, China
CHEN S.	Beijing Theoretical Physics Institute, Beijing, China
CHENG F.	University of Science and Technology of China, Hefei, Anhui, China
CHENG F.-Z.	University of Science and Technology of China, Hefei, Anhui, China
CHINCARINI G.	Osservatorio Astronomico Brera-Merate, Merate, Italy
CHU Y.	University of Science and Technology of China, Hefei, Anhui, China
CLOWES R.	Royal Observatory, Edinburgh, Scotland, U.K.
COWLEY A.P.	Arizona State University, Tempe, Arizona, U.S.A.
COYNE G.V.	Vatican Observatory, Rome, Italy
CRAMPTON D.	Dominion Astrophysical Observatory, Victoria, B.C. Canada
CUI Z.-X.	Beijing Astronomical Observatory, Beijing, China
DA COSTA L.A.	Observatório Nacional, Rio de Janeiro, Brasil

DAVIES R.D.   N.R.A.L., Jodrell Bank, Cheshire, England, U.K.
DAVIES R.L.   Kitt Peak National Observatory, Tucson, Arizona,
                     U.S.A.
DEKEL A.   Weizmann Institute, Rehovot, Israel
DENG Z.-G.   University of Science and Technology of China,
                     Hefei, Anhui, China
DUNCAN D.   Space Telescope Science Institute, Baltimore,
                     Maryland, U.S.A.
ELLIS R.S.   Durham University, Durham, England, U.K.
FALL M.   Space Telescope Science Institute, Baltimore,
                     Maryland, U.S.A.
FANG D.-P.   Beijing Normal University, Beijing, China
FANG L.-Z.   University of Science and Technology of China,
                     Hefei, Anhui, China
FANTI C.   Istituto di Radioastronomia, Bologna, Italy
FANTI R.   Istituto di Radioastronomia, Bologna, Italy
FAN X.-M.   Beijing Astronomical Observatory, Beijing, China
FILIPPENKO A.V.   University of California, Berkeley, California,
                     U.S.A.
FLIN P.   Jagiellonian University Observatory, Krakow, Poland
GAO J.G.   Xin Jiang Institute of Technology, China
GARILLI B.   Istituto di Fisica Cosmica, Milano, Italy
GELLER M.   Center for Astrophysics, Cambridge, Massachusetts
                     U.S.A.
GIANNONE P.   Observatory of Monte Mario, Rome, Italy
GIOIA I.M.   Center for Astrophysics, Cambridge, Massachusetts,
                     U.S.A.
GIOMMI P.   EXOSAT Observatory, ESOC, Darmstadt, F.R.G.
GIRAUD E.   European Southern Observatory, Munich, F.R.G.
GIURICIN G.   Osservatorio Astronomico, Trieste, Italy
GONG S.M.   Purple Mountain Observatory, Nanjing, China
GOTT J.R.   Princeton University, Princeton, New Jersey, U.S.A.
HALPERN M.   University of British Colombia, Vancouver, B.C.,
                     Canada
HAMMER F.   Observatoire de Meudon, Meudon, France
HARTWICK F.D.A.   University of Victoria, Victoria, B.C., Canada
HAWKINS, M.R.S.   Royal Observatory, Edinburgh, Scotland, U.K.
HE X.T.   Beijing Normal University, Beijing, China
HEWETT P.C.   Institute of Astronomy, Cambridge, England, U.K.
HEWITT A.   University of California, San Diego, La Jolla,
                     California, U.S.A.
HEWITT J.   Massachusetts Institute of Technology, Cambridge,
                     Massachusetts, U.S.A.
HU F.-X.   Purple Mountain Observatory, Nanjing, China
HUANG K.-L.   Nanjing University, Nanjing, China
HUANG W.-L.   Institute of High Energy Physics, Beijing, China
HU J.-Y.   Beijing Astronomical Observatory, Beijing, China
HUNSTEAD R.W.   University of Sydney, Sydney, Australia
ILLINGWORTH G.D.   Space Telescope Science Institute, Baltimore,
                     Maryland, U.S.A.

JIANG S.-D.	University of Science and Technology of China, Hefei, Anhui, China
JUGAKU J.	Tokyo Astronomical Observatory, Tokyo, Japan
KAPAHI V.K.	TIFR Centre, Bangalore, India
KARACHENTSEV I.D.	Special Astrophysical Observatory, Arkhyz, U.S.S.R.
KAYSER R.	Hamburger Sternwarte, Hamburg, F.R.G.
KACHIKIAN K.	Byurakan Observatory, Yerevan, U.S.S.R.
KIANG T.	Dunsink Observatory, Dublin, Ireland
KOO D.	Space Telescope Science Institute, Baltimore, Maryland, U.S.A.
KRISTIAN J.	Mount Wilson and Las Campanas Observatories, Pasadena, California, U.S.A.
LACHIEZE-REY M.	CEN-Saclay, Gif-sur-Yvette, France
LI J.	Beijing Astronomical Observatory, Beijing, China
LIU C.	Hua Zhong Normal University, Wuhan, China
LIU J.-Y.	Beijing Astronomical Observatory, Beijing, China
LIU M.-C.	Hefei Teachers College, China
LIU Y.-Z.	University of Science and Technology of China, Hefei, Anhui, China
LIU R.-L.	Purple Mountain Observatory, Nanjing, China
LIU Y.-L.	Beijing Astronomical Observatory, Beijing, China
LI Z.-W.	Beijing Normal University, Beijing, China
LOH E.	Princeton University, Princeton, New Jersey, U.S.A.
LONGAIR M.	Royal Observatory, Edinburgh, Scotland, U.K.
LU R.-W.	Yunnan Observatory, China
LUKASH V.N.	Space Research Institute, Moscow, U.S.S.R.
LU T.	Nanjing University, Nanjing, China
MACCACARO T.	Center for Astrophysics, Cambridge, Massachusetts, U.S.A.
MACCAGNI D.	Istituto di Fisica Cosmica, Milano, Italy
MACCHETTO F.	Space Telescope Science Institute, Baltimore, Maryland, U.S.A.
MACHALSKI J.	Mazowiecka 36/33, Krakow, Poland
MA E.	Beijing Astronomical Observatory, Beijing, China
MANDOLESI N.	Istituto TESRE/CNR, Bologna, Italy
MAO X.-J.	Beijing Normal University, Beijing, China
MARANO B.	Dipartimento di Astronomia, Bologna, Italy
MATHEZ G.	Observatoire de Toulouse, Toulouse, France
MATSUMOTO T.	Nagoya University, Nagoya, Japan
MAZURE A.	Observatoire de Meudon, Meudon, France
MILEY G.K.	Space Telescope Science Institute, Baltimore, Maryland, U.S.A.
NARLIKAR J.	TIFR, Bombay, India
NETZER H.	Tel Aviv University, Tel Aviv, Israel
NORMAN C.A.	Space Telescope Science Institute, Baltimore, Maryland, U.S.A.
OKAMURA S.	Kiso Observatory, Nagano-Ken, Japan
OLOWIN R.P.	University of Oklahoma, Norman, Oklahoma, U.S.A.
PAN R.-S.	Shanghai Astronomical Observatory, Shanghai, China
PARTRIDGE R.B.	Institute of Astronomy, Cambridge, England, U.K.
PECKER J.C.	Collège de France, Paris, France

PENG W.   Beijing Astronomical Observatory, Beijing, China
PREUSS E.   Max-Planck-Institut für Radioastronomie, Bonn, F.R.G.
PRITCHET C.   University of Victoria, Victoria, B.C., Canada
RAMELLA M.   Osservatorio Astronomico, Trieste, Italy
ROWAN-ROBINSON M.   Queen Mary College, London, England, U.K.
RUBIN V.C.   Carnegie Institute of Washington, Washington, D.C.
                         U.S.A.
RUFFINI R.   University of Rome I, Rome, Italy
SANCISI R.   Kapteyn Laboratorium, Groningen, The Netherlands
SANDAGE A.   Mt. Wilson and Las Campanas Observatories,
                         Pasadena, California, U.S.A.
SARGENT W.L.W.   California Institute of Technology, Pasadena,
                         California, U.S.A.
SAVAGE A.   United Kingdom Schmidt Telescope, Coonabarabran,
                         N.S.W., Australia
SCHMIDT M.   California Institute of Technology, Pasadena,
                         California, U.S.A.
SCHULTZ G.V.   Max-Planck-Institut für Radioastronomie, Bonn, F.R.G.
SETTI G.   European Southern Observatory, Munich, F.R.G.
SHAVER P.A.   European Southern Observatory, Munich, F.R.G.
SILK J.R.   University of California, Berkeley, California,
                         U.S.A.
SONG D.J.   University of Rome I, Rome, Italy
SPINRAD H.   University of California, Berkeley, California,
                         U.S.A.
SUBRAHMANYA C.R.   TIFR Centre, Bangalore, India
SUN K.   Beijing University, Beijing, China
SWARUP G.   TIFR Centre, Bangalore, India
SWINGS J.P.   Université de Liège, Cointe-Ougrée, Belgium
SZALAY A.   Eötvös University, Budapest, Hungary
TAMMANN G.A.   St. Alban-Ring 172, Basel, Switzerland
TANG Z.-M.   Changsha Railway Institute, Changsha, China
TSURUTA S.   Montana State University, Bozeman, Montana, U.S.A.
TULLY R.B.   University of Hawaii, Honolulu, Hawaii, U.S.A.
TURNER E.L.   Princeton University, Princeton, New Jersey, U.S.A.
ULMER M.P.   Northwestern University, Evanston, Illinois, U.S.A.
VERON P.   Observatoire de Haute-Provence, Saint-Michel-
                         l'Observatoire, France
VETTOLANI G.   Instituto di Radioastronomia, Bologna, Italy
WALL J.V.   Royal Greenwich Observatory, E. Sussex, England,
                         U.K.
WALSH D.   N.R.A.L., Jodrell Bank, Cheshire, England, U.K.
WAMPLER E.J.   European Southern Observatory, Munich, F.R.G.
WANG G.   Beijing Astronomical Observatory, Beijing, China
WANG R.-J.   Beijing Astronomical Observatory, Beijing, China
WANG S.G.   Beijing Astronomical Observatory, Beijing, China
WANG Y.-J.   Changsha Railway Institute, Changsha, China
WANG Z.-R.   Nanjing University, Nanjing, China
WEBB J.   Sterrewacht Leiden, Leiden, The Netherlands
WESTPFAHL D.   Montana State University, Bozeman, Montana, U.S.A.
WILLS B.J.   University of Texas, Austin, Texas, U.S.A.

WILLS D.	University of Texas, Austin, Texas, U.S.A.
WINDHORST R.A.	Mount Wilson and Las Campanas Observatories, Pasadena, California, U.S.A.
WORRALL D.M.	Center for Astrophysics, Cambridge, Massachusetts, U.S.A.
WRIGHT J.P.	National Science Foundation, Washington, D.C., U.S.A.
WU X.-D.	Beijing Astronomical Observatory, Beijing, China
XIA C.L.	Purple Mountain Observatory, Nanjing, China
XIA X.-Y.	Tianjin Normal University, Tianjin, China
XIANG S.-P.	University of Science and Technology of China, Hefei, Anhui, China
XIAO X.-H.	Beijing Normal University, Beijing, China
XIE G.-Z.	Yunnan Observatory, China
XU C.-M.	Fudan University, China
YAHIL A.	State University of New York, Stony Brook, New York, U.S.A.
YANG L.-T.	Hua Zhong Normal University, Wuhan, China
YANG P.-B.	Hua Zhong Normal University, Wuhan, China
YATES M.G.	Royal Observatory, Edinburgh, Scotland, U.K.
YE Z.-C.	University of Science and Technology of China, Hefei, Anhui, China
YEE H.K.	University of Montreal, Montreal, P.Q., Canada
YIN Q.-F.	Beijing University, Beijing, China
YOU J.-H.	University of Science and Technology of China, Hefei, Anhui, China
YU D.-W.	Shanghai No. 3 Steel Plant College, Shanghai, China
YU X.	Beijing Astronomical Observatory, Beijing, China
YU Y.-Q.	Beijing University, Beijing, China
ZAMORANI G.	Space Telescope Science Institute, Baltimore, Maryland, U.S.A.
ZHANG J.-L.	University of Science and Technology of China, Hefei, Anhui, China
ZHANG S.-W.	Shanghai Chemical Engineering School, Shanghai, China
ZHANG Y.	Yunnan Observatory, China
ZHOU Z.-H.	Shanghai Observatory, Shanghai, China
ZHOU Y.	Yunnan Observatory, China
ZHOU Y.-Y.	University of Science and Technology of China, Hefei, Anhui, China
ZHU J.-Z.	Northwest College of Tele-Communication Engineering, Xian, China
ZHU S.-C.	Shanghai Normal University, Shanghai, China
ZHU X.-F.	University of Science and Technology of China, Hefei, Anhui, China
ZITELLI V.	Dipartimento di Astronomia, University of Bologna, Bologna, Italy
ZOU Z.-L.	Beijing Astronomical Observatory, Beijing, China

INTRODUCTORY LECTURE

OBSERVATIONAL COSMOLOGY 1920-1985:   AN INTRODUCTION TO THE CONFERENCE

Allan Sandage
Center for Astrophysical Sciences
Department of Physics and Astronomy
The Johns Hopkins University
Baltimore, Maryland 21218, USA

ABSTRACT.   The history of modern cosmology has been divided into two periods. The primary discovery phase extended from 1920 to $\sim$1950 when most of the classical tests were developed. The subsequent consolidation period from 1950 to $\sim$1980 saw a deeper understanding of the tests and the introduction of concepts of stellar populations that led to estimates of evolutionary trends with look-back time. The only test not affected by evolutionary effects is the comparison of the Hubble time, H_o^{-1}, with independent estimates of the age of the Universe. The $N(m)$, $m(z)$, $\theta(z)$, $\theta(m)$, and $H_o T_o(q_o,\Lambda)$ tests are reviewed as they have been discussed in the archive literature. The problem of observational bias is emphasized, as is the difficult recent search for secular evolutionary effects. Current problems which have prospects for solutions in the next decades are set out in the final section.

PREVIEW

Cosmology is now a dominant field of modern astronomy, carrying a principal theme in its own right, but also being the background motivation for much of the present work in even traditional astronomical fields from astrometry, to stellar atmospheres and interiors, to galactic structure, and the interstellar medium. The present cosmological thrust is to devise methods for calibrating various cosmological parameters such as H_o, the time scale, the 3°K flux, etc., but ultimately to reach back to the origins of everything from stars, to galaxies, to the Universe itself; the borders of cosmology and cosmogony here being blurred.

The connection of cosmology to nearly all other parts of astronomy, and the fact that a good fraction of the young astronomers now take cosmology to be their subject, has not always been. As late as 1938, the last year before the war, there were probably less than 30 cosmologists in the world, and of these perhaps only $\sim$15 had the status of builders of the foundations. Today there are $\sim$1000 cosmologists, of which $\sim$30% are at this conference. This rapid growth corresponds to a doubling time of $\sim$10 years, or a 7% growth per year. At this rate, in

1

A. Hewitt et al. (eds.), Observational Cosmology, 1–27.
© *1987 by the IAU.*

100 years there will be a million cosmologists, giving hope that the cosmological problem can, in fact, be solved. At the moment it is not.

We know the questions now better than at any previous time, and, fortunately, the powerful instrumentation, either just operating or on the horizon to be built, signals the start of the observational survey effort required to truly sample the World. It is in this sense that our conference here, I think, marks the final beginning of the subject.

Looking over the program, we shall see discussions this week of the enormous progress and the high expectations for the massive data collection that is required. To be sure, there <u>has</u> been much exploratory work over the past half century which has led us to this starting place. My plan for this introduction is to review the past so as to trace the origins of the major themes we shall hear repeatedly throughout the meeting. In one way or another, each speaker will return to the same set of basic topics, centered on the four classical cosmological tests. The history of the development of these tests forms the organizing frame of this introduction.

It is convenient to divide the past into two segments, separated at 1950 with the beginning of observations with the Hale 200-inch reflector. This divides the time from 1920 to 1980 into two equal intervals - the foundational period from 1920 to 1950 when the elements of the subject were laid down, and 1950 to 1980 which was a period of consolidation, punctuated, to be sure, by the unexpected major discoveries of radio astronomy and the 3°K radiation.

The last 30 years have been a period of understanding the tests, set out early, primarily by Hubble, but sharpened by the interpretative insight of Tolman and Robertson. At the very start of the Hale telescope period in 1950, high simplistic hopes were held that only a few years would be required to find such things as the expansion rate and the deceleration, to actually measure the curvature of space by Gauss's experiment with surfaces, generalized to volumes in Riemannian manifolds, and to test if the expansion is real or if the redshift is from some unknown new principle. But at every step, each of the tests has been blocked or delayed until a great deal had to be understood about the details of the test objects. This now has taken cosmologists about 30 years to accomplish.

Hubble was primarily interested in the large scale geometry of space. He looked upon galaxies as markers, either of the geometry directly, or as a means to measure the deceleration of the redshift with time (i.e., to determine the "second term" in the redshift - distance relation). This second term - $q_0 z$ - leads to the space curvature through the postulate of general relativity that the components g_{ij} of the metric tensor are related to the mean mass density, which itself controls the deceleration.

Baade, Hubble's colleague at Mount Wilson, was always opposed to this direct path. Each of the primary tests [the $m(z)$, $N(m)$, $\theta(z)$, and $SB(z)$ relations] uses some measured parameter of the galaxy such as apparent magnitude, angular diameter, or surface brightness. These obviously are subject to changes with time, and therefore with redshift in an evolving Universe. Baade's dictum was always

> "You must understand the galaxies before you
> can get the geometry right."

Hubble clearly understood this, but rather than be stopped because this part of his subject was 30 years before its time, he pushed ahead with an abandonment known to pioneers in any milieu who try to reach Everest without proper equipment. Some, indeed, succeed. Without Hubble's attempt, the foundations of our subject would not now be in place for us and for the next generation.

Meanwhile, while Hubble was emphasizing both the redshift - magnitude relation and the galaxy number counts as ways to determine the spatial curvature, Baade was indeed trying to understand the galaxies themselves. From the late 1930's until $\sim$1950 two parallel, nearly non-intersecting flows in cosmology were occurring in Pasadena - one using galaxies as markers; the other with galaxies as keepers of various stellar populations. This second avenue - Baade's - hardly existed in its present recognized form until its beginnings in $\sim$1940. The modern ideas of stellar evolution, age dating the stars, the change of the HR diagram with time, color and luminosity evolution of the total light, etc., is a product of the decade from 1950 to 1960 in its foundations, and 1960 to 1970 in the working out of its details. Hence, Baade's dictum could not have led to any reasonable actions until quite recently. It has only been a few years now since these two great separate streams of inquiry met to form the modern formulation of the work.

This review is centered on the tests. I propose to introduce each test as Hubble first set it out, to discuss how each was treated and what we learned in the consolidation period, and finally to look at what is possible concerning the completion of each test in the future with the new instrumentation.

THE TESTS

a) <u>The N(m) count distribution</u> was thought to be a direct attack on the experimental geometry following the method of Gauss for surfaces. In curved space the circumference of a small circle of radius d on, for example, the surface of a sphere of radius R is not $2\pi d$ but is

$$L(d) = 2\pi d \left(1 - \frac{1}{6}\frac{d^2}{R^2} + \text{---}\right), \tag{1}$$

providing a means to determine the radius of curvature R by measuring L and d <u>along the surface</u> of the sphere.

Likewise, the area of a spherical cap of radius d on the surface is not Πd^2 as in Euclidean space, but is

$$A(d) = \pi d^2 \left(1 - \frac{1}{12}\frac{d^2}{R^2} + \text{---}\right), \tag{2}$$

again affording an experimental determination of R by measuring the deviation of the measured area from πd^2.

These two examples are easy to visualize because the curved surface is the familiar two dimensional space of a sphere which is embedded in a Euclidean three dimensional enveloping manifold. The visualization of three dimensional Riemannian space is more difficult because it is outside direct experience. Nevertheless, it forms the essence of Einstein's replacement of action at a distance (force without contact via a field) where, in the presence of a mass, a test particle moves along a straight path in the warp of curved space rather than on a curved path in straight space. By 1930 the idea of curved space was commonplace enough to attempt a direct measurement of volumes, similar to the earlier experiment by K. Schwarzschild on stellar parallaxes.

In the presence of a three dimensional Gaussian spatial curvature of strength k/R^2, volumes out to a measured "distance" (properly defined) do not increase as $\frac{4\pi d^3}{3}$ but rather as

$$V(d) = \frac{4}{3}\pi d^3 \left(1 - \frac{1}{5}\frac{d^2}{R^2} + \text{---}\right), \tag{3}$$

again affording a test <u>if proper volumes could be measured</u>.

Hubble proposed galaxy counts to do that, using apparent magnitudes as indicators of distance in the mean, and determining the number, $N(m)$, of galaxies brighter than apparent magnitude m.

After correcting the apparent magnitudes for the effects of redshift, Hubble assumed that the correct distance to use was a luminosity distance based on

$$\ell \sim \frac{L}{\pi d^2 (1 + z)^2} \tag{4}$$

if the redshift z was a true expansion, or

$$\ell = \frac{L}{4\pi d^2 (1 + z)} \tag{5}$$

if the redshift was not due to expansion, where ℓ and L are the apparent and absolute luminosities over a fixed proper wavelength band pass, correcting the observed ℓ for the selective K term due to the redshift of the spectrum through the band pass. (He forgot the stretching factor of 1 + z in the bandpass due to redshift in his total K term).

One can show that, given any non-divergent distribution of absolute luminosities $\phi(L)$, i.e. the luminosity function, an integration over $\phi(L)$ for the Euclidean case of $R = \infty$ (in the closed analytical form of equation 3 rather than a series expansion), gives

$$N(m) \sim 10^{0.6m} \tag{6}$$

for all m. The spatial curvature flattens the $N(m)$ curve in a way which Hubble expressed as

$$\log N(m) = \text{const} + 0.6\,(m - \Delta m), \tag{7}$$

where Δm is a function of distance (i.e. redshift) that depends on the spatial curvature R. His task was to determine the $\Delta m = f(z)$ function and to infer the spatial curvature from it using the closed form of equation (3), assuming either equation (4) or equation (5).

His data and analysis (Hubble 1936) are shown in Figure 1 here, taken from his Figure 16 in The Realm of the Nebulae, together with the caption he gave the diagram.

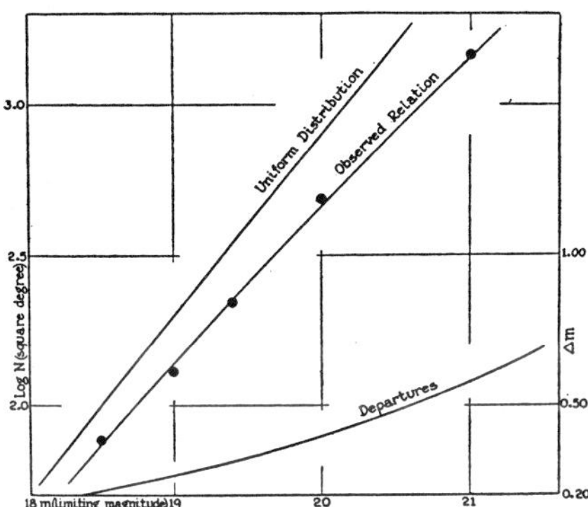

FIG. 16. *Apparent Distribution of Nebulæ in Depth.*

Each point on the Observed Relation represents the average number (actually log N) of nebulæ per square degree which are equal to, or brighter than, a particular apparent magnitude, as determined by an entire survey. The line through the points (Observed Relation) is a least square solution in the form, log $N = 0.6\,(m - \Delta m) +$ constant, derived from the assumption that Δm is a linear function of distance.

In the vicinity of the galactic system, Δm should be negligible, and the Observed Relation should coincide with a Uniform Distribution indicated by the straight line. As the surveys are extended to greater distances (fainter limiting magnitudes), Δm increases, and the Observed Relation departs from the line representing Uniform Distribution. The departures, Δm (horizonal displacements between the two lines), are plotted against m (limiting magnitudes of the surveys) in the lowest curve. The departures are interpreted as effects of red-shifts.

Fig. 1   Hubble's (1936) final analysis of his count data as a function of blue magnitude, corrected according to his precepts of the K correction. His caption from Realm of the Nebulae is reprinted.

His conclusion from the analysis was that the deviation of the observed log N(m) curve from a slope of 0.6 was so great that either R was very small and that curved space had indeed been found, or that equation (5) was correct and that the redshift was not due to expansion.

We know now that he could not have found the answer by this route, even in principle because the theory of the redshift - distance relation for any spatial curvature had not yet been worked out. He assumed that the d to be used in equations (4) or (5) was simply proportional to the redshift z for all geometries (i.e. for an value of R). However, the correct distance to use (Mattig 1958, Sandage 1961a,) is known to be

$$r = \frac{c}{R_o H_o q_o^2 (1+z)} \left\{ q_o z + (q_o - 1)[-1 + (1 + 2q_o z)^{1/2}] \right\} \qquad (8)$$

which differs from

$$d = \frac{cz}{R_o H_o} \qquad (9)$$

assumed by Hubble (note that d in eqs. 4 and 5 is the same as $R_o r(1+z)$ here). The correct equation for the volume contained out to the co-moving coordinate r is

$$V(r) = \frac{4\pi R^3 r^3}{3} [\frac{3}{2} \frac{\sin^{-1} r}{r^3} - \frac{3}{2} r^2(1-r^2)^{1/2}] \qquad (10)$$

which can be put in terms of the observed redshift z by substituting equation (8) in (10). A series approximation, correct to order z in the correction term is

$$V(z) = \frac{4\pi z^3}{3H_o^3} [1 - \frac{3}{2}(1-q_o)z + ---]. \qquad (11)$$

This can be expressed in apparent luminosity by using equation (4) to finally give the predicted N(m) relation for different values of q_o which is the intrinsic geometry parameter, related to the expansion rate H_o, and the spatial curvature R_o by

$$\frac{kc^2}{R_o^2} = H_o^2(2q_o-1). \qquad (12)$$

When this is done to produce the predicted N(m) relation, it turns out that N(m) is independent of q_o to first order in z entering only at order z^2 (Mattig 1959, eq. 9; Sandage 1961a; Brown and Tinsley 1974). Only at very high redshifts is the spatial curvature effect felt. Hence the N(m) test is quite insensitive to q_o and the test essentially fails even if there would be no luminosity evolution of the galaxies in the look back time.

On the contrary, N(z) rather than N(m) is sensitive to q_o to first order in the correction term, as shown by equation (11), and this will be the form of the count test to find q_o in the future by counting

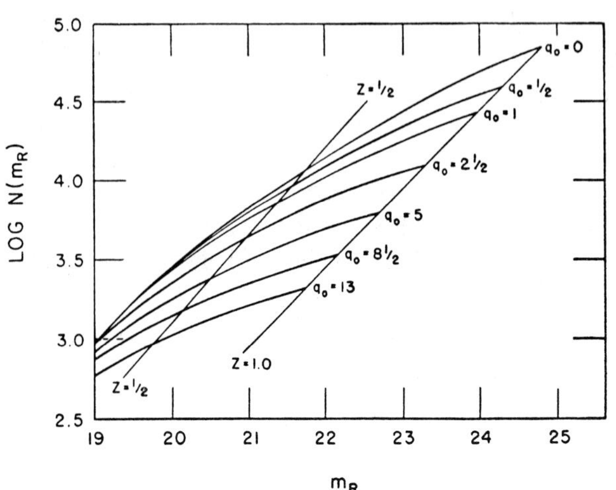

Fig. 2   Predicted $N(m,q_o)$ from the standard Friedman model using fully corrected red magnitudes m_r.

objects in a complete sample to a given z (see the review by Sargent in
this volume of an application of the test using L_α absorption lines in
the Lyman forest of quasar spectra where line of sight distances are
used rather than volumes).

The insensitivity of the N(m) test to variations in q_o for $z \lesssim 1$ is
shown by idealized calculations made as if $\phi(L) = \delta$ (i.e. no dispersion
in L)(Sandage 1961a, Figure 5, shown as Figure 2 here).
Reasonable values of q_o are between 0 and 1/2, giving very little handle
for the N(m) test according to Figure 2. The actual difference in the
family of curves is even less than shown in the diagram when an
integration over $\phi(L)$ is made.

The N(m) test has been applied in recent years to test primarily
luminosity evolution of the galaxies rather than intrinsic geometry.
Counts by Karachentsev and Kopylov (1978), Kron (1980), Tyson and Jarvis
(1979), and Peterson et al. (1979) are examples. Figure 3 shows the
final analysis of Peterson et al. (their Figure 4), using proper
integrations over luminosity functions for different galaxy types. A
model for galaxy luminosity evolution has also been folded in, being the
major effect compared with differences of q_o.

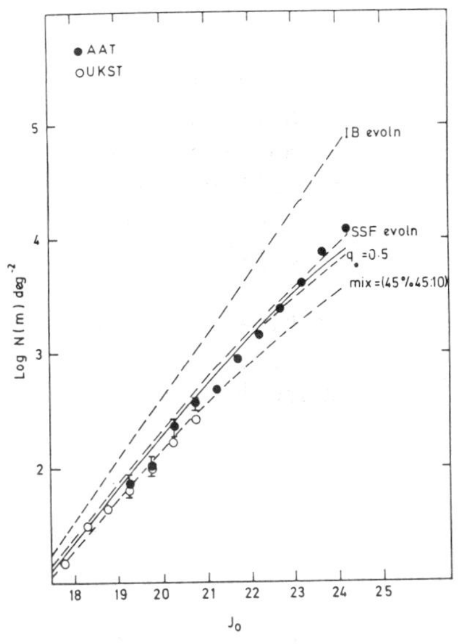

Fig.3   Observed N(m) from
Peterson et al. (1979, their
Fig. 4). Various calculated
theoretical curves are shown.

The conclusion from Figure 3 is that the N(m) distribution is very close
to what is predicted by the standard model, but that it tells us nothing
about the intrinsic geometry; its main use is to test models of galaxy
luminosity evolution, on the assumption that the Friedman standard model
is correct.

During this conference we will hear of new N(m) and especially N(z)
data from several different experiments (e.g. Yee, Loh, Tytler and also
Murdock, et al. 1986 in Sargent) in attempts to find q_o.

b) the m(z) or Hubble diagram test is the most famous of the
relations. Hubble's first formulation of the linear redshift-distance
relation in 1929 is shown in Figure 4, together with his figure caption
from Figure 9 of Realm of the Nebulae.

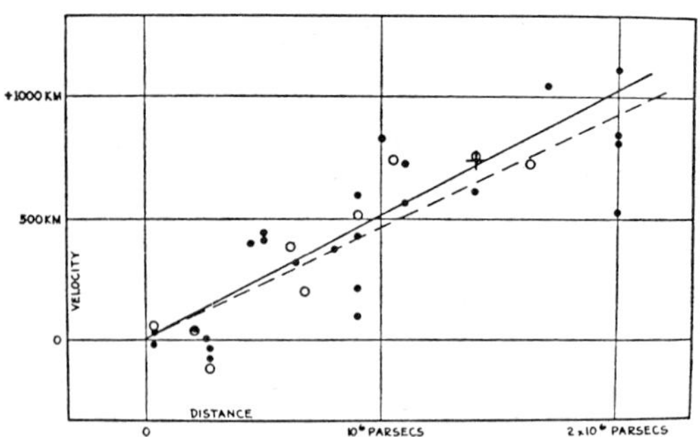

FIG. 9. *The Formulation of the Velocity-Distance Relation.*

The radial velocities (in km/sec.), corrected for solar motion, are plotted against distances (in parsecs) estimated from involved stars and, in the case of the Virgo cluster (represented by the four most distant nebulæ), from the mean luminosity of all nebulæ in the cluster. The black disks and full line represent a solution for the solar motion using the nebulæ individually; the circles and dashed line, a solution combining the nebulæ into groups.

Fig. 4 Hubble's original 1929 discovery plot of the velocity-distance relation, reproduced from Figure 9 of Realm of the Nebulae.

In his last discussion, which was the Darwin Lecture, Hubble (1953) used the m(z) observations of brightest cluster galaxies, measured in the first years of the Hale telescope operation. He fit the preliminary data well with an $m_c \alpha 5 \log z$ relation, where m_c was the apparent magnitude corrected by <u>only one factor</u> of $1+z$ as in equation(5), plus that part of the K term which accounts for the shifting of the non-flat galaxy spectrum through the stationary band pass of the detector (but again he forgot that 1+z passband stretching term due to the redshift). He discussed the data as if $\log z \, \alpha \, 0.2 \, m_c$ represented the true linear velocity-distance relation, and that deviations from it would indicate a change of z with time (i.e. acceleration or deceleration). He stated that if the additional 1+z term (of equation 4) was applied to the bolometric magnitude correction, then the upward deviation (i.e. a brighter m) of the resulting m(z) function relative to the $z \sim 0.2 \, m$ "linear case" would give a non-linear velocity-"distance" relation <u>in the sense of an accelerated expansion</u>. (italics are Hubble's words).

However, Hubble's supposition that $z \sim 0.2 \, m_c$ is the non-decelerated case is based on equation (9), which, although it is the intuitive guess for the linear velocity - "distance" relation, it is wrong even if $R_o r(1+z)$ or $R_o r(1+z)^{1/2}$ (equations 4 or 5) are taken to be the "luminosity distance". The correct expression to all orders of z in the expanding case (eq. 4) for the luminosity distance is given by equation (8), whose series expansion is

$$R_o r(1+z) = \frac{cz}{H_o} \left[1 - \frac{z}{2} (1-q_o) + O(z^2)\right]. \qquad (13)$$

This differs from Hubble's supposition (eq. 9) except when $q_o = 1$; a highly decelerating case.

Mattig (1958) first called attention to this difference by deriving the exact formulation of the m(z) relation as

$$m_{bol} = 5 \log \frac{1}{q_o^2} \{q_o z + (q_o - 1)[-1 + (1 + 2q_o z)^{1/2}]\} - 5\log H_o + M \qquad (14)$$

whose series expansion, known to Heckmann (1942, which he derived from a model-independent Taylor series in $\dot{R}_o$ and $\ddot{R}_o$ using the light travel time as the independent variable) is

$$m = \mathrm{const} + 5 \log z + 1.086 (1 - q_o) z + \text{---}. \qquad (15)$$

Note that in the zero acceleration case of $q_o = 0$, the expected $m(z)$ relation is $\sim 5\log z + 1.086 z$, rather than merely $5\log z$.

The chief cosmological program for the 200 inch telescope, aside from correcting the distance scale, developed into obtaining reliable magnitude and redshift data to test equation (14), with the (unrealized) hope of finding q_o. The data and analysis, covering the 25 years from 1950 to 1975, is described in a series of papers (Sandage 1968, 1972b,c, 1975a,b; Sandage and Hardy 1973; Gunn and Oke 1975; Kristian et al. 1978, Sandage and Tammannn 1983, 1986) whose final conclusion has been that the $m(z)$ test is spoiled by the lack of precise knowledge of the evolutionary change, dM/dt, of the absolute magnitude of the standard candles with time (Sandage 1961b; Tinsley 1980 with references to her many detailed earlier papers on the effect). Again, Baade's dictum applies -- one cannot obtain q_o until we know how m changes with time, i.e. until we understand how the stellar content of galaxies evolves.

It is fair to say that the present status of the work is that the velocity -- distance relation fits the linear case well to $z \sim 0.5$, and that deviations from $\log z \sim 0.2m$ for $z > 0.5$ can be interpreted either as due (1) to a value of q_o in the range $0.5 \lesssim q_o \lesssim 2$, or (2) to true small departures from linearity for small z, for example as if $1+z \sim \exp Hr$ (see Pecker in this volume), or (3) to the combined effects of $q_o \neq 1$ and a luminosity evolution, $\ell(t)$, with time.

In this week we shall hear of the modern studies of the $m(z)$ relation by Lilley and Longair (1984), Lilley et al. (1985), and by Spinrad (1986) and Djorgovski and Spinrad (1986) where these effects are discussed anew in the near infrared bandpass.

Finally, it is of interest to inquire into how universal is the redshift-distance relation, what is its scatter (intrinsic dispersion), and how close it can be traced to the Local Group.

To the extent it can be measured, the effect is universal, and isotropic to within small limits. Gross deviations from the Hubble flow, although occasionally claimed (Arp 1967, 1980; Burbidge 1981) are controversial -- not yet established.

These are, however, claims of "large" perturbations on the ideal Hubble flow (cf. Davies, this conference), but no claim of this kind has been made for systemic deviations larger than $\Delta v/v \sim 15\%$. This is the limit that is still possible from the observed scatter in the $m(z)$ relation for first ranked cluster and group galaxies (Sandage 1972b, 1975b; Sandage and Hardy 1973). Therefore, even if the recent claims

(cf. Davies et al., opt. cit.) of deviations of $\sim$600 km s^{-1} for $\langle cz \rangle$
$\sim$4000 km s^{-1} are confirmed, the linear Hubble velocity-distance relation
(with then <u>local</u> perturbations of $\lesssim$ 15%) holds.

Finally, the linear expansion can be traced to within $\sim$2 Mpc of the
Galaxy using precise radio redshift data and distances determined
largely from Cepheid variables. The relation is shown in Figure 5. The
dispersion about the linear relation is small at $\sigma(v) \lesssim 60$ km s^{-1},
indicating a very quiet local Hubble flow.

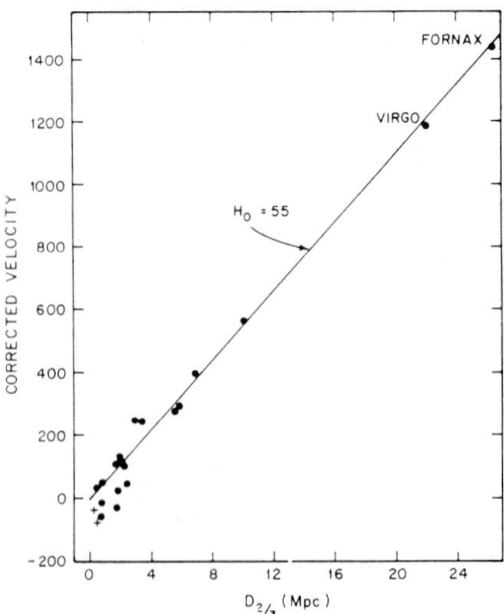

Fig. 5   Modern data for the
very local velocity-distance
relation (Sandage 1986).

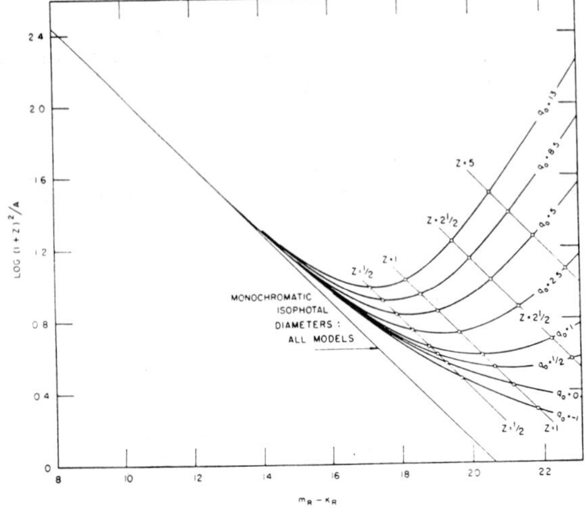

Fig. 6   Theoretical
expectation of the
$\theta(m, q_o)$ relation (Sandage
1961a, Fig. 7).

c) <u>The angular diameter-redshift test</u> was not proposed by Hubble, but by Hoyle (1959), where he showed that $\theta(z)$ has a dependence on q_o. The dependence of the corresponding $\theta(m)$ relation on q_o for standard candles is shown in Figure 6. The $\theta(z)$ diagram for standard meter sticks follows from the $m(z,q_o)$ relation of equation (14). Figure 6 shows that the <u>metric</u> diameters are predicted to decrease more slowly than $\theta \sim z^{-1}$ (depending on q_o), whereas isophotal diameters (i.e. angular sizes to a given isophote) are degenerate in q_o. However, the $\theta(z)$ isophotal relation is not so degenerate,   but forms a family, giving, in principle another way to find q_o, provided that θ does not evolve with time.

The $\theta(z)$ isophotal relation for first ranked cluster galaxies is known to vary closely as z^{-1}. Early optical data using eye estimates of θ (isophotal) are shown in Figure 7.

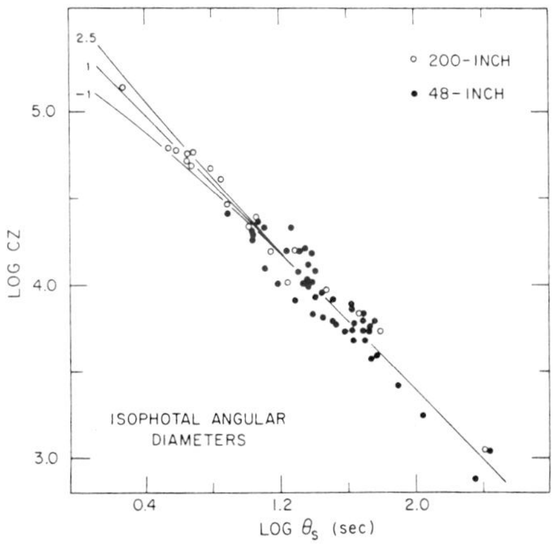

Fig. 7.   Observed $\theta(z)$ relation for a first ranked cluster galaxies (Sandage 1972a).

This is, of course, a very strong test of the linearity of the velocity-distance relation, but it would be equivalent to the $m(z)$ Hubble diagram test if the surface brightness was strictly constant among first ranked cluster galaxies.

Fig. 8   Measured relation between metric diameter and redshift for first ranked cluster galaxies (Djorgovski and Spinrad 1981, their Fig. 3).

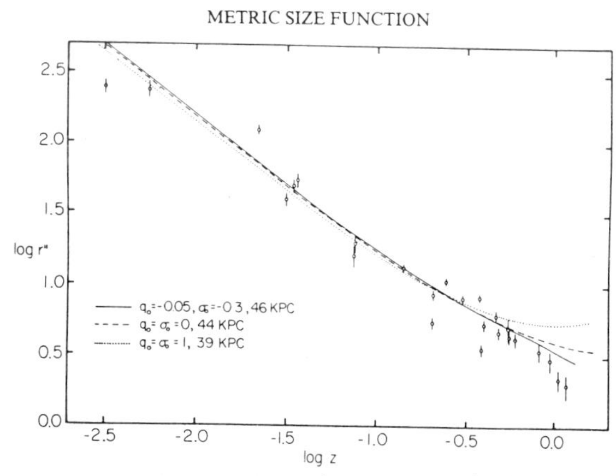

The current status of the metric diameter test is shown in Figure 8 from measurements of Djorgovski and Spinrad (1981) using θ (metric) for first ranked cluster galaxies as defined by Petrosian (1976). There is no indication of the expected turn-up in θ for large z. This can either be interpreted as an evolutionary change in the metric size with time, or as a proof that the theory is simply wrong. The point is crucial because it shows the circularity of the argument in each of the tests discussed so far. The observational fact is that the predictions of the Friedman models are <u>not</u> confirmed in detail in either the N(m), the m(z), or the $\theta(z)$ tests <u>using the data as they are directly measured</u>. In each case, to save the phenomenon, one imposes evolution of m or θ with time to force agreement and says, thereby, that one sees direct evidence for cosmic evolution with time. But it is clear that the conclusion would be correct only if we had independent knowledge that the standard Friedman cosmology <u>is</u> correct. This we do not have, being the object of the exercise itself. The arguments for each test, taken singly, are therefore circular. However, the case for the validity of the Friedman standard model and the consequent need for evolution corrections can be made in principle if all the available tests are used in concert and if the resulting evolution required by each makes a consistent evolutionary picture.

An important discussion of the consistency from such an approach is the parallel study of the angular size — flux density [$\theta(S)$] relation (Swarup 1975) and the number count — angular size [$N(\theta)$] relation (Kapahi 1975) for radio sources over a wide range of flux density, suggesting indeed that evolutionary has occurred.

The $\theta(z)$ test was applied early to radio sources (Miley 1968, 1971; Legg 1970) using the angular separation of the double radio lobes. The test had been used earlier to show that, if radio galaxies were at their redshift distance, so then were quasars (Ryle 1968). Figure 9 shows Miley's 1971 $\theta(z)$ diagram where the lack of the expected turn-up in θ for large z is evident, violating the model predictions.

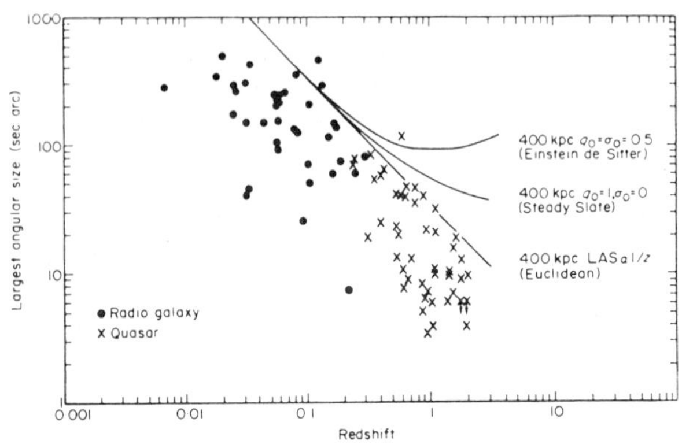

Fig. 9   The $\theta(z)$ relation for double lobed radio galaxies and quasars (Miley 1971).

However, in this case there is a very strong selection effect caused by the flux limited radio sample. This particular case serves as a good

example by which to introduce the problem of selection effects in
general, which we now discuss.

SELECTION EFFECTS

This week we shall repeatedly hear discussions of selection effects
as they bias various results. Many of the effects are variations of the
well known Malmquist bias which, in ·its simplest form, states that in a
flux (apparent magnitude) limited sample, the average absolute
luminosity of the nearby members of the sample is fainter than that
which holds for more distant members. Stated that way, the effect is
obvious, but the consequences of this $\langle M(z)\rangle$ variation in flux limited
samples are profound. In every data sample where the objects have a
non-zero dispersion, $\sigma(M)$, in absolute luminosity, the bias must be
accounted for. The purpose of this section is merely to illustrate the
effect with two examples.

The bias as it relates specifically to H_o will often be discussed
in the sessions that follow. This most crucial case is not discussed
here as a third example, except to say the obvious that each side of the
H_o controversy believes that their data have been properly corrected for
selection effects, while those of their opponents have not. It remains
to be seen if the explanation of the present discrepancy by a factor of
two in H_o is, indeed, caused by selection effects.

a)   Bias in the radio data:

When large numbers of redshifts became available for identified
radio galaxies and quasars, and when the absolute radio power radiated
between the rest-frequencies of 10^7 and 10^{10} Hz was calculated for each
(assuming $q_o=+1$), it was found that the radio power L_R increased nearly
precisely as z^2 for the available sample. The conclusion was not, of
course, that the average radio power for the radio sources did, in fact,
increase as z^2, but rather that the sample was severely restricted by
the apparent flux limit of the catalogs, which was between $\sim$ 9 and 3 Jy.
Because the radio luminosity function $\phi(L)$ is so broad, covering at
least a factor of 10^6 in L_R, the observed effect is entirely the
Malmquist bias.

Figure 10 shows the data, using mostly 3C sources. The 3C catalog
has an apparent flux limit 9 Jy at 178 MHz. Lines of constant apparent
flux at 10, 3, and 1 Jy are sloped with a $\sim z^2$ dependence in Figure 10.
The distance modulii, m-M, are calculated from the redshifts, using
H_o=50. The lack of data points in the upper left is due to the volume
effect, coupled with the very steep $\phi(L_R)$ at the bright absolute radio
power end.

Figure 10 shows the method by which any sample can be tested for a
bias of this kind. One plots absolute magnitude (or log L) vs log
distance for objects which have had these values calculated by <u>any</u>
algorithm (eg. brightest stars, redshift, total apparent magnitude,

Tully-Fisher etc. as the distance indicator). If there is a trend of
<M> with distance, then either the effect <u>is</u> real, or one is sampling
only a decreasing fraction of the total spread in L as the distance is
increased. The test is infallible. It should be applied to all samples
as an indication of either freedom from Malmquist, or, very rarely, a
real effect.

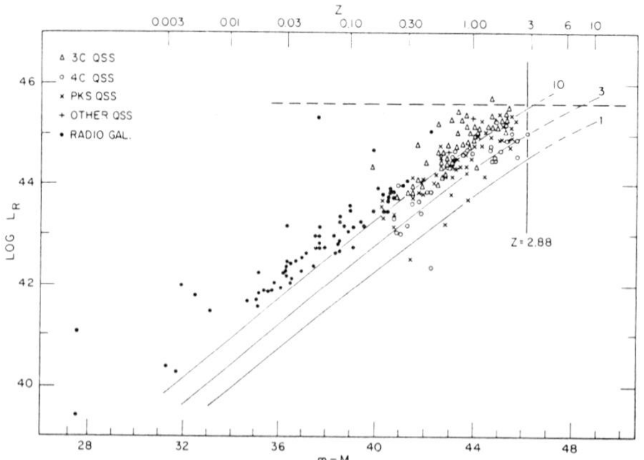

Fig. 10 Correlation
between the absolute
radio power, L_R,
between 10^7 and 10^{10}
Hz in proper
wavelength, and the
distance modulus
calculated from the
redshift-distance
relation. Lines of
constant apparent
flux density at
10,3, and 1 Jz are
shown (Sandage
1972c).

     Figure 10 is an extreme example because the 3C catalog contains
only the <u>apparently</u> brightest radio sources. Only very nearby radio
galaxies of low absolute radio power can enter the catalog. For the
same reason, there are no M dwarf stars in the Bright Star Catalog,
reaching as it does only to apparent magnitude of V $\sim$ 7. The
correlation in Figure 10 shows that the galaxies in Miley's diagram
(Figure 9 here) at large redshift have much higher <u>absolute</u> radio power
than those at low redshift, by about $(z_1/z_2)^2$. If, then, the <u>linear</u>
separation of double lobed radio sources were to be a function of L_R
(from the unknown physics), one would not be comparing a standard meter
stick at different distances in Figure 9, but rather one that itself
stretched or shrunk in linear diameter, with z. This would then provide
no cosmological test. Much will be said this week about ways to test
for such true physical effects (eg. especially Kapahi in this volume).

<u>b) Example of bias in optically selected data:</u>

     Figure 11 is equivalent to Figure 10 but is a simulation adapted
from Spaenhauer (1978), showing in the upper panel the distribution of
absolute magnitudes in a volume limited sample, assuming <M> = -18.0 and
$\sigma(M)$ = 2 mag. The lower panel is that part of the upper panel that is
available in a magnitude limited sample, cut at apparent magnitude m =
13. The mean absolute magnitude at any distance in such a biased sample
is clearly a function of distance.
     The effect of more luminous galaxies at larger distances in a
magnitude limited sample is strongly present in the Shapley-Ames galaxy
sample, shown in Figure 12, which shows the absolute magnitude

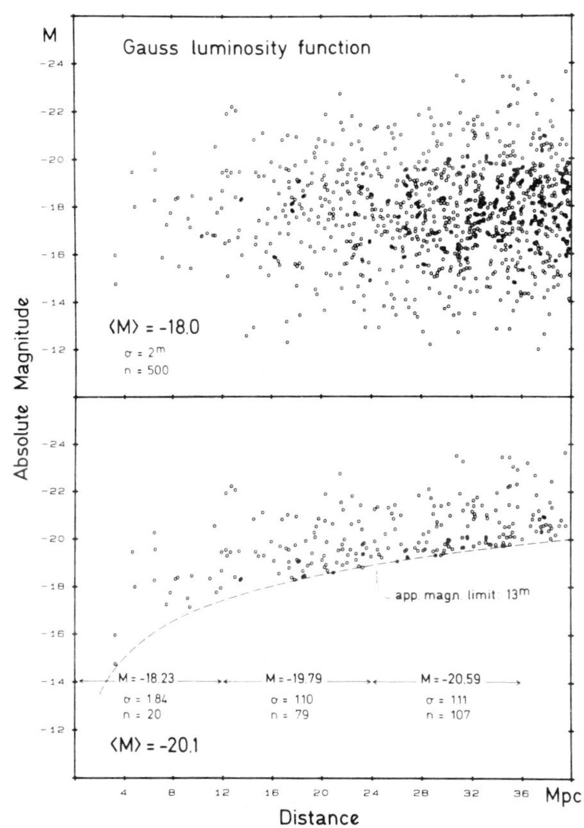

Fig. 11 Spaenhauer's simulation illustrating the Malmquist bias for a sample of galaxies whose volume limited mean magnitude is ⟨M⟩ = -18, with an intrinsic dispersion of $\sigma(M)$ = 2 mag. The apparent magnitude limit of the catalog at m=13 is shown in the lower panel.

(calculated from a non-perturbed Hubble flow using H_o=50, together with the measured redshift) vs log redshift for all E and SO Shapley-Ames galaxies with $|b|$> 30°, excluding the Fornax and the Virgo Cluster fields. Figure 13 shows that the variation of $\langle M_B\rangle$ with distance is the expected Malmquist bias. The lower limit line is for a cut-off magnitude of m_B=13.2. The upper curved envelope is the calculated volume effect using the standard galaxy luminosity function. The fact that the observed data are contained within the two calculated limit lines shows directly the Malmquist bias.

The data for spirals (Tammann, et al. 1979) in Figure 14 again shows the strong Malmquist effect, indicating that these galaxies have a broad luminosity function. It has been claimed that part of the effect in Figures 11, 12, 13, and 14 is caused by an error due to the assumption of an unperturbed Hubble flow locally — to distances of $\sim$2000 km s^{-1}. However, calculations using any reasonable local perturbation on the Hubble flow (Kraan-Korteweg et al. 1984) show the same pattern, indicating that $\phi(M)$ is broad, and that the Malmquist effect biases all discussions of field galaxy correlations in magnitude limited spaces.

Without correction for this bias, the apparent Hubble constant

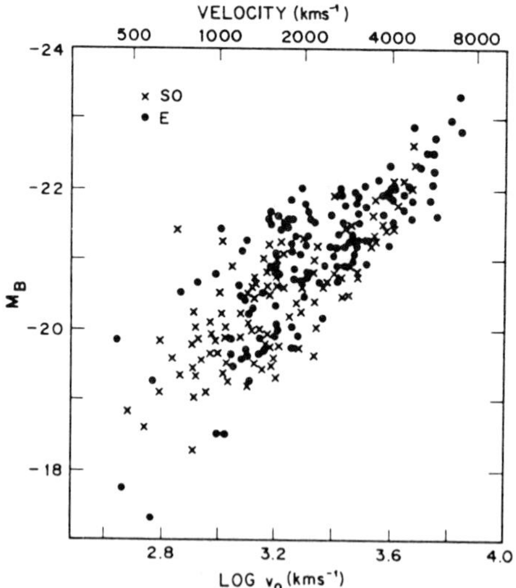

Fig. 12   The observed correlation of M_B and redshift for E and SO galaxies in the Shapley-Ames catalog. The strong increase of mean luminosity with increasing velocity is a direct indication of the Malmquist bias in the Shapley-Ames bright galaxy list. (Diagram from Sandage <u>et al</u>. 1979).

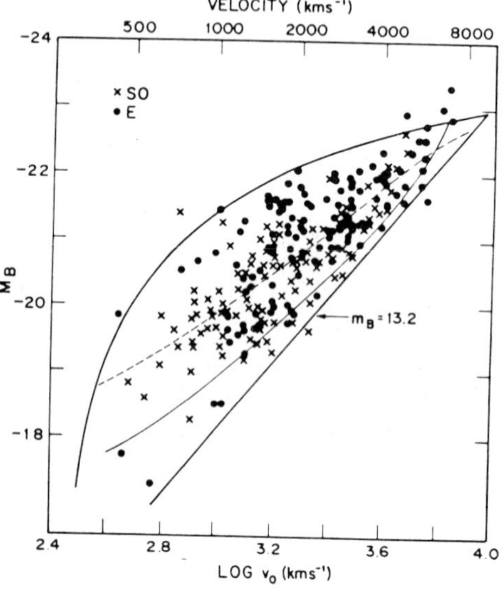

Fig. 13   Same as Figure 12 but with the calculated limit lines superposed as if the entire correlation was due to magnitude limited sample with m_{limit}=13.2 and a normal luminosity function with which to calculate the upper volume effect curved envelope (same reference as for Fig. 12).

<u>would appear to increase</u> with distance, simply because incorrect faint values for <M> would be assumed for the more distant galaxies, making their calculated distances smaller than their real distances — the effect increasing with increasing distances (de Vaucouleurs and Peters 1986, their Figs 2a and 2b).

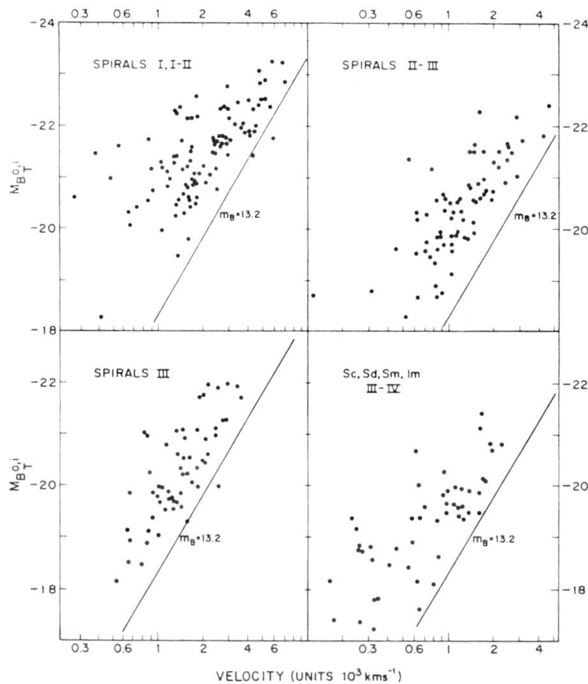

Fig. 14 The correlation of absolute blue magnitude with velocity for Shapley-Ames spirals of various van den Bergh luminosity classes illustrating the presence of the strong Malmquist bias in this sample. (Diagram from Tammann _et al._ 1979).

THE TIME SCALE TEST

The most powerful test concerns the various time scales; powerful because it is independent of the evolutionary problems that effect the $N(m)$, $N(z)$, $m(z)$, $\theta(m)$, and $\theta(z)$ relations. The test is to compare the time found from H_o^{-1} with an independent determination of the age of the Universe from the age of globular clusters, increased by the gestation period of the Galaxy after the creation event.

The age of Galactic globular clusters, based on the luminosity of their main sequence turn-offs, determined by the requirements of the Oosterhoff effect for the relative level of the horizontal branches (Sandage _et al._ 1981; Sandage 1981, 1982), is

$$T_{\otimes} = 17\pm3 \times 10^9 \text{ years.} \tag{16}$$

This assumes $M_V(RR,M3) = 0.6$ and $M_V(RR, M15) = 0.8$, consistent with the requirements of the Oosterhoff effect, and zero-pointed by trigonometric parallaxes of field subdwarfs of low metal abundance (Sandage 1970, 1982).

The intrinsic geometry of the Universe, characterized by the value of q_o _could_ be found by comparing the value of H_o^{-1} with the age of the Universe T_o, if both were known, by using the $T_o = H_o^{-1}(q_o, \Lambda)$ relations (Sandage 1961a, Eqs. 61,62,65).

The present situation is that H_o^{-1} is not agreed upon to within a factor of 2. The consequence of this disagreement are so well known as to require no additional comment here, except to set out that, if $H_o = 50$ km s^{-1} Mpc^{-1}, then $H_o^{-1}(50) = 19.5 \times 10^9$ years, and that if $H_o = 100$ then $H_o^{-1}(100 = 9.8 \times 10^9$ years.

If $q_o = 0$, then $T_o = H_o^{-1}$. Hence equation (16) could be satisfied if $H_o = 50$ and $q_o \sim 0$. However, if $q_o = 1/2$ (or $\Omega_o = 1$) then $T_o = 2/3 \, H_o^{-1}$ $= 13 \times 10^9$ years for $H_o = 50$, and 6.5×10^9 years for $H_o = 100$. Neither case would satisfy equation (16) if $\Lambda = 0$.

A consistent solution is for $\Lambda = 0$, $H_o = 50$ (or $H_o^{-1} = 19.5 \times 10^9$ years), and $T_o = T_\circledast$ (from eq. 16) $+ 2.5 \times 10^9$ years for the age of the Universe, assuming 2.5×10^9 yr as the gestation time for galaxies, This puts the epoch of galaxy formation at $z \sim 7$. My own belief is that $H_o = 42$. All time-scale problems would then be solved if $q_o \sim 0.1$ and $\Lambda = 0$.

Because the time scale test is independent of the types of evolutionary corrections that are necessary for the $m(z)$, $\theta(m)$, and $\theta(z)$ tests, its importance cannot be overemphasized. It is for this reason that such a major effort must be expended to determine its value using powerful enough methods that the problem is placed beyond controversy. This has not yet happened. I doubt if any of the present methods are powerful enough to instill that kind of universal acceptance today, and I believe that no amount of further argumentation by any of the present groups, using the same methods that are yet so controversial, will solve the problem. It was in this vain that I suggested to Longair (quoted in his summary) that similar argumentation in the future along the same paths as used in the past, will not relieve the dilemma, and hence that a moratorium on the present arguments be called. But, not for the moment do I believe the solution to the problem should be called off — only that new methods are needed by new groups whose advocacy positions are not involved.

EVOLUTION IN THE LOOK BACK TIME

We shall hear very much about attempts to find evidence for evolution of galaxy properties in the look-back time. Such effects must be present in an evolving Universe. Failure to detect them would lead to the important conclusion that a creation event did not occur. Unequivocal evidence that the effects are present would be the Ur proof that a creation event did occur.

At present, the evidence for the existence of such evolutionary differences (with z) is marginal. Direct evidence for changes of spectral energy distributions (Wilkinson and Oke 1978) and of colors (Kristian et al. 1978) show no measurable evolution to redshifts of ~ 0.5, but none is expected over this short look-back time.

The most direct test for some sort of evolution of galaxies in clusters is that of Butcher and Oemler (1984, 1985) where they show (with rather high statistical significance) that the % of blue cluster galaxies increases with z. This suggests a conversion process of star-

forming galaxies into benign systems in the very short look-back times
of z<0.5. Dressler and Gunn (1982) found evidence of a similar kind in
the increased spectral peculiarity of cluster galaxies with increasing
redshift. Neither of these results can be understood as passive
evolution of stars in a given galaxy over so short a look-back time.
The effects, if proved, must then be interpreted as some type of
"interactive" evolution caused by the cluster environment. This type of
work is expected to provide a dominant theme in the next decade.

Evidence for evolution of the Universe with time was found soon
after work on the surface density of quasars as a function of apparent
magnitude was begun. From even the first surveys it was found that $N(m)$
increased more steeply than 0.6m (Sandage and Luyten 1967, 1969).
Detailed work in the intervening years by the Bologna group (Braccasi et
al.), by the Wills (cf. this conference), by Usher and collaborators,
Osmer and Smith, Arp, Schmidt and Green (1983 for a recent summary), and
by many others has strengthened the data. There now seems little doubt
that either density or luminosity evolution is required to explain the
steepness of the $N(m)$ quasar counts brighter than B=21 and its rapid
flattening at fainter levels. These data are generally considered to
be the strongest evidence now available for look-back evolution.

Evidence for the direct evolution of the stellar content has been
sought from changes in the spectra of E and SO galaxies over the
redshift range of 0<z<1. Although Wilkinson and Oke failed to find
evidence in their sample to $z \sim 0.5$, Spinrad (1986) and Hamilton (1985)
have searched for the effect on the H and K break near $\lambda 4000$ A in
spectra to $z \sim 0.8$. Models by Bruzual (1983) predict a $\sim 15\%$ effect at
$z \sim 0.8$. Models by Hamilton predict essentially no effect for $z \lesssim 0.8$. The
observations are contradictory. Spinrad observes the predicted $\sim 15\%$.
Hamilton observes none.

As this is a most direct test, the result is fundamental.
Selection effects, operating differently on the different samples, are
thought to be the explanation for part of the difference between Spinrad
and Hamilton. It is expected that a large effort will be put on this
and similar tests in the next decade to test for evolution.

LARGE SCALE STRUCTURE

Hubble (1934) was among the first to suggest large scale clustering
properties. He did not observe the tendency to clump into sheets and
nodes, as has now been done starting in ~ 1975, but rather he obtained a
distribution of galaxy counts per unit area, reduced to standard
observing conditions, that was a Gaussian normal distribution in log
$N(m)$ rather than $N(m)$. This is the distribution to be expected in
contagion situations, rather than in conditions of independence.
Hubble's data were convincing. Because the log normal form with the
same dispersion was observed over the entire area of his survey, and
because he obtained log $N(m) \sim 0.6m$ for $m \lesssim 19$, he concluded that the
Universe was homogeneous on the largest scale that he could measure
(over $\sim 2\pi$ steradians and to a depth of $m \sim 22$), but that it was clumped

(i.e. clustered) on an intermediate scale. Hubble's evidence for the
log normal distribution is shown in Figure 15, together with the figure
caption to his diagram from Realm of the Nebulae.

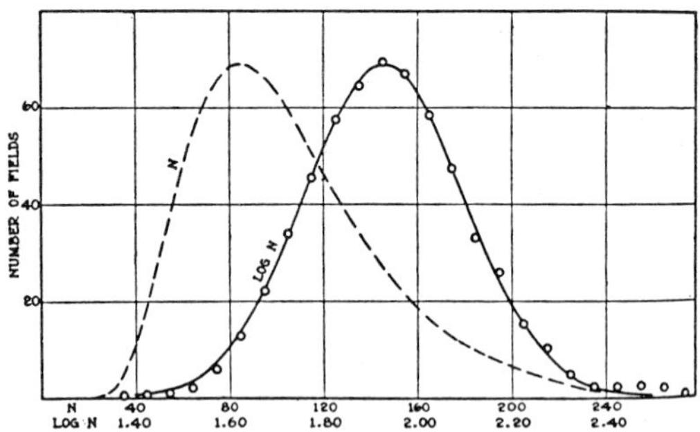

FIG. 7. *Frequency-Distribution of Samples of Various Sizes.*

The circles indicate numbers of samples in which the logarithms of the numbers of nebulæ have various values. The smooth curve drawn through the circles is a normal error-curve. The close fit of the error-curve indicates that the log N per sample are distributed at random around the mean log N for all the samples.

When simple numbers of nebulæ per sample are substituted for the logarithms, the frequency distribution follows the unsymmetrical curve indicated by the dashes.

Fig. 15   Hubble's (1934) distribution of galaxy number counts from his survey of the number of galaxies per unit sky area in many areas. The distribution is a Gaussian in log N rather than N, indicating contagion. The diagram is Figure 7 in The Realm of the Nebulae.

A modern study by Crane and Saslaw (1986), using the Zwicky et al.
catalogs, obtained the same result as Hubble, and drew the same
conclusion.

The discovery of the filaments along which the galaxies concentrate
is the major advance during the past decade. Much will be discussed
during the week on this topic, with the review of the recent work by
Geller (this volume). The subject began, in two dimensions, with a
reanalysis of the classical galaxy catalogs in a series of papers by
Peebles and his collaborators, with a parallel development by Einasto
and his collaborators. The most striking result from the series of
papers by Peebles et al. was an analysis of the Shane-Wirtanen counts
which showed the filamentary nature of the distribution of galaxies
brighter than m∿19 (Seldner et al. 1977). The conclusions were
unexpected, but confirmation from a different direction soon came using
the addition of the third dimension as redshift space. The convincing
data and powerful visualization was by Gregory and Thompson (1978, their
Fig. 2). This paper marks the discovery of voids, which have become
central to the subject. Prior work by Einasto et al. (1980 with earlier
references), Tifft and Gregory (1976), and Chincarini and Rood (1976),
foreshadowed the development, but the Gregory and Thompson discovery is
generally recognized as the most convincing early demonstration.

Rapid developments in the mapping of various filaments and voids
include the studies of Tarenghi et al. (1979), Gregory et al. (1981),
Kirshner et al. (1981), Gregory and Thompson (1982), Chincarini et al.
(1983), and Huchra et al. (1983). A general review is given by Oort
(1983).

Illustrations of the very rapid recent progress in deep mapping of
particular regions will be given by Ellis and by Koo at this conference.

Figures 16 and 17, from Haynes and Giovanelli's (1986) mapping of the
Perseus region, show the richness of the present data.

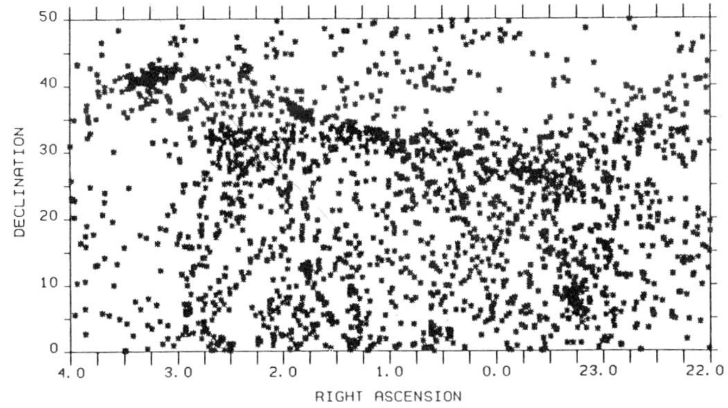

Fig. 16   The
distribution in RA
and Dec of galaxies
in the Perseus
region from the UGC
catalog showing the
projection of the
Perseus filament on
the plane of the
sky. Diagram from
Haynes and
Giovanelli (1986).

Fig. 17   The π
diagram for galaxies
in the same general
direction as Figure
16. The presence of
filaments and voids
is typical of all
other regions
surveyed to date.
The data and diagram
is from Haynes and
Giovanelli (1986).

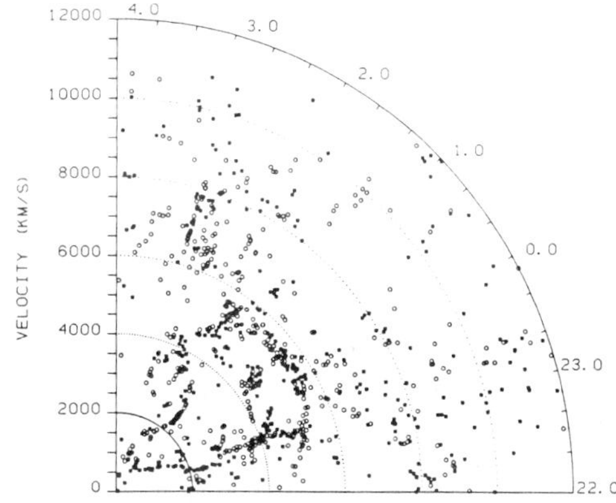

The π diagram of Figure 17 is typical of the recent evidence for sheets
and voids over scales of ∿200 Mpc. It is the projection of such
structures on the plane of the sky that explains the filaments that are
very clearly visible in the Shane-Wirtanen Lick counts (Seldner et al.
1977) and, for example, in Figure 16.

Theories of the production of the voids and sheets, following the
original ideas of Zeldovich for pancake structures, have dominated the
recent theoretical literature on galaxy and supergalaxy formation.
These are to be reviewed by Dekel and others at this meeting.

OTHER TOPICS

A principal subject of the conference is galaxy formation (Silk)
and the related topics of the galaxy luminosity function and the
properties of galaxies along the Hubble sequence. Hubble began both of
the latter subjects in his papers between 1920 and 1936. The work was
later systematized by the concept of stellar populations within his
scheme of galaxy classification.

Little of that will be discussed at this conference, although the variation of galaxy properties along the Hubble classification sequence undoubtedly holds clues to galaxy formation and evolution.

THE NEXT TWO DECADES

Finally, one can set out a list of unsolved problems for which the new large instrumentation will be adequate for solution. My list includes:

1)   Proof or not that the redshift is a true expansion via the surface brightness test that $SB \sim (1+z)^{-4}$ for a true expansion, $SB \sim (1+z)^{-1}$ for some form of tired light. For this, it is crucial that a metric diameter be used, probably adopting Petrosian's (1976) suggestion for its form. Although the tired light hypothesis is unlikely theoretically on the basis of the non-observed spreading of the images with increasing redshift, and the non-widening of the spectral lines, also with increasing z, the test should be made. Until now, attempts have been unsuccessful due to lack of a properly defined metric diameter.

2)   Proof or not of evolution of the galaxy stellar content in the look-back time. The variation of the H and K, $\lambda 4000A$ break with z would seem to be the likely experiment, following Spinrad and Hamilton using larger data samples and a larger redshift range.

3)   The most important experiment to test the Friedman model and/or the value of q_0 (with a possible need for $\Lambda \neq 0$) is the value of H_0 compared with the globular cluster time scale.

4)   The compatibility of the clustering properties (filaments and "small" scale voids and inhomogeneities) with possible variations of the Hubble flow, following tne bias introduced in the Hubble diagram by such inhomogeneities (Sandage and Tammann 1975, Appendix; Fall and Jones 1976). The necessary large scale redshift surveys that will be completed in this period should provide the data by which to model the problem.

5)   Studies of the galaxy luminosity functions for different galaxy types and for the extremes of the density environments such as in the rich cluster cores and in the voids. Do dwarf galaxies exist in the voids such as is expected in certain models of biased galaxy formation?

6)   The necessity to detect the $\delta T/T$ fluctuations of the 3°K radiation at a level of one part in $\sim 10^{5.5}$ on small angular scales. A failure to detect such fluctuations (presumed to be necessary for galaxy formation) would cause one to wonder if the Friedman standard model is correct. If not, the subject might again be back near its first beginnings.

ACKNOWLEDGEMENTS

Part of the work on galaxies and observational cosmology at Mount Wilson was supported by National Science Foundation grant AST 82-15063 for which I am grateful. It is a pleasure to thank Janet Krupsaw for the preparation of the many drafts of this review with skill and patience.

DISCUSSION

Burke: In the spoken version of your talk, you set out a list of 12 great cosmologists of the 1930's, six observers and six theoreticians. The observers were Lundmark, Lindblad, Hubble, Humason, Baade, and Mayall. The theoreticians were Friedman, Lemaitre, de Sitter, Tolman, Robertson, and Gamow. To bring your list to 15 would you accept Einstein and Weyl on the theorist list, and V. M. Slipher among the observers?

Sandage: My list of 12 was meant to be evocative rather than definitive. Your choices are, of course, fascinating. Why are they not on my list? With hindsight, Slipher is often mentioned as having found the expansion. It has never been clear that this was understood in $\sim$1925. He did not publish his redshift values, nor did he discuss them. He sent them to Strömberg who discussed them circa 1924 and to Eddington who published them in $\sim$1925. Without a discussion, are isolated measurements a basis to change the World which the discovery of the linear velocity-distance relation for all galaxies has, in fact, done?

   Einstein: In the presence of God one does not mention God. It was his equations to which Friedman found a particular expanding solution. Nevertheless, we all learned cosmology not by reading Friedman but rather we teach our students from the Newtonian analogs of the equations first set out by Milne and McCrea. Their influence and that of G. C. McVittie has been paramount in making the subject understandable to more ordinary minds. Their names should have been added to the first hurried list.

Rubin: You discussed the effect of the Malmquist bias on the velocity studies, but are there not equivalent effects on the spatial plots, i.e. the strings in the distributions that are now showing up. Won't a bias due to such local non-homogeneities be present from that effect?

Sandage: Indeed. Both Fall and Jones (1976) and Tammann and I (1975) suggested that apparent deviations from the Hubble flow need not be real but would <u>appear</u> in the analysis of the $m(z)$ relation for ScI galaxies (for example), caused by the bias effect on the data sample by a non-homogeneous spatial distribution of galaxies such as you suggest.

Pecker: What about the age of the globular clusters? Quite possibly, turbulent diffusion, as suggested by Maeder and Schatzman might slow down the stellar evolution, and push to $\sim$25 Gyr. the age of globular clusters. This would add a severe difficulty to the modeling of the universal cosmology.

Sandage: The Friedman cosmology is already in difficulty (if $\Lambda=0$) unless $H_0 \sim 50$ km s^{-1} Mpc^{-1}, or less. Tammann and my current value from the supernovae observations of $m(z)$ and with $M_B = -20.0$ at maximum gives $H_0 = 42$ km s^{-1} Mpc^{-1}. This can be fit with a $\Lambda=0$, $q_0 \sim 0$, $T_U = 20 \times 10^9$ year data set. But, of course, I must agree with you that the longer the age scale of globular clusters is set and the shorter one puts H_0^{-1} (i.e. the larger one puts the value of H_0), then the more you squeeze yourself out of the Friedman box.

Narlikar: I think that correct scientific practice requires that those who measure H_0 are not influenced (indirectly) by what answers the ages of globular clusters give. Likewise those who estimate globular cluster ages do not look over their shoulders to see what answers the cosmologists want. After the two determinations have been made independently then the answers can be compared to check whether our cosmological picture is right. I hope this procedure is followed in the next twenty years.

Sandage: I believe, in fact, that it has been scrupulously followed in the past 20 as well.

Giraud: The variation of the Hubble constant that you have shown in a diagram by de Vaucouleurs and Peters is in the first few Mpc. How can you explain that by a Malmquist bias?

Sandage: The claim by de Vaucouleurs and Peters of a local variation of H_0 is over $\sim$40 Mpc -- more precisely it is over at least two times the distance to the Virgo cluster (i.e. for observed velocities of $0 \lesssim v \lesssim 2000$ km s^{-1}). This is precisely the range over which the very strong Malmquist bias exists shown in Figures 12, 13 and 14 of the written text for the Shapley-Ames spirals.

Audouze: In your introduction you did not mention the nucleochronology techniques to determine the age of the Universe. I am aware that these methods are still very uncertain (maybe more uncertain than the globular cluster technique), but we may hope in the future to obtain the age of the Universe from "chemical" (meteoritical) determinations.

Sandage: Is it true that the present range of nucleosynthesis ages is
           from $\sim$9 to 22 Gyr, depending on the assumptions of how the
           heavy element yield has changed with time — i.e. whether
           "sudden synthesis" or "gradual" enrichment of the ISM has
           occurred? I believe the globular cluster ages are accurate to
           $\sim\pm$25% and that the time between the creation event and the
           formation of globular clusters is accurately known from the
           high redshift cut-off of the quasar distribution at $z \sim 4$,
           signalling the first creation of the galaxies.

Norman: If one finds non-evolving galaxies at say $z\sim$1 (for example as
           indicated by Hamilton's work, although the following statement
           is not dependent on this result), then they will be the oldest
           objects. Surely they should be the focus of our efforts to
           find T_o, <u>not</u> the globular clusters. One unambiguous non-
           evolving galaxy at redshift greater than $z=1$ would be a
           fundamental discovery and within it would contain the value of
           T_o.

Sandage: The problem then becomes, how do you age date that galaxy?
           The only technique is to divine the stellar content by some
           type of spectral synthesis, and from it determine the main
           sequence termination point of the oldest stellar component.
           This is precisely the same technique as used for globular
           clusters except, for them, the main sequence termination
           luminosity is observed directly, and is therefore a more
           precise determination.

REFERENCES

Arp, H. C. 1967, <u>Ap. J.</u> <u>148</u>, 321.
Arp, H. C. 1980 in Ninth Texas Symposium on Relativistic Astrophysics,
     ed. J. Ehlers, J. Perry, M. Walker: New York Academy of Sciences,
     p. 94.
Brown, G. S., and Tinsley, B. M. 1974, <u>Ap. J.</u> <u>194</u>, 555.
Bruzual, A. G. 1983, <u>Ap. J.</u> <u>273</u>, 105.
Burbidge, G. R. 1981 in Tenth Texas Symposium on Relativistic
     Astrophysics, ed. R. Ramaty and F. C. Jones: New York Academy of
     Sciences, p. 123.
Butcher, H. R., and Oemler, A. 1984, <u>Ap. J.</u> <u>285</u>, 426.
Butcher, H. R., and Oemler, A. 1985, <u>Ap. J. Suppl.</u> <u>57</u>, 665.
Chincarini, G. L., Giovanell, R., and Haynes, M. P. 1983, <u>Ap. J.</u> <u>269</u>,
     13.
Chincarini, G., and Rood, H. J. 1976, <u>Ap. J.</u> <u>206</u>, 30.
Crane, P., and Saslaw, W. C. 1986, <u>Ap. J.</u> <u>301</u>, 1.
de Vaucouleurs, G., and Peters, W. L. 1986, <u>Ap. J.</u> <u>303</u>, 19.
Djorgovski, S., and Spinrad, H. 1981, <u>Ap. J.</u> <u>251</u>, 417.

Djorgovski, S., and Spinrad, H. 1986, Ap. J., in press.
Dressler, A., and Gunn, J. E. 1982, Ap. J. 263, 533.
Einasto, J., Joeveer, M., and Saar, E. 1980, M.N.R.A.S., 193, 353.
Fall, S. M., and Jones, B. J. T. 1976, Nature, 262, 457.
Gregory, S. A., and Thompson, L. A. 1978, Ap. J., 222, 784.
Gregory, S. A., and Thompson, L. A. 1982, Ap. J. 286, 422.
Gregory, S. A., and Thompson, L. A., and Tifft, W. G. 1981, Ap. J., 243,
     411.
Gunn, J. E., and Oke, J. B. 1975, Ap. J. 195, 255.
Hamilton, D. 1985, Ap. J., 297, 371.
Haynes, M. P., and Giovanelli, R. 1986, Ap. J. Lett., 306, L55.
Hoyle, F. 1959 in Paris Symposium on Radio Astronomy, ed: R. N.
     Bracewell, Stanford Univ. Press, p. 529.
Hubble, E. 1934, Ap. J., 79, 8.
Hubble, E. 1936, Realm of the Nebulae, Yale University Press.
Hubble, E. 1953, M.N.R.A.S., 113, 658.
Huchra, J., Davis, M., Latham, D., and Tonry, J. 1983, Ap. J. Suppl.,
     52, 89.
Kapahi, V. K. 1975, M.N.R.A.S., 172, 513.
Karachentsev, I. D., and Kopylov, A. I. 1978, IAU Symp., 79, 339.
Kirshner, R. P., Oemler, A., Schechter, P. L., and Schectman, S. A.
     1981, Ap. J. Lett., 248, L57.
Kraan-Korteweg, R. C., Sandage, A., and Tammann, G. A. 1984, Ap. J.,
     283, 24.
Kristian, J., Sandage, A., and Westphal, J. A. 1978, Ap. J., 221, 383.
Kron, R. G. 1980, Ap. J. Suppl., 43, 305.
Legg, T. H. 1970, Nature, 226, 65.
Lilley, S. J., Longair, M. S. 1984, M.N.R.A.S., 211, 833.
Lilley, S. J., Longair, M. S., and Allington-Smith, J. R. 1985,
     M.N.R.A.S., 215, 37.
Mattig, W. 1958, Astron. Nach., 284, 109.
Mattig, W. 1959, Astron. Nach., 285, 1.
Miley, G. 1968, Nature, 218, 933.
Miley, G. 1971, M.N.R.A.S., 157, 477.
Murdock, H. S., Hunstead, R. W., Pettini, M., and Blades, J. C. 1986,
     Ap. J., 309, October 1.
Oort, J. H. 1983, Annual Rev. Astron. Astrophys., 21, 373.
Peterson, B. A., Ellis, R. S., Kibblewhite, E. J., Bridgeland, M. T.,
     Hooley, T., and Horne, D. 1979, Ap. J. Lett., 233, L109.
Petrosian, V. 1976, Ap. J. Lett., 209, L1; 210, L53.
Ryle, M. 1968 in Highlights of Astronomy, Vol. 1, ed. L. Perek, IAU
     publications: Reidel, p. 33.
Sandage, A. 1961a, Ap. J., 133, 313.
Sandage, A. 1961b, Ap. J., 134, 916.
Sandage, A. 1968, Observatory, 88, 91.
Sandage, A. 1970, Ap. J., 162, 841.
Sandage, A. 1972a, Ap. J., 173, 485.

Sandage, A. 1972b, Ap. J., 178, 1.
Sandage, A. 1972c, Ap. J., 178, 25.
Sandage, A. 1975a in Galaxies and the Universe: ed. A. Sandage, M.
     Sandage, and J. Kristian: Univ. Chicago Press, p. 761.
Sandage, A. 1975b, Ap. J., 202, 563.
Sandage, A. 1981, Ap. J., 248, 161.
Sandage, A. 1982, Ap. J., 252, 553.
Sandage, A. 1986, Ap. J., 307, 1.
Sandage, A., and Hardy, E. 1973, Ap. J., 183, 743.
Sandage, A., Katem, B., and Sandage, M. 1981, Ap. J. Suppl., 46, 41.
Sandage, A., and Luyten, W. J. 1967, Ap. J., 148, 767.
Sandage, A., and Luyten, W. J. 1969, Ap. J., 155, 913.
Sandage, A., and Tammann, G. A. 1975, Ap. J., 197, 265.
Sandage, A., and Tammann, G. A. 1983 in Large Scale Structure of the
     Universe, Cosmology and Fundamental Physics. First ESO-CERN
     Conference ed. G. Setti and L. van Hove, p. 127.
Sandage, A., and Tammann, G. A. 1986 in Inner Space Outer Space, ed. E.
     W. Kolb, M. S. Turner, D. Lindley, K. Olive, and D. Steckel: Univ.
     of Chicago Press, p. 41.
Sandage, A., Tammann, G. A., and Yahil, A. 1979, Ap. J., 232, 352.
Schmidt, M., and Green, R. F. 1983, Ap. J., 269, 352.
Seldner, M., Siebers, B., Groth, E. J., and Peebles, P. J. E. 1977, A.
     J., 82, 249.
Spaenhauer, A. M. 1978, Astron. Astrophys., 65. 313.
Spinrad, H. 1986, Pub. A.S.P., 98, 269.
Swarup, G. 1975, M.N.R.A.S., 172, 501.
Tarenghi, M., Tifft, W. G., Chincarini, G., Rood, H. J., and Thompson,
     L. A. 1979, Ap. J., 234, 793.
Tammann, G. A., Yahil, A., and Sandage, A. 1979, Ap. J., 234, 775.
Tifft, W. G., and Gregory, S. A. 1976, Ap. J., 205, 696.
Tinsley, B. M. 1980 in Physical Cosmology: Les Houches summer school;
     session XXXII: ed. R. Balian, J. Audouze, and D. Schramm: North
     Holland, P. 161.
Tyson, J. A., and Jarvis, J. F. 1979, Ap. J. Lett., 230, L153.
Wilkinson, A., and Oke, J. B. 1978, Ap. J., 220, 376.

CHAPTER I

THE MICROWAVE BACKGROUND RADIATION

The Cosmic Microwave Background

R. B. Partridge
Haverford College
Haverford, PA 19041, U.S.A.
and
The Institute of Astronomy
Madingley Road, Cambridge, England

ABSTRACT.   This review summarizes recent observational work on the
cosmic microwave background, or 3 K radiation.   Recent measurements
of its spectrum and large-scale angular distribution are described, as
well as searches for small angular scale fluctuations on arcsecond to
degree scales.   A few of the consequences of these measurements and
upper limits for cosmology, astrophysics, and theories of galaxy
formation are touched on here.

Like many others, I am here in China for the first time.   In my
three days here so far, I have already seen enough hospitality and
energy to hope this trip will not be my last.   I would like particularly
to thank the Local Organizing Committee and my Chinese colleagues
for making me feel so welcome.

My task today is to review observations of the cosmic microwave
background (CBR).   This radiation, discovered 22 years ago by Arno
Penzias and Robert Wilson (1965), was immediately interpreted by
Robert Dicke and his colleagues as radiation left over from a hot Big
Bang of the Universe (Dicke, Peebles, Roll and Wilkinson, 1965).   In
this model, the radiation should be approximately isotropic and have a
blackbody spectrum.

There are, however, a number of astrophysical processes which
may have operated in the redshift interval $10^6 \gtrsim z \gtrsim 3$ to perturb
either the spectrum or the isotropy of the radiation.   These small
epartures from perfect isotropy or an exact Planckian spectrum, if
detected, would provide important information about epochs in the
Universe not accessible by other observational techniques.   I will
mention some of these astrophysical processes as I summarize the
recent observational results, but my main emphasis will be on the
observations.   These fall essentially under three main headings:--the
spectrum of the CBR, its large-scale anisotropy (especially the dipole
component in its intensity), and small-scale anisotropies or
fluctuations in the observed temperature of the radiation.

A. Hewitt et al. (eds.), Observational Cosmology, 31–53.
© 1987 by the IAU.

I intend this to be a very general overview of recent observational results. Several more detailed papers, on measurements of the dipole anisotropy, the spectrum, and intermediate and fine-scale anisotropies, follow this review. I should also mention that much of the material I present here has been drawn more or less directly from a long review paper I prepared at about the same time as this IAU symposium (Partridge, 1987). Since that review and other recent overviews of the field are available (e.g., Wilkinson, 1986; Partridge, 1986), I will omit in this written version of my remarks much of the detail I presented orally in Beijing. In particular, I will devote considerably less attention here to measurements of the angular distribution of the radiation.

1. SPECTRUM

Early measurements of the spectrum of the microwave background (as reviewed, for instance, by Wilkinson, 1980; Richards, 1980; or Weiss, 1980) showed that the spectrum was on the whole consistent with a Planck curve of temperature 2.5-3.0 K. However, measurements made by Woody and Richards (1981), using a bolometric detector flown above much of the earth's atmosphere, suggested a departure from a blackbody curve near the peak of a 2.5-3.0 K spectrum. The nature of the departure is shown in the cartoon diagram below. Considerable theoretical energy (e.g., Negroponte et al, 1981; Bond et al, 1986) went into offering explanations for the ~ 10 percent increase in temperature or intensity at wavelengths of a few millimeters.

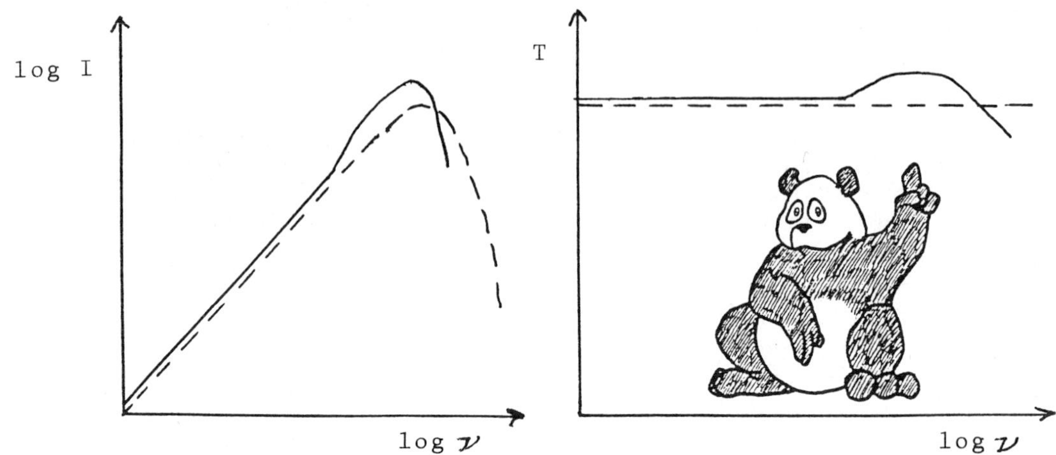

Figure 1. Schematic representation of the intensity and of the thermodynamic temperature of the CBR as determined by observations prior to 1982. The panda indicates the apparent departure from a Planck curve.

By the early 1980's, improved experimental techniques had made
refined measurements of the spectrum possible and had allowed us to
test the findings of Woody and Richards. In the past five years, a
number of groups have worked to improve measurements of the
spectrum of the background radiation over a wavelength range of
more than 100, from λ = 12 cm to $\lambda \simeq$ 1 mm.

1.1.   Sampling the CBR temperature using interstellar molecules.

Interstellar CN (cyanogen) molecules have low-lying rotational energy
levels which can be excited by photons at wavelengths of 2.64 and
1.32 mm. Measurements of the relative populations of these levels can
thus tell us the thermodynamic temperature of the microwave radiation
field in space at these two wavelengths. The relative populations, in
turn, may be determined by measuring the equivalent width of optical
lines originating on these low-lying levels---see figure 2. As the
figure suggests, these observations are made in absorption, using a

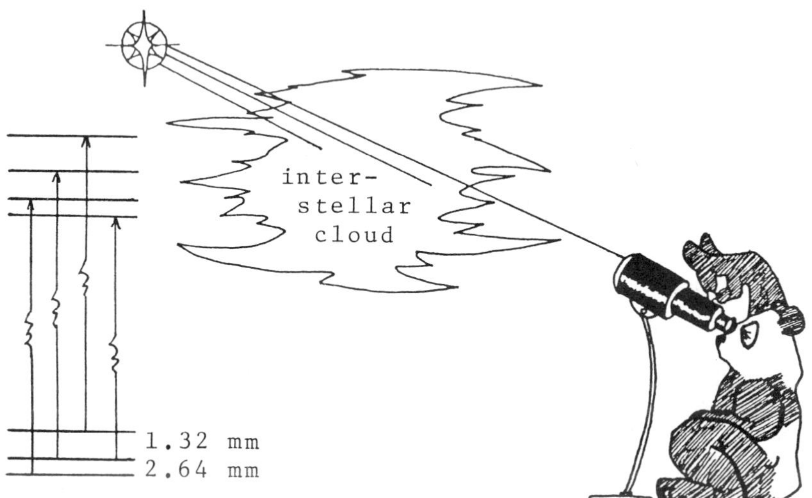

Figure 2.   Sketch of the means used to measure the CBR temperature
using the excitation of interstellar CN molecules.

bright star as a background source. Two groups, Meyer and Jura
(1984, 1985) and Crane et al (1986) have made use of this technique to
redetermine the temperature of the microwave background at 2.64 and
1.32 mm. The work of the latter group is described in more detail
here by one of the members of the collaboration, Dr. Mandolesi. Here,
I merely summarize the observational results in Table 1. Note that
the measurement at 2.64 mm, near the peak of the 2.5-3.0 K blackbody
spectrum, is the single most accurate measurement we have of the
temperature of the CBR.

Observers	T, λ = 2.64mm	T, λ = 1.32 mm
Meyer and Jura (1984)	2.73 ± 0.04	2.8 ± 0.3
Crane et al (1986); see Mandolesi here	2.74 ± 0.05	2.75 ± 0.3

Table 1. Measurements of the CBR temperature using CN molecules.

1.2. Long wavelength measurements.

More precise measurements of the spectrum in the Rayleigh-Jeans region were undertaken in 1982 and 1983 by an international collaboration of astronomers from Bologna, Milan, Padua, Berkeley and Haverford. The results are fully described in Smoot et al (1985) and in review papers that I have prepared (1986, 1987). The final results of this collaborative work are displayed in both Table 2 and Figure 3. I note here that the experiment was specifically designed to ensure good intercomparability of measurements made at different wavelengths. For instance, the same antenna design was used for

Wavelength, cm	Atmos. Temp.	Galactic Emission	T CBR
12	0.95	0.15-0.20	2.77 ± 0.13
6.3	1.00	~ .04	2.70 ± 0.08
3.0	1.14	< .01	2.75 ± 0.08
0.9	4.5-4.9	~ 0	2.81 ± 0.12
0.33	10-13	~ 0	2.57 ± 0.14

Table 2. Radiometric measurements of the CBR temperature (Smoot et al, 1985). The weighted means of 1982 and 1983 measurements are shown. The atmospheric temperature at 0.9 and 0.33 cm depended on the water vapor content of the atmosphere--approximate values are indicated.

measurements at all five wavelengths, and the cold-load calibrator was common to all. In addition, we devoted considerable attention to the measurement of sources of systematic error, such as emission from the Galaxy, the ground and the earth's atmosphere. The magnitude of some of these are indicated in Table 2. While the emphasis was on relative accuracy within our set of five measurements, it is gratifying to see how well they mesh with the CN measurements referred to above.

Because the equipment employed was cumbersome, this experiment was necessarily ground-based. We were thus faced with the problem

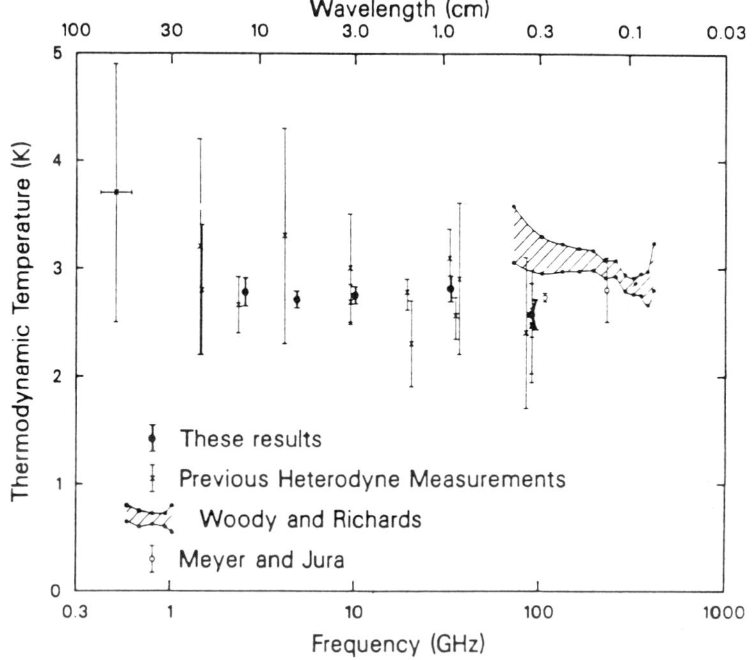

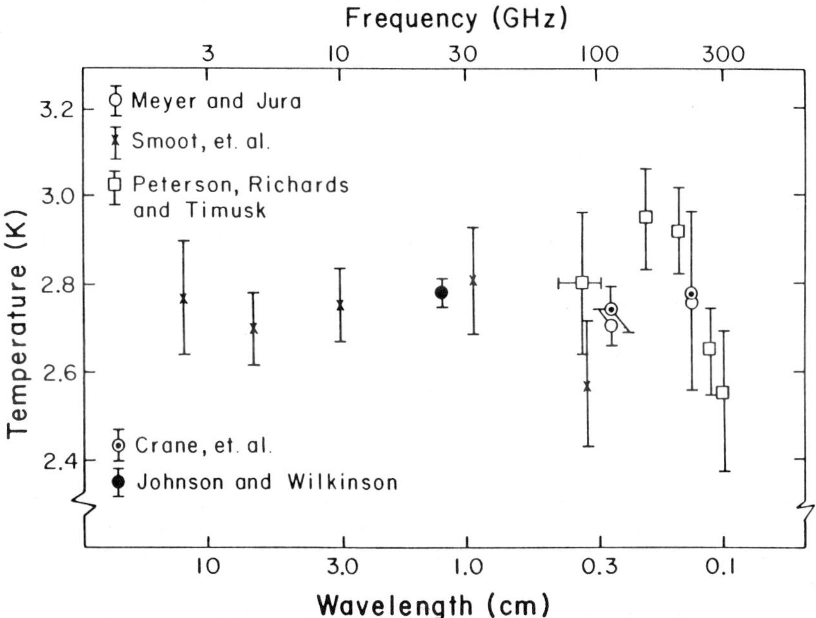

Figure 3. (Top) Measurements of the spectrum of the CBR available by early 1985 ("These results" = Smoot et al, 1985). (Bottom) Recent spectral measurements discussed here.

of measuring accurately and subtracting microwave emission from the
earth's atmosphere (Partridge et al, 1984; Danese and Partridge,
1987). Another approach to the problem presented by emission from
the earth's atmosphere is literally to rise above it, as Johnson and
Wilkinson (1986) did by flying their instrument on a balloon. At an
altitude of ~ 25 Km, the residual emission of the earth's atmosphere is
easily corrected for. Their measurement was made at a wavelength of
1.2 cm. That experimental result is presented in poster form here;
the authors give two possible errors, a formal statistical error and a
more conservative error. Using the latter, they find at a wavelength
of 1.2 cm, T = 2.78 ± 0.08 K, once again in excellent agreement with all
the recent results described so far.

1.3.   Measurements near the peak of the spectrum.

Finally, Richards and his colleagues have repeated their observations
near the peak of a 2.5-3.0 K blackbody spectrum, in the wavelength
range $5 \gtrsim \lambda \gtrsim 1$ mm. They again used a broad-band, bolometric
detector. For the most recent observations, however, they employed
filters to define five wavelength bands in the range of wavelengths
noted above, and the instrument was calibrated in flight. The results
of these new observations (Peterson, Richards and Timusk, 1985) are
also shown in Figure 3. While there is still some barely significant
evidence for a slight increase in the temperature of the CBR just at
the peak, that is, around 2 mm, the overall results are now in
excellent agreement with the cyanogen measurements which fall in the
same wavelength range.

1.4.   Summary of spectral measurements.

To summarize, over a wavelength range extending from 0.1 to 12 cm,
right across the peak and well into the Rayleigh-Jeans region of a
2.5-3.0 K spectrum, there is no evidence for spectral distortions.
Taken together, the measurements establish the temperature of the
CBR at 2.75 K with an accuracy of ~ 1 percent--in my view an
important baseline in astrophysics.

1.5.   Consequences.

Let us now look briefly at some of the consequences of the recent
measurements summarized above.
    First, the fact that the observed spectrum of the CBR is so close
to blackbody is hard to explain in models which invoke re-emission by
hot dust to explain some or all of the microwave background (e.g.,
Negroponte et al, 1981). Very special optical properties of the dust
would have to be assumed to make the observed temperature constant
over so wide a range of wavelength.
    The observational results also place interesting limits on any sort
of energy release into the microwave background over a redshift
range of order $10\text{-}10^{6}$ (see review by Danese and De Zotti, 1977).
Adding energy to the radiation field would perturb the spectrum; the

absence of evident spectral distortion limits the energy release, ΔE, to $\lesssim$ 5 percent of the energy density of the CBR. For instance, the energy released in the process of galaxy formation cannot exceed this limit. Likewise, the energy released by the decay of possible exotic particles cannot exceed this limit, hence placing limits on the physical parameters of such particles (e.g., Silk and Stebbins, 1983). I could continue with a catalog of other ways in which the absence of evident spectral distortion has been used to set limits on astrophysical processes occurring at these large redshifts; instead I will just remark that I feel the power of this constraint has not been fully recognized or employed. I hope some of you present here will take up the challenge that this remark implies.

Finally, a value for the present temperature of the CBR is one of the parameters needed to calculate the relative abundances of light nuclei synthesized in the Big Bang, as reviewed in this volume by Prof. Audouze. The error in the value of the CBR temperature is now negligible compared with other uncertainties in this calculation. A slightly subtler point is that the absence of spectral distortions permits us to assume with more certainty that the temperature we now measure can be directly related to the temperature prevailing at the much earlier epoch of nucleosynthesis.

2.   LARGE-SCALE ISOTROPY

Early measurements of the large-scale angular distribution of the CBR showed it to be isotropic to a few percent, as expected in the Big Bang model of Dicke et al (1965). These observations, reviewed by Weiss (1980) and elsewhere by the present author (1987), were thus in agreement with the predictions of the Big Bang model. But it has been known for 20 years that large-scale anisotropy in the radiation may be introduced in two general ways. The first of these is motion of the observer; the Doppler effect ensures that in the direction of motion the intensity of the CBR, and hence its measured temperature, will be slightly increased. The Doppler effect introduces a dipole component into the measured temperature of the CBR. A second cause of large-scale anisotropy is anisotropic expansion of the Universe as a whole. In the simplest case, a pure quadrupole component would result, but other possibilities including temperature anisotropy on an angular scale of $\sim \theta \simeq \Omega$ radians are also possible in open Universes with $\Omega \equiv \rho_0/\rho_c < 1$. Some of these possibilities are mentioned briefly in the review by Dr. Lukash here and also in Barrow et al (1983).

2.1.   Recent observational results.

Just as in the case of spectral measurements, measurements of the large angular scale distribution of the CBR have been improved in the past few years. Four groups have worked on these measurements. The results obtained by three of these groups are summarized in Table 3 below. The Berkeley and Princeton groups employed radiometers carried aloft by balloons; the results of the Soviet group

were obtained from the first satellite devoted to measurements of the
CBR, the Relict experiment on the Prognoz 9 satellite.   The
fourth group that has made such measurements is based at MIT.
Since their measurements are made at quite short wavelengths, they
cannot be directly compared to measurements made at wavelengths
longer than the peak of a 2.5–3.0 K blackbody unless the temperature
of the CBR is known accurately.   Therefore, the MIT group has used
their results to obtain estimates of both the dipole component and the
temperature of the CBR, and these results are described in detail in
the following paper by Dr. Halpern.

Group	Berkeley	Princeton	Moscow*
Reference	Lubin et al (1983)	Fixsen et al (1983)	Strukov and Skulachev (1984)
λ, cm	0.3	1.2	0.8
dipole amplitude	3.4 ± 0.2 mK	3.1 ± 0.2 mK	3.16 ± 0.12 mK
direction of sun's velocity	$11^h\!.5$, -6°	11^h, -10°	$11^h\!.3 \pm 0^h\!.15$, $-7.5^o \pm 2.5^o$

*Updated results from the paper of Lukash here.

Table 3.   Results of three recent measurements of the dipole
component of the CBR (see the following paper by Halpern for the
fourth such measurement).   The three measurements above imply a
velocity of the local group of galaxies of $V_{LG} \sim 600$ km/sec towards
$\ell = 270^o$, b $= 30^o$ in Galactic coordinates.

Returning to the longer wavelength measurements summarized in
Table 3, note the excellent agreement on the dipole amplitude.   In
particular, the re-analysis of the Prognoz 9 results reported here by
Dr. Lukash gives a larger value for the dipole amplitude than that
reported earlier by Strukov and Skulachev (1984), one in better
agreement with the other measurements.   If we interpret the dipole
amplitude as a consequence of the Doppler effect, we can determine
the direction of the motion of the sun (neglecting the small motion of
the earth around the sun); we can then correct this value for the
known motion of the sun around the center of the Galaxy, and the
motion of the Galaxy in the local group, to provide an estimate of the
velocity of the local group itself.   That value is something like
600 Km/sec in the direction 270° galactic longitude and 30° galactic
latitude.   While there is some uncertainty in the sun's motion relative
to the local group (somewhat larger than the experimental uncertainty
in the CBR dipole), the qualitative conclusions of these measurements
are not affected.   They show that the local group is moving with a
substantial velocity, ~ 0.002 c, and that the direction of this motion is
at $\sim 45^o$ to the direction to the center of the nearest large
concentration of mass, the Virgo cluster.

The same observational programs that have determined the dipole component of the CBR may also be used to establish upper limits on the quadrupole component. Earlier reports (e.g., Fabbri et al, 1980; Boughn et al, 1981) of a quadrupole component of amplitude ~ 1 mK appear to be in error; the present upper limits on the quadrupole component are $\Delta T/T \lesssim 3 \times 10^{-5}$, the most sensitive upper limit resulting again from a reanalysis of the Soviet satellite observations (additional details provided in the paper here by Lukash).

Since most of these observational results are now several years old, and in any case are discussed in review articles (Wilkinson, 1986; Partridge, 1987), I will turn now to a discussion of some of the consequences of these findings. In the remainder of section 2, I want to focus on three points:--a comparison of the velocity inferred from CBR measurements with that derived from optical measurements; a discussion of the possible causes of the dipole component; and consequences of the apparent absence of true, cosmological, anisotropy.

2.2.   Comparison to optical results.

The microwave dipole component provides, as we have seen, a measure of the velocity of the local group relative to matter at large distances. The velocity of the local group may also be determined by making observations of the distribution of redshifts of more local objects. A decade ago, Dr. Rubin and her colleagues (Rubin et al, 1976) began a program of optical measurements of the redshifts of a shell of galaxies centered on us. By determining the systematic variations of redshift across the sky, they were able to deduce the velocity of the local group. The _speed_ they determined was roughly comparable to the 600 Km/sec figure mentioned above, but the direction was approximately orthogonal to the microwave result. Since those results were announced, there has been considerable discussion and even controversy about the optical results and further measurements (see, e.g., Fall and Jones, 1976; Schechter, 1977; deVaucouleurs and Peters, 1984; Yahil et al, 1986; and Meiksin and Davis, 1987). I have elected to skip over most of that material and move to a very recent--and also controversial--paper on this subject (Collins et al, 1986). Using a different technique to determine the distance to some of the galaxies used originally by Rubin and her colleagues, these authors find results in good agreement with the original 1976 work. They therefore make the suggestion that both the optical _and_ the CBR results are experimentally correct. If so, we must infer that the shell of galaxies observed by Rubin and her colleagues is in rapid motion with respect to matter at large distances. This indeed is the claim of Collins et al--velocities of 1000 km/sec on megaparsec scales are present in the Universe. As Collins et al (1986) note, such large-scale and substantial velocities are very difficult to reconcile with models in which the density of the Universe and the process of galaxy formation are dominated by cold dark matter. These issues are discussed later in this volume by Drs. Dekel and Silk, among others.

2.3.   Gravity as a cause of the microwave dipole.

A further consequence of the assumption that both the microwave and
the optical velocity measurements are correct is that it is quite
difficult to say exactly what produces either of them. The simplest
assumption is that a large, localized, lump of matter induces the
velocity of the local group by gravitational attraction. If that were
true, and if the local lump lay well inside the shell of galaxies used
for the optical observations, then the optical and microwave results
should agree. Under the assumption that the microwave and optical
results are both correct, therefore, the situation must be more
complex. For now, however, I wish to put this reservation aside to
sketch out some of the consequences of the simpler assumption that a
single large lump induces gravitational acceleration, producing the
observed dipole component.

If the observed dipole component of the microwave background is
gravitationally induced, we may use its amplitude to tell us something
about the large-scale distribution of matter in the Universe and even
to estimate the mean mass density, ρ_O. As an example of this sort of
calculation, I will assume that the accelerating lump has spherical
symmetry, and that its center is located a distance r away from us.
If we neglect nonlinear effects, the gravitationally induced velocity as
a fraction of the recession velocity is given simply by:

$$V/H_O r \;=\; \frac{1}{3}\, \bar{\delta}\; (\rho_O/\rho_C)^{0.6}$$

where $\bar{\delta}$ is the overdensity of the local lump. To a first
approximation, we may determine $\bar{\delta}$ simply by counting up galaxies.
For instance, the overdensity in numbers of the Virgo cluster is ~ 2.
If we use this value, and set r equal to the distance to the Virgo
cluster, we obtain $\rho_O/\rho_C \approx 0.3 \pm 0.2$. But there are good theoretical
reasons for believing that the matter in the Universe may be more
smoothly distributed than the light, as measured by counts of
galaxies. If that is true, the overdensity $\bar{\delta}$ is smaller than the
value inferred from galaxy counts alone, and higher values of ρ_O are
possible; indeed the possibility that ρ_O is equal to the critical density
ρ_C can by no means be excluded.

Some of these issues are discussed in this volume by Drs. Yahil
and Rowan-Robinson and in more detail in a review paper by Davis in
IAU Symposium 117 (Davis, 1987). Here I want only to emphasize that
one must use some caution in interpreting the microwave results until
the apparent discrepancy (if any) with the optical results is
resolved. This, too, is an area where further work, both
observational and theoretical, is needed. The microwave observers
have told us with a precision of a few percent how we are moving
relative to matter at large distances; what can we make of this result?

2.4.   Intrinsic anisotropy of the Universe.

If the observed dipole component of the CBR is interpreted as a
result of the Doppler effect, any residual dipole moment is very small,
like the quadrupole component. Taken together, these results imply
that the expansion of the Universe has been isotropic and shear-free
for the vast majority of the history of the Universe, from the time
that the microwave photons last interacted with matter until the
present. As pointed out nearly 20 years ago, originally by Thorne
(1967) and Hawking (1969), limits on large-scale anisotropies in the
CBR can be used to fix quite stringent limits on the anisotropic
expansion of the Universe for many classes of spatially homogeneous,
but anisotropic, cosmological models (see a recent discussion by
Barrow et al, 1983). One possible escape from these conclusions is
opened up by some low density anisotropic cosmological models in
which the anisotropy is effectively "squeezed" into a small solid
angle of the sky of angular dimension $\theta \sim \Omega$ radians (Novikov, 1968;
see also the paper by Lukash here). In this connection, the upper
limit of $0.004(mK)^2$ placed on $[\Delta T/T]^2$ over the whole sky by the
Prognoz 9 experimenters (and reported here by Lukash) is particularly interesting. This limit applies for angular scales $\gtrsim 20^\circ$, and
effectively constrains the anisotropy in any cosmological model with
$\rho_0 \gtrsim 0.3 \, \rho_C$. Fixsen et al (1983) have reported somewhat less
sensitive upper limits, but on scales reaching down to 10°, corresponding to $\rho_0 \simeq 0.16 \, \rho_C$.

2.5.   Summary.

Provided we interpret the dipole component as the result of
gravitationally induced velocity, there is no evidence in any of the
recent observations for any true, cosmological, anisotropy in the CBR
on angular scales of tens of degrees or larger. These results, taken
in conjunction with the spectral measurements discussed in section 1,
show that the cosmic background radiation is isotropic and blackbody
to high precision, just the properties expected if it is a relic of a hot
Big Bang phase in the history of the Universe.

3. SMALL ANGULAR SCALE FLUCTUATIONS IN THE CBR TEMPERATURE

In addition to possible large-scale anisotropy in the CBR, smaller
angular scale anisotropies or fluctuations may be present. To
understand their origin, we must ask where the photons of the CBR
we detect originate. While the CBR photons are produced early in the
hot Big Bang explosion of the Universe, we cannot see all the way
back to arbitrarily early epochs. Instead, the photons we detect here
on earth originate at some surface of last scattering, as indicated in
the cartoon below. A useful physical analogy is looking at a cumulus
cloud. Inside the cloud, optical photons are frequently scattered, so
the cloud is opaque and we have no information about its internal

Figure 4. The role of the surface of last scattering. We get no
information about the angular distribution of the radiation beyond
that surface (i.e., at larger redshifts).

structure. Once optical photons reach the edge of the cloud, they
may travel freely towards us through the transparent atmosphere.
Hence we see only the cloud's surface, that is the surface of last
scattering. In the case of the CBR, the primary scattering mechanism
is Thomson scattering from free electrons. Before the matter of the
hot Big Bang cooled to a temperature of ~ 3000 K at z ~ 1000, the
material contents of the Universe were ionized, and the plentiful free
electrons ensured that Thomson scattering was strong. Hence one
well defined surface of last scattering was present at a redshift of
~ 1000, when free electrons combined with protons to form neutral
hydrogen. For the moment, let us adopt that epoch as the time of
last scattering.

Now if the matter of the Universe was inhomogeneously
distributed at the epoch of last scattering, small temperature
fluctuations in the CBR will result. Unfortunately the connection
between the degree of density inhomogeneity $\Delta\rho/\rho$ and the
temperature fluctuations $\Delta T/T$ is both complex and model dependent.
The very model dependence, of course, makes the observational
results so interesting--we have in principle a means of checking our
theories of the formation of large-scale structure in the Universe by
making straightforward, if difficult, radio astronomical observations
from earth.

In this overview, I will focus on three angular scales. The first
is an angular scale of a few degrees or so, corresponding to the scale
of the causal horizon at z = 1000, which for the moment I take to be

the redshift of the surface of last scattering. The second is an angular scale of a few arcminutes. Detailed calculations (for instance by Bond and Efstathiou, 1984; and Vittorio and Silk, 1984) suggest that the maximum amplitude of $\Delta T/T$ fluctuations should be on roughly this angular scale if the surface of last scattering is at about 1000. These same calculations indicate that the amplitude of temperature fluctuations should fall sharply at angular scales below a few arcminutes. If fluctuations are in fact found on angular scales $\lesssim 1'$, they are then most likely due to reionization of the material contents of the Universe which shifts the surface of last scattering to substantially lower redshifts than 1000 (Hogan, 1980, 1982, 1984; Ostriker and Vishniac, 1986).

3.1.   Measurements on scales of a degree or more.

A recently reported search for fluctuations in the CBR at a wavelength of 3 cm (Mandolesi et al, 1986) has established an upper limit of $\sim 5 \times 10^{-4}$ on $\Delta T/T$ on scales $\gtrsim 2^O$. Earlier work on angular scales larger than a degree or so, the causal scale for last scattering at $z \simeq 1000$, are discussed briefly by me elsewhere (1987). Newer observational results, obtained at a wavelength of 3 cm, are described in a paper in this volume by Dr. Davies of Jodrell Bank. The results he reports are at an angular scale of 8°. All of these results show that the Universe is clearly quite homogeneous on these scales. That fact requires some explanation. In the case of the classical Big Bang picture, the only plausible explanation is that the homogeneity of the Universe was ensured by initial conditions, since no causal process could have produced homogeneity on scales larger than the light horizon at $z = 1000$, corresponding in scale to a few degrees. A period of inflationary expansion early in the history of the Universe, however, does provide a convincing physical explanation for the observed homogeneity on these scales, as first noted by Guth (1981, see also his 1986 review). Parenthetically, changing the redshift of last scattering to a smaller value does not change the thrust of these arguments; the angular scale corresponding to the causal horizon becomes larger, say 10°, but an explanation for the observed homogeneity of the Universe is still required.

3.2.   Upper limits on $\Delta T/T$ on scales of a few arcminutes.

Over the past decade, most observational and theoretical attention has been focused on anisotropies in the CBR on scales of a few arcminutes. The most interesting and sensitive upper limit yet published is the one set at a wavelength of 1.5 cm by Uson and Wilkinson (1984, 1985). They used the 140-ft. telescope of the National Radio Astronomy Observatory in Green Bank, West Virginia, which has a beam size of 1.5' and a beam switch angle of 4.5'. Their experimental technique permitted them to sample temperature differences between pairs of points separated by 4.5'. A dozen such pairs were sampled. Their statistical analysis of the data established an upper limit of $\Delta T/T \leqslant 2.5 \times 10^{-5}$ on this angular scale. Subsequent

analyses by others have suggested that the data can more reasonably
set upper limits of $\Delta T/T \lesssim 5 \times 10^{-5}$. In any case, these results set
extremely stringent limits on temperature fluctuations in the CBR, and
hence on density perturbations on the surface of last scattering.

A somewhat similar experimental program, at the same wavelength,
is being conducted by Dr. Readhead and his colleagues at the
California Institute of Technology. No results have yet been
published, although some preliminary results are being discussed
privately, and the first published results should soon be appearing.
The sensitivity is comparable to or perhaps better than that attained
by Uson and Wilkinson. One result that has now made the transition
from "rumor" to "lore" is a single sample difference
$\Delta T/T = (2.5 \pm 8) \times 10^{-6}$. This is the difference in temperature
between the north celestial pole and a concentric circle a few minutes
of arc away. Although it is a bit like asking Zen question, "What is
the sound of one hand clapping?", attempts have been made to use
this single difference to establish limits on the statistical properties
of temperature fluctuations in the CBR. The resulting upper limits, of
course, depend very much on one's assumption about the spectrum of
CBR fluctuations, but upper limits in the range of a few times 10^{-5}
are being quoted.

Perhaps the most useful thing I can say is that rigorous and
sensitive upper limits on $\Delta T/T$ are now available and may soon be
pushed even lower, and that these limits are on precisely the angular
scales where the maximum fluctuation amplitude in $\Delta T/T$ is expected in
models in which the surface of last scattering is at $z \simeq 1000$.

3.3. Search for fluctuations on scales $\lesssim 1'$.

Let us now drop the assumption that the surface of last scattering
corresponds to the epoch of recombination, at $z \simeq 1000$. If the matter
contents of the Universe are reionized at lower redshifts, free
electrons will again be plentiful, and Thomson scattering will result.
The surface of last scattering of the CBR can be shifted to redshifts
as low as about 15 (at lower redshifts, even complete re-ionization will
produce an optical depth in Thomson scattering well below unity). If
the surface of last scattering is shifted to lower redshifts, all
information about anisotropy introduced at earlier times is lost. In
particular, the arcminute scale fluctuations suggested by the
theoretical work of Bond and Efstathiou (1984) or Vittorio and Silk
(1984) will be erased. On the other hand, matter in the Universe
cannot be exactly homogeneously distributed at these more recent
epochs either, and hence new temperature fluctuations will be
imprinted on the CBR (see Hogan, 1980 and 1982; Ostriker and
Vishniac, 1986; for example). One of the intriguing predictions of
these authors is that fluctuations on angular scales less than an
arcminute may predominate, unlike the case discussed above.

To make observations of the CBR on scales below an arcminute
requires a new observational technique, the use not of a single radio
antenna, but an array of antennas. Signals received by an array of
radio telescopes are coherently combined to produce a two-dimensional

map of the sky with angular resolution corresponding to the size of
the array, not the diameter of a single antenna. The advantages and
disadvantages of this observing technique are discussed by Fomalont
et al (1984a), Knoke et al (1984) and Partridge (1987). Observations
made using an array of 27 telescopes, the Very Large Array of the
National Radio Astronomy Observatory in New Mexico, operating at
6 cm wavelength, have reached sensitivities in $\Delta T/T$ of 10^{-4} or better
on angular scales $< 1'$.

Briefly, the technique employed (Fomalont et al, 1984a; Knoke
et al, 1984) was to use the array first to make a map of the sky in a
region free of bright discrete sources. The remaining visible discrete
sources were then removed from the map, and the noise properties of
the remaining, nominally source-free, map were then examined.
Essentially, we compared the rms noise level at the center of such a
map to the noise level near its edges. At the center of the map,
where the response of the individual elements of the array is large,
the variance contributed by fluctuations in the CBR (or, of course, by
faint discrete sources) will be at a maximum. On the other hand, near
the edges of the map where the diffraction power patterns of the
individual elements of the array have fallen to zero, the noise will be
completely dominated by instrumental and atmospheric contributions.
Thus a subtraction provides an estimate of the excess variance
produced by the sky. This value, in turn, can be corrected for the
contribution due to sources too weak to detect individually. The
result is an estimate of, or upper limit on, fluctuations in the CBR.
Table 4 shows the results of such an analysis. The results are
presented as upper limits on $\Delta T/T$. Our own most recent work (Martin
and Partridge, 1986) <u>suggests</u> the presence of true sky fluctuations at
a level of $1-2 \times 10^{-4}$ on angular scales $\theta = 18''-60''$. This value has
been corrected for weak discrete sources by extrapolating direct

Reference	Angular Scale	Upper Limit on $\Delta T/T \times 10^{-4}$
Fomalont	18"	10
et al (1984a)	30"	8
	60"	5
Knoke et al	6"	32
(1984)	12"	17
	18"	12
Martin and	18"- 80"	~ 2
Partridge (1987)	36"-160"	~ 1.5

Table 4. Upper limits on small-scale fluctuations in the CBR. All
results are from the VLA at $\lambda = 6$ cm. As noted in the text, the
results of Martin and Partridge (1987) are tentative; Fomalont and his
colleagues also have new measurements (see Wall's paper here).

source counts at $\lambda \simeq 6$ cm and 21 cm (Fomalont et al, 1984b; Partridge et al, 1986; and Condon and Mitchell, 1984; Windhorst et al, 1985, respectively). On the other hand, I must caution you that Dr. Martin and I have not yet eliminated all possible sources of instrumental error. It happens that Dr. Fomalont and his colleagues, including Jasper Wall, have now made more extended observations of a different region of the sky and see roughly comparable results. Those results will be reported here by Dr. Wall, but I believe it is true to say that the analysis of their results is not complete either. Nevertheless, we have here a tantalizing hint that anisotropies in the CBR may have been detected. What is more interesting, if these results hold up, is that fluctuations in the CBR are apparently most visible at the "wrong" angular scale, a factor of 10 smaller than expected from models of temperature fluctuations induced by density perturbations at the epoch of recombination (Bond and Efstathiou, 1984; Vittorio and Silk, 1984).

3.4.   Summary.

Nearly two decades of effort have gone into searches for anisotropies in the CBR on angular scales of a few arcseconds to many degrees. As the papers by Davies and Wall in this volume suggest, we may be on the verge of detecting CBR fluctuations for the first time. Even if these observational results hold up, however, the main outcome of these two decades of effort has been to show how smooth and featureless the microwave background truly is. The upper limits on $\Delta T/T$ have already had a decisive influence on theories of the formation of large-scale structure in the Universe (see Silk's review here). I turn to some of the consequences of these observations next.

3.5.   Consequences of upper limits on $\Delta T/T$.

If the CBR is observed to be isotropic on angular scales larger than the causal horizon scale, as it appears to be, then either large-scale homogeneity has to be assumed as an initial condition, or some physical cause for it is required. As noted above, inflationary expansion of the Universe can produce just the kind of large-scale homogeneity we see. Thus the observations summarized under 3.1 above support the introduction of an inflationary phase into standard Big Bang cosmology.

The upper limits on arcminute scales can tell us much more. To interpret these upper limits, however, requires an assumption about the surface of last scattering. If we take it to correspond to the epoch of recombination, $z \simeq 1000$, then the absence of detectable CBR fluctuations allows us to set limits on the density inhomogeneity at that crucial epoch. The most straightforward theoretical possibility for the formation of galaxies and other large structures is that they form from adiabatic perturbations in pure baryonic matter. This possibility is effectively ruled out by the CBR results, especially the measurement of Uson and Wilkinson (1985). As a consequence, most aspects of this simplest picture have been modified in one way or

another to attempt to fit the observational constraints. First, the
notion that the inhomogeneities are adiabatic, that is that both the
matter and the radiation content of the Universe are perturbed, has
been dropped in favor of isothermal perturbations, in which the
radiation, to first order, is not perturbed. This change can reduce
the predicted amplitude of $\Delta T/T$ fluctuations by approximately an
order of magnitude (in baryonic matter models [Davis and Boynton,
1980], but see Efstathiou and Bond, 1986). Second, both to make
models of galaxy formation conform with the CBR upper limits and for
other more direct reasons, additional forms of matter have been
introduced into cosmology, known generically as "dark matter." These
include neutrinos with nonzero rest mass and various forms of cold
dark matter, possibilities touched on by other speakers at this
symposium. Since these nonbaryonic forms of matter couple less
strongly to the radiation field, it is possible to accommodate larger
density perturbations in them for a given upper limit on $\Delta T/T$.
Galaxy formation then occurs when the more tightly coupled baryons
decouple from radiation and fall into the gravitational potential wells
established by the neutrinos, cold dark matter, or what have you.
Nevertheless, as Figure 5 shows, even some of these models find
themselves in trouble with the observational upper limits. As a

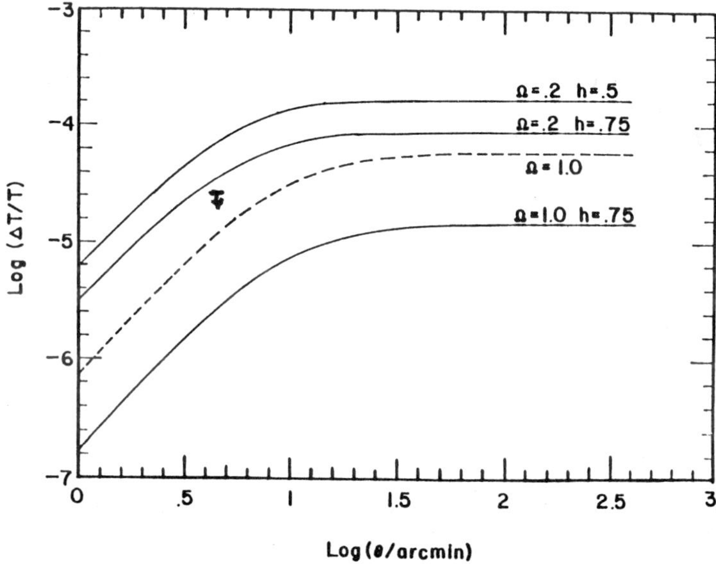

Figure 5. Adapted from Bond and Efstathiou (1984). The predicted
amplitude of CBR fluctuations is shown for various cosmological models
under the assumption that last scattering occurred at z ~ 1000. Solid
curves--models with ρ_0 dominated by cold dark matter. Dashed
curve--a model with $\rho_0 = \rho_c$ of hot dark matter, such as massive
neutrinos. The observational upper limit is that of Uson and
Wilkinson (1985).

consequence, yet another modification to the standard theory of
galaxy formation has been introduced--biased galaxy formation, in
which galaxies form only in regions of particularly high overdensity,
e.g., 2.5σ or 3σ deviations (Bardeen et al, 1986). Since the theoretical
predictions of the amplitude of $\Delta T/T$ fluctuations are calibrated by
using the present distribution of galaxies (see Bond and Efstathiou,
1984; or Vittorio and Silk, 1984), including bias in effect lowers the
predicted amplitude of $\Delta T/T$ by a like factor of ~ 3.

Despite their importance, I have moved rapidly over the
consequences of the upper limits on $\Delta T/T$ because they are discussed
much more fully by others (e.g., Bond and Efstathiou, 1987). In
addition, you will be hearing more about the constraints on theories
of galaxy formation from other speakers at this symposium.

Let me now move on to the exciting possibility that fluctuations
in the CBR have been detected on angular scales below 1'. If these
results are confirmed, they suggest that the material contents of the
Universe have been reionized at some redshift below z = 1000, shifting
the surface of last scattering to more recent epochs. Although the
results are still extremely tentative, it is interesting to note that the
amplitude of the fluctuations on scales 0.1-1' are roughly at the level
predicted by the preliminary calculations of Ostriker and Vishniac
(1986). These predicted amplitudes are derived from simple models of
the heating of the intergalactic plasma. At the moment, these
arguments and the models included in them are quite general, and
both the amplitude and the predicted spectrum of $\Delta T/T$ fluctuations
may eventually be modified by more detailed calculations.
Nevertheless, it is interesting that the predicted amplitudes in the
models which include reionization are roughly comparable to or
perhaps slightly larger than the amplitude of fluctuations expected
from density perturbations present at z = 1000. Of more interest to
the observers, perhaps, is the fact that the predicted amplitudes are
within our immediate experimental reach.

4.   OTHER OBSERVATIONS AND CONCLUDING REMARKS

Since this is intended to be a general overview of CBR observations, I
would like to mention two other sorts of measurements, at least briefly.

4.1.   Polarization of the CBR.

This topic will not be mentioned elsewhere in this volume, so let me
introduce it briefly. It has been recognized for some time
(Negroponte and Silk, 1980; Basko and Polnarev, 1980; see also Tolman,
1985) that large angular scale linear polarization can be introduced
into the CBR by anisotropic expansion at the epoch of recombination
(or, more generally, at any epoch of last scattering). Observations of,
or upper limits on, the linear polarization of the CBR thus provide
constraints on possible anisotropic cosmological models which
complement direct measurements of $\Delta T/T$. The best upper limits to
date are those established by Lubin et al (1983b). The limit on linear

polarization is $\leq$ 3 x 10^{-4}. That same paper reports weaker limits also on the circular polarization of the CBR, though these appear to be of less theoretical interest.

There are as yet no sensitive upper limits on small-scale linear polarization which is expected on angular scales slightly less than the ~ 10' angular scale where intensity fluctuations are largest according to many models of galaxy formation (Bond and Efstathiou, 1986). In addition to having slightly smaller angular scales, the polarization fluctuations are expected to have an amplitude roughly one tenth that of the intensity fluctuations--but they may still be measurable because some of the instrumental problems in making searches for intensity fluctuations in the CBR may be avoided.

4.2.   Searches for the Sunyaev-Zel'dovich effect

The Sunyaev-Zel'dovich effect (1972) is a small perturbation in the spectrum of the CBR produced by inverse-Compton scattering of photons by hot electrons. If the hot electrons are localized, say in the intergalactic plasma of a cluster of galaxies, a localized temperature perturbation, ΔT, will be produced. In the Rayleigh-Jeans region, this will have a negative sign (see, for instance, the paper by Birkinshaw here). Several groups have searched for this effect in clusters of galaxies (see recent papers by Birkinshaw et al, 1984; Meyer et al, 1983; Radford et al, 1986; Partridge et al, 1987; and references therein). The most reliable of the observational results are those obtained by Dr. Birkinshaw and his colleagues, and those will be discussed by him in this volume. Note also that Dr. Xie will discuss in this volume a novel method for searching for the same effect, this time using an infrared background rather than the CBR. While the small temperature perturbation produced by this effect in clusters of galaxies is not, strictly speaking, a cosmological effect like others described earlier, I mention it for three reasons. First, as pointed out independently by several authors including Gunn (1978), a careful measurement of the Sunyaev-Zel'dovich effect in a cluster of galaxies, combined with X-ray observations of the same cluster, can in principle provide a measurement of Hubble's constant independent of all the steps in the optical distance ladder. This hope has not yet been realized, in part because the microwave background observations are insufficiently accurate, and in part because the best studied clusters do not have good X-ray fluxes or temperature determina- tions. A second reason for mentioning the Sunyaev-Zel'dovich effect is that the small temperature fluctuations produced by this effect in distant clusters may set limits on our ability to measure cosmological temperature fluctuations from the surface of last scattering (Rephaeli, 1981). A rough calculation I have done suggests that the Sunyaev- Zel'dovich effect may be a more important source of background "fluctuations" than discrete sources, at least at angular scales below an arcminute, but that is a rough calculation which needs refinement.

A final reason for mentioning the Sunyaev-Zel'dovich observa- tions is that they provide a good illustrative example of both the importance and the difficulties of CBR observations. In the case of

the Sunyaev-Zel'dovich effect, we can use the observations to tell us
about the properties of intergalactic gas in clusters, and possibly also
to provide an independent measure of Hubble's constant. Likewise,
other observations of the CBR may be used as tools to solve a number
of cosmological problems ranging from nucleosynthesis to galaxy
formation. As is true in the case of the Sunyaev-Zel'dovich
observations, the utility of the CBR observations is largest when they
can be combined with other astronomical measurements. One salient
example is the power of the constraints set by observations of the
abundance of light elements (to be described here by
Professor Audouze). As an observer, I would like to add one other
common feature--searches for the Sunyaev-Zel'dovich effect, like most
of the other observations of the CBR summarized here, are difficult
experiments. Some of the difficulties will be summarized in papers
following this one by, among others, Halpern, Mandolesi, Davies,
Lukash, Birkinshaw and Xie.

   I hope this brief overview has given you some idea of the
present observational status of the CBR and of the many ways in
which the observations may be used in cosmology. I also hope we will
be able to meet the challenges over the next few years of improving
the measurements and of making further use of the ones we now have.

   This paper was prepared while I was a guest of King's College,
Cambridge and of the Institute of Astronomy, also in Cambridge. I
would like to thank both for their hospitality, and for the opportunity
to work both in concert with colleagues and undisturbed by myself.
My appearance at this symposium was made possible by a travel grant
from the American Astronomical Society, and by funds provided to
Haverford College by Bettye and Howard Marshall. The panda
cartoons are by Nicholas Bruel, Haverford '87. I would especially like
to thank the Local Organizing Committee and the Scientific Organizing
Committee for making this symposium so pleasant and fruitful for us
all, and to thank the members of the Chinese delegation to this
symposium for making me feel so much at home in their midst at the
Jia Li Hotel.

REFERENCES

Bardeen, J. M., Bond, J. R. Kaiser, N. and Szalay, A. 1986, Ap. J.,
    304, 15.
Barrow, J. D. Juszkiewicz, R., and Sonoda, D. H. 1983, Nature, 305,
    397.
Basko, M. M. and Polnarev, A. G. 1980, Monthly Notices Royal Astr.
    Soc., 191, 207.
Birkinshaw, M., Gull, S. F., and Hardebeck, H. 1984, Nature, 309, 34.
Bond, J. R. and Efstathiou, G. 1984, Ap. J. (Letters), 285, L45.
___________________ 1986, submitted to Monthly Notices Royal Astr. Soc.
___________________ 1987, in preparation for Annual Reviews of Astron. and
    Astrophys.
Bond, J. R., Carr, B. J., and Hogan, C. J., 1986, Ap. J., 306, 428.

Boughn, S. P., Cheng, E. S., Wilkinson, D. T. 1981, Ap. J. (Letters), 243, L113.

Collins, C. A., Joseph, R. D. and Robertson, N. A. 1986, Nature, 320, 506.

Condon, J. J., and Mitchell, K. J., 1984, A. J., 89, 610.

Crane, P., Hegyi, D. J., Mandolesi, N., and Danks, A. C. 1986, submitted to Ap. J.

Danese, L., and De Zotti, G. 1977, Riv. Nuovo Cimento, 7, 277.

Danese, L., and Partridge, R. B. 1987, in preparation.

Davis, M. 1987, I.A.U. Symposium 117 (Reidel Publish. Co., Dordrecht, Holland).

Davis, M., and Boynton, P. 1980, Ap. J., 237, 365.

deVaucouleurs, G., and Peters, W. L. 1984, Ap. J., 287, 1.

Dicke, R. H., Peebles, P. J. E., Roll, P. G., and Wilkinson, D. T., 1965, Ap. J., 142, 414.

Efstathiou, G., and Bond, J. R. 1986, Monthly Notices Royal Astr. Soc., 218, 103.

Fabbri, R., Guidi, I., Melchiorri, F., and Natale, V. 1980, Phys. Rev. Letters, 44, 1563.

Fall, S. M., and Jones, B. J. 1976, Nature, 262, 457.

Fixsen, D. J., Cheng, E. S., and Wilkinson, D. T. 1983, Phys. Rev. Letters, 50, 620.

Fomalont, E. B., Kellermann, K. I. and Wall, J. V. 1984a, Ap. J. (Letters), 277, L23.

Fomalont, E. B., Kellermann, K. I., Wall, J. V., and Weistrop, D. 1984b, Science, 225, 23.

Gunn, J. E. 1978, in Observational Cosmology, eds. A. Maeder, L. Martinet and G. Tammann (Geneva, Geneva Observatory).

Guth, A. H. 1981, Phys. Rev., D23, 347.

__________ 1986, in Inner Space/Outer Space, ed. E. W. Kolb et al (Univ. Chicago Press).

Hawking, S. W. 1969, Monthly Notices Royal Astr. Soc., 142, 129.

Hogan, C. J. 1980, Monthly Notices Royal Astr. Soc., 192, 891.

__________ 1982, Ap. J., 252, 418.

__________ 1984, Ap. J. (Letters), 284, L1.

Johnson, D. G., and Wilkinson, D. T. 1986, submitted to Ap. J. (Letters).

Knoke, J. E., Partridge, R. B., Ratner, M. I. and Shapiro, I. I. 1984, Ap. J., 284, 479.

Lubin, P. M., Epstein, G. L., and Smoot, G. F. 1983a, Phys. Rev. Letters, 50, 616.

Lubin, P. M., Melese, P., and Smoot, G. F. 1983b, Ap. J. (Letters), 273, L51.

Mandolesi, N., Calzolari, P. Cortiglioni, S., Delpino, F., Sironi, G., Inzani, P., De Amici, G., Solheim, J.-E., Berger, L., Partridge, R. B., Martenis, P. L., Sangree, C. H., and Harvey, R. C. 1986, Nature, 319, 751.

Martin, H. M., and Partridge, R. B. 1986, submitted to Ap. J. (Letters).

Meiksin, A., and Davis, M. 1986, A. J., 91, 191.

Meyer, S. S., Jeffries, A. D., and Weiss, R. 1983, Ap. J. (Letters), 271, L1.

Meyer, D. M., and Jura, M. 1984, Ap. J. (Letters), **276**, L1.
————————— 1985, in The Cosmic Microwave Background and
   Fundamental Physics, ed. F. Melchiorri (Editrice Compositori, Bologna).
Negroponte, J., and Silk, J. 1980, Phys. Rev. Lett., **44**, 1433.
Negroponte, J., Rowan-Robinson, M., and Silk, J. 1981, Ap. J., **248**, 38.
Novikov, I. D. 1968, Soviet Astron., **12**, 427 (Ast. Zh., **45**, 538).
Ostriker, J. P., and Vishniac, E. T. 1986, Ap. J. (Letters), **306**, L51.
Partridge, R. B. Cannon, J., Foster, R., Johnson, C., Rubinstein, E., and
   Rudolph, A., 1984, Phys. Rev. D, **29**, 2683.
Partridge, R. B. 1986, Highlights of Astronomy, ed. J.-P. Swings.
Partridge, R. B., Hilldrup, K. C., and Ratner, M. I. 1986, Ap. J., **308**, 46.
Partridge, R. B. 1987, submitted to Reports Prog. Physics.
Partridge, R. B., Perley, R. A., Mandolesi, N., and Delpino, F. 1987
   submitted to Ap. J.
Penzias, A. A. and Wilson, R. W. 1965, Ap. J., **142**, 419.
Peterson, J. B., Richards, P. L. and Timusk, T. 1985, Phys. Rev.
   Letters, **55**, 332.
Radford, S. J. E. Boynton, P. E., Ulich, B. L., Partridge, R. B.,
   Schommer, R. A., Stark, A. A., Wilson, R. W., and Murray, S. A., 1986,
   Ap. J., **300**, 159.
Rephaeli, Y. 1981, Ap. J., **245**, 351.
Richards, P. L. 1980, Physica Scripta, **21**, 610.
Rubin, V. C., Thonnard, N., Ford, W. K. and Roberts, M. S. 1976, A. J.,
   81, 719.
Schechter, P. L. 1977, A. J., **82**, 569.
Silk, J., and Stebbins, A. 1983, Ap. J., **269**, 1.
Smoot, G. F., De Amici, G., Friedman, S., Witebsky, C., Sironi, G.,
   Bonelli, G., Mandolesi, N., Cortiglioni, S., Morigi, G., Partridge,
   R. B., Danese, L., and De Zotti, G. 1985, Ap. J. (Letters), **291**, L23.
Strukov, I. A. and Skulachev, D. P. 1984, Sov. Astron. Letters, **10**, 1.
Thorne, K. S. 1967, Ap. J., **148**, 51.
Tolman, B. W. 1985, Ap. J., **290**, 1.
Uson, J. M., and Wilkinson, D. T. 1984, Ap. J. (Letters), 277, L1.
————————— 1985, Nature, **312**, 427.
Vittorio, N., and Silk, J., 1984, Ap. J. (Letters), **285**, L41.
Weiss, R. 1980, Annual Rev. Astron. and Astrophys., **18**, 489.
Wilkinson, D. T. 1980, Physica Scripta, **21**, 606.
————————— 1986, Science, **232**, 1517.
Windhorst, R. A., Miley, G. K., Owen, F. N., Kron, R. G. and Koo, D. C.
   1985, Ap. J., **289**, 494.
Woody, D. P. and Richards, P. L. 1981, Ap. J. **248**, 18.
Yahil, A., Walker, D., and Rowan-Robinson, M. 1986, Ap. J. (Letters),
   301, L1.

DISCUSSION

TURNER: Concerning the comparison of optical and microwave background
peculiar velocity determinations, the optical studies are subject to
a bias, related to the Malmquist bias, such that spatial inhomogeneities in the galaxy distribution appear as a peculiar velocity of
the observer. This effect discussed by Fall and Jones in 1976 is
likely to be particularly severe given the remarkable spatial
structures apparent in the CFA "Slice of the Universe" survey and
other studies.

SILK: If fluctuations of $\Delta T/T \sim 10^{-4}$ are present on sub arc minute
scales in the microwave background, then they are probably too large
to be due to fluctuations associated with reionization and galaxy
formation. A more plausible-interpretation would invoke a new
population of radio sources at high frequency. Could you comment on
this?

PARTRIDGE: Let me re-emphasize that the measured values I mentioned
are very tentative. Likewise, the predictions of Ostriker and
Vishniac (Ap. J. Letters, 306, 1986) are approximate. If we accept
both, however, they are consistent if we assume the fluctuations we
(may) see are dominated by the smallest angular scales we observe.
This is because Ostriker and Vishniac predict values of $\Delta T/T$ which
depend on large negative powers of Θ (e.g. $\propto \Theta^{-3}$).

CHEN: Will the interaction between host intergalactic clouds (IGC)
and the CBR through the inverse-compton scattering influence the
isotropic nature if the IGC distribution is not uniform?

PARTRIDGE: Yes, in addition to possibly perturbing the spectrum of
the radiation. This could take two forms - the usual Sunyaev-Zeldovich
effect of hot gas concentrated in clusters of galaxies (investigated
about five years ago by Rephaeli) and, for larger optical depths,
something like the effect discussed in more recent papers by Hogan
and by Ostriker and Vishniac. Either could add to purely "cosmic"
temperature fluctuations. The problem, of course, is that we
haven't reliably detected fluctuations of any sort.

HIGH SENSITIVITY OBSERVATIONS OF THE MICROWAVE BACKGROUND RADIATION

R.D. Davies
Nuffield Radio Astronomy Laboratories, Macclesfield, Cheshire

A.N. Lasenby, Mullard Radio Astronomy Observatory, Cambridge

INTRODUCTION

Angular fluctuations in the cosmic microwave background radiation are indicators of the seed structure at recombination ($z \simeq 1000$) which subsequently evolves into the present-day Universe ($z = 0$). Such measurements provide stringent tests of the cosmological theories of universal evolution and have implications for allowed values of the density and composition of matter in the Universe.

Observations on different angular scales give a direct indication of the amplitude of structure on different mass scales. The angular scale θ of adiabatic fluctuations of mass M is given by

$$\theta\,(\text{arcmin}) = 8.7\ \Omega^{2/3}\ h^{1/3}\ (M/10^{15}\ M_\odot)^{1/3}$$

where h is the ratio of the present value of Hubble's constant to an assumed value of 50 $\text{kms}^{-1}\text{Mpc}^{-1}$ and Ω is the ratio of the present density of the Universe to that required for closure. Galaxies with masses of 10^{11} to 10^{12} $M_\odot$ will give rise to angular structure on scales of 10-20 arcsec while the largest scale structures known with dimensions of 100 Mpc and masses of 10^{18} $M_\odot$ will have dimensions on the scale of degrees. The adiabatic scenario requires that the temperature fluctuations at recombination are related to the matter density fluctuations by

$$\Delta T/T = \frac{1}{3}\ \Delta\rho/\rho \simeq 3\text{x}10^{-4}\ \Omega^{-1}$$

This fluctuation amplitude applies to angular scales greater than a few tens of arcminutes; on progressively smaller scales the fluctuations are reduced by the "random walk" scattering in the finite depth of the recombination zone so that at 10 arcsec $\Delta T/T$ may be as low as 10^{-5}. We describe results for observations on the 10-20 arcsec scale and the 5°-15° scale.

A. Hewitt et al. (eds.), Observational Cosmology, 55–58.

OBSERVATIONS ON A SCALE OF 10-20 ARCSEC

The broadband (400 MHz) interferometer operating between the MkIA (76-m) and MkII (35-m x 25-m) telescopes on a 430-m baseline (Padin & Davis 1986) was used to make a deep survey for fluctuations in the CMB at 5 GHz. The region chosen for study was adjacent to the cluster A576 which had been extensively studied at high sensitivity at Jodrell Bank (Lasenby & Davies 1983) and at the VLA. The field centre was RA=07^{h}17^{m}55^s; Dec=55°51'30"; the only source expected in the field was detected at the correct level on the edge of the primary beam. At 5.0 GHz, the area synthesized is the MkIA beam 2.6 arcmin in diameter while the interferometer baseline provided a 20" resolution. There were accordingly some 50 independent resolution elements sampled in the survey region from which an estimate could be made of the fluctuation amplitude. This interferometer is a simple and clean system free from spurious responses and whose sensitivity is entirely limited by receiver noise. The amplitude of receiver noise was determined from the higher delay components of the interferometer which correspond to positions outside the main beam of the MkIA. Repeated 24-hr tracks were made on the field from which 7.7 x 10^5 secs of data were accumulated.

Any fluctuations in the sky (i.e. the CMB) observed with the interferometer will have an angular distribution attenuated by the primary beam. From a study of the radial autocovariance, the set noise and sky noise contributions can be separated. This leads to a limit on the power spectrum of background fluctuations of 3.74 x 10^{-8}K with 95% confidence at a wavenumber corresponding to an angular scale of 35". For fluctuations containing a realistic blend of wavenumbers our sensitivity maximizes for those with an equivalent Gaussian FWHM of 15". At 95% confidence level the upper limit to the amplitude of such fluctuations is 1.6 mK which corresponds to $\Delta T/_T$ < 5.8 x 10^{-4}.

We consider that this is a "clean" result from a system with a low level of spurious effects. Although this upper limit is still a factor 10 higher than may be expected from primordial fluctuations arising at recombination, it is of relevance to scenarios which envisage the reionization of low-redshift matter at z = 3-10.

OBSERVATIONS ON A SCALE OF 5° TO 15°

A high sensitivity experiment to measure CMB fluctuations at 10.4 GHz on a high site has been running over the last two years. The observing system consists of a dual channel receiver recording continuously the difference between a central 8° (FWHM) beam and two adjacent beam positions displaced by $\pm$8°. The cryogenic receivers have system noises between 80 and 100K and a bandwidth of 500 MHz. The observing site at Izana, Tenerife, lies at a height of 2300 metres and is above the inversion layer for most of the year. In one day the fixed beams sweep a 24-hr RA scan through the sky.

The programme has two aims, firstly to survey a large area of the sky, namely all Right Ascensions at Declinations between Dec= -15° and $+55^\circ$; this will allow a detection of any rare high amplitude components of the fluctuation spectrum. The second is a deep survey of limited areas at Declinations 0° and 40°. The observations are continuously calibrated from a switched noise source and have frequent primary calibrations via the Moon and the Sun.

We will discuss here the analysis of the deep survey made at Dec=40°. The equivalent of 10 days of observations were stacked and used to estimate the amplitude of the true sky auto-covariance function (ACF). The observing system has sensitivity to structure on angular scales between about 5° and 15°; it can be shown analytically that for our system the maximum sensitivity is on a scale of $10^\circ.6$ (FWHM). A maximum likelihood analysis was made of the data at Dec=40° between RA = 12^h and 17^h, a region clear of any obvious confusing effects from the galactic plane. The analysis showed that a non-zero variance is preferred over a zero one. The equivalent standard deviation on the $8^\circ.3$ scale of the observing beam is 0.10 mK corresponding to $\Delta T/_T = 3.7 \times 10^{-5}$. This is for a Gaussian-shaped true sky ACF with FWHM = $10^\circ.6$ corresponding to the scale of our maximum sensitivity and with height adjusted for what we would observe in our $8^\circ.3$ beam. The $\Delta T/_T$ of the unsmeared intrinsic ACF is 0.16 mK corresponding to $\Delta T/_T = 5.8 \times 10^{-5}$. The likelihood analysis showed that virtually all the probability is contained in the interval 0 to 0.20 mK in ΔT, leading to a firm upper limit on the sky fluctuations of $\Delta T/_T < 7.3 \times 10^{-5}$ as seen in the $8^\circ.3$ beam.

The fluctuations apparently detected in this experiment could be intrinsic to the CMB or due to emission from irregular emission patches at high galactic latitude or a combination of both. We are pursuing observations to separate these contributions to the observed fluctuations.

The results of this experiment on a scale of $\sim 8^\circ$ have direct consequences for theories of the evolution of the Universe for two reasons. Firstly, the intrinsic fluctuations on this scale cannot be smoothed out by subsequent reionization. Our result rules out most baryon-dominated models (with adiabatic fluctuations) in which $\Omega < 0.1$ and places strong constraints on pictures involving weakly interacting dark matter.

We acknowledge the contributions to the above work of our collaborators S. Padin, R.J. Davis, D. Waymont, R.Watson, E.J. Daintree, J. Hopkins, J.E. Beckmann, J. Sanchez-Almeida and R. Rebolo.

REFERENCES

Lasenby, A.N. & Davies, R.D.: 1983, Mon.Not.R.astr.Soc., 203, 1137.
Padin, S. & Davis, R.J.: 1986. Radio Science, 21, 437.

DISCUSSION

SILK: When the galactic emission is modelled and subtracted out, as
you presumably can do with an all sky map, what is the residual signal,
if any, that you find for $\delta T/T$?

DAVIES: Our observations can be corrected for the (lumpy) emission
from the Milky Way using published 408 and 1400 MHz surveys by
adopting a value for the spectral index. However the spectral index
of this emission around 10 GHz is not well-known (somewhere between
2.7 and 3.0) and the wavelength baseline is large, so there is a
major uncertainty in making the correction. This is why we are
making high sensitivity observations of the galactic emission at 5.0
GHz. If for example we assume a spectral index of 2.8 the galactic
contribution to $\delta T/T$ for the $\delta = 40^{\circ}$ scan is 1.8×10^{-5}, and then
the intrinsic CMB fluctuations would be $\delta T/T \simeq 2 \times 10^{-5}$.

MEASUREMENT OF THE MICROWAVE BACKGROUND TEMPERATURE FROM CN ABSORPTION

N. Mandolesi
Istituto TESRE-CNR
Via de' Castagnoli, 1
40126 Bologna, Italy

P. Crane
European Southern Observatory
Karl Schwarzschildstrasse, 2
D-8045 Garching, Western Germany

D.J. Hegyi
Department of Physics
University of Michigan
Ann Arbor, Michigan USA

ABSTRACT. Observations of the interstellar lines of CN near 3875Å toward ζ Oph provide precise measurements of the Microwave Background Radiation (MBR) temperature and determine the intrinsic linewidth necessary for the saturation correction. We report $T_{MBR}=2.74\pm0.05$ K at $\lambda=2.64$ mm and $T_{MBR}=2.75$ (+0.24, -0.29) K at $\lambda=1.32$ mm.

INTRODUCTION

The CN absorption line technique for determining the temperature of the MBR can provide results with precision better than 2% at $\lambda=2.64$ mm. This can be compared to the best microwave radiometer measurements so far reported (Smoot et al., 1985; Mandolesi et al., 1986) which have an accuracy of 5% in the millimeter region.
The present results were obtained in the summer of 1984 and 1985 at the ESO 1.5 m CAT coupled to a Coude' spectrograph. The data were recorded on a Reticon detector with a resolution of up to 156,000.
The interstellar cloud in the direction of ζ Oph is particularly well suited to these observations because it is a single cloud of relatively large column density and has been well studied (Morton, 1975) in the past; thus many potential sources of error are known to be unimportant.

DATA AND ANALYSIS

Of the 22 nigths assigned to this project, the weather allowed only seven nights data of sufficient quality for our purposes. These data were analysed in the standard way to remove instrumental effects and then wavelength calibrated. Non stochastic events limited our ability to reduce noise by adding data from several nights.
The 1984 observations were obtained with a resolution of 156,000 allowing us to determine the intrinsic width of the CN lines. The observed width of the R(0) line was measured to be 31.8±0.3 mÅ (FWHM)

A. Hewitt et al. (eds.), Observational Cosmology, 59–61.

Table I
Summary of CN Results at λ=2.64 mm.

Lines	Equivalent Widths (mÅ)	T (K)	ΔT_{sat} (K)	ΔT_{elec} (K)	T_{MBR} (K)
$\dfrac{R(1)}{R(0)}$	$\dfrac{2.420\ (51)}{7.646\ (91)}$	2.956 (31)	−0.159 (3)	−0.060 (40)	2.737 (50)
$\dfrac{P(1)}{R(0)}$	$\dfrac{1.254\ (67)}{7.646\ (91)}$	3.015 (71)	−0.197 (3)	−0.060 (40)	2.758 (82)

Table II
Summary of CN Results at λ=1.32 mm

Lines	Equivalent Widths (mÅ)	T (K)	ΔT_{sat} (K)	T_{MBR} (K)
$\dfrac{R(2)}{R(1)}$	$\dfrac{0.072\ (26)}{2.420\ (51)}$	2.78 (+0.26; −0.30)	−0.03	2.75 (+0.25; −0.29)
$\dfrac{R(2)}{P(1)}$	$\dfrac{0.072\ (26)}{1.254\ (67)}$	2.75 (+0.27; −0.31)	−0.01	2.74 (+0.27; −0.30)

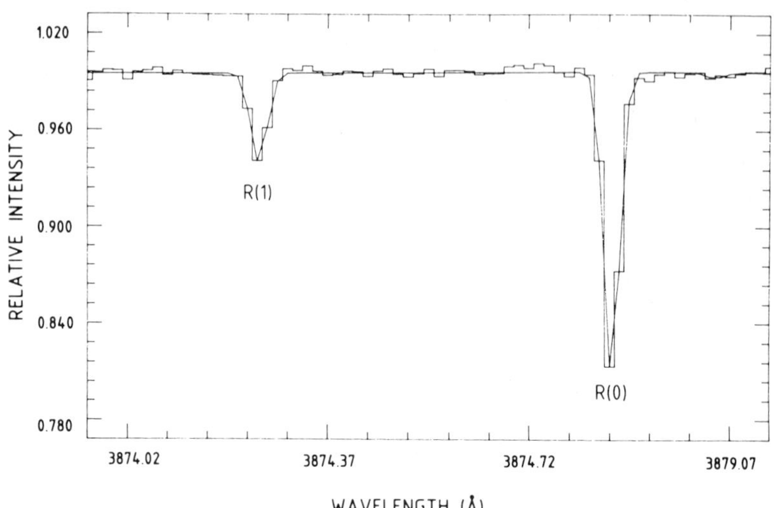

Figure 1: Plot of CN R(0) and R(1) lines. Solid line is a fit to the data. Wavelengths are observed not rest wavelengths.

(Fig. 1), which is assumed to be the convolution of an intrinsic line profile with a FWHM of 19.0 ± 0.5 mÅ and the instrumental profile of 24.8 ± 0.4 mÅ.

The CN linewidth of 19.0 mÅ implies a velocity spread in the cloud of 0.88 ± 0.02 km/s; this value should be compared with a value of 1.3 km/s reported by Hegyi et al. (1972). For a number of reasons (Crane et al., 1986) we prefer our measured value. We note however that the choice of 1.3 km/s, in making saturation corrections, would increase our final T_{MBR} at 2.64 mm by only 0.05 K.

Our final equivalent widths are given in Table I and II together with errors. Also shown in the Tables are the CN excitation temperatures T at $\lambda=2.64$ and 1.32 mm before saturation corrections.

The Microwave Background Radiation temperature is derived from T (Table I and II) by making the saturation correction ΔT_{sat} and a correction for local electron excitation ΔT_{elec} (Meyer and Jura, 1985). At $\lambda=1.32$ mm the correction for electron excitation is not necessary.

CONCLUSIONS

At $\lambda=2.64$ mm the CN technique gives a result already accurate to 2%. However this error is dominated by the uncertainty in the local excitation correction (Table I); hence further optical observations will not improve this result substantially.

At $\lambda=1.32$ mm the precision in T_{MBR} is dominated by the uncertainty in the R(2) equivalent width. In this case more precise optical observations could improve the situation.

Our analysis has, so far, not considered one possible source of systematic error in the final T_{MBR} values. We have assumed that the width of the R(0) line is dominated by the velocity spread in a single cloud and not by the superposition of several clouds each with a low internal velocity spread. If this were the case then our saturation correction would be too small because the lines would be intrinsically narrower. We note that any effect of this sort would reduce the final T_{MBR} values.

However we have some confidence that our hypothesis is correct since our saturation corrected temperatures give very nearly the same values by using R(1) and P(1) lines.

References
Crane, P., Hegyi, D.J., Mandolesi, N., and Danks, A.C., 1986, Ap. J., 309.
Hegyi, D.J., Traub, W., and Carleton, N.P., 1972, Phys Rev. Letters, 28, 1541.
Mandolesi, N., Calzolari, P., Cortiglioni, S., Morigi, G., Danese, L., and De Zotti, G., 1986, Ap. J., 310.
Meyer, D.M., and Jura, M., 1985, Ap. J., 297, 119.
Morton, D.C., 1975, Ap. J., 197, 85.
Smoot, G.F., De Amici, G., Friedman, S.D., Witebsky, C., Sironi, G. Bonelli, G., Mandolesi, N., Cortiglioni, S., Morigi, G. Partridge, R.B., Danese, L., and De Zotti, G., 1985, ap. J. (Letters), 291, L23.

Measurement of the Dipole Moment of the Cosmic Background Radiation at mm and Sub-mm Wavelengths

Mark Halpern

Department of Physics, UBC, Vancouver, Canada

Abstract

Sensitive measurements of a dipole anisotropy of the Cosmic Background Radiation at wavelengths of 1.7 and 0.8 mm are presented. Under the assumption that the dipole moment is caused by a doppler shift, these measurements are of sufficient precision to constrain spectral distortions of the CBR at short wavelengths. Measurements of the brightness and shape of diffuse galactic emission are also presented.

Introduction

This report contains some of the results of a balloon borne observing program undertaken at MIT to measure large angular scale anisotropy of the Cosmic Background Radiation.[1] The program started in 1974 and has included six flights of differential radiometers which operate at mm and sub-mm wavelengths. They are sensitive to the bulk of the energy in the CBR, well into the Wein tail. At these wavelengths atmospheric emission seen from the surface of the earth is brighter than the CBR and sensitive measurements, even differential measurements, are not possible from the ground. Therefore, we have used scientific balloons to carry our radiometers to an altitude of roughly 40km., above 99% of the atmosphere.

The Experiment

The sophistication of the radiometers has improved over the years as allowed by improvements in detector technology and as demanded by our deepening understanding of emission by competing sources, primarily diffuse galactic emission. The early radiometers were all two channel instruments. One channel extended from $\lambda = 1$cm to $\lambda = 1$mm, embracing the peak of the CBR. The second channel was sensitive to wavelengths as short as 330 μm and was included to monitor atmospheric emission. However, thermal emission by dust lying in the galactic plane was clearly evident in the short wavelength channel and we understood that better spectral resolution and galactic modelling were both needed in order to interpret our results. The final version of the radiometer had four channels centered at $\lambda = 1.7$mm, 830μm, 380μm and 290μm respectively.

All of the radiometers involved in this program measure the difference in brightness of two regions in the sky 45° from the vertical and 180° apart in azimuth. The radiometers are set into constant rotation about their vertical axis after launch, thus surveying a circle on the sky. This circle "drift scans" as the earth turns and as the package moves across the earth. The field of view varied slightly from flight to flight but was always quite large, approximately 17° FWHM. The data stream, consisting of detector signals, the output of an instrumented bubble level and three axis magnetometer to determine package orientation, and several temperatures and other housekeeping data, were tape recorded on board as well as being telemetered to the ground station. All of the flights were launched from the NSBF in Palestine, Texas, so our sky coverage is limited to the region -20° $\leq \delta \leq$ 85°.

63

A. Hewitt et al. (eds.), Observational Cosmology, 63–68.

Because of our differential geometry, we do not measure brightness as a function of position on the sky directly. To make physical sense of our data we have fit the measured brightness differences to the model of sky brightness given in Table 1. The dipolar (T_n) and quadrupolar (Q_n) terms have become standard in expressing measurements of anisotropy of the CBR. The galactic model, G(b), is appropriate if the galaxy looks like a translucent disk with little dependence of brightness on longitude.

TABLE (1)

Model Components:

α is right ascension, δ is declination and b is galactic lattitude.

$\mathbf{T}_x\cos\alpha\cos\delta \quad \mathbf{T}_y\sin\alpha\cos\delta \quad \mathbf{T}_z\sin\delta \quad \mathbf{Q}_1\left(\tfrac{3}{2}\sin^2\delta - \tfrac{1}{2}\right)$

$\mathbf{Q}_2\sin 2\delta\cos\alpha \quad \mathbf{Q}_3\sin 2\delta\sin\alpha \quad \mathbf{Q}_4\cos^2\delta\cos 2\alpha \quad \mathbf{Q}_5\cos^2\delta\sin 2\alpha$

$\mathbf{G}(b)\!:\dfrac{1}{\sin(b)}$ convolved with our beam profile, for $b \geq 1°$

Galactic Emission

The galactic polar antenna temperature inferred from our four channel data is plotted in Figure 1. Also shown are the estimates obtained by others at both higher and lower frequencies. There are two features of this graph which have a bearing on continued measurements of the CBR. First, there is no frequency at which one should expect to find a spot in the sky from which galactic emission is less than several times 10^{-5} of the CBR. Second, at the most promising wavelengths from the point of view of low galactic signal, $\lambda \approx$ several mm, measured galactic brightness lies above reasonable extrapolations from higher and lower frequencies, making proper subtraction difficult. We have made maps of sky brightness which seem to show that this excess emission does not have the same spatial distribution as the dust seen at shorter wavelengths, compounding the problem.

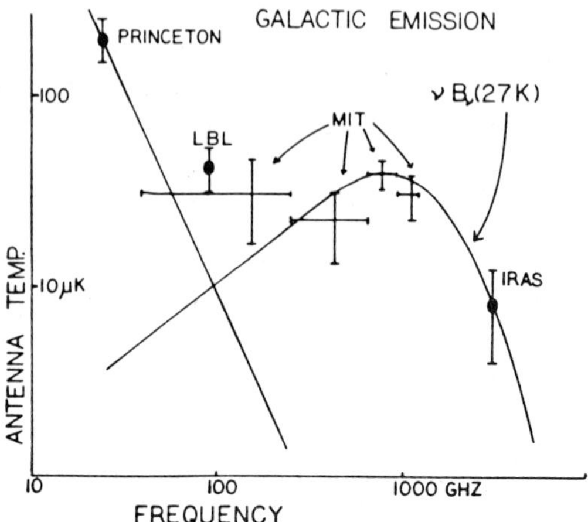

Figure 1. Galactic Polar Antenna Temperature, G(b). The thermal curve at high frequencies has been constrained to pass through our data at 380 μm and the IRAS data at 100 μm reported by Hauser *et al.* The straight line passing through the Princeton data at 25 GHz falls as $1/\nu^2$. Reasonable emission processes fall more rapidly than this with increasing frequency.

The Dipole Moment

The best fit dipole moment for each of the two low frequency channels is listed in Table 2. The higher frequency fit uses data from the only flight of the four channel radiometer whereas the lower frequency fit includes data from all but one of the six balloon flights. In the flight excluded the sky coverage is such that galactic emission is very strong, rendering the flight useless for determining properties of the CBR.

TABLE (2)
Dipole Moment in $m°K$

	1-8cm.$^{-1}$			5-18cm.$^{-1}$	
	ΔT_{RJ}	ΔT_{CBR}		ΔT_{RJ}	ΔT_{CBR}
T_x	-1.41±.19	-3.13±.41		-.25±.11	-3.14±1.43
T_y	-.03±.09	-.o6±.20		.15±.09	1.96±1.13
T_z	-.59±.13	-1.32±.29		-.22±.11	-2.85±1.33
$G(b)$	.053±.013			.022±.009	
	$\chi^2_\nu = .9571$ (4167/4354)			$\chi^2_\nu = .9809$ (1155/1178)	

The stated errors include a 10% uncertainty in calibration.

The data are listed as antenna temperatures and also as equivalent thermal perturbations of the CBR assuming a temperature of $2.74°K$. Under the assumption that the dipole moment is due to a doppler shift (or in fact any other mechanism which gives rise to a *thermal* perturbation), the brightness of the dipole moment is given by

$$\Delta I(\nu) = \frac{dB_\nu(T)}{dT}\Delta T_{dipole} = \frac{1.72 \times 10^{-12}\,\nu^4\,e^{hc\nu/kT}}{T^2\left(e^{hc\nu/kT}-1\right)^2}\Delta T_{dipole}\frac{watts}{cm.^2 str.cm^{-1}} \qquad (1)$$

At short wavelenghts this expression depends very strongly upon the temperature of the CBR. A proper comparison of our measurements to low frequency results[2,3,4] requires using Equation 1 to determine ΔT_{dipole} and T_{CBR} simultaneously from the data. The result is:

$$\Delta T_{dipole} = 3.35 \pm .12\,mK \qquad T_{CBR} = 2.87 \pm .28°K \qquad 2\,to\,8\,cm^{-1} \qquad (2)$$

$$\Delta T_{dipole} = 3.34 \pm .11\,mK \qquad T_{CBR} = 2.97 \pm .24°K \qquad 5\,to\,18\,cm^{-1} \qquad (3)$$

The inferred vaules of T_{CBR} are plotted in Figure 2, along with some recent direct measurements of the brightness of the CBR[5,6,7] and the indirect measurements obtained from the temperature of CN clouds.[8,9]

Conclusions

Notice that the resulting values of T_{CBR} are reasonable. This means that the dipole moment does in fact have the spectrum expected from the doppler shift of a $2.7°K$ blackbody. We are moving at $v = 1.2 \times 10^{-3} c$ with respect to a frame in which the CBR would appear to be isotropic. This velocity is 1000 times too large to be the remnant of a primordial peculiar velocity. Also notice that these results are competitive in accuracy with the direct

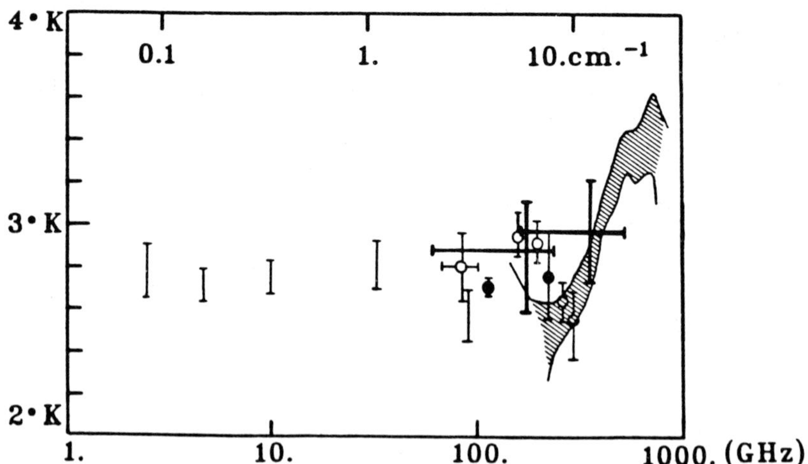

Figure 2.   Inferred Temperature of the CBR as a function of frequency. Our data are plotted as bold crosses where the vertical bar shows the meaured value $\pm\sigma$ and the horizontal bar shows the spectral response. The vertical bars, open circles, shaded region and solid circles are direct and indirect measurements of the CBR spectrum taken from references 5,6,7 and 8 and 9 respectively.

measurements in this difficult spectral region even though we are only sensitive to 10^{-3} of the power in the CBR. This is because spatial chopping is easier than absolute radiometry and also because of the sensitivity of the dipole moment to T_{CBR} at short wavelengths. In fact, in the higher frequency channel a 30% measurement of a dipole moment has led to an 8% constraint on T_{CBR}, a pleasing shrinkage of uncertainty.

More measurements are needed to understand the spectrum of the CBR at high frequencies, especially since it is possible to imagine mechanisms which might distort the spectrum in this region. A new radiometer has been built at MIT using improved detectors which in turn allow higher spectral resolution. At the same time, an experiment is underway to measure the short wavelength spectrum directly. The Physics Department at the University of British Columbia has built a liquid helium cooled differential fourier transform spectrometer which will be carried aloft on a sounding rocket this spring.[10] If successful it will attain an accuracy of $\delta T/T = 1\%$ in the wavelength range from $\lambda = 3$ mm to 300 μm.

References.

1 M. Halpern, R. Benford, S. Meyer, D. Muehlner, and R. Weiss In Peparation.
2 D. Fixsen, E. Cheng, and D. Wilkinson Phys. Rev. Lett. **50**,620, 1983
3 P. Lubin, T. Villela, G. Epstein and G. Smoot Ap. J. **298**:L1, 1985
4 R. Weiss Ann. Rev. Astron. and Astrophys. **18**,489, 1980
5 G. Smoot *et alia* Soc. Ital. di Fisica Conf. Proc. **1**,27, 1985
6 J. Peterson, P. Richards and T. Timusk Phys. Rev. Lett. **55**,332, 1985
7 H. Gush Phys. Rev. Lett. **47**,745, 1981
8 D. Meyer and M. Jura Ap. J. **297**,119, 1985
9 P. Crane, D. Hegyi, N. Mandolesi and A. Danks Ap. J. 15 October 1986.
10 H. Gush Proc. of 1983 Space Helium Dewar Conference. Hendricks, Karr eds.p.99, 1984

BIRKINSHAW: What Quadrupole moment limits can you place from your measurements?

HALPERN: We can not place any interesting quadrupole moment limit. We have been unable to model diffuse galactic emission well enough to remove it from our signal. At lower frequencies as well as in our own data at high frequencies where the CBR no longer has any power, the necessary modelling is comparatively straightforward. This is what first convinced us that diffuse galactic emission has different shapes at different frequencies. When we include the quadrupolar terms in our model along with the dipole and the galaxy we get a significant non-zero result. However, this result looks like the galactic plane and therefore cannot be attributed to the CBR. Please see the figure.

As sensitivity increases there will certainly be reports of lumps in the sky. To attribute them to structure of the CBR they should be seen at several frequencies. One should ask if the spectrum is reasonable and if the shape is reasonable. Does it look like any competing sources? The dipole moment clearly passes these tests. The quadrupole moment clearly fails.

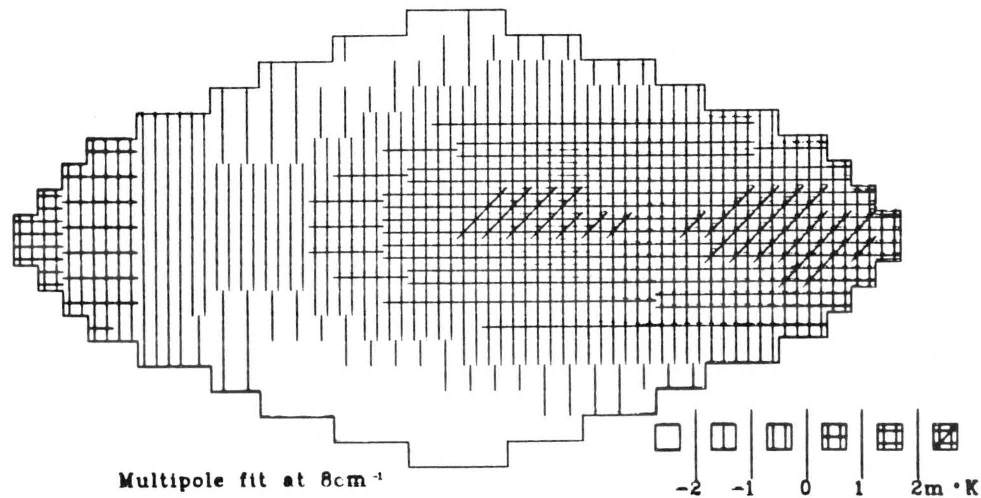

The brightness assigned to the Dipole (T_n) and Quadrupole (Q_n) terms when all of our data is fit to the model in Table 1. (Q_1 was not included in the fit.) The entire night sky is shown in a $cos(b)$ projection of galactic coordinates. The galactic center is at the center of the drawing. The galactic plane runs horizontally through the middle and the north and south poles are at the top and bottom respectively.

The galactic plane is evident. The plane does not appear to be uniformly bright because the effect of the dipole moment of the CBR has not been removed from the data.

Because of the strong galactic residue which is assigned to the Q_n terms in preference to the galactic model, $G(b)$, our search for a quadrupole moment of the CBR has not been successful.

ULMER: Do your measurements give the same direction as previous measurements of a dipole moment?

HALPERN: Yes, roughly. Our results are not as precise as the others. The directions of each of the dipole moments which we have measured are listed below along with those of Fixsen et al. and Lubin et al.

Author	Frequency	Right Ascension	Declination
Fixsen et al.	25 GHz.	$11.08 \pm .04$ Hours	$-8.1^{\circ} \pm .62$
Lubin et al.	90 GHz.	$11.2 \pm .1$ Hours	$-6.^{\circ} \pm 1.5$
MIT CH.1	180 GHz.	$12.1 \pm .24$ Hours	$-23^{\circ}\ ^{+5}_{-4}$
MIT CH.2	360 GHz.	$9.9\ ^{+1.7}_{-1.1}$ Hours	$-38^{\circ}\ ^{+13}_{-21}$

ROCKET EXPERIMENT TO SEARCH FOR THE NEAR-INFRARED EXTRAGALACTIC
BACKGROUND LIGHT

T. Matsumoto, M. Akiba and H. Murakami
Department of Astrophysics, Nagoya University
Chikusa-ku, Nagoya, 464
Japan

ABSTRACT. A rocket experiment was carried out to search for the extragalactic background light at 1-5 μm. After subtracting the foreground
radiation, there still remains an appreciable amount of isotropic
diffuse radiation with a complex spectral feature which is possibly
attributed to extragalactic origin.

1. OBSERVATION

The photometer consisted of 2 parts both of which had $4°$ beam. One was
a wide band photometric channel, covering the following standard filter
bands : J(1.27 μm), K(2.16 μm), L(3.8 μm) and M(5.0 μm), each of which
had a lens of 14 mm dia. Another was a narrow band photometric channel
consisting of two 26 mm dia. lenses and a filter wheel which rotated
intermittently and covered 1-5 μm region with 10% spectral resolution.
A Ge detector for J band and InSb detectors for others were used. The
zero signal level was confirmed by closing the cold shutter every 35
seconds. The whole system was cooled below 60 K by solid nitrogen for
better performance. The instrument was installed at the top of the
rocket body with the optical axis parallel to the rocket axis.
     The sounding rocket, K-9M-77, was launched on 14 Jan. 1984 at 04:30
JST (19:30 UT, 13 Jan.) from the Kagoshima Space Center of Institute of
Space and Astronautical Science. At 282 sec after launch, the rocket
reached the apogee at an altitude of 316 km. The lid of the cryostat
was opened at 80 sec after launch, and the survey started according to
the precession of the rocket.

2. DATA ANALYSIS

During the flight, bright stars such as γLeo were observed, whose fluxes
were consistent with those expected from the pre-flight calibration
with 10% accuracy.
     The contribution of the atmospheric emission and the contamination
of the environmental emission due to the rocket engine fuel exhaust etc.

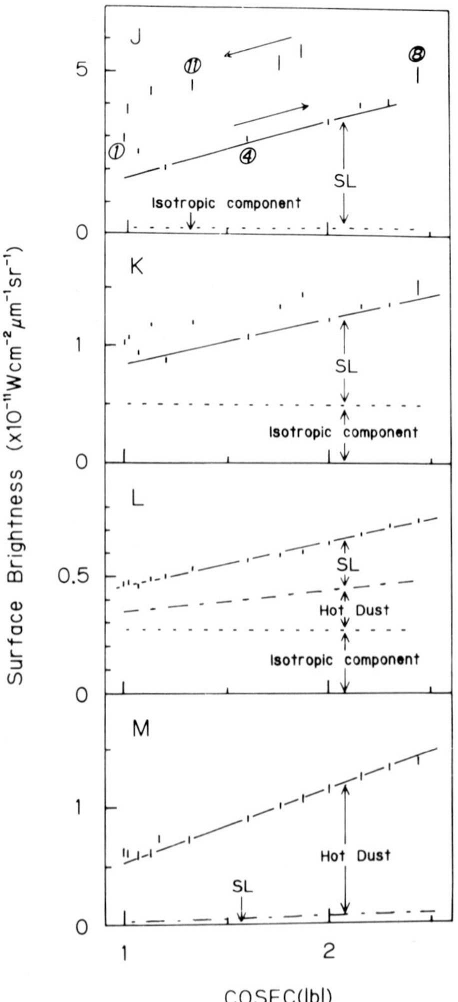

was estimated from the altitude and temporal dependence of signals, and it was found to be negligible above 300 km altitude even for M band.

The earthshine was also observed at the longer wavelength bands when the optical axis approached to the earth limb. But the observed earth shine is consistent with off-axis response measured in the lab and can be neglected at the large elevation angle.

Stars brighter than 3 mag in K band were identified by the wide band channels and confirmed by reference to the IRC catalogue. The surface brightnesses for 13 regions where no such a bright star was observed are shown as functions of cosec (b) in Fig.1.

The J and K band data show a loop feature such that the sky at a higher ecliptic latitude has a lower surface brightness. This is attributed to the spatial distribution of the zodiacal light (ZL). We tried to look for the best fit combination for J and K which separates the observed brightness to three components, ZL, star light (SL) and isotropic radiation. Here, we assumed that color of ZL, (J-K), is constant and SL should be consistent with the model Galaxy[1,2,3]. The result indicates an upper limit for J but considerable amount of isotropic radiation for K.

A little different but a simple feature is seen at L and M. SL extrapolated from J and K can't explain observed galactic component and the new galactic component appeared (designated as "hot dust"). Again the isotropic component is clearly seen for L, but only upper limit was

Fig.1. The observed surface brightnesses for the wide band channels are plotted versus cosec (b). The solid lines show the result of fittings. Breakdown of the observed surface brightness to the star light (SL), the isotropic radiation, and the new galactic component (Hot dust) is also shown.

obtained for M. The break down to individual components is also indicated in Fig.1.

3. DISCUSSION

The spectrum observed at the galactic pole region is shown in Fig.2, where the systematic errors are included. The data of the narrow band channels are consistent with those of the wide band channels, and the component found at M can be represented by 310 ± 70 K blackbody. The isotropic component is also shown in Fig.2. The upper limit of EBL at the visible region[4] is also presented. In Fig.2 calculated EBL for two extreme cases is also shown[5]. Model 1 assumes no evolution for galaxies, while model 4 assumes that all He were synthesized in stars during early era of galaxy formation. The observed level is somewhat lower than model 4, but still considerably higher than model 1. Provided that the observed isotropic component is really extra-galactic origin, some activities at the early universe is required. Comparing our data with the recent observation of the smoothness of the sky at K[6], the fluctuation of EBL is so small that the observed isotropic radiation is hardly explained by the integrated light of the primeval galaxies. It may be worthy to mention that Carr et al.[7] predicted a similar line feature at the redshifted wavelengths for Lymann α due to the pregalactic pop. III objects.

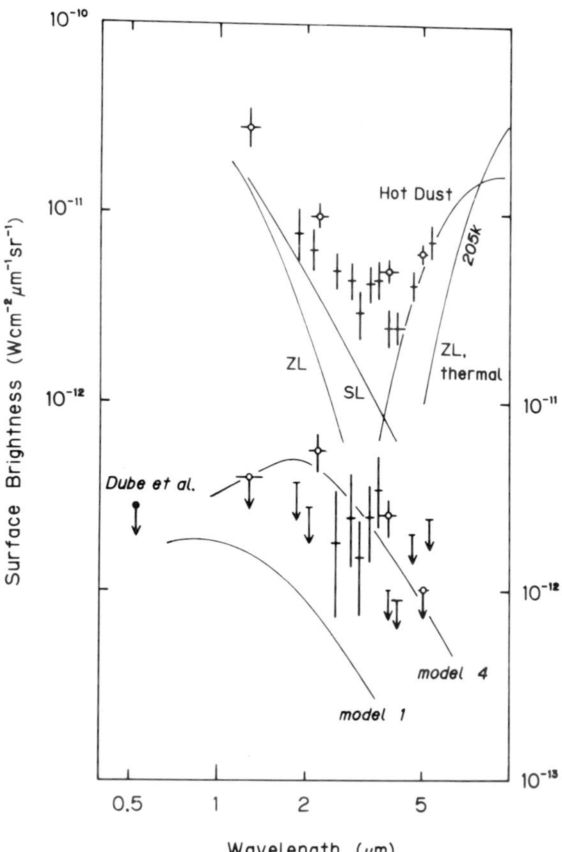

Fig.2. The spectrum of the observed surface brightness at the galactic pole region. Open circles and crosses are data of the wide and narrow band channels, respectively. The lower part shows the spectrum of the isotropic component. The unit of the surface brightness is indicated at the right side.

REFERENCES

1. Bahcall, J.N. and Soneira, R.S.: 'Comparisons of a Standard Galaxy Model with Stellar Observations in Five Fields', 1984, Ap.J. Suppl. 55, 67.

2. Mamon, G.A. and Soneira, R.M.: 'Stellar Luminosity Functions in the R, I, J, and K Bands Obtained by Transformation from the Visual Band' 1982, Ap. J. 255, 181-190.

3. Ishida, K. and Mikami, T.: 'Surface Brightness and Colors of the Galactic Disk' 1982, Publ. Astron. Soc. Japan 34, 89.

4. Partridge, R.B., and Peebles, P.J.E.: 'Are Young Galaxies Visible? II. The Integrated Background' 1967, Ap. J. 148, 377.

5. Dube, R.R. et al.: 'Extragalactic Background Light at 5100Å' 1977, Ap. J. Lett. 215, L51.

6. Boughn, S.P. et al.: 'The Smoothness of the 2.2 Micron Background: Constraints on Models of Primeval Galaxies' 1986, Ap. J. 301, 17.

7. Carr, B.J. et al.: 'Cosmological consequences of Population III Stars' 1984, Ap. J. 277, 445.

LARGE DISTANCES, HIDDEN MASS AND FLUCTUATIONS OF THE RELIC RADIATION

V. N. Lukash and I. D. Novikov
Space Research Institute
Academy of Sciences of the U.S.S.R.
Profsoyuznaja 84/32
Moscow 117810, U.S.S.R.

ABSTRACT. The problem of determining the fundamental constants of the Universe – H_o, q_o, Ω, Λ, hidden matter, and the primordial perturbation parameters is considered. The likelihood of finding new evidence for large-scale $\Delta T/T$-fluctuations is discussed.

1. INTRODUCTION

In the past observational cosmology was a science concerned with the measurement of two numbers – the Hubble constant and the deceleration parameter. Now we can investigate more parameters, because of successes in investigations of the large-scale structure of the Universe. There are only a few observations in cosmology which crucially influenced the development of the science. They are first and foremost:

- The discovery of the recession of the galaxies by Hubble;
- The discovery of the relic microwave background radiation.

These discoveries confirmed the Friedmann model of the universe and the hot big bang. The next experiment on the list is now on the agenda. It is:

- The measurement of the large-scale angular $\Delta T/T$ fluctuations of the microwave background.

Theoretical predictions and current observational possibilities are of the same order ($\Delta T/T \sim 10^{-5}$). So, it may well be that the relic fluctuations on scales $\Theta \gtrsim 20'{}^*$ will be discovered soon. They could tell us a lot not only about the origin of galaxies but also about the processes in the very early Universe when the primordial cosmological perturbations (galactic embryos) were only forming.

One of the future experiments important for cosmology is:

- The use of Space radio interferometers with baselines much

A. Hewitt et al. (eds.), Observational Cosmology, 73–81.
© 1987 by the IAU.

longer than the Earth's diameter. These may allow us to determine
directly distances to other galaxies. This follows from a simple
estimate for the maximum distance R_{max} to be measured by a interferometer with baseline L at a wavelength[*] λ:

$$R_{max} = f \cdot L^2 / \lambda \tag{1}$$

where $f \sim S/N$ is the numerical coefficient. For $\lambda \sim 1$ cm, $L \sim 2$a.u.
$\simeq 3 \cdot 10^{13}$ cm, and $f \sim 10$, $R_{max} \sim 10^{28}$ cm, which is the cosmological
horizon length scale.

   Before we consider the current status of the $\Delta T/T$ problem we
consider this latter argument in more detail.

2. SPACE INTERFEROMETERS

   There are two proposals associated with large distances.
   The first is to determine the curvature of the signal wave-front
from a distant object which in fact, means to determine the parallax
π. This may be done with the help of three or more satellites by
measuring the delay time of the wave signal.
   In the simple case in which it is assumed that heavy particles
make up the missing mass (Novikov 1977)

$$\pi = \frac{H_o L}{2 c} (4 - \Omega + \Omega^2 \cdot (\sqrt{1+\Omega z} - 1)^{-1} (1 - \frac{1 - \Omega}{\sqrt{1+\Omega z}})^{-1}) \tag{2}$$

where Ω is the total density of the Universe in critical density units.
For the large redshift objects ($\Omega z \gg 1$)

$$\pi_\infty = \frac{H_o L}{c} (2 - \Omega/2) \tag{3}$$

π_∞ is independent of the redshift and allows us to determine Ω when
the Hubble constant H_o is known. In the general case the parallax

---

* See previous page.
   According to conventional theories the scales $\Theta \sim 20'$ in $\Delta T/T(\Theta)$ are
the scales most likely to be responsible for the formation of clusters
and superclusters. Fluctuations on a scale of degrees are related to
the structural correlations ~ 100–300 Mpc in the distribution of rich
clusters (see Kopylov et al. 1984; Bahcall and Burgett 1986; Koo and
Ellis in these Proceedings). Also a very important scale is $\sim 5^o$ which
is the cosmological horizon at $z \sim 10^3$ (the hydrogen recombination
epoch). The large-scale fluctuations ($\Theta > 5^o$) are of particular
interest since they are expected to be the biggest. They are directly
coupled to the primordial gravitational perturbations, and they are not
smoothed by the secondary ionization and the hydrogen recombination
dynamics.

depends on the deceleration and the structural filling factor as well.
Thus the shape of the diagram $\pi \leftrightarrow z$ gives us direct information about
the fundamental constants of the universe.

The second proposal was recently discussed by Kardashev (1986),
and he assumes that a long baseline interferometer can detect the
motion of distant galaxies on a background of very distant objects.
The relic radiation fixes the universal coordinate system in space-time,
and any object moves relative to it. For example, our Sun has the
absolute velocity v = 350 km/s in the direction $l \simeq 253^o$, $b \simeq 47^o$.
Thus, relative to our laboratory system any remote object moves
systematically (i.e. in addition to its random proper motion) in the
direction of the antiapex with an angular shift depending on the
distance and direction to the object. In a year the Sun travels $\sim 10^{15}$
cm giving a gain of a factor ~ 35 compared with the annual parallax
(cf. L = 2 a.u. $\simeq 3 \cdot 10^{13}$ cm).* This effect grows linearly with time.
While investigating angular shifts and redshifts we can extract information about the universal constants.

We do not discuss here statistical, absorption, gravitational and
other problems (see Kardashev et al. 1973). Our aim is just to
emphasize the possibility of determining the main cosmological parameters with help of space interferometers.

However, for now the most precise tool for the exploration of
the universe is the study of fluctuations of the microwave background.
Let us first recall the contemporary upper limits for $\Delta T/T$ in large
scales, and then proceed to discuss the theoretical implications and
the whole problem.

3. OBSERVATIONAL DATA FOR $\Delta T/T$

Up to now, and after 21 years of investigation of the relic
radiation, no fluctuations of its temperature have been found although
they ought to exist on the level $\sim 10^{-5}$ because of the existence of
superclusters. The best up-to-date limits on $\Delta T/T$ on scales $\theta > 7^o$
are from the "RELIC" experiment that was carried out on the satellite
Prognoz two years ago. This produced a sky-map at $\lambda = 8$ mm with a
sensitivity $\sigma \simeq 0.2$ mK, which covered ~ 8.8 steradians. The final
analysis of all of the observations is now complete (see Strukov and
Skulachev 1986, Strukov et al. 1986, Klypin et al. 1986) and we
summarize below that part which is of main interest here.

According to Klypin et al. (1986) the cosmological information
can be listed under three headings:

1. Harmonic analysis. No $\Delta T/T$ signal was found at the 95%
confidence level for the multipole components with the spherical

---

* In a year one could measure systematic angular shifts for objects
at $z \lesssim 0.01$ with angular resolution $\sim 2 \cdot 10^{-6}$ arc sec (say, using as
a baseline the Earth-Lagrange point $\sim 1.5 \cdot 10^6$ km (Kardashev 1986).

harmonic index $1 = 2,3,\ldots, 10$. This leads to the following upper
limits for the quadrupole and octupole anisotropies

$$(\Delta T/T)_2 < 3\cdot10^{-5} \, , \, (\Delta T/T)_3 < 7\cdot10^{-5} \tag{4}$$

and $(\Delta T/T)_\ell < 5\cdot10^{-5}$ for the higher harmonics ($\ell = 4,5,\ldots, 10$). The
octupole amplitude ($\ell = 3$) is less well-determined since the dipole
component was subtracted in every scan before the analysis and harmonics
with $1 > 10$ are reduced by the antenna beamwidth and the method of
analysis. Note that the quadrupole limit ($\ell = 2$) is a factor of 2 less
than that in the previous experiment (Fixen et al. 1983) and the large-
scale background pattern is now outlined.

    2. Dispersion analysis was based on the background signal
dependence

$$(\Delta T/T)_\ell \sim \frac{2\ell+1}{1(\ell+1)} \, , \, \ell \geq 2, \tag{5}$$

which is valid for the flat primordial perturbation spectrum and the
assumption that $\Omega = 1$ with massive weakly interacting particles making
up the missing mass. For the quadrupole amplitude this yields

$$(\Delta T/T)_2 < 2\cdot10^{-5} \quad \text{(95\% confidence limit)} \tag{6}$$

Note, that the upper limit in eq. (6) is model dependent (see eq. (5))
whilst the numbers in eq. (4) come purely from observation.

    3. The correlation function calculated from the noise sky map
displays no statistically significant peaks:

$$< \Delta T_1 \cdot \Delta T_2 > \quad < \quad 4\cdot10^{-3} \, \text{mK}^2 \tag{7}$$

for angular scales $20^\circ < \Theta < 160^\circ$. This result adds no information
to that obtained from the harmonic analysis.

    These observational data put severe limits on the "market" of
cosmological models which we now discuss.

4. COSMOLOGICAL MODELS

    In Table I we present theoretical predictions for the quadrupole
$\Delta T/T$ – amplitude ($\ell = 2$) as measured by a 6° antenna and those for the
small-scale structure function $\Delta T/T$ ($\Theta = 20'$) to be detected by the
ideal antenna ($\Theta < 5'$). The numbers are calculated for $H_\circ = 50$ km/s/Mpc,
$T_{BR} = 2.7^\circ$K, $\Omega =_a 1$ and a flat spectrum for the primordial$^\circ$ perturbations.
The results are given for four different assumptions concerning the
nature of the hidden mass.

    The stable particle column (see Lukash et al. 1984) assumes the
hot missing matter variant (ν – type particles). Here the primordial
perturbations are normalized by assuming an r.m.s. density contrast
~ 1 at $z = 3$. The same numbers hold for the cold variant (axions or
very heavy particles) if the correlation scale of the dynamical mass
at $z = 0$ is ~ 30 Mpc.*

TABLE I.   Large and Small Scale Microwave Background Anisotropy
            Expected for Different Carriers of Hidden Matter

Δ T/T	Stable Particles	Decaying Particles	$\Lambda \neq 0$	Strings
Quadrupole	2×10^{-5}	2×10^{-5}	10^{-5}	10^{-5}
20' dispersion	10^{-4}	2×10^{-5}	8×10^{-5}	4×10^{-6}

In the second column we use the calculations of Kofman et al.
(1986) and Doroshkevich and Khlopov (1985) for the decaying particle
models. For the $\Lambda \neq 0$ model, the dynamical matter density in the form
of axions was supposed to be 20% of the critical one (see Kofman and
Starobinskii 1985). The data for the cosmological string model are due
to Brandenberger and Turok (1985).

The large-scale Δ T/T fluctuations in all the models are seen to
be of the same order of magnitude, and they are on the verge of the
current observational limits (see Section 3). Detection of these
fluctuations would be a crucial test of cosmological theory, since
they bear direct information about the primordial gravitational
perturbations on large scales which are not smoothed out by the
recombination and secondary ionization dynamics.

A subtle point about the fluctuations in angles $\Theta \gtrsim 5^{o}$ is that
the related density perturbations are still in the linear regime of the
evolution now, i.e. they do not manifest themselves as a space structural pattern yet and we do not know their amplitude. Another parameter
which the Δ T/T ($\Theta > 5^{o}$) fluctuations are sensitive to is the overall
matter density Ω (see discussion on the spottiness in Section 5). Thus
the large scale Δ T/T fluctuations can tell us primarily something
about the primordial spectrum and Ω and secondarily something about the
nature of the missing mass.

A different situation applies for the small scales ($10' < \Theta < 5^{o}$).
These are related to the large-scale structure of the Universe which
we now observe. However, the Δ T/T level at $\Theta \sim 20'$ drastically
depends on the recombination and ionization processes. The numbers in
Table I are given for the conventional recombination dynamics. If some
unstable particles decay with small probability through the γ-channel
at $z < 10^3$, they can quite easily reionize or draw out recombination of
the cosmic hydrogen. This, in turn, decreases the small-scale Δ T/T

* If the correlation length is three times smaller (the Peebles
correlation scale), then the quadrupole amplitude for the axion model
is $\sim 5 \cdot 10^{-6}$; with biasing at $\sim 2\sigma$ (Kaiser 1984) the last number is less
by a factor ~ 1.5 (see Lukash 1987).

- anisotropy by a factor ~50 (Dorosheva and Naselskij 1986). Thus
Δ T/T fluctuations on small scales are a good test of the recombination
and ionization dynamics at large redshifts.

5. DISCUSSION

In Table II we present the results of calculations of Lukash et al.
(1984) for the dipole (in scales ~100 Mpc) and quadrupole anisotropies
in the hot and cold missing matter models (without biasing) for different
assumed power-law spectra of the primordial perturbations. It can be
seen that the flat spectrum is a corner stone of the Δ T/T problem from
the observational point of view. The white noise spectrum growing to
larger scales is practically forbidden by the large-scale observations.
On the other hand, decaying spectra which are most naturally predicted
by the parametric amplification theory (Lukash 1980, Kompaneets et al.
1982) or in multi-inflationary universes (Kofman et al. 1985, Kofman
1986), decrease the expected value of Δ T/T on large-scales which would
postpone the Δ T/T discovery.

TABLE II. Dipole (ℓ = 1) and Quadrupole (ℓ = 2) Δ T/T Amplitudes
           Expected for Different Primordial Spectra

Hidden Matter	ℓ	Flat Spectrum	White Noise	Decaying Spectrum
Hot	1	3×10^{-3}	5×10^{-3}	3×10^{-3}
	2	2×10^{-5}	4×10^{-4}	$< 10^{-5}$
Cold	1	10^{-3}	10^{-2}	10^{-3}
	2	4×10^{-6}	7×10^{-5}	$< 10^{-6}$

Another way to change the structure of the anisotropy at $5^{\circ}<\Theta<180^{\circ}$
is to put $\Omega \neq 1$. In open universes a qualitatively new effect arises.
It is "spottiness": the large-scale Δ T/T pattern over the sky for
$\Omega < 1$ represents an arbitrary superposition of cold and hot spots with
angular dimensions $\Theta_{0} \sim \Omega$ independent of the perturbation wavelength
(Novikov 1968, Lukash 1977, Lukash and Novikov 1985). As $\Omega \rightarrow 1$ the
phenomenon disappears since the spot shape converts into the quadrupole
structure, and any superposition of quadrupoles results in quadrupole
structure again, i.e. the statistical effect fails. The spottiness
brings about a relative gain in the multipole Δ T/T amplitude with
$\ell_{0} \sim \pi/\Theta_{0} \sim \Omega^{-1}$. In other words the following result is found. If
we fix the flat primordial spectrum and vary Ω from unity to lower
values than the maximum expected, the Δ T/T anisotropy will shift from
the quadrupole to a higher harmonic with $\ell_{0} \sim \Omega^{-1} > 1$. The normalization assumes that the Δ T/T amplitude will grow with Ω decreasing,

which allows for the lower limit $\Omega \gtrsim 0.3$ established by the high harmonic upper $\Delta T/T$ limits ($\ell \sim 10$, see eq. (4)).

Of course, both factors, spottiness and primordial spectrum as well as the cosmological hidden mass model determine the large-scale $\Delta T/T$ structure. Only after we know the whole sky map for $\ell \lesssim 15$ can we answer all of the questions about the main cosmological parameters.

What are the observational possibilities as far as the detection of large-scale $\Delta T/T$ fluctuations are concerned?

First, we have the exciting result of Dr. R.D. Davies for $\Theta \sim 5^{\circ}$ in these Proceedings. This is in good agreement with the balloon data of Dr. Melchiorri (private communication) which are obtained on the same angular scales but at a much shorter wavelength. Both results still need verification, but let us assume for a moment that the background signal exists on the level $\Delta T/T \sim 2 \cdot 10^{-5}$ at $\Theta \sim 5\text{-}10^{\circ}$. Then we can normalize the primordial perturbations using this value in the correlation function of $\Delta T/T$. This is much more precise than the normalization by the density contrast since $\Delta T/T$ calculations are made using linear perturbation theory, and we do not need such poorly known quantities as the correlation radius for the dynamical mass, or the type of particle making up the missing mass. In doing this for the flat spectrum and $\Omega = 1$, we find from eq. (5) the following estimate for the quadrupole, $(\Delta T/T)_2 \sim 10^{-5}$ which is on the verge of detection. This problem was also discussed at the SAO Conference (USSR) in September 1986.

Another line of evidence comes from the large-scale motions (Collins et al. 1986, Burstein et al. 1986, and also the paper by Roger Davies in these Proceedings). If we assume the streaming velocity $v/c \sim 2 \cdot 10^{-3}$ on scales ~ 100 Mpc this will not contradict the dipole distribution of IRAS galaxies (Meiksin and Davis 1985, Yahil et al. 1986) if the density contrast in 100 Mpc is ~ 0.1 and the velocity vector correlates with the IRAS dipole. In this way we conclude that the microwave dipole anisotropy on these scales is totally determined by the cosmological "growing" mode of perturbations and $(\Delta T/T)_1 \sim 2 \cdot 10^{-3}$ purely from the velocity observations (Yudin 1986, cf. Table I). Again recalculating this result for the quadrupole $\Delta T/T$ anisotropy on the assumption of a flat spectrum and $\Omega = 1$ we get the same number $(\Delta T/T)_2 \sim 10^{-5}$.

We note that the two points at the diagram $\Delta T/T \leftrightarrow \Theta$ (at $\Theta_1 = 5^{\circ}$ and $\Theta_2 = 180^{\circ}$) give us the same estimate $\sim 10^{-5}$ for the quadrupole$_1$ independent of any assumptions about the cosmological model. Thus the discovery of only quadrupole structure at this level would strongly confirm a flat spectrum for the primordial cosmological perturbations.

6. CONCLUSIONS

Our main conclusions are briefly summarized as follows.

1. Baryon models and $\Lambda \neq 0$ models with baryons or light massive particles as a dynamical hidden matter contradict the large-scale $\Delta T/T$ limits.

2. Large-scale $\Delta T/T$ fluctuations may be detected in the satellite experiment with a sensitivity $\sim 3-5 \times 10^{-6}$.

3. The flat primordial perturbation spectrum is a crucial test of the $\Delta T/T$ problem. The optimal range to search for $\Delta T/T$ is $20' < \Theta < < 180^{\circ}$.

If $\Delta T/T$ fluctuations are discovered at $\Theta > 5^{\circ}$ the spectrum of primordial perturbations as well as the total matter density in the Universe Ω could be determined.

REFERENCES

Bahcall, N.A., and Burgett, W.S., _Ap. J. (Lett.)_ 300, L35 (1986).
Brandenberger, R.N., and Turok, N., Preprint NSF-ITP-85-88 (1985).
Burstein, P., Davies, R.L., Dressler, A., Faber, S.M., Lynden-Bell, D.,
    Terelevich, R., and Wegner, J., _Galaxy Distances and Deviations
    from Universal Expansion_, eds. Madore, B.S., and Tully, R.B.
    (Reidel), p.123 (1986).
Collins, C.A., Joseph, R.D., and Robertson, N.A., _Nature_ 320, 506 (1986).
Doroshkevich, A.G., and Khlopov, M. Yu., _Astron. Zh. Lett._ 11, 563 (1985).
Dorosheva, E.I., and Naselskij, P.D., _Astrophysica_ 24, 561 (1986).
Fixen, D., Cheng, E., and Wilkinson, D., _Phys. Rev. Lett._ 50, 620 (1983).
Kaiser, N., _Ap. J. (Lett.)_ 284, L9 (1984).
Kardashev, N.S., _Astron. Zh._ 63, 845 (1986).
Kardashev, N.S., Parijskij, Yu.N., and Umarbaeva, N.D., _Astrof.
    Issledovaniya_ 5, 16 (1973).
Klypin, A.A., Sazhin, M.V., Skulachev, D.P., and Strukov, I.A., _Astron.
    Zh. Lett., in press_ (1986).
Kofman, L.A., _Phys. Lett._ 173B, 400 (1986).
Kofman, L.A., Linde, A.D., and Starobinskii, A.A., _Phys. Lett._, 157B,
    361 (1985).
Kofman, L.A., Pogosyan, D.Yu., and Starobinskii, A.A., _Astron. Zh. Lett._,
    12, 419 (1986).
Kofman, L.A., and Starobinskii, A.A., _Astron. Zh. Lett._, 11, 643 (1985).
Kompaneets, D.A., Lukash, V.N., and Novikov, I.D., _Astron. Zh._ 59, 424
    (1982).
Kopylov, A.I., Fetisova, T.S., and Shvartsman, V.F., _Astron. Tsyrkulyar_,
    1344, 1 (1984).
Kopylov, A.I., Kuznetsov, D.U., Fetisova, T.S., and Shvartsman, V.F.,
    Astron. Tsyrkulyar 1347, 1 (1984).
Lukash, V.N., _Contr. Papers GRG-8 Conference_, Waterloo University,
    237, (1977).
Lukash, V.N., _JETP_ 79, 1601 (1980).
Lukash, V.N., Proc. Symp. IAU 117 _'Dark Matter in the Universe'_,
    Princeton, eds. Kormendy, J., and Knapp, G.R., (Reidel) (1987).
Lukash, V.N., Naselskij, P.D., and Novikov, I.D., Proc. _'Quantum
    Gravity-3'_, Moscow, eds. Markov, M.A., Beresin, V.A., and Frolov,
    V.P., (World Scientific), 675 (1984).
Lukash, V.N., and Novikov, I.D., _Nature_ 316, 46 (1985).
Meiksin, A., and Davis, M., _Ap. J._, 291, 191 (1985).

Novikov, I.D., Astron. Zh. 45, 538 (1968).
Novikov, I.D., Astron. Zh. 54, 722 (1977).
Strukov, I.A., and Skulachev, D.P., Astron. Zh. Lett., in press (1986).
Strukov, I.A., Skulachev, D.P., Boyarskij, M.N., and Tkachev, A.N.,
     Astron. Zh. Lett., in press (1986).
Yahil, A., Walker, D., and Rowan-Robinson, M., Ap. J. Lett. 301, L1
     (1986).
Yudin, V.M., Astron. Tsyrkulyar, 1428, 1 (1986).

DISCUSSION

PARTRIDGE: In simple, low density, anisotropic cosmological models,
there is a "hot spot" of scale $\Theta \sim \Omega$ radians. Has the map from the
RELIC satellite been inspected for such a "hot spot?"

LUKASH: Individual hot or cold spots can exist if there were some
peculiarities in the primordial spectrum. For example, if some wave
vector in the initial perturbation amplitude was large in comparison
with all of the others, then a spot will form with angular scale
$\sim \Omega_{tot} < 1$ and with the large scale structure totally described by the
Bianchi type V model. (The RELIC experiment was proposed in order to
search for such spots in the sky.)

        However, for example, for the Gaussian perturbation field,
the sky pattern in the open universe represents an arbitrary superposition of the spots. In this case you cannot find separate spots,
but instead the angular correlation scale, $\sim \Omega_{tot} < 1$. This is what
we call the spottiness effect. It can be detected from the shape of the
correlation function.

DEKEL: Could you please elaborate on the way you've normalized the
spectrum of density fluctuations to obtain the various theoretical
predictions for $\delta T/T$? In particular, shouldn't it depend on the type
of dark-matter, the initial fluctuations, and on whether the galaxies
trace the mass?

LUKASH: Thank you for the question since I did not have time enough
to dwell on these points. The numbers for the stable particles were
given for the hot scenario. Here the normalization is done by putting
the r.m.s. density fluctuation equal to unity at $z = 3$, and $H_0 = 50$
km/s/Mpc. For the very heavy particles the predicted $\Delta T/T$ values in
different normalization procedures are a bit lower (see refs.) but
still the $\Delta T/T$-fluctuations in large scales will be detected if the
sensitivity is several units times 10^{-6}.

THE SUNYAEV-ZEL'DOVICH EFFECT AND H_0

M. Birkinshaw
Department of Astronomy, Harvard University,
60 Garden Street,
Cambridge, MA 02138, U.S.A.

ABSTRACT. Observational data on the Sunyaev-Zel'dovich effect for three clusters are used to obtain limits to the value of the Hubble constant. Only a wide bound on the value of H_0, $10 \ \mathrm{km\,s^{-1}\,Mpc^{-1}} < H_0 < 160 \ \mathrm{km\,s^{-1}\,Mpc^{-1}}$, can be derived, mostly because of the large uncertainty in the measured temperature of the intracluster medium in each cluster. Better information on this temperature and on the distribution of the intracluster medium are needed for this method to yield a good measurement of H_0.

1. THE USE OF THE SUNYAEV-ZEL'DOVICH EFFECT TO MEASURE H_0

The Sunyaev-Zel'dovich effect is a change of brightness of the microwave background radiation produced by its inverse-Compton scattering by electrons in a hot intracluster medium (Sunyaev & Zel'dovich 1972). The X-ray surface brightness of the intracluster medium is usually approximated by

$$l_{\mathrm{X}} \approx l_0 \left(1 + \frac{\theta^2}{\theta_{\mathrm{c}}^2}\right)^{\frac{1}{2}-3\beta}$$

with $\beta \approx 0.5$. If the intracluster medium is isothermal, the predicted microwave background decrement is

$$\Delta \mathrm{T_{RJ}} = \Delta \mathrm{T_0} \left(1 + \frac{\theta^2}{\theta_{\mathrm{c}}^2}\right)^{\frac{1}{2}-\frac{3}{2}\beta}$$

where $\Delta \mathrm{T_0}$, the central decrement, and l_0, the central X-ray surface brightness, are related by

$$\frac{\Delta \mathrm{T_0}^2}{l_0} = 55 \frac{\mathrm{T_r^2}}{\Lambda(\mathrm{T_g})} \left(\frac{\mathrm{k\,T_g}}{\mathrm{m_e\,c^2}}\right)^2 \sigma_{\mathrm{T}}^2 \, \mathrm{D_A} \, \theta_{\mathrm{c}}$$

for $\beta = 0.5$. $\Delta \mathrm{T_0}$, l_0, θ_{c} and β can be fitted from the observed Sunyaev-Zel'dovich data and X-ray images, the gas temperature $\mathrm{T_g}$ can be derived from X-ray spectra, the X-ray emissivity $\Lambda(\mathrm{T_g})$ is a known function of $\mathrm{T_g}$, and $\mathrm{T_r}$ (the radiation temperature of the microwave background radiation), k, $\mathrm{m_e}$, c, and σ_{T} are known constants. This relation can, therefore, be used to determine the angular diameter distance $\mathrm{D_A}$ of a cluster of galaxies, and hence the Hubble constant. This method has been described by a number of authors (Gunn 1978; Silk & White 1978; Cavaliere *et al.* 1979; Birkinshaw 1979).

83

A. Hewitt et al. (eds.), Observational Cosmology, 83–86.
© *1987 by the IAU.*

2. THE DATA

The Sunyaev-Zel'dovich data are taken from the results of OVRO observations made over the period 1982-6, which add the data taken during the 1985-6 season to those summarized in Birkinshaw & Moffet (1986). These data have been corrected for the presence of weak non-thermal radio sources lying near the clusters, and they have been adjusted to make an allowance for the maximum likely residual systematic errors. The central microwave background decrements, ΔT_0, (deconvolved for the telescope beam) are given in Table 1.

Table 1. Central Sunyaev-Zel'dovich effects $\left(\beta = \tfrac{1}{2}\right)$

cluster	$\Delta T_0/\mathrm{mK}$
0016+16	-0.92 ± 0.19
A 665	-1.01 ± 0.16
A 2218	-0.63 ± 0.13

The values of θ_c needed in the deconvolution were taken from the X-ray data of White *et al.* (1981), Forman & Jones (private communication) and Boynton *et al.* (1981). The central X-ray surface brightnesses, l_0, were derived from the fluxes given in these papers, and approximate gas temperatures were derived from the *Einstein* X-ray data for 0016+16 ($T_g > 6$ keV; White *et al.* 1981) and A 2218 ($T_g = 11^{+\infty}_{-4}$ keV; Perrenod & Henry 1981). For A 665 and A 2218, estimates of T_g may also be derived from the cluster velocity dispersions, σ_v (subject to the assumed value of β), since $\beta = \mu m_p \sigma_v^2 / kT_g$ (Cavaliere & Fusco-Femiano 1976). The results for an assumed $\beta = \tfrac{1}{2}$ are 29 ± 16 keV and 24 ± 7 keV, respectively, and are probably overestimates (Mushotzsky 1984). The uncertainties in the temperatures derived from the velocity dispersions are too large for these estimates to be useful at present, but T_g may be derived in this way in the future if more accurate values of σ_v can be measured and if β can be determined independently.

3. RESULTS

The gas temperature, the central Sunyaev-Zel'dovich decrement, the central X-ray surface brightness, and the structural data can now be combined to produce an independent estimate of the distance (and hence the Hubble constant) for each cluster. The values obtained for H_0 are > 11 km s^{-1} Mpc^{-1} from the data for 0016+16 and 70 ± 90 km s^{-1} Mpc^{-1} from the data for A 2218 (using the X-ray inferred gas temperatures). The large errors in these estimates reflect mostly the uncertainties in the values of T_g, and indicate that accurate estimates of the Hubble constant can be derived from this method only with better (error less than 1 keV) measurements of T_g. At present it is more useful to invert the argument and predict the gas temperatures that would be inferred from good X-ray spectra for these clusters for different assumed values of the Hubble constant. These temperatures are given in Table 2.

Table 2. Predicted T_g for two values of H_0

cluster	$H_0 = 50$ km s^{-1} Mpc^{-1} T_g/keV	$H_0 = 100$ km s^{-1} Mpc^{-1} T_g/keV
0016+16	8 ± 3	12 ± 4
A 665	10 ± 2	14 ± 2
A 2218	10 ± 2	13 ± 3

It can be seen that *overall* the predicted gas temperatures are significantly higher for $H_0 = 100$ km s^{-1} Mpc^{-1}. The difference between the gas temperatures for the two cases suggests that a test for the value of H_0 between the bounds of 50 and 100 km s^{-1} Mpc^{-1} may be possible with X-ray spectra with high signal/noise near 10 keV. The values of T_g predicted in Table 2 were derived for an isothermal model with $\beta = \frac{1}{2}$, but are found to depend only weakly on β.

4. THE FUTURE

This method of limiting the value of the Hubble constant may become important if good X-ray data on these clusters of galaxies, or good Sunyaev-Zel'dovich data on nearer clusters of galaxies (for which extensive X-ray data are already available) can be obtained. High quality data should allow the systematic uncertainties in the H_0 estimate caused by the unknown equation of state of the gas (assumed here to be isothermal) and the uncertain value of β (taken to be $\frac{1}{2}$) to be resolved.

5. ACKNOWLEDGEMENTS

The Sunyaev-Zel'dovich data on which these calculations are based were obtained in collaboration with A.T. Moffet and S.F. Gull on the OVRO 40-m telescope.

6. REFERENCES

Birkinshaw, M., 1979. *Mon. Not. R. astr. Soc.*, **187**, 847.
Birkinshaw, M. & Moffet, A.T., 1986. *Highl. Astron.*, **7**, 321.
Boynton, P.E., Radford, S.J.E., Schommer, R.A. & Murray, S.S., 1981. *Astrophys. J.*, **257**, 473.
Cavaliere, A., Danese, L. & De Zotti, G., 1979. *Astr. Astrophys.*, **75**, 322.
Cavaliere, A. & Fusco-Femiano, R., 1976. *Astr. Astrophys.*, **49**, 137.
Gunn, J.E., 1978. In *Observational Cosmology*, ed. Maeder, A., Martinet, L. & Tammann, G., Geneva Observatory.
Mushotzsky, R.F., 1984. *Phys. Scr.*, **T7**, 157.
Perrenod, S.C. & Henry, J.P., 1981. *Astrophys. J.*, **247**, L1.
Sunyaev, R.A. & Zel'dovich, Ya.B., 1972. *Comm. Astrophys. Sp. Phys.*, **4**, 173.
Silk, J.I. & White, S.D.M., 1978. *Astrophys. J.*, **226**, L103.
White, S.D.M., Silk, J.I. & Henry, J.P., 1981. *Astrophys. J.*, **251**, L65.

DISCUSSION

ELLIS: Why in the case of 0016+16 were you only able to determine an upper limit to the angular diameter distance? Is the X-ray data inadequate? I thought it was one of the strongest distant X-ray clusters.

BIRKINSHAW: The limit to D_A arises from the lower limit to the gas temperature ($T_g > 6$ keV) for this cluster. Although 0016+16 is very X-ray <u>luminous</u>, its flux is small, and it is difficult to extract any temperature information from the Einstein data.

CANIZARES: How sensitive is your determination of H_0 to the assumption that the X-ray gas is isothermal?

BIRKINSHAW: Quite sensitive - but in principle this assumption, like
the β-parameter assumption, can be tested by comparing X-ray and
Sunyaev-Zel'dovich effect maps of the clusters, by using a spatially-
resolved X-ray spectrum, or by doing a better job of modelling the
X-ray emitting gas.

MANDOLESI: Have you taken into account the emission from weak, unresolved radio sources?

BIRKINSHAW: Yes. The emission from these sources was measured in a
multi-frequency VLA survey of the fields of each of these three
clusters.

CHAPTER II

THE ORIGIN AND ABUNDANCES OF THE LIGHT ELEMENTS

ON THE ORIGIN OF THE LIGHT ELEMENTS
(D, ^{3}He, ^{4}He AND ^{7}Li)

Jean AUDOUZE

Institut d'Astrophysique du CNRS – Paris, France

and Laboratoire René Bernas – Orsay Campus, France

ABSTRACT. The abundances of the very light elements (D, ^{3}He, ^{4}He and ^{7}Li) constitute indeed one of the most powerful constraints in cosmology : they are known to fix very interesting limits on the baryonic density of the Universe and on the maximum number of neutrino (lepton) families in the frame of the simplest canonical models. Given the importance of these predictions, these models should be analysed very cautiously at the light of recent developments in the observations of these elements. In order to make the simplest models consistent with the observations, it is argued that a thorough destruction of D should occur during the galactic evolution. Moreover this review deals also with some models invoking the possible existence of massive unstable neutrinos, gravitinos or photinos which would decay into high energy photons or of quark nuggets which could be created during the quark–hadron phase transitions. Such models have been designed in an attempt to overcome the limitation on the Universe density coming from these abundance determinations. Although the simple canonical models are especially attractive such models cannot be disregarded .

1. INTRODUCTION

It is known since quite a long time (see e.g. Peebles 1966, Wagoner *et al.* 1967, Wagoner 1969 and 1973, Reeves *et al.*, 1973) that the lightest elements D, ^{3}He, ^{4}He and ^{7}Li are likely to be formed during the very early phases of the Universe i.e. about 100 sec after the Big Bang. It is fair to say that the nucleosynthesis of these elements constitutes an argument in favour of this cosmology theory as important as the recession of galaxies and the discovery of the 2.7 K background radiation. A very large number of papers (see e.g. the most recent reviews of Boesgaaard and Steigman, 1985 and Audouze 1987) recall that the simplest (canonical) Big Bang models seem to account for the observed abundances of these elements if the present baryonic density of the Universe ρ_B is such that the baryonic cosmological parameters ρ_B ~ 0.1

89

A. Hewitt et al. (eds.), Observational Cosmology, 89–117.

(corresponding to a open Universe if there is no others significant constituent in the Universe). Moreover such models would predict the existence of a limited number (3 or 4) of different families of neutrinos (leptons) consistent with the three lepton families predicted by Grand Unification Theories GUT.

Given the importance of what can be inferred from the observed abundances of these elements, this symposium devoted to observational cosmology is indeed the place where the recent developments concerning that type of research should be analyzed. Section 2 contains a discussion of the observations relevant to these elements and their resulting primordial abundances. Section 3 summarizes the consequences on the baryonic density and the maximum number of neutrinos deduced from the simple (canonical) Big Bang models. From the current observations of D and ^{4}He especially it seems that primordial D should be largely destroyed during the Galaxy evolution. This cannot be achieved by the simplest galactic models such as those used and analyzed e.g. by Audouze and Tinsley (1974). Section 4 presents a brief account of the chemical evolution proposals which are currently made to lead to such D destruction. ^{7}Li which should (could) originate from many different astrophysical sources is also an interesting species for such evolutionary studies. Finally some of the possibilities offered by recent developments in particle physics are also considered presently in an attempt to overcome the limitation on the total density of the Universe are briefly presented in section 5. Since the bulk of the material which has been presented at Beijing has been published elsewhere, I will mainly concentrate on the latest developments and refer the reader to the list of references such as the two reviews quoted above.

2. THE "PRIMORDIAL" ABUNDANCES OF THE VERY LIGHT ELEMENTS

Let us consider in turn the four nuclear species D, ^{3}He, ^{4}He and ^{7}Li concerned by the primordial nucleosynthesis.

2.1 Deuterium

The most recent and thorough review concerning the abundance determinations of that isotope is due to Vidal–Madjar (1987) which provides an exhaustive list of references relative to this problem. This author points out the extreme variability on very short scales (down to a few parsecs) of the D/H abundances deduced from far UV observations of the nearby interstellar medium. From them he deduces that in our vicinity the "average" D/H $\sim$ $(1 \pm 0.3)\ 10^{-5}$ (following in particular the reanalysis of the Copernicus data performed by Gry *et al.* (1984). Concerning the Solar System D/H value, the best estimate remains that proposed by Geiss and Reeves (1972) from solar wind $^3\mathrm{He}/^4\mathrm{He}$ determinations : (D/H) solar system $= 2 \pm 1 \; 10^{-5}$.

A most exciting new information comes from the work of Carswell *et al.* (1987) who analyzed the higher order Lyman lines of a metal line absorption system at Z=3.08571 in the quasar Q0420-388. This system has a metallicity Z=0.2 $Z_\odot$ (where $Z_\odot$ is the solar metallicity). These authors argue that it is in principle possible to measure a D/H ratio in such systems where the column H density $N(H) \sim 10^{18}$ cm^{-2} and where the velocity dispersion $\Delta v \leq$ 20 km s^{-1}. In that specific case they determine $\log[n(DI)/n(HI)]$=-4.43$^{+0.45}_{-0.26}$ which leads to $\frac{D}{H} \sim \left(4^{+4}_{-2.5}\right) 10^{-5}$. They do not exclude entirely a possible confusion of this D determination with a low H density with high velocity although such an interpretation is quite unlikely. With their work one has the first report of a D abundance determination in astrophysical sites formed just after the birth of the Universe. We will come back in Section 4 on the importance of such measurements concerning an element like D especially affected by stellar (galactic) evolution process.

2.2 Helium 3

Nothing new has appeared in the literature since the presentation of previous reviews. The interstellar $^3\mathrm{He}/H$ abundance is very ill defined due to very large variations between the $^3\mathrm{He}^+$ radio measurements concerning different HII regions and performed by Rood *et al.*, 1984 : the corresponding interstellar $^3\mathrm{He}/H$ range from less than $2\;10^{-5}$ up to $5\;10^{-4}$. In the Solar System $\frac{^3He}{H} \sim (1.4 \pm 0.4)\;10^{-5}$ according e.g. to Geiss and Reeves (1972) and Black (1972).

2.3 Helium 4

There is an excellent book edited by Shaver *et al.*, 1983 which constitutes a compendium of all the abundance determinations of $^4\mathrm{He}$ in any possible astrophysical site. Blue compact galaxies (BCG) appear to be the most appropriate objects to obtain the primordial abundance of $^4\mathrm{He}$ (Y) ; Pagel *et al.*, 1986 have reviewed three independent accurate Y determinations due respectively to Lequeux *et al.*, 1979, Kunth and Sargent 1983 and themselves *.

---

* As pointed out recently by Vigroux (1986), one should be cautious not to consider together BCG results with galactic HII region ones in order to deduce $Y_{primordial}$. Any analysis mixing these two sets of determinations should be in error because of the quite different evolution behaviour of BCG and spirals.

— From Lequeux *et al.*, 1979, $Y = (0.230 \pm 0.004) + (3.3 \pm 1.1)Z$ (where Z is the metallicity of BCG).

— From Kunth and Sargent, 1983, $Y = (0.243 \pm 0.010) + (1.4 \pm 3.8)Z$, but when IIZw40 is removed because of a ^{4}Helium confusion with galactic Na absorption ** their result becomes $Y = (0.234 \pm 0.008) + (4.6 \pm 5.9)Z$

— Finally Pagel *et al.*, 1986 find $Y = (0.236 \pm 0.005) + (5.7 \pm 2.7)Z$ and $Y = (0.238 \pm 0.005) + (2.9 \pm 1.5)\, 10^3 \frac{N}{H}$ (this last correlation between Y and the N abundance has also been noted by Vigroux *et al.*, 1986).

Depending of ones own optimism or pessimism regarding these correlations

$$Y_p = (0.235 \pm 0.005) \text{ or } (O.24 \pm 0.01)).$$

2.4 Lithium 7

The great breakthrough regarding the primordial abundance of this element has been achieved first by Spite and Spite (1982) who showed convincingly that $\left(\frac{^7Li}{H}\right)_{primordial} = 10^{-10}$ by measuring the Li abundance in a large set of halo stars (if ^{7}Li is not destroyed subsequently in the atmosphere of these stars). Subsequent works by Spite *et al.* 1984 and Spite and Spite 1986 corroborate this discovery. The fact that in their 1986 paper they report that N rich halo stars have a similar (Li/H) atmospheric abundance seems to exclude any significant mixing between the surface and the deep layers of the star and give some weight to the idea that the observed Li has not been significantly destroyed during the evolution of such stars. It is worth mentioning that the subsequent speaker Doug Duncan (who has reported on quite similar but independent observed Li abundances in pop. II metal poor stars) has argued in favour of some significantly higher primordial Li abundance such that $2\ 10^{-10} < \frac{Li}{H} < 8\ 10^{-10}$ (Duncan and Hobbs, these proceedings). Should they found to be right, the cosmological implications of such higher Li abundance are evoked in the next sections.

In order to follow the evolution of this nuclear species one should recall that $\frac{Li}{H} \sim 10^{-9}$ and $\frac{^7Li}{^6Li} \sim 12.5$ in the Solar System (see e.g. Reeves, 1974). The Li abundance has also been determined in a series of open clusters the Pleiades (Duncan and Jones, 1983) the Hyades (Cayrel *et al.*, 1984, Boesgaard and Tripico 1986), NGC 752 (Hobbs and Pilackowski 1986) and M67 (Hobbs and Pilackowski 1987). The age of these open clusters is respectively $0.1, 0.7, 1.7$ and $5\ 10^9$ years). All these measurements seem to indicate that

---

** D. Kunth (private communication) concurs with B.E.J. Pagel that the measurement of IIZw40 should be removed from the Kunth and Sargent, 1983, BCG sample.

$\frac{Li}{H}$ has roughly remained constant i.e. $\sim 10^{-9}$, within a factor 2, during the last $5\ 10^9$ years (Hobbs and Pilackowski, 1987).

Another important piece of information is the isotopic $\frac{^7Li}{^6Li}$ ratio which is 12.5 in the Solar System but is significantly larger in the line of sight : $25 < \frac{^7Li}{^6Li} < 150$ with an average value of ~ 40 (Ferlet and Dennefeld, 1984).

2.5 Summary of the available abundances of the light elements

From this review one can deduce the following table where all these determinations are assembled and listed.

TABLE 1

Abundances of the light elements D, ^{3}He, ^{4}He
and ^{7}Li/^{6}Li isotopic ratio

	Primordial abundances	QSO abundances	Solar System	Interstellar medium
	$t \simeq 0$ Gyr	$t \simeq 1$ Gyr	$t \simeq 10$ Gyr	t=15 Gyr
D/H (by number)	$4\ 10^{-5}$–$4\ 10^{-4}$	$(2$–$8)^*\ 10^{-5}$ *(if confirmed)	$(1$–$3)\ 10^{-5}$	$(0.7 - 1.3)\ 10^{-5}$
^{3}He/H (by number)	$2\ 10^{-5}$–$4\ 10^{-5}$		$(1.4 \pm 0.4)\ 10^{-5}$	$<2\ 10^{-5}$–$5\ 10^{-4}$
^{4}He/H (by mass)	0.235 ± 0.005 (O) 0.24 ± 0.01 (P)		0.17–0.28	0.22–0.30
^{7}Li/H (by number)	$(1 \pm 0.3)\ 10^{-10}$ (S^2) $(2$–$8)\ 10^{-10}$ (DH)		10^{-9}	$(0.5$–$1)\ 10^{-9}$
^{7}Li/^{6}Li			12.5	24–(40)–150

(O) : optimistic – (P) : pessimistic – (S^2) : F. and M. Spite –
(DH) : Duncan and Hobbs.

The lower limit of the Solar System Y value comes from Gautier (1983) while the interstellar Y values are extracted from various contributions published in Shaver *et al.* (1983).

The primordial D and ^{3}He values are especially uncertain because, as we will discuss it at some length in §4, they depend much on galactic evolution.

3. THE LIGHT ELEMENTS AND THE CANONICAL BIG BANG NUCLEOSYNTHESIS

The current literature contains many papers publicizing the success of the standard (canonical) model with respect to the early nucleosynthesis of D, ^{3}He, ^{4}He and ^{7}Li. Those which come to my mind are Yang *et al.* (1979 and 1984), Olive *et al.*, 1981, Boesgaard and Steigman, 1985, Steigman 1986, Steigman *et al.*, 1986, Audouze (1982, 1984, 1987, 1987) and Matzner 1986. Let me recall that the standard (canonical) model adopts the following set of hypotheses (i) the Universe was born from a very hot and dense phase (Big Bang) ensuring statistical equilibrium between the existing particles (ii) the Universe which is homogeneous and isotropic expands according the Einstein General Relativity (GR) laws where the cosmological constant Λ is equal to O. (iii) the neutron life time is equal to $\tau_{1/2} = 10.5 \pm 0.1$ minutes according to the recent analysis of Steigman *et al.*, 1987 (iv) the baryon density $\eta = n_B/n_\gamma$ (relative to that of photons) ranges between 1 to 10 10^{-10} (we will use also in the sequel $\eta_{10} = 10^{10}\eta$). One should recall at this point the relations existing between this baryon parameter η, the baryonic ρ_B density and the baryonic cosmological parameter Ω_B

$$\rho_B = 6.64 \ 10^{-32}\eta_{10}(T/2.7)^3 \ g \ cm^{-3} \tag{1}$$

$$\Omega_B = \rho_B/\rho_c = 3.53 \ 10^{-3}\eta_{10}h^{-2}(T/2.7)^3 \tag{2}$$

where T is the temperature of the background radiation, $h = (\frac{H_o}{100})$ where H_o is the Hubble constant expressed in km s^{-1} Mpc^{-1}. v) the existing particles are non degenerate which is another way to state that the neutrino chemical potentials are negligible.

Early nucleosynthesis starts at a time short enough i.e. where the neutrons which are decaying since they are more in equilibrium with protons and leptons have still a large density but long enough in order for the temperature to be such that $10^8 < T < 3 \ 10^9$ K : above such values D photodisintegrates into protons and neutrons. Therefore the current time–scale for the early nucleosynthesis is ~ 100 sec.

The most critical physical parameters which govern the outcome of this nucleosynthesis are :

1– The expansion rate which can be deduced from GR and which increases with the number of boson or fermion flavors : the characteristic time of expansion is given by

$$t(s) = 2.4 g_{eff}^{-1/2} T_{MeV}^{-2} \qquad (3)$$

where g_{eff} is the effective number of relativistic degrees of freedom

$$g_{eff} = \frac{43}{4}[1 + \frac{7}{43}(N_\nu - 3)] \qquad (4)$$

and N_ν is the equivalent number of neutrino families

$$N_\nu = \sum_F (\frac{g_F}{2})(\frac{T_F}{T_\nu})^4 + \frac{8}{7} \sum_B (\frac{g_B}{2})(\frac{T_B}{T_\nu})^4 \qquad (5)$$

In this last formula, F and B relate to the fermion and boson families which are relativistic at nucleosynthesis, T_F and T_B are their temperature relative to the neutrinos temperature T_ν. This time of expansion determines the temperature T_* at which the equilibrium between protons and neutrons freeze out, which in turn fixes the initial n/p ratio such that

$$\frac{n}{p} = \exp -(\frac{\Delta m}{T_*}) \qquad (6)$$

The resulting Y is

$$\sim \frac{2n/p}{1 + n/p} \qquad (7)$$

and therefore depends very much on the expansion rate : an increase of N_ν by one unit corresponds to an increase of Y by 0.01. The neutron lifetime is also a parameter which plays some role in fixing the resulting abundances of these light elements. An increase of the neutron lifetime by 0.1 mn has a quantitative effect on Y_p about ten times lower than the addition of a new neutrino family.

2 – The baryon density which determines the fusion reactions rates like those which transform D and ^{3}He into ^{4}He and ^{7}Li (^{7}Be at high densities).

The classical figure 1 extrated from Yang *et al.*, 1984 displays the resulting abundances of D, ^{3}He, ^{4}He and ^{7}Li obtained with $\tau_{1/2}=10.6$ mn and where the dependence of Y with respect to N_ν is clearly apparent.

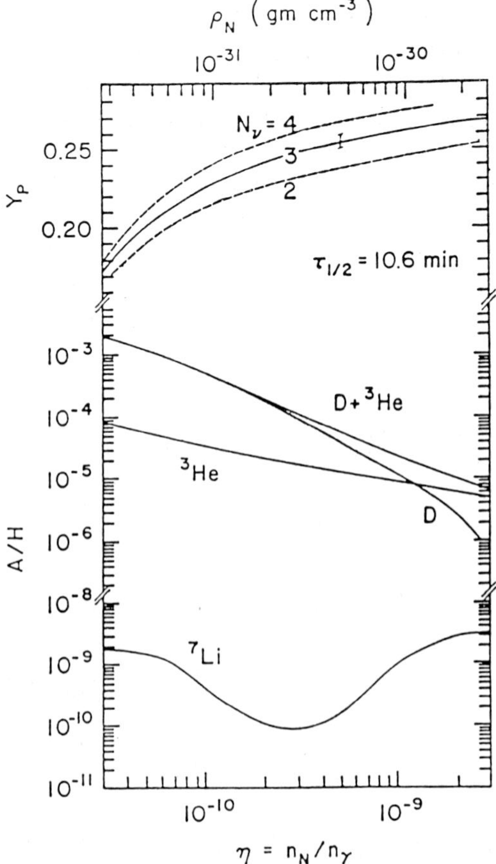

Figure 1

Resulting abundances of D, ^{3}He, ^{4}He and ^{7}Li predicted by the canonical Big Bang model against the baryon density parameter η (for a neutron lifetime $\tau_{1/2} = 10.6$ minutes). The dependence of Y (the ^{4}He abundance by mass) with N_ν the number of neutrino families is shown on this figure coming from Yang *et al.*, 1984.

By using such abundance dependences with the baryon density η, Yang *et al.*, 1984 (repeated by Boesgaard and Steigman, 1985) argue (figure 2a) that there is a range for the baryon density parameter η_{10} where an agreement can be found between the predictions coming from the four different nuclear species. According to these authors 3–$4 < \eta_{10} < 7$–10. The upper limit comes from Y which is limited by the observations (from figure 2a one can see that $\eta_{10} < 10$ corresponds to $Y < 0.26$, $\eta_{10} < 7$ to $Y < 0.254$ The lower limit is constrained by the $(D + {}^3He)$ primordial abundance : $\eta > 3$ corresponds to $(D + {}^3He) < 9.10^{-5}$. In this analysis they assume that D is not destroyed during the galactic evolution more than a factor 2–3 as specified by the classical models of galactic evolution such as those with infall of unprocessed material considered by Audouze and Tinsley, 1974.

From the recollection of Table 1 the alert reader realizes that now we are going to challenge somewhat this optimistic reasoning. We will not be however as extreme as e.g. Vidal–Madjar and Gry (1984) and Gautier (1983) who claim that there is absolutely no agreement between a very low $\eta(Y)$ and a very high $\eta(D)$ obtained in the case where D is destroyed by a modest factor during the galactic evolution.

Before doing this comparison we can make use of the instrumental relations between $Y(D+{}^3He)$ abundances, N_ν and $\tau_{1/2}$ designed in Boesgaard and Steigman (1985) and especially in Steigman $et\ al.$, 1987

$$Y_p = 0.23 + 0.011 \ln \eta_{10} + 0.013(N_\nu - 3) + 0.014(\tau_{1/2} - 10.6) \qquad (8)$$

and

$$Y_p = 0.243 + 0.014[(N_\nu - 3) + (\tau_{1/2} - 10.6)] - 0.018 \log(10^4 \frac{(D + {}^3He)}{H}) \quad (9)$$

From relation (7) one can draw the dependence between Y_p and $\left(\frac{D+{}^3He}{H}\right)_p$ shown on figure 3.

For $N_\nu = 3$, Table 2 indicates the $\left(\frac{D+{}^3He}{H}\right)_p$ calculated by Yang $et\ al.$, 1984 for critical values of Y (Table 1).

TABLE 2

Comparison between the primordial abundances of ${}^4He$ and $(D+{}^3He)$ with $N_\nu = 3$ (from Yang $et\ al.$, 1984).

Y_p	$(D+ {}^3He)/H$
0.25	$5\ 10^{-5}$
0.24	$1.8\ 10^{-4}$
0.235	$3\ 10^{-4}$
0.23	$4\ 10^{-4}$

For most of the values contained in this Y_p range coming from our discussion of §2 (Table 1) $\frac{D+{}^3He}{H} > 9\ 10^{-5}$ which is the upper limit for $\frac{D+{}^3He}{H}$

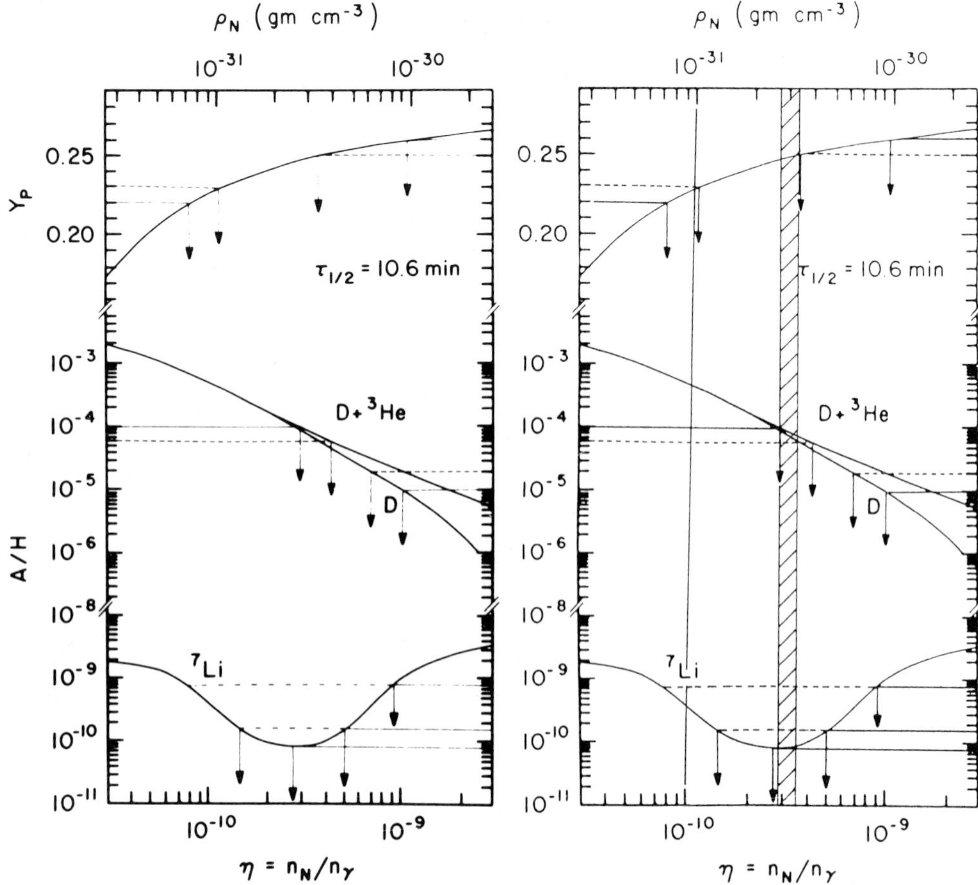

Figure 2a

Same figure as figure 1 from Yang *et al.*, 1984 for N_ν=3 and $\tau_{1/2}$=10.6 minutes. The arrows come from the "optimistic" analysis made by these authors of the ranges for the primordial abundances of the light elements which allow them to deduce $0.01 < \Omega_B < 0.14$–0.19.

Figure 2b

Same as figure 1 and 2a where I indicate the $\eta(Y,D)$ range compatible both with Y_p=0.24 $\pm$ 0.01 (Kunth 1986) and with the Yang *et al.*, 1984 prescription in $\frac{D}{H} < 10^{-4}$. One sees that the resulting η range is quite narrow η=3.2$\pm$0.2 making such comparison fairly contrived. One notes also that $\frac{^7Li}{H})_p$ must be as low as the Spite and Spite (1982) prescriptions which means that population II stars cannot have destroyed their initial Li to keep this picture marginally consistent.

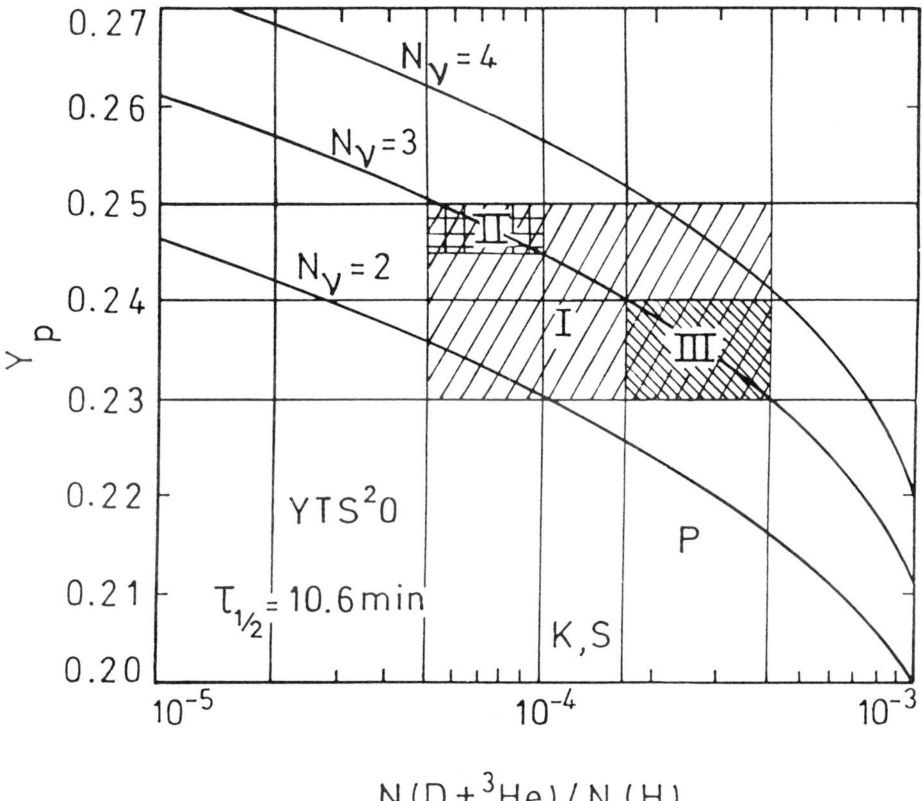

Figure 3

Dependence between Y_p (the primordial abundance of ^{4}He) and that of $(D +{}^3\text{He})$ computed from relation (9) [from Yang *et al.*, 1984]. Assuming that $N_\nu=3$, box 1 corresponds to the Kunth (1986) and Shields (1986) prescription on Y_p. Box 2 corresponds to the domain where these prescriptions agree with those of Yang *et al.*, 1984. Box 2 is clearly separate from box 3 corresponding to $Y_p=0.234 \pm 0.005$. One realizes on this diagram the importance of determining Y_p as precisely as possible.

adopted by Yang *et al.*, 1984 and subsequent publications of this group. It appears therefore that <u>at best</u> there is a quite marginal agreement between $\eta_{10}(D)$ and $\eta_{10}(^4\text{He})$ as indicated by these authors for a much narrower range $\eta_{10} \approx 3.2 \pm 0.2$ (figure 2b) leading to $1.1\ 10^{-2} \leq \Omega_b \leq 0.048\ 10^{-2}$ (by taking the two extremes for the Hubble constant H_o).

A similar difficulty was noted already by Pagel, 1986, who found that $3 < \eta_{10} < 3.6$ corresponding to typically the same range of Ω_B. One should note here that <u>no agreement</u> between $\eta_{10}(D)$, $\eta_{10}(^4\text{He})$ and $\eta_{10}(^7\text{Li})$ can be found if the Duncan and Hobbs (1987) conclusion, regarding $(^7Li)_p (2\ 10^{-10} \leq$

$\frac{^7Li}{H} \leq 9 \ 10^{-10})$, is found to be correct. This is why in our group we have found useful to analyze chemical evolution models leading to a more thorough D destruction during the galactic life without any overproduction of ^{3}He.

4. SPECIFIC MODELS OF GALACTIC EVOLUTION ALLOWING A LARGER D DESTRUCTION

For that purpose we have considered three possible scenarios (a) infall or inflow of processed (D poor) material in the considered zone (here the solar vicinity) (b) destruction of D by stellar winds released during the premain sequence phase. These two types of scenarios have been analyzed in some details by Delbourgo–Salvador $et\ al.$, 1985, (c) evolution models assuming bimodal star formation processes (see e.g. Gusten and Mezger 1983, Larson 1986, Wyse and Silk 1987). These models have inspired us (Audouze $et\ al.$, 1987) to perform some preliminary computations leading to the expected large D destruction.

I am not writing again here the classical equations of chemical evolution which can be found e.g. in Delbourgo–Salvador $et\ al.$, 1985. These equations become most simple in the <u>case of Deuterium</u> which is destroyed inside the stars. The general equation describing the classical evolution of the D abundance by mass X_D is quite simple :

$$\frac{d(\sigma X_D)}{dt} = -\nu \sigma X_D + \delta(X_D) \tag{10}$$

where σ is the gas density (relative to the total mass of galactic matter), ν the rate of astration and δ the rate of infall or inflow.

(i) In models without infall

$$X_D = (X_D)_{primordial}\, e^{-\nu t} \tag{11}$$

(ii) In models with infall of unprocessed material

$$\frac{X_D}{(X_D)_{primordial}} = \frac{1}{\sigma}\left(1 - \frac{\delta}{\nu}\right)e^{-\nu t} + \frac{\delta}{\nu} \tag{12}$$

(iii) With infall (or inflow) of processed material

$$\sigma X_D = (X_D)_{primordial}e^{-\nu t} + (X_D)_{process}\frac{\delta}{\nu}(1 - e^{-\nu t}) \tag{13}$$

(iv) In models where stellar mass loss from pre main sequence stars is taken into account (in these stars the only nucleosynthetic process which could occur is the D transformation into ^{3}He) : equation (10) is replaced by (14) :

$$\frac{d(X_D \sigma)}{dt} = -\nu(1 + f)\sigma X_D + \delta(X_D)_{process} \tag{14}$$

where f is the mass fraction of premain sequence stars lost by stellar winds during these early phases. The resulting X_D abundance is

$$\frac{X_D}{(X_D)_{primordial}} = \frac{1}{\sigma}(1 - \frac{\delta}{\nu(1 + f)})e^{-\nu(1+f)t} + \frac{\delta}{\nu} \tag{15}$$

In the <u>case of Helium 3</u>, classical models of stellar evolution teach us that low mass stars can release significant fractions of ^{3}He during the red giant phases. According to Iben and Truran (1978), the resulting $(\frac{^3He}{H})$ at the end of this phase is :

$$\frac{^3He}{H} = 2.6\ 10^{-4}(\frac{M}{M_\odot})^{-2} + 0.7(\frac{^3He}{H})_{init} \tag{16}$$

when one takes into the IMF the rate of ^{3}He released after that phase is $\propto K(M/M_\odot)^{-4}$. As discussed in Delbourgo–Salvador *et al.*, 1985 the rate of ^{3}He released during the Asymptotic Giant Branch is less important. Finally ^{3}He should mainly come from low mass stars (stars with m>5 $M_\odot$ should not eject any ^{3}He which is itself destroyed inside such stars.

The general equation describing the ^{3}He abundance by mass X_3 evolution is

$$\frac{d(\sigma X_3)}{dt} = -\nu\sigma X_3 + \int_{m_L}^{m_u} E_3(M)\phi(M)\nu\sigma(t - \tau_m)dM + \delta X_3(\inf) + \nu f\sigma X_D \tag{17}$$

where $E_3(M)$ is the ^{3}He fraction released from a star of mass m, τ_m being its lifetime and $\phi(M)$ the usual initial mass function IMF.

The galactic evolution of D and ^{3}He computed through equations (13), (15) and (17) in the frame of models a) and b) are shown in figures 4a, 4b and 5 coming from Delbourgo–Salvador *et al.*, 1985.

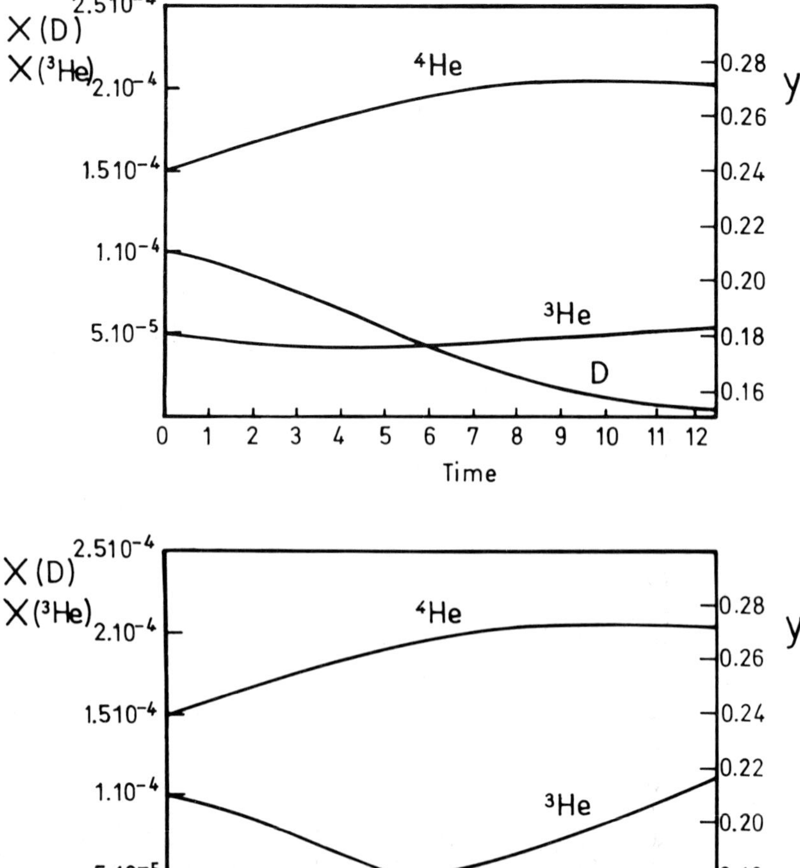

Figures 4a and 4b

Resulting abundances by mass of ^{4}He, ^{3}He and D as a function of time (in Gyr) in chemical evolution models with infall of processed material such that the ^{3}He production rate is a) $5\ 10^{-5}\ (M/M_\odot)^{-4}$. b) $5\ 10^{-4}(M/M_\odot)^{-4}$. The infall rate is $\delta = 0.012$ with $\nu = 0.45$ (from Delbourgo–Salvador *et al.*, 1985).

These two types of models are able to account for a D destruction such that $D/D_p \sim 1/15$ sufficient to reconcile $\eta_{10}(D)$ with $\eta_{10}(^4$He$)$. With the prescriptions of these two types of models the baryonic density η_{10} range inside

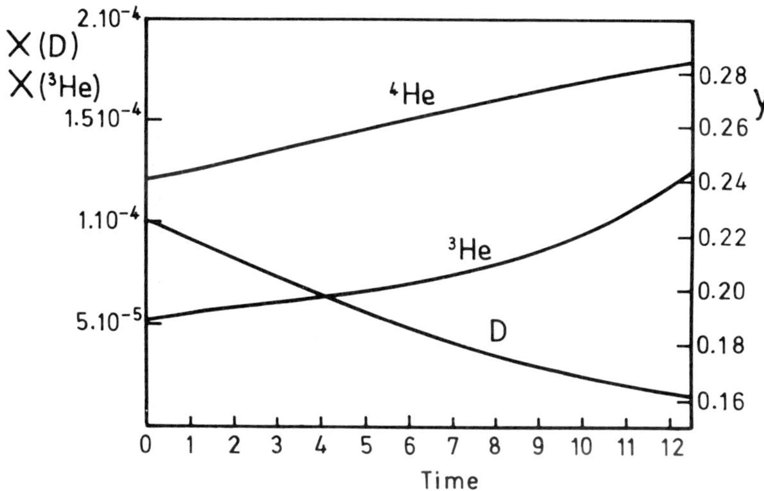

Figure 5
Abundances by mass of ^{4}He, ^{3}He and D as a function of time (in Gyr) with f=0.2 (see text). The production rate of ^{3}He is 5 10^{-4} $(M/M_\odot)^{-4}$, the infall rate δ=0 here (from Delbourgo–Salvador *et al.*, 1985).

which there is such an agreement is $1.2 < \eta_{10} < 4.5$ leading to $4\ 10^{-3} \leq \Omega_B \leq 6\ 10^{-2}$. It should be noted that this resulting Ω_B parameter is significantly lower than that proposed by Yang *et al.*, 1984.

The third type of model c) called the bimodal star formation model has been proposed and considered by several authors (see e.g. Larson 1986). The aim of such models where the rate of star formation especially those of low mass is increased to provide a viable solution for many problems like the G dwarf problem and the evolution of metallicity with time (Pagel and Patchett, 1975), an increase of the resulting M/L ratios and a better account of the colors of the solar vicinity and spiral galaxies than it is achieved in simpler and therefore more classical models. For instance, Wyse and Silk, 1987, consider a bimodal star formation model such that the massive star mode has a star formation rate (SFR) decreasing exponentially with time while the mode which contains stars of all mass has a constant SFR. By doing it these authors have constructed galactic evolution models able to account for the age metallicity relation of the thin disk and the halo. After hearing the oral presentation of my Princeton paper (Audouze, 1987), Larson (1986) noted that this type of model should lead to high (D_{prim}/D_{pres}) ratios.

In Audouze *et al.*, 1987, following a suggestion of J. Silk we have computed the D evolution in a model of that type : In this model the SFR is supposed to be constant with time scale : $\tau_{*1} \propto \nu_1$ such that $\nu_1 = 1$ for times $t \leq 10^9$ years, then when $t > 10^9$ years, the SFR is exponentially decreasing with time such that the time scale $\tau_{*2} \propto \nu_2 \sigma$ where $\nu_2 = 0.3$.

With such models at times t=12.5 Gyr, σ=0.04 $D_{pres}/H=10^{-5}$, $\frac{^3He}{H}pres \sim 10^{-4}$, Y $\sim$ 0.30, Z=0.025 and $\frac{D_{prim}}{D_{pres}}=15$.

That type of model then fulfills the two requirements that we seek to reconcile any baryonic density parameter η, i.e. a large D destruction and a not too large present ^{3}He/H abundance.

The merit of these galactic models is that they can be proved or disproved by three different ways (i) one could expect an improvement of the radio search of interstellar ^{3}He$^+$ abundance as currently pursued by R.T. Rood and his associates. Any viable galactic evolution cannot predict a present ^{3}He abundance higher than the largest ^{3}He measured abundances in the interstellar medium ; (ii) more determinations of the D abundances in high redshift QSOs : these models are only valid if $(\frac{D}{H})_{QSO} > (5\text{-}10)\ 10^{-5}$ for lower values $D/H \leq 5\ 10^{-5}$ the more classical models in which D is not as destroyed would be favoured. One notes that the present (D/H) abundance proposed by Carswell *et al.*, 1987 is quite ambiguous in that respect for the z=3.09 QSO 0420–388. (iii) the third method suggested by Delbourgo–Salvador *et al.*, 1987 is to look for any D abundance variation between different galactic sites where the gas density can vary. Remember that

$\sigma < 0.01$ in central region of our galaxy while

$\sigma \sim 0.05$ in the solar vicinity.

If, for simplicity one uses a model with the constant recycling approximation, the gas density is given by

$$\frac{d\sigma}{dt} = -\nu\sigma(1 - \alpha) + \delta \qquad (18)$$

with $\alpha \simeq 0.2$ being the integrated mass fraction coming back from stars to ISM. From (18) it is easy to show that the relation between the infall (inflow) rate and the astration rate depends significantly on σ_{pres} (the present gas density) :

$$\delta = \nu(1 - \alpha)\frac{\sigma_{pres} - \exp(-\nu(1 - \alpha)t_{pres})}{1 - \exp(-\nu(1 - \alpha)t_{pres})} \qquad (19)$$

In the case of inflow of processed material,

$$\frac{(X_D)_{prim}}{(X_D)_{pres}} = \frac{\delta}{\nu(1 - \alpha)}e^{\nu t} + \left(1 - \frac{\delta}{\nu(1 - \alpha)}\right)e^{\alpha \nu t} \qquad (20)$$

and if D is destroyed during the stellar premain sequence phase

$$\frac{(X_D)_{prim}}{(X_D)_{pres}} = e^{\nu t(\alpha + f)} \qquad (21)$$

Figures 6a and 6b show how the resulting $(X_D)_{prim}/(X_D)_{pres}$ ratios are much dependent on σ_{pres} for these two types of model.

Future work will then tell us who is right between those like us who favor large D destruction during the galactic history or those like Yang *et al.*, 1984 who trust the classical galactic evolution models leading to moderate D destructions.

Before closing this section devoted to the interplay between galactic evolution models and the early nucleosynthesis, I would like to make the following remark regarding ^{7}Li. The present debate between those like Spite and Spite (1982) who argue in favour of a primordial ^{7}Li/H abundance as low as $\sim 10^{-10}$ and Duncan and Hobbs (1987) for whom $(^7\text{Li/H})_{prim} > 2\,10^{-10}$ has the following consequences : in favour of the Spite and Spite (1982) view is the fact that one deduce from it a η_{10} (^7Li) value consistent with $\eta(\text{D})$ and $\eta(\text{Y})$ but one has to explain why ^{7}Li has increased dramatically during the first billion years of galactic evolution and remain constant afterwards as pointed out by Hobbs and Pilackowski (1987). If $(\text{Li/H})_p$ is higher, then one has difficulty to reconcile the resulting $\eta(\text{Li})$ with $\eta(\text{D})$ and $\eta(\text{Y})$ and if one does so, one is led to very low $\Omega_B \ll 0.01$. On the other hand the chemical evolution of ^{7}Li in that case is far easier to be explained. We (i.e. J. Silk, E. Vangioni–Flam, R. Wyse and myself) are currently investigating this problem in the Spite and Spite (1982) terms within the frame of bimodal star formation which should lead to a rapid Li increase at the beginning of the galactic history if the ^{7}Li production is related to massive stars.

5. EARLY NUCLEOSYNTHESIS AND PARTICLE PHYSICS

The simple (canonical) Big Bang models with specific galactic evolution hypotheses if one follows the french school or without if one trusts the Chicago–Ohio–Minesotta group seems to imply that the number of neutrino (lepton)

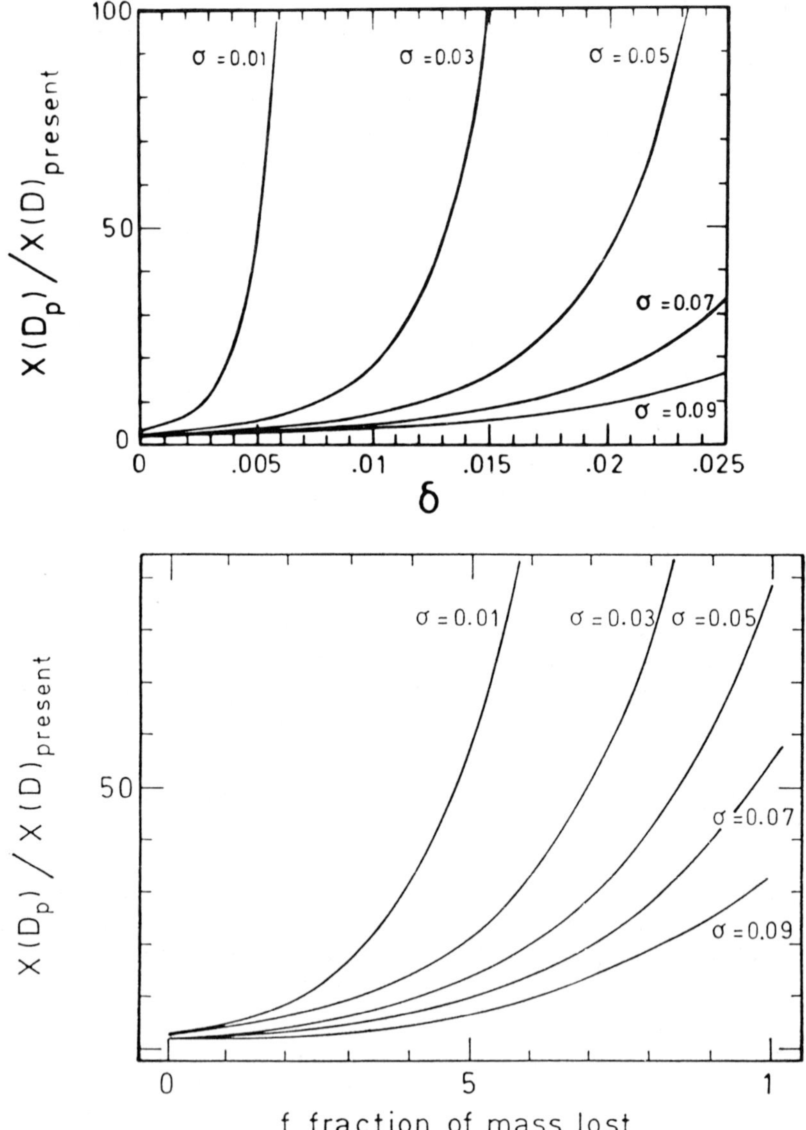

Figure 6a
Relation between the ratio $X(D)_{prim}/X(D)_{pres}$ and the rate of infall δ for different values of the <u>present</u> gas density σ ranging from 0.01 to 0.09 (from Delbourgo–Salvador *et al.*, 1987).

Figure 6b
Relation between the ratio $X(D)_{prim}/X(D)_{pres}$ and the fraction of mass f lost during the pre main sequence phase for different values of the present gas density σ ranging from 0.01 to 0.09 (from Delbourgo–Salvador *et al.*, 1987).

families is limited to the three already known species (in agreement with GUT theories) and that the baryonic cosmological parameter $\Omega_B \sim 0.1$. In this section, I would like to make a brief account of some models based on recent particle physics developments which have been built in order to see if the light element abundances resulting from the Big Bang can be consistent with $\Omega=1$ values. These high Ω values have been found to be a consequence of current inflationary models. There are two ways to achieve that goal (i) either Ω_B is large, then the Big Bang nucleosynthesis would result in ^{4}He and ^{7}Li overabundances and D and ^{3}He underabundances ; *. these discrepancies with respect to the observed "primordial" abundances can be solved if photodisintegration processes ,triggered by high energy photons coming from the decay of massive particles destroy in part ^{4}He and ^{7}Li and restore the D and ^{3}He observed abundances from them. This type has been considered e.g. by Audouze *et al.*, 1985 who analyzed such processes occuring after the decay of massive neutrinos and gravitinos while Salati *et al.*, 1987 conducted a similar analysis with decaying photinos (ii) the other alternative is to have Ω large because of the existence of non baryonic particles which act as mere spectators of the early nucleosynthesis : the literature and these proceedings contain many different proposals advocating in favour of hot dark matter (neutrinos with mass of 50 eV ...) or of cold dark matter (axions, strings ...). In our work we have considered only stable photinos (Salati *et al.*, 1987) or stable quark nuggets (Schaeffer *et al.*, 1985). We will summarize only here the few points which deserve to be made on that question.

5.1 Early photodisintegration processes affecting the light element abundances

Three possibilities have been considered up to now although any other massive decaying particle able to produce high energy gamma rays with the appropriate time scale is fitted for this specific purpose. There are (i) massive neutrinos with masses of ~ 500 MeV and life-time $\sim 10^5$–10^6 sec ; (ii)

---

  * After the Beijing conference I became acquainted with the proposal currently made by Applegate *et al.*, 1987 according whom the quark–hadron phase–transition which should be a first order phenomenon could lead to huge inhomogeneities of proton and especially neutron distribution : neutrons are especially favoured in the evolution of such Big Bang plasmas because their diffusion mean free path is much larger than that of protons. With such a proposal interesting, nucleosynthetic byproducts can be synthetized not only the very light elements under consideration here but also heavier elements. In their so–called segregated model, this new type of early nucleosynthesis could be consistent with the D, ^{3}He, ^{4}He and ^{7}Li primordial values for an Ωh^2 value of 1 which is the main aim of the models described here.

are (i) massive neutrinos with masses of $\sim$ 500 MeV and life–time $\sim 10^5$–10^6 sec ; (ii) gravitinos, the $3/2$ spin particles which are the supersymmetric (SUSY) counterparts of the 2 spin gravitons. Their mass range from 20 GeV and 1 TeV and their lifetime is $\sim 10^5$–10^6 sec ; these two first possibilities have been considered by Audouze *et al.*, 1985 and (iii) unstable photinos of comparable lifetime (Salati *et al.*, 1987).

Lindley (1985) has shown very convincingly why these particles should have a lifetime as long as 10^5–10^6 sec. The physical reason is the following : there exists an energy threshold E_γ for these gamma rays such that

$$E_\gamma kT = \frac{1}{50} MeV^2 \qquad (22)$$

where kT is the thermal energy of the Universe background radiation.

Above this threshold, the high energy photons scatter on the thermal photons and produce $(e^+ e^-)$ pairs while below the threshold they can Compton scatter on electron, induce $(e^+ e^-)$ pair production by scattering on the nuclei or induce photofission. The energy E_γ of the photofission gamma rays should be high enough to be above the photofission reaction threshold i.e. $E_\gamma > 20$ MeV. On the other hand $E_\gamma kT < \frac{1}{50}$ MeV^2 for this photofission to take place : For $E_\gamma > 20$ MeV, $kT < 10^{-3}$ MeV, $t \geq 10^5$–10^6 sec. Therefore photofission processes cannot occur at times shorter than these 10^5–10^6 sec during which the light elements are protected against photofission by $(e^+ e^-)$ pair production.

The minimum energy for the gamma rays to induce photofission and the conditions on the time scale define a fairly precise set of decay life–time–mass conditions for these particles. As examples Figure 7 extracted from Audouze *et al.*, 1985, shows how part of ^{4}He can be transformed into D and ^{3}He by high energy gamma rays coming from the decay of gravitinos. Figure 8 from Salati *et al.*, 1987 is a map of the lifetime–mass of decaying photinos defining 5 different regions for a Universe such that $\Omega=0.02$.

In region I the photinos are not massive enough to induce any photofission : they only act as another neutrino family and lead therefore to an overproduction of ^{4}He. Region II corresponds to photinos with a lifetime too short to affect the nucleosynthesis. In region III D is dissociated , in Region IV all three elements D, ^{3}He and ^{4}He are dissociated while in region V a partial photodisintegration of ^{4}He leads to an overproduction of ^{3}He. The results of the early nucleosynthesis can then constrain significantly the mass–lifetime relation for unstable photinos and have interesting predictions on the physics of the different possible decay models of photinos (see Salati *et al.*, 1987 for more details).

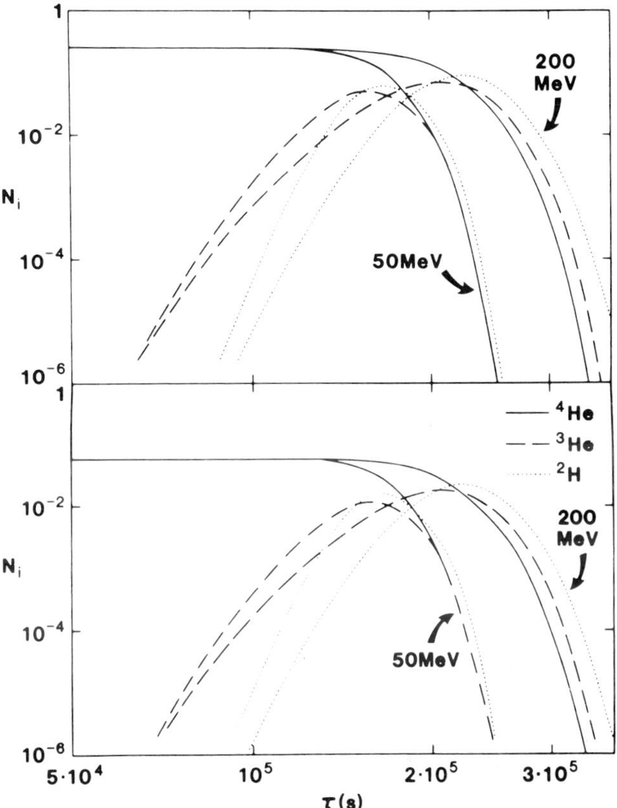

Figure 7
Effects of decaying gravitinos on light element abundances : $Y_p=1$ for the upper panel and $Y_p=0.24$ for the lower one ; $X_p(D)=X_p(^3He)=0$. The two sets of curves have been calculated for 50 MeV and 200 MeV electrons (respectively 100 MeV and 400 MeV photons), τ_s is the gravitino lifetime (from Audouze *et al.*, 1985).

To sum up any massive ≥ 500 MeV long lived ($\tau \geq 10^5$–10^6 sec) particle decaying in high energy gamma rays can affect the outcome of early nucleosynthesis by photofission reactions.

5.2. Early nucleosynthesis in the presence of particles acting as spectators

We have envisaged so far two possible types of particles which are non baryonic, which could act as spectators of the early nucleosynthesis and induce large cosmological parameter value $\Omega=1$. The first possibility is <u>stable</u> photinos of mass in the fairly narrow range $3 \leq M_{\tilde{\gamma}} \leq 8$ GeV which could close the Universe and still be consistent with the primordial abundances

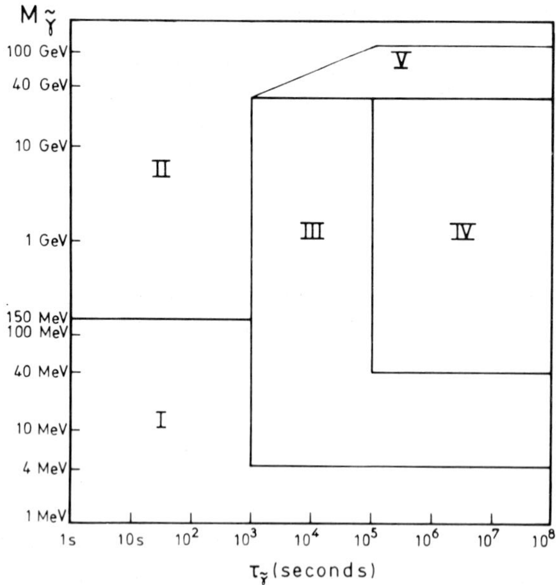

Figure 8

Photino mass $(M_{\tilde{\gamma}})$ and lifetime $(\tau_{\tilde{\gamma}})$ domains defined by the nucleosynthesis constraints for $\Omega=1$. Each domain is defined in the text. The only allowed domain is region I (low lifetime – high mass) and the border of this region with domain III (from Salati *et al.*, 1987).

of the light elements (Salati *et al.*, 1987). The second possibility are quark nuggets. This term designates particles with atomic mass $10^2 < A < 10^{57}$ and $Z \sim 5\,A^{1/3}$ made up of 3 A quarks (u, s and d quarks in equal amounts) and which would fill up the "nuclear desert" between the heaviest nuclei and the neutron stars (de Rujula, 1986). These quark nuggets could have been produced during the quark–hadron transition (Witten, 1984). Schaeffer *et al.*, 1985 have suggested that an $\Omega \geq 1$ Universe can be consistent with the results of the early nucleosynthesis if such nuggets exist, if they are stable and if they have an appropriate atomic mass : $10^{15} < A < 10^{17}$. This A range may have to be changed after a better account of the nuclear absorption and emission rates from these nuggets. However one should remind that Alcock and Fahri, 1985 claim that quark nuggets of any atomic mass disappear by transforming themselves into nucleons before the time of nucleosynthesis.

In short it is not impossible to build up models with low Ω_B and large Ω coming from some specific types of particles like stable photinos and quark nuggets but it may not be the most likely hypothesis.

6. CONCLUSION

To sum up the discussion presented above, the very light elements D, ^{3}He, ^{4}He and ^{7}Li remain still very powerful and handy cosmological tools.

1 – Regarding the observations, one can expect in a foreseable future significant progress on several important issues :
— determination of D abundances both in various galactic locations (where the relative gas densities are different from each other) and in the absorption line systems of quasars at high redshifts. Such observations should allow to distinguish between the galactic evolution model preferred in our group (see e.g. Delbourgo–Salvador *et al.*, 1987) which lead to a huge D destruction during the galactic history and therefore large variations of D with the considered galactic locations and significantly large D abundances in high redshifts QSOs. If such variations are not found, one should come back to the views expressed by the Chicago group (see e.g. Yang *et al.*, 1984). In this case, one may have some difficulty to reconcile the baryon densities predicted by ^{4}He and D primordial abundances respectively.

— determination of ^{3}He abundances : we anticipate forthcoming exciting data on the interstellar ^{3}He abundances coming from R.T. Rood and his associates. A significant decrease of the interstellar ^{3}He concerning especially the high value found for the W3 HII region would be most useful in constraining effectively the rate of ^{3}He formation and therefore of D destruction during the galactic evolution.

— after the discussions concerning the ^{4}He abundance determinations in blue compact galaxies (Pagel *et al.* 1986, Kunth 1986, Shields private communication) one might expect to know soon Y_p with an accuracy of 5 %.

— after the works of different groups pionnered by F. and M. Spite and their associates and followed by those of Duncan and Hobbs, Boesgaard *et al.*, one should have soon a clear–out view on the primordial abundance of ^{7}Li.

I have argued here that the standard (simple or canonical) Big Bang model still stands very well to account for the "observed" primordial abundances of the very light elements if D is largely destroyed during the galactic evolution. In that respect new and successful galactic evolution models like the bimodal star formation model of Larson, 1986 considered also by Wyse and Silk (1987) are especially suited to achieve that type of D behaviour. In that case as noted by Delbourgo–Salvador et $al.$, 1985, the resulting Ω_b would be somewhat smaller than the value quoted e.g. by Boesgaard and Steigman (1985) $4 \ 10^{-3} < \Omega_b < 6 \ 10^{-2}$ instead of $0.01 < \Omega_B < 0.14$. In this situation the Universe should be filled by large density of non baryonic particles in order to account for the dynamics of its large scale structures.

I have also considered also models derived from recent developments in particle physics. Most of them have been designed in our group, but see also Scherrer and Turner, 1987, like those of Audouze et $al.$, 1985, or of Salati et $al.$, 1987 which take advantage of the possible existence of massive, long lived particles like massive neutrinos, gravitinos or photinos which could decay in releasing high energy photons able to partially photodisintegrate ^{4}He and ^{7}Li produced in a large Ω_B universe (the dark matter would be baryons in that case). Other models like Schaeffer et $al.$, 1985 attempt to take advantage of the possible existence of quark nuggets which are nuclearites which could appear during the quark–hadron phase transition (Witten, 1984) and assist like spectators to the early nucleosynthesis while filling the Universe with their mass. In that respect it is interesting to mention the recent proposal of Applegate et $al.$ (1987) who take advantage of the possible existence of large baryon density fluctuations which could occur in the QCD or electroweak phase transition and lead to the formation of very neutron rich regions which in turn would trigger some interesting processes pertaining to primordial nucleosynthesis.

Although these problems have not been evoked in this presentation, interesting works have still to be pursued in three directions. a) study of models with non zero cosmological constant Λ which should be the signature of fast expanding Universes, b) study of models where the chemical potential of neutrinos can be high and then pursue the discussion initiated by David and Reeves, 1980 and by Steigman, 1985, c) analysis of models corresponding to inhomogeneous and/or anisotropic Universes and pursue the work of e.g. Rothman and Matzner (1984), Delbourgo–Salvador (1986) and Matzner (1986).

Because of the many progresses achieved both in the observation of faint far and/or old objects in the overall geometry of the Universe, in the understanding of particle physics and in the construction of new unification schemes, one expects that the formation of the very light elements (D, ^{3}He, ^{4}He and ^{7}Li) will inspire for sometime exciting new ideas which may be useful to cosmology.

Acknowledgements

The work which is summarized here comes from a long–standing collaboration with several colleagues : P. Delbourgo–Salvador, D. Lindley, G. Malinie, P. Salati, J. Silk, E. Vangioni–Flam, A. Vidal–Madjar with whom I have much pleasure to work. Let me express my appreciation to professors Li Zhi Fang, G.R. Burbidge, and their colleagues from China who have arranged such an outstanding and memorable symposium. This contribution has been written up when I was visiting Miller Professor at the University of California, Berkeley. I thank Professors G. Shields, J. Silk and Dr R.G. Wyse for enlightening discussions. I would like to thank the Miller Institute and its director R. Ornduff and Professor J. Silk for their hospitality. I thank also M.C. Pelletan for her skilled typing of this contribution and E. Vangioni–Flam for a careful reading and criticism of its draft. Part of the research reported here has been supported by PICS No 18.

REFERENCES

Alcock C. and Fahri E., 1985, Phys. Rev. D, **32**, 1273

Applegate J., Hogan C. and Scherrer R.J., 1987, Phys. Rev. D, in press

Audouze, J., 1987, in Nucleosynthesis and Chemical Evolution, A. Hauck and A. Maeder ed., Saas Fee, in press

Audouze J., 1987, in Dark matter in the Universe, eds. J. Kormendy and G. Knapp, Reidel–Dordrecht

Audouze J., Lindley D and Silk J., 1985, Ap. J. Letters, **293**, l53

Audouze J., 1982, in Astrophysical cosmology, eds H.A. Brück *et al.*, Pontificiae Academie Scientarum Scripta Varia, p. 395

Audouze J., 1984, in Large Scale Structure of the Universe, Cosmology and Fundamental Physics, eds G. Setti and L. Van Hove, CERN, Geneva p.293

Audouze J., Delbourgo–Salvador, P. and Vangioni–Flam, E., 1987, in Advances in Nuclear Astrophysics, J. Audouze *et al.*, eds. Frontières, to be published

Audouze J. and Tinsley B.M., 1974, Ap. J., **192**, 487

Black D.C., 1972, Geochim. Cosmochim. Acta, **36**, 347

Boesgaard A.M. and Tripicco, M.J., 1986, Ap. J. (Letters), **302**, L49

Boesgaard A.M. and Steigman G., 1985, Ann. Rev. Astron. Astrophys. , **23**, 319

Carswell R.F., Irwin M.J., Webb J.K., Baldwin J.A., Atwood B., Robertson J.G. and Shaver P.A., 1987, to be published

Cayrel, R., Cayrel de Strobel, G., Campbell, B., Dappen, W., 1984, Astrophys. J., **283**, 205

David Y. and Reeves H., 1980, in Physical cosmology, R. Balian , J. Audouze, D.N. Schramm eds., North Holland Amsterdam, p.433

Delbourgo–Salvador P., Gry C., Malinie G. and Audouze J., 1985, Astron. Astrophys.,**150**, 53

Delbourgo–Salvador P., Audouze J. and Vidal–Madjar A., 1987, Astron. Astrophys., in press

Delbourgo–Salvador P., 1986, Thèse de Doctorat, Paris 7 University, unpublished

De Rujula A., 1985, Nucl. Phys., **A434**, 605

Duncan D.K.. and Hobbs L., 1987, in Observational Cosmology, eds G.R. Burbidge, A. Hewitt and L.Z. Fang, Reidel–Dordrecht

Duncan, D.K., Jones, B.F., 1983, Astrophys. J., **271**, 663

Ferlet R. and Dennefeld M., 1984, Astron. Astrophys., **138**, 303

Hobbs L.M. and Pilackowski C., 1986a, Ap. J. Letters, **309**, L17

Hobbs L.M. and Pilackowski C., 1987, Ap. J. Letters, to be published

Gautier D., 1983, in Primordial helium, P.A. Shaver *et al.* eds, ESO Garching, p. 139

Geiss J., Reeves H., 1972, Astron. Astrophys., **18**, 126.

Gry C., Lamers H.J.G.L.M. and Vidal–Madjar A., 1984, Astron. Astrophys., **137**, 29

Iben Jr I. and Truran J.W., 1978, Ap. J., **220**, 980

Kunth D., 1986, P.A.S.P., in press

Kunth D. and Sargent W.L.W., 1983, Ap. J., **273**, 81

Larson R.B., 1986, M.N.R.A.S., **218**, 409

Lequeux J., Peimbert M., Rayo J.F., Serrano A. and Torres–Peimbert S., 1979, Astron. Astrophys., **80**, 155

Lindley D., 1985, Ap. J., **294**, 1

Matzner R.A., 1986, P.A.S.P. in press

Olive K.A., Schramm D.N., Steigman G., Turner M.S. and Yang J., 1981, Ap. J., **246**, 557

Pagel B.E.J. and Patchett B.E., 1975, M.N.R.A.S., **172**, 13

Pagel B.E.J., 1986, in Material Content of the Universe, published by the Astronomical Royal Soc.

Pagel B.E.J., Terlevich R., Melnick J., 1986, Pub. Astr. Soc. Pacific, in press

Peebles P.J.E., 1966, Astrophys. J., **146**, 542

Reeves H., Audouze J., Fowler A. and Schramm D.N., 1973, Ap. J., **179**, 909

Reeves H., 1974, Ann. Rev. Astron. Astrophys., **12**, 437

Rood R.T., Bania T.M. and Wilson T.L., 1984, Ap. J., **280**, 629

Rothman T. and Matzner R., 1984, Phys. Rev. D, **30**, 1649

Salati P., Delbourgo–Salvador J. and Audouze J., 1987, Astron. Astrophys.,
        in press
Schaeffer R., Delbourgo–Salvador P. and Audouze J., 1985, Nature, **317**,
        6036
Scherrer R.J. and Turner M.S., 1987, Ap. J., in press
Shaver P.A., Kunth D. and Kjär K. eds, 1984, Primordial Helium, ESO
        Garching publication
Spite F. and Spite M., 1986, Astron. Astrophys., **163**, 140
Spite M., Maillard J.P. and Spite F., 1984, Astron. Astrophys., **141**, 56
Spite F. and Spite M., 1982, Astron. Astrophys., **115**, 357
Steigman G., 1985, Nucleosynthesis and its implications on nuclear and par-
        ticle physics, eds J. Audouze and N. Mathieu, NATO ASI Series, **163**,
        p.45
Steigman G., Olive K.A., Schramm D.N. and Turner M.S., 1987, Phys. Let-
        ters B, to be published
Steigman G., 1985, in Nucleosynthesis : challenges and new developments
        W.D. Arnett and J.W. Truran ed., U. of Chicago Press, p. 48
Vidal–Madjar, A., Gry, C., 1984, Astron. Astrophys., **138**, 285.
Vidal–Madjar, A., 1987, in Space Astronomy and Solar System Exploration,
        ed. W.R. Burke, ESA–SP 268
Vigroux L., Stasinska G. and Comte G., 1986, Astron. Astrophys., in press
Vigroux, L., 1986, Comment made during the Japan–France seminar at
        Sendai (Japan), November 1986
Wagoner R.V., Fowler W.A. and Hoyle F., 1967, Ap. J., **148**, 3
Wagoner R.V., 1969, Ap. J. Suppl., **18**, 247
Witten E., 1984, Phys. Rev. D **30**, 272
Wyse R.F.G. and Silk J., 1987, Ap. J. Letters, in press
Yang J., Turner M.S., Steigman G., Schramm D.N. and Olive K.A., 1984,
        Ap. J., **281**, 493
Yang J., Schramm D.N., Steigman G., Rood R.T., 1979, Astrophys. J.,**227**,
        697

DISCUSSION

PECKER: 1) The He^4 may not be as "primaeval" as you think it should
be, if diffusion of H vs. He, under the influence of radiation pressur
acting on H-Lyα, but not on He Lyα or HeII-Lyα, is acting in objects
where a large number of ionizing photons ionize H and He locally. In
other terms, loss of H may be a way to see an apparent enhancement
of He abundance.

2) About the Sourian baryonic-symmetry model, based on
distribution of quasars. Do you mean you do not accept his
statistical analysis of such a distribution?

AUDOUZE: 1) There are ample discussions among the specialists about
ionization effects on He in blue compact galaxies. In a sample of
about 100 objects, only 10 of them which show sufficiently ionization
effects were kept in order to avoid any improper correction. Their
method seems very safe in that respect.

2) About symmetric models. I do not believe that the Sourian models belong to the class I am criticizing, in which the amount of matter is equal to that of antimatter. The only thing I said is that the relative number of photons relative to baryon prevents equality between matter and antimatter densities.

NORMAN: As we heard this morning, we may well need a significant value for the cosmological constant. What will this do to your standard model. For example, will this change your limit on the number of neutrinos. Possibly the extra push in expansion due to λ will then reduce the allowed number of ν's to below three. An interesting way to look at this, perhaps, is to use the ^{4}He abundance and $N_\nu = 3$ to limit the cosmological constant.

AUDOUZE: I should say that both the Chicago and the Paris groups have not yet investigated properly models with non-zero cosmological constant. My bet is that you are right in saying that the ^{4}He abundance puts a strong limit to near zero values for this constant. For non-zero values the expansion is accelerated and would lead to much too high He abundance. Therefore the limits that we could put on this constant should be quite stringent and close to zero.

NARLIKAR: If we assume that Y_p lies in the range 0.24 ± 0.01, how do we account for very low values of Y ($\lesssim 0.20$) in certain objects?

AUDOUZE: At present the only real difficulty would come from the Jupiter atmosphere observations. Some chemical fractionation could affect the He/H ratio deduced from this site. But it is true that if $Y < 0.22$ the simple Big Bang model is in trouble.

BURKE: Recently Kardashev has proposed that the baryon asymmetry that we observe locally is a fluctuation of finite extent in a larger universe with a mean baryonic density of zero. What implication would this have for your model?

AUDOUZE: The simple early nucleosynthesis model works only with specific expansion in rates and baryonic density contents. I am not aware of the Kardashev model. I can only say that in order to explain the light element abundances in the frame of our hypothesis any model should come with the same type of expansion and baryonic density content. To pursue this question, I have heard in informal conversations that Dr. Hogan is developing some models in which he would predict the appropriate amount of ^{4}He in models emphasizing the role of the quark-hadrun first order transition.

SETTI: A couple of years ago it was said that primordial nucleosynthesis models would constrain the number of neutron types to no more than 4-5. What has changed now to bring the limit down to 3?

AUDOUZE: It is the noticeable decrease of the He abundance quoted by Pagel (1986) and independently supported by Kunth (1986) which makes me feel that N_ν = 3 is the actual limit to the number of neutrino families. This conclusion would also be reinforced with higher primordial Li abundances as suggested by Duncan (this conference).

SILK: The Chicago models of primordial nucleosynthesis allowance of Ω_M require baryonic dark matter, but your standard models (at least some of them) do not. Could you comment on the origin of this difference?

AUDOUZE: The main difference between the Chicago school models and ours comes from the galactic evolution of D (and possibly also of ^{3}He): we presently favour significant decreases of D by factors of about 10 during the galaxy history which would lead to large ($\sim 10^{-4}$) primordial D abundance in agreement with the primordial ^{4}He abundance (and also larger ^{7}Li abundances). This would lead to lower values of η (Ω) such that $4 \times 10^{-3} < \Omega < 0.06$ which would limit in a more stringent way the baryonic dark matter present in the Universe. As said in the talk, there is an observational test (the D abundance/interstellar gas density correlation) which should support (rule out) our models relative to theirs.

PRIMORDIAL NUCLEOSYNTHESIS OF ^{7}Li

Douglas K. Duncan
The Space Telescope Science Institute
3700 San Martin Drive, Baltimore, MD 21218 USA

L. M. Hobbs
Yerkes Observatory, University of Chicago
Box 258, Williams Bay, WI 53191 USA

ABSTRACT. We have observed 23 halo stars with space velocities $|\vec{v}_{LSR}| \geq 100$ km s^{-1} and metallicities [Fe/H] ≤ -0.6. Twelve of these 23 show the more extreme properties $|\vec{v}_{LSR}| \geq 160$ km s^{-1} and [Fe/H] ≤ -1.4 and should therefore constitute an especially old, homogeneous subgroup. The principal results for these 12 extreme halo stars and 5 similar stars observed in previous studies are that (1) a single, well defined relation, previously discovered and discussed by Spite and Spite, exists without exception between the atmospheric Li/H ratio and T_e, and (2) at $T_e \geq 5600$ K the average lithium abundance is $<$Li/H$> = 1.2 \pm 0.3 \times 10^{-10}$. The latter value constitutes a lower limit on the ^{7}Li fraction produced in primordial nucleosynthesis and thereby significantly constrains the cosmic ratio of baryons to photons.

1. INTRODUCTION

Accurate measurements of the abundance of lithium in the surface layers of both old and young stars and in the interstellar medium can increase our knowledge of stellar structure, Galactic element production, and big-bang nucleosynthesis. The latter subject recently has been reviewed by Boesgaard and Steigman (1985). Observational efforts to measure the lithium abundance in stars sufficiently old to test theoretical estimates of ^{7}Li production in the pre-stellar universe were pioneered by Spite and Spite (1982; hereinafter SS) and by Spite, Maillard, and Spite (1984; hereinafter SMS). This paper presents new observations which confirm and extend their work.

In conventional theoretical models of the sun, temperatures hot enough to destroy lithium are reached somewhat below the base of the convective envelope. The present abundance of lithium in the solar atmosphere is about 100 times smaller than its initial main-sequence value. The discovery by SS that lithium is present in halo stars of approximately solar temperature in amounts at least ten times the solar value was therefore surprising and important. An explanation of

119

A. Hewitt et al. (eds.), Observational Cosmology, 119–126.
© *1987 by the IAU.*

how the lithium abundance in such halo stars can exceed the value
observed in the younger sun may be related to their lower
metallicities. Calculations by Däppen (1984) and by D'Antona and
Mazzitelli (1984) show that convective zones become cooler at the base
and thinner, as the metal abundance is reduced at a fixed stellar
effective temperature. The rate of lithium destruction, which depends
strongly on temperature, may therefore be severely reduced in the halo
stars.

Stimulated by the initial results of SS, we independently set out
in 1983 to extend those results to a larger number of subdwarfs chosen
to be as homogeneously old as possible. Our separate results are
combined here, where measurements of the Li I $\lambda6707$ doublet at a
resolution typically of 0.25Å are reported for a primary group of 23
subdwarfs with iron abundances [Fe/H] $\leq$ -0.6 and space velocities
$v = |\vec{v}_{LSR}| \geq 100$ km s^{-1}. Twelve of these stars in fact are extreme
halo stars which show [Fe/H] $\leq$ -1/4 and $v \geq 160$ km s^{-1}.

2. OBSERVATIONS AND RESULTS

2.1. The Spectra

Spectra of 16 of the stars were obtained at an instrumental resolution
(FWHM) of 0.26 Å, using the Digicon detector and grating B of the
coude spectrograph at the 2.7 m reflector of McDonald Observatory.
The entrance slit corresponded to 0.6 arcsec and yielded a projected
slit two diodes wide. A useful spectral interval of about 120 Å was
recorded in each exposure. The typical S/N ratio of 80 which was
achieved corresponds to a 2 σ detection limit of about 6 mÅ. For a
few of the fainter stars a slit width of 1.2 arcsec and a consequent
resolution of 0.52 Å were chosen instead. Data for 9 of the stars
were obtained in 1983 and 1984 at a resolution of 0.3 Å with the 3.0 m
reflector of Lick Observatory, using the bare reticon detector at the
coude camera with a 40-inch focal length. The agreement of the $\lambda6707$
equivalent widths for the five stars in common between McDonald and
the Lick observations is excellent, the average difference being 0 $\pm$ 4
mÅ. Finally, observations of five of the faintest stars were acquired
in October 1985 at a resolution of 0.2 Å, using the TI-2 CCD and the
echelle spectrograph at the 4 m Mayall reflector of Kitt Peak National
Observatory.

The agreement of our results with the data of SS and SMS, which
were recorded at generally similar resolution and slightly better
photometric accuracy, is generally excellent. The average difference
in the Li I equivalent widths is 1 $\pm$ 3 mÅ for the 12 stars for which
positive $\lambda6707$ detections were obtained in both observing programs.

2.2. The Adopted Temperatures

Li I $\lambda6707$ is the resonance line of the neutral species of an atom
which is almost completely ionized in these stars. The equivalent
width of the line is therefore very temperature sensitive, and

accurate stellar temperatures are needed in order to derive accurate
lithium abundances from model atmospheres. Peterson and Carney (1979)
and Carney (1983) derived temperatures for most of these stars from R-
I and V-K colors and by matching spectrophotometric scans in the
region 5000 to 8400 Å to surface fluxes calculated from ATLAS6 model
atmospheres. From photometry available in the literature, we have
calculated temperatures anew by using their two photometric methods.
The methods are accurate at $4500 \lesssim T_e \lesssim 7000$ K, a range which includes
all of our halo stars. After initial calibration, these three kinds
of temperature determination are independent, so random errors should
be reduced by averaging them. In an effort to get the best possible
temperatures, one more set of colors, the Stromgren photometry
tabulated by Hauck and Mermilliod (1979), was also examined. In a few
cases where there was disagreement by more than 100 K among the
various temperatures discussed above, the Stromgren colors were used
to indicate which temperature is more nearly correct. An
intercomparison of the three sets of temperatures indicates a random
error of approximately 80 K in the temperature estimated from any
color or scan, and we conclude that the random component of the
standard (1 σ) error in the final temperatures adopted here is about
60 K.

2.3. The Derived Abundances

Curves of growth for Li I λ6707 were computed using model atmospheres
constructed by Bell (1984) and the spectrum-synthesis program WIDTH6
(Kurucz 1983). Calculations were carried out for T_e = 4500 K, 5000 K,
5500 K, and 6000 K, log g = 4.5 and 3.75, and [Fe/H] = 0, -1, and
-2. The method described by Duncan and Jones (1983) was used to
extend the calculations to T_e = 6500 K as well. An isotope ratio
^{6}Li/^{7}Li = 0 was adopted. Errors in the relative abundances from star
to star arise from uncertainties in both the equivalent widths and the
temperature differences. A typical random error of 60 K at T_e = 5800
K and W_λ = 30 mÅ causes an abundance error of ±15% in our results; a
measurement error of ±20% in the equivalent width contributes a
further abundance error of ±20%. Quadratically combining these
independent errors yields a total error of ±25%, or 0.10 dex, which we
take as the representative standard (1 σ) error of the _relative_
abundances. Our derived abundances also can be compared directly with
those of SS and SMS, who used a different set of model atmospheres.
For 11 stars in which λ6707 is definitely detected in both sets of
observations the mean logarithmic difference is 0.02 ± 0.12 dex.

3. DISCUSSION

The lithium abundances of the 23 halo stars investigated here are
plotted as a function of effective temperature in the Figure. The
most important results of the present study are seen to be (1) a
confirmation of the discovery by SS and SMS that lithium is present at
an average abundance near $\langle N(Li)\rangle = \langle 12 + \log (Li/H)\rangle = 2.05$ in nearly

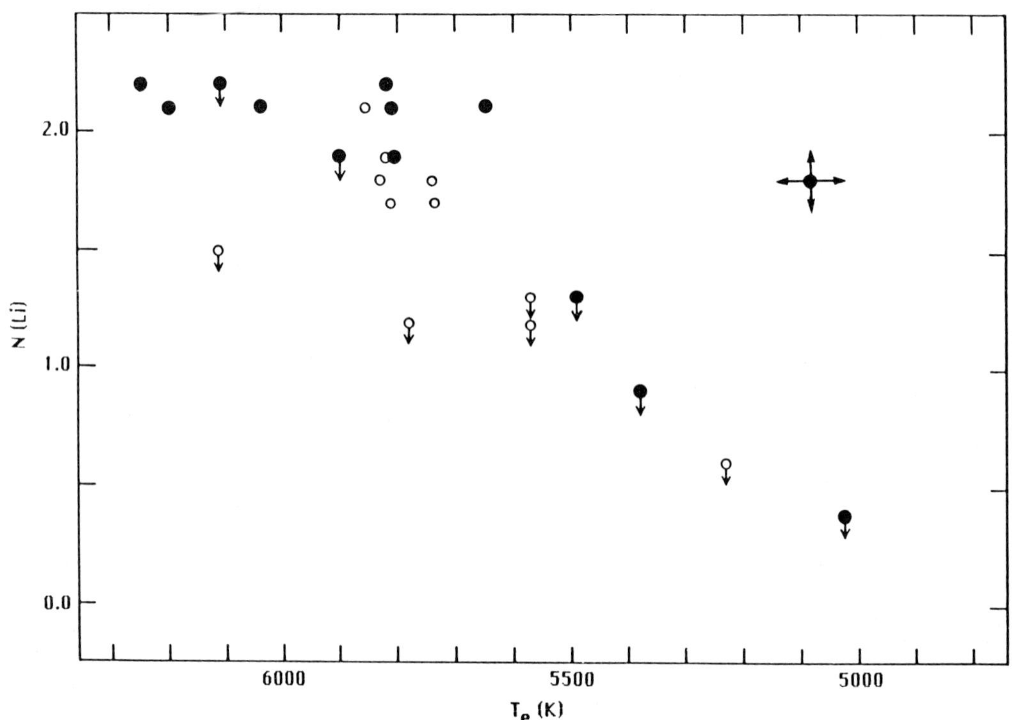

Figure. The variation of N(Li) = 12 + log(Li/H) with T_e for the 23 halo stars of Table 4. The 12 extreme halo stars, defined as those with $|\vec{v}_{LSR}| \geq 160$ km s^{-1} and [Fe/H] $\leq$ -1.4, are shown by filled circles; the other 11 stars, by open circles. Representative 1 σ errors of ±0.10 dex in N(Li I) and ±60 K in T_e are indicated in the upper right hand corner.

all halo stars with $T_e \geq 5600$ K and (2) an extension of these
spectroscopic results to nine additional halo stars. The main
conclusion of SS, that lithium in the halo stars probably was produced
in the Big Bang, is therefore immediately supported by our results.
We now further ask how certain it is that this lithium is a product of
primordial nucleosynthesis alone and whether the lithium observed in
the halo stars is an unmodified sample of the lithium fraction
produced in the Big Bang.

3.1. The Extreme Halo Stars

Twelve of the 23 stars which have the properties $v \geq 160$ km s^{-1} and
[Fe/H] ≤ -1.4 are indicated by filled circles in the Figure. These 12
stars should constitute a selectively very old, homogeneous subset of
the 23 halo stars observed here, and they will therefore be referred
to as extreme halo stars, or extreme subdwarfs. The nine hotter
extreme halo stars have atmospheric lithium abundances, or upper
limits on the abundances, in the narrow range $1.9 \leq N(Li) \leq 2.2$. The
average value derived from the seven definite detections of $\lambda 6707$ is
$\langle N(Li) \rangle = 2.11 \pm 0.09$. In contrast, the abundances of the three
cooler, extreme subdwarfs are $N(Li) \leq 1.3$. Five stars from the list
of SMS which are not among our program stars also conform to the
respective limits just noted. For this selectively old group of 17
extreme halo stars, we conclude that (1) a single, very well defined
relation exists *without exception* between Li/H and T_e, and (2) $\langle N(Li) \rangle =$
2.07 ± 0.11 for 10 stars with $T_e \geq 5600$ K in which the Li I has been
definitely detected.

3.2. Variations in the Lithium Abundance

Since the fundamental details of lithium destruction are unknown, even
in the sun, it is impossible to predict reliably from theory which of
the observed halo stars have destroyed lithium. Michaud, Fontaine,
and Beaudet (1984) suggest that some lithium destruction is likely to
have occurred in all halo stars.
   Our observations may provide direct evidence as to whether the
relatively hot, extreme halo stars in the present study have destroyed
some of their original atmospheric lithium. These stars show a very
small dispersion in lithium abundance, as emphasized above. We have
performed chi-square tests to determine whether the dispersion is
explained by the estimated errors in the abundances, or whether it is
real. We conclude that, while there is slight evidence for real
differences, the apparent star-to-star variations in lithium abundance
are probably accounted for by the estimated errors of measurement.

3.3. Other Sources of Lithium

Lithium is observed in many Population I objects, such as the Hyades
(Cayrel *et al.* 1984), the Pleiades (Duncan and Jones 1983), NGC 752
(Hobbs and Pilachowski 1986), early-F field stars (Boesgaard and
Tripioco 1986b), and late-F and G stars in the solar neighborhood

(Duncan 1981). In all these objects, the maximum abundance observed is N(Li) $\approx$ 3.0 (cf. Boesgaard and Steigman 1985, Figure 10). In type I carbonaceous chondrites the abundance is N(Li) $\approx$ 3.3, but there is some evidence of chemical enrichment (Nichiporuk and Moore 1974). Interstellar abundances are typically 2.5 $\lesssim$ N(Li) $\lesssim$ 3.6 and are appreciably more uncertain than the stellar ones (Hobbs 1984; Ferlet and Dennefeld 1984). This fairly uniform upper limit to N(Li) makes it unlikely that the lithium fraction produced in the Big Bang exceeded N(Li) $\approx$ 3.0.

Several sources of lithium other than the Big Bang have been suggested, such as novae or red giants (cf. Audouze $et\ al.$ 1983). It seems unlikely that these objects could explain the ^{7}Li seen in the halo stars, however. If the ^{7}Li were the result of stellar processing, a relation between lithium abundance and other elements such as iron might be expected but is not observed. As [Fe/H] varies by a factor of about 13 over the range -1.4 to -2.5, N(Li) varies by a factor of about 3 in an unrelated fashion. We conclude that the ^{7}Li in the halo stars in almost certainly primordial.

3.4. Cosmological Implications

In agreement with the discussion of Boesgaard and Steigman (1985), the two limiting possibilities appear to be that either the Big Bang produced an abundance N(Li) $\approx$ 3.0 and all sufficiently hot halo stars have suffered lithium destruction amounting uniformly to almost an order of magnitude, or it produced an amount N(Li) $\approx$ 2.1 and the halo stars have suffered little lithium destruction, as argued by SS. The extreme halo stars studied so far show no positive evidence of the destruction required by the former hypothesis, which is perhaps simpler requiring only one significant source of ^{7}Li. In the latter case, Galactic sources not yet conclusively identified must have produced somewhat more ^{7}Li than the Big Bang.

Whether or not additional lithium has been produced by galactic sources, the present investigation corroborates the conclusion of SS and SMS in indicating that the primordial production of ^{7}Li was 2.1 $\lesssim$ N(Li) $\lesssim$ 3.0. Compared with standard models of light element production in the Big Bang (Yang $et\ al.$ 1984), which assume a neutron half life of t = 10.6 minutes and the existence of three types of neutrinos, this lithium abundance restricts the baryon-to-photon ratio r essentially to 1 x 10^{-10} < r < 10 x 10^{-10}. If the primordial lithium abundance was N(Li) $\approx$ 2.1, the ratio of baryons to photons is rather narrowly restricted to r $\approx$ 3 x 10^{-10}. These constraints appear to be consistent with those imposed by the abundances of D, ^{3}He, and ^{4}He (SS; SMS; Boesgaard and Steigman 1985). All indicate a low baryon density, $\Omega_B \approx$ 0.1.

REFERENCES

Bell, R. A. 1984, private communcation.
Boesgaard, A. M., and Steigman, G. 1985, $Ann.\ Rev.\ Astr.\ Ap.$, **23**, 319.

Boesgaard, A. M., and Tripicco, M. 1986a, *Ap. J. (Letters)*, **302**, L49.
__________. 1986b, *Ap. J.*, **303**, 724.
Carney, B. W. 1983, *A. J.*, **88**, 623.
Cayrel, R., Caryel de Stroebel, G., Campbell, B., and Dappen, W. 1984,
    Ap. J., **283**, 205.
D'Antona, F., and Mazzitelli, I. 1984, *Astr. Ap.*, **138**, 431.
Däppen, W. 1984, private communication.
Duncan, D. K. 1981, *Ap. J.*, **248**, 651.
Duncan, D. K., and Jones, B. F. 1983, *Ap. J.*, **271**, 663.
Ferlet, R., and Dennefeld, M. 1983, *Astr. Ap.*, **138**, 303.
Hauck, B., and Mermilliod, J. 1979, *uvbyβ Photoelectric Photometric
    Catalogue*, (Centre de Donnes Stellaires, Strasbourg).
Hobbs, L. M. 1984, *Ap. J.*, **286**, 252.
Hobbs, L. M., and Pilachowski, C. 1986, *Ap. J. (Letters)*, **309**, Lxxx.
Kurucz, R. L. 1983, private communcation.
Michaud, G., Fontaine, G., and Beaudet, G. 1984, *Ap. J.*, **282**, 206.
Nichiporuk, W., and Moore, C. B. 1974, *Geochim. Cosmochim. Acta.*, **38**,
    1691.
Peterson, R. C., and Carney, B. W. 1979, *Ap. J.*, **231**, 762.
Spite, F., and Spite, M. 1982, *Astr. Ap.*, **115**, 357 (SS).
Spite, M., Maillard, J. P., and Spite F. 1984, *Astr. Ap.*, **141**, 56
    (SMS).
Yang, J., Turner, M. S., Steigman, G., Schramm, D. N., and Olive, K.
    A. 1984, *Ap. J.*, **281**, 493.

DISCUSSION

AUDOUZE: If the resulting primordial lithium abundance is higher than
what is presently quoted by F. and M. Spite, I claim that it would
favour lower values of baryonic density and not higher ones in order
to agree with the present limits on the ^{4}He abundances. This would
strengthen the conclusion raised by Delbourgo-Salvador et al. 1985
according to whom the baryonic density parameter Ω_N should be low.

DUNCAN: I agree that this is a possibility. It would increase the
agreement with ^{4}He results, at the expense of those from D and ^{3}He, and
the latter two are less certain.

FILIPPENKO: Since even your most metal-poor stars contain a substantial
quantity of metals, they are probably not Population III objects. The
gas they contain has presumably been processed by supernovae. Could
these supernovae have significantly depleted the ^{7}Li abundance? After
all, their internal temperatures are extremely high. Alternatively,
could the supernovae have ejected some "extra" ^{7}Li into the interstellar
medium from which your stars formed?

DUNCAN: The strongest argument that supernovae have not significantly increased or decreased the primordial ^{7}Li abundances is that the metal abundances of our true halo stars range over about two orders of magnitude, and the ^{7}Li abundances show no correlation with metallicity over that range.

CHAPTER III

THE CLASSICAL QUANTITIES OF COSMOLOGY

THE STATUS OF THE HUBBLE DIAGRAM IN 1986

Hyron Spinrad
Astronomy Department
University of California
Berkeley, CA 94720, USA.

S. Djorgovski
Harvard–Smithsonian Center for Astrophysics
60 Garden St.
Cambridge, MA 02138, USA.

ABSTRACT.   We present visual and near-infrared Hubble diagrams for first-ranked cluster galaxies, and moderate-flux and high-flux radio galaxies. The photometric improvements and the extension of the diagrams to large redshifts (up to $z \simeq 1.8$) for both 3CR and "1 Jy" class radio galaxies are highlighted. The absolute luminosities of these three types of "standard candles", where their redshifts overlap, agree adequately. The near-IR (2μm) Hubble diagrams may be used to determine the global deceleration parameter, q_0, and the current data favor $q_0 \sim +0.2$ to $+0.3$. The sensitivity to evolutionary changes in the giant galaxies is quite modest at these long wavelengths. On the other hand, the visual regime (BVR) shows a dramatic dependence on differential evolution effects, which dominate over the cosmological model differences. Most radio galaxies apparently have had active star formation, with a continually declining rate, but a few are quite faint and red; they may have had a relatively passive evolution.

I. Introduction. The New Observational Capabilities

The principal goal of cosmology is to determine the global geometric structure and evolutionary history of the Universe. The Hubble diagram, a plot of the (m, z) relation for a given type of luminous galaxies, has been one of the classic methods in this quest ($cf.$ Sandage 1975). But in the real world these goals, in particular the determination of the deceleration parameter q_0 and galaxy luminosity evolution, have been sufficiently intertwined so that neither have been determined with any certainty so far. [We will interchangeably use $2q_0 = \Omega_0$, and assume $\Lambda_0 = 0$ in this paper.] This research is difficult: we are still beset with sample selection biases that are probably functions of redshift; there remain some problems with the faint end of the galaxy photometry, and to a lesser degree, with the bright end aperture corrections. The most referenced works on the subject in the 1970's (Gunn and Oke 1975, Kristian et $al.$ 1978) were inconclusive and somewhat contradictory in their determinations of q_0. This was at least partly due to the galaxy evolution uncertanties. Kristian et $al.$ (1978) and Sandage (1972) found from the brightest

A. Hewitt et al. (eds.), Observational Cosmology, 129–141.

cluster members (BCM's) Hubble diagrams an *apparent* (evolution uncorrected) value of $q_0 \geq 1$. Application of the simple E-galaxy evolution models by Tinsley (1972, 1977) suggested q_0(corrected) $\simeq$ +0.3 to +0.5, again in the provocative, near–closure range.

However, because of the recent advances in the solid state detectors (CCD's) now available to the observational astronomers of the 1980's, we can probably do better. Area photometry with CCD's now permits identification and photometry of radio galaxies and optically-selected BCM's down to $V \geq 24$. This procedure is also superior to the old aperture photometry, because apertures of any size can be synthesized, unwanted companions can be excised, and, most important for good faint object photometry,   the dominant contribution of the night sky foreground

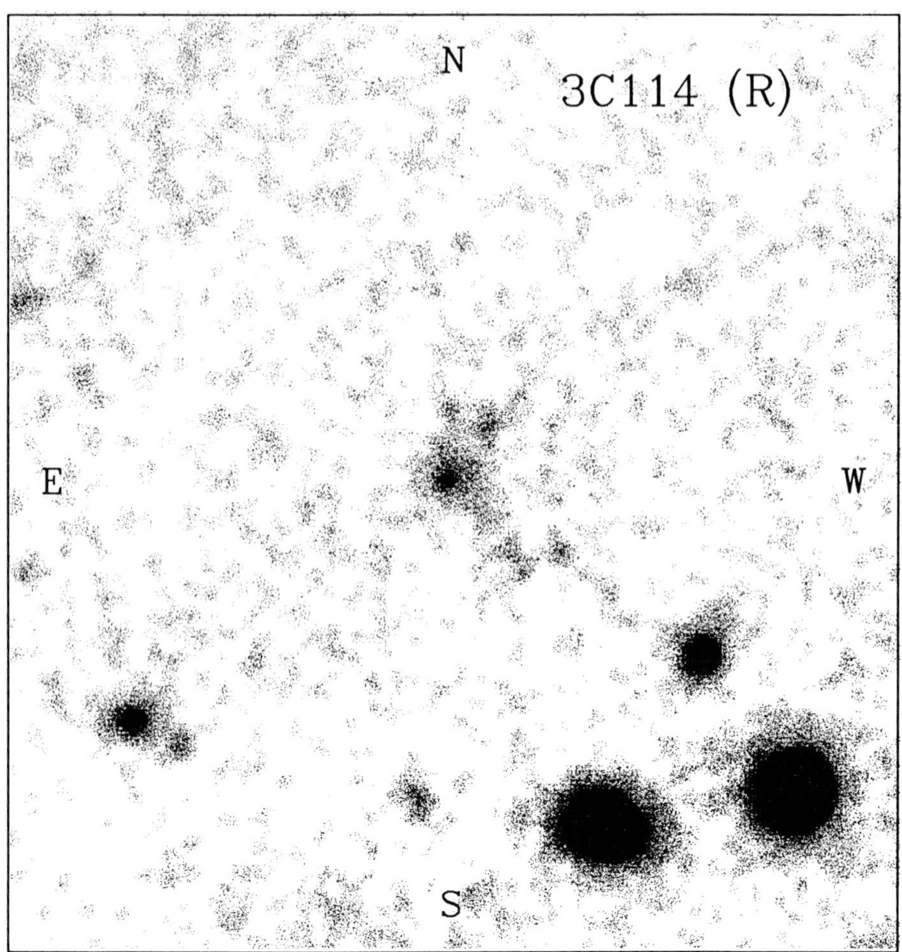

Figure 1.   A 1-arcmin square section of a red CCD frame centered on 3C114. This distant radio galaxy ($z = 0.81$, $R = 22.2$) was identified only recently, by using a modern radio map. This image, and the one shown in Figure 2, were obtained with a TI CCD camera on the KPNO 4-m telescope in December 1985.

can be measured with considerable precision. Figure 1 shows a field centered on 3C 114, a $R = 22.2$ radio galaxy at $z = 0.815$. Figure 2 shows the field of 3C 210, which is at $z = 1.17$. Photometry of distant galaxies like these in the visual region is now fairly routine. These 3CR radio galaxies are quite extended even at $z > 1$, and very rarely show bright or blue nuclei; they are mainly radiating starlight, even in the observed V band ($\sim$ emitted UV).

Our ability to measure galaxy magnitudes in the near–IR, especially in the K-band (2.2μm) has also steadily improved through the 1980's. Early reviews were given by Lebofsky (1980) and Grasdalen (1980). The new data, mainly compiled by Lilly and Longair (1982, 1984; hereafter LL), Lebofsky (1981), Lebofsky and Eisenhardt (1986; hereafter LE)   and Lilly *et al.* (1985b) are of high quality and

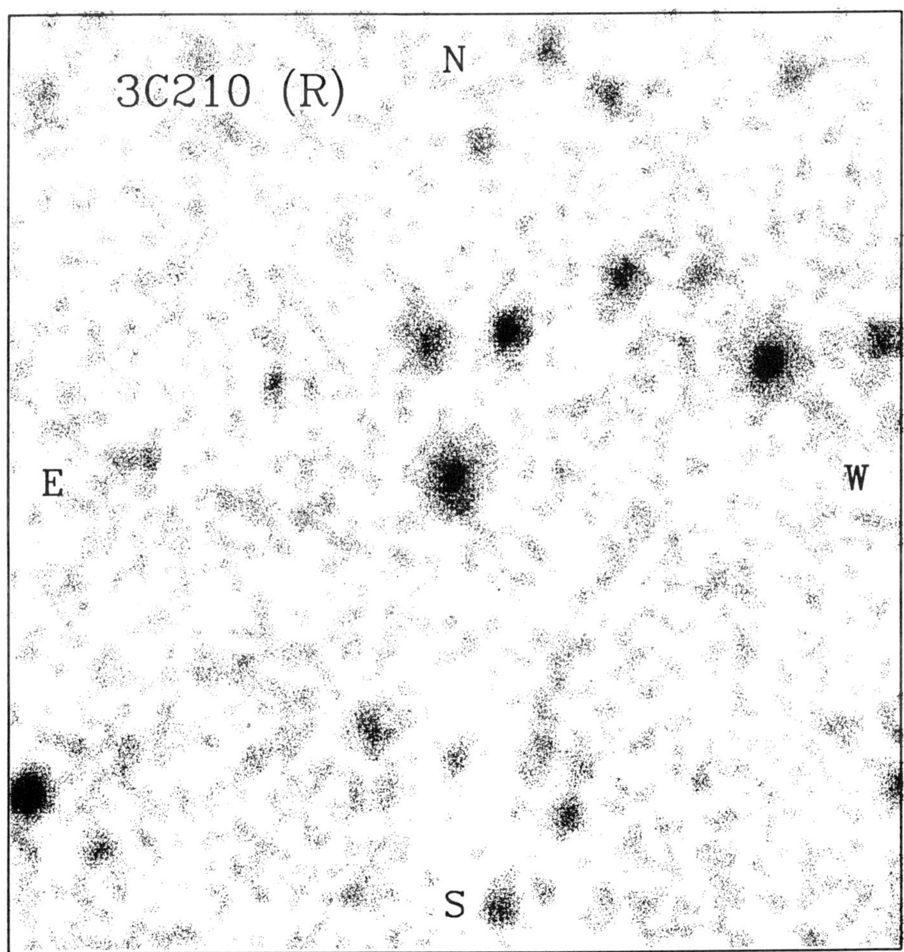

Figure 2.   A red CCD frame of 3C 210, a faint ($R = 21.5$) radio galaxy at $z = 1.17$. The field is 1 arcmin square. Note the obvious resolution of the object, and a substantial elongation in the N–S direction. The galaxy has moderately weak emission lines indicative of a very low ionization state.

enable us to sample the brightness of active and quiescent high-redshift galaxies near their spectral energy distribution maxima. Even in most powerful radio galaxies the stellar radiation from old stars dominates any non–thermal radiation that may be present at these long wavelengths (Lilly *et al.* 1985a).

Finally, and most importantly, the availability of fast CCD slit spectrographs has yielded new faint galaxy redshifts, e.g., for the Gunn, Hoessel, and Oke (1986) clusters, for 24 intermediate-power "1 Jy" galaxies (Allington-Smith *et al.* 1985), and for 21 powerful 3CR galaxies with $z > 1$ (Spinrad and Djorgovski 1984, and in preparation). Of course, the presence of emission lines in the radio galaxy spectra greatly simplifies the measurements of their redshifts, and study of their physical properties (*cf.* Spinrad and Djorgovski 1984, Spinrad 1986, and Perryman *et al.* 1984). Not all of these objects have good photometry as yet. Progressing out to $z \simeq 1.8$ for both classes of radio galaxies is critical to a new interpretation of the Hubble diagram; recall that the papers from the 1970's had available redshifts for the samples reaching only to about $z \simeq 0.5$.

Another interpretative tool of the 1980's, which we use intensively, is the set of galaxy evolution models of Bruzual (1983). These models are by intent simple, yet sufficiently predictive. With some reasonably justified assumptions on the IMF and the star formation rate (SFR) history, we can predict luminosities and colors as functions of redshift for evolving galaxies.

While these strong radio sources may not be ideal "standard candles" from all points of view, they do have one redeeming virtue: since the 3CR sample is now virtually complete in its identification content, and at the 97% completeness level in redshift determinations for the "safe" extragalactic identifications (Spinrad *et al.* 1985; Spinrad 1986), we can virtually ignore the standard statistical problem plaguing many samples, *viz.* the faint-end Malmquist bias. The galaxies we use are radio-selected, with no apparent correlation between the optical and the radio luminosities. Low optical luminosity *might* only matter for perhaps the remaining 3% of the 3CR galaxies so far unidentified, or without measured redshifts. The problem of luminosity selection is thus not very serious.

II. Results in the Visible Regime

Djorgovski, Spinrad, and Dickinson (1987; hereafter DSD), following the early presentation by Djorgovski, Spinrad, and Marr (1985), have used new CCD photometry of 3CR galaxies, combined with a little of spectrophotometry of slightly lower precision, to extend the visual (BVR) Hubble diagrams to $z \simeq 1.82$. Only the NLRG were used (this meant discarding a few broad-line N-galaxies, such as 3C 297, and 208.1). The DSD 3CR galaxy photometry was combined with other photometric data on the lower redshift BCM's by the various Mt. Palomar groups (Gunn and Oke 1975, Sandage *et al.* 1976, Kristian *et al.* 1978, Hoessel *et al.* 1980, and Schneider *et al.* 1983). The photometric systems were standardized with a fair success to the BVR Mould-Bessel system, though it is probably true that the full DSD sample is not on a perfectly homogenous photometric system throughout the visible regime. The differential zero-point and small color terms which probably still remain are small when compared to our internal faint end photometric uncertainties, and are much smaller than the evolutionary effects which we discuss below. We present the photometry without the cluster richness corrections, or anything beyond the aperture corrections for $z < 0.5$; the K-corrections are incorporated in the model curves.

Figure 3 illustrates the contemporary Hubble diagram for all galaxy classes in the V-band, with several evolutionary models overlaid. The Bruzual models shown are characterized either by an early single 1-Gyr burst of star formation (c-models), or an exponentially decreasing SFR rate (μ-models; $\mu = 0.5$ implies an e-folding time of about 1.4 Gyr). Also shown are the changes in models for an open ($\Omega_0 = 0 = q_0$) and the critical ($\Omega_0 = 1$, or $q_0 = \frac{1}{2}$) universes.   A fictitious

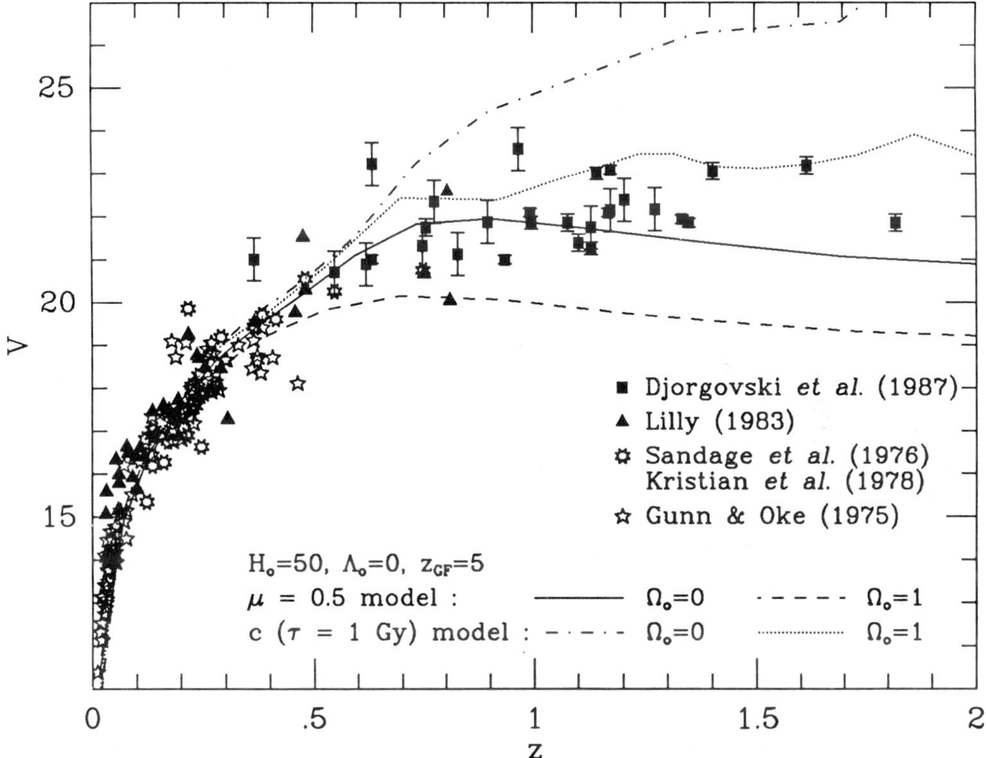

Figure 3.   The visual region Hubble diagram for 3CR galaxies and BCM's. The data include cluster gE's at the lower redshift ($z \leq 0.76$), and 3CR radio galaxies at the higher redshifts. Superposed are the simple cosmological models described by $q_0 = \Omega_0 = 0$ and $q_0 = \frac{1}{2}$, combined with two galaxy evolution models by Bruzual: one is a c-model with a 1 Gyr burst duration, the other is for $\mu = 0.5$ (exponential SFR decay, with an e-folding time of $\simeq 1.4$ Gyr), both with a Salpeter IMF. Note the relatively large sensitivity to galactic evolution and the modest dependence on the cosmological model; in the V-band one cannot satisfactorily separate the two effects. A non-evolving model (just a K-corrected present-day E-galaxy energy distribution) at $z \geq 1.5$ lies $\sim 2^m$ above the upper c-model curve, and $\sim 5^m - 6^m$ above the μ-model curves and the data.

"No evolution" model (not shown), incorporating the K-correction alone would be consistently fainter than the data at the high redshifts. The effect of the global geometry is considerably smaller than the range of evolutionary changes indicated by the data.

Even a casual glance at Fig. 3 assures the reader that Tinsley's (1972, 1977) conclusions are basically correct; at redshifts where the cosmological leverage becomes fairly large, the evolutionary effects due to possible different histories of early galactic star formation become much larger. Model details, such as the exact shape of the giant branch and the stellar luminosity function slope near $M = 1M_\odot$ are of a minor significance. What matters at large redshifts in the visible regime is the length of any star formation burst, and to some extent, the epoch of the last strong burst (that is, the z_{GF}) in these giant galaxies. Other data, mainly relevant to non-radio galaxies (see Hamilton 1985, Spinrad 1986) suggests $z_{GF} \geq 4$ for many large gE galaxies, and so we use $z_{GF} = 5$ as a convenient guess value for the models.

The main conclusion of DSD is that the size of the detected evolutionary effects, and the variety of plausible evolutionary histories which affect the emitted UV spectra of galaxies are large at $z > 1$, and prevent us from making any firm conclusions about cosmology. Thus, the data and models do not allow sharp cosmological decisions in the visual region. We can say that completely passively evolving galaxies (no large starbursts after the initial one) are rare in our sample of 3CR galaxies, but we will mention a few 3CR candidates for quiet evolution in the Section IV below. This is consistent with O'Connell's (1986) conclusions from population synthesis of nearby E galaxies. An important future project will be to select optically a sample of very distant clusters ($z \geq 1$) to complement the 3CR data set shown here. We conclude that these luminous 3CR galaxies are characterized by μ-models with e-folding times of $\sim 1 - 1.5$ Gyr; such models can also provide enough ionizing radiation (UV starlight) at $\lambda < 912$ Å to produce the observed emission line intensities, even the far-UV lines (e.g., Ly-α or the carbon lines) in 3C 256 and 3C 239 ($z \sim 1.8$). The differences between the passive c-models and the more active μ-models fade as the galaxies age, so that at $z \leq 0.5$ no substantive differences remain. In summary, a separation of cosmology and galaxy evolution is quite difficult in the observed BVR bands.

III. The IR Hubble Diagram: A Chance to Determine q_0 ?

As previously stated, the 2μm photometry samples mainly the light from old stars, and thus it should be much less sensitive to the fraction of young stars in these distant galaxies. The photometry of 3CR sources by LL and LE contains a sufficient overlap to permit a comparison of magnitudes for 5 radio galaxies with $K > 15$; the photometric agreement is good with a slight zero point offset. We have corrected the LL and Lilly *et al.* (1985b) K-band data by $-0^m.15$ to place everything on the LE system. There have been no aperture, richness, or K-corrections to any of the data; these corrections were applied to the models instead.

Figure 4 shows a K-band Hubble diagram, and some of the Bruzual evolutionary models. In this and the other figures showing the K-band Hubble diagrams, we have normalized the models at small ($z < 0.15$) redshifts; aperture corrections were applied for that purpose, but are not shown on the plotted data points. This Figure illustrates the lack of sensitivity to the galaxy formation redshift, z_{GF}. This insensitivity to evolutionary variables is typical for the models in the near-IR, as opposed to our conclusions from the V band, where the galaxy evolution dominated the Hubble diagram and the conclusions drawn from it. In the K-band we mainly see the mild evolution of the red giant branch, and the relatively slow-changing K-corrections.

The models in the near-IR are also insensitive to the SFR history: whether the SFR is prolonged or in a single burst. Figure 5 shows all currently available K-band photometry for distant galaxies, compared to the same c and μ models shown for our V-band Hubble diagram of Fig. 3. The reader should note that at 2μm the evolutionary model differences are very small, and there is not much sensitivity to the adopted H_0, which enters in the conversion of model timesteps to redshifts. The evolutionary differences are now overshadowed by the cosmological curvature differences between open and closed world models. This may be the cosmologically sensitive type of diagram that many investigators have long hoped for; a choice between global geometries should be possible from data of this quality at redshifts this large. Recall that the previous investigations often showed the scatter of galaxy magnitudes in excess of the differences between decelerations of

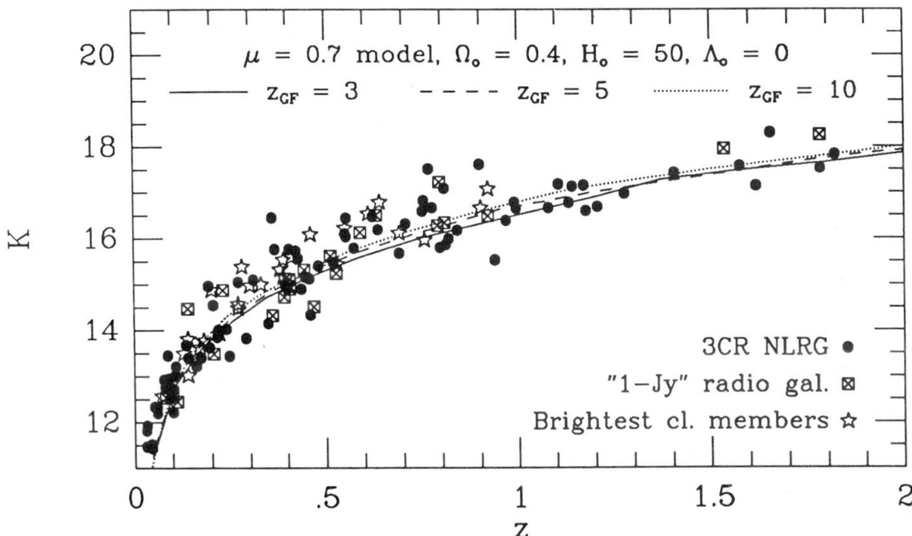

Figure 4. An infrared K-band Hubble diagram, used to illustrate the insensitivity to evolution models at long emitted wavelengths, and in particular the insensitivity to the redshift of galaxy formation, z_{GF}. The data points are BCM's from LE, the "1 Jy" radio galaxies of Allington-Smith *et al.*, and the narrow–line 3CR galaxies of LL and LE. The data are quite homogenous, and *no corrections* of any kind have been applied. Thus, at $z \lesssim 0.1$, the models appear slightly too bright ($K \leq 13$) because the necessary aperture corrections to the galaxy photometry have not been explicitly applied to the data.

$q_0 = 0$ and $q_0 = 1$. The 3CR galaxies have a fairly small luminosity dispersion in the near-IR, roughly $\sigma(m) \simeq 0.5$, independent of the redshift.

But before we feel too optimistic about the deceleration parameter which the eye would chose in Figure 5 ($q_0 \simeq 0.25$, say), we ought to consider any possible sources of systematic errors. Hardest of all is the evaluation of the quality and appropriateness of the red giant branch (RGB) spectra adopted by Bruzual (1983), and the post-RGB evolution ($cf.$ Renzini and Buzzoni 1986). All we can do empirically is to note that the predicted broad-band colors, $(V - K)$ of the models, which are sensitive to the giant branch shape and the ratio of main sequence stars and subgiants to the luminous giants, fit the available data at $z \leq 0.45$ quite well (Bruzual 1983, 1986). This is a fair, but obviously an incomplete check. New models are needed, with improved post-RGB tracks, and with a more complete library of stellar spectra, including more metal-rich giants; that work could take some time, and is obviously beyond the scope of this review.

The second worry is the quality of our three types of "standard candles" in

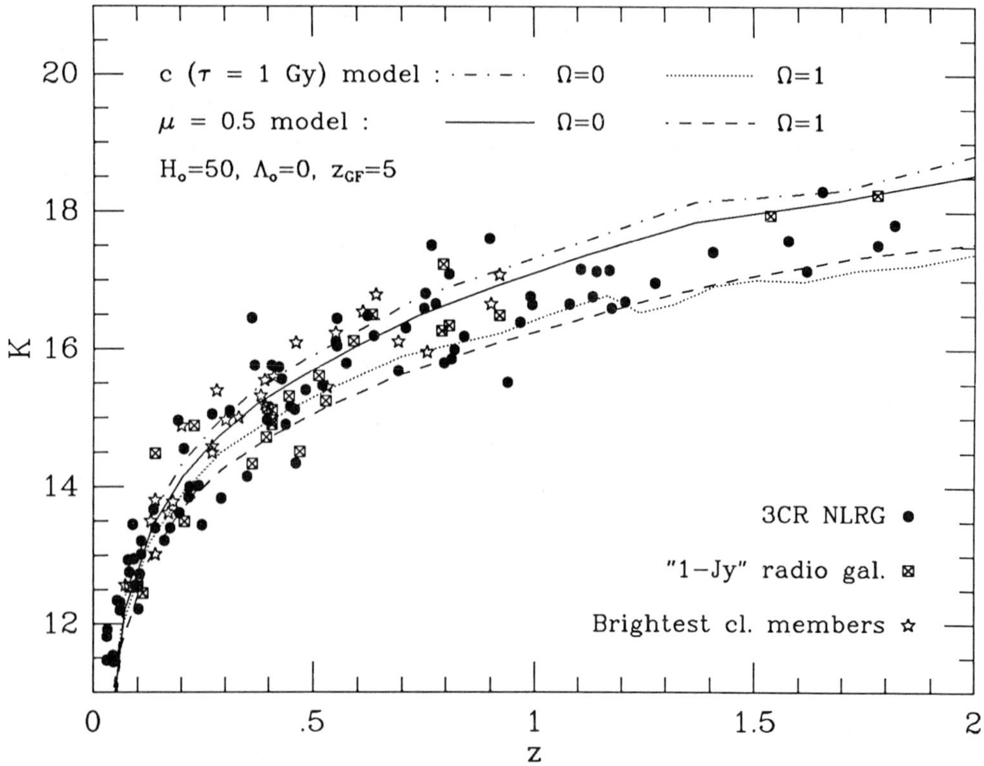

Figure 5.   The complete K-band photometry sample is compared to the Bruzual evolution and standard cosmology models shown before in Fig. 3. There is now a good sensitivity to cosmological parameters. The q_0 probably lies between 0.2 and 0.3, judging from this data set. The models are also quite insensitive to the adopted H_0. Galaxy evolution differences here are much less important than what we see in the visual region.

the near-IR. A careful comparison of the K-band luminosities of the BCM's ($0.05 \leq z \leq 0.92$), "1 Jy" radio galaxies ($0.1 \leq z \leq 1.78$), and the 3CR galaxies ($0.04 \leq z \leq 1.82$) in their overlap regions suggests that at small redshifts ($z < 0.4$) the first-ranked cluster gE's are about $0^m.3 \pm 0^m.2$ fainter than the 3CR galaxies; the "1 Jy" galaxies have about the same M_K as the 3CR galaxies at low redshifts, -26 or so for $H_0 = 50$ km s^{-1} Mpc^{-1}. However, at $z > 1$, the two available "1 Jy" galaxies are slightly fainter ($\leq 0^m.4$) than 3CR galaxies with the similar redshifts. Can these zero-point M_K differences sway our world-model conclusions? Not entirely, since $\Delta q_0 = 0.5$ corresponds to $\Delta K \approx 1^m$ at $1 < z < 1.8$. Any possible luminosity differences which are *systematic with redshift* can be much more serious. Recently Yates, Miller, and Peacock (1986) have suggested that the most powerful 3CR galaxies, at $z \geq 1$, are more luminous than the lower redshift 3CR's and the "1 Jy" galaxies at the highest redshifts. A correlation with radio power and optical luminosity cannot be pervasive in our samples, since the "1 Jy" and BCM luminosities are virtually identical in M_K at $z \sim 0.8$, and their radio powers are vastly different at those redshifts. But if we restrict our comparison to the most powerful radio galaxies with $P_{178} > 10^{27} W Hz^{-1}$ sr^{-1}, a correlation between M_K and P_{178} may well be present. Recall that at $z > 1$, the "1 Jy" galaxies appear slightly fainter than 3CR galaxies shown in Fig. 5. However, we do not find any correlation between cluster membership and radio power. To explore the tentative Yates *et al.* correlation somewhat further, we have tried to put together an IR Hubble diagram at approximately equal radio power (log $P \sim 27$), using the high-flux 3CR sample at $z < 0.5$ and the "1 Jy" sample at $z > 0.7$. Figure 6 shows this composite diagram. The model line for $q_0 \simeq +0.2$ is a good fit to this data set. Using only the 3CR data set, we obtain a best fit with $q_0 \simeq +0.5$; these differences may be caused in part by the data systematics.

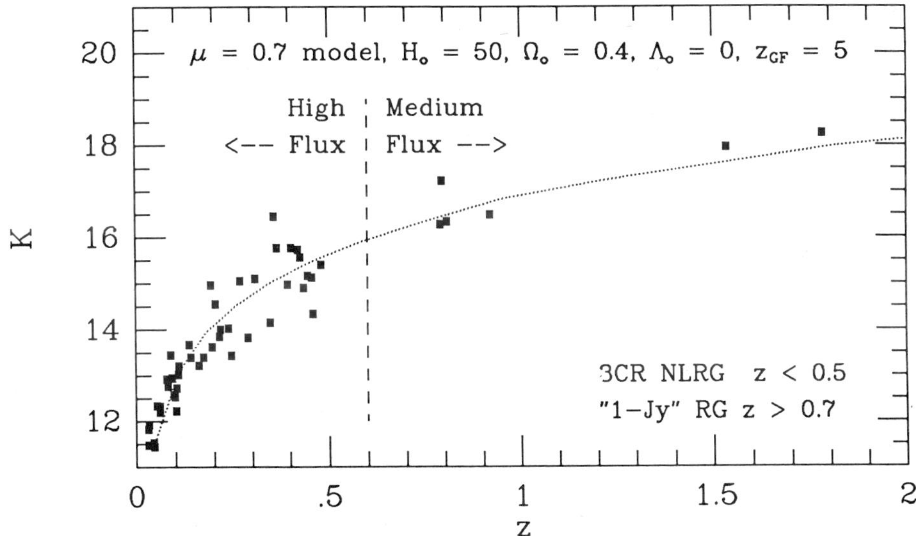

Figure 6. An infrared Hubble diagram combining the data points from the 3CR set ($z < 0.5$) and the "1 Jy" galaxies ($z > 0.70$). The idea is to equalize roughly the radio power at all observed cosmic epochs; the effect is to decrease slightly q_0 — our model is now quite satisfactory for $q_0 = +0.2, \Lambda_0 = 0$. This probably is a very sensible compromise solution.

Finally, there is another possible bias: if galaxy mergers or cannibalism are important in the early formation of giant elliptical galaxies (especially in proto-clusters) as originally suggested by Ostriker and Hausmann (1977), and more recently by Davis (1986) from N-body simulations, then our q_0 (apparent) will be a systematic underestimate of q_0 (true). This is because gE or cD galaxies become brighter as they age. But recall that most of the weight of our samples is at $z > 0.7$, where 90% of the galaxies which we use are strong radio emitters. Considerable recent work in this field has indicated a *qualitative* connection between the presence of radio emission and galaxy interactions. These 3CR and "1 Jy" galaxies are, then, the systems that evolved dynamically very early on, and they may have completed their merger growth at the epoch we observe them.

So, it appears that little or no systematic correction is needed in the comparison of $z < 0.8$ galaxies with their distant counterparts. Thus, we conclude by suggesting a value of $q_0 = 0.2 \pm 0.3$ as a best guess from the 1986 Hubble diagrams in the 2μm region. This is an *estimate*, rather than a *measurement*, but we believe that this value of q_0 is more reliable than the previous measurements using similar methods, and it is consistent with other modern estimates of the density parameter Ω_0 (Peebles 1986).

IV. Consistency Checks Over the Whole Spectrum

It is possible to use the visual and the near-IR color and magnitude data together to check whether the same μ-models/cosmology combinations are indicated over the entire available wavelength range. Following LL, we note that the $(R - K)$ color provides our best broad-brush coverage of the UV–optical spectrum, but by itself it cannot constrain the q_0 or the evolution models very well. If we adopt $q_0 \simeq 0.25$, we would conclude, as do LL, that the galaxies mainly follow a μ-model, but some are sufficiently faint and red as to suggest a passive c-model evolutionary history. Among these distant red galaxies are 3C 65, 3C 68.2, and 3C 241. Their mean redshift is $\langle z \rangle = 1.45$. The data in the visual regime indicate the more active models, $\mu \sim 0.6$, but the data in near-IR are consistent with the quieter models, $\mu \sim 0.8$, or even the passive c-models. The interpretation of this moderate discrepancy is unclear; it could be due to some unknown model flaw, but it can also be understood simply by assuming a *two-component* population model.

This scenario suggests that the majority of the stars are fairly old, but a small fraction of the mass is now undergoing a starburst (associated also with the presence of a radio source?). That would make the composite starlight at short wavelengths resemble a μ-model (decaying SFR) for a few Gyr. A sequence of such sporadic starbursts declining in time is what the μ-models were intended to represent anyway. On the other hand, the few faint and red galaxies may have lead a quiet existence since their initial collapse. A future check on this "color accord problem" could come from new data on more distant cluster or "1 Jy" red galaxies at large redshifts. At the moment the inconsistency in our derived μ as a function of wavelength leads to only some mild doubts about our quantitative application, and should not affect our q_0 measurements appreciably. We note also that the correlation of "UV-excess" with the radio galaxy emission line strengths has weakened as more spectroscopic data have become available.

V. Desiderata for the Future, and the Concluding Remarks

The progress of the 1980's is very encouraging, and we expect the field to advance by extension of observations on these three types of standard candles toward even greater lookback times. We listed some of the possible problems which need more attention, but new problems will arise as well. For example, what is the effective formation epoch of clusters? Will their initial density contrast be a major factor (and a bias) in their collapse time and later recognition, by either optical or X-ray techniques?

There is hope that some of the "1 Jy" radio galaxies whose observations have just really begun, may prove to be very distant and secure cosmological probes, perhaps at $z \leq 2$, if the (K, z) relations we show here may be extrapolated. Future *imaging photometry* in the near-IR will greatly improve the quality of the raw K photometric data at faint levels ($K \geq 18$, say), and we look forward to that capability. Finally, some independent method to check the validity of evolutionary models has to be established. Then we should be able to capitalize on the distant galaxy data sets like those presented here in attacking one of the cosmology's oldest and most important problems.

We thank Drs. M. Rieke, J. Silk, R. Wyse, and M. Davis for their comments and encouragement, and for the use of unpublished data. Mark Dickinson provided valuable assistance with our new CCD photometry, and Patrick McCarthy and Michael Strauss helped with related work. H. S. wishes to acknowledge continued support form the U.S. National Science Foundation, and S. D. acknowledges partial support from Harvard University. We again thank the staffs of Kitt Peak, Lick, MMT, and CFHT observatories for their assistance in gathering much of the observational data presented here.

REFERENCES:

Allington-Smith, J. R., Lilly, S. J., and Longair, M.S. 1985, *M.N.R.A.S.* **213**, 243.
Bruzual, G. 1983, *Ap.J.* **273**, 105.
Bruzual, G. 1986, in *Spectral Evolution of Galaxies*, C. Chiosi and A. Renzini (eds.), p. 263. Dordrecht: D. Reidel.
Davis, M. 1986, private communicaton.
Djorgovski, S., Spinrad, H., and Dickinson, M. 1987, Ap.J. (in press). [DSD]
Djorgovski, S., Spinrad, H., and Marr, J. 1985, in *New Aspects of Galaxy Photometry*, J.L. Nieto (ed.), p. 193. New York: Springer Verlag.
Grasdalen, G. 1980, in *Objects of High Redshift*, IAU Symposium #92, G. O. Abell and P. J. E. Peebles (eds.), p. 269. Dordrecht: D. Reidel.
Gunn, J. E., and Oke, J. B., 1975, *Ap.J.* **195**, 255.
Gunn, J. E., Hoessel, J., and Oke, J. B. 1986, *Ap.J.* **306**, 30.
Hamilton, D. 1985, *Ap.J.* **297**, 371.
Hoessel, J., Gunn, J. E., and Thuan, T. X. 1980, *Ap.J.* **241**, 486.
Kristian, J., Sandage, A. R., and Westphal, J. 1978, *Ap.J.* **221**, 383.
Lebofsky, M. 1980 in *Objects of High Redshift* IAU Symposium #92, G. O. Abell and P. J. E. Peebles (eds.), p. 257. Dordrecht: D. Reidel.
Lebofsky, M. J. 1981 *Ap.J.* **245**, L59.
Lebofsky, M. J., and Eisenhardt, P. R. M. 1986, *Ap.J.* **300**, 151. [LE]

Lilly, S. J., and Longair, M. S. 1982, *M.N.R.A.S.* **199**, 1053. [LL]

Lilly, S. J. 1983, Ph. D. Thesis, University of Edinburgh, UK.

Lilly, S. J., and Longair, M. S. 1984, *M.N.R.A.S.* **211**, 833. [LL]

Lilly, S. J., Longair, M. S., and Miller, L. 1985a, *M.N.R.A.S.* **214**, 109.

Lilly, S. J., Longair, M. S., and Allington-Smith, J. R. 1985b, *M.N.R.A.S.* **215**, 37.

O'Connell, R. 1986, to appear in *STScI Conference on Stellar Populations*, M. Tosi and C. Norman (eds.), Cambridge University Press.

Ostriker, J., and Hausman, M. 1977, *Ap.J.* **217**, L125.

Peebles, P. J. E. 1986, *Nature* **321**, 27.

Perryman, M. A. C., Lilly, S. J., Longair, M. S., and Downes, A. J. B. 1984, *M.N.R.A.S.* **209**, 159.

Renzini, A., and Buzzoni, A. 1986 in *Spectral Evolution of Galaxies*, C. Chiosi and A. Renzini (eds.), p. 195. Dordrecht: D. Reidel.

Sandage, A. R. 1972, *Ap.J.* **178**, 1.

Sandage, A. R. 1975, in "Galaxies and the Universe", Vol. *IX* of *Stars and Stellar Systems*, A. Sandage, M. Sandage, and J. Kristian (eds.), p. 761. Chicago: University of Chicago Press.

Sandage, A. R., Kristian, J., and Westphal, J. A. 1976, *Ap.J.* **205**, 688.

Schneider, D. P., Gunn, J. E., and Hoessel, J. B. 1983, *Ap.J.* **264**, 337.

Spinrad, H. 1986, *Publ. Astr. Soc. Pacific* **98**, 269.

Spinrad, H., and Djorgovski, S. 1984, *Ap.J.* **285**, L49.

Spinrad, H., Djorgovski, S., Marr, J., and Aguilar, L. 1985, *Publ. Astr. Soc. Pacific* **97**, 932.

Tinsley, B. T. 1972, *Ap.J.* **173**, L93.

Tinsley, B. T. 1977, *Ap.J.* **211**, 621.

Yates, M. G., Miller, L., and Peacock, J. A. 1986, *M.N.R.A.S.* **221**, 311.

DISCUSSION

MILEY: I would like to report evidence that indicates that the high-luminosity radio galaxies are indeed different from the lower-luminosity radio galaxies and that plotting their magnitudes on the same diagram may well introduce selection effects. First, a survey of lower-z matched samples of high-luminosity and low-luminosity radio galaxies by Heckman, Illingworth, van Breugel, Bothun, Baum, Smith and myself shows that the higher powered sources have preferentially peculiar optical morphologies. Second, Golombek, Neugebauer and I have shown that they have IRAS (60-micron) excesses compared with low-luminosity radio galaxies. Both these results suggest that the high-powered radio galaxies are undergoing mergers and are definitely not normal ellipticals as presumed by conventional astronomical folk-lore.

SPINRAD: Yes, there is little doubt that some radio galaxies have disturbed outer isophotes. Whether that would matter in a centered-aperture light measurement that includes the inner 60-80 kpc (D), is, to me, unclear.

I also believe that the IR is more immune to these "disturbances", if the latter are young-object dominated.

The Hubble-diagrams used here, are, in fact, for very luminous sources - radio galaxies well above the slope-change in the RLF.

WINDHORST: First, an addition to George Miley's comment. At much fainter radio fluxes, milli Jansky levels, the radio source population also consists of normal giant ellipticals and weird looking beasts. So we can always construct a subsample of "normal looking" giant ellipticals for the purpose of a Hubble diagram. My question is twofold. I can imagine two effects that could brighten the K-magnitude of galaxies at $z \sim 1$ with respect to the model predictions. First, the Paschen or other lines might enhance the K-luminosity and second, AGB stars or carbon stars might produce more light at $\lambda \gtrsim 1\mu$. I am sure you share this worry, but could you give us some feeling how important these effects might be for the Hubble diagram in K.

SPINRAD: I doubt if any infrared emission lines would compete with the stellar energy maxima and the very broad-band IR spectral response(s). Emission lines should hardly modulate $\underline{K}$ magnitudes.

The giant branch details now follow the Yale evolutionary tracks. The current Bruzual models are a satisfactory match in color $(V-K)_0$ for redshifts $0 < z \leq 0.5$; this gives me some limited confidence that the stellar giant branch is fairly realistic for recent cosmic epochs. At longer look-backs (e.g. - younger turnoffs and more-massive giant precursor stars), I cannot make a comparison of theory with the E galaxy data. So the question will have to be answered indirectly, presumably by tests Dr. Bruzual is now assembling. Good question!

LONGAIR: It is a great pleasure to have the opportunity to express my admiration for the remarkable achievement which Dr. Spinrad has made in measuring the redshifts and spectra of the 3C and 1-Jy radio galaxies. I believe the studies of the faintest of the radio galaxies in the 1-Jy sample are important for understanding when these galaxies first formed. Simon Lilly, Jeremy Allington-Smith and I showed that, among the faintest and presumably most distant of the galaxies in the 1-Jy sample, there are objects with the optical-infrared colors of passively evolving elliptical galaxies. Since their redshifts are presumably in the range $1.5 \lesssim z \lesssim 3$, this means that they must have formed their stellar populations at redshifts greater than 3.5. The study of these galaxies is very important in setting astrophysical constraints upon when these massive galaxies first formed.

THE RADIO AND INFRARED LUMINOSITIES OF 3CR RADIO GALAXIES -ARE THEY CORRELATED?

M.G. Yates
Department of Astronomy, University of Edinburgh, U.K.

L. Miller & J.A. Peacock
Royal Observatory, Edinburgh, U.K.

1. INTRODUCTION

The infrared photometric study of a sample of 90 3CR radio galaxies by Lilly & Longair (1984, hereafter LL) has demonstrated that the high redshift objects are brighter in the infrared than their low redshift counterparts; this has been interpreted as being entirely due to the evolution of their constituent stellar populations. There is however a great difference between the radio luminosities of the high and low redshift objects in this flux limited sample and we have therefore examined statistically the possibility of a correlation between the infrared and radio luminosities of these galaxies, the presence of which could bias our interpretation of the infrared Hubble diagram. We find that the radio and infrared luminosities do indeed correlate for the most powerful radio galaxies.

2. THE ANALYSIS

The sample of LL is complete in both K magnitude and redshift and we have defined a subset of this sample by including in our analysis only classical double FR II sources and excluding broad-lined radio galaxies. Infrared magnitudes (already corrected for galactic extinction) were taken from LL, and flux densities and spectral indices at 178 MHz from Laing et al. (1983). The infrared magnitudes were corrected to a standard metric diameter of 43kpc, assuming $H_0 = 50$ km s^{-1} Mpc^{-1} and $\Omega_0 = 1$. Lilly & Longair (1982) showed that the K-z Hubble diagram is well described by the passively evolving C model of Bruzual (1983). This is the most natural null hypothesis to consider, although the exact amount of passive evolution is model dependent, and the observations of LL indicate that a reasonable set of parameters to choose is $q_0 = 0.5$, an IMF slope x = 1.35 and a formation redshift of 3.5. If the 3CR galaxies were a homogeneous set at all redshifts, subject to no other evolutionary changes the resulting corrected magnitudes would be constant for all redshifts. Figure 1 shows absolute infrared magnitude plotted against redshift for the sample.

A. Hewitt et al. (eds.), Observational Cosmology, 143–146.
© 1987 by the IAU.

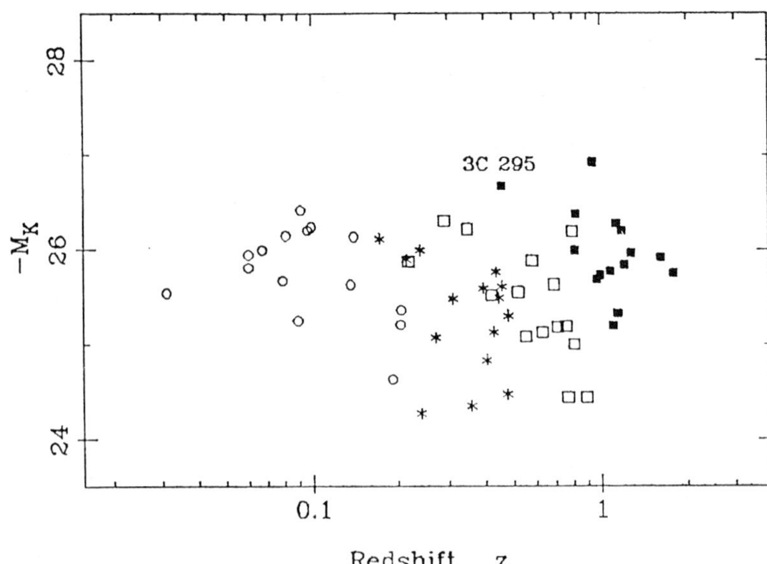

Redshift   z

Figure 1.   Absolute K magnitude against redshift.   The radio galaxies
have been divided into four radio luminosity bins, each plotted with a
different symbol.   The filled squares are the most radio luminous
galaxies, followed in decreasing radio luminosity by the open squares,
the asterisks and the open circles respectively.

The relationships between absolute magnitude and radio power (M-P) and absolute
magnitude and redshift (M-z) are related by the implicit relation between radio
power and redshift (P-z) in a flux limited sample such as this, and so we have
used a Spearmann partial rank correlation analysis to disentangle the underlying
relationships, as described by Macklin (1982);   this allows us to test the correlation
between M and P in the presence of the correlation of either with z.   The null
hypothesis is that any M-P correlation is entirely due to a combination of separate
P-z and M-z relations.   If this null hypothesis can not be accepted, there must be
some additional real correlation between radio luminosity and absolute magnitude,
independent of redshift.   A more comprehensive discussion of the partial rank
analysis as applied to this data set is given in Yates et al. (1986).   Table 1
presents the first and second-order (i.e. partial) coefficients for the correlated
quantities, the significance calculated in terms of the Student's t statistic.

Table 1.

	r	t
Mz	0.050	0.379
MP	-0.050	0.379
Pz	0.950	23.096
MP,z	0.309	2.456 (~1.5%)
Mz,P	-0.309	2.456 (~1.5%)

Thus the galaxies' absolute infrared magnitudes correlate with radio luminosity (MP,z) at the 1.5% significance level, and anti-correlate with redshift (Mz,P) at the 1.5% level. The similar sizes of the two correlations (although of opposite sign) are as we should expect since the magnitudes have been corrected using the C model which LL empirically found to be a good fit to the data - these two correlations conspire to produce a near-zero first order correlation between M and z.

3. DISCUSSION

The correlation of infrared luminosity with radio luminosity has important implications for our understanding of the evolution of stellar populations in these objects since it has previously been assumed that these galaxies are a homogeneous set of standard candles. A likely mechanism for the correlation is hinted at by the bright absolute magnitude of 3C 295 (z=0.459), a cD galaxy in a rich cluster and thus a galaxy which has probably undergone cannibalism (Hausman & Ostriker 1978). At low redshifts (z<0.2) classical-double radio galaxies of the sort considered here do not appear to lie in rich clusters (Longair & Seldner 1979), however, that work did not extend to high enough redshifts to encompass the top decade of radio power which we have been able to include in our analysis here. We suggest therefore that the most powerful radio galaxies known are those rare sources which are associated with rich clusters of galaxies and which are subject to processes such as cannibalism; the most radio luminous galaxies occur at high redshift and have stellar luminosities enhanced by dynamical evolution. The first results from a deep imaging survey of the most distant 3CR radio galaxies by Spinrad (these proceedings) tend to lend support to the hypothesis that the most powerful radio galaxies do indeed inhabit rich environments.

An encouraging aspect of this study concerning the use of radio galaxies to probe evolution on the Hubble diagram is that at low radio powers, classical double sources do not lie in dense environments and the effects of dynamical evolution are small. One can thus reduce the effect dynamical evolution is likely to have on the form of the Hubble diagram by only employing galaxies of moderate radio power.

REFERENCES

Bruzual, G.A., 1983. Rev. Mexicana Astron. Astrofis., 8, 63.
Hausman, M.A. & Ostriker, J.P., 1978. Astrophys. J., 224, 320.
Laing, R.A., Riley, J.M. & Longair, M.S., 1983. M.N.R.A.S., 204, 151.
Lilly, S.J. & Longair, M.S., 1982. M.N.R.A.S., 199, 1053.
Lilly, S.J. & Longair, M.S., 1984. M.N.R.A.S., 211, 833.
Longair, M.S. & Seldner, M., 1979. M.N.R.A.S., 189, 433.
Macklin, J.T., 1982. M.N.R.A.S., 199, 1119.
Yates, M.G., Miller, L. & Peacock, J.A., 1986. M.N.R.A.S., 221, 311.

DISCUSSION

WAMPLER: One can create a correlation between the absolute radio luminosity and the absolute optical luminosity by using a wrong value for q_0. Since your correlation is rather weak could a big change in q_0 remove it?

YATES: The partial rank analysis we have employed allows us to examine the correlation of absolute magnitude with radio luminosity independent of redshift - any magnitude redshift correlation is removed and thus a big change in q_0 will not significantly change the strength of the effect seen here.

A STUDY OF THE HUBBLE FLOW

E. Joseph Wampler
European Southern Observatory
Karl-Schwarzschild-Str. 2
D-8046 Garching bei München, F.R.G.

ABSTRACT. It is found that the Hubble diagrams for quasars and giant radio galaxies are very similar. Either the evolution of giant branch stars in the galaxies closely matches the evolution of the quasar non-thermal emission or the cosmological deceleration parameter q_0 is close to 3.

1. INTRODUCTION

Lilly and Longair (1984) have shown that a Hubble diagram that uses 2.2 μ photometry of a complete sample of 3CR radio galaxies either shows strong galaxy luminosity evolution at high redshift or indicates a very high value for q_0. For this sample there is expected to be little contribution to the 2.2 μ magnitudes either from a non-thermal nuclear continuum or from strong emission lines. If the observed effect is due to luminosity evolution it must result from a decrease in giant branch luminosity as the Universe ages.

Quasar data also show strong evolutionary effects (Schmidt and Green, 1983) or indicate a high value for q_0 (Wampler and Ponz, 1985). Baldwin (1977) showed that the equivalent width of the CIVλ1548 emission line in quasars is luminosity sensitive allows quasar magnitudes to be corrected for luminosity effects. Since the "Baldwin relationship" seems to be independent of redshift (Wampler et al., 1984), corrected quasar magnitudes should be largely free of luminosity bias.

2. ANALYSIS

Complete samples of radio loud quasars from Wampler et al. (1984), radio quiet quasars from Schmidt and Green (1983) and from Osmer and Smith (1980) were supplemented with a few additional quasars to form a quasar Hubble diagram with magnitudes corrected for luminosity spread using the "Baldwin relation". Since the quasar data overlapped the galaxy data in redshift the quasar magnitudes were adjusted by 6.75 mag. so that the two sets of data could be used together in a plot of

A. Hewitt et al. (eds.), Observational Cosmology, 147–150.
© *1987 by the IAU.*

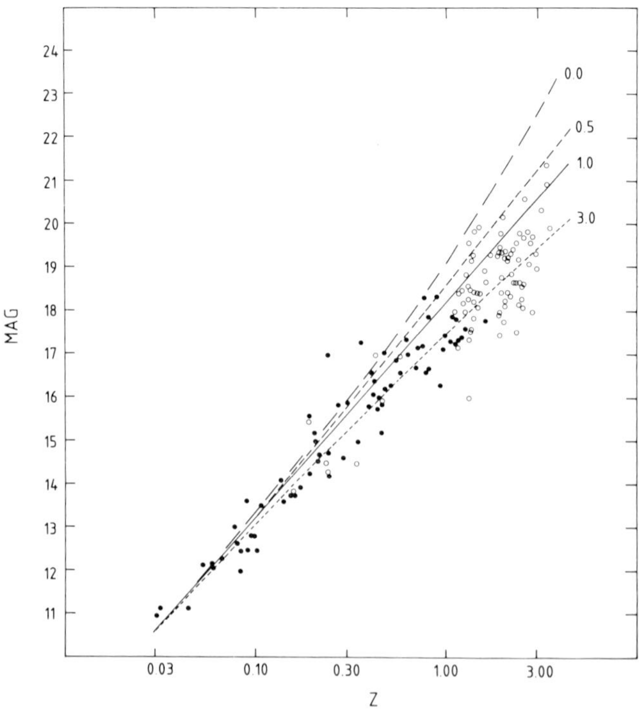

Fig. 1. The Hubble diagram for 3CR radio galaxies (●) and quasars (○).

the composite Hubble diagram. This is shown in Fig. 1. It is clear
from Fig. 1 that both sets of data show a similar smooth departure
from the predictions of low q_0 Friedmann models. The data are
reasonably well fit by Friedmann models with $q_0 \approx 3$. Taking the two
sets of data separately the squares of the residuals from Friedmann
curves were calculated as a function of q_0. These are plotted in
Fig. 2. It can be seen that both sets of data indicate similar, high
values for q_0. The minimum r.m.s. scatter in the quasar data is about
0.8 mag while for the galaxy data it is about 0.6 mag.

3. DISCUSSION

At present it is not possible to decide if the apparent brightening of
the magnitudes at high z is due to luminosity evolution or indicates a
high value for q_0. If it is due to evolution then the amount required
for giant branch stars in luminous radio galaxies must approximately
equal the evolution of the UV non-thermal evolution in quasars. In
both cases the rate of evolution approximately agrees with the amount
expected from geometrical effects in a Friedmann cosmology with a high
value of q_0. For quasars the evolution must shift the zero-point of
the "Baldwin relationship" but not its slope since a universal
"Baldwin relationship" would remove the luminosity evolution.

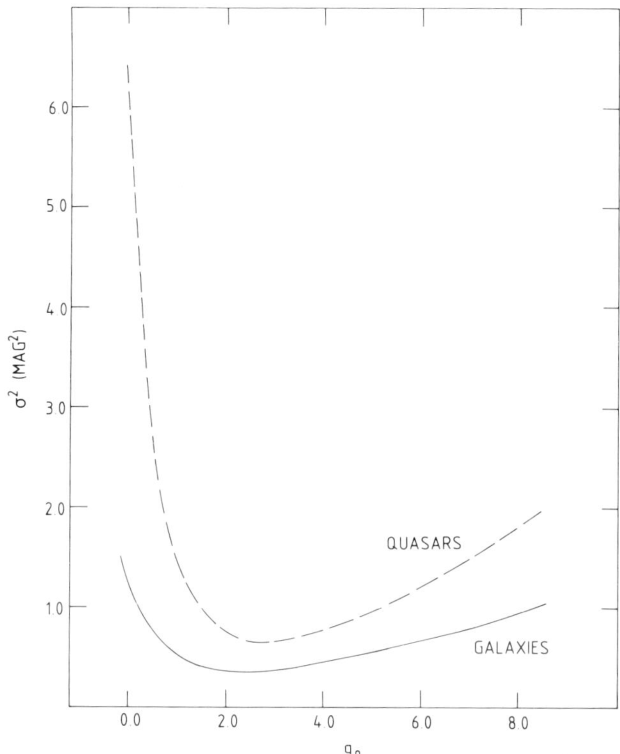

Fig. 2. The square of the residuals as a function of q_0.

If $q_0 \approx 3$ there is an age problem unless H_0 is very low or unless the cosmological constant Λ is positive. For currently accepted values of H_0 (40 km/sec/Mpc $\lesssim H_0 \lesssim$ 80 km/sec/Mpc) if $\Lambda = 0$ and $q_0 = 3$ the age of the Universe is less than the age of pop. II stars in the Galaxy. A positive value for Λ is required to increase the age of the Universe. A very slight positive vacuum energy would give a high enough Λ to satisfy the data. The inflationary model of the Universe requires that in the past the vacuum energy was > 0.

The results given here are discussed in more detail in Wampler (1987).

REFERENCES

Baldwin, J.A. 1977, Astrophys. J., **214**, 679.
Lilly, S.J., and Longair, M.S. 1984, Mon. Not. R. astr. Soc., 211, 833.
Osmer, P., and Smith, M.G. 1980, Astrophys. J. Suppl., **42**, 333.
Schmidt, M., and Green, R.F. 1983, Astrophys. J., **269**, 352.
Wampler, E.J., Gaskell, C.M., Burke, W.L., and Baldwin, J.A. 1984, Astrophys. J., **276**, 403.
Wampler, E.J., and Ponz, D. 1985, Astrophys. J., **298**, 448.
Wampler, E.J. 1987, Astron. Astrophys., submitted.

DISCUSSION

RUBIN: I think I do not understand what you mean by a complete
sample. Isn't there an observational selection which means that at
any z the velocities will be available for the brighter galaxies, but
not for the fainter galaxies? Could your distribution be incomplete
for fainter galaxies at each z?

WAMPLER: It is believed that powerful radio sources are bright
galaxies. If this is not so then the 5-10 blank fields (out of ~80
radio galaxies) in the 3CR catalogue may be nearby, faint galaxies.
If they were at just the right place in the Hubble diagram they could
change the conclusions for galaxies a little, but not much. The
quasars are all much brighter and the Baldwin correction removes most
of the distortion in the Hubble diagram caused by selection effects.

THE COSMIC DISTANCE SCALE

G.A.Tammann
Astronomisches Institut der Universitaet Basel
European Southern Observatory, Garching

ABSTRACT. The zero-point of the extragalactic distance
scale, defined by about two dozens of nearby, late-type
galaxies, has remained nearly unchanged for the last decade,
in spite of the advent of new techniques and great efforts.
The distances are essentially tied to trigonometric parallax
stars and hence independent of the Hyades modulus; they are
consistent with RR Lyr stars. The mean zero-point is
therefore probably secure to better than 10%.
All known secondary distance indicators are still
affected by zero-point errors, by problems in the definition
of their relation between distance indicator and absolute
magnitude (or linear size), and/or by selection bias. The
effect of the very important selection bias (Malmquist
effect), which causes a seemingly non-linear expansion
field, is illustrated by two examples. To test for any true
deviations from a linear expansion the Hubble diagram of
nearly bias-free first-ranked cluster galaxies and
supernovae Ia is shown; this imposes stringent limits on any
non-linearity of the Hubble flow within $v < 5000$ km s^{-1}.
After freeing the available distances of field galaxies
from selection bias and after reducing them to a common
zero-point, one finds $H_0 = 55$-65. Several distance indicators
require a best Virgo cluster modulus of $(m-M) = 31.60$, which
implies for the Coma cluster $(m-M) = 35.38$ and, with $v(Coma) =$
7217 km s^{-1}, $H_0 = 60$. Supernovae Ia and first-ranked cluster
galaxies out to large distances give $H_0(global) = 53$. Thus the
evidence from clusters <u>and</u> field galaxies is best satisfied
by $H_0 = 55$; the assigned mean error of ± 7 is to indicate a 3σ
range of $35 < H_0 < 75$.
Purely physical methods to determine extragalactic
distances have modest weight yet; they will contribute
eventually much to the determination of H_0.
If H_0 were as large as 100, several paradoxa would
arise. The Milky Way would have a very high supernova
frequency, our Galaxy and M31 would be oversized, the baryon

A. Hewitt et al. (eds.), Observational Cosmology, 151–185.

density would fall short to bind clusters, and Friedman
universes were excluded.

Because all systematic errors have conspired and
probably still conspire to measure H_O too high, the true
value could well be 40. Until new, decisive evidence becomes
available, it is suggested for all practical purposes to use
$H_O=50$.

I. INTRODUCTION

Distance determinations outside our Galaxy have still to
rely on the true luminosity or on the true size of
extragalactic objects, which have "identical" counterparts
close enough that their distances are known by some method
or another.

In spite of considerable efforts, the history of the
extragalactic distance scale is beset by errors and
setbacks. Much of the reason lies in the term "identical"
counterparts. A priori they may not exist outside our Galaxy
due to differences in the evolutionary history, metallicity
etc. Even if they exist in first approximation, their
properties must have some intrinsic scatter and the term
"identical" must be replaced by "similar". This makes a
fundamental difference, because as soon as distance
indicators have intrinsic scatter of their luminosity or
size, they are subject to selection bias: at large distances
only overluminous or oversized objects tend to enter our
catalogues. This tendency distorts <u>systematically</u> the
distance scale. The problem becomes particularly severe for
the determination of the global value of the Hubble constant
H_O [km s^{-1}Mpc^{-1}], because to find its true value freed of
any local peculiar and streaming motions, one has to go to
distances where the recession velocities are larger than at
least $v_O=5000$ km s^{-1}. Such distances are presently difficult
to bridge in one step (the exception may be supernovae of
type Ia [SNe Ia] at maximum light), and by the time one has
patched up the distance ladder one may have accumulated
several systematic errors. Because all systematic errors due
to selection bias tend to <u>underestimate</u> the distances, the
resulting, typically non-linear distance scale may be
seriously compressed with a correspondingly, unrealisticly
high value of H_O. The reliability of any distance indicator
must therefore be thoroughly studied before it can be put to
use. In many cases this can be achieved by consistency
checks as shown below.

The development of extragalactic astronomy has been
accompanied by only a slow grasp of the true size of the
universe and its brightest/largest constituents. This is
reflected in a continuous reduction of H_O. Hubbles original
value of $H_O=513$ (1929) or 526 (1936) had to be halved after

Baade's (1952) 1.5 mag. correction of the zero-point of the
P-L relation of Cepheids. Sandage's (1958) reanalysis of the
P-L relation, his application of the new magnitude scale by
Stebbins, Whitford, and Johnson (1950) to the brightest
stars in external galaxies, and his proof that Hubble's
brightest stars were actually HII regions, led him to
conclude that $50 < H_O < 100$. In particular he reasoned that if
the brightest blue stars have a - presently generally
accepted - absolute magnitude of "M_{pg}(stars)$=-9.5$, then H$=$
55". The first suggestion that the best value of H_O is as
low as 50 is due to de Vaucouleurs (1970). Later work by
many researchers has led to a biforcation with values either
near $H_O = 100$ (sometimes called the short distance scale) or
near $H_O = 50$ (long distance scale). This discrepancy,
corresponding to 1.5 mag. in the distance moduli, is <u>not</u>
mainly due to uncertainties of the distances to very nearby
galaxies, but it accumulates gradually as one goes to larger
distances. The steady divergence of the two distance scales
suggests that their main difference lies in the treatment of
the already mentioned selection effects. The demonstration
of this is in fact the main goal of the present review.

In Section II a brief discussion is given of the
primary distance indicators. They provide distances to 24
nearby galaxies, which can be used to calibrate additional
distance indicators which reach to larger distances; modern
results for these galaxies are compiled in Section III and
they are shown to have changed on average surprisingly
little during the last twelve years. A selection of proposed
secondary distance indicators is listed in Section IV and it
is stressed that they must be subjected to a severe scrutiny
of their reliability; in particular their possible
vulnerability against the Malmquist effect is illustrated.
As an example, the internal consistency of the distances
derived from the luminosity index Λ_C (Section V) and from
21cm-line widths combined with infrared (IR) magnitudes
(Section VI) is checked. The possibility of an overall non-
linearity of the expansion field out to $v_O \sim 5000$ km s^{-1} is
rejected in Section VII. The calibration of the luminosity
of SNe Ia at maximum light is discussed in Section VIII.
Section IX compiles the evidence for the distance of the
Virgo cluster. A compromise global value of H_O is derived in
Section X. Conclusions are drawn in Section XI.

II. PRIMARY DISTANCE INDICATORS

Classical Cepheids are ideal extragalactic distance
indicators because as supergiants they are luminous enough
to be observed from the ground out to ~ 7 Mpc, and because
they follow the period-luminosity-color (P-L-C) relation
with very little scatter (Sandage and Tammann, 1971), which

makes them insensitive to selection bias. They show
considerably more scatter in a simplified P-L relation, but
the known intrinsic width of the relation offers also here a
handle against selection bias. A disadvantage of the optical
P-L-C and P-L relation is that they yield too large
distances for low-metallicity Cepheids. This difficulty is
minimized by the IR P-L relation (for references see Table
1). The IR relation has also the advantage that it has a
reduced width and that it is insensitive to internal
absorption. - The zero point of the optical and IR P-L
relation rests on the Cepheids which are members of galactic
clusters. The distances of the latter are generally tied to
the Hyades modulus. When the classical Hyades modulus of
$3^{m}_{.}03$ was revised upwards by $\sim 0^{m}_{.}25$ the question arose whether
<u>all</u> Cepheid distances had to be increased by the same
amount. Van den Bergh (1977) warned that the correction may
be compensated by the over-metallicity of the Hyades.
Indeed, tying the Cepheid-bearing clusters M25 and NGC 6087
to parallax stars, which are reduced to the same
metallicity, yields cluster moduli (Pel 1985; Cameron 1985a,
b; Turner 1986) only $0^{m}_{.}05$ larger on average than those
derived on the basis of a Hyades modulus of $3^{m}_{.}03$ and no
metallicity correction (Sandage and Tammann, 1969). The two
cases suggest that the revision of the Hyades modulus
affects the extragalactic distance scale by a barely
significant amount of only $0^{m}_{.}05$. It has therefore been
decided here to fit the Cepheid-bearing clusters, whose
metallicity is yet unknown, to a <u>formal</u> Hyades main-sequence
with (m-M)=$3^{m}_{.}03$. The remaining zero-point error of the
Cepheids is estimated to $\pm 0^{m}_{.}10$. Eventually it will become
possible to base the zero-point of the P-L relation on
Cepheid distances derived from the purely physical Baade-
Becker-Wesselink method, but at present its accuracy is not
compatible; the main reason is that no models of <u>moving</u>
atmospheres are yet available (cf.Gautschy 1986).

RR Lyr stars are powerful distance indicators, because
they follow a tight P-L relation (Sandage 1981). Their zero-
point of $\langle M_{V} \rangle = 0^{m}_{.}6 \pm 0^{m}_{.}1$ is widely accepted. This zero-point
cannot yet be based on the Baade-Becker-Wesselink method,
because the problem of non-static atmospheres is here still
more severe than for Cepheids, as evidenced by the
appearance of emission lines and shock fronts. The
disadvantage of RR Lyr stars is that they are intrinsically
much fainter than Cepheids, and that their luminosity is a
function of metallicity. The internal accuracy of
statistical parallaxes is not yet sufficient to determine
the sign of the metal correction (Strugnell, Reid, and
Murray, 1986; Barnes and Hawley, 1986), but from cluster
variables it follows that they become fainter with
decreasing metallicity (Sandage 1970; Carney 1980).

Spectral types of sufficiently bright stars in external galaxies have been used to determine distances. The assumption here is, of course, that the empirical relation between spectral type and absolute magnitudes, derived in our Galaxy, holds also in other galaxies. In view of the width of the relation, the sampling of the extragalactic stars poses a severe problem. Sampling from bright magnitudes into a galaxian population tends to distort the sample and to underestimate the mean stellar magnitudes and hence the distances. This trend can be traced historically through the literature for instance in the case of LMC.

Color-magnitude diagrams (CMD) of brightest stars in external galaxies are subject to the same selection problems as spectral types. Evolved and untypically bright stars enter the sample first. In addition blends with very faint stars tend to brighten and redden the sample stars (unless the "sky" is measured near to the stars), which leads to an overestimate of their absorption and underestimate of their distance.

If only the brightest part of the CMD can be observed, brightest blue stars, Hubble-Sandage variables, and brightest red stars can be used as distance indicators. Unfortunately the luminosity of the brightest blue stars correlates with the size of the parent galaxy, yet the latest calibration of the mean of the three brightest blue stars shows a useful plateau at $M_B(3)=-10^m\!.0$ for galaxies with $M_B<-20^m$ (Sandage 1986c). The same source shows the dependence of the brightest red stars on the galaxy size to be shallower; they have correspondingly higher weight as distance indicators.

Mira variables have been shown by M.Feast and his collaborators at the Cape to follow a P-L relation. The uniqueness of this relation is, however, not yet established (Menzies and Whitelock, 1985). The applicability of Miras is anyhow yet restricted to LMC (Feast 1984, 1986).

Since Novae exhibit a correlation between absolute magnitude and decline rate they may be useful distance indicators. In principle their Galactic zero-point can be found from expanding nova shell parallaxes, which makes them independent of all other distance scales. An old zero-point was provided by T.Schmidt-Kaler, but the results were not very encouraging (van den Bergh 1977; Tammann 1977; de Vaucouleurs 1978). A new start has been made by Cohen (1985) and van den Bergh and Pritchet (1986a). The method is of course also vulnerable to selection effects due to the still not well known intrinsic scatter of the absolute magnitude-decline rate relation. Moreover in distant galaxies there is the danger that the nova sample is biased by objects whose decline rate is so steep that they fall below the detection limit before their decline rate can be determined. This would result in a sample of novae which are overly bright at

discovery <u>and</u> at the reference time (or magnitude) where
their decline rate is determined.
    Supernovae of type Ia as distance indicators are
discussed in Section VII and VIII.

Table 1:   True Distance Moduli to Nearby Calibrating
Galaxies as Adopted by Sandage and Tammann (1974) and from
New Evidence.

	ST (1974)	new	(m-M)	Sources
LMC	18.59	18.50	-0.09	1
SMC	19.27	18.85	-0.42	2
M31	24.16	24.2	+0.04	3
M33	24.56	24.4	-0.16	4
NGC 6822	23.95	23.4	-0.55	5
IC 1613	24.43	24.1	-0.33	6
NGC 300	–	26.0	–	7
NGC 247	–	26.7	–	8
NGC 253	–	27.5	–	9
NGC 7793	–	27.5	–	9
NGC 2403*	27.56	27.8	+0.24	10
NGC 3031	27.56	28.7	+1.14	11
M 101 **	29.2	29.2	0	12

$$\overline{-0.01\pm0.17}$$

* Other group members: NGC 2366, NGC 4236, IC 2574, HoII,
HoI, HoIX
** Other group members: NGC 5204, NGC 5474, NGC 5477, NGC
5585, HoIV

The adopted "new" distances are based on the following
determinations:

1. Feast (1986) gives as the best mean from optical and
    infrared observations of Cepheids, from RR Lyr stars,
    Mira variables, and clusters, including the extensive
    work at the Cape, 18.5$\pm$0.1 (cf. also Stothers, 1983).
    Infrared observations of Cepheids alone give 18.45
    (Welch et al., 1985) and 18.50$\pm$0.07 (McAlary and Welch,
    1985). Aperture photometry at 1.05 μm of Cepheids yields
    18.56$\pm$0.07 (Visvanathan, 1985, after reducing the value
    by $0\overset{m}{.}26$ to agree with the presently adopted Hyades zero
    point for Cepheids in Galactic clusters). ZAMS fitting
    of LMC clusters tends to yield small moduli of
    18.1-18.42$\pm$0.2 (Andersen et al., 1985; Schommer et al.,
    1984; Walker, 1986). However, numerical experiments
    moving Galactic clusters to roughly the LMC distance and
    calculating the effect of resulting blends shows that
    the method yields systematically too low moduli

(Brodbeck, 1986). The outlying cluster NGC 1841 at 18.7 (Andersen et al., 1985) may not be a member (or may suffer less blends?). Chiosi and Pigatto (1986) have shown that the inclusion of convective overshooting increases the ZAMS fitting distances by typically $0\overset{m}{.}2$. Conti and Garmany (1986) found from O,B stars 18.3 ± 0.3, as compared to 18.63 ± 0.2 by Crampton (1979). Shobbrock (1986) finds from $H\beta$ photometry of B supergiants $18.8+0.3$. The adopted value of 18.50 ± 0.10 appears to be a good compromise.

2. A review by Feast (1986) gives 18.8. Eggen (1977) derived from optical observations of Cepheids 18.95. BVI (Caldwell and Coulson, 1985) and JHK (Laney and Stobie, 1986) photometry of Cepheids shows SMC to be more distant than LMC by $0\overset{m}{.}30\pm0\overset{m}{.}06$ and $0\overset{m}{.}32\pm0\overset{m}{.}04$, respectively. Infrared photometry of Cepheids by Welch et al. (1985) and McAlary and Welch (1985) shows this difference to be $0\overset{m}{.}38_5$, while Visvanathan (1985) obtained $0\overset{m}{.}29$ at $0.05\ \mu$m and Madore (1986) $0\overset{m}{.}50$ in the infrared. RR Lyr stars give 0.35 (Graham, 1975). Assuming (m-M)=0.35 (cf. Tammann et al., 1980), a true modulus of 18.85 ± 0.15 is adopted.

3. Infrared observations of Cepheids give 24.26 ± 0.08 (Welch et al., 1986). From the photometry of giant branch Pop.II stars follows 24.4 ± 0.25 (Mould and Kristian, 1986). Assuming M_B(RR Lyr)=1.02 and $A_B=0.32$ van den Bergh and Pritchet (1986) found from three RR Lyr stars 24.16 ± 0.18. Novae in M31 yield, with the same value of A_B, 24.03 ± 0.20 (Cohen, 1985). Baade and Swope's (1963) value of 24.20 from Cepheids was revised by Sandage (1983a) to 24.11, if $(m-M)^O_{Hyades}=3.03$ and $A_B=0.64$ for Cepheids is assumed. A value of 24.2 is consistent with all of these determinations.

4. Blue photometry of Cepheids gives $(m-M)_{AB}=25.35$ (Sandage and Carlson, 1983; Sandage 1983a), or $0\overset{m}{.}08$ less if $(m-M)^O_{Hyades}=3.03$. With $A_B=0.8$ from Freedman (1986) a true modulus follows of 24.47 ± 0.15. Infrared Cepheids yield 24.1 ± 0.1 (Freedman, 1986) or 24.17 ± 0.15 (McAlary and Welch, 1985). Halo giants in M33 indicate 24.8 ± 0.3 (Mould and Kristian, 1986). The value of 24.4 is the best compromise.

5. Infrared observations of Cepheids give 23.47 ± 0.11 (McAlary et al., 1983), 23.30 ± 0.13 (McAlary and Welch, 1985), 23.4 ± 0.2 (Freedman, 1986), and 23.5 (Madore, 1986). The adopted mean value of 23.4 is $0\overset{m}{.}55$ smaller than the value adopted from blue Cepheid magnitudes

(Sandage and Tammann, 1974) presumably due to intrinsic
absorption and/or low metallicity of NGC 6822.

6. Cepheids in the infrared yield 24.08±0.14 (McAlary and
    Welch, 1985), 24.0±0.2 (Freedman, 1986), and 24.2
    (Madore, 1986). Conservatively a mean value of 24.1
    is adopted, rather than the considerably higher value of
    24.43 from blue Cepheid magnitudes (Sandage, 1971),
    which may be similarly affected as in NGC 6822.

7. Graham (1982) found from old red giants 25.8±0.5. Blue
    magnitudes of Cepheids give 26.09±0.2 (Graham, 1984) and
    infrared magnitudes 25.9±0.2 (Freedman, 1986). Planetary
    nebulae suggest 25.85±0.34 (Lawrie and Graham, 1983).

8. From resolution into brightest stars, which is
    intermediate between NGC 300 and NGC 253/NGC 7793
    (Sandage, 1986b). The value agrees within $0^{m}_{.}2$ with that
    given by de Vaucouleurs (1975).

9. From resolution into brightest stars (Sandage, 1986a).

10. Infrared magnitudes of Cepheids yield 28.15±0.20
    (McAlary and Madore, 1984), 28.09±0.21 (McAlary and
    Welch, 1985), and 27.5±0.2 (Freedman, 1986). Sandage
    (1984b) obtained 27.66 (with $A_{B}=0^{m}_{.}24$) from Cepheids and
    the C-M diagram of the brightest stars.

11. Brightest blue and red stars, blue irregular Hubble-
    Sandage variables, and Cepheids require (with $A_{B}=0^{m}_{.}10$)
    28.7 (Sandage, 1984a).

12. Brightest blue and red stars, blue irregular Hubble-
    Sandage variables, and the absence of Cepheids require $\geqslant$
    29.2 (Sandage, 1983b). This is compatible with 28.9±0.3
    from M supergiants (Humphreys and Strom, 1983). The
    discovery of Cepheids at very faint levels suggests 29.3
    (Cook, Aaronson, and Illingworth, 1986).

III. DISTANCES TO NEARBY CALIBRATING GALAXIES

The distance indicators, which can be calibrated in our
Galaxy and which are briefly discussed in the previous
Section, have been used by many authors to determine
distances to galaxies within ~7 Mpc. In Table 1 in the
column headed "new" the mean distance moduli from many such
determinations since about 1980 are given. The multitude of
sources, as discussed in the footnotes to Table 1, should
minimize any personal bias, and the entries - although of

different weight - are intended to represent generally
acceptable compromises.

The distance moduli available in 1974, essentially
based on observations of Cepheids in optical wavebands, are
also listed in Table 1 (Sandage and Tammann, 1974a; 1974b,
cf. also Tammann, Sandage, and Yahil, 1980, Table 9). A
comparison of the 1974 and "new" distances shows for
individual galaxies revisions by up to $1^m.1$, but the
impressive result is that the <u>mean</u> zero-point of the
calibrating galaxies has hardly changed at all.

The fact that the zero-point of the extragalactic
distance scale has essentially remained unchanged during an
interval of a decade, in spite of the advent of new
techniques and the enormous efforts of many researchers,
gives considerable confidence that the mean zero-point as
defined by the galaxies in Table 1 can be trusted at the
$0^m.1-0^m.2$ level.

Because several recent distance determination methods
have based their results relative to an adopted LMC modulus
of $18^m.5$, the latter value carries considerable weight.
Recent infrared work on LMC Cepheids has shown that this
value may be low by $0^m.2-0^m.3$ (Laney and Stobie, 1986). This
suggests that the adopted "new" zero-point is, if anything,
somewhat faint.

The number of 13 calibrating galaxies in Table 1 can be
increased by members of the NGC 2403 group and of the M101
group. They are listed at the foot of Table 1; they bring
the total number of calibrators to 23. There are in addition
a number of small and very small irregulars whose distances
are known from Cepheids and/or brightest stars (Sandage
1986d). We will return to these nearby objects in Section X,
but so far dwarf galaxies have not been applied as stepping
stones of the extended distance ladder.

It should be noted that the calibrating galaxies are
either of spiral or later type. Some distances of dE
galaxies in the Local Group are known from RR Lyr stars, but
a real E galaxy is missing among the calibrators, unless one
wants to accept M32, an exceptionally faint E-type companion
of M31, to be representative for its much brighter fellow
galaxies.

IV. PROPOSED DISTANCE INDICATORS BEYOND ~7 MPC AND THE
EFFECT OF SELECTION BIAS

The calibrating galaxies in Table 1 offer a data base from
which either the intrinsic size/luminosity of particularly
conspicuous objects or the global properties of galaxies can
be determined. The calibration can then, hopefully, be
applied to more distant galaxies. A multitude of objects and

global parameters have been proposed for this purpose. Some
of the more important ones are listed in the folowing.

Among individual objects, HII regions can be identified
out to large distances. Their size (Sandage and Tammann,
1974a), Hα flux (Kennicutt 1981), and velocity dispersion
(Melnick et al., 1986) have been applied for deriving
distances. Another example is the peak of the luminosity
function of globular clusters (van den Bergh, Pritchet, and
Grillmair, 1986).

Among the observable global parameters of galaxies
which correlate with the galaxian luminosity, and which
hence may be distance indicators, are
- the surface brightness/luminosity relation of spirals
  (Holmberg 1958) and of dE galaxies (Binggeli, Sandage, and
  Tarenghi, 1984),
- the luminosity classification of spiral galaxies according
  to the "beauty" of the spiral structure (van den Bergh
  1960; Sandage and Tammann, 1974c; however also Kraan-
  Korteweg, Sandage, and Tammann, 1984),
- the luminosity index Λ_C as a derivative of the luminosity
  classification (de Vaucouleurs 1979),
- the color/luminosity relation of E (Visvanathan and
  Sandage, 1977) and spiral (Visvanathan and Griersmith,
  1977) galaxies,
- the 21cm line width/luminosity relation of spiral galaxies
  in optical (Tully and Fisher, 1977) and infrared
  (Aaronson, Huchra, and Mould, 1979) wavelengths,
- the velocity dispersion/luminosity relation of E galaxies
  (Faber and Jackson, 1976) and of the spheroidal components
  of spirals (Whitmore, Kirshner, and Schechter, 1979;
  1981),
- the brightness distribution/luminosity relation of E
  galaxies (Kormendy 1977),
- look-alike galaxies ("sosies") as standard candles
  (Paturel 1984; Bottinelli et al., 1985),
- the metallicity/luminosity relation of E galaxies,
  particularly the Mg_2 index (Dressler 1984),
- the velocity dispersion/magnitude-related diameter
  relation of E galaxies (Burstein et al., 1986; Djorgovski
  and Davis, 1986), and
- first-ranked cluster galaxies as standard candles (Sandage
  1972).

Before these or any other distance indicators can be
relied upon it is essential to gain <u>full control</u> of the
following points:
(1) The <u>shape</u> of the distance indicator/luminosity (or size)
relation must be known, unless the distance indicator is
treated as a standard candle. The number of calibrating
galaxies and their absolute-magnitude range prove frequently
insufficient to yield a reliable shape. The shape must
therefore typically be based on members of a cluster, whose

number is necessarily also limited, or on the assumption of
an ideal (or corrigible) Hubble flow of field galaxies.
(2) The zero-point of the relation must be well known, which
requires a sufficient number of applicable calibrators.
(3) The intrinsic scatter of the relation between observable
and luminosity (or size) is of overwhelming importance for
the distance scale and will therefore be discussed below.
(4) The parameter space in which the distance indicator can
be applied must be well tested. A distance indicator may be
sensitive to the type of galaxy, it may be reliable only
within a certain luminosity interval etc. Of particular
importance is here the question of field (or group) and
cluster galaxies. Given the possibility that the intrinsic
properties change between the two kinds of galaxies, one
normally depends on an assumption when the calibrating
galaxies, which are typical for the field and group
population, are directly compared with cluster galaxies.

So far none of the listed secondary distance indicators
satisfies all four points. Either the shape of the distance
indicator/luminosity relation is uncertain or no zero-point
is available. The latter problem is particularly severe for
E galaxies as discussed above. While these difficulties are
hardly controversial, there is considerable disagreement on
the importance of the intrinsic scatter. Because I believe
this problem to be the main reason for the present existence
of two distance scales, the short scale with $H_o \sim 100$ and the
long scale with $H_o \sim 50$, a somewhat more extended explanation
is in place.

To illustrate the last point the Spaenhauer diagram is
repeated here in Fig.1 (cf. Sandage 1987, Fig.11). It
represents in the upper panel the distribution in absolute
magnitude of galaxies which scatter by an intrinsic value σ_M
about a mean value. Because the number of objects increases
with the volume surveyed, one reaches more and more objects
as the distance increases. The eventually large number of
objects causes the appearance of "improbable" objects, i.e.
their absolute magnitudes deviate from the mean by $2\sigma_M$, $3\sigma_M$
or even $4\sigma_M$ in either direction. The character of the
diagram is universal if only $\sigma_M \neq 0$; it holds equally for E
galaxies, for galaxies of given luminosity parameter, or for
any other subsample of galaxies.

The lower panel of Fig.1 repeats the upper panel, but
now an apparent-magnitude limit is introduced (here at $m=$
13^m). The devastating effect can be clearly seen, which the
observational limit causes on the sample. At larger
distances only "improbably" luminous objects are retained,
while their underluminous counterparts are eliminated.
Therefore the mean luminosity of the objects increases with
distance, and one must not expect that a distant sample
complies with the zero-point determined from nearby objects.
There is an additional effect, which enhances the problem.

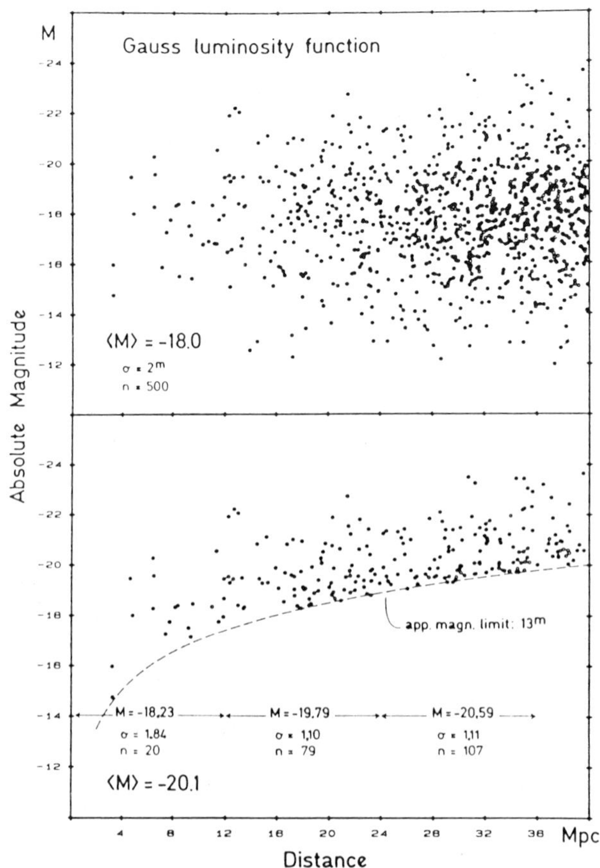

<u>Fig.1</u>. Upper panel: Monte Carlo distri-
bution in distance and absolute magnitude
of 500 galaxies within 38 Mpc. Constant
space density and a mean absolute magnitude
of ⟨M⟩=-18^m with a Gauss standard deviation
of σ_M=2^m are assumed. Lower pannel: The
same sample cut by an apparent-magnitude
limit of m=13^{m}0. Note the increase of the
galaxian luminosities with increasing
distance and the small effective (observable)
scatter σ_M within individual distance
intervals.

The observable "effective" scatter is much smaller than the
intrinsic scatter. One can therefore easily be misled to
conclude from the observed scatter that the bias is

negligible. In fact the bias is determined by the intrinsic scatter, which in most practical cases is very difficult to determine.

The lower panel of Fig.1 is an illustration of a very general case. Almost all galaxy catalogs are limited by apparent magnitude and are therefore subject to the "Malmquist bias". An exception are first-ranked cluster galaxies, which are identified, irrespectively of any overluminosity, in independently discovered clusters. Also all SNe Ia, due to their extreme luminosity, can be found within a reasonable redshift limit; they should carry therefore a minimum bias. In general the bias is the less severe the smaller σ_M(intrinsic).

In practice it is not possible to correct analytically for the bias. Its size does not only depend on σ_M, but also on the shape of the true luminosity distribution of the objects considered, and on their space density ϱ , which realisticly is variable and unknown as long as no correct distances are known. Malmquist (1920) has derived an explicit correction for the idealized case of a Gaussian luminosity distribution and ϱ = const; this correction, however, implies an infinite sample in space, which in the presence of the K-term is highly unrealistic for recessing galaxies.

There are tactics to compensate the Malmquist in first approximation. A possibility is to shift the limiting magnitude to fainter values as one goes to larger recession velocities. Strictly speaking this requires still a well understood expansion field and ϱ =const. The method has been applied for distances from luminosity classes (Sandage and Tammann, 1975), but for most purposes the distant galaxies must be sampled to unpractically faint limits. Another possibility is to rely on an empirical and in each case newly determined correlation between luminosity and recession velocity (Sandage, Tammann, and Yahil, 1979). Yet another possibility is not to treat the luminosity indicator (e.g. the 21-cm line width) as the independent variable, but rather the absolute magnitude, which can be predetermined except for a constant term from any arbitrary value of H_0, if a perfect Hubble flow is assumed (Schechter 1980). This requires, however, that the sample is unbiased with respect to the distance indicator, an assumption which is not generally fulfilled (Kraan-Korteweg, Cameron, and Tammann, 1986). Finally the Malmquist bias can be excluded if one samples by a fixed magnitude interval into the population of clusters at different distances.

Neglect of the Malmquist bias always leads to too short a distance scale. However, the error is not easily recognizable, because the artificially decreased luminosity scatter and the a priori unknown space density of galaxies will still allow a seemingly consistent model. To test for

the presence of a Malmquist bias somewhat more sophisticated
methods have to be applied. They involve the fundamental
linearity of the expansion field. How these tests can be
performed is illustrated in the next two Sections for two
different distance indicators, i.e. the luminosity index Λ_c
and the infrared Tully-Fisher relation.

V. THE LUMINOSITY INDEX Λ_c AS DISTANCE INDICATOR

Distances derived from the luminosity index Λ_c were
published for 309 non-Local Group, Shapley-Ames spiral galaxies (de Vaucouleurs 1979). Their recession velocities v_{220}
corrected for the Virgocentric infall component following
Kraan-Korteweg (1985), are plotted in a double-logarithmic
diagram (Fig.2). De Vaucouleurs has concluded from these
data that $H_O=100$ (e.g. de Vaucouleurs and Corwin, 1986), and
he has stressed that the diagram proved the Λ_c distance
scale to be linear. However, the eye is insensitive against

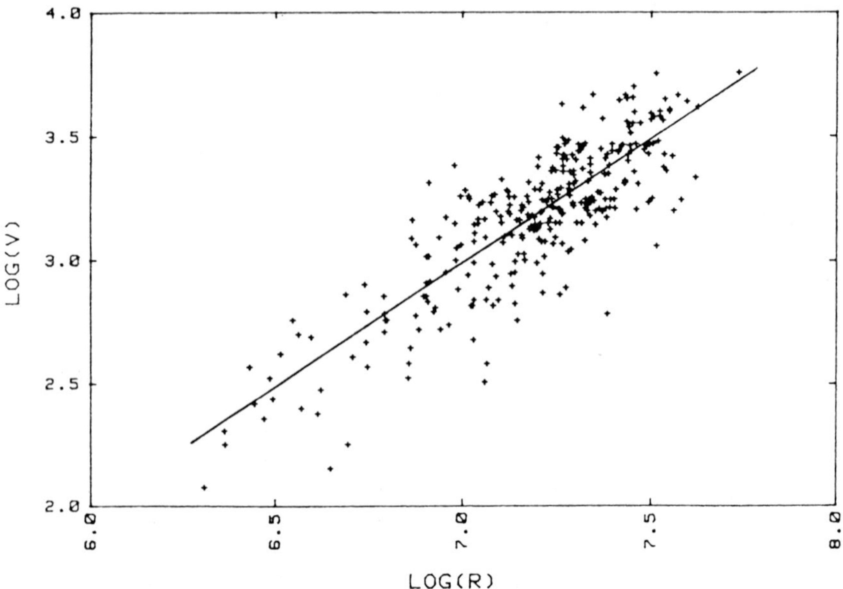

Fig.2. A plot of the logarithm of the corrected recession
velocity versus the logarithm of the published Λ_c distance
(de Vaucouleurs 1979) of field spiral galaxies. The data
seem compatible with $H_O=100$ (full line).

systematic effects in a log-log plot. Therefore the data of
Fig.2 are repeated in Fig.3, but now the numeric values of
distance R and velocity v_{220} are plotted. The diagram
reveals a strong non-linearity with the velocities
increasing faster than the Λ_c distances. With an ansatz of
$R=a \cdot v_{220}^b$ one finds from a least-squares solution a=0.227 and
b=0.589. This seems to require an increasing value of the
Hubble constant, viz. H_o=57 at v_{220}=500 km s^{-1}, H_o=100 at
v_{220}=2000 km s^{-1} and H_o=146 at v_{220}=5000 km s^{-1}; the
asymptotic (global) value of H_o would then have to be >150.
A much more attractive interpretation is, however, to
explain the non-linearity by the _expected_ Malmquist bias.

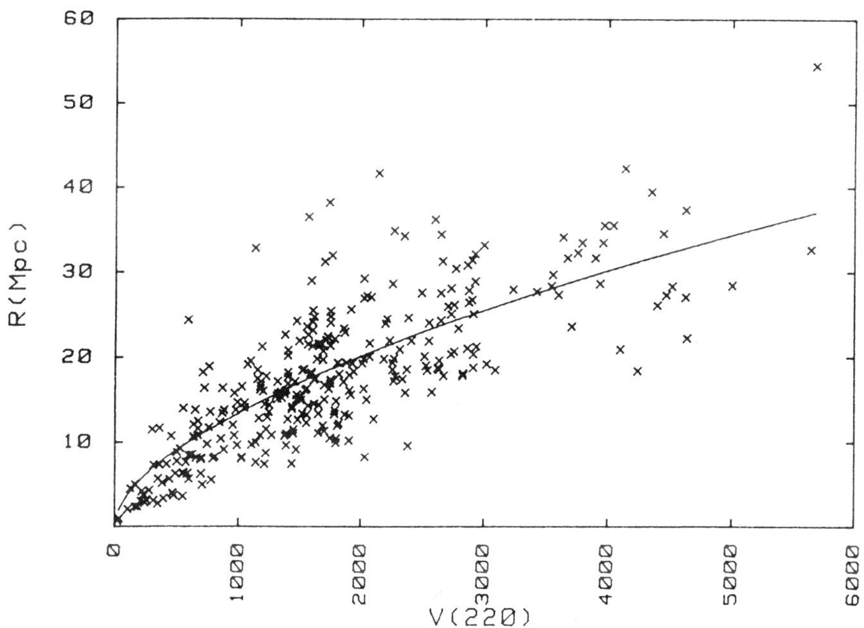

<u>Fig.3</u>. A plot of the Λ_c distances, as published by de
Vaucouleurs (1979), against the recession velocity
(corrected for a Virgocentric infall model with a local
infall velocity of v_{vc}=220 km s^{-1}). Note the non-linearity
of the relation.

     The question is then, how the best, unbiased value of
H_o can be derived from the Λ_c distances. A possibility is to
determine a Hubble constant $H_i=v_{220}(i)/R_i$ for each galaxy i.
The resulting values are plotted against the velocity v_{220}
in Fig.4. Here the increase of H_i with velocity (i.e.

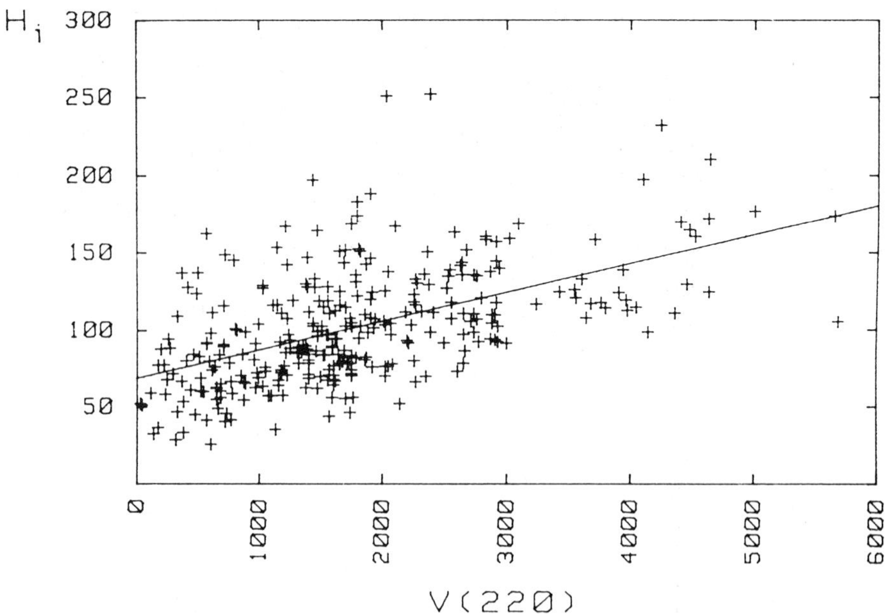

Fig.4. Hubble ratios H_i for individual galaxies with published Λ_c distances plotted against the corrected velocity v_{220}. Due to the Malmquist effect the values of H_i increase with velocity/distance.

Table 2: de Vaucouleurs' Calibrators

Galaxy	$(m-M)^o_{deV}$	$(m-M)^o_{new}$	$(m-M)$
LMC	18.31	18.50	−0.19
M31	24.07	24.2	−0.13
M33	24.30	24.4	−0.10
NGC 2403	27.10	27.8	−0.70
NGC 3031	27.7	28.7	−1.00
NGC 4236	27.7	27.8	−0.10
M101	28.5	29.2	−0.70
NGC 5474	28.5	29.2	−0.70
NGC 5585	28.5	29.2	−0.70
NGC 247	27.0	26.7	+0.30
NGC 253	27.0	27.5	−0.50
NGC 300	27.0	26.0	+1.00
NGC 7793	27.0	27.5	−0.50
			−0.31±0.15

distance) becomes very clear, in fact a least-squares
solution gives

$$H_i = 68.0 + 0.019 v_{220}. \tag{1}$$

From eq.(1) the best value of H_o is judged to be $H_o=68$. This
should not be interpreted as the Hubble constant at zero
velocity/distance, because eq.(1) is still well determined
if one excludes the nearby galaxies with $v_{220} < 1000$ km s^{-1}.
Rather the value $H_o=68$ should be taken as the best estimate
at zero bias. The conclusion from this is that the Λ_c
distances do not require $H_o=100$, but much more likely H_o
$=68$.

The value of $H_o=68$ is still based on de Vaucouleurs'
(1979) calibrators. They define, as Table 2 shows, a
somewhat fainter zero-point than the "new" calibrators of
Table 1. The zero-point difference of $-0^m.31$ requires an
increase of the Λ_c distances by a factor of 1.15. Including
this correction, the best estimate of the Hubble constant,
derived from Λ_c distances, is $H_o=59$. It is impossible to
assign a formal error to this value. The Λ_c distances
contain internal inconsistencies, e.g. one obtains different
values of H_o if the galaxies are subdivided into absolute
magnitude or type bins (Tammann and Sandage, 1983).
Moreover, the numerical result depends on the linear
correlation between H_i and velocity, which was assumed in
eq.(1). Any other, non-linear relation would have led to a
different, zero-bias value of H_o. These remaining
uncertainties introduce an estimated error of about ± 20 km
s^{-1} Mpc^{-1}, i.e. $H_o=59\pm20$ from the Λ_c distance indicator.

VI. 21cm-LINE WIDTHS AND INFRARED MAGNITUDES AS DISTANCE
    INDICATORS

21cm-line widths Δv_{21} of spiral galaxies, corrected for
inclination, and (nearly absorption-free) IR magnitudes m_{IR}
have been used to derive a Hubble constant of $H_o=90$
(Aaronson and Mould, 1986, and references therein). In order
to test whether this distance scale is compatible with a
linear expansion field, the same procedure has been employed
as in Section V. For this purpose, IR magnitudes and incli-
nation-corrected 21cm-line widths are used of 308 field
spirals outside the Local Group (17 of which are certain or
probable Virgo cluster members) from Aaronson et al. (1982).
A quadratic relation between absolute magnitude M_{IR} and Δv_{21}
has been prescribed by Aaronson et al. (1986). Following
strictly these precepts, the distance moduli $m_{IR}-M_{IR}$ and the
corresponding linear distances R have been calculated for
all 308 spirals. Their observed recession velocities have
been corrected for the effect of a selfconsistent Virgo-

centric infall model with a Local Group infall of v_{vc}=220 km s^{-1} (following Kraan-Korteweg, 1985). If one plots log R versus log v_{220} one finds a seemingly linear relation (analogously to Fig.2). A more revealing plot of the numeric values of R versus v_{220}, however, discloses a definite non-linearity (Fig.5), which can be expressed in the form R= $a \cdot v_{220}^{b}$ with a=0.0430 and b=0.811. This non-linearity, requiring an asymtotic value of H_O>110, is naturally explained by the Malmquist effect. The bias is here somewhat milder than for the Λ_C distances, as can be seen from a comparison of the exponents b. The procedure followed here to correct in first approximation for the bias is the same as in Section V. Individual Hubble ratios H_i are determined for each galaxy and these are plotted versus v_{220} (Fig.6). A linear regression then yields

$$H_i=78.8 + 0.0117v_{220}. \tag{2}$$

This suggests a zero-bias value of H_O=79, which is based, however, on only three calibrating galaxies, viz. M31, M33,

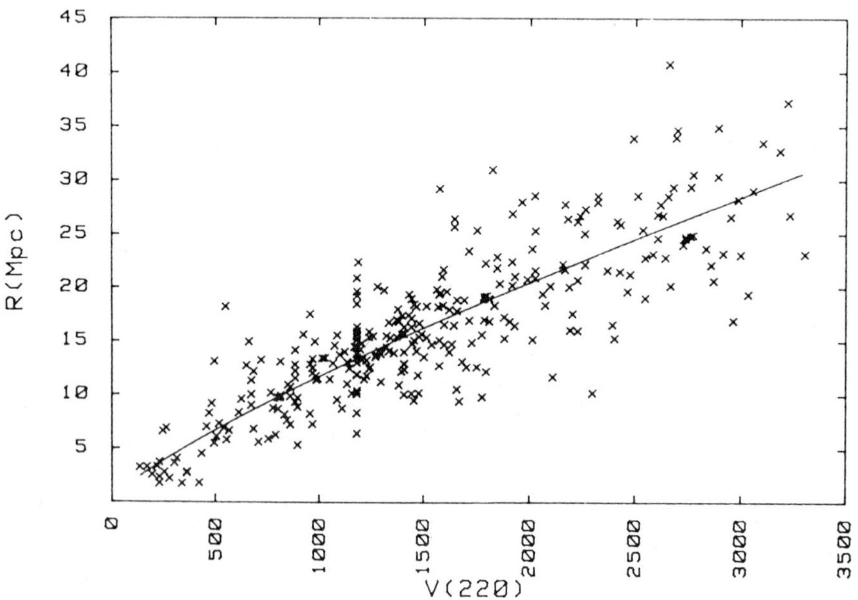

Fig.5. A plot of the IR 21cm-line width distances of 308 field and Virgo cluster spirals against the corrected recession velocity v_{220}. The distances are calculated following the precepts of Aaronson et al. (1986). Note the non-linearity of the relation.

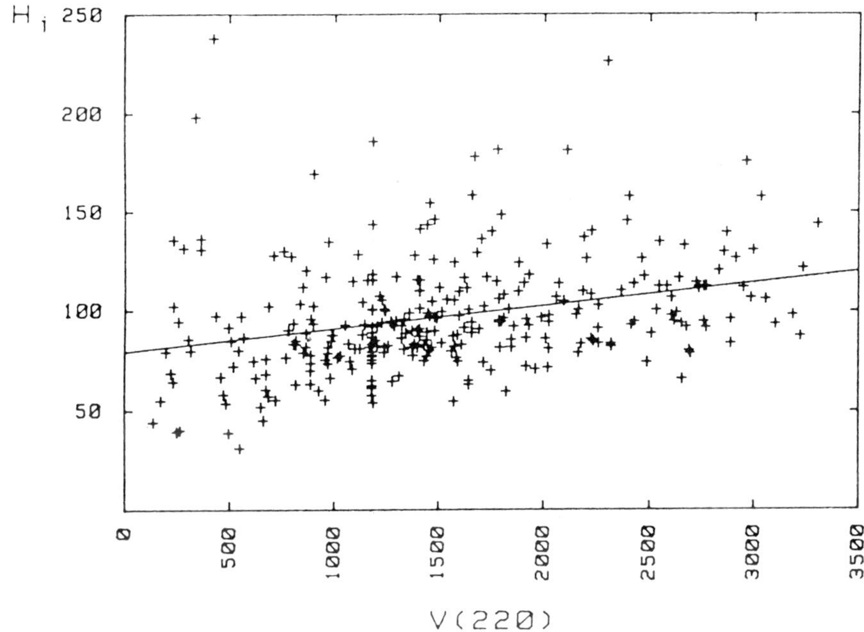

<u>Fig.6.</u> The Hubble ratios H_i for individual galaxies with known distances from IR magnitudes and 21cm-line widths. Due to the Malmquist effect the values of H_i increase with velocity/distance.

and NGC 2403 (Aaronson et al., 1986). One of these calibrators, M31, may have an unrealiable IR magnitude (Manousoyanniki and Chincarini, 1986). A safer procedure is to use as many calibrators as possible. In Table 3 the distance moduli $(m-M)_{Aa}$ are listed as obtained from the precepts of Aaronson et al. (1986) and they are compared to the new distance moduli repeated from Table 1. The mean difference of $(m-M)=-0^m.75$ requires a stretching factor of 1.4, i.e. $H_O=79:1.4=56$.

The details of this discussion may be debated. For instance, one could criticize that the quadratic $M_{IR}/\Delta v_{21}$ relation should not be used for the intrinsically fainter galaxies but rather the linear relation of Aaronson et al. (1982a). If this second possibility is followed up one finds a zero-bias value of $H_O=72$ and after correction to the "new" zero-point $H_O=60$. Another objection might be that only calibrators should be used with $\Delta v_{21}>200$ km s^{-1}; the nine remaining calibrators would then give $H_O \sim 70$.

Table 3: The Calibrators of Aaronson et al.(1986)

Galaxy	$(m-M)_{Aa}$	$(m-M)_{new}$	$\Delta(m-M)$
M31	<u>24.12</u>	24.2	-0.08
M33	<u>24.17</u>	24.4	-0.23
NGC 247	27.05	26.7	+0.35
NGC 253	27.22	27.5	-0.28
NGC 7793	27.61	27.5	+0.11
NGC 2366	26.20	27.8	-1.60
NGC 2403	<u>27.57</u>	27.8	-0.30
NGC 3031	<u>27.57</u>	28.7	-1.13
IC 2574	26.22	27.8	'-1.58
NGC 4236	27.58	27.8	-0.22
NGC 5204	27.16	29.2	-2.04
HoIV	26.69	29.2	-2.51
NGC 5585	28.92	29.2	-0.28
			-0.75 ± 0.25

Note: Only the underlined distances are used by
Aaronson et al. (1986) as calibrators, but the
distances of the remaining galaxies are implied by
their calibrated eq.4.

These remarks already show that the $M_{IR}/\Delta v_{21}$ data leave
considerable margin for interpretation. A compromise value
may be $H_o=65\pm15$. If anything, this margin is increased by
the impossibility to apply a rigid correction for the Malmquist bias; that the optical and IR Tully-Fisher relations
have considerable intrinsic scatter (viz. $\sigma_M=0\overset{m}{.}7$ according
to Rubin et al., 1985), and must therefore be vulnerable to
the Malmquist effect, is not new (cf. Bottinelli et al.,
1986; Kraan-Korteweg et al., 1986; Giraud, 1985, 1986).
Other unsolved problems of the Tully-Fisher relation concern
its quadratic or linear shape and its observed type
dependence (Roberts, 1978; Giraud, 1985; Rubin et al., 1985;
Kraan-Korteweg et al., 1986). To take the Tully-Fisher relation as the one reliable distance indicator would therefore be a misrepresentation. The method will have to be
replaced eventually by physically more meaningful kinematical para-meters (Persic and Salucci, 1986).

The IR Tully-Fisher relation has also been applied to
ten clusters beyond the Virgo cluster (Aaronson et al.,
1986). The resulting distances relative to Virgo carry less
of a Malmquist bias, i.e. they define a more nearly linear
velocity field beyond Virgo, because the distant clusters
were sampled to fainter magnitudes than the Virgo cluster.
It was mentioned in Section IV that a <u>consistent</u> change of
the apparent limiting magnitude is indeed a powerful remedy
against Malmquist bias.

VII. THE LINEARITY OF THE LOCAL EXPANSION FIELD

In the two previous Sections the demonstration of the
Malmquist effect relied on the systematic deviations from a
linear expansion field. This made the tacit assumption that
the true expansion field _is_ linear. The evidence for a truly
linear Hubble flow at a scale of v<5000 km s^{-1} is given
here.

   The linearity of the expansion field can be tested with
relative distances only, i.e. with standard candles for
which an arbitrary absolute magnitude can be adopted. SNe Ia
at maximum light and first-ranked galaxies in groups are
ideal for this purpose, because they have not only small
intrinsic scatter σ_M, but their discovery is also nearly
independent of their luminosity as argued in Section IV,
making them insensitive to Malmquist bias.

   For 49 SNe Ia with v_{220}<5500 km s^{-1} blue maximum
magnitudes are known (Cadonau 1986). They are corrected for
Galactic absorption and, in the case of the 38 SNe Ia in
S/Im galaxies, for intrinsic absorption in the parent
galaxy. For the latter correction a preliminary absorption
law of A_B =E(B-V), as evidenced by IR data, was adopted. No
absorption correction within the parent galaxy was applied
for the 11 SNe Ia in E/S0 galaxies. Plotted into a Hubble
diagram the resulting values of m_B^O(max), combined with their
respective corrected recession velocities v_{220}, define a
Hubble line of the form

$$m_B^O(\max)=5\cdot\log\ v_{220}-3.32. \tag{3}$$

It was seen before that the (logarithmic) Hubble diagram is
insensitive to a linearity test. An arbitrary absolute
magnitude of SNe Ia at maximum of M_B(max)=$-20^m\!.0$ was therefore adopted, and the corresponding linear distances R were
calculated. They are plotted versus v_{220} in Fig.7.

   Corrected magnitudes m_V are published for 38 first-
ranked galaxies in groups and clusters with v_{220}<5500 km s^{-1}
(Sandage 1975). They define a Hubble line of

$$m_V(1)=5\cdot\log\ v_{220}-6.82 \tag{4}$$

A comparison of eq.(3) and (4) shows that M_V(1) is $3^m\!.50$
brighter than M_B(max) for SNe Ia, for which an arbitrary
value of $-20^m\!.0$ was adopted. If one therefore adopts for
first-ranked galaxies M_V(1)=$-23^m\!.50$, it will put them on the
same (arbitrary) distance scale. With this precept linear
distances were calculated for the first-ranked galaxies and
they were added to Fig.7.

   Inspection of Fig.7 puts stringent limits on any
systematic deviations from a linear expansion field. This
proves that the local Hubble flow _is_ linear and justifies

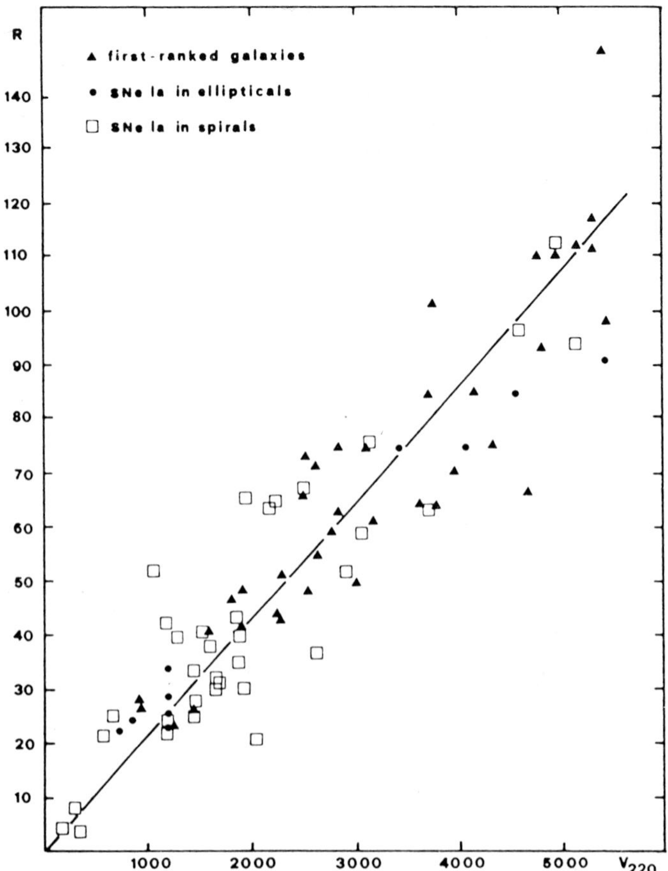

<u>Fig.7</u>. The distance-velocity diagram of SNe Ia and
first-ranked cluster galaxies in groups and
clusters. The full drawn line corresponds to an
arbitrary Hubble constant of 46.1, which follows
from the choice of $M_B(max)=-20^m0$ for SNe Ia and
correspondingly of $M_V(1)=-23^m50$ for first-ranked
galaxies. The diagram puts strong limits on any
non-linearity of the local expansion field.

the procedure in Sections V and VI, where the deviations
from linearity were charged to the Malmquist bias.
     In passing it is noted that the random scatter in Fig.7
is considerable. This scatter can be interpreted as a
scatter in distance or in velocity or as a combination of
both. If the scatter is read in distance it corresponds to a
mean error of $\Delta r/r \sim 0.25$ or $\sigma(mag)=0^m5$, which can be due to

luminosity scatter of the standard candles and/or to random photometric errors. If on the other hand, one assumes (unrealistically) that the total scatter is caused by peculiar motions of the individual objects, it follows $\sigma(v_{220})=550$ km s^{-1}. This value is an upper limit of the mean one-dimensional peculiar motion of field galaxies, groups and clusters. The corresponding three-dimensional value of $\sigma(v_{220})<950$ km s^{-1} can be compared with the absolute peculiar velocity of 600 km s^{-1} of the Local Group with respect to the microwave background. This suggests then, that our peculiar motion may be quite typical.

VIII. THE ABSOLUTE MAGNITUDE OF SNe Ia AT MAXIMUM

The evidence for the absolute magnitude M_B(max) of SNe Ia has been recently compiled by Cadonau, Sandage, and Tammann (1985). Therefore only a brief, slightly updated review is given here.

For two SNe, presumably of type Ia, distances are known from the brightest stars of their parent galaxies (Sandage and Tammann, 1982). Their somewhat revised data are given in Table 4.

Table 4: Data for two SNe Ia with known distances of their parent galaxies

SN	Galaxy	(m–M)	m_B(max)	(B–V)	E_{B-V}	m_B^o(max)	M_B^o(max)
1937c	IC 4182	28.21	8.79	+0.02	0.29	8.50	−19.71
1954a	N 4214	28.92	9.79	−0.03	0.24	9.55	−19.37
							−19.54

The apparent magnitudes m_B(max) and colors (B–V) are taken from Cadonau (1986), as well as the intrinsic color $(B-V)_{max}=-0^m\!.27$. As in Section VII it was assumed that the resulting reddening E_{B-V} is equal to the blue absorption A_B from within the parent galaxy. Note that this absorption correction is very conservative; a conventional law of $A_B=4 E_{B-V}$ would bring the mean absolute magnitude from M_B(max)= $-19^m\!.54$ to $-20^m\!.34$. In addition, if one or both SNe in Table 4 had not been of type Ia, this type, as the brightest known, could only be still brighter.

The absolute magnitude of two historical SNe, Tycho and Kepler, was estimated to be $-20^m\!.0\pm0^m\!.6$ (Cadonau et al., 1985). This implies distances of the SN remnants of 4.4 and 3.6 kpc, respectively, as can be seen from data given elsewhere (Tammann 1982). These distances lie within the limits of current determinations, which disagree however by factors of almost 2. The calibration from historical data is therefore still weak. However, in principle this calibration sets a lower limit, because if the Tycho and/or Kepler SNe had not been of type Ia, they could yield only too low a luminosity.

From the angular extent of the supernova remnant of the type II SN 1979c, measured with VLBI radio techniques, and the optical expansion velocity Bartel et al. (1985) have determined a Virgo modulus of 31.4 ± 0.8. The large error is due to the difficulty to combine the radio and optical data, but the distance has been closely confirmed by Chevalier und Fransson (1985) who studied the SN interaction with a circumstellar wind. Because the seven SNe Ia in Virgo have $<m_B(max)>=12^m_.24$, a value of $M_B(max)=-19^m_.2\pm0^m_.8$ follows.

A Baade-Becker-Wesselink parallax of a SN Ia yields $M_B(max)=-20^m_.5$ with an internal error of $\pm0^m_.6$ (Branch et al., 1983). This value should be made brighter by $\sim0^m_.5$, because the canonical time between explosion and B maximum of 15 days should now be revised to >20 days (Cadonau, 1986). From a type II SN a Virgo cluster distance of 23 ± 3 Mpc was found consistently by different authors (Panagia et al., 1980; Branch et al., 1981; Kirshner 1985), which implies from the argument in the foregoing paragraph $M_B(max)=-19^m_.6$ for SNe Ia. The quantitative exploitation of SN expansion parallaxes poses still considerable difficulties, but a mean value and error of $M_B(max)=-20^m_.2\pm0^m_.7$, stemming from two types of SNe, may be realistic (cf. also Branch 1985).

Physically determined distances of SNe Ia follow also from deflagrating C/O White Dwarf models producing radioactive ^{56}Ni. The models of several authors give $M_B(max)=-19^m_.6$ with larger margins toward brighter magnitudes than vice versa (Sutherland and Wheeler, 1984; Arnett, Branch, and Wheeler, 1985; Nomoto 1986). Converting the bolometric luminosity at maximum of $2.5\cdot10^{43}$ erg s^{-1} by Woosley, Taam, and Weaver (1986) into B magnitudes according to Arnett et al. (1985) gives $M_B(max)=-20^m_.2$ ($-0^m_.3$; $+0^m_.8$). Thus the deflagration models seem to require $M_B(max)=-19^m_.8\pm0^m_.6$.

The five routes to $M_B(max)$ of SNe Ia are all still inflicted by considerable uncertainties, but it is encouraging in view of their complete autonomy that they are fully consistent with a single value of $M_B(max)=-19^m_.7\pm0^m_.4$. We can now return to eq. (3) to find that a SN Ia has, e.g. at $v_{220}=5000$ km s^{-1} an apparent magnitude of $m_B(max)=15^m_.17$, corresponding to a distance modulus of $(m-M)=34.87$ or 94.4 Mpc. All existing evidence of SNe Ia, including Baade-Becker-Wesselink parallaxes of SNe II, points therefore to a global (i.e. large-scale) value of $H_o=53\pm11$.

IX. THE DISTANCE OF THE VIRGO CLUSTER

Simple arguments suggest that the Virgo cluster is roughly
three times more distant than the M101 group (Sandage and
Tammann, 1976), i.e. at a distance of ~20 Mpc. The arguments
involve, besides 21cm-line widths, the brightest stars, H II
region sizes, luminosity classes, galaxian magnitudes and
diameters, as well as the velocity ratio. Indeed from more
modern data the ratio v_{220}(Virgo)/v_{220}(M101 group) becomes
2.9 (Kraan-Korteweg 1985). This points to a Virgo distance
of 16-26 Mpc if allowance is made for a peculiar group
velocity of ±100 km s^{-1}.

On the basis of recent data the blue-magnitude/21cm-
line width relation gives a Virgo modulus of 31.70 (Richter
and Huchtmeier, 1984) and 31.40 (Sandage and Tammann, 1984).
These values may be low by ~0^{m}.25, because the Virgo spirals
may have been overcorrected for intrinsic absorption (van
den Bergh,1984). Visvanathan (1983), however, confirms the
conventional absorption corrections by using Tully-Fisher
relations for optical magnitudes and for a magnitude
measured at 1.05 μm. He finds a Virgo modulus of 31.33. -
The three distance moduli in this paragraph are all reduced
to the new zero-point defined in Table 1. The resulting mean
of $(m-M)_{\mathrm{Virgo}}$=31.5 is in principle a lower limit because no
Malmquist correction has been applied.

A bias-free distance of Virgo can be derived in the
following way from IR magnitudes and 21cm-line widths. For
all 15 Sbc, Sc galaxies with inclinations i>45° from the
Revised Shapley-Ames Catalog (Sandage and Tammann, ...;
RSA), which are certain Virgo cluster members, IR magnitudes
and 21cm-line widths Δv_{21} are available. The sample is
therefore unbiased in Δv_{21} (but of course not in M_{IR}).
Treating M_{IR} as the independent variable should therefore
lead to an unbiased Virgo distance, which in particular
should be free of any Malmquist bias. The field galaxies
with IR magnitudes (Aaronson et al. 1982) and 13 of the
calibrators of Table 1 determine the following <u>inverse</u>
Tully-Fisher relation

$$\log\Delta v_{21}=-0.0780\ M_{\mathrm{IR}} + 0.820\ (\pm0.017). \qquad (5)$$

This relation is not strictly correct, because the field
galaxies constitute a sample which is still biased in Δv_{21}
(Kraan-Korteweg et al., 1986), and this will influence the
slope of eq. (5). With a blind eye for the remaining bias,
we derive with the 15 Virgo spirals a cluster modulus of
31.57$\pm$0.25. This somewhat crude application of the inverse
Tully-Fisher relation shows, that the IR magnitudes and
21cm-lin widths do <u>not</u> lead to low Virgo cluster distances,
but rather that they yield a marginally higher modulus than
the blue Tully-Fisher relation.

It was stated before that the seven SNe Ia in the Virgo cluster have $\langle m_B(max)\rangle=12^m\!.24\pm0^m\!.19$. Combining this with the calibration in the previous Section gives a cluster modulus of 31.94 ± 0.44.

Globular clusters have resisted for a long time to yield a meaningful Virgo distance. However, the recent discovery of the turnover of the magnitude distribution of the globulars in M87 at $B=25^m\!.0\pm0^m\!.3$ leads, together with the assumption that the turnover lies at the same luminosity as in the Local Group, to a Virgo modulus of 31.43 ± 0.3 (van den Bergh, Pritchet, and Grillmair, 1986).

Very recently Dressler (1986) has discovered that the relation between a photometric diameter and the central velocity dispersion, known for ellipticals, can be extended to the bulges of early-type spirals. From this relation he finds a Virgo modulus of 31.64 ± 0.3 if M31 is used as the sole calibrator, or $31.5<(m-M)<32.2$ if M31 and M81 are used for the calibration. In the latter case the lower modulus corresponds to the conventional assumption that M81 lies at the distance of NGC 2403, while the higher modulus follows from Sandage's (1984a) larger M81 distance.

The foregoing distance determinations are in excellent agreement with an adopted Virgo cluster modulus of $(m-M)=31.6\pm0.3$ or $r=20.9\pm3.0$.

The unquestionable Virgo members define a mean velocity of $v_0=976\pm67$ km s^{-1}; higher values in the literature are due to outlying galaxies which belong to the background as judged from their morphological appearance (Binggeli, Sandage, and Tammann, 1987). Increasing this value by our Virgocentric infall velocity of $v_{VC}=220\pm50$ km s^{-1} (Tammann and Sandage, 1985) we obtain v_{220}(Virgo)$=1196\pm84$ km s^{-1}. From this follows a Hubble constant at the distance of the Virgo cluster of $H_0=57\pm9$.

The Virgo cluster is quite rigidly tied into the global expansion field by relative distance indicators. From this it was concluded that $H_0=(50\pm7)(21.6/r_{Virgo})$ (Tammann and Sandage, 1985). With the present distance of $r_{Virgo}=20.9$ one finds $H_0=52\pm10$.

To illustrate this point, one may use for instance the well determined relative distance between the Virgo cluster and the Coma cluster (Table 5) to obtain $r_{Coma}=119\pm18$ Mpc and with v_{220}(Coma)$=7217\pm400$ km s^{-1} a large-scale Hubble ratio of $H_0=61\pm10$. The error of the Coma cluster velocity is to allow for a possible one-dimensional peculiar velocity of the cluster of 400 km s^{-1}.

Distances relative to Virgo are also available for ten clusters with IR and Δv_{21} data (Aaronson et al. 1986). With the present Virgo distance they require $H_0=65$. This is rather an upper limit to H_0 because the clusters were not sampled to the same <u>absolute</u> magnitude, as mentioned before, and some remaining Malmquist bias is to be expected.

Table 5: The Modulus Difference Coma-Virgo

(m-M)	Author(s)	Remarks
3.87 ± 0.13	Dressler (1984)	1
3.90 ± 0.10	Tammann and Sandage (1985)	2
3.69 ± 0.12	Aaronson et al. (1986)	3
3.80 ± 0.06	Vader (1986)	4
3.60 ± 0.10	Lynden-Bell (1986)	5
3.81 ± 0.12	Lucey (1986)	6
3.78 ± 0.05		

Remarks:
1. Mean of luminosity/velocity dispersion and luminosity/Mg index relations.
2. Mean from various authors and methods (excluding Dressler 1984).
3. From IR Tully-Fisher relation.
4. From multivariate analysis of E galaxies.
5. From velocity dispersion/diameter relation.
6. From central-luminosity/velocity dispersion relation.

Along still another route the adopted Virgo and Coma distances define the absolute magnitude of first-ranked cluster galaxies to $M_V(1)=-23\overset{m}{.}37$ and $-23\overset{m}{.}56$, respectively, using apparent magnitudes from Sandage (1975). Combined with eq. (4) the mean value of $M_V(1)=-23\overset{m}{.}56\pm0\overset{m}{.}4$ yields $H_O=45\pm9$. Although the value relies on only two calibrators, it is in principle a very powerful determination, because eq. (4) is defined out to $v\sim10\ 000$ km s^{-1} (Sandage 1975) and must describe the truly global expansion.

X. THE GLOBAL VALUE OF H_O

Field galaxies out to $v\sim3000-5000$ km s^{-1} give an estimate of the Hubble constant of 59 ± 20 and 65 ± 15 from the Λ_c distance indicator and from the IR Tully-Fisher relation, respectively, if one allows in first approximation for Malmquist bias (cf. Sections V and VI).

An impartial test for the presence of a Malmquist bias is provided by the inverse Tully-Fisher relation, as mentioned before. It is uncontested that the regression of Δv_{21} on M is independent of the sample selection in M; it requires, however, that the sample is unbiased in Δv_{21}. The practical difficulty is to define objectively a sample of _field_ galaxies with complete Δv_{21} data. An approximation is provided by the 147 Sbc, Sc galaxies with $m_B<12\overset{m}{.}0$ and inclination $i>45^O$ from the RSA; for 136 of these galaxies 21cm-line widths are available and they form hence an almost complete sample, which is hopefully unbiased in Δv_{21} (cf. Kraan-Korteweg et al., 1986). The 136 galaxies define a regression

$$\log \Delta v_{21} = -0.082 M_B + 0.88, \tag{6}$$

$\underline{if}$ $H_O=55$ is assumed. Ten calibrating RSA galaxies from Table 1 require a constant term of 0.862 ± 0.025, which translates into $H_O=50\pm8$. However, five of the calibrators are of type Sd; if they should be excluded, the constant term becomes 0.906 ± 0.022 and $H_O=64\pm8$. These results are in satisfactory agreement with the de-biased values above.

SNe Ia in field galaxies out to $v \sim 5000$ km s^{-1} give an independent and unbiased determination of $H_O=53\pm11$ (cf. Section VIII).

The combined evidence from field galaxies is therefore well satisfied by $H_O=55\pm10$.

The distance scale building on clusters yields a Hubble constant at the Virgo cluster distance of $H_O=57\pm9$ and a value of $45<H_O<65$ for clusters out to $v \sim 10\ 000$ km s^{-1}.

The agreement of the distance scales from field galaxies and from cluster galaxies is very satisfactory. It suggests that no major systematic errors dominate the solution. Errors due to selection effects propagate differently through the field and cluster distance scales, because field galaxies (mainly from the RSA) and cluster galaxies (from specific catalogs) are typically sampled to different apparent-magnitude limits.

The impossibility of the short distance scale with $H_O \sim$ 100 becomes most obvious just from the discrepancy between clusters and field galaxies. Even if the cluster scale may appear as linear, and to define a single value of H_O, the value of H_O from field galaxies agrees at one specific distance only, being lower at smaller distances and higher at larger distances.

With a consistent, linear distance scale at hand, which reaches out to $v=5000-10\ 000$ km s^{-1} and defines H_O(global)= 55 ± 10, a rather unexpected result emerges. If one determines the Hubble constant from the calibrating galaxy M101 alone (Table 1; $v_{220}=415$ km s^{-1}), one finds $H_O=60$. In a detailed analysis of the local expansion field, including several additional dwarf galaxies with known distances, Sandage (1986d) has drawn the same conclusion, viz. H_O(local)=55. It is surprising that in the presence of large galaxy clusterings and of considerable streaming velocities the local value of H_O is the same as H_O(global) within the observational errors.

XI. CONCLUSION

The available data, involving various methods and different
authors, impose consistently a Virgo cluster modulus of
31.6$\pm$0.3 and a global value of the Hubble constant, derived
independently from clusters and field galaxies, of H_O=55$\pm$7.
The mean errors quoted here are estimates, because in the
presence of possibly remaining systematic errors they cannot
be calculated in a formal way. The error of the Hubble
constant defines a probably realistic 99% confidence
interval of 35<H_O<75.

The so-called short distance scale with H_O ~100 implies
discordant scales for clusters and field galaxies, which is
obviously the result of a well understood selection effect
(Malmquist bias). A first-order correction for this bias
removes the discrepancy from the above results.

The short distance scale requires all distant extragalactic objects to be fainter by ~$1\overset{m}{.}5$ than presently
assumed. This difference is <u>not</u> caused by major differences
of the adopted local calibrators. The bulk of the difference
builds up between the local calibrators and the Virgo
cluster, which the short distance scale locates almost 1^m
nearer than in the case of H_O=55. Taken at face value the
short distance scale would create a number of artificial
problems, e.g.:
- It would imply M_B(max)=$-18\overset{m}{.}3$ for SNe Ia. In that case at
  least four historical SNe would have occurred within 2.5
  kpc during the last millennium (cf. Tammann 1982, Table
  4), which corresponds for the whole Galaxy to an un-
  parallelled frequency of 1 SN per <10 years.
- The Galaxy and M31 would have the largest diameters among
  the galaxies in the Local Supercluster (van der Kruit
  1986).
- Primordial nucleosynthesis is most easily understood with
  a baryon density of Ω_O ~ 0.1 $(50/H_O)^2$ (Rees 1986). With H_O
  =100 one obtains Ω_O=0.025, which is about four times less
  than required to bind clusters of galaxies (Sandage and
  Tammann, 1984b). In that case the clusters were bound by
  mainly non-baryonic matter.
- The maximum Friedman time would be $1/H_O$=10^{10} years, i.e.
  less than the age of globular clusters (Sandage 1982;
  Renzini 1986). Therefore H_O ~100 excludes a priori all
  Friedman models.

No such problems arise with H_O~50. It seems to require
superluminal velocities up to ~10c in some radio sources
(Porcas 1985), but a value of 5c in the case of H_O~100 poses
in principle similar constraints on the models.

As the historical evidence shows, it is much easier to
measure too high a value of H_O than too low a value. Also
future corrections will probably tend to lower H_O even
further. Very little work has been done to determine an

absolute lower limit of H_O, but if $H_O < 40$ our present
understanding of galaxian parameters and SN luminosities
would rather be hindered than helped. For all practical
purposes the convenient number of $H_O = 50$ km s^{-1}Mpc^{-1} is
recommended; it will probably take a long time before a
<u>significantly</u> different value becomes necessary.

ACKNOWLEDGEMENTS. Because this Conference coincides within a
few weeks with Allan Sandage's 60th birthday it is
appropriate to recollect here how much he has contributd to
our present topic. I could accompany him for a part of his
work on the distance scale for almost 25 years; for this
privilege I am deeply grateful.
    I thank Miss A.Schroeder for computational help and
Mrs.M.Saladin for preparing the manuscript. Support from the
Swiss National Science Foundation is gratefully acknowledged.

REFERENCES

Aaronson, M., Bothun, G., Mould, J., Huchra, J., Schommer,
        R.A., and Cornell, M.E. 1986, Ap.J.<u>302</u>, 536.
Aaronson, M., Huchra, J., and Mould, J. 1979, Ap.J.<u>229</u>, 1.
Aaronson, M., Huchra, J., Mould, J.R., Tully, R.B., Fisher,
        J.R., van Woerden, H., Goss, W.M., Chamaraux, P.,
        Mebold, U., Siegman, B., Berriman, G., and
        Persson, S.E. 1982, Ap.J.Suppl.<u>50</u>, 241.
Aaronson, M., Huchra, J., Mould, J., Schechter, P.L., and
        Tully, R.B. 1982a, Ap.J.<u>258</u>, 64.
Aaronson, M., and Mould, J. 1986, Ap.J.<u>303</u>, 1.
Andersen, J., Blecha, A., and Walker, M.F. 1985, Astron.
        Astrophys.Letters <u>150</u>, L12.
Arnett, W.D., Branch, D., and Wheeler, J.C. 1985, Nature
        <u>314</u>, 337.
Baade, W. 1952, Trans.I.A.U.<u>8</u>, 398.
Baade, W., and Swope, H.H. 1963, Astron.J.<u>68</u>, 435.
Barnes, T.G., and Hawley, S.L. 1986, Ap.J.Letters <u>307</u>, L9.
Bartel, N., Rogers, A.E.E., Shapiro, I.I., Gorenstein, M.V.,
        Gwinn, C.R., Marcaide, J.M., and Weiler, K.W.
        1985, Nature <u>318</u>, 25.
Binggeli, B., Sandage, A., and Tammann, G.A. 1987, A.J., in
        press.
Binggeli, B., Sandage, A., and Tarenghi, M. 1984, A.J.<u>89</u>,
        64.
Birkinshaw, M. 1987, this volume.
Bottinelli, L., Fouqué, P., Gouguenheim, L., Paturel, G.,
        and Teerikorpi, P. 1986, in: <u>Galaxy Distances and
        Deviations from Universal Expansion</u>, eds. B.F.
        Madore and R.B.Tully, Dordrecht: Reidel, p.73.

Bottinelli, L., Gouguenheim, L., Paturel, G., and de
        Vaucouleurs, G. 1985, Ap.J.Suppl.59, 293.
Bottinelli, L., Gouguenheim, L., Paturel, G., and
        Teerikorpi, P. 1986, Astron.Astrophys.156, 157;
        166, 393.
Branch, D. 1985, in: Supernovae as Distance Indicators, ed.
        N.Bartel, Berlin: Springer, p.138.
Branch, D., Falk, S.W., McCall, M.H., Rybski, P., Uomoto,
        A.K., and Wills, B.J. 1981, Ap.J.244, 780.
Branch, D., Lacy, C.H., McCall, M.L., Sutherland, P.G.,
        Uomoto, A., Wheeler, J.C., and Wills, B.J. 1983,
        270, 123.
Brodbeck, K. 1986, in preparation.
Burstein, D., Davies, R.L., Dressler, A., Faber, S.M.,
        Lynden-Bell, D., Terlevich, R., and Wegner, G.
        1986, in: Galaxy Distances and Deviations from
        Universal Expansion, eds. B.F.Madore and R.B.
        Tully, Dordrecht: Reidel, p.123.
Cadonau, R. 1986, Ph.D.Thesis, Univ.Basel.
Cadonau, R., Sandage, A., and Tammann, G.A. 1985, in:
        Supernovae as Distance Indicators, ed. N.Bartel,
        Berlin: Springer, p.151.
Caldwell, J.A.R., and Coulson, I.M. 1985, Mon.Not.R.astr.
        Soc.212, 879; 214, 639.
Cameron, M.L. 1985a, Astron.Astrophys.147, 39.
Cameron, M.L. 1985b, Astron.Astrophys.152, 250.
Carney, B.W. 1980, Ap.J.Suppl.42, 481.
Chevalier, R.A., and Fransson, C. 1985, in: Supernovae as
        Distance Indicators, ed. N.Bartel, Berlin:
        Springer, p.123.
Chiosi, C., and Pigatto, L. 1986, Ap.J.308, 1.
Cohen, J.G. 1985, Ap.J.292, 90.
Conti, P.S., and Garmanay, C.D. 1986, Astron.J.92, 48.
Cook, K.H., Aaronson, M., and Illingworth, G. 1986, Ap.J.
        Letters 301, L45.
Crampton, D. 1979, Ap.J.230, 717.
de Vaucouleurs, G. 1970, Ap.J.159, 435.
de Vaucouleurs, G. 1975, in: Galaxies and the Universe, eds.
        A. and M.Sandage and J.Kristian, Chicago:
        University of Chicago Press, p.557.
de Vaucouleurs, G. 1978, Ap.J.223, 730.
de Vaucouleurs, G. 1979, Ap.J.227, 380.
de Vaucouleurs, G., and Corwin, H.G. 1986, Ap.J.308, 487.
Djorgovski, S., and Davis, M. 1986, in: Galaxy Distances and
        Deviations from Universal Expansion, eds. B.F.
        Madore and R.B.Tully, Dordrecht: Reidel, p.135.
Dressler, A. 1984, Ap.J.281, 512.
Dressler, A. 1986, preprint.
Eggen, O.J. 1977, Ap.J.Suppl.34, 1.
Faber, S.M., and Jackson, R. 1976, Ap.J.204, 668.
Feast, M.W. 1984, Mon.Not.R.astr.Soc.211, 51P.

Feast, M.W. 1986, in: Galaxy Distances and Deviations from
        Universal Expansion, eds. B.F.Madore and R.B.
        Tully, Dordrecht: Reidel, p.7.
Freedman, W.L. 1986, in: Galaxy Distances and Deviations
        from Universal Expansion, eds. B.F.Madore and R.B.
        Tully, Dordrecht: Reidel, p.21.
Gautschy, A. 1986, preprint.
Genzel, R., Reid. M.J., Moran, M., and Downes, D. 1981,
        Ap.J.244, 884.
Giraud, E. 1985, Astron.Astrophys.153, 125.
Giraud, E. 1986, ESO Preprint No.473.
Graham, J.A. 1975, Pub.A.S.P.87, 641.
Graham, J.A. 1982, Ap.J.252, 474.
Graham, J.A. 1984, Astron.J.89, 1332.
Holmberg, E. 1958, Medd.Lund Obs., Ser.2, No.136, p.102.
Hubble, E. 1929, Proc.Nat.Acad.Sci.15, 168.
Hubble, E. 1936, Ap.J.84, 270.
Humphreys, R.M., and Strom, S.E. 1983, Ap.J.264, 458.
Kayser, R. 1986, Astron.Astrophys.128, 156.
Kayser, R., and Refsdal, S. 1983, Astron.Astrophys.128,
        156.
Kennicutt, R.C. 1981, Ap.J.247, 9.
Kirshner, R.P. 1985, in: Supernovae as Distance Indicators,
        ed.N.Bartel, Berlin: Springer, p.171.
Kormendy, J. 1977, Ap.J.218, 333.
Kraan-Korteweg, R.C. 1985, Basel Preprint No.18.
Kraan-Korteweg, R.C., Cameron, L., and Tammann, G.A. 1986,
        in: Galaxy Distances and Deviations from Universal
        Expansion, eds. B.F.Madore and R.B.Tully,
        Dordrecht: Reidel, p.65.
Kraan-Korteweg, R.C., Sandage, A., and Tammann, G.A. 1984,
        Ap.J.283, 24.
Laney, R.T., and Stobie, R.S. 1986, in press.
Lawrie, D.G., and Graham, J.A. 1983, Bull.Am.Astron.Soc.15,
        907.
Lucey, J.R. 1986, Mon.Not.R.astr.Soc.222, 417.
Lynden-Bell, D. 1986, Q.Jl.astr.Soc.27, 319.
McAlary, C.W., and Madore, B.F. 1984, Ap.J.282, 101.
McAlary, C.W., Madore, B.F., McGonegal, R., McLaren, R.A.,
        and Welch, D.L. 1983, Ap.J.273, 539.
McAlary, C.W., and Welch, D.L. 1985, in: Cepheids: Theory
        and Observations, ed. B.F.Madore, Cambridge:
        Cambridge Univ.Press, p.228.
Madore, B.F. 1986, in: Galaxy Distances and Deviations from
        Universal Expansion, eds. B.F.Madore and R.B.
        Tully, Dordrecht: Reidel, p.29.
Malmquist, K.G. 1920, Medd.Lund Obs., Ser.2, No.22.
Manousoyanniki, I., and Chincarini, G. 1986, Astron.Astro-
        phys.160, 331.
Melnick, J., Moles, M., Terlevich, R., and Garcia-Pelayo,
        J.-M. 1986, ESO Preprint No.440.

Menzies, J.W., and Whitelock, P.A. 1985, Mon.Not.R.astr.Soc.
          212, 783.
Mould, J., and Kristian, J. 1986, Ap.J.305, 591.
Nomoto, K. 1986, Ann.New York Acad.Sci 470, 294.
Panagia, N., et al. 1980, Mon.Not.R.astr.Soc.192, 861.
Paturel, G. 1984, Ap.J.282, 382.
Pel, J.W. 1985, in: Cepheids: Theory and Observations, ed.
          B.F.Madore, Cambridge, p.1.
Persic, M., and Salucci, P. 1986, Mon.Not.R.astr.Soc.223,
          303.
Porcas, R.W. 1985, in: Active Galactic Nuclei, ed. J.E.
          Dyson, Manchester: Manchester Univ.Press, p.22.
Rees, M.J. 1986, in: Cosmology, Astronomy and Fundamental
          Physics, eds. G.Setti and L.Van Hove, Garching:
          ESO, p.227.
Renzini, A. 1986, in: Galaxy Distances and Deviations from
          Universal Expansion, eds. B.F.Madore and R.B.
          Tully, Dordrecht: Reidel, p.177.
Richter, O.-G., and Huchtmeier, W.K. 1984, Astron.Astro-
          phys.132, 253.
Roberts, M.S. 1978, A.J.83, 1026.
Rubin, V.C., Burstein, D., Ford, W.K., and Thonnard, N.
          1985, Ap.J.289, 81.
Sandage, A. 1958, Ap.J.127, 513.
Sandage, A. 1970, Ap.J.162, 841.
Sandage, A. 1971, Ap.J.166, 13.
Sandage, A. 1972, Ap.J.178, 1.
Sandage, A. 1975, Ap.J.202, 563.
Sandage, A. 1981, Ap.J.248, 161.
Sandage, A. 1982, Ap.J.252, 553.
Sandage, A. 1983a, Astron.J.88, 1108.
Sandage, A. 1983b, Astron.J.88, 1569.
Sandage, A. 1984a, Astron.J.89, 621.
Sandage, A. 1984b, Astron.J.89, 630.
Sandage, A. 1986a, preprint.
Sandage, A. 1986b, unpublished.
Sandage, A. 1986c, A.J.91, 496.
Sandage, A. 1986d, Ap.J.307, 1.
Sandage, A. 1987, this volume.
Sandage, A., and Carlson, G. 1983, Ap.J.Lett.267, L25.
Sandage, A., and Tammann, G.A. 1969, Astrophys.J.157, 683.
Sandage, A., and Tammann, G.A. 1971, Ap.J.167, 293.
Sandage, A., and Tammann, G.A. 1974a, Ap.J.190, 525.
Sandage, A., and Tammann, G.A. 1974b, Ap.J.194, 223.
Sandage, A., and Tammann, G.A. 1974c, Ap.J.194, 559.
Sandage, A., and Tammann, G.A. 1975, Ap.J.197, 265.
Sandage, A., and Tammann, G.A. 1976, Ap.J.210, 7.
Sandage, A., Tammann, G.A., and Yahil, A. 1979, Ap.J.232,
          352.
Sandage, A., and Tammann, G.A. 1981, A Revised Shapley-Ames
          Catalog of Bright Galaxies, Wash.: Carnegie Inst.

Sandage, A., and Tammann, G.A. 1982, Ap.J.$\underline{256}$, 339.
Sandage, A., and Tammann, G.A. 1984, Nature $\underline{307}$, 326.
Sandage, A., and Tammann, G.A. 1984a, in: Large-Scale
        Structure of the Universe, Cosmology and
        Fundamental Physics, eds. G.Setti and L.Van Hove,
        Garching: ESO, p.127.
Schechter, P.L. 1980, A.J.$\underline{85}$, 801.
Schommer, R.A., Olszewski, E.W., and Aaronson, M. 1984,
        Ap.J.Lett.$\underline{285}$, L53.
Stebbins, J., Whitford, A.E., and Johnson, H.L. 1950, Ap.J.
        $\underline{112}$, 469.
Stothers, R.B. 1983, Ap.J.$\underline{274}$, 20.
Strugnell, P., Reid, N., and Murray, C.A. 1986, Mon.Not.R.
        astron.Soc.$\underline{220}$, 413.
Sutherland, P.G., and Wheeler, J.C. 1984, Ap.J.$\underline{280}$, 282.
Tammann, G.A. 1977, Mitt.Astron.Ges.$\underline{42}$, 42.
Tammann, G.A. 1982, in: Supernovae: A Survey of Current
        Research, eds. M.J.Rees and R.J.Stoneham,
        Dordrecht: Reidel, p.371.
Tammann, G.A., and Sandage, A. 1983, Highlights of Astronomy
        6, 301.
Tammann, G.A., and Sandage, A. 1985, Ap.J.$\underline{294}$, 81.
Tammann, G.A., Sandage, A., and Yahil, A. 1980, in: Physical
        Cosmology, eds. R.Balian, J.Audouze, and D.N.
        Schramm, Amsterdam: North-Holland, p.53.
Tully, R.B., and Fisher, J.R. 1977, Astron.Astrophys.$\underline{54}$,
        661.
Turner, D.G. 1986, A.J.$\underline{92}$, 111.
Vader, J.P. 1986, Ap.J.$\underline{306}$, 390.
van der Kruit, P.C. 1986, Astron.Astrophys.$\underline{157}$, 230.
van den Bergh, S. 1960, Publ.David Dunlap Obs.$\underline{2}$, No.6.
van den Bergh, S. 1977, in: I.A.U.Coll.$\underline{37}$, 13.
van den Bergh, S. 1984, A.J.$\underline{89}$, 608.
van den Bergh, S., and Pritchet, C. 1986, in: Galaxy
        Distances and Deviations from Universal Expansion,
        eds. B.F.Madore and R.B.Tully, Dordrecht: Reidel,
        p.35.
van den Bergh, S., and Pritchet, C.J. 1986a, Publ.A.S.P.$\underline{98}$,
        110.
van den Bergh, S., Pritchet, C., and Grillmair, C. 1986,
        preprint.
Visvanathan, N. 1983, Ap.J.$\underline{275}$, 430.
Visvanathan, N. 1985, Ap.J.$\underline{288}$, 182.
Visvanathan, N., and Griersmith, D. 1977, Astron.Astrophys.
        $\underline{59}$,317.
Visvanathan, N., and Sandage, A. 1977, Ap.J.$\underline{216}$, 214.
Walker, A.R. 1985, Mon.Not.R.astr.Soc.$\underline{217}$, 13P.
Welch, D.L., McAlary, C.W., McLaren, R.A., and Madore, B.F.
        1985, in: Cepheids: Theory and Observtions, ed.
        B.F.Madore, Cambridge: Cambridge Univ.Press,
        p.219.

Welch, D.L., McAlary, C.W., McLaren, R.A., and Madore, B.F.
          1986, Ap.J.305, 583.
Whitmore, B.C., Kirshner, R.P., and Schechter, P.L. 1979,
          Ap.J.234, 68; 1981, Ap.J.250, 43.
Woosley, S.E., Taam, R.E., and Weaver, T.A. 1986, Ap.J.301,
          601.

DISCUSSION

M.Aaronson: I have a comment and a question. First the
comment: It might have been fair for you to point out that
our reason for collecting the nearby field galaxy data was
not to study H_o but to analyze deviations from uniform
Hubble flow. I believe our analysis of potential bias in the
latter problem was correct. As far as H_o goes, we have
always emphasized working with clusters, as this avoids
difficulties in knowing how to properly treat magnitude bias
effects or bias effects from streaming motions within the
Local Supercluster. (Nevertheless, one would expect the
field and cluster samples to yield consistent results. In
fact they do! Both sets of data give a similar value for
Virgocentric motion). Now the question: Could you tell me
what paper of mine you purported to have taken distances
from in the table you showed?

G.A.Tammann: It is, of course, interesting for me to learn
about the reason for your collecting the data. But since you
have published them, and since you have published formulae
to derive absolute magnitudes from them, I deemed it fair to
do just this. I chose your quadratic equation (Aaronson et
al., 1986, Ap.J.302, 536, eq.4); I will try in addition your
earlier linear version in the written version of my talk. I
hope it will not make too much difference, because otherwise
the IR Tully-Fisher method would depend on details which are
difficult to control. - You cannot test the linear distance
scale by the Virgocentric motion because it depends only on
relative distances. Finally you mention a central point of
our disagreement: I cannot see how nearby calibrating
galaxies and distant Virgo cluster galaxies, all obeying the
same apparent-magnitude limit of the Shapley-Ames catalog,
could have the same mean properties.

M.Aaronson: The reason for the big discrepancies with the
low luminosity galaxies in the table you showed was a result
of your extrapolation of the quadratic form of the relation
down to the low luminosity regime. As we discussed in our
last paper, where we introduced this quadratic form, this
extrapolation should not be made. (One should also confine
the calibrating galaxies to the same velocity width range as
the more distant cluster galaxies, as we have done).

A DISTANCE SCALE FROM THE IR MAGNITUDE/HI VELOCITY WIDTH RELATION

M. Aaronson
Steward Observatory
University of Arizona
Tucson, Arizona   85721
U.S.A.

ABSTRACT.   A summary is given of recent work using the IR/H I method, arguably the best global distance indicator presently available. Reflex motion toward the microwave dipole anisotropy has been seen relative to a sample of nearby galaxy clusters; this result is contrasted with the somewhat divergent conclusions obtained by Burstein et al. from a similar study of elliptical galaxies.   A best guess calibration of the IR/H I zero point continues to lead to a high value for the expansion rate.   This finding cannot be explained by appealing to Malmquist bias, as demonstrated by a straightforward linearity test of the velocity-distance relation.   However, all current estimates of the Hubble constant are plagued by the large uncertainties to the distances of nearby calibrating galaxies, a problem whose full solution probably must await the launch of the Hubble Space Telescope.

1.   INTRODUCTION -- ARE ANY GALAXY DISTANCES SECURE?

It is widely acknowledged that current estimates of the Hubble constant are uncertain by a factor perhaps as large as two.   The underlying reasons for this situation are usually obscured by claims and counter-claims involving magnitude bias effects on samples of distant galaxies. It is the author's belief, however, that the principal problem in all current determinations of H_0 starts right in our own backyard with the rubbery state of the nearby distance scale.   I would like to begin this contribution by illustrating the point with three cases: the Galaxy, the LMC, and M33.

Turning first to the Milky Way, the last few years has seen a dramatic reduction in measures of the galactic center distance R_0.   As a result, the old recommended IAU value of 10 kpc was changed at the 1985 General Assembly to 8.5 kpc   (Kerr and Lynden-Bell 1986).   Even so, the best recent determinations of R_0 from four independent methods indicate a smaller value still.   From RR Lyraes, Blanco and Blanco (1985) obtain $R_0 = 7.3 \pm 0.5$ kpc.   Using globulars, Frenk and White (1982)   have   found $R_0 = 6.8 \pm 0.8$ kpc.   Reid et al. (1987)   have

A. Hewitt et al. (eds.), Observational Cosmology, 187–195.
© *1987 by the IAU.*

measured $R_0 = 7 \pm 1$ kpc from maser expansion velocities. Finally, kinematic modeling of field Cepheids calibrated via main sequence fitting leads Caldwell and Coulson (1987) to propose $R_0 = 7.8 \pm 0.7$ kpc. The straight mean of these four results is ~ 30% less than the (not so) old IAU value!

Well, perhaps we should not be too surprised that it is difficult to obtain an accurate distance to a place hidden by 30 magnitudes of visual extinction, so let us consider instead the Large Cloud, our next nearest companion galaxy. In one way or another, modern distance estimates to most other galaxies scale directly with the LMC modulus, giving it a high significance. This fact should be born in mind by the reader when confronted with what is often the "blind" averaging of discrepant distances to farther systems. The most oft-quoted LMC modulus is based on Cepheid work of the South African group. Martin, Warren, and Feast (1979) gave 18.69 $\pm$ 0.15, based on a zero-point calibration again obtained from galactic main sequence fitting. Little changed between this and the more recent Cepheid study of Caldwell and Coulson (1986), where a value of 18.65 $\pm$ 0.07 was offered.

For the last few years, however, there has been continually mounting evidence that the Cepheid distance was problematic, first coming from the RR Lyraes (see the discussion following Graham 1984), then from the reduced galactic cluster distances found using Stromgren photometry by Schmidt (1984) and subsequently by Balona and Shobbrook (1984), from differences in the Mira P-L relations (see Menzies and Whitelock 1985), and from direct main sequence fitting to the LMC itself (Schommer, Olszewski, and Aaronson 1984). A shorter LMC modulus had of course been advocated for a long time by Eggen (1977, see also de Vaucouleurs 1980).

Recently, the South African work seems to have come into better agreement, with the latest result being a modulus of 18.45 $\pm$ 0.05 (Caldwell and Coulson 1987), a 3σ change from their earlier effort! The difference is largely due to the number of very fine CCD color-magnitude diagrams for galactic calibrating clusters that have emerged in the last few years which have tended to yield systematically smaller moduli by ~ 0.2 mag (in accord with the Stromgren findings). For comparison, recent RR Lyrae studies by Walker (1985) and Reid and Strugnell (1986) give LMC moduli of 18.42 $\pm$ 0.10 and 18.37 $\pm$ 0.15, respectively. The luminosity calibration of OB stars lead Conti, Garmany, and Massey (1986) to propose 18.3 $\pm$ 0.3. Within the errors, all these values are in agreement with 18.2 $\pm$ 0.2 obtained by Schommer et al. (1984) via Cloud main sequence fitting. (Convective overshooting does not alter this last result, since fits are made to the unevolved main sequence. As discussed by Schommer 1986, the luminosity function method advocated by Chiosi and Pigatto 1986 cannot independently determine age and distance, and is itself subject to the uncertainties of convective effects, so is therefore not to be preferred.) Finally, we note that application of stellar evolutionary models to the clump luminosities of Cloud clusters was found by Seidel, Da Costa, and Demarque (1986) to yield consistent results only for the "short scale."

The dust has by no means settled on the LMC distance, though a best guess modulus would seem to fall in the range 18.3 - 18.4. Once again, within a few years time, a "firm" distance has been modified by ~ 15 - 20%. It should be remembered that this has in part come about because the data base of material for the Magellanic Clouds is far more extensive than for any other extragalactic systems.

Finally, we turn to M33. Two recent studies involving Cepheids, one based on photographic photometry by Sandage and Carlson (1983), and another on multi-color CCD observations by Freedman (1985), yield moduli of 25.35 and 24.1, respectively, a difference of 1.25 magnitudes! Although the former value is an apparent modulus, it is nonetheless roughly the same one used by Sandage and Tammann (1984) in their attempt to derive a Hubble constant from the IR/H I relation. (These authors also adopt a large M81 apparent modulus of 28.8, which Sandage 1986 has indicated to be based primarily on novae. Use of the latter as reliable distance indicators, though, has been called into question by the recent work of Pritchet and van den Bergh 1986 on novae in the Virgo cluster.)

It is illuminating to examine the origin of this gross difference in M33 modulus. Reddening is part of the problem; Freedman's multicolor data convincingly demonstrates the presence of a <u>primarily internal</u> extinction of A_V ~ 0.6 mag. A difference in Cepheid calibrating precepts also seems involved. The relative LMC-M33 distance quoted by Sandage and Carlson (1983) implies an LMC modulus of 19.0, whereas Freedman (1985) adopts 18.5 mag. This leaves a scale difference of only ~ 0.2 mag, which presumably rests almost entirely with the less precise photographic work. (A larger scale error appears to be present for Hubble's Cepheids as corrected by Sandage and Carlson; in this regard, a similar conclusion has been reached by Christian and Schommer 1986.)

Well, what is one to make of the above. First of all, the answer to the question posed in the section heading should be obvious. Secondly, it is also clear that all of the older photographic work is in desperate need of checking via multi-wavelength CCD observations and modern point-spread-function fitting reduction techniques. Furthermore, it is interesting to note that the distance scale within the Local Group appears to be undergoing a considerable shrinkage, which will certainly have important implications with regard to future efforts to measure H_0.

2. DEVIATIONS IN THE HUBBLE FLOW

Departures from uniformity in the Hubble flow can provide important information about the value of Ω and the processes involved in galaxy formation. For a number of years, the author and collaborators have been involved in the study of large scale flow deviations using the IR/H I relation. This method satisfies most of the desirable qualities one would like to have in a distance indicator based on the global properties of galaxies: The technique connects with physics via Newton's law of gravity, all of the measurables involved are quantitative, extinction effects are not problematic, the method can be employed over a wide distance range, and the scatter in the relation is demonstrably small (e.g. Aaronson and Mould 1986).

Our initial interest in mapping the Hubble flow lay with determining the Local Group infall velocity toward Virgo. For this purpose a flow-model analysis was made of ~ 300 galaxies within the Local Supercluster (Aaronson et al. 1982, hereafter AHMST). An infall in the range 250 - 300 km s^{-1} was found, along with a significant random component of Local Group motion. More recently, Lilje, Yahil, and Jones (1986) have reanalyzed the AHMST data set by also allowing for the presence of a quadrupolar tidal velocity field. Most of the peculiar motion seen by AHMST can be accounted for with the presence of such a tidal field. It is interesting that while Lilje et al. argue the form of the tidal field indicates the existence of more than a single attractor, they find the expansion axis to point in the general direction of the Hydra-Centaurus Supercluster, and further suggest the size of the field implies a perturber distance roughly three times greater than the Virgo cluster. We come back to these points below.

Our most recent work with the IR/H I relation has involved the study of ten nearby galaxy clusters in the redshift range 4,000 - 11,000 km s^{-1} (Aaronson et al. 1986, hereafter ABMHSC). Two results on large scale velocities have emerged from this effort. First, a Virgocentric motion of ~ 300 km s^{-1} is indicated, in good agreement with the earlier analysis of the Local Supercluster sample. Second, the reflex motion of the Local Group relative to the cluster sample yields a vector that agrees well in both magnitude and direction with the velocity inferred from the dipole anisotropy of the microwave background.

The difference between the dipole vector and Local Group motion within the Local Supercluster yields an estimate of bulk Supercluster motion as a whole. The direction of this bulk motion is again found to be toward Hydra-Centaurus; its magnitude, however, is not so well determined, and can be as low as 300 or as high as 500 km s^{-1}, depending on precisely what is assumed for the tidal/peculiar motion of the Local Group (see ABMHSC, Lilje et al. 1986). In any event, these results and those of the tidal field work point toward Virgo and Hydra-Centaurus as the two principle attractors giving rise to the dipole anisotropy.

A further result to follow from the ABMHSC study concerns relative cluster-cluster motions. After correcting the velocities for the dipole effect, the observed scatter of the ten cluster sample about purely uniform Hubble flow is only ~ 270 km s^{-1}. An amount larger than this would be expected from the formal errors in velocity and distance alone. The latter, though, were overestimated since they did not make any allowance for subclustering. If zero error is assumed in the observables, then it would seem that 270 km s^{-1} was a firm upper limit to the 1-d bulk motion of a typical cluster, or 470 km s^{-1} for the 3-d bulk motion, comparable with our measure for the Local Supercluster.

The above findings stand in somewhat sharp contrast to those recently reported by Burstein et al. (1986, hereafter BDDFLTW), who have applied an improved version of the Faber-Jackson relation to an all-sky sample of elliptical galaxies. The resulting reflex vector (V = 470 km s^{-1} toward ℓ = 183°, b = 29°) points in an orthogonal direction to the dipole velocity (V = 600 km s^{-1} to ℓ = 268°, b = 27°), such that when these two components are subtracted from one another, a residual bulk motion of 600 km s^{-1} still remains, pointing once again roughly towards Hydra-Centaurus. However, the bulk motion now involves a scale that extends outwards from the Local Group some 4000 km s^{-1}, actually encompassing Hydra-Centaurus itself. That is, the latter is claimed to be moving with a velocity comparable to the Local Group <u>and in the same direction</u>, so that the attractor must lie well beyond. (A similar, though apparently less statistically significant result, has also been reported by Collins, Joseph, and Robertson 1986.)

Although the conclusions of ABMHSC and BDDFLTW look very different, the results are not formally inconsistent, as the two data sets overlap very little spatially or in redshift. In fact, for the few clusters in common between the two samples, the agreement in relative modulus appears to be excellent. Most of the power in the BDDFLTW solution comes from E galaxies in the redshift range 2000 – 4000 km s^{-1}, whereas the cluster sample of ABMHSC does not start until 4000 km s^{-1}. The latter is confined to a broad ring around the sky defined by the Arecibo declination range, and is least sensitive to the direction of the BDDFLTW motion, which is roughly perpendicular to the Arecibo band. Nevertheless, the implication is that on average most structures beyond ~ 4000 km s^{-1} would be co-moving with respect to the microwave frame.

The BDDFLTW results do appear to differ with those of AHMST. The former authors claim that their 0 – 2000 km s^{-1} shell yields the same reflex vector as the whole sample, i.e., they find no signal indicating strong Virgocentric motion. However, they have not fit a proper flow model to the data; because of the way the velocity vectors then cancel, this will lead to a result biased toward small infall values (Yahil 1986). Furthermore, while the 600 km s^{-1} bulk velocity found by BDDFLTW is perhaps just marginally consistent with the typical 1-d bulk motions indicated by the ABMHSC study, it is definitely a problem for theoretical models of biased galaxy formation with cold dark matter (e.g. Vittorio, Juszkiewicz, and Davis 1986). Such models imply that large scale bulk velocities should not be greater than ~ 150 km s^{-1}. There have been suggestions that the BDDFLTW result and the earlier Rubin-Ford effect could be biased by spatial inhomogeneities in the samples (see Vittorio <u>et al</u>. 1986), and it is interesting to note in this regard the recent study of the Centaurus cluster by Lucey, Currie, and Dickens (1986). These authors find Centaurus to be composed of two distinct clumps separated in redshift by ~ 1500 km s^{-1}. They argue, though, that the clumps lie at the same distance, and are just now merging. The cosmological implications of such large scale velocity flows are clearly very substantial, and the whole problem is obviously deserving of much greater study.

3. A HUBBLE CONSTANT FROM THE IR/H I METHOD

As discussed in the first section, the greatest difficulty in trying to derive a value for H_0 using the IR/H I relation (or any other global distance indicator) arises from knowing what moduli to select for nearby calibrating galaxies. A judicious choice of moduli can lead to an expansion rate that is near just about anyone's preferred value. Nevertheless, the author believes it is possible to lay down a sensible set of ground rules that should be followed in treating the problem.

The first of these rules is to work with galaxy clusters. This avoids the thorny issue of how to properly treat the Malmquist effect, because the galaxies in a cluster are generally all at the same distance. In fact, because the cluster objects are ultimately H I selected, it is likely that any sort of magnitude bias effects can be dispensed with entirely. Support for this contention can be seen in Figure 3 of Aaronson and Mould (1986), which shows no significant departure from a linear velocity-distance relation <u>over</u> <u>a</u> <u>factor</u> <u>of</u> <u>eight</u> <u>in</u> <u>radial</u> <u>velocity</u>. (The test is precisely that advocated by Tammann at this Symposium to check for Malmquist bias.) We also note in this regard the excellent agreement in relative Virgo-Coma modulus measured by ABMHSC from the IR/H I relation [Δ(m-M) = 3.69 mag] and that obtained by BDDFLTW from their improved Faber-Jackson method [Δ(m-M) = 3.65 mag]. Sandage and Tammann (1984) themselves apparently accept this latter approach as bias free.

In choosing a proper set of calibrating objects, one should also place restrictions on inclination (we have followed i > 45°, which may be too optimistic), morphology (types no earlier than Sab or later than Sdm), and corrected velocity width (200 < ΔV < 600 km s^{-1} probably being a good choice). The last of these criteria is driven by the desirability both to cover the same range in width sampled by the more distant cluster spirals, and to avoid the regime of small linewidth objects where the IR/H I relation is not well defined. The relation may in fact break down entirely for dwarfish galaxies, since inclination as judged from the optical structure often bears little resemblance to that indicated from detailed H I maps (Bothun 1986).

Finally, the calibrators should be confined to those systems having modern multi-color CCD (or good IR) Cepheid photometry, with no assumptions about group membership being introduced. This last criteria is particularly restrictive, and reduces the number of acceptable objects to just three: M31, M33, and NGC 2403. As discussed by ABMHSC, these galaxies yield a high value for the expansion rate, $H_0 \sim 90$ km s^{-1} Mpc^{-1}.

Tammann at this Symposium (see also Kraan-Korteweg, Cameron, and Tammann 1986) has attempted to analyze the biases in the nearby field sample of AHMST, claiming that a smaller value of H_0 thereby follows. There are three possible ways to reconcile his results with the above: either (1) cluster and field spirals differ in their IR/H I properties; or (2) a different set of calibrating distances and/or calibrating precepts has been employed (involving, for instance, the inappropriate extrapolation of the quadratic form of the relation down to the low-luminosity regime, as was indicated in the oral version of

Tammann's talk); or (3) an incorrect analysis of bias in the field sample. There is considerable evidence to suggest that the first of these possibilites has no basis. In particular, the clusters studied by ABMHSC cover a wide range of type, from dense, spiral-poor concentrations such as Coma, to loose, spiral-rich systems like Pegasus; and there is no hint of any systematic dependence of the ABMHSC results on cluster type or H I deficiency. This is not really too surprising, since the cluster samples tend to be weighted by objects outside the central cores, and are therefore expected to be more akin to field galaxies in nature. A further indication that there is no dichotomy between the field and cluster samples is the fact that <u>both independent sets of data yield the same value for Virgocentric motion</u>. The details of Tammann's analysis are not available at the time of writing, so it is provisionally concluded that his lower expansion rate results from some linear combination of the second and third reasons mentioned above.

If there is a weakness in the ABMHSC value for H_0, it surely rests with the fact that there are only three trustworthy calibrators at our disposal (whose Cepheid CCD photometry is still in a preliminary stage). When the number of reliable calibrators has grown into double digits, which will hopefully be the case a few years after the Hubble Space Telescope has commenced operation, we shall perhaps have a value for the expansion rate that is truly believable.

It is a great pleasure to thank the organizers of this Symposium for bringing us all to such a marvelous and stimulating country as China. Preparation of this article was partially supported with funds from United States National Science Foundation grant AST83-16629.

REFERENCES

Aaronson, M., Bothun, G., Mould, J., Huchra, J., Schommer, R. A., and Cornell, M. E.: 1986, <u>Ap. J.</u>, **302**, 536 (ABMHSC).
Aaronson, M., Huchra, J., Mould, J., Schechter, P. L., and Tully, R. B.: 1982, <u>Ap. J.</u>, **258**, 64 (AHMST).
Aaronson, M., and Mould, J.: 1986, <u>Ap. J.</u>, **303**, 1.
Balona, L. A., and Shobbrook, R. R.: 1984, <u>M.N.R.A.S.</u>, **211**, 375.
Blanco, V. M., and Blanco, B. M.: 1985, <u>Mem. Soc. Astron. Ital.</u>, **56**, 15.
Bothun, G.: 1986, private communication.
Burstein, D., Davies, R. L., Dressler, A., Faber, S. M., Lynden-Bell, D., Terlevich, R., and Wegner, G.: 1986, in "Galaxy Distances and Deviations from Universal Expansion," ed. B. F. Madore and R. B. Tully (NATO: Brussels); and private communication (BDDFLTW).
Caldwell, J. A. R., and Coulson, I. M.: 1986, <u>M.N.R.A.S.</u>, **218**, 223.
__________: 1987, in preparation.
Chiosi, C., and Pigatto, L.: 1986, <u>Ap. J.</u>, **308**, 1.
Christian, C. A., and Schommer, R. A.: 1986, preprint.
Collins, A., Joseph, R. D., and Robertson, N. A.: 1986, <u>NATURE</u>, **320**, 506.

Conti, P. S., Garmany, C. D., and Massey, P. 1986, A. J., 92, 48.
de Vaucouleurs, G.: 1980, Pub.A.S.P., 92, 579.
Eggen, O. J.: 1977, Ap. J. Suppl., 34, 1.
Freedman, W. L.: 1985, in "IAU Colloqiuium 82, Cepheids: Theory and
    Observation," ed. B. F. Madore (Cambridge: Cambridge University
    Press), p. 225.
Frenk, C. S., and White, S. D. M.: 1982, M.N.R.A.S., 198, 173.
Graham, J. A.: 1984, in "Structure and Evolution of the Magellanic
    Clouds," ed. S. van den Bergh and K. S. de Boer (Dordrecht:
    Reidel), p. 207.
Kerr, F. J., and Lynden-Bell, D.: 1986, M.N.R.A.S., 221, 1023.
Kraan-Korteweg, R. C., Cameron, L. M., and Tammann, G.: 1986, in
    "Galaxy Distances and Deviations from Universal Expansion," ed. B.
    F. Madore and R. B. Tully (NATO: Brussels).
Lilje, P. B., Yahil, A., and Jones, B. J. T.: 1986, Ap. J., 307, 91.
Lucey, J. R., Currie, M. J., and Dickens, R. J.: 1986, M.N.R.A.S.,
    221, 453.
Martin, W. L., Warren, P. R., and Feast, M. W.: 1979, M.N.R.A.S., 188,
    139.
Menzies, J. W., and Whitelock, P. A.: 1985, M.N.R.A.S., 212, 783.
Pritchet, C. J., and van den Bergh, S.: 1986, preprint.
Reid, I. N., and Strugnell, P. R.: 1986, M.N.R.A.S., 221, 887.
Reid, M. J., et al.: 1987, in "Star Formation," ed. M. Peimbert and
    and J. Jugaku (Dordrecht: Reidel).
Schmidt, E. G.: 1984, Ap. J., 285, 501.
Sandage, A.: 1986, private communication.
Sandage, A., and Carlson, G.: 1983, Ap. J. (Letters), 267, L25.
Sandage, A., and Tammann, G. A.: 1984, NATURE, 307, 326.
Schommer, R. A.: 1986, in "Galaxy Distances and Deviations from
    Universal Expansion," ed. B. F. Madore and R. B. Tully (NATO:
    Brussels),
Schommer, R. A., Olszewski, E. W., and Aaronson, M.: 1984, Ap. J.
    (Letters), 285, L53.
Seidel, E., Da Costa, G. S., and Demarque, P.: 1986, preprint.
Vittorio, N., Juszkiewicz, R., and Davis, M.: 1986, NATURE, 323, 132.
Walker, A. R.: 1985, M.N.R.A.S., 212, 343.
Yahil, A.: 1986, private communication.

DISCUSSION

SILK: You used only three galaxies as distance calibrators for the infrared Tully-Fisher relation. If you instead used Tammann's calibration, revised to 1986 distance moduli as he just presented, what global value of H_o would you obtain?

AARONSON: First, let me say that Tammann's "revised" moduli are his own, and not what I would necessarily agree with. I do not subscribe to the notion of blindly averaging together a wide variety of results. Second, by a judicious choice of calibrators I can give you an H_o anywhere between 65 and 100. This is my whole point - the nearby distances are so poorly known, it is not too difficult to argue things up or down, which ever suits your bias.

ULMER: If you reduce distance to center of the Milky Way Galaxy, how does this affect the supernova rate/ brightness "problem" created by choosing $H_o = 100$?

AARONSON: I believe it helps to alleviate this "problem."

LOCAL CALIBRATORS AND H_0:   THE DISTANCE TO M31 USING RR LYRAE STARS

C.J. Pritchet
Physics Department
University of Victoria
P.O. Box 1700
Victoria, B.C.   V8W 2Y2

ABSTRACT. We have used the CFH 3.6 m telescope in excellent seeing to
observe RR Lyrae variables in a halo field in M31. From 32 variables
found, we obtain $(m - M)_0 = 24.34 \pm 0.15$ for M31.

1   INTRODUCTION

In another paper in this volume, Marc Aaronson points out how
important the determination of accurate distances to nearby galaxies is
to the precise measurement of the Hubble parameter H_0. With modern
detectors and consistent subarcsecond seeing, it is now possible to
greatly improve our confidence in the local distance scale, and hence
in the zeropoint calibration of secondary distance indicators, by
observing RR Lyrae variables. These stars are well-respected distance
indicators locally, and it is now within the reach of some ground-based
telescopes to observe these objects in most, if not all, Local Group
galaxies. As part of an on-going program to improve the Local Group
distance scale from observations of RR Lyrae stars, Sidney van den
Bergh and I would like to report on the first observations of RR Lyrae
stars in the nearby spiral galaxy M31.

2   OBSERVATIONS AND REDUCTIONS

Our observations were obtained on the Canada-France-Hawaii
Telescope during a 6 night observing run with an RCA CCD and B filter.
The final data set consists of 16 1 hour exposures of an area
$2'.2 \times 3'.3$ located 40' SE of M31 (along the minor axis). Average
seeing for the observations was $0".9$. Reduction procedures were
standard, and will be described elsewhere (Pritchet and van den Bergh
1987).
Variables were found by blinking independent pairs of frames.
Magnitudes were measured both on individual 1 hour exposures, and also
on a frame constructed by averaging all 16 1 hour exposures, using the
crowded field photometry algorithm DAOPHOT (Stetson 1987). The mean

A. Hewitt et al. (eds.), Observational Cosmology, 197–198.

frame magnitude provided an estimate of the phase-averaged, intensity-mean, blue magnitude $\langle B \rangle$ for each object, without needing to know the magnitude at minimum light on individual frames.

3   RESULTS

A total of 32 variables were found, of which 2 may be eclipsing systems. This number of objects implies a relatively high RR Lyrae star frequency, comparable to that found in the Galactic globular cluster M3. In turn this suggests a relatively blue horizontal branch, which is in contrast to expectations based on the relatively high metallicity of the M31 halo (Mould and Kristian 1986).

Our best estimate of $\langle B \rangle$ for all RR Lyrae stars is 25.68 ± 0.06 (error in the mean). Taking the recent determination of $\langle M_V(RR) \rangle = +0.77$ from statistical parallaxes, corrected for magnitude scale errors, reddening scale errors, and put on an intensity-weighted system (Barnes and Hawley 1986), and using $\langle B-V \rangle = 0.26$ and $A_B = 0.31$ (Burstein and Heiles 1984), we obtain $(m - M)_0 = 24.34 \pm 0.15$ for M31. The principal uncertainty in this result is in the calibration of $\langle M_V \rangle$ for RR Lyrae stars.

This result is in good agreement with a number of recent determination of the M31 distance modulus. For example, Welch et al (1986) obtain $(m - M)_0 = 24.26 \pm 0.08$ from IR observations of Cepheids, Cohen (1985) finds $(m - M)_0 = 24.04 \pm 0.20$ from novae, and Mould and Kristian (1986) obtain 24.4 ± 0.2 from Population II stars in a halo field near ours.

Details of this work will be published in the Astrophysical Journal.

REFERENCES

Barnes, T.G., and Hawley, S.L. 1986, Ap. J. (Letters), in press.
Burstein, D., and Heiles, C. 1984, Ap. J. Suppl., 54, 33.
Cohen, J. 1985, Ap. J., 292, 90.
Mould, J., and Kristian, J. 1986, Ap. J., 305, 591.
Pritchet, C.J., and van den Bergh, S. 1987, in preparation.
Stetson, P. 1987, in preparation.
Welch, D.L., McAlary, C.W., McLaren, R.A., and Madore, B.F. 1986, Ap.J., 305, 583.

DISTANCE MODULI FROM THE TULLY-FISHER RELATION

E. Giraud
European Southern Observatory
Karl-Schwarzschild-Strasse 2
D-8046 Garching bei München, F.R.G.

ABSTRACT. When the Bottinelli et al. version of the Tully-Fisher
relation is applied to derive distances, and when the observed
parameters are used to predict the Malmquist bias at a given distance,
the observed variation of the Hubble ratio as a function of kinematic
distance is about 2.5 times the predicted variation.

1. INTRODUCTION

It is now well-known that the sample used by de Vaucouleurs and his
collaborators in 1981 to derive the mean Hubble ratio through the
Tully-Fisher method or the tertiary distance indicators, is such that
the Hubble ratio is an increasing function of the velocity. A similar
deviation was reported many years ago by de Vaucouleurs at the IAU
Symposium 44 in 1972.

Because H is a velocity/distance ratio, some dependence between H
and the velocity would be obtained in the case of a poor representation of the velocity field. However, 1) if the distances are well
estimated, 2) if after correction for the perturbation caused by the
overdensity of Virgo, the velocity field is quiet and 3) if there is
no bias in the data, then a [velocity - Hubble ratio] diagram must be
random. We know that there is a similar increase in H, even after
correction for infall velocities toward Virgo through Peeble's
spherical infall model.

I recall the selection rules of the data. The set consisted of
254 galaxies from the catalogue of Bottinelli and her collaborators
selected such that:
a)   the inclination of a galaxy is larger than $35°$
b)   internal errors are lower than 0.45 mag
c)   de Vaucouleurs' morphological types are between 2 and 7.
The HI velocities were corrected to the centroid of the Local Group
and then were converted to kinematic distances. The parameters are:
velocity of Virgo: 1000 km s^{-1}; infall velocity of the Local Group
toward Virgo: 330 km s^{-1}.

The main difference between this sample and the sample of

A. Hewitt et al. (eds.), Observational Cosmology, 199–205.

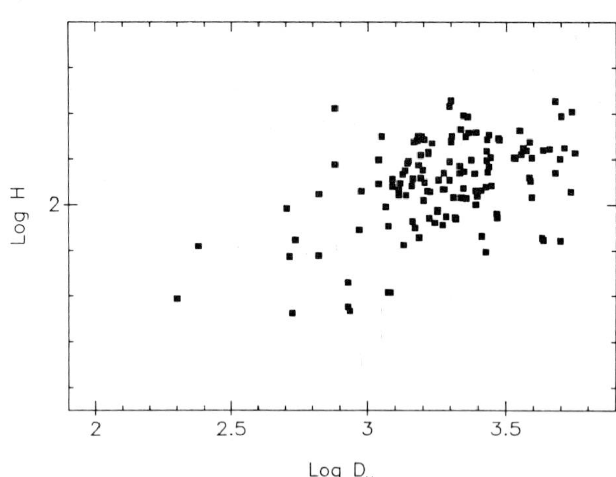

Fig. 1a.   The logarithm of the Hubble ratios H vs. the logarithm of
the kinematic distances D_v (in km s^{-1}) for Sbc–Sc galaxies. The
distance moduli are derived from the B-band Tully-Fisher relation with
slope of 5.

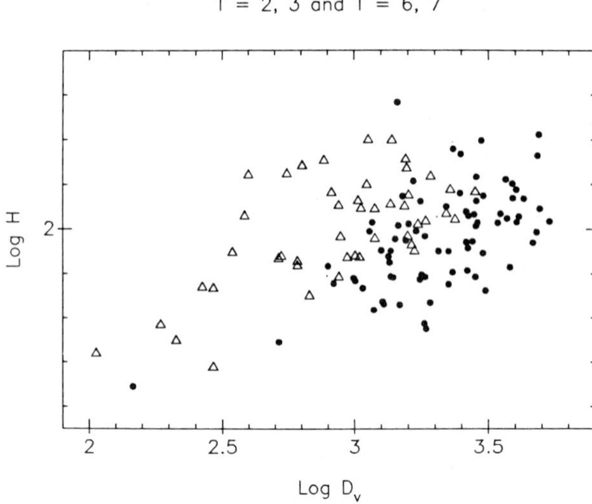

Fig. 1b.   Same as 1a for Sab-Sb and Scd-Sd galaxies.

Bottinelli and her collaborators is that they do not exclude data with
large internal errors. So they have many galaxies with known errors
larger than 1 mag.
     The distances are derived from the version of the Tully-Fisher
relation adjusted on the de Vaucouleurs distance indicators. In this

method the slope of the relation between the maximum rotational
velocity of a spiral galaxy and its absolute magnitude in the B-band
has a value of 5 and, at a given rotation velocity, the luminosity
does not depend on Hubble type.

With these hypotheses we obtain the relation between the
kinematic distances and the Hubble ratios shown in figures 1a and 1b
presented separately for the Sbc-Sc, the Sab-Sb and the Scd-Sd
galaxies. I use a logarithmic scale because the distances are in
magnitudes.

These figures show two main features:
1) the Hubble ratio increases with respect to the velocities
2) in the shell $\log D_V$ between 2.8 and 3.3 the mean Hubble ratio for
   the Scd-Sd is larger than that of the Sab-Sb's suggesting that
   there is a small type offset.

2. MALMQUIST BIAS

The Malmquist bias has played a crucial role in the value of the
Hubble constant and apparent non-linearities in the Hubble flow. In a
magnitude-limited sample, when the distances become larger than some
critical value, intrinsically faint galaxies are progressively lost.
Sandage and Tammann pointed out that neglect of the Malmquist effect
gives higher values of H in the biased part of the sample in which the
fainter objects are absent and gives only an apparent correlation of H
with velocities.

Bottinelli and her collaborators have investigated the effect of
the Malmquist bias using substantially the same plot of $\log H$ versus
$\log D_V$ as in the last figures. The main objection to their analysis
may follow from what they call normalized kinematic distances. In

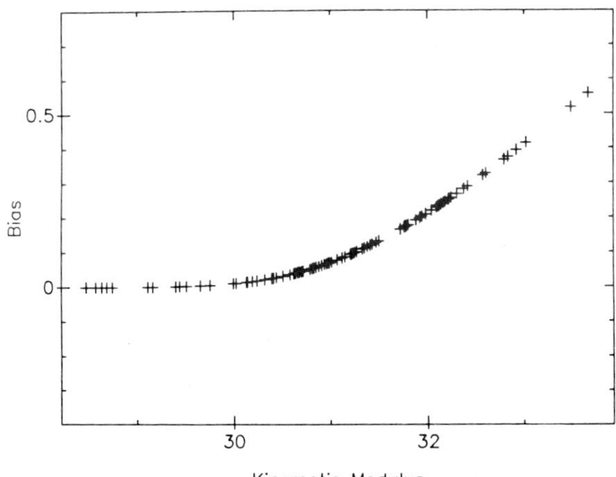

Fig. 2.   The bias (in mag) in the determination of absolute
magnitudes through the Tully-Fisher relation plotted as a function of
the kinematic modulus $\mu = 5(\log D_V + 3)$ (in mag).

particular they use a dispersion of 0.5 mag of classes of galaxies
instead of the observed dispersion of the luminosity function of
spirals which is about 1 mag for the sample. I have remade the
analysis of Malmquist bias with the observed parameters. We need to
know the mean absolute magnitude M_0 of the sample, the dispersion and
the shape of the luminosity function, the magnitude-limit and the
dispersion of the Tully-Fisher relation. We do not know the true mean
absolute magnitude but we can find the biased mean absolute magnitude
and dispersions to determine approximately the maximum value of
distance within which the sample is not much biased. Then we assume
that the parameters in that range are good estimates of the true mean
absolute magnitude and dispersion. The variation in the mean absolute
magnitude is the Malmquist effect but is not the bias in the value of
absolute magnitudes estimated through the Tully-Fisher relation. This
bias depends on the accuracy of the Tully-Fisher relation and is
illustrated by figure 2 in the case of the observed parameters of the
sample. When the observed parameters are used the variation of the
Hubble ratio as a function of the kinematic distance in the case of
Bottinelli's version of the Tully-Fisher relation is 2.5 to 3 times
the expected variation due to the Malmquist bias at a given distance.

   After correction for Malmquist bias the remaining variation is
shown in Figure 3. Here the zero point is adjusted in the value of the
mean Hubble ratio at small distance in the short scale (see Giraud
1986c). We note in passing that the low value of the Hubble ratio at
small distance is not due to the Local Group deceleration.

   We know that the problem that we have to face is more difficult
than a Malmquist bias because:

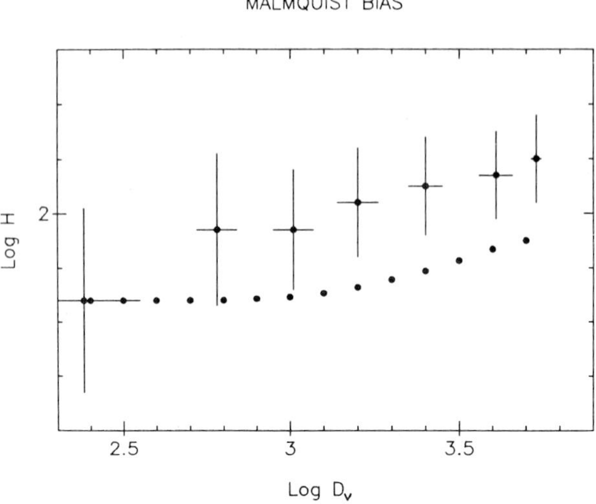

Fig. 3.   The observed and the predicted variation in <log H> binned
as a function of kinematic distance (the bars are the dispersions).
The adjustment is made at <log H> = 1.84 for log D_V ≤ 2.7.

1) The dispersion of the absolute magnitudes derived from the field is
larger than that derived from the Tully-Fisher relation with a slope
of 5 and without type offset.
2) If we select galaxies with large rotation velocities, the variation
of H as a function of absolute magnitudes derived from the velocity
field remains almost the same.

3. TYPE EFFECT

A source of uncertainty in the B-band comes from the unknown amount of
scatter introduced by the wide range in morphological types. In particular we have seen that there is a small type offset between Sab-Sb
and Scd-Sd in the shell $\log D_V$ between 2.8 and 3.3. Part of this
offset is due to a differential Malmquist bias because the sample of

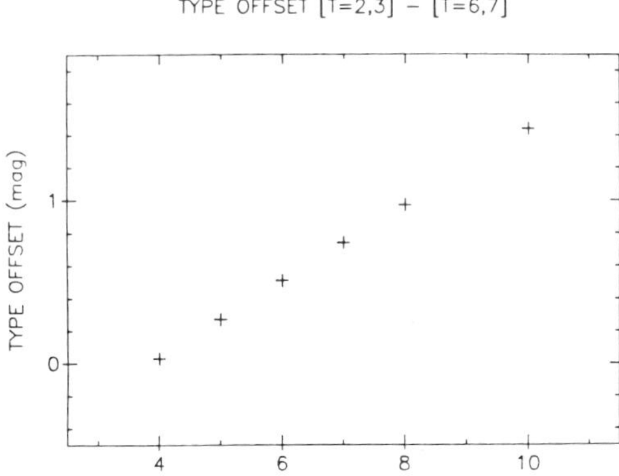

Fig. 4a.   The type offset in the zero-point of the Tully-Fisher
relation between Sab-Sb and Scd-Sd plotted as a function of the slope
of that relation (after correction for differential Malmquist bias).

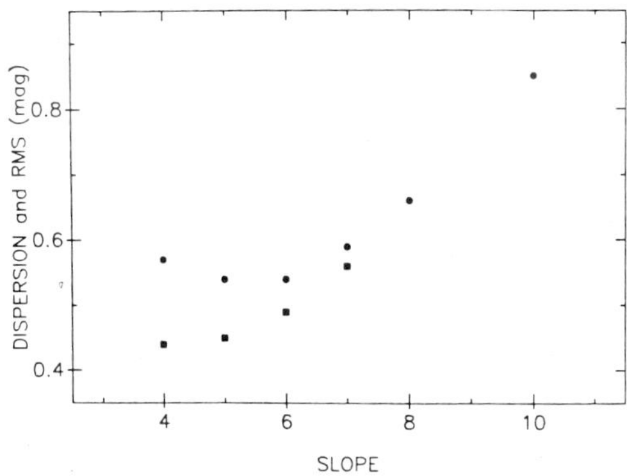

Fig. 4b.   The weighted mean dispersion of the Tully-Fisher relation
for various types plotted as a function of the slope. The filled
squares are for a non-linear Hubble flow.

Scd-Sd galaxies being less luminous is more biased. The differential
Malmquist bias has been found to be of 0.25 mag in the shell. The rest
is a true type offset.

Many authors have found values of the slope of the B-band Tully-
Fisher relation in the range 6 to 10. I have measured the dispersion
by type and the offset between Sab-Sb and Scd-Sd corrected for
differential Malmquist bias for various values of the slope (fig. 4).
The values of the dispersion favour low values of the slope of the
Tully-Fisher relation between 5 and 6 and a true type offset of 0.3 to
0.5 mag. A slope of 10 would give very poor results but we note that
the values of the type offset and of the dispersions in that case are
more or less in agreement with the results obtained by Rubin and her
collaborators. Finally if we assume that the Hubble flow is not linear
we obtain the dispersions marked by filled squares in the figure.

CONCLUSIONS

1.   The observed variation in log H with kinematic distance is not
      due to the Malmquist effect.

2.   The values of the dispersion favour slopes of the B-band Tully-
      Fisher relation between 5 and 6 and a true type offset of 0.3 to
      0.5 mag.

3.   The low value of the Hubble ratio at small distance is not due to
      the Local Group deceleration.

REFERENCES

Bottinelli, L., Gouguenheim, L., Paturel, G., and de Vaucouleurs, G.,
      1983, Astron. Astrophys. **118**, 4.
Bottinelli, L., Gouguenheim, L., Paturel, G., and Teerikorpi, P.,
      1986, Astron. Astrophys. **156**, 157.
Giraud, E., 1986a, Astrophys. J. **301**, 7.
Giraud, E., 1986b, Astrophys. J. **309**, in press.
Giraud, E., 1986c, Astron. Astrophys. in press.
Giraud, E., 1986d, Astron. Astrophys. in press.
Teerikorpi, P., 1984, Astron. Astrophys. **54**, 661.
de Vaucouleurs, G., 1972, in IAU Symposium 44, External Galaxies and
      Quasi-Stellar Objects, ed. D.S. Evans (Dordrecht: Reidel) p. 353.

DISCUSSION

ARP: With the most accurate possible correction for Malmquist effect you get a clear difference between a "local" value of H_o and the value of H_o at greater distances. What is the H_o you get at close distances and what value further away?

GIRAUD: 75 kms^{-1} Mpc at small distance; 90 kms^{-1} Mpc at the location of Virgo (depending on the infall velocity); and 100-110 kms^{-1} Mpc further away.

ESTIMATES OF Ω BASED ON MOTIONS WITHIN THE LOCAL SUPERCLUSTER

R. Brent Tully
Institute for Astronomy, University of Hawaii
2680 Woodlawn Drive
Honolulu, HI   96822
U.S.A.

ABSTRACT. Three ways to estimate Ω from the motions of nearby galaxies
are reviewed: (i) comparisons can be made between observed local
streaming motions and observed density fluctuations, (ii) a timing
argument can give a value of $f(\Omega) = t_0 H_0$, and (iii) the motions in
groups provide solid evidence regarding the amount of matter that is
distributed like galaxies on a scale of 0.5 Mpc. In each case, the
evidence suggests the Universe is open, but only by a factor of 3-5
and, given the uncertainties, a closed Universe definitely is not
precluded.

1. LOCAL DENSITY FLUCTUATIONS AND STREAMING MOTIONS

Silk (1974) made the first serious attempt to measure Ω, the ratio of
the density of the Universe ρ to the critical density for closure ρ_c,
from the kinematic effects of density fluctuations on the scale of the
Local Supercluster. Peebles' (1980) description in the linear regime
can be approximated by:

$$u \propto H_0 \ r \ \Omega^{0.6} \ \delta \quad . \tag{1}$$

It might be possible to estimate Ω by making measurements of streaming
velocities u on a scale r, if it is possible to estimate the local
overdensity of matter $\delta = \Delta\rho/\rho$.

Yahil, Sandage, and Tammann (1980), Davis and Huchra (1982), and
Lahar (1986) have compared the distribution of nearby galaxies with
evidence for infall toward the Virgo Cluster of ~300 km s^{-1} and a
motion of a similar amplitude toward the apex defined by the cosmic
microwave background dipole anisotropy and concluded $\Omega \sim 0.15$, 0.4, and
0.3, respectively. Higher values were deduced from the observed
coincidence in the anisotropy of IRAS sources and the cosmic microwave
background (Meiksin and Davis 1986: $\Omega \sim 0.5$; Yahil, Walker, and Rowan-
Robinson 1986: $\Omega \sim 1.0$).
These calculations require an extremely important assumption: that
blue or far-infrared light accurately traces the distribution of

A. Hewitt et al. (eds.), Observational Cosmology, 207–215.
© 1987 by the IAU.

matter on supercluster scales. This assumption is always identified,
but it is routinely ignored that the assumption **must** be invalid if
$\Omega \sim 1$, at least, in the case of blue light. The mass-to-light ratio
for the Universe at large is (Davis and Huchra 1982 but adjusted to
include internal and galactic absorption):

$$M/L \sim 1600 \; \Omega \; h_{100} \; M_{\odot}/L_{\odot} \; . \tag{2}$$

The value $\Omega \sim 1$ is larger than values associated with galaxies on
scales <1 Mpc. Hence, if the Universe is closed then dark matter is
more widely spread than galaxies or distributed differently.

Hoffman, Olson, and Salpeter (1980) and Hoffman and Salpeter
(1982) have pointed out that there is another constraint in the environment of the Local Supercluster, namely, the central density of the
mass concentration in the Virgo Cluster. If the assumption that mass
follows light is dropped, then the consequence of an established central density is _lower_ peculiar velocities in a _higher_ Ω universe, since
the overdensity is lower.

In summary, the tenet of this approach has been that larger peculiar velocities imply greater Ω. Yet, if peculiar velocities were
observed large enough to imply $\Omega \sim 1$, then there would have to be mass
concentrations in excess of what is seen in clusters. The assumption
that mass follows blue light is so dubious that the present formulation
of the method probably only has value as a lower limit on Ω. Far-
infrared emission might be a fairer tracer of the mass since it is
less concentrated, as it is preferentially associated with spiral
galaxies, but the very difference between the blue and infrared results
cautions us to mistrust either result.

If these criticisms are not enough, the conventional interpretation is thrown in extreme doubt by recent supercluster collapse simulations that reveal biases in "observed" Ω in models with closure density
(Villumsen and Davis 1986; Melott 1986). In the mean, an observer in a
region with $\delta \sim 2-3$ would estimate $\Omega \sim 0.8-0.9$, with systematically
lower values of Ω estimated in regions of higher δ. Worse, the _range_
of "observed" Ω values is very large: $\Omega \sim 0.3$ is a typical measurement
of an observer situated on the major axis of a quadrupole feature.
There is evidence in the Local Supercluster for the sorts of motions
seen in the n-body simulations: rotation or shear (Aaronson _et al._
1982), bulk transverse motion (Aaronson _et al._ 1986), and quadrupole
distortion (Lilje, Yahil, and Jones 1986). The evidence, though, is
that we are orthogonal to the major axis of the quadrupole distortion
of the Local Supercluster.

Recently, Burstein _et al._ (1986) and Collins, Joseph, and
Robertson (1986) have claimed to detect streaming motions on scales
larger than the Local Supercluster. What these motions imply with
regard to the density of the Universe remains to be seen. It need not
be exceptional if it should prove that there is a substantial density
enhancement within the Hydra-Centaurus complex (which we are part of)
that lies in the zone of obscuration.

2. A TIMING ARGUMENT

Tully and Shaya (1984) did some preliminary work on the idea that will
be presented, but Shaya (1986a,b) has developed the concept more fully.
The aim is to measure the quantity

$$f(\Omega) = t_0 \, H_0 \, , \tag{3}$$

where

$$f(\Omega) = (1 - \Omega)^{-1} - 1/2\Omega(1 - \Omega)^{-3/2} \cosh(2\Omega^{-1} - 1), \tag{4}$$

so that $f(0) = 1$ and $f(1) = 2/3$.

It will obviously be difficult to constrain Ω in a situation where
the difference in the product $t_0 H_0$ between empty and closed models is
only 33%. However, to our distinct advantage, the product does <u>not</u>
require a knowledge of the distance scale. The product is a function
of accurately measurable quantities only.

We must deal with the nonlinear collapse regime, which requires
numeric integration of the equations that describe a Freidmann universe. This situation is sufficiently difficult to describe that it is
worthwhile to start by presenting essentially a dimensionality
argument.

A galaxy that is being pulled into a cluster will have a streaming
velocity toward the cluster that is dependent on the accelerating force
and the age of the Universe. At the present epoch, the infall velocity
of a galaxy falling toward a mass $M(<r)$ at a distance r will have the
approximate dependencies

$$v_r \propto \frac{M(<r)}{r^2} \, t_0 \, . \tag{5}$$

Suppose that the only mass interior to r is the cluster mass which can
be estimated by an application of the virial theorem. Then equation (5)
can be rewritten:

$$v_r \propto \frac{\sigma_v^2 \, \Theta_v \, D}{\Theta_r^2 \, D^2} \, t_0 \, , \tag{6}$$

where σ_v is the rms velocity dispersion of the cluster, Θ_v is the angular virial radius (related to a harmonic radius), Θ_r is the angular
distance of the infalling galaxy from the center of the cluster, and D
is the distance of the cluster. If the cluster would have a systemic
velocity if it were freely expanding of $V_c = H_0 D$, then

$$t_0 H_0 \propto \frac{v_r \, \Theta_r^2 \, V_c}{\sigma_v^2 \, \Theta_v} \, . \tag{7}$$

Hence, if galaxies can be identified that are just falling into clusters, it is possible to derive an estimate of Ω that only depends on measurements of velocities and positions on the sky.

Shaya has developed this concept more formally with application to a substantial group of galaxies that Tully and Shaya identified to be falling into the Virgo Cluster today. There is a large group of spiral and irregular galaxies, well contained both spatially and in redshift, that distance estimates place at, or just beyond, the cluster distance but that are substantially blueshifted with respect to the cluster. It is argued that this group is just at the edge of the Virgo Cluster and will soon be merged into it.

Shaya uses the formalism developed by Schechter (1980) to describe the collapse of spherical shells in terms of dimensionless expansion and density parameters that depend only on the familiar arc-parameter, η. He calculates the value of the arc-parameter, η_6, for the shell at $6°$ radius that we surmised is just entering the Virgo Cluster, where the mass interior to this shell is the virial estimate of the mass of the cluster. The arc-parameter at infinity, η_∞, is given by Schechter's expansion formula and an estimate of the free-expansion velocity of the Virgo Cluster, V_c (the observed redshift of the cluster adjusted for the infall and peculiar velocities of the Galaxy). Then Ω is found from Schechter's density formula:

$$\Omega = \frac{2}{1 + \cos \eta_\infty} . \tag{8}$$

In his initial development of the idea, Shaya (1986a) concluded $\Omega \sim 0.1$, but this result is quite uncertain. In particular, there is a gap between most of the galaxies that are claimed to be falling into Virgo (they are at a typical radius of $10°$) and the $6°$ core that contains the mass estimated by the virial analysis. Then, there are uncertainties in the virial mass estimate, in the deprojected infall velocities, and the free-expansion Virgo redshift.

In the subsequent work, Shaya (1986b) is placing further constraints, such as knowledge regarding the position of the turn-around radius, and allowing himself the assumption that the mass falloff in the region of the supercluster outside the central cluster but within our radius is the same as the distribution of light. He has developed an overconstrained set of equations that should only be satisfied by a restricted range of Ω.

This work is still in progress. Ultimately, more sophisticated models that do not necessarily assume spherical symmetry can be used. In spite of the complexity of the problem, and the need for great accuracy, there is some hope because, with time, the method can be applied to a large number of clusters and circumstances.

3. NEARBY GROUPS

This topic deviates from the spirit of the title of the talk because it does not lead to a global Ω estimate. However, it is relevant to application of the "cosmic virial theorem" which could be construed as doing so. Also, on the scale of groups, ~1 Mpc, we now have very good estimates of the density and these estimates are remarkably high. Large M/L values for groups have been claimed before, but the new data is especially convincing.

The present analysis is based on a soon-to-be-published group survey (Tully 1987). It has the following features: (i) there is uniform, all-sky coverage, (ii) low-luminosity galaxies are well represented, (iii) velocity measurement errors are insignificant, (iv) galaxies are linked into groups by an algorithm that mimics gravity and includes the tidal effects of adjacent groups, (v) galaxies were accepted as group members if they satisfied a specific local density (not overdensity) requirement.

Within a distance of reasonable completion of 25 h_{75}^{-1} Mpc (corresponding to velocities <1900 km s^{-1} but with a Virgocentric retardation model incorporated) 178 groups of two or more galaxies were defined, and there are 50 groups with five or more members. There is no evidence of an interloper problem in that faint candidate members do not tend to have larger redshifts than bright candidate members. The median virial radius for the 50 largest groups is $\langle r_v \rangle = 340\ h_{75}^{-1}$ kpc and $r_v < 500\ h_{75}^{-1}$ kpc for 80% of these groups. There is evidence that the groups are bound in that (i) crossing times are substantially less than the age of the Universe (the median for the 50 large groups is $t_x H_0 = 0.2$), and (ii) collapse times, defined as $t_c = 8.5 \times 10^6$ $(r_v^3/M_v)^{1/2}$ years (r_v is the virial radius in Mpc and M_v is the virial mass in $M_\odot$), are typically less than a Hubble time (the median for the 50 large groups is $t_c H_0 = 0.6$ and 94% satisfy $t_c < 2\ H_0^{-1}$). This evidence strongly suggests the groups are bound, though not necessarily well virialized.

Figure 1 presents data that some might find contentious. It is a histogram of unweighted group velocity dispersions, corrected for measurement errors in velocities. The corrections are usually small but 29 pairs to quadruples have velocity dispersions less than the uncertainties in the measurements. The median one-dimensional dispersion for the 50 largest groups is 100 km s^{-1}. Eighty-eight percent of the groups have velocity dispersions below 130 km s^{-1}, although there is a long tail that extends out to 715 km s^{-1} (the Virgo Cluster).

Rivolo and Yahil (1981) found the similarly small characteristic velocity dispersion of 100 km s^{-1} for pairs of galaxies, but Huchra and Geller (1982) derived a median value of 180 km s^{-1} for groups with more than 5 members identified in the Shapley-Ames sample and Davis and Peebles (1983) found a characteristic one-dimensional velocity dispersion of 300 km s^{-1} at a radius of 300 h_{75}^{-1} kpc between correlated pairs in their application of the cosmic virial theorem. The studies that get larger velocity dispersions (i) tend to include a larger percentage of rich groups than my very local sample, and (ii) may suffer a greater interloper problem because insufficient account might be taken of the filamentary nature of large-scale structure and the consequence that the background density is high in the vicinity of groups.

Given the distribution of velocity dispersions in Figure 1 (most below 130 km s^{-1} and a long tail to high velocities), the cosmic virial

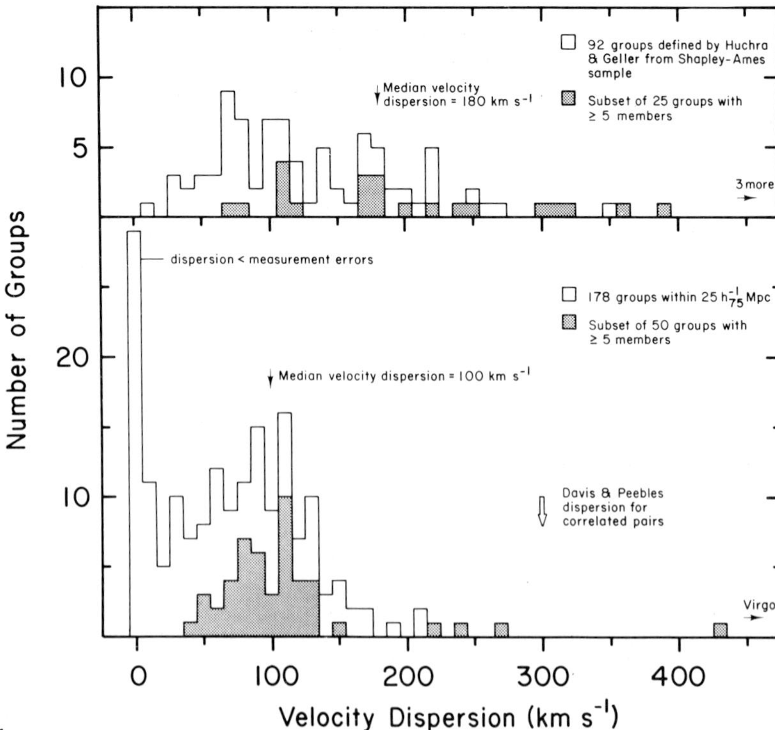

Figure 1

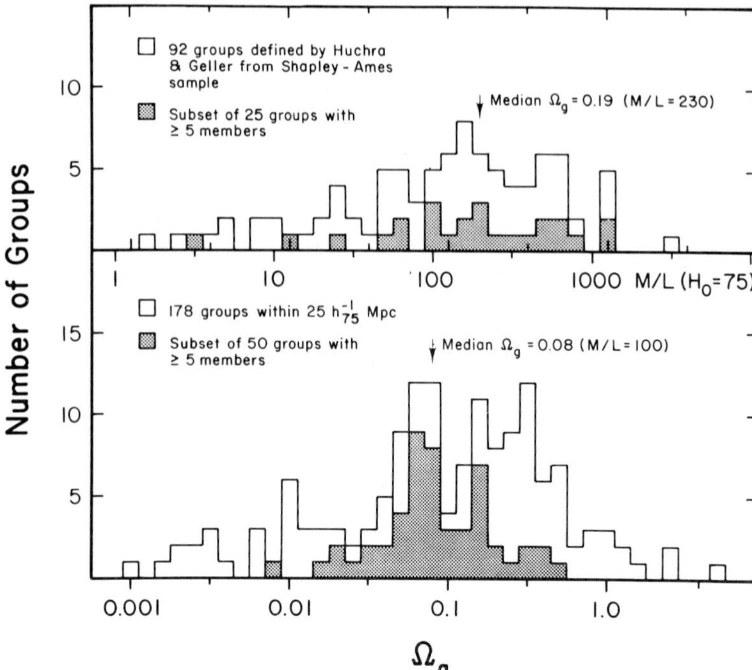

Figure 2

dispersion of 300 km s^{-1} is not easily interpreted. It is not representative of the "field" outside of the principal clusters. Davis and Peebles estimated $\Omega \sim 0.2$ from the cosmic virial theorem, but Rivolo and Yahil concluded that this estimate should be reduced.

If velocity dispersions are modest, M/L values are nonetheless substantial. Figure 2 is a histogram of mass-to-light values for the individual groups, ratioed to the value required for a closed universe (to eliminate the dependency on the distance scale). The median value for the 50 largest groups corresponds to $\Omega = 0.08$ ($M/L_B{}^{b,i} = 100$ h_{75} $M_\odot/L_\odot$). My typical groups have large mass-to-light ratios in spite of low velocity dispersions because they also have low integrated luminosities.

4. FINAL COMMENTS

Comparison of velocity streaming motions with fluctuations in the distribution of light leads to estimates of $\Omega \sim 0.4$. However, standard methodology requires the assumption that mass is distributed like light, which is obviously dangerous. A result of the group analysis is that <1% of galaxies lie outside of the filamentary clouds that dominate large-scale structure, a circumstance that strongly implies biased galaxy formation. The present velocity streaming studies do not preclude $\Omega = 1$ and, hence, do not yet strongly constrain the density parameter.

In principle, the timing argument could give us a value of Ω relatively free of assumptions (of course, the entire discussion is pointless if $\Lambda = 0$ is not assumed). It is premature to make much of present results, but the method should not be written off because it eventually could be applied to a large number of situations.

The group analysis provides the firmest results, though the implications are restricted. There is strong evidence that $\Omega_g \sim 0.1$ on a scale of 0.5 Mpc about galaxies. It is to be noted that the 178 groups of pairs and larger include 69% of the galaxies within 25 $h_{75}{}^{-1}$ Mpc and 77% of the light (single galaxies tend to be low-luminosity systems). Hence, I argue that the value $\Omega_g \sim 0.1$ ($M/L_B{}^{b,i} \sim 100$ h_{75} $M_\odot/L_\odot$) characterizes the distribution of mass on a scale of ~ 0.5 Mpc around the average galaxy and, hence, represents a rather firm <u>lower limit</u> to the universal value of Ω.

ACKNOWLEDGMENTS

Section 2 is essentially a discussion of Ed Shaya's work, and I have profited from extensive discussions with him. Support for this research has been provided by grant AST 83-19951 from the U.S. National Science Foundation.

REFERENCES

Aaronson, M., Huchra, J. P., Mould, J. R., Schechter, P. L., and Tully, R. B. 1982, Ap. J., 258, 64.

Aaronson, M., Bothun, G., Mould, J. R., Huchra, J. P., Schommer, R. A., and Cornell, M. 1986, Ap. J., 302, 536.

Burstein, D., Davies, R. L., Dressler, A., Faber, S. M., Lynden-Bell, D., Terlevich, R., and Wegner, G. 1986, in: Galaxy Distances and Deviations from Universal Expansion, eds. B. F. Madore and R. B. Tully (Dordrecht: Reidel), p. 123.

Collins, C. A., Joseph, R. D., and Robertson, N. A. 1986, in: Galaxy Distances and Deviations from Universal Expansion, eds. B. F. Madore and R. B. Tully (Dordrecht: Reidel), p. 131.

Davis, M., and Huchra, J. P. 1982, Ap. J., 254, 437.

Davis, M., and Peebles, P. J. E. 1983, Ap. J., 267, 465.

Hoffman, G. L., Olson, D. W., and Salpeter, E. E. 1980, Ap. J., 242, 861.

Hoffman, G. L., and Salpeter, E. E. 1982, Ap. J., 263, 485.

Huchra, J. P., and Geller, M. J. 1982, Ap. J., 257, 423.

Lahar, O. 1986, preprint.

Lilje, P. B., Yahil, A., and Jones, B. J. T. 1986, Ap. J., 307, 91.

Meiksin, A., and Davis, M. 1986, A.J., in press.

Melott, A. L. 1986, in: Galaxy Distances and Deviations from Universal Expansion, eds. B. F. Madore and R. B. Tully (Dordrecht: Reidel), p. 225.

Peebles, P. J. E. 1980, in: The Large-Scale Structure of the Universe (Princeton: Princeton University Press), Ch. 14.

Rivolo, A. R., and Yahil, A. 1981, Ap. J., 251, 477.

Silk, J. 1974, Ap. J., 193, 525.

Shaya, E. J. 1986a, in: Galaxy Distances and Deviations from Universal Expansion, eds. B. F. Madore and R. B. Tully (Dordrecht: Reidel), p. 229.

Shaya, E. J. 1986b, in preparation.

Schechter, P. L. 1980, A.J., 85, 801.

Tully, R. B. 1987, Ap. J., to be submitted.

Tully, R. B., and Shaya, E. J. 1984, Ap. J., 281, 31.

Villumsen, J. V., and Davis, M. 1986, Ap. J., in press.

Yahil, A., Sandage, A., and Tammann, G. A. 1980, Ap. J., 242, 448.

Yahil, A., Walker, D., and Rowan-Robinson, M. 1986, Ap. J. (Letters), 301, L1.

DISCUSSION

GELLER: (1) Could you clarify your procedure for selecting groups? (2) How do you choose the cutoff in your selection parameter?

TULLY: I use a dendogram method. Galaxies are merged into units on the basis of the "force" measure and groups are defined as those units

that exceed a certain threshold in density. The threshold was chosen
after considerable experimentation and taking into account the core-
halo nature of observed groups. If the cutoff is increased or decreased
a factor of 2 or 3, either the crossing times of outlying objects
become long compared with a Hubble time or qualitatively evident groups
are frequently broken up.

FALL: When discussing estimates of Ω from the Virgocentric infall, you
emphasized the uncertainty caused by the possible segregation of mass
and light. It seems to me that the same considerations apply to your
virial analysis of small groups, which is based on the assumption that
mass and light are distributed in the same way. One might even suspect
that the segregation of mass and light is <u>greater</u> on small scales
although the sign of the effect is far from clear.

TULLY: While it is possible that dark matter might be more centrally
concentrated than the galaxies and, hence, virial masses might be overestimated, it would seem considerably more likely that, if dark matter
deviates from the light, it would be <u>less</u> centrally concentrated and
then my mass estimates represent lower limits.

KIANG: I suppose your results do not exclude the possibility of a much
larger mean density of the universe. If dark matter is distributed
more uniformly than luminous matter, then both the results from Virgo
cluster infall analyses and your application of the cosmic virial
theorem will be consistent with a much larger value of Ω.

TULLY: Definitely, yes.

PECKER: Does the hierarchical structure of the universe, as postulated
years ago by de Vaucouleurs (after others!) affect your results? In
a sense, the ρ_o value has (if such a "fractal" structure applies to the
whole universe) no real meaning; and when the extrapolation of the
observed data is done, one can go as well to $\rho_o \sim 2 \times 10^{-31}$ g cm^{-3} or
much lower values of ρ_o at the large scale, the latter giving rise to
a definitely open universe, - if I am not mistaken! How do you react
to this difficulty? Does not your methodology imply a well-defined
ρ_o?

TULLY: Yes, it assumes both a well-defined closure density and that
a reasonable estimate has been made of the mean luminosity density.
Given the apparent existence of extremely large scale structures,
even the simpler second assumption involves considerable uncertainty.

A NEW MEASUREMENT OF THE GEOMETRY OF SPACE

Edwin D. Loh
Physics Department
Princeton University
P. O. Box 708
Princeton, New Jersey 08544
USA

ABSTRACT. This paper discusses the recent measurement of the number of galaxies *vs.* redshift and flux and presents new results pertaining to the two dimensionless geometrical quantities that describe the geometry of the conventional big-bang cosmology, the density parameter Ω and the dimensionless form $\lambda = \Lambda/(3H_0^2)$ of the cosmological constant. In contrast to the classical redshift-magnitude test as applied to the brightest galaxies in clusters, this new method is able to separate the effects of evolution from geometrical effects and is therefore able to measure the geometry of space. The 95% confidence limits are $\Omega - \lambda = 0.9^{+0.7}_{-0.5}$ and $-1.5 < \Omega + \lambda < 7.1$. The principal conclusions are these: (1) For both $\lambda = 0$ and inflationary models of the universe, this measurement and primordial nucleosynthesis imply a large density of nonbaryonic matter. (2) Hubble's constant H_0 and the age of the universe τ are constrained by $0.60 < H_0\tau < 0.88$ (95% confidence).

1. INTRODUCTION

Measuring the geometry of space has been hindered because comparing quantities over large distances $\sim H_0^{-1}$ to measure the geometry entails comparing them over long times $\sim H_0^{-1}$ and uncertain histories. Recently Loh and Spillar (1986b) measured the joint distribution of redshift and flux for a flux-limited sample of galaxies. They argue that the geometrical effects and the effects of luminosity evolution are separable and the evolution of the number of galaxies is small, and therefore they have measured the volume element as a function of redshift, a geometrical quantity. Here I review these results briefly and present the current constraints on the geometry of space.

2. NUMBER OF GALAXIES *vs.* REDSHIFT AND FLUX

One measures the flux and redshift z of every galaxy with a greater than minimum flux in a pencil beam of solid angle $d\omega$. The number of galaxies with a given redshift and flux is the product of the volume and the density.

$$\text{Volume} = H_0^{-3}a^3\frac{r^2\,dr}{\sqrt{(1 - Kr^2)}}\,d\omega = H_0^{-3}a^3 A(z;\Omega,\lambda)z^2\,dz\,d\omega,$$

217

A. Hewitt et al. (eds.), Observational Cosmology, 217–221.
© 1987 by the IAU.

$a = (1+z)^{-1}$, and r is the comoving coordinate. The density in space and luminosity L is the Schechter function

$$\phi(L)\, dL = a^{-3}\phi^* e^{-x} x^\alpha\, dx, \quad x = L/L^*, \quad \text{and } \alpha \approx -1.25.$$

From the measured number of galaxies one extracts the quantity $A\phi^* H_0^{-3}$. The quantity $A(z;\Omega,\lambda)$, the volume element relative to $a^3 H_0^{-3} z^2\, dz\, d\omega$, contains the geometrical information. The volume element and hence the observed number of galaxies is a strong function of Ω; $eg.$, $A(0.75;0,0)/A(0.75;1,0) = 1.9$.

The technical advance that enables this measurement is the photometric method (Loh and Spillar 1986a) for measuring redshifts, which Baum invented 30 years ago. In approximate terms one determines the redshift of every galaxy in a field by finding the wavelength of the 400 nm break with six broad-band filters. In 3 hours on the Wyoming 2.3 m telescope, we obtain the redshifts of 200 galaxies, for which the median redshift is 0.5. Spectroscopy yields redshifts with higher precision, but photometry is faster.

It is instructive to compare this measurement of geometry with the classical redshift-magnitude test of the brightest galaxy in rich clusters (Sandage 1961). For the redshift-magnitude test, a 10% difference in Ω implies a 3% difference in the luminosity at $z = 0.75$. Stellar evolution can shift the apparent value of Ω by $\Delta\Omega \geq 2.8$ (Tinsley and Gunn 1975). Mergers of galaxies can shift Ω by $\Delta\Omega \approx -2.2$ (Ostriker and Tremaine 1975). Therefore the systematic corrections are large and uncertain.

For this the redshift-volume test, a 10% difference in Ω implies a 5.5% difference in the number of galaxies. Loh and Spillar (1986b) measured the correction for luminosity evolution; $\Delta\Omega < 0.2$ (95% confidence). The correction for mergers is $\Delta\Omega \approx 0.05$ from Toomre's (1977) estimates of merging NGC galaxies.

With the present amount of data, the systematic errors are smaller than the random errors. The largest systematic error in $\phi^* A$ at $z = 0.75$, due to possible spectral evolution, is less than 11% (2σ). For comparison, the random errors are 21%. See Loh and Spillar (1986b) for a more complete discussion.

Several comments on the redshift-volume test are important:
1. Numbers of galaxies are dimensionless as are Ω and λ. To make this measurement one need not transfer distant times and lengths to laboratory standards as is the case for measuring dimensional quantities such as Hubble's constant.
2. For the interpretation of this measurement to be secure, the number of galaxies in a comoving box must be conserved. That this assumption is close to the truth can be seen from two observations. (a) It is difficult to prevent large galaxies from forming since the formation time is approximately the dynamical time, which is 0.2 Gyr for the Galaxy. (b) Galaxies live a long time because stars do.
3. 'First-order' evolution, where all galaxies brighten by the same factor, does not affect the measurement of $\phi^* A$.
4. Even though galaxies are observed, they are used solely as benchmarks of space. Therefore this experiment is sensitive to all kinds of matter, be it luminous or dark.

3.   RESULTS

Fitting 1000 galaxies in a solid angle $24\,\mu$sr from Loh and Spillar (1986b), one finds these limits for the independent linear combinations of Ω and λ

$$\Omega - \lambda = 0.9^{+0.7}_{-0.5} \quad \text{and} \quad -1.5 < \Omega + \lambda < 7.1$$

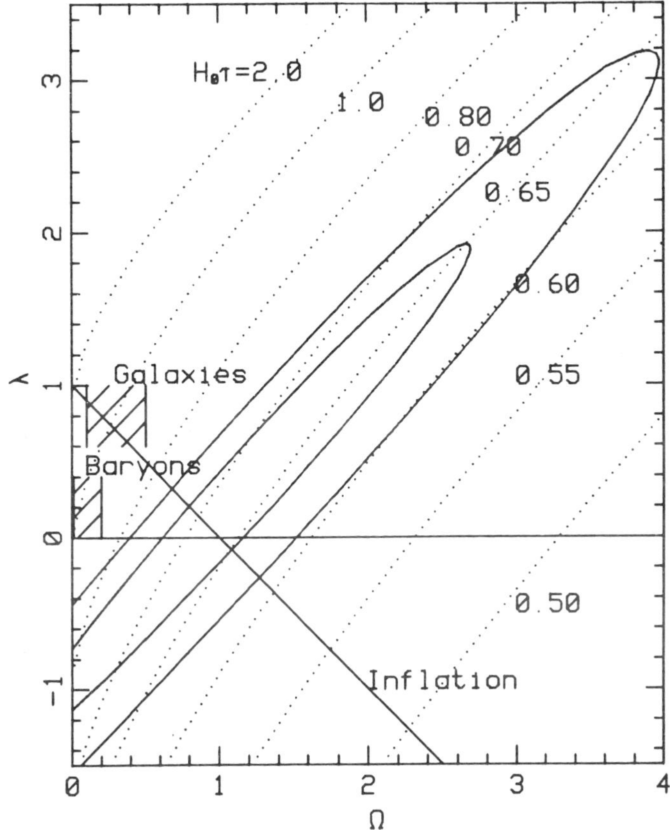

Fig 1.— The 95% and 67% confidence regions in the Ω, λ plane. Other measurements (at the 95% confidence level) of the mass density are shown as hatched regions: Yang *et al.* (1984) find that $0.01 < \Omega_{\rm baryon} < 0.2$. Peebles (1984) finds that the mass associated with galaxies is bound by $0.1 < \Omega_{\rm gal} < 0.5$. The prediction of the inflationary theory is shown as the diagonal line. Loci (dotted lines) of models with equal $H_0\tau$ are shown.

for a homogeneous and isotropic universe with pressureless matter. 95% confidence limits are used here and in the following. See Fig. 1.

3.1   Mass Density

Nonbaryonic matter exists in abundance if $\lambda = 0$ or space is flat ($K = \Omega + \lambda - 1 = 0$). From the abundances of the light elements and the theory of nucleosynthesis, Yang *et al.* (1984) conclude that $0.01 < \Omega_{\rm baryon} < 0.2$. This experiment implies that $\Omega = 0.9^{+0.7}_{-0.5}$ if $\lambda = 0$ and $\Omega = 0.9^{+0.4}_{-0.2}$ if $K = 0$.

Dynamical measurements of galaxies, which are sensitive only to the matter associated with galaxies, yield $0.1 < \Omega_{\rm gal} < 0.5$ (Peebles 1984). If $K = 0$, then galaxies do not trace the mass.

3.2   Hubble's Constant H_0 and the Age of the Universe τ

This measurement implies

$$0.60 < H_0\tau < 0.88, \quad \text{or} \quad 6.0\,\text{Gyr} < h\tau < 8.8\,\text{Gyr},$$

$H_0 = h\,100\,\text{km/sec/Mpc}$. The 95% confidence limits span a range of 47%. The age as measured by nuclear clocks is accurate to a factor of 2 (Fowler and Meisl 1985). For the age of globular clusters the 95% confidence limits, which account for measurement errors only, span a factor of 2, and systematic corrections amounting to 40% are possible (Iben and Renzini 1984). There is a well known disagreement in the value of Hubble's constant of a factor of 2. Therefore this measurement of $H_0\tau$ is more accurate and precise than the separate measurements of H_0 and τ.

3.3   Cosmology

At the present this measurement is not precise enough to choose between $\Omega = 1$ and $\Omega = 0$ models of the universe without assumptions about the cosmological constant. However if one accepts the prediction $K = 0$ of the inflationary theory because it provides an unforced explanation for the isotropy of the cosmic background radiation, then this experiment yields $\Omega = 0.9^{+0.4}_{-0.2}$ and $\lambda = -0.1^{+0.2}_{-0.4}$. This experiment with the theory of inflation is consistent with the Einstein-deSitter model ($\Omega = 1$ and $\lambda = 0$), which requires no fine tuning at large z.

4.   REFERENCES

Fowler, W. A. and Meisl, C. C. 1985, Caltech preprint.
Iben, I. and Renzini, A. 1984, *Phys. Rep.*, **105**, 329.
Loh, E. D. and Spillar, E. J. 1986a, *Ap. J.*, **303**, 154.
Loh, E. D. and Spillar, E. J. 1986b, *Ap. J. Lett.*, **307**, L1.
Ostriker, J. P. and Tremaine 1975, *Ap. J. Lett.*, **202**, L113.
Peebles, P. J. E. 1984, *Ap. J.*, **284**, 439.
Sandage, A. 1961, *Ap. J.*, **133**, 355.
Tinsley, B. M. and Gunn, J. E. 1975, *Ap. J.*, **203**, 52.
Toomre, A. 1977, in *Evolution of Galaxies and Stellar Populations*, ed. B. M. Tinsley and
   R. Larson (New Haven: Yale University Observatory), p. 401.
Yang, J., Turner, M. S., Steigman, G., Schramm, D. N., and Olive, K. A. 1984, *Ap. J.*,
281, 493.

DISCUSSION

ELLIS: The validity of this method rests ultimately on the accuracy of photometric redshifts. Spectroscopy shows most faint galaxies have weak or non-existent 4000 Å breaks. Thus I imagine significant errors could arise in their estimation from multi-colour methods. Can you check the values, or even the redshift distributions with those obtained spectroscopically?

LOH: We have compared spectroscopic and photometric redshifts for the cluster 0024+1654 (Ap. J. 303, 154, 1986), which contains about 20 blue galaxies. It is true that the errors in the photometric redshifts are larger for the blue galaxies. However, we have shown that there is no bias in the photometric redshifts, and that is all that is needed for the measurement of the volume. We also show that the errors in the photometric redshifts are estimated accurately. Furthermore, the photometric redshifts are sufficiently accurate that any systematic errors that they introduce will be small compared to the random errors (Ap. J. (Letters) 307, L1, 1986).

NON-UNIFORMITIES IN THE HUBBLE FLOW: RESULTS FROM A SURVEY OF
ELLIPTICAL GALAXIES

Roger L. Davies, Kitt Peak National Observatory, NOAO.
David Burstein, Dept. of Physics, Arizona State University.
Alan Dressler, Mt. Wilson and Las Campanas Observatories.
S. M. Faber, Lick Observatory, U. C. Santa Cruz.
Donald Lynden-Bell, Institute of Astronomy, Cambridge, England.
Roberto Terlevich, Royal Greenwich Observatory, Sussex, England.
Gary Wegner, Dept. of Physics and Astronomy, Dartmouth College.

ABSTRACT. We have used a new distance estimator for elliptical
galaxies to determine the peculiar velocities, with respect to a
uniform Hubble flow, of approximately 400 galaxies. The relative
distances of five clusters in common with those of Aaronson et al.
(1981, 1986), based on the infrared Tully-Fisher relation for spirals,
are in good agreement.
     We do not see the reflex of the Local Group motion with respect
to the microwave background out to recession velocities of 6000 km s^{-1}.
Rather, the frame of elliptical galaxies appears to be moving with
respect to the microwave background with a velocity of 600 km s^{-1}
towards $l = 312°$, $b = +6°$. This motion is consistent with a reanalysis of the Rubin et al. (1976) data on the magnitude-diameter
relation for ScI galaxies and with the nearby and cluster samples of
Aaronson et al. (1982, 1986).

1. INTRODUCTION

We have carried out a survey of elliptical galaxies, measuring both
spectroscopic and photometric parameters (Davies et al. 1987, Burstein
et al. 1987a), that is 84% complete to B = 13.0 magnitude. Fainter
galaxies were studied in clusters and a few in the field. Our aim was
to discover the nature of the second parameter in the Faber-Jackson
relation and use this to form a more accurate luminosity indicator to
determine relative distances that can be compared with those predicted
on the basis of a uniform Hubble flow model.
     Sandage and Tammann (1984) have suggested that the motion of the
Local Group (LG) with respect to the microwave background (MWB),
600 km s^{-1} toward $l = 268°$, $b = +27°$ (Smoot & Lubin 1979), is caused
by the combination of the gravitational pull on the LG towards the
center of the Virgo Supercluster, together with that on the
Virgo Supercluster towards the next nearest supercluster in
Hydra-Centaurus. Aaronson et al. (1986), using the infrared Tully-
Fisher relation (IRTF) for spirals, claim to have detected this bulk

A. Hewitt et al. (eds.), Observational Cosmology, 223–227.

motion of the Local Supercluster towards the microwave anisotropy, in
their survey of 10 clusters.

We discuss here the non-Hubble velocities that we derive using
our new distance indicator for ellipticals and refer the reader to the
other papers in this series for more details of the method and survey.

2. RESULTS

We have used the relationship between the diameter enclosing a mean
surface brightness of 20.75 with central velocity dispersion,
$D_n \propto \sigma^{1.33}$ to predict the relative distances of galaxies and
aggregates of galaxies (Dressler et al. 1987a).

We have determined the motion of the Local Group with respect to
the frame of ellipticals by fitting a model of a uniform Hubble flow,
plus a dipole motion for the LG, and maximizing the likelihood of
obtaining the observed peculiar velocities. We find that the LG is
moving at 480 km s^{-1} towards l = 193°, b = +28° with respect to the
frame of ellipticals with recession velocities less than 6000 km s^{-1}.
Thus, the elliptical galaxies do not show the reflex of the LG motion
with respect to the MWB, indicating that they are not at rest with
respect to it. In fact, we find that the frame of elliptical galaxies
is moving at 600 km s^{-1} towards l = 312°, b = +6° with respect to the
MWB (Dressler et al. 1987b). We have not reached a shell of galaxies
that is at rest with respect to the MWB and cannot identify the mass
causing the LG microwave motion, if indeed it is of gravitational
origin. However, we can eliminate the Hydra-Centaurus supercluster as
a candidate for such a mass because its peculiar motion is directed
away from the LG, not towards it, as would be the case if the two
superclusters were falling towards each other.

This large velocity of the local universe, roughly 120h Mpc in
diameter, is a somewhat surprising result. In the following section
we compare the result obtained here with those from other independent
methods.

3. COMPARISON WITH OTHER METHODS

The IRTF method for spirals has been used to determine the relative
distance moduli to Virgo of 10 clusters (Aaronson et al. 1986) and the
Fornax-Virgo relative distance modulus (Aaronson et al. 1981). Of
these 11 clusters there are 5 in common with the elliptical galaxy
survey and for these the two estimates of relative distance modulus
agree to within the errors. These two methods are completely
independent and suffer quite distinct systematic errors, so their
mutual agreement is encouraging.

We have looked for evidence of large scale coherent motions in
both the Aaronson et al. nearby and cluster samples (1982, 1986) and
the ScI galaxies of Rubin et al. (1976). Our re-analysis of the
magnitude-diameter data from Rubin et al. has used the Peterson and
Baumgart (1986) photometry and a new treatment of the Malmquist bias;

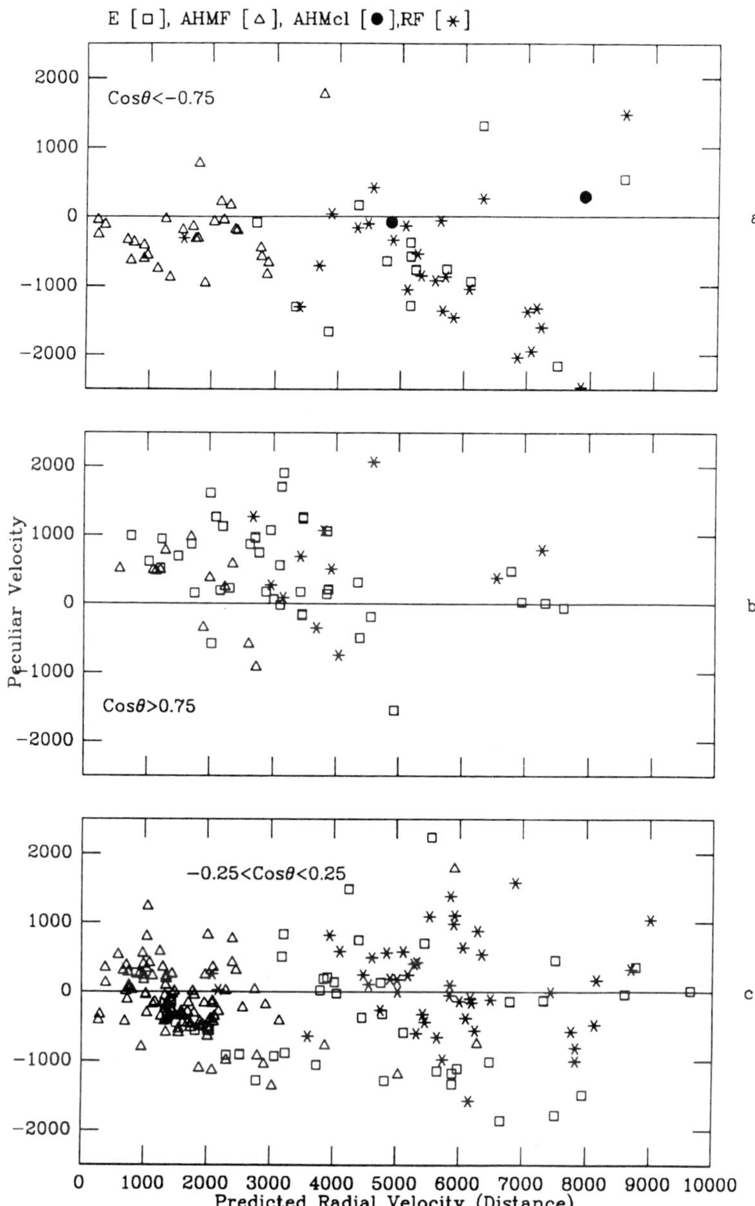

Figure 1. The peculiar velocity of galaxies taken from the four samples plotted as a function of predicted distance expressed as a recession velocity. The elliptical galaxies from this survey are plotted as open squares, the Aaronson et al. (1982) field spirals as open triangles, the Aaronson et al. (1986) clusters as filled circles and the Rubin et al. (1976) ScIs as asterisks.

this analysis is reported more fully in Burstein et al. (1987b).
Preliminary results are shown in Figure 1 (a) - (c) which plot the
peculiar velocity against predicted distance, expressed as a recession
velocity, for the four samples. Fig. 1(a) is for the direction of the
antapex of the motion and shows that the field spirals from Aaronson
et al., the Rubin et al. ScIs, and the ellipticals all show a net
negative peculiar velocity in this direction. Fig 1(b) shows the apex
direction where all the samples have a net positive peculiar velocity
for recession velicities less than 5000 km s^{-1}, beyond which there are
few galaxies in our sample. Figure 1(c) shows that there is no net
peculiar velocity in the directions perpendicular to the motion. Note
that in neither the apex nor antapex direction is there any indication
of a reduction in the peculiar velocity as the galaxies become more
distant.

4. CONCLUSIONS

1. We have used a new distance estimator for elliptical galaxies to
show that the local volume of the universe 120 Mpc across is moving at
600 km s^{-1} towards l = 312°, b = +6° with respect to the MWB.

2. We do not see the mass causing the Local Group's microwave motion
out to recession velicities of 5000 km s^{-1}, beyond which there are few
galaxies in the direction of the apex of the motion. We can eliminate
the possibility that it is caused by a combination of the
gravitational pulls of the Virgo Supercluster and the next nearest,
Hydra-Centaurus Supercluster, because the peculiar motion of Hydra-
Centaurus is away from, not towards, the LG.

3. Our method gives good agreement with the IRTF method applied by
Aaronson et al. (1986) to 5 clusters we have in common. Further, the
pattern of peculiar velocities across the sky that indicates a large
scale coherent motion is present in both a re-analysis of the Aaronson
et al. (1982) field data and the Rubin et al. (1976) data using the
ScI magnitude-diameter relation. The detailed agreement among
clusters in common, and the use of three independent methods to
confirm the presence of a large coherent motion, suggest that the
motion is real rather than an artifact of the changing properties of
ellipticals from place to place in the universe.

REFERENCES

Aaronson, M., Dawe, J. A., Dickens, R. J., Mould, J. R. and Murray, J.
     B., 1981, M.N.R.A.S., **195**, 1p.
Aaronson, M., Huchra, J., Mould, J., Schechter, P. L. and Tully, R.
     B., 1982, Ap. J., **258**, 64.
Aaronson, M., Bothun, G., Mould, J. R., Huchra, J., Schommers, R. A.,
     Cornell, M., 1986, Ap. J., **302**, 536.

Burstein, D., Davies, R. L., Dressler, A., Faber, S. M., Stone, R. P.
    S., Lynden-Bell, D., Terlevich, R. J. and Wegner, G., 1987a, in
    preparation.
Burstein, D., Davies, R. L., Dressler, A., Faber, S. M., Lynden-Bell,
    D., Terlevich, R. J. and Wegner, G., 1987b, in preparation.
Davies, R. L., Burstein, D., Dressler, A., Faber, S. M., Lynden-Bell,
    D., Terlevich, R. J. and Wegner, G., 1987, Ap. J. Suppl., in
    press.
Dressler, A., Lynden-Bell, D., Burstein, D., Davies, R. L., Faber, S.
    M., Terlevich, R. J. and Wegner, G., 1987a, Ap. J., in press.
Dressler, A., Faber, S. M., Burstein, D., Davies, R. L., Lynden-Bell,
    D., Terlevich, R. J. and Wegner, G., 1987b, Ap. J., in press.
Peterson, C. and Baumgart, C. W., 1986, A. J., **91**, 530.
Rubin, V. C., Ford, W. K., Thonnard, N. and Roberts, M. S., 1976, A.
    J., **81**, 719.
Sandage, A. and Tammann, G. A., 1984, in "Large Scale Structure of the
    Universe, Cosmology and Fundamental Physics," ed. G. Setti and L.
    Van Horn (Geneva: CERN), p.127.
Smoot, G. and Lubin, P. M., 1979, Ap. J. Lett. **234**, L83.

DISCUSSION

R.D. DAVIES: L. Staveley-Smith and I have investigated the local
velocity field out to 4000 kms^{-1} for 350 Sb, Sbc and Sc galaxies
using the Tully-Fisher relation for blue magnitude and diameter.
We find that the apparent direction of the local group motion varies
with the redshift range taken for the reference shell. It agrees
with the CMB at 2000 km s^{-1} but progressively moves away up to 4000
km s^{-1} by 60°; implying large-scale streaming motions of about 200-300
km s^{-1} on scales of 20-30 Mpc or larger. Local departures on scales of
10-20 Mpc are of similar magnitude. These values are significantly
less than the values you quote. Do you think that the apparent
increased streaming motion at your larger distances is a consequence of
the distance uncertainty (which increases linearly with distance)?

R.L. DAVIES: No. We have divided our sample into $0 < v < 2000$,
$2000 < v < 4000$ and $4000 < v < 6000$ km s^{-1} and find no significant
differences in the vector. I believe that apparent discrepancies
can only be resolved by comparing the relative distances to clusters
and groups in common between different methods and samples. When we
do this with the few Aaronson _et al._ clusters in common the agreement
is good.

DISTRIBUTION OF IRAS GALAXIES

Michael Rowan-Robinson
Theoretical Astronomy Unit,
Queen Mary College, Mile End Road,
London E1 4NS

ABSTRACT   Infrared wavelengths are free of several of the problems that plague optical galaxy surveys.   At high galactic latitude $\gtrsim$99% of 60μ sources in the IRAS Point Source Catalog, after deletion of obvious stars, are galaxies.   At lower latitudes care has to be taken to avoid confusion with emission from interstellar dust (the 'cirrus').   IRAS galaxies have been used to determined the direction of the gravitational acceleration acting on the Local Group due to galaxies and clusters within about 200 Mpc.   This agrees well with the direction of the microwave background dipole.   The density of matter in the universe, distributed like IRAS galaxies, needed to account for the observed velocity of the Local Group, corresponds to $\Omega_0 = 1.0 \pm 0.2$.   In the standard hot Big Bang model, 90–95% of this matter would have to be non–baryonic.

IRAS galaxies are significantly less clustered than optically selected galaxy samples.

1.   INTRODUCTION.

Galaxy surveys at optical wavelengths have so far failed to derive a convincing value for the velocity of the Local Group of galaxies with respect to the cosmological reference frame.   While some determinations are in reasonable agreement with the value derived from the dipole component of the microwave background anisotropy (Hart and Davis 1982, Aaronson et al 1986), which we take here to be

$$v = 600 \text{ km sec}^{-1} \text{ towards } (l,b) = (277, 29) \tag{1}$$

(Lubin et al 1983, Fixsen et al 1983, Yahil et al 1986),
others are wildly discrepant from this value (Rubin et al 1976, de Vaucouleurs 1978, Sandage et al 1979, de Vaucouleurs and Peters 1981, de Vaucouleurs et al 1981, Aaronson et al 1982).   Attempts to explain these discrepancies in terms of large–scale streaming motions seem premature.   A more likely explanation of the discordant optical results is a combination of the known problems of current galaxy surveys:

(i) there is no all–sky galaxy survey with a well–calibrated homogeneous magnitude scale deeper than the Revised Shapley Ames Catalogue (Sandage & Tammann 1981), which has a completeness limit of about $m_B = 12$ mag.   For a typical galaxy with absolute magnitude $M_B = -21$, this corresponds to a survey depth of only 40 Mpc, a volume clearly dominated by the Virgo Supercluster.

(ii) the strength of extinction by interstellar dust is a matter of lively controversy.   Estimates of polar extinction range from 0 to 0.3

A. Hewitt et al. (eds.), Observational Cosmology, 229–245.

magnitudes of visual extinction (Sandage 1973, de Vaucouleurs et al 1976, Burstein & Heiles 1978) and at lower galactic latitudes the uncertainty can be 1 magnitude or more. The neutral hydrogen column-density (Burstein & Heiles 1978, 1982) and the IRAS 100μ background intensity (Rowan-Robinson 1986a) offer useful indicators of the column-density of interstellar dust associated with diffuse neutral gas, but there is little prospect of an effective indicator of the column-density of dust in lines of sight with significant concentrations of molecular gas.

(iii) for studies of late-type galaxies, the strength of extinction by internal dust and its dependence on the orientation of the galaxy is even less well known than that by interstellar dust in our Galaxy.

The IRAS Point Source Catalog provides us with a database for cosmological studies free of the above problems. Interstellar and internal extinction are neglible at far infrared wavelengths, the calibration is carried out in a homogeneous way over the whole sky (Neugebauer et al 1984) and outside the Galactic plane, regions of strong 'cirrus' emission and the 4% of the sky which lies in coverage gaps, the survey is 98% complete and 99.9% reliable (Rowan-Robinson et al 1984). Lawrence et al (1986) have shown that at high galactic latitudes, $\gtrsim$99% of 60μ sources (after exclusion of sources which are obviously stellar) can be identified with galaxies and that the median depth of the IRAS 60μ survey is 200 $(50/H_0)$ Mpc, where H_0 is the Hubble constant in km s^{-1} Mpc^{-1}. Known systematic errors can be shown to be small (see section 4 below). There remains the problem of emission from interstellar dust (the infrared 'cirrus'), which has to be carefully controlled to achieve reliable results (see section 3 below).

In this paper I review the work carried out by my group at Queen Mary College, London, in collaboration with Amos Yahil of Stony Brook University, N.Y., and colleagues at the Royal Greenwich Observatory. We have used the unique qualities of the IRAS 60μ survey to derive the direction of the gravitational acceleration acting on the Local Group due to galaxies and clusters within about 200 Mpc. This direction agrees well with that of the microwave background dipole. From the 60μ luminosity function for galaxies derived from a redshift survey of IRAS sources we estimate the density of matter in the universe, distributed like the IRAS galaxies, required to accelerate our Galaxy to its observed velocity today (eqn (1)). I shall also comment on the parallel work of Meiksin and Davis (1986) and Lahav (1986).

2. USING IRAS GALAXIES TO MAP THE LOCAL GRAVITATIONAL FIELD

We assume that IRAS 60μ galaxies trace the matter in the universe and that there exists a universal luminosity function $\Phi(L)$, so the number of galaxies in luminosity range dL, volume element d^3r, can be written

$$dN = D(\underline{r})\ d^3r\ \Phi(L)\ dL \tag{2}$$

where $D(\underline{r})$ is the local relative density function ($D=1$ corresponds to the mean density of the universe).

Then the smoothed surface brightness due to sources is (Yahil et al 1986)

$$4\pi\sigma(S,\omega) = 4\pi S \frac{dN}{dSd\omega} = \int D(\underline{r})dr \int L\Phi(L) \; \delta \left[S - \frac{L}{4\pi r^2}\right] dL. \tag{3}$$

The peculiar gravitational acceleration $\underline{g}$ acting on the Local Group is proportional to the density moment

$$\underline{G} = \frac{3}{4\pi} \int D(\underline{r}) \left[\frac{\underline{r}}{r^3}\right] d^3r = 3g/4\pi G\rho_0 \; . \tag{4}$$

Now the QMC-RGO redshift survey (Lawrence et al 1986) yields a 60μ luminosity function which can be well represented by (Fig 1)

$$\Phi(L) = CL^{-2} (1 + L/\beta L_*)^{-\beta} \tag{5}$$

with $C = (5.75 \pm 0.2) \times 10^6 (H_0/50) \; L_\odot \; \mathrm{Mpc}^{-3}$

$$\beta = 2.4 {+ 1.5 \atop - 0.8}.$$

Substitution into eqn (3) gives

$$4\pi S\sigma(S,\omega) = C \int D(\underline{r})(1 + r^2/\beta r_*^2)^{-\beta} \; dr \tag{6}$$

where $r_* = (L_*/4\pi S)^{\frac{1}{2}}$.

The dipole moment of this

$$4\pi S\underline{\sigma}(S) = \int 3 \, S \, \sigma(S,\omega) \, \hat{\underline{r}} \; d\omega$$

$$= \frac{3C}{4\pi} \int D(\underline{r}) \left[\frac{\underline{r}}{r^3}\right] (1 + r^2/\beta r_*^2)^{-\beta} \; d^3r \tag{7}$$

is now of the same form as $\underline{G}$ except for the cutoff factor $(1+r^2/\beta r_*^2)^{-\beta}$, which has only a small effect.

3. THE IRAS DIPOLE

To evaluate eqn (7) we have used IRAS 60μ sources brighter than the completeness limit of 0.6 Jy (Rowan-Robinson et al 1986a). Stars have been excluded by omitting sources with $S(25\mu)>3S(60\mu)$: virtually all such sources are identified with catalogued stars. Lawrence et al (1986) have shown that $\geqslant 99\%$ of the remaining sources at $b>60^0$ are galaxies.

Fig 1: 60μ luminosity function for IRAS galaxies, compared with 2 simple analytical models. The dashed curve is eqn (5). (Filled circles: Lawrence et al 1986, crosses: Soifer et al 1986, x's: Rieke and Lebofsky 1986).

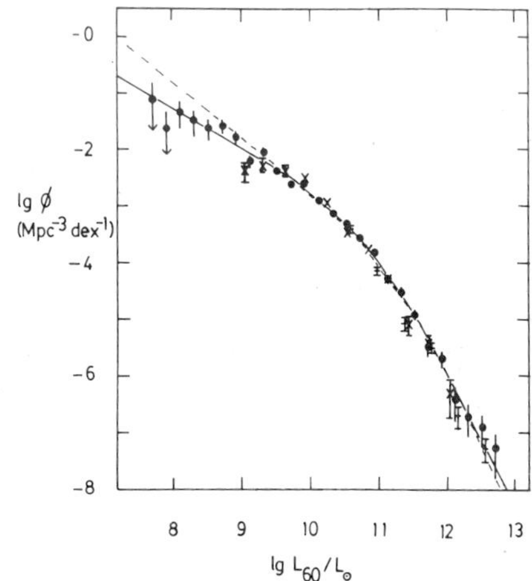

Table 1: IRAS dipole moment

Mask	%of sky excluded	l	b	θ_{CBR}	$\|4\pi S\sigma\|$ (Jy sr^{-1})	Ω_0 (linear theory)
n=0	66%	237±11	33±12	34±12	1290±190	1.15 ± 0.29
n=1	53%	248±9	40±8	26±10	1550±170	0.85 ± 0.16
n=2	46%	256+8	37+8	19+9	1590+160	0.81 ± 0.14
n=1						
0.5–1Jy		286	50	22		
1–2 Jy		232	30	39		
2–4 Jy		257	20	20		
4–8 Jy		255	37	20		
8–16Jy		240	43	33		
16–32Jy		240	33	32		
n=1, plus 20°x20° area centred on Virgo	54%	242	36	30		
Meiksin and Davis (1986)	24%	235	43.5	36		
Lahav (merged diameter limited optical galaxy catalogue, $\|b\|>20°$	34%	220	44	47		

To avoid contamination by 'cirrus' and regions of very high source-density like the Milky Way and the Magellanic Clouds, we exclude $|b|<5^\circ$ and any one degree square bin which has been flagged in the IRAS Point Source Catalog as a region of exceptionally high source-density at 60μ. We also exclude any one square degree bin in which the cirrus flag CIRR1 (the number of 100μ only sources in a 1 sq deg area centred on a source) has been set $>$ n. After investigating the sensitivity of our results to the value of n, and the values of n found in known clouds of 'cirrus' (eg Rowan-Robinson et al 1986a), we adopt the conservative value $n = 1$.

The excluded areas, including the 4% of the sky in coverage gaps, define a mask, which is illustrated in Fig 2 a and b. Little of the sky remains unmasked below $|b|=30^\circ$. The spherical harmonic components have been calculated over the un-masked area only (this modifies the orthogonality matrix for the harmonics). We are essentially assuming that the un-masked area is representative of the whole sky. Fig 3a,b show the distributions of IRAS 60μ sources in the unmasked area.

Fig 5 of Yahil et al (1986) shows the 3 components of $4\pi S\underline{\sigma}(S)$, eqn (7), as a function of S, together with the average over all S. The amplitude of the dipole component is 20% of the mean surface brightness at 2 Jy and 10% at 0.6 Jy. Table 1 gives the direction of the average dipole and the angular displacement from the microwave background dipole, Θ_{CBR}, for 3 values of the cirrus mask parameter n = 0, 1, 2. The direction of the IRAS dipole agrees well with that of the microwave background dipole. This suggests that we have now identified the cause of our motion with respect to the microwave background,

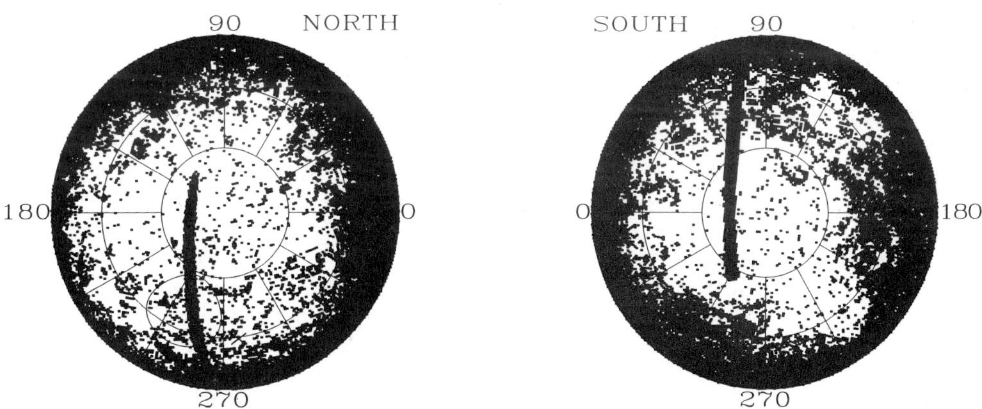

Fig 2: Equal area projections of the north and south Galactic hemispheres showing the mask n=1.

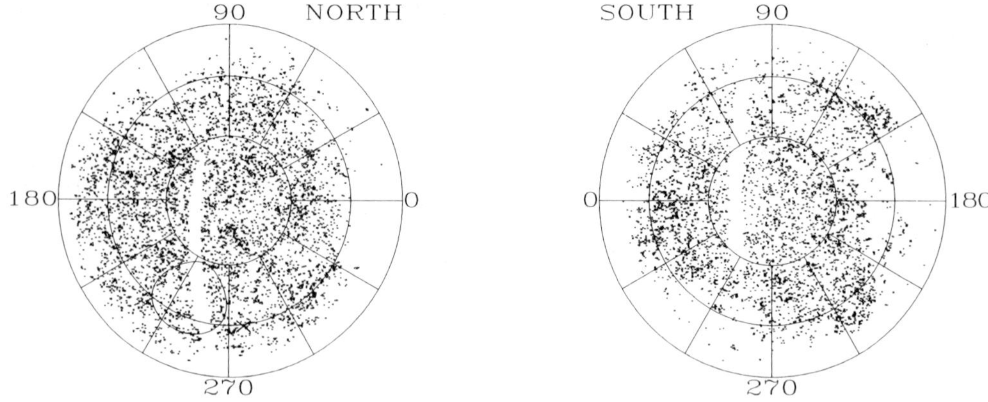

Fig 3:   Equal area projections of the north and south Galactic hemispheres showing the IRAS 60μ sources in the unmasked region. Also shown are the 90% confidence limit of the IRAS dipole moment, and the direction of the microwave background dipole moment (large cross).

namely the net gravitational attraction of galaxies and clusters within about 200 Mpc.   Table 1 also shows how the direction of the dipole depends on the flux bin: the dependence appears to be slight over the flux range 0.6 - 32 Jy.

4.   POSSIBLE SOURCES OF SYSTEMATIC ERROR IN THE IRAS SURVEY

The known systematic errors in the IRAS data have been extensively discussed in the IRAS Explanatory Supplement (1984) and their implications for cosmological studies have been discussed by Rowan-Robinson et al (1986a), Yahil et al (1986) and Rowan-Robinson and Needham (1986).   The major effects which could contribute to the IRAS dipole are:

(A) photon-induced responsivity enhancement ('hysteresis') as the detector field of view crosses the Galactic plane and other bright sources.   The magnitude of this effect is <1% at 60μ at high Galactic latitudes, but great care would have to be taken when working within 20° of the Galactic plane.

(B) responsivity changes due to particle hits on orbits passing near the edge of the South Atlantic Anomaly, but where bias boost was not applied to the detectors.   This affects about 15% of orbits and the magnitude of the effect is <7% at 60μ, so the net effect on the 60μ fluxes is <1%.   Moreover the effect would have opposite signs at 60 and 100μ and the 60 and 100μ source-counts at $|b|$>60° are well correlated with each other.   The effects of the Polar Horns of the radiation belts on fluxes in the IRAS Point Source Catalog are likely to be even smaller than that of the SAA.

To explain the observed dipole, a dipole calibration error at 60μ would have to have a 7% amplitude at 0.6 Jy and 13% amplitude at 2 Jy.

5.   CALCULATION OF Ω_0

We can now calculate how much matter there needs to be in the universe, distributed like the IRAS galaxies, to accelerate the Local Group to its observed velocity with respect to the microwave background (eqn (1)).   In linear perturbation theory, our peculiar velocity would be (Peebles 1980):

$$\underline{u} = \frac{1}{3} \Omega_0^{0.6} H_0 \underline{G} \tag{8}$$

$$= \frac{1}{3} \Omega_0^{0.6} H_0 \langle 4\pi S\underline{\sigma}\rangle/C, \text{ using eqns (4) and (7)}.$$

Using the value for C of Lawrence et al (1986), eqn (5) above, and correcting for the fact that the area they surveyed has a source-density 18% above the average for the unmasked sky, we find

$$\Omega_0 = 0.85 \pm 0.16 \;. \tag{9}$$

Applying a small correction for the effects of non-linearity (see Yahil et al 1986) our final result is

$$\Omega_0 = 1.0 \pm 0.2 \;. \tag{10}$$

In the standard hot Big Bang picture, this would imply that 90–95% of the matter in the universe would be non-baryonic, since consistency with the observed primordial helium and deuterium abundances requires $\Omega_{b,0} \approx 0.05$–0.1.

6. MEIKSIN AND DAVIS STUDY

Meiksin and Davis (1986) have also studied the IRAS dipole and the direction they derive, $(l,b) = (235, 43.5)$, is similar to ours. However there are some key differences of approach and also some puzzling discrepancies between their work and ours (Rowan-Robinson 1986b).   Figure 4 shows the mask used by Meiksin and Davis.   They have been much less severe than us in excluding regions affected by cirrus and in fact exclude only 24% of the sky, compared with 53% excluded by the n=1 mask.   They have used a different colour condition to exclude stars, $S(60\mu)>3S(12\mu)$, but this should have led to the exclusion of only a further 6% of the sample.   However I find (i) Meiksin and Davis appear to have used less than half of the sources in the IRAS catalog satisfying their constraints, (ii) the amplitude of their 2-dimensional covariance function for IRAS galaxies is more than twice that found by Rowan-Robinson and Needham (1986).   (i) may be caused by using upper limits to the 12μ flux

Fig 4: Meiksin and Davis (1986) mask. The excluded area consists of the boxed regions plus $|b|<10^{\circ}$.

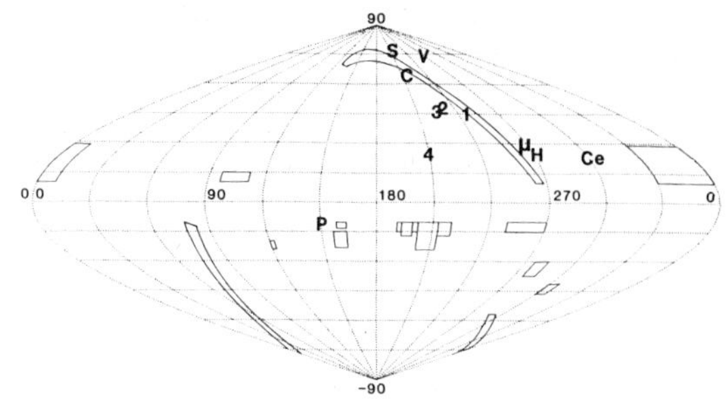

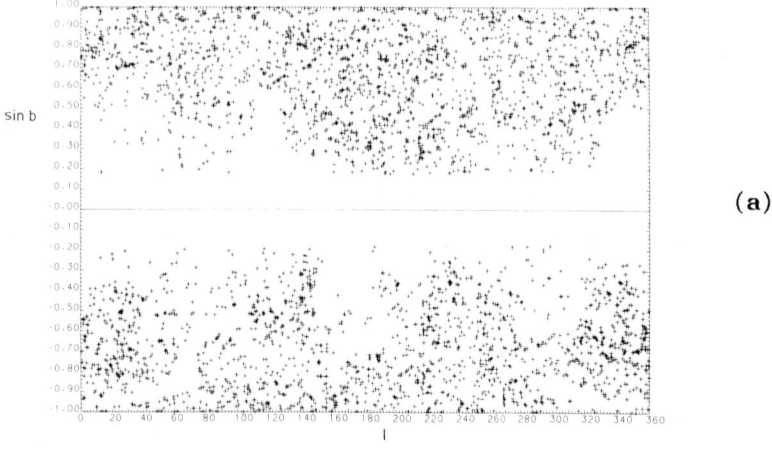

(a)

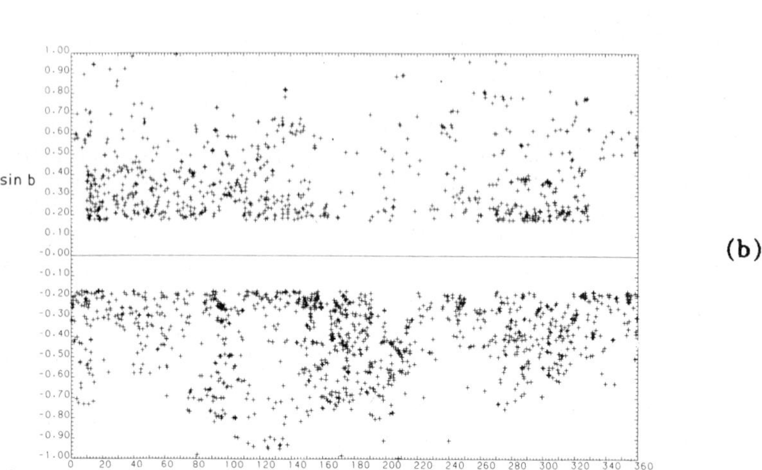

(b)

Fig 5: sin b – l distribution for IRAS 0.6 –1Jy 60μ sources outside the Meiksin and Davis mask and satisfying S(60)>3S(12), (a) CIRR1 = 0, (b) CIRR1 > 2.

in place of the actual 12μ flux, $S(12\mu)$, in their colour condition. (ii) is almost certainly due to a substantial contamination of their sample by cirrus.

Figure 5 shows the sin b – l distribution of 0.6 – 1.0 Jy IRAS 60μ sources outside the Meiksin and Davis mask, satisfying $S(60) > 3S(12)$, (a) with CIRRI = 0, (b) with CIRRI > 2. The striking difference in the degree of clustering between Fig 5b and Fig 5a is clearly due to strong contamination by cirrus amongst the unidentifed sources with CIRRI > 2. The contamination by cirrus in the flux range $1 < S(60) < 2$ Jy is still marked though it is weaker (but not absent) for $S(60) > 2$Jy. Even in this brighter flux range care has to be exercised. Figure 6 shows the same distribution as Fig 5b for IRAS 60μ sources brighter than 2 Jy, satisfying the Meiksin and Davis colour condition, and with $|b| > 10^{\circ}$ (no additional masking). Strong contamination by cirrus and other local sources can still be seen in Fig 6 in the Orion–Taurus and Ophiucus regions (as well as the SMC and LMC). In such regions not only are spurious sources generated but the accuracy of the fluxes of the detected sources can no longer be relied on, nor will the IRAS Point Source Catalog be complete. The very high figures for completeness and reliability on which the present cosmological study relies apply only outside regions affected by cirrus.

7. LAHAV STUDY

Lahav (1986) has attempted to carry out an analysis similar to that of Yahil et al (1986) using optical galaxy catalogues. Merging together the Uppsala, ESO and MCG Catalogues, he has created a diameter – limited $(\theta > 1.3)$ galaxy catalogue covering the whole sky (Figure 7). Replacing the flux in the expression for the surface brightness by (diameter)^2, he calculates the dipole component of the surface brightness, analogously to Yahil et al. The masks he investigates are very simple, of the form $|b| > b_0$ for example, and he calculates the effect of these masks on the dipole components analytically. For $b_0 = 20^{\circ}$, for example, he finds dipole direction (l,b) = (220,44), similar to that of the IRAS dipole (Table 1). I have also used our software to calculate the dipole for Lahav's galaxy catalogue using the IRAS n = 1 mask, which should delete all area significantly affected by interstellar extinction. The resulting direction is (l,b) = (218,41), not very different from that for the much cruder $|b| > 20^{\circ}$ mask.

Lahav's approach is clearly a very interesting one and we plan to carry out further intercomparison of the optical and IRAS samples. Although Lahav's merged galaxy catalogue is not a homogeneous or complete one, the use of diameters rather than magnitudes may have overcome some of the problems of optically selected samples.

Fig 6: sin b - l distribution
for IRAS 60μ sources with
$S(60) > 2$ Jy, $|b| > 10°$,
$S(60) > 3S(12)$, and CIRR1 > 2.

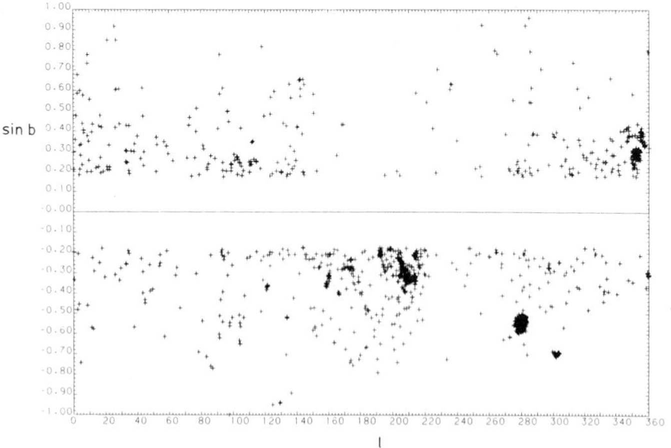

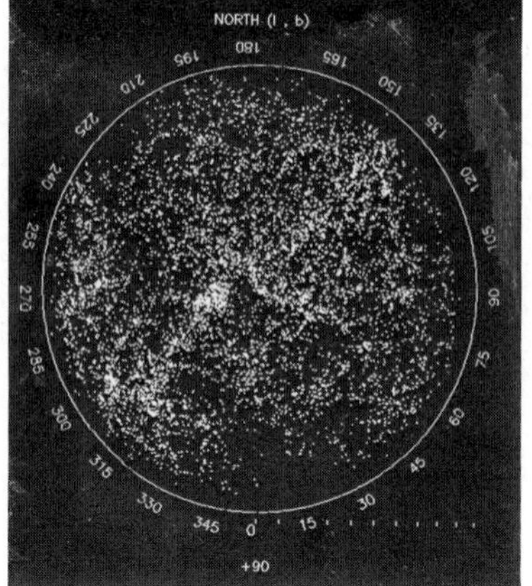

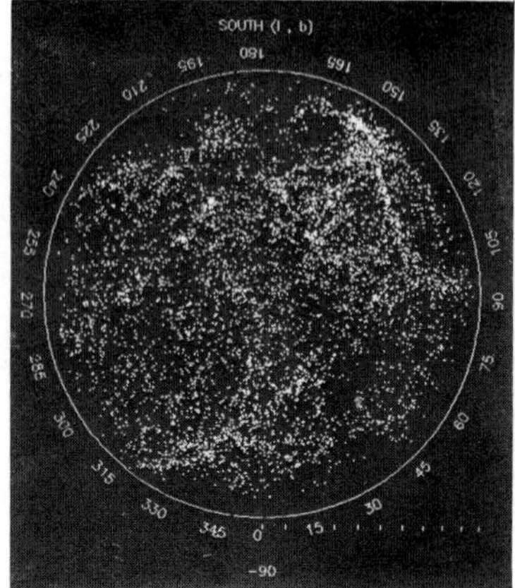

Fig 7: Equal area plot of galaxies
in Lahav's (1986) diameter limited
merged galaxy catalogue

8. CLUSTERING OF IRAS SOURCES AND THE DEPTH OF THE IRAS SURVEY

The distribution of IRAS galaxies (Fig 3) shows clear evidence of clustering and many of these concentrations of IRAS sources correspond to known clusters of galaxies at distances ranging from $20 - 200$ ($50/H_o$) Mpc. Rowan-Robinson et al (1986) gave a list of IRAS clusters with $\geqslant 7$ members in a 1° radius circle for $|b| > 60^o$. I have now extended this analysis to the whole sky outside the n = 1 mask. Fig 8 shows the sin b - l distribution of these IRAS clusters, with selected identifications with known galaxy clusters (and their distances) indicated. 57 out of a total of 161 IRAS clusters can be identified in this way.

Figure 8 also shows a contour plot of the surface density of IRAS sources in the flux range $0.6 < S < 30$ Jy, made with a resolution of 0.05 in sin b $\times$ 10° in l. Here we can see the regions of above average surface density, many identifiable with prominent superclusters like Virgo, Coma, Perseus-Pisces, Hydra-Centaurus, and the regions of below average surface density (voids?), responsible for the IRAS dipole. Virgo itself does not play a major role in our acceleration, since adding the Virgo area to the mask hardly changes the dipole direction (Table 1).

The depth of the IRAS survey has been investigated by Lawrence et al (1986) and Fig 9 shows the redshift distribution found in their survey for sources with S(60) > 0.85 Jy. We can also use the survey of Lawrence et al to derive the radial density distribution for IRAS galaxies in this part of the sky ($0^o < 1 < 110^o, b > 60^o$), using the method of Kirschner et al (1978). The result is shown in Fig 10: the Coma cluster appears as a strong peak at d ~ 140 Mpc.

9. DISCUSSION

The calculations of this paper are based on a number of assumptions, which I now discuss in turn.

(i)   The 60μ luminosity has to be a measure of the mass of a galaxy. In support of this is the good correlation of 60μ luminosity with optical luminosity (Rowan-Robinson et al (1986b), which in turn is well correlated with galaxy mass. The minority of IRAS galaxies with exceptionally high infrared-to-optical ratios do not invalidate this assumption.

(ii)   IRAS galaxies have to be a good tracer for the overall matter distribution, most of which has to be dark and non-baryonic. It is known that ellipticals and lenticulars are biassed towards rich clusters of galaxies, but IRAS galaxies are predominantly spirals and are the best available candidate for an unbiassed tracer of matter. If IRAS galaxies too are biassed tracers of matter then the density estimate (10) would be a lower limit.

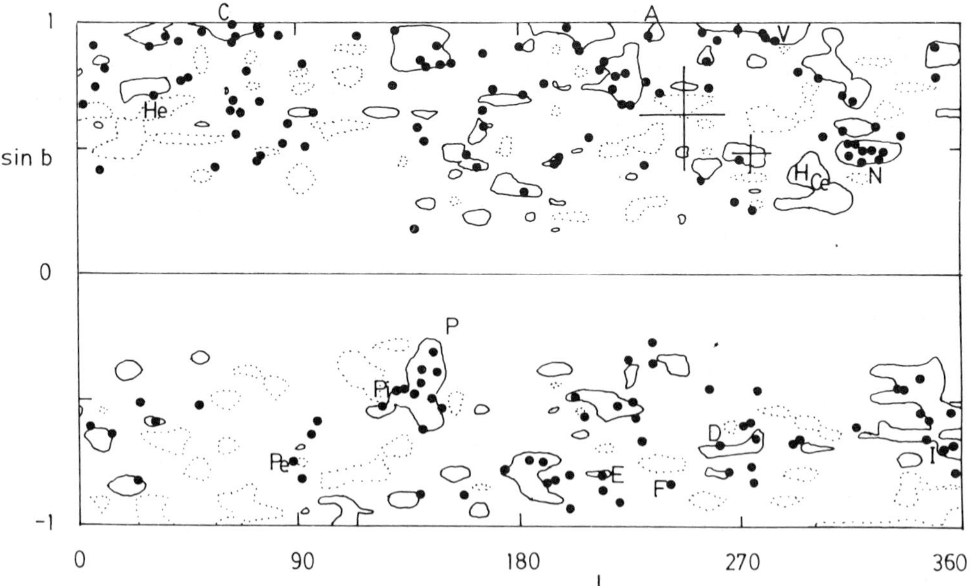

Fig 8: Sin b - l distribution for IRAS clusters (filled circles) superposed on contour plot of surface density of IRAS 60μ sources. Solid curves: source-density 20% higher than average, dotted curves: source-density 20% lower than average. Large cross: direction of IRAS dipole, small cross: microwave background dipole.

Letters denote prominent clusters, with distances in Mpc given in brackets (calculated as $v_0/50$) : C = Coma (133), A = A1367 (122), V = Virgo (20), He = Hercules (220), H = Hydra (60), Ce = Centaurus (60), N = N5419 (83), P = Perseus (108), Pi = Pisces (105), Pe = Pegasus I (77), E = Eridanus (31), F = Fornax I (30), D = Dorado (20), I = Indus II (45).

It is clear that the distribution of IRAS sources (Fig 3) is markedly less clustered than the corresponding distribution of optically selected galaxies (Fig 7). Rowan-Robinson and Needham (1986) have calculated the 2-dimensional covariance function for IRAS 60μ sources outside the n=1 mask (Fig 11). While the power-law form is similar to that found in optical studies, the amplitude is considerably lower. Using the redshift survey of Lawrence et al (1986), Rowan-Robinson and Needham (1986) have calculated the 3-D covariance function corresponding to the model fit in Fig 11 and find

$$\xi(r) = (r/r_0)^{-1.7} \tag{11}$$

where $r_0 = 4.1 \pm 0.7$ (50/H_0) Mpc.

which can be compared with a typical value of 10(50/H_0) Mpc found in optical studies.

We have also now calculated the 3-D covariance function directly for the Lawrence et al sample (Needham, Lawrence and Rowan-Robinson 1986). The result is shown in Fig 12, compared with the power-law of eqn (11). The agreement is excellent.

(iii) we have to assume that there are no clusters (or voids) behind the mask which are so pronounced as to invalidate the assumption that the unmasked sky is representative of the whole sky. We plan to carry out a careful search for IRAS galaxies at low galactic latitudes to try to check this.

I conclude that the IRAS survey has provided us with an exceptionally rich database for cosmological studies. We have determined the direction of the gravitational acceleration due to galaxies and clusters within about 200 Mpc and find it to agree well with the direction of the microwave background dipole. We have therefore, I believe, found the cause of our motion with respect to the cosmological frame. The density of matter in the universe, distributed like the IRAS galaxies, required to explain our observed velocity with respect to the microwave background corresponds to $\Omega_0 = 1.0 \pm 0.2$. Most of this matter would in the standard model, have to be non-baryonic.

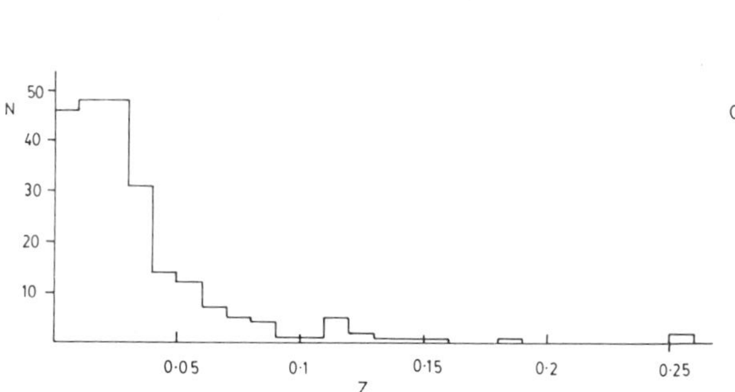

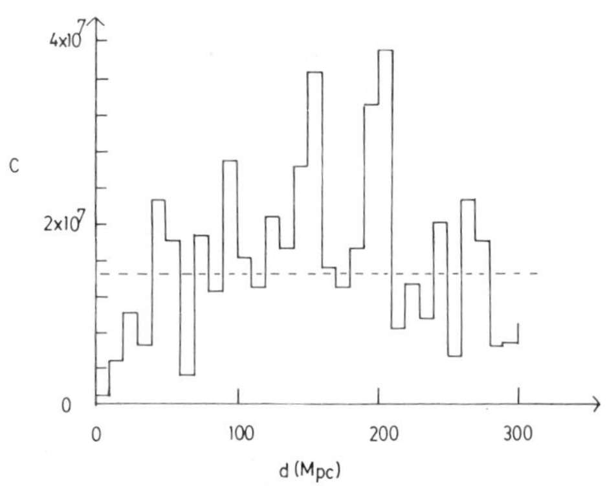

Fig 9: Redshift distribution for IRAS 60μ sources brighter than 0.85 Jy (from Lawrence et al 1986).

Fig 10: Variation of spatial density of IRAS galaxies with radial distance. The vertical scale is the constant C in eqn (5). (from Rowan-Robinson and Lawrence 1986)

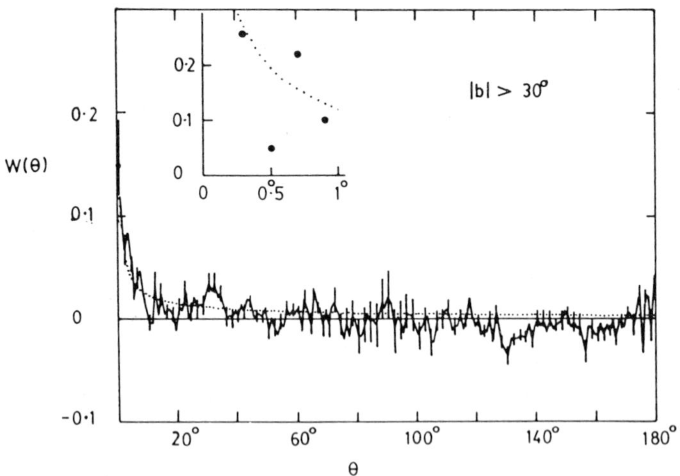

Fig 11: Angular correlation function for IRAS sources at $|b|>30^o$, with 1° bins. Inset is shown data for $12^1<\theta<1^o$. The dotted curve is $\omega(\theta)=0.12\ \theta^{-0.7}$. (from Rowan-Robinson & Needham 1986).

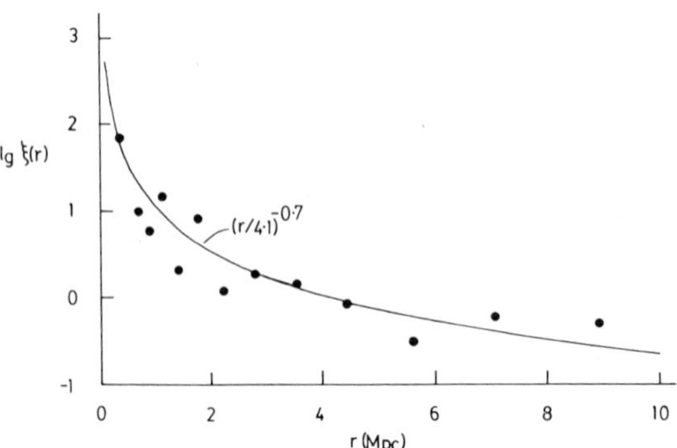

Fig 12: 3-D covariance function for IRAS galaxies (Needham, Lawrence and Rowan-Robinson 1986).

REFERENCES

Aaronson M., Huchra J., Mould J., Schecter P.L. & Tully R.B.,
    Ap.J. 258, 64
Aaronson M., Bothum G.D., Mould J.R., Huchra J., Schommer R.A.
    & Cornell M.E., 1986, Ap.J. 302, 536
Burstein D. & Heiles C., 1978, Ap.J. 225, 40
Burstein D. & Heiles C., 1982, A.J. 87, 1165
Fixsen D.J., Cheng E.S. & Wilkinson D.T., 1983, Phys.Rev.Letters 50,
    620
Hart L. & Davis R.D., 1982, Nature 297, 191
IRAS Explanatory Supplement, 1984, eds, C.A. Beichman, G. Neugebauer,
    H.J. Habing, P.E. Clegg & T.J. Chester, JPL D-1855
Lahav O., 1986, MNRAS, (in press).
Lawrence A., Walker D., Rowan-Robinson M., Leech K.J. & Penston M.V.,
    1986, MNRAS 219, 687
Lonsdale C.J., Helou G., Good J.C. & Rice W., 1985, Catalogued
    Galaxies and Quasars observed in the IRAS survey, JPL D-1932
Lubin P.M., Epstein G.L. & Smoot G.F., 1983, Phys.Rev.Letters 50, 616
Meiksin A. & Davis M., 1986, A.J. 91, 191
Needham, G. Lawrence A. and Rowan-Robinson M., 1986, in preparation
Neugebauer G., Wheelock S., Gillett F., Aumann H.H., Gautier N.,
    Low F.J., Hacking P., Hauser M., Harris S. & Clegg P., 1984, IRAS
    Introductory Supplement, Ch. VI
Peebles P.J.E., 1980, The Large-scale Structure of the Universe
    (Princeton University Press), §14
Rieke G.H. and Lebofsky M.J., 1986, A   304, 326.
Rowan-Robinson M., 1986a, M.N.R.A.S. 219, 737
Rowan-Robinson M., 1986b, A.J. (to be submitted)
Rowan-Robinson M., Clegg P., Beichman C., Chester T., Conrow T.,
    Habing H., Helou G., Neugebauer G., Soifer T. & Walker D.,
    1984, IRAS Introductory Supplement, Ch. VIII
Rowan-Robinson M., Walker D., Chester T., Soifer T. & Fairclough J.,
    1986, MNRAS 219, 273
Rowan-Robinson M., Walker D., & Helou G., 1986b, MNRAS (submitted)
Rowan-Robinson M. & Needham G., 1986, MNRAS, 222, 611.
Rowan-Robinson M. and Lawrence A., 1986, in preparation
Rubin V.C., Thonnard N., Ford W.K. & Roberts M.S., 1976, A.J. 81, 719
Sandage A., 1973, Ap.J. 183, 711
Sandage A., Tammann G.A. & Yahil A., 1979, Ap.J. 232, 352
Sandage A. & Tammann G.A., 1981, Revised Shapley-Ames Catalog of
    Bright Galaxies (Carnegie Institute of Washington)
Soifer B.T. et al, 1986, Ap.J., 283, L1.
de Vaucoulers G., de Vaucouleurs A. & Corwin H.G.Jr, 1976, 2nd
    Reference Catalogue of Bright Galaxies (University of Texas Press)
de Vaucouleurs G., 1978, IAU Symposium 79, ed. M.S. Longair and
    J. Einasto (Reidel) p. 205
de Vaucouleurs G., Peters W.L., Bottinelli L., Gouguenheim L. &
    Paturel G., 1981, Ap.J. 248, 408
Yahil A., Walker D. & Rowan-Robinson M., 1986, Ap.J. 301, L1

DISCUSSION

BOLDT: What is the number of sources used for the dipole determination?
Is it sufficient to minimize statistical fluctuations?

ROWAN-ROBINSON: There are 9908 IRAS 60 μ sources, after exclusion of
stars, with 0.5 Jy $\leq$ S < 30 Jy, outside the N = 1 mask.   The uncertainties I have shown are the statistical uncertainties.

DEKEL: I wish to challenge the assumption that we move with a velocity
of 600 km/s relative to the IRAS sample, based on the finding discussed
by Dr. Davies (Burstein et al.) that we do not move at all relative to
the sample of ellipticals which is only twice smaller.

ROWAN-ROBINSON: Although the depth of the IRAS survey may exceed that
of the elliptical sample by rather more than a factor of two, I accept
that there is probably an inconsistency between the two results.   My
first comment is that Lahav's study, based on an optically selected
galaxy sample covering the whole sky, is consistent with the IRAS
result.   My second comment is that the Faber-Jackson distance method
(and this applies also to the Tully-Fisher method) does not have a
real physical basis, so the uncertainty in the distance estimates may
be larger than the purely statistical estimates.   The fact that you
find similarly sloped correlations between observed quantities in
different clusters does not prove that the galaxies in two different
clusters are identical.

ELLIS: The reduced amplitude of the IRAS galaxy correlation function
can largely be explained as due to the dominant spiral mixture.   The
best estimate of the morphological variation in correlations comes from
the Durham/AAT Survey where I find ξ_{spiral} ~ 1.8 times lower than
$\xi E/S\emptyset$.   Thus your result is not necessarily evidence for biassed
galaxy formation.

ROWAN-ROBINSON: Davis and Geller (1976) also found that the clustering
amplitude for spirals was lower, by a factor ~ 1.5, than that for
galaxies as a whole. However the factor by which the clustering
amplitude for IRAS galaxies lies below that found for optical galaxy
samples appears to be larger than this.   My interpretation of this,
and of the higher value of Ω we find compared with that from
dynamical studies of optical galaxy samples, is that while the spirals
predominantly found by IRAS trace the general matter distribution,

ellipticals form preferentially in regions of high density contrast, which define the cores of rich clusters.

CHINCARINI: In your detection of clusters using the IRAS survey, how do Coma, A1367 and Virgo compare in number density?

ROWAN-ROBINSON: The numbers of IRAS sources found in clusters associated with Coma, A1367 and Virgo, by the method described in section 8 (1^O radius search area) are 10, 7 and 48 respectively. The peak surface-density of sources seen with a pixel size of 0.05 in sin b and 10° in L (section 8, Fig. 8) is 0.89 per sq degree for Coma and Virgo and 0.55 for A1367, compared with a mean surface-density of 0.39 per sq degree.

CONTRIBUTIONS TO THE LOCAL GRAVITATIONAL FIELD FROM BEYOND THE VIRGO SUPERCLUSTER

Amos Yahil
State University of New York
Stony Brook, NY 11794-2100, USA

In the last decade, the inter-relation between the density structure and velocity perturbations in the local Virgo Supercluster (VSC) has provided a major means of determining the cosmological density parameter Ω_o. The underlying dynamics, the so-called Virgocentric Flow Model (VFM), assumes that the relevant peculiar gravitational field in the VSC is dominated by the monopole term, due to material inside the VSC. Over the past year, however, this premise has been called into question, on both observational and theoretical grounds. It has become clear that the contributions to the local gravitational field from beyond the VSC cannot be neglected. The major new developments are summarized here. A more detailed review is given elsewhere (Yahil 1987).

In order to make the distinction between the internal and external contributions more concrete, imagine the demarcation of the VSC to be a spherical surface. The peculiar gravitational potential inside this volume, due to material outside it, can be written as a multipole expansion:

$$\phi = \Sigma \ a_{lm} Y_{lm} r^l \quad , \tag{1}$$

with a corresponding gravitational field

$$g = d + Q \cdot r + ... \tag{2}$$

The first term in the expansion is a homogeneous dipole field. (By Newton's theorem there is no monopole term inside the cavity due to material outside it.) The second term is the quadrupolar tidal field, with Q a symmetric and traceless tensor.

In linear perturbation theory the peculiar velocity u is proportional to g:

$$u = \tfrac{2}{3} H_o^{-1} \Omega_o^{-0.4} \ g = u_\mu + \sigma \cdot r + ... \tag{3}$$

The first (dipole) term is the constant bulk velocity of the VSC relative to the Robertson-Walker frame. It can not be measured within the VSC since it cancels for all relative velocities, and therefore needs to be measured relative to the microwave background radiation (MBR), or a sufficiently distant frame of

A. Hewitt et al. (eds.), Observational Cosmology, 247–250.
© *1987 by the IAU.*

galaxies. The quadrupolar tidal velocity field is measurable in principle, and a detection has, in fact, been claimed (Lilje, Yahil, and Jones 1986). Recently, a study of a large sample of elliptical galaxies with $v \lesssim 6000$ km s^{-1} (Dressler *et al.* 1986) has concluded that the velocity of the Local Group (LG) relative to that frame is different from the one relative to the MBR (Lubin, Epstein, and Smoot 1983; Fixsen, Cheng, and Wilkinson 1983). The new data have not yet been available for independent evaluation, but the plots presented show that the velocity field is turbulent. In fact, for $v \lesssim 3000$ km s^{-1}, it resembles the tidal one found by Lilje *et al.*

A complete dynamical picture, and a means of determining Ω_o, is possible only when the kinematic data are coupled to the density structure. The coefficients of the multipole expansion, eq. (1), are given by

$$a_{lm} = \frac{4\pi}{2l+1} \int \rho(\mathbf{r}) r^{-l-1} Y^{*}_{lm}(\Omega) \, d^3 r \tag{4}$$

The dipole term is simply the vector sum of the gravitational r^{-2} attractions of all the structure outside the VSC. The quadrupole term measures the r^{-3} tidal force.

Complete surveys are needed in order to measure the density of galaxies. The initial RSA catalog was used for the volume inside $|b| > 30°$ and $v \lesssim 4000$ km s^{-1} (Yahil, Sandage, and Tammann 1980; Yahil 1981). The CfA catalog extends to the greater depth of 8000 km s^{-1}, but is confined to $b > 40°$, and only limited coverage in the South Galactic Hemisphere (Davis *et al.* 1980; Davis and Huchra 1982). These studies were unable to find a dipole term in the direction of the MBR. It was therefore concluded that the structure responsible for the bulk velocity of the VSC either lay beyond the observed distances, or in the zone of avoidance, or both.

The advent of the IRAS catalog has made it possible to search further for the illusive density perturbations. The catalog is ideal for this purpose, because it is calibrated homogeneously over almost the entire sky, ranges in distance far beyond the limits of present redshift surveys, and is unaffected by extinction. Two independent teams quickly discovered a dipole anisotropy in the sky distribution of the IRAS galaxies, aligned within the errors with the velocity of the LG relative to the MBR (Yahil, Walker, and Rowan-Robinson 1986; Meiksin and Davis 1986). The interpretation has been that the density structure responsible for the bulk motion of the VSC has been identified.

Most of the IRAS galaxies do not yet have redshifts, but initial surveys have provided the infrared luminosity function (e.g., Lawrence *et al.* 1986). This can be used to work out the dipole moment of g (Yahil *et al.* 1986). When coupled via eq. (3) to the peculiar velocity, assumed to be equal to the velocity of the LG relative to the MBR, the estimate $\Omega_o \approx 1$ is obtained. This contradicts the previous observational "consensus", based on the VFM, the cosmic virial theorem, and N-body calculations, that $\Omega_o = 0.1$-0.2 (e.g., the review by Yahil 1984).

The theoretical disposition in favor of $\Omega_o = 1$, now augmented by the observational contradiction, has led to the proposal that galaxy formation is more efficient in regions that have higher densities, even though at the epoch of galaxy formation their overdensity may still have been very small (Kaiser 1984; Schaeffer and Silk 1985; Bardeen *et al.* 1986). As a result of this biased formation process, galaxies are no longer fair tracers of the total mass-energy density. Details of the biasing physics are still sketchy (e.g., Rees 1987).

Biasing is difficult to establish observationally, because mass can only be traced dynamically. It is possible, however, to check whether one type of galaxy has the same density structure as another one. It is well known that the frequencies of early and late type galaxies are functions of density (Dressler 1980), but this has not seemed to bias the determination of the largescale density structure in the VSC (Yahil *et al.* 1980). It is important to see whether a similar effect might result in different determinations of density from IR (spiral) and optical (spiral and elliptical) galaxies. A comprehensive comparison of the two is not yet possible, because of the lack of redshifts for the IRAS galaxies. However, in the small area of the sky surveyed by Lawrence *et al.* (1986), 90% of the IRAS galaxies with $S_{60} \geq 0.85$ and $v < 3000$ km s^{-1} are already included in the CfA catalog. (Of course the CfA catalog also contains many galaxies which are not IRAS galaxies, e.g., elliptical galaxies.) If this is true in all the overlap region between the IRAS and CfA surveys, then their density structures can be compared directly within the limited redshift range $v < 3000$ km s^{-1}. Such a comparison clearly shows the two density structures to be similar (Yahil 1986). It also shows that the mean cosmological density baseline defined by the IRAS galaxies is higher than that determined in the optical studies. A possible interpretation might be that the existing optical surveys do not extend deep enough, and the volume covered is not a fair sample of the universe. This possibility needs to be examined carefully, both by re-evaluating the existing density determinations, and by extending both the optical and IR redshift surveys.

A full understanding of the density structure of the IRAS galaxies can come only from a complete redshift catalog, which will enable analyses which are not possible with only fluxes and a luminosity function. First, the dipole gravitational field can be determined using individual distances for each galaxy. This will greatly increase the reliability of the result, and minimize biases. Secondly, one can evaluate higher moments of the gravitational field. The quadrupole moment, in particular, should be compared with the quadrupolar tidal velocity field (Lilje *et al.* 1986). Thirdly, other quantitative descriptions of the density structure around us can be sought, such as correlation functions, hole distribution, and continuous largescale structures. Fourthly, the density structure of IR selected galaxies may be compared with optical galaxies, to obtain some understanding of possible biased galaxy formation. Finally, biased galaxy formation can be tested using the principle of superposition, by determining the effect of introducing density-dependent weighting on the resultant gravitational field (Gunn **1986**).

A collaboration (Davis, Huchra, Strauss, Tonry, Yahil) is already underway to obtain a redshift catalog complete to $S_{60} \geq 2$ Jy for $|b| > 10^{\circ}$. The flux limit was chosen in order to make the project manageable, with the aim of completing it within two years. The broad luminosity function guarantees that the sample will extend to large distances. The eighty percentile of the redshift distribution will be at 10,000 km s^{-1}, but there will be adequate sampling even beyond that range. A British team (Efstathiou, Ellis, Frenk, Hewitt, Kaiser, Rowan-Robinson) is undertaking the complimentary approach of sparse sampling the catalog down to $S_{60} \geq 0.5$. Many of the problems posed above should be answered shortly

This research was supported in part by USDOE grant DE-AC02-80ER10719 at the State University of New York. The hospitality of the Research Institute for Fundamental Physics, Kyoto University, where this work was completed, is gratefully acknowledged.

References

Bardeen, J.M., Bond, J.R., Kaiser, N., and Szalay, A.S. 1986, *Astrophys. J.*, **304**, 15.

Davis, M., Tonry, J., Huchra, J., and Latham, D. 1980, *Astrophys. J. (Letters)*, **238**, L113.

Davis, M., and Huchra, J. 1982, *Astrophys. J.*, **254**, 437.

Dressler, A. 1980, *Astrophys. J.*, **236**, 351.

Dressler, A., Faber, S.M., Burstein, D., Davies, R., Lynden-Bell, D., Terlevich, R., and Wegner, G. 1986, *Astrophys. J.*, submitted.

Fixsen, D.J., Cheng, E.S., and Wilkinson, D.T. 1983, *Phys. Rev. Letters*, **50**, 620.

Gunn, J.E. 1987, in *Dark Matter in the Universe*, eds. G. Knapp and J. Kormendy (Dordrecht: Reidel).

Kaiser, N. 1984, *Astrophys. J. (Letters)*, **284**, L9.

Lilje, P.B., Yahil, A., and Jones, B.J.T. 1986, *Astrophys. J.*, **307**, 91.

Lawrence, A., Walker, D., Rowan-Robinson, M., Penston, M., and Leech, K. 1986, *Mon. Not. Roy. Astr. Soc.*, **219**, 687.

Lubin, P.M., Epstein, G.L., and Smoot, G.F. 1983, *Phys. Rev. Letters*, **50**, 616.

Meiksin, A., and Davis, M. 1986, *Astr. J.*, **91**, 191.

Rees, M. 1987, in *Nearly Normal Galaxies: From the Planck Time to the Present*, ed. S.M. Faber (Berlin: Springer). in press.

Schaeffer, R., and Silk, J. 1985, *Astrophys. J.*, **292**, 319.

Yahil, A. 1981, *Ann. N.Y. Acad. Sci.*, **375**, 169.

________. 1984, in *The Virgo Cluster of Galaxies*, ed O.G. Richter and B. Binggeli (Munich: ESO) pp. 359-373.

________. 1986, in *Galaxy Distances and Deviations from Universal Expansion*, eds. B.F. Madore and R.B. Tully (Dordrecht: Reidel) pp. 151-158.

________. 1987, in *Nearly Normal Galaxies: From the Planck Time to the Present*, ed. S.M. Faber (Berlin: Springer), in press.

Yahil, A., Walker, D., and Rowan-Robinson, M. 1986, *Astrophys. J. (Letters)*, **301**, L1.

Yahil, A., Sandage, A., and Tammann, G.A. 1980, *Astrophys. J.*, **242**, 448.

THE ANGULAR SIZE - REDSHIFT RELATION AS A COSMOLOGICAL TOOL

V.K.Kapahi
Tata Institute of Fundamental Research
Post Box 1234, Bangalore 560012
India

1. INTRODUCTION

The angular size - redshift (θ - z) relation can in principle be used to discriminate between world models because the angular size subtended by a rigid rod is quite a sensitive function of cosmology, specially at $z > 0.5$. The test is simpler to apply to objects for which a metric diameter is measured than to objects with isophotal diameters (Sandage 1961). It was first suggested by Hoyle (1958) at the Paris symposium on Radio Astronomy, that the separation between the two lobes of extragalactic radio sources such as Cyg-A, could be used for performing such a test. In an Einstein-de Sitter Universe sources like Cyg-A cannot have angular sizes < 15 arcsec (the minimum occuring at $z = 1.25$) whereas in the Steady State Universe their sizes should asymptotically approach a value near 4 arcsec at large redshifts. It was not until the early seventies that the test was actually applied to samples of radio quasars with redshifts of upto ~ 2 (Legg 1970; Miley 1971; Wardle & Miley 1974). The angular sizes were found to show a large scatter due to a wide distribution of physical sizes and the projection effects associated with the essentially linear radio structures. The upper envelope to the θ -values (which would be expected to show much less scatter) nevertheless appeared to fall off monotonically with increasing z, more or less like the Euclidean relation $\theta \propto z^{-1}$. The θ - z test thus appeared to be incompatible with the predictions of uniform world models in which the linear sizes of quasars are independent of epoch.

In the Friedmann models of the universe the straightforward interpretation of the θ - z plots is that linear sizes (ℓ) of quasars were systematically smaller at earlier epochs. If a simple epoch dependence of the form ℓ (z) $\propto \ell(z=0) (1+z)^{-n}$ is assumed, values of the evolution parameter, n, of 1 to 2 are necessary to fit the data. Such evolution can closely mimic an Euclidean relation between θ and z. The evolutionary interpretation was not unique, however, because flux density limited samples of radio sources show a strong correlation between radio luminosity (P) and redshift (the ubiquitous Malmquist bias), so that a possible inverse correlation between P and ℓ cannot be easily distinguished from size evolution (eg. Jackson 1973; Stannard & Neal 1977; Wardle & Potash 1977; Hooley et al. 1978; Wills 1979; Saikia

A. Hewitt et al. (eds.), Observational Cosmology, 251–266.
© *1987 by the IAU.*

& Kulkarni 1979; Masson 1980; Macklin 1982).

In this review I will discuss the observed θ - z relation for large samples of quasars and radio galaxies that are currently available, and the recent attempts to disentangle the dependence of linear sizes on radio luminosity and redshift. I will also briefly consider the present situation with regard to the angular size - flux density (θ - S) relation, whose interpretation has been somewhat controversial. The conclusion now seems inescapable that the maximum linear sizes of extragalactic radio sources have undergone fairly strong evolution with cosmic epoch.

I will not discuss here the application of the θ - z test to clusters of galaxies (Hickson 1977; Bruzual & Spinrad 1978) which also appears to be strongly affected by evolution of cluster sizes with epoch (see Hickson & Adams 1979 for a discussion).

2. THE θ - z RELATION FOR QUASARS AND RADIO GALAXIES

The θ - z plots for 296 quasars and 215 radio galaxies are shown separately in Figures 1 and 2 respectively. The plot for quasars includes essentially all those with known redshift that have a steep high frequency spectral index (generally 1.4 to 5 GHz; $\alpha > 0.5$; $S \propto \nu^{-\alpha}$) and for which the angular sizes have been measured with sufficient angular resolution with aperture synthesis telescopes. Our starting point was the compilation of quasar data by Veron-Cetty & Veron (1985). Spectral and structural information was compiled

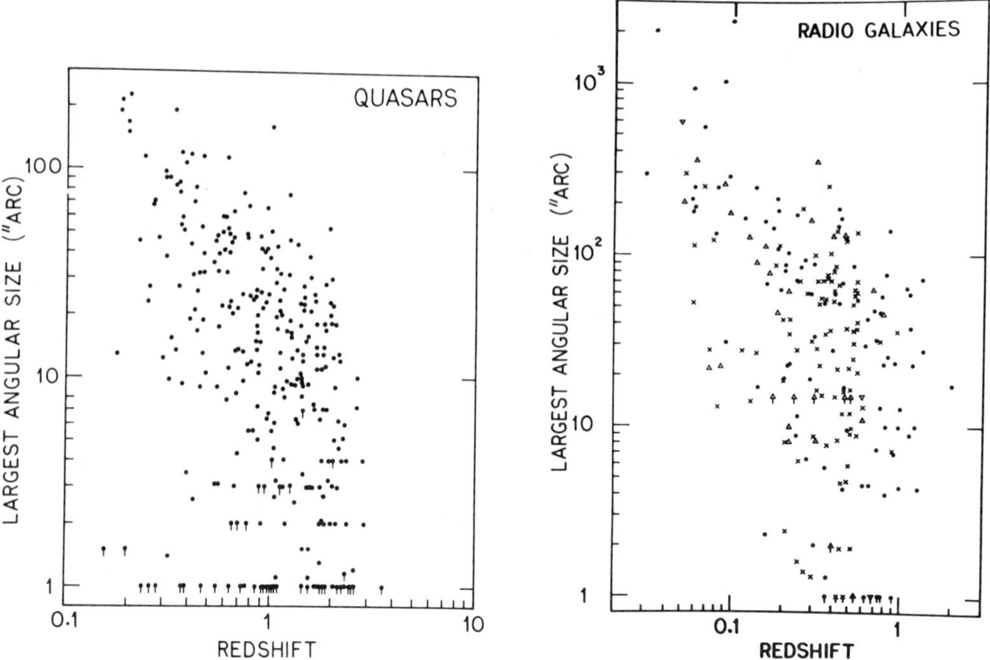

Figs. 1 & 2. The θ - z plots of quasars and radio galaxies respectively.

from a large number of references. As the number of radio galaxies with large measured redshifts is much smaller than quasars, the galaxy sample was put together as follows. We restricted the search to the following complete samples for which good optical identification and structural information is available; the 3CR sample (Laing et al. 1983); BDFL sample (Bridle et al. 1972); 2.7 GHz northern hemisphere Catalogue (Peacock & Wall 1981) and the GB/GB2 samples of radio galaxies (Machalski & Condon 1985). Spectroscopic redshifts are not available for most of the galaxies in the last sample; we have however used the redshifts estimated from their optical magnitudes by Machalski & Condon. We exclude galaxies with radio spectral indices <0.5 and those with estimated radio luminosities $P(408$ MHz$) < 10^{25}$ WHz^{-1} ster^{-1}(H_0 = 50 km s^{-1} Mpc^{-1}; q_0 = 1/2) because on the Fanaroff & Riley (1974) classification most such objects do not have the edge brightened structure typical of quasars.

As can be seen from Figures 1 and 2 there is a large scatter in θ values at all redshifts, but the upper envelope shows a monotonic decrease of θ with increasing redshift. The upper envelope is however not a good parameter for quantitative analysis (there is in fact the possibility that one or two objects that strongly influence the estimate of the upper envelope may not have been plotted correctly, either because their redshifts, estimated from just a couple of emission lines, may be grossly in error, or their angular sizes may have been overestimated due to the chance presence of an unrelated radio source in a direction close to their radio axes). We shall therefore use the median angular size (θ_m) as the more appropriate parameter to characterise the angular sizes. Although the samples we have used are not strictly complete, it is very unlikely that selection or instrumental effects would have introduced a significant systematic dependence of θ on z of the form apparent in Figures 1 and 2.

A word about the steep-spectrum compact (SSC) sources. Many of these are unresolved and appear in the plots mostly at high redshifts and are more common among quasars than radio galaxies, as is known from complete surveys at high frequencies (Kapahi 1981; Peacock & Wall 1982; Kapahi et al. 1986). Their angular sizes do not appear to form a smooth continuation of the distribution of sizes of the more extended sources (particularly apparent in Figures 1 and 2 at lower redshifts). Most of the SSC sources (generally < 20 kpc in extent) are known to show low frequency turnovers in their spectra due to synchrotron self absorption or thermal absorption effects (eg. van Breugel et al. 1984). As they may form a class quite distinct from the extended objects that we are essentially interested in for the θ - z tests, we shall exclude such objects (all those with linear sizes < 20 kpc) from consideration, but note that their inclusion only strengthens our conclusions regarding evolution in linear sizes.

The median angular sizes for radio galaxies and quasars in different ranges of z are shown in Figure 3. There appears to be a very good agreement in the values of θ_m for radio galaxies and quasars in the region of overlap in redshift. Also, the sizes of radio galaxies at low redshifts show a smooth continuity with those of quasars at higher redshifts. So whatever their other differences, radio galaxies and quasars appear to be indistinguishable as far as the linear sizes of their extended radio structure is concerned. It is also worth noting that the excellent continuity between radio galaxies and quasars, together with the decrease in angular sizes with redshift, provides

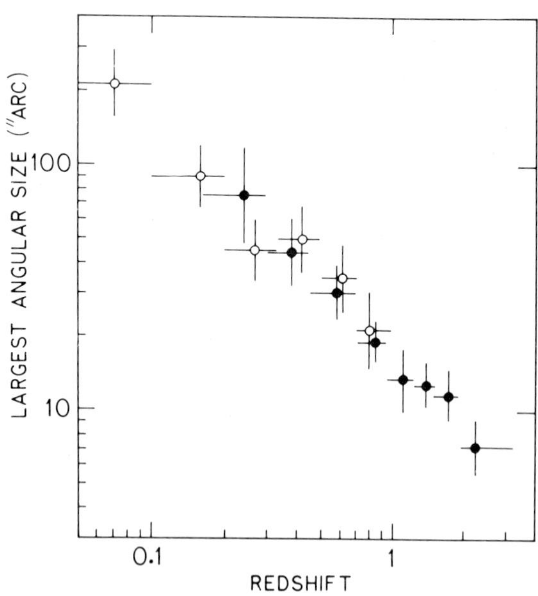

Fig. 3. The median angular size in different ranges of z for quasars (filled circles) and radio galaxies (unfilled circles).

strong support to the cosmological nature of redshifts. Any non-cosmological interpretation of quasar redshifts must explain these features of Figure 3.

Before comparing the observed θ - z relation with predictions of different world models we shall first examine the possible dependence of angular sizes on radio luminosity.

3. DEPENDENCE OF LINEAR SIZES ON RADIO LUMINOSITY

The strong correlation between z and P in the data for quasars due to the Malmquist bias is shown in Fig.4. A similar correlation exists for the radio galaxies. In order to investigate the epoch dependence of linear sizes it is therefore first necessary either to determine the luminosity dependence of linear sizes by comparing the sizes of samples with widely different luminosities at the same redshift or to determine the epoch dependence from samples of similar luminosity at different redshifts. With the availability of deep optical photometry on some samples, only recently has it become possible to carry out such a comparison with reasonable statistical significance (Kapahi 1985, 1986; Eales 1985; Kapahi & Kulkarni, in preparation). We will review this briefly.

Kapahi (1986) has considered four complete samples of radio galaxies selected from 1.4 GHz surveys to have similar luminosities at different redshifts as follows. i) BDFL; S(1.4 GHz) $\geq$ 2 Jy; 0.075 < z < 0.2, ii) GB/GB2; S(1.4) $\geq$ 0.55 Jy; 0.15 < z < 0.4, iii) GB/GB2; S(1.4) > 0.2 Jy; 0.25 < z < 0.6 and iv) LBDS; S(1.4) > 10 mJy; z $\gtrsim$ 0.8. The choice of these samples and redshift ranges is based on the availability of good angular size information and the need to have a reasonably large number of sources for statistics. The relati-

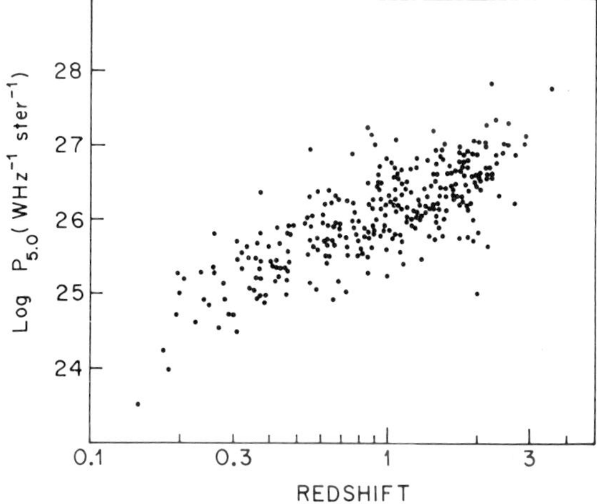

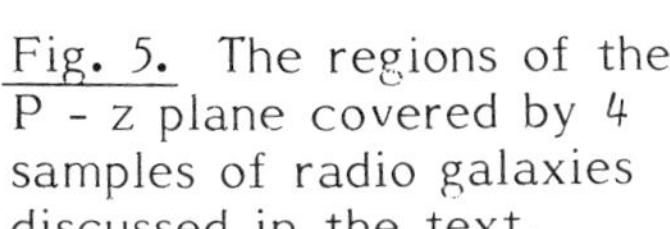

Fig. 4. The correlation between redshift and radio luminosity for quasars.

vely small range of P covered by the samples is shown in Fig.5. Note that spectroscopic redshifts are known for most of the sources only in the brightest BDFL sample and have been estimated (by Machalski & Condon 1985) from the optical magnitudes for the GB/GB2 samples. The Leiden Berkeley Deep Survey (LBDS) sample (Windhorst et al. 1985a) consists of all those sources in the mentioned flux density range that are either identified with galaxies of F magnitude > 22 or have no optical counterparts (implying F $\gtrsim$ 22.75) down to the plate limit (Windhorst et al. 1985b). Most of these are expected to be galaxies at z $\gtrsim$ 0.8. The angular sizes of these sources have now been determined with high angular resolution (~ 1.2 arcsec) using the VLA at 21 cm (Kapahi & Kulkarni, in preparation). The median angular sizes for these four samples are shown in Figure 7 (using crosses) for comparison with the other data for radio galaxies and quasars. Although there is little direct

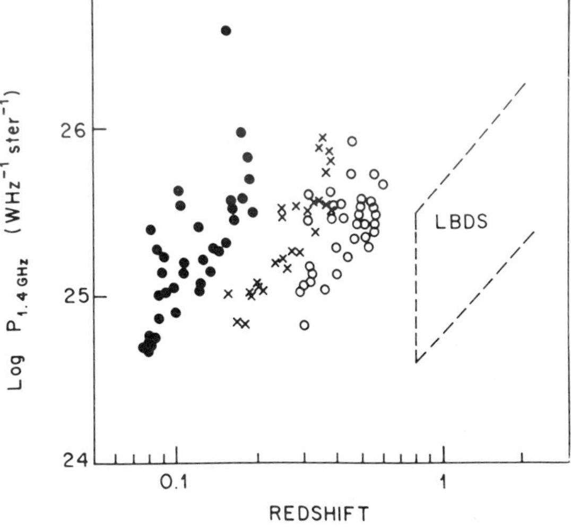

Fig. 5. The regions of the P - z plane covered by 4 samples of radio galaxies discussed in the text.

information about the redshifts of sources in the LBDS sample, it seems unlikely that the redshifts for radio galaxies extend much beyond ~ 2 or so (Windhorst 1986; Downes et al. 1986). We have assumed a median redshift of 1.25 for this sample but the conclusions are not likely to be seriously affected even if the actual z_m is considerably different ($0.8 \lesssim z \lesssim 2$). Note also that the values of θ_m for these samples should be directly comparable with the other data in Figure 7 since the number of flat spectrum ($\alpha' < 0.5$) sources is negligible (<10%; see Figure 6) in all the four samples.

The value of θ_m for the LBDS sample is seen in Fig.7 to be considerably smaller than for quasars in a comparable redshift range. As the quasars come mostly from much stronger radio surveys than the LBDS, they are of higher radio luminosity. This means that the angular sizes are likely to be directly correlated with radio luminosity rather than being smaller at higher luminosities. This can be seen more clearly from Figure 6 in which we have compared the linear size distributions in the GB/GB2 samples and in the LBDS sample with complete samples of radio galaxies from the strong source 3CR sample (Laing et al. 1983) in the same redshift ranges. The comparison is thus between sources of different luminosities in three different redshift ranges. The 3CR galaxies compared in Fig.6 with the LBDS sample consist of all those that are either known or estimated (from optical magnitudes) to have $z > 0.8$ together with the few sources that still do not have optical counterparts. If these two samples have a similar distribution of redshifts, their median luminosities must differ by about 2 orders of magnitude. There is clearly no evidence in Fig.6 of a $P - \ell$ anticorrelation at any redshift. The higher luminosity 3CR galaxies in general appear to have

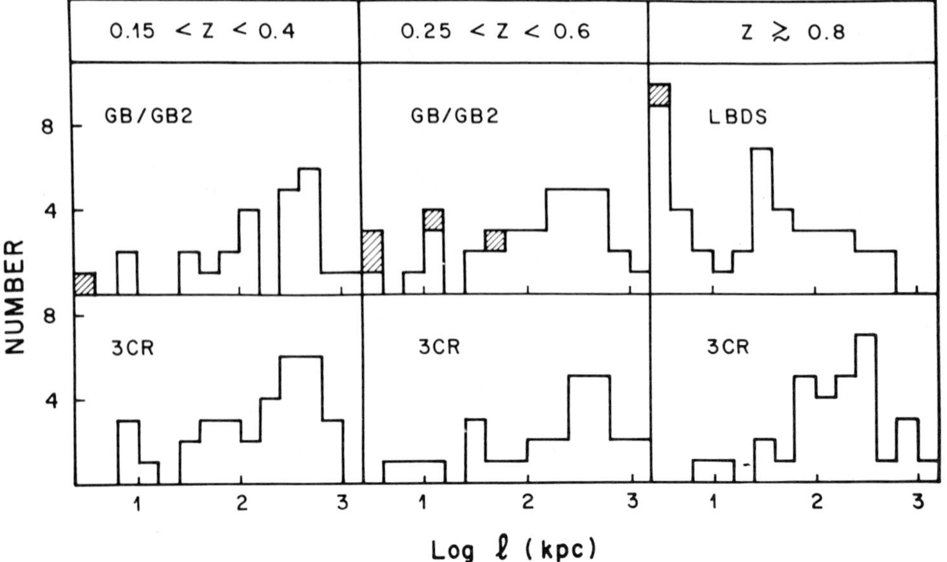

Fig. 6.   Comparison of linear size distributions for three samples of radio galaxies discussed in the text with 3CR sources. Hatched area indicates flat spectrum sources.

larger linear sizes, the difference in sizes being largest in the highest redshift samples. The distributions are consistent with a weak $P - \ell$ correlation of the form $\ell \propto P^{0.25 \pm 0.1}$, assuming the dependence to be similar at different redshifts. A more accurate determination of the relation between P and ℓ requires the determination of many more redshifts for weak radio samples. It is interesting to note that a weak correlation between P and ℓ is already known to exist (Gavazzi & Perola 1978; Ekers et al. 1981) for nearby (z < 0.1) galaxies of lower radio luminosity ($P(1.4) \lesssim 10^{25.5}$ WHz^{-1} ster $^{-1}$).

It is seen from Fig.6 that the LBDS sample has a much larger fraction of SSC sources than the corresponding 3CR sample. This is likely to be due partly to the difference in the frequencies of the two surveys, as such sources are well known to be much more common in higher frequency surveys. It is also possible that several of these sources are optically faint quasars because most of the SSC sources in strong high frequency surveys are known to be quasars at high redshifts. In fact over 50% of all sources in the Peacock & Wall (1981) 2.7 GHz catalogue with known or estimated redshifts $\gtrsim$ 0.8 are found to be of the SSC variety and over 60% of these are identified with quasars (Kapahi et al. 1986).

A comparison of the properties of a complete sample of 67 sources (mostly galaxies) from a 151 MHz 6C survey (with 2.2 < S (151 MHz)< 4.4 Jy) with those in the 3CR sample has been reported recently by Eales (1985). Spectroscopic redshifts are available for about 24% of the sample and have been estimated for the rest either from the optical magnitudes ($\sim$ 42%) or from the absence of optical identifications. It is found that on average the 6C sources have larger redshifts but similar radio luminosities compared to the 3CR sample. The angular sizes of the 6C sources are however considerably smaller than for 3CR. The data provide a statistically significant evidence for the dependence of linear sizes on epoch at constant luminosity (n = 1.45 ± 0.4 for q_0 = 0.5 and n = 1.1. ± 0.5 for q_0 = 0), but none for a correlation between size and luminosity at fixed redshift. A weak correlation between P and ℓ cannot however be ruled out by Eales' data as the difference in the mean luminosities of the samples compared is not very large.

It is harder to investigate the $P - \ell$ relation for quasars alone because of the difficulty of defining complete samples. Not only are identifications with quasars and spectroscopic observations incomplete, particularly in low flux density samples, the larger spread in optical luminosities makes it difficult to estimate redshifts from apparent magnitudes. We assume therefore in what follows that $P - \ell$ relation for quasars is similar to that for galaxies.

4. COMPARISON WITH COSMOLOGICAL MODELS AND SIZE EVOLUTION

The predictions of several cosmological models, including the standard Friedmann type models with q_0 = 0 & 0.5, and some 'non-standard' models are compared with the observed θ - z relation in Figure 7. The predicted relations have all been normalized to a constant value of θ_m at low redshifts. Apart from the static Euclidean relation which fits the data rather well, all the other models show different amounts of departure from the observed data. Although the 'tired light' model (a non expanding universe in which the redshift arises because photons loose energy with distance; e.g. La Violette 1986) cannot be ruled out by the present data, its prediction does appear to depart from observations

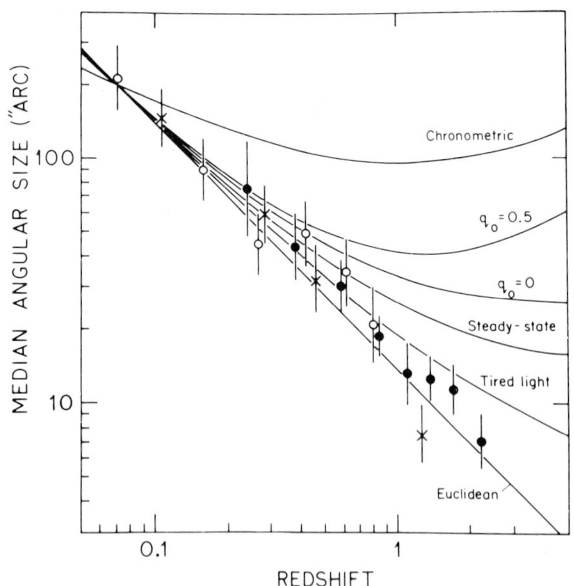

Fig 7. Predictions of several world models compared with the observed θ - z relation. θ_m values for the 4 samples of galaxies of similar luminosity are shown by crosses. The curves have all been normalized to $\theta_m = 200''$ arc at z = 0.07.

at the highest redshifts, particularly with respect to the data for galaxies of constant luminosity. Note that the model with $q_0 = 0$ has the steepest possible relation between θ and z among the 'standard' models. The prediction of the Chronometric model (Segal 1976) appears to be grossly discrepant with the observations.

It is clear from Fig.7 that in the Friedmann models the linear sizes must depend more and more strongly on cosmic epoch the larger the value of q_0. But since the nature and amount of evolution is far from being understood at the present state of our knowledge about the formation and evolution of radio sources, the θ - z relation cannot be used to determine a value of q_0. It is more meaningful therefore to determine the amount of evolution needed for a given value of q_0 in order to compare with possible models for the physical nature of the evolution.

The amount of evolution with epoch could in general depend on the radio luminosity as well. There is insufficient data however to consider this general case at present. We shall assume that the epoch dependence is the same over the range of luminosity covered by the data. We also assume the simplest mathematical form of epoch dependence, viz. $\ell(z) \propto (1+z)^{-n}$. This refers of course to the maximum linear sizes attained by the populations of radio sources at different epochs and not to the temporal evolution of individual sources because the life times of radio sources are likely to be much smaller than the Hubble time.

The expected θ - z relations for different values of n in the $q_0 = 0$ and 0.5 models are shown in Figures 8 and 9 respectively. Good fits to the data are obtained for values of n = 1.5 ± 0.5 for $q_0 = 0$ and n = 2 ± 0.5 for $q_0 = 0.5$. The fits shown in the figures correspond to local median linear

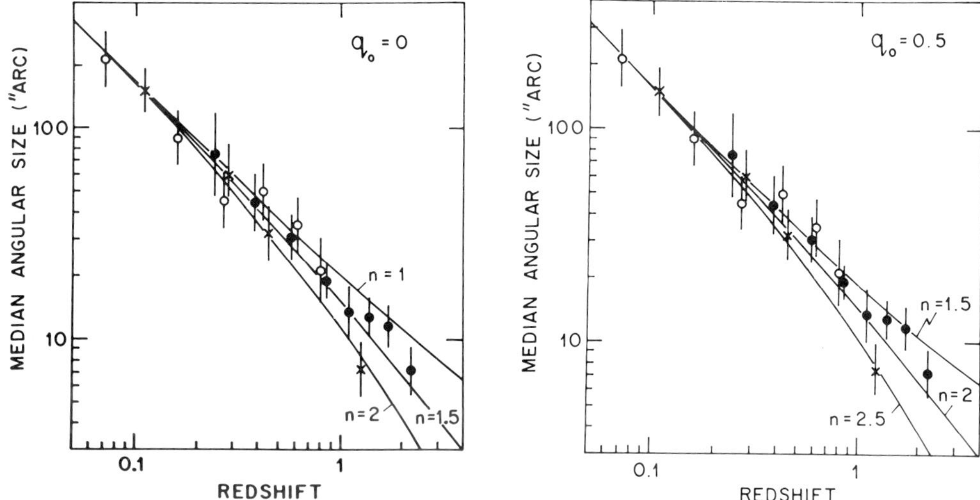

Figs. 8 & 9. Predicted relations for different amounts of size evolution for q_0 = 0 and 0.5 respectively.

sizes $\ell(z = 0)$ of about 450 to 475 kpc.

In deriving the evolution parameter we have ignored a possible inhomogeneous distribution of matter in the Universe which can have the effect of steepening the θ - z relation (Roeder 1975). The effect is, however, applicable only for $q_0 > 0$ and as noted by Katgert (1977) has a relatively small effect on the determination of n even for $q_0 = 1$ and the extreme assumption that the entire mass density of the Universe is condensed into galaxies.

5. θ - z RELATION FOR 'HOTSPOTS'.

The θ - z test has been applied also to the sizes of hotspots (typically $\lesssim 5$ kpc) generally found at the extremities of the outer lobes of extragalactic sources by estimating their sizes from observations of interplanetary scintillations (IPS) at 81 MHz (Hewish et al. 1974; Readhead & Hewish 1976). The results were interpreted to imply either an increase in the physical sizes of hotspots with increasing radio luminosity or a rather high value for q_0 (0.5 < q_0 < 2). Although hotspots sizes may be less likely to show strong projection effects, the determination of their sizes from IPS observations may suffer from effects of blending of different source components (eg. Swarup & Bhandari 1976). Several subsequent observations with higher angular resolution (eg. Kapahi & Schilizzi 1979) have infact found structure in the radio lobes on a scale considerably smaller than the IPS estimates. Synthesis maps of the hotspots in quasars do not appear to show any strong dependence of their physical sizes on redshift or radio luminosity (Swarup et al. 1984). High resolution (0.1 to 0.3 arcsec) maps of much larger samples covering a wide range in radio luminosity and redshift are however needed to investigate the θ - z relation for hotspots reliably.

6. THE θ - S RELATION

A useful variant of the θ - z test is the angular size - flux density (θ - S) relation, first investigated by Swarup (1975) and Kapahi (1975). This has the advantage that large unbiased samples of radio sources can be used without having to measure redshifts. It is necessary, however, in interpreting the relation to relate flux density to redshift through models for the epoch dependence of the radio luminosity function (RLF) that have been developed in order to fit the extensive source count data and other available informat- ion on redshifts, optical identifications, spectral index distributions etc. at various flux levels.

In the last ten years the θ - S relation has been confirmed and extended to lower flux levels (~ 0.1 Jy at 408 MHz and ~ 1 mJy at 1.4 GHz) using high resolution observations of many complete samples (eg. Ekers & Miley 1977; Downes et al. 1981; Kapahi & Subrahmanya 1982; Fielden et al. 1983; Windhorst et al. 1985a). For samples selected at or near 408 MHz (where the compact flat-spectrum sources make a negligible contribution at all flux levels) it is now well established that θ_m reaches a constant value of ~ 8 to 10 arcsec at low flux levels (see Fig.10). The observed relation can be compared with predictions (for an assumed world model) based on an evolving RLF and a local size function derived from the observed linear sizes of nearby sources in a well observed sample such as 3CR. On the assu- mption that the size function is independent of radio luminosity (there is no evidence of a P - ℓ relation among the nearby 3CR sources) the predicted values of θ_m are found to be considerably higher than observed at lower flux levels and can be made to fit the data only by invoking evolution in linear sizes (Kapahi 1975, 1977; Swarup & Subrahmanya 1977; Katgert 1977). Narlikar & Chitre (1977) have attempted to explain the observations without evolution but their claims have been questioned by Subrahmanya (1977).

An improvement over the above method (Downes et al. 1981) is to use each source in a well observed parent sample of bright sources individually to estimate the contribution of sources of that luminosity to different bins of flux density and angular size by a V/V_m type of analysis in which the known evolution in the RLF is taken into account. The method assumes that the type of sources found in weaker samples are well represented in the parent sample and any relation between P and ℓ that may be present in the parent sample is automatically taken into account. Predictions using this technique and the 3CR parent sample are also found to fit the data (including distributions of θ at different flux levels) only when size evolution is included in the computations (Kapahi & Subrahmanya 1982), contrary to the conclusions of Downes et al. (1981).

The choice of the parent sample in predicting θ - S relations has also created some controversy. It was pointed out by Downes (1982) that if week sources in low frequency surveys have large redshifts the observed emission would correspond to much higher emitted frequencies so that the observed θ_m values may be too low because of the presence of a large number of steep-spectrum compact (SSC) sources which constitute a substantial fraction of the sources found in high frequency surveys. Parent samples selected at high frequencies may therefore be more appropriate. Predictions based on the Peacock & Wall catalogue of strong sources at 2.7 GHz and the multi- frequency evolutionary models of the RLF investigated by Peacock & Gull

(1981) have led Downes (1982), Fielden et al. (1983) and Allington-Smith (1984) to conclude that in at least one or two RLF models the observed θ_m values at low flux densities at 408 MHz can be explained without need for any evolution in linear sizes. This has been examined recently by Kapahi et al. (1986), who point out that predictions based on high frequency surveys overestimate the contribution of SSC sources to low frequency samples at all flux levels because models of the evolving RLF do not take into account the low frequency curvature in the spectra of such sources. Because SSC sources occur at fairly high redshifts even in bright source samples at high frequencies the use of such samples to predict θ_m values in deep low frequency samples is inappropriate unless the latter have extremely high redshift of $\gtrsim 10$. Furthermore, recent high resolution observations of weak 5C sources do not show an abnormally large number of SSC sources (Goapl-Krishna et al. 1986). By correcting partially for the spectral curvature in SSC sources, new predictions based on parent samples selected at 1.4 and 2.7 GHz and a wide range of possible evolutionary schemes for the RLF suggested in the literature (Peacock & Gull 1981; Subrahmanya & Kapahi 1983; Condon 1984) it is found (Kapahi et al. 1986) that the θ - S data cannot be fitted without invoking evolution in linear sizes, even for $q_0 = 0$. Predictions based on the 2.7 GHz sample are reproduced in Fig.10. Values of n between about 1 and 2.5 are indicated depending on q_0, the parent sample

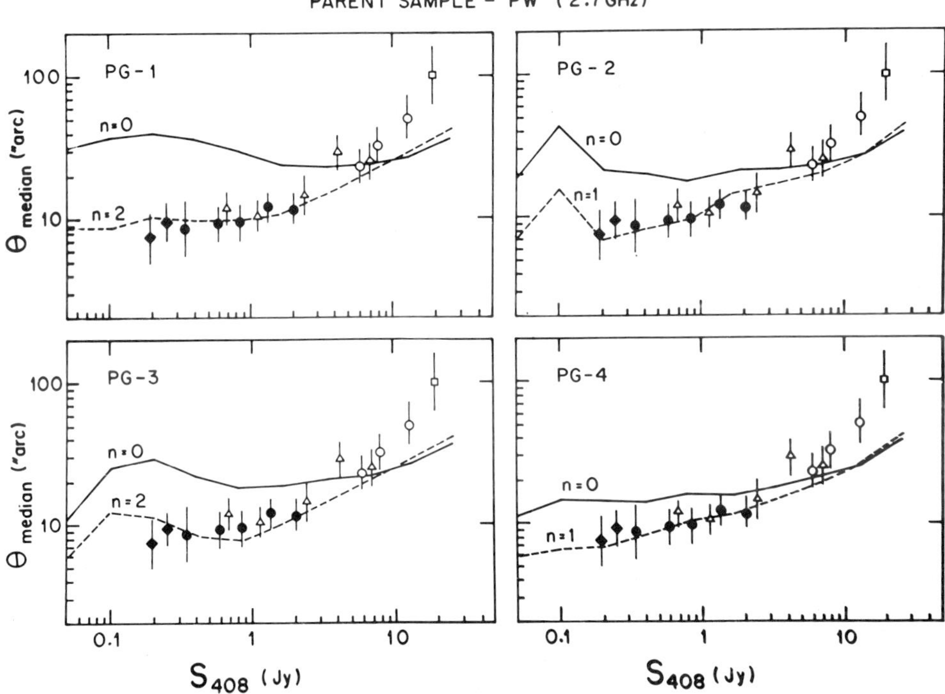

Fig. 10. Predicted θ - S relations for the 2.7 GHz parent sample and the four evolutionary RLF models of Peacock & Gull (1981), with and without size evolution. RLF models 1a & 2 are for $q_0 = 0.5$ and 3 & 4 for $q_0 = 0$.

and the evolutionary model. This is in good agreement with the evolution inferred from the θ - z relation discussed earlier.

From a high resolution deep VLA survey, Coleman & Condon (1985) have recently extended the θ - S relation to extremely low flux levels of ~0.1 mJy at 1400 MHz. They find that the value of θ_m drops again below ~1 mJy, reaching $\lesssim 3$ arcsec at the lowest flux levels. They interpret this as arising from the faintest sources residing in spiral or Seyfert galaxies unlike the stronger ones which are produced largely by elliptical galaxies. Using a simple scheme based on the BDFL parent sample and the evolutionary model of Condon (1984) they have fitted the entire θ - S relation without requiring size evolution. While the change in population from elliptical to spiral galaxies may indeed be responsible for the fall off in θ_m at very low flux levels, their conclusion about not requiring size evolution is, however, misleading because their scheme for predicting the θ - S relation is over simplified. Instead of treating each source in the BDFL sample individually in order to estimate its contribution to different flux density and angular size bins they have summed the number of BDFL sources in different ranges of linear size (as calculated from their redshifts and angular sizes) to obtain a distribution of intrinsic sizes assumed to be independent of radio luminosity. But since many sources in BDFL are seen at high redshifts, the effect of evolution in linear sizes is already present in the derived size distribution which is then used to predict values of θ_m at different flux levels.

7. WHAT CAUSES SIZE EVOLUTION ?

Two possible causes for the evolution of linear sizes have been in the literature for quite some time, namely the increase with epoch in the energy density of the microwave background radiation and in the density of an intergalactic medium.

The increased radiative losses at high redshifts due to inverse Compton scattering off the microwave background photons can snuff out radio sources (Rees & Setti 1968) before they have reached large physical sizes. Unless the magnetic fields in radio sources are much weaker than implied by equipartition arguements, however, inverse Comption losses do not appear to dominate over synchrotron losses even at fairly high redshifts (Scheuer 1977). The importance of the inverse Compton losses has been investigated by Allington-Smith (1984) and Eales (1985), who find that their effect on linear size evolution, for realistic models of radio sources, is rather small, although it may be important for some of the largest sources. The similar range in observed angular sizes at different redshifts (Figures 1 and 2), also argues against this effect being important in determining the distribution of linear sizes.

The ram pressure confinement of radio sources due to an intergalactic medium, whose cosmological density should vary as $(1+z)^3$, can slow down the expansion of the hot spots so that sources would attain smaller physical sizes at earlier epochs, assuming that the activity in their central engines lasts for a fixed period of time. But the existence of an intergalactic medium of sufficient density is uncertain. Two recent developments are however noteworthy. First, analysis of X-ray data from the Einstein Observatory has indicated (Forman et al. 1985) that hot gaseous coronae (extending to a radius of ~ 100 kpc in some cases) are perhaps an ubiquitous feature of bright early type galaxies, including isolated ellipticals. Second, the possibility that the 5-200 kev X-ray background arises from a hot intergalactic medium has been revived recently by Guilbert & Fabian (1986). The propagation of the twin beams of radio sources as they plough their way first through the hot coronae of the parent galaxies and then through the hot intergalactic medium has been investigated recently by Gopal-Krishna & Wiita (1986), by assuming that the two media are pressure matched at their interface which moves closer to the parent galaxy at higher redshifts due to the rising pressure of the intergalactic medium. This hastens the onset of rapid deceleration of the beam heads whose advance velocity becomes subsonic relative to the IGM at earlier times at higher redshifts, and the growth of linear size can subsequently be ignored inspite of continuing nuclear activity. Their model appears to provide a possible explanation for the strong evolution of linear sizes with epoch.

The evolution of galactic haloes with epoch is at present unknown. Even on the conservative assumption that galactic halos do not evolve significantly with epoch, the interface between them and the IGM would be expected to take place close to or within the confines of the parent galaxy at high redshifts ($z \gtrsim 2$). This may be responsible (Gopal-Krishna & Wiita 1986) for the subgalactic dimensions of a large number of quasars (SSC sources) and for the observed increase in distorted radio structures at high redshifts (Barthel 1986; see also Miley's contribution in the present symposium) which may be caused by a substantial heating of and the consequent bulk flows of the interstellar gas due to the discharge of large amounts of momentum flux by the twin beams in such small volumes (e.g. Lonsdale & Barthel 1986).

8. CONCLUSION

The angular size - redshift relation is important for observational cosmology as it provides valueable information on th evolution with epoch of the linear sizes of extragalactic radio sources and can be used for discriminating against some models of the Universe. The physics of the evolution of individual sources and the various factors responsible for the inferred evolution with epoch are, however, far from being understood. There is therefore little hope at this stage of using the relation as a test of the geometry of the Universe.

REFERENCES

Allington-Smith, J.R. 1984, M.N.R.A.S., 210, 611.
Barthel, P.D. 1986, in Quasars, IAU Symp.119, eds. G.Swarup & V.K.Kapahi, D.Reidel, Dordrecht, p.181.
Bridle, A.H., Davis, M.M., Fomalont, E.B. & Lequeux, J. 1972, A.J., 77, 405.
Bruzual, G.A. & Spinrad, H. 1978, Ap.J., 220, 1.
Coleman, P.H. & Condon, J.J. 1985, A.J., 90, 1431.
Condon, J.J. 1984, Ap.J., 287, 461.
Downes, A.J.B. 1982, in Extragalactic Radio Sources, IAU Symp. 97, eds. D.S.Heeschen & C.M.Wade, D.Reidel, Dordrecht, p.393.

Downes, A.J.B., Longair, M.S. & Perryman, M.A.C. 1981, M.N.R.A.S., **197**, 593.
Downes, A.J.B., Peacock, J.A., Savage, A. & Carrie, D.R. 1986, M.N.R.A.S., **218**, 31.
Eales, S.A. 1985, M.N.R.A.S., **217**, 179.
Ekers, R.D. & Miley, G.K. 1977, in Radio Astronomy and Cosmology, IAU Symp.74., ed. D.L.Jauncey, D.Reidel, Dordrecht, p.109.
Ekers, R.D., Fanti, R., Lari, C. & Parma, P. 1981, A & A., **101**, 194.
Fanaroff, B.L. & Riley, J.M. 1974, M.N.R.A.S., **167**, 31P.
Fielden, J., Downes, A.J.B., Allington-Smith, J.R., Benn, C.R., Longair, M.S. & Perryman, M.A.C. 1983, M.N.R.A.S., **204**, 289.
Forman, W., Jones, C. & Tucker, W. 1985, Ap.J., **293**, 102.
Gavazzi, G. & Perola, G.C. 1978, A & A., **66**, 407.
Gopal-Krishna, Saripalli, L., Saikia, D.J. & Sramek, R.A. 1986, in Quasars, IAU Symp.119, eds. G.Swarup & V.K.Kapahi, D.Reidel, Dordrecht, p.193.
Gopal-Krishna & Wiita, P.J. 1986, submitted to M.N.R.A.S.
Guilbert, P.W. & Fabian, A.C. 1986, M.N.R.A.S., **220**, 439.
Hewish, A., Readhead, A.C.S. & Duffett-Smith, P.J. 1974, Nature, **252**, 657.
Hickson, P. 1977, Ap.J., **217**, 964.
Hickson, P. & Adams, P.J. 1979, Ap.J., **234**, L91.
Hooley, T.A., Longair, M.S. & Riley, J.M. 1978, M.N.R.A.S., **182**, 127.
Hoyle, F. 1959, in Paris Symp. on Radio Astronomy, ed. R.N.Bracewell, Stanford Univ. Press, Stanford, p.529.
Jackson, J.C. 1973, M.N.R.A.S., **162**, 11P.
Kapahi, V.K. 1975, M.N.R.A.S., **172**, 513.
Kapahi, V.K. 1977, in Radio Astronomy and Cosmology, IAU Symp.74, ed. ed. D.L.Jauncey, D.Reidel, Dordrecht, p.119.
Kapahi, V.K. 1981, A. & A. Suppl., **43**, 381.
Kapahi, V.K. 1985, M.N.R.A.S., **214**, 19P.
Kapahi, V.K. 1986, in Highlights of Astronomy, Vol 7, ed. J.P.Swings, D.Reidel, Dordrecht, p.371.
Kapahi, V.K., Kulkarni, V.K. & Subrahmanya, C.R. 1986, submitted to J.Astrophys. Astron.
Kapahi, V.K. & Schilizzi, R.T. 1979, Nature, **277**, 610.
Kapahi, V.K. & Subrahmanya, C.R. 1982, in Extragalactic Radio Sources, IAU Symp. 97, eds. D.S.Heeschen & C.M.Wade, D.Reidel, Dordrecht, p.401.
Katgert, P. 1977, Ph.D. Thesis, University of Leiden
La Violette, P.A. 1986, Ap.J., **301**, 544.
Laing, R.A., Riley, J.M. & Longair, M.S. 1983, M.N.R.A.S., **204**, 151.
Legg, T.H. 1970, Nature, **226**, 65.
Lonsdale, C.J. & Barthel, P.D. 1986, Ap.J., **303**, 617.
Machalski, J. & Condon, J.J. 1985, A.J., **90**, 973.
Macklin, J.T. 1982, M.N.R.A.S., **199**, 1119.
Masson, C.R. 1980, Ap.J., **242**, 8.
Miley, G.K. 1971, M.N.R.A.S., **152**, 477.
Narlikar, J.V. & Chitre, S.M. 1977, M.N.R.A.S., **180**, 525 and **181**, 799.
Peacock, J.A. & Gull, S.F. 1981, M.N.R.A.S., **196**, 611.
Peacock, J.A. & Wall, J.V. 1981, M.N.R.A.S., **194**, 331.
Peacock, J.A. & Wall, J.V. 1982, M.N.R.A.S., **198**, 843.
Readhead, A.C.S. & Hewish, A. 1976, M.N.R.A.S., **176**, 571.
Rees, M.J. & Setti, G. 1968, Nature, **219**, 127.
Roeder, R.C. 1975, Nature, **255**, 124.

Saikia, D.J. & Kulkarni, V.K. 1979, M.N.R.A.S., **189**, 393

Sandage, A. 1961, Ap.J., 133, 355.

Scheuer, P.A.G. 1977, in Radio Astronomy & Cosmology, IAU Symp.74, ed. D.L.Jauncey, D.Reidel, Dordrecht, p.343.

Segal, I.E. 1976, Mathematical Cosmology and Extragalactic Astronomy, Academic Press, New York.

Stannard, D. & Neal, D.S. 1977, M.N.R.A.S., **179**, 719.

Subrahmanya, C.R. 1977, Ph.D. Thesis, University of Bombay.

Subrahmanya, C.R. & Kapahi, V.K. 1983, in Early Evolution of the Universe and Its Present Structure, IAU Symp. 104, eds. G.Abell & G.Chincarini, D.Reidel, Dordrecht, p.47.

Swarup, G. 1975, M.N.R.A.S., **172**, 501.

Swarup, G. & Bhandari, S.M. 1976, Astrophys. Lett., **17**, 31.

Swarup, G., Sinha, R.P. & Hilldrup, K. 1984, M.N.R.A.S., **208**, 813.

Swarup, G. & Subrahmanya, C.R. 1977, in Radio Astronomy & Cosmology, IAU Symp.74, ed. D.L.Jauncey, D.Reidel, Dordrecht, p.125.

van Breugel, W., Miley, G. & Heckman, T. 1984, A.J., **89**, 5.

Veron-Cetty, M.P. & Veron, P. 1985, E.S.O. Scientific Report No.4.

Wardle, J.F.C. & Miley, G.K. 1974, A. & A., **30**, 305.

Wardle, J.F.C. & Potash, R. 1977, Ann.N.Y.Acad.Sci., **302**, 605.

Wills, D. 1979, Ap.J.Suppl., **39**, 291.

Windhorst, R.A. 1986, in Highlights of Astronomy, Vol.7, ed. J.P.Swings, D.Reidel, Dordrecht, p.355.

Windhorst, R.A., van Heerde, G.M. & Katgert, P. 1985a, A. & A. Suppl., 58, 1.

Windhorst, R.A., Kron, R.G. & Koo, D.C. 1985b, A. & A. Suppl., **58**, 39.

DISCUSSION

NARLIKAR: I have two comments. (1) For the θ-z plot at high redshifts you took a tentative value of 1.5 for redshift since there were no identifications in your sample (z > 0.8). Taking this tentativeness into account the steady state model appears consistent with your θ-z plot. (2) The non-evolving luminosity function computed by DasGupta et al. was used by DasGupta and me together with weak-inverse correlation between size and power in your luminosity range log L between 24.5 to 26 to reproduce your θ-z plot. Whether such inverse correlation exists can be decided by looking at large samples of radio sources which are nearby (to avoid confusion with evolution).

KAPAHI: (1) Although there is indeed some uncertainty about the median redshift of the deep LBDS sample (z_m ~ 1 to 3?), this is not crucial to my argument. The observed median size for this sample cannot in fact be made to fill the steady-state prediction whatever the actual value of z_m is for this sample. (2) In this luminosity range there is just no evidence in the data for an inverse relation between size and power. The data in fact indicate a weak direct correlation between the two.

WALSH: Do you use the same evolution factor for compact steep-spectrum sources as for more extended ones? There is some evidence that they are qualitatively different: possibly the SSC sources have not burst out of their parent galaxy whereas the more extended ones have, so that they may evolve in different ways.

KAPAHI: Yes, models of the evolving RLF do not distinguish between steep-spectrum compact (SSC) sources and more extended ones as far as evolution is concerned. The data on quasars shows quite clearly that SSC sources are generally of high radio luminosity and are much more likely to be at high redshifts, which is consistent with strong evolution.

MACHALSKI: Do the medians shown contain sources larger than 20 Kpc, or all sources including steep spectrum compact sources? I am not sure that using medians is the best method to investigate any distribution since one can imagine <u>completely</u> different distributions with the same median. A non-parametric test would be much more appropriate for this purpose.

KAPAHI: Sources with $\ell < 20$ Kpc ($q_o = \frac{1}{2}$, $H_o = 50$) have indeed been excluded from the θ-z plots in estimating values of θ_{median}. But as I pointed out, including them would only strengthen the case for size evolution. The observed distributions of log θ are quite well behaved and median values are simple to estimate. The estimated value of θ_m and the errors are indeed non-parametric because no particular form for the distribution of θ is assumed.

RADIO BENDING AT HIGH REDSHIFTS - A NEW PROBE OF PROTOGALAXIES?

G. K. Miley
Space Telescope Science Institute, Baltimore, USA.
Affiliated to Astrophysics Div., Space Science Dept., ESA,
Noordwijk, Netherlands.
On leave from Sterrewacht, Postbus 9513, 2300 RA Leiden,
Netherlands.

ABSTRACT. Evidence is presented that radio sources are more bent at
larger redshifts. This effect is interpreted as due to interaction of
the radio source with an ambient circumgalactic medium which must then
have been clumpier at early epochs. It is suggested that the
increased clumpiness could be associated with the late stages of
galaxy formation and possibly also with CIV absorption line systems.

1. INTRODUCTION

Studies of the redshift dependance of radio source structure have been
part of the cosmological scene for almost two decades (Miley, 1967).
As Dr. Kapahi has pointed out, one of the main problems in making such
studies has been entangling effects due to the geometry of the
Universe from those due to physical evolution of the radio sources and
of their surroundings.

Here I shall use radio sources as a probe of their environment
and provide evidence that this environment is epoch-dependant. In
particular, I shall discuss the results of a project to map a large
sample of high-redshift quasars with the VLA (Barthel and Miley 1986),
which show that the high redshift objects are systematically smaller
and more distorted than similar quasars at low redshift.

Until now such statistical studies have been hampered severely at
high redshifts by (i) inadequate resolution and (ii) insufficient
sensitivity to map objects of low enough luminosity to be comparable
with those seen at low redshifts. Our observations with the VLA were
designed to overcome both these problems. The data consist of maps of
80 steep spectrum (extended) quasars with z > 1.5, observed with ~0.4"
resolution at λ6cm. We compared the structures with data taken from
the literature for similar sources having z < 1.5 and constructed
subsamples of each of these matched in radio luminosity.

A. Hewitt et al. (eds.), Observational Cosmology, 267–271.

2. RESULTS

2.1. Smaller Sizes.

We reach similar conclusions to Dr. Kapahi concerning the decrease of
median linear size as a function of redshift for currently reasonable
values of q_o and Ω. However, we are still unable to establish that
this effect is unambiguously one of redshift dependance rather than
luminosity dependance. As Dr. Kapahi has stated, the dependance of
size on redshift can be naturally interpreted as due to interaction of
the radio sources with a denser confining medium at early epochs
(Miley,1971; Wardle and Miley, 1972).

2.2. Bending.

The most striking and important new result from our data concerns the
appearance of the radio morphologies. Whereas at small redshifts,

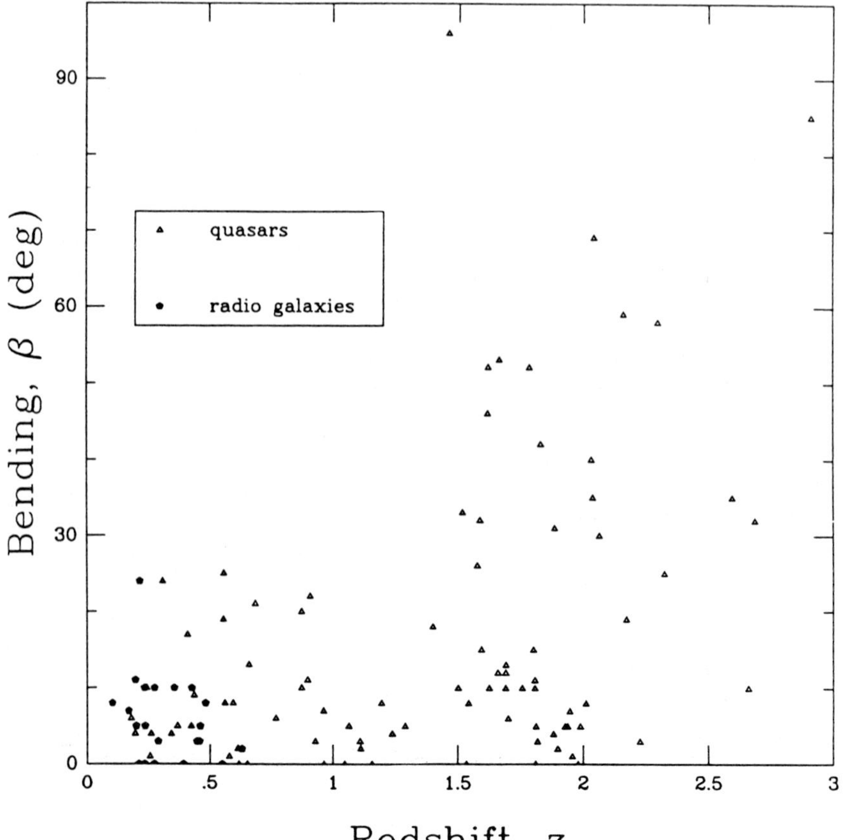

Figure 1. Bending of quasar radio sources as a fuction of redshift.
$\beta = 0$ refers to a straight source. All sources have steep spectra
($\alpha < -0.6 : S \propto \nu^{\alpha}$) and high luminosity ($P_{1.4GHz} > 10^{27}$ W Hz^{-1}).

high luminosity radio sources associated with quasars typically have
linear structures resembling that of Cygnus A, there appears to be a
significant excess of bent sources at high redshifts. We have
partially quantified this effect using a "bending angle", β, defined
using the source extremities and the nucleus. For linear (straight)
sources, $\beta = 0$. About 50 high-redshift sources were resolved
sufficiently to allow β to be estimated. Figure 1 shows β as a
function of redshift. There is a striking increase in the curvature
of the sources with increasing redshift. For $z < 1.4$, no objects have
$\beta > 30°$, whereas 30% of the high redshift sources have $\beta > 30°$.

3. DISCUSSION

Our results that high-redshift radio sources are both crookeder and
smaller than lower redshift sources can be best understood as caused
by interaction with a denser and clumpier ambient medium at earlier
epochs. At small redshifts we have a direct demonstration that such
interaction can cause bending in radio sources. There are now several
cases where bends in radio sources are observed to occur close to
extranuclear emission-line gas (e.g. Miley 1984).
        Perhaps the most dramatic example is 3C277.3 (Coma A) at $z = 0.09$
(van Breugel et al 1984) where a radio jet is observed to bend and
decollimate close to a bright optical-line knot. Collision with the
line-emitting gas cloud appears to cause the bending. For 3C277.3
simple Newtonian arguments show that collision with a mass of
7×10^5 $M_\odot$ would deflect the jet (assuming minimum energy), whereas
the observed mass of the ionized gas cloud is $\sim 2 \times 10^6$ $M_\odot$.
Extrapolating to the larger luminosities and redshifts of typical
sources in our sample, similar arguments require gas clouds of
$>10^8$ $M_\odot$ to produce corresponding deflections.
        It is interesting to enquire how many preexisting clouds would be
needed to intercept a typical jet, thereby explaining the observed
statistics.

$$\begin{array}{c} \text{No. of Clouds} \\ \text{per Galaxy} \end{array} \sim \left(\frac{4\pi}{\text{Radio Opening Angle}} \right) (\text{Observed Fraction Bent})$$

This argument would imply typically a few hundred of these 10^8 M
clouds per galaxy and a total mass per galaxy of $>5 \times 10^{10}$ $M_\odot$.
        As Dr. Silk has described, current models of galaxy formation
invoke collapse of the galaxy nucleus before the rest of the galaxy
(Silk and Norman 1979). Since the radio bending takes place typically
within about several tens of kiloparsecs of the QSO, it seems
reasonable to attribute the increased bending at high redshift as
being caused by interaction with a clumpy protogalactic disk. If this
interpretation is correct, radio sources provide a unique new probe of
the late stages of galaxy formation. Also, significant clumpiness
would have to persist in the disks of protogalaxies until an epoch
corresponding to about $z = 1.5$.
        A 10^8 $M_\odot$ cloud with a size comparable to the observed emission-
line knot in 3C277.3 (~ 3 kpc) would have a column density of

10^{21} cm^{-2}. Since this is comparable with the column densities
observed from absorption line systems in QSOs, it is tempting to
associate the clouds responsible for bending the radio sources with
quasar absorption lines. A possible culprit may be provided by the
CIV systems, which are reported to occur excessively frequently with
redshifts close to the emission redshifts of QSOs (Weymann et al 1979,
Foltz et al 1986).

4. FUTURE PROSPECTS

We are at present pursuing several follow-up projects related to this
work:

First, we are mapping several of the more bizarre sources at
higher resolution with the VLA (Barthel, Lonsdale, Miley). There is
the possibility that a study of the bending as a function of distance
from the nucleus will provide information about the mass distributions
in protogalactic disks. Also, the statistics of polarization
distributions will be another tool for such an investigation.

Secondly, a comparison of the statistics of absorption lines
between QSOs having straight radio sources and those having crooked
ones is being made with the 200" telescope at Mount Palomar (Barthel,
Miley, Tytler). A significant difference would be unambiguous proof
that there is a connection between gas clouds that bend the radio
sources and those responsible for the absorption.

Thirdly, we are engaged in a search for optical emission lines
occurring close to the radio bends (Barthel, Heckman, Macchetto,
Miley).

I wish to thank Drs. P. Shaver and J. Webb for useful discussions,
and my various collaborators, particularly Dr. P. Barthel. It is also
a pleasure to thank Mr. D. Golombek for drawing up the figure.

REFERENCES

Barthel, P. and Miley, G. K.,1986, in preparation.
Foltz, C. B., Weymann, R. J., Peterson, B. M., Sun, L., Malkan, M. A.,
     Chafee Jr., F. H., 1986, preprint.
van Breugel, W. J. M., Miley, G. K., Heckman, T. M., Butcher, H. R.,
     and Bridle, A., 1984, Ap. J., **290**, 496.
Miley, G. K., 1967, Nature, **218**, 933.
Miley, G. K., 1971, Mon. Not. Roy. Astro. Soc., **152**, 477.
Miley, G. K., 1984, Proc. ESA Workshop on Quasat, ESA publ. p. 131.
Silk, J. and Norman, C. 1979, Ap. J., **234**, 86.
Wardle, J. F. C. and Miley, G. K., 1974, Astron. and Astrophys., **30**,
     305.
Weymann, R. J., Williams, R. E., Peterson, B. M., and Turnshek, D. A.,
     1979, Ap. J., **234**, 33.

DISCUSSION

WAMPLER: You find a sudden "turn on" of sharp bending at z ≈ 1.5.
Is this consistent with the observed gradual departure of the largest
angular size from the expected cosmological models?

MILEY: Yes, there does appear to be a sudden "turn on" in the bending.
However, even above z = 1.4 only about 30% of the sources are bent by
more than 30°. In view of the large dispersion in linear sizes, the
resultant effect on the LAS - z relation is not great.

TURNER: It might be worth noting that the gravitational field of an
intervening object could distort and bend the image of a distant radio
source. The cross section for such a distortion is substantially
larger than for multiple imaging. This process might also play some
role in explaining the effect you report.

MILEY: Yes it might be important in a few cases but in general the
strange morphologies observed would be difficult to explain as the
effect of gravitational lenses.

SWARUP: Do you exclude flat spectrum sources for which relativistic
beaming can enhance apparent bending?

MILEY: Yes. The sources in the sample were all selected to have steep
spectra.

CHAPTER IV

THE LARGE SCALE DISTRIBUTION OF GALAXIES

THE DISTRIBUTION OF BRIGHT GALAXIES

G. Chincarini (1), G. Vettolani (2)
(1) Osservatorio di Brera, Milano, Italy
(2) Istituto di Radioastronomia CNR, Bologna, Italy

ABSTRACT   Following a brief review of the early developments in the
field, we discuss a few new advances in the distribution of bright
galaxies which occured after the extensive review paper by Oort (1983)
We show that the large scale structure, shells, filaments and/or sheets,
may bias the determination of the velocity dispersion in clusters of
galaxies and that the boundaries of the voids may often be biassed by
the clusters and groups velocity dispersion.

Of interest are the very large structures selected on catalogs of
clusters of galaxies. The "Local Structure" claimed by Tully seems to
be somewhat flattened and about parallel to the plane of the Local
Supercluster.   If the structure is real the alignement is relevant in
relation to the physical mechanisms at work at the time of formation.
Noticeable progress has been done in the measurement of the large scale
velocity field. Large scale motions may somewhat bias the study of the
topology of the Universe.

Relevant work has be done on the shape of the boundaries of the voids
and observational work is progressing to detect faint galaxies in voids
and determine their charactestics. This is important also in relation
to the theory of biassed galaxy formation.   To better focus the
observational problem and eventually related biasses, we give some
statistics on dwarf galaxies in the Virgo cluster.

1.   INTRODUCTION

The study of our environment, the Cosmos, dates back to the time   the
human being took consciousness of himself and of its surroundings.   The
increase of scientifical knowledge and technology expanded, during the
centuries of our evolution, the orizon of the physical world.

In most modern times and after the theoretical work on cosmological
models it became clear that the knowledge of the distribution of
galaxies, taken by most scientists as a mark of the distribution of
matter in the Universe, would let us understand not only the.present
status of the Universe, but also give us some observational data to be
used as test for models about the origin and evolution of the Universe
as a whole.

275

In the last few years a lot has been written on the large scale structures and varius review papers have been published, e.g. among others Chincarini (1982) and the comprehensive review by Oort (1983). We will therefore only schetch some of the early development and focus our attention on some of the recent work which we believe particularly important.

Sandage (these proceedings) mentioned that in the data of Hubble (1936) we already have the notion of clustering. Shapley and his collaborators (e.g. Shapley and Ames, 1932; Shapley, 1935) , however, pioneered the two fundamental aspects of the large scale distribution of galaxies. They called attention to the existence of a local large scale structure, the Local Supercluster, and detected other distant structures

Later and in a different form Abell (1958) and Zwicky (e.g. Zwicky, 1957) called attention on other agglomerates and it became fairly clear that not only the distribution of galaxies is clumpy but that also the distribution of clusters of galaxies is not random (Abell 1961).

De Vaucouleurs (1958) is probably the one who had an early understanding on the significance of Shapley's work on the local distribution of galaxies and devoted important years of research in understanding this structure.

Shane and Wirtanen (1967) undertook a monumental observational work in a very accurate survey on the distribution of galaxies while Neymann and Scott (see e.g. their 1959 review) developed foundamental statistical tools to allow their interpretation.

The discovery of the microwave background (Penzias and Wilson 1965 and Dicke et al, 1965) gave confidence to a theoretical framework around which to develop numerical models and marked the birthday of the observational cosmology in all its branches and complexity.

In the early seventies Peebles and collaborators (see Peebles 1980) were heavily involved in developing the autocorrelation function machinery and in using it to understand the large scale clustering using two dimensional catalogs. Earlier a pioneer analysis of the Shane-Wirtanen catalog was carried out by Totsuji and Kihara (1969).

In the same period it became feasible, thanks to the new image intensifiers and availability of telescopes, to observe a fair number of redshifts of galaxies in a reasonable amount of time. It is worth stressing that good telescopes in good sites and with advanced instrumentation always allow good progress in research. This may be important for those countries which plan a growth in observational Astronomy.

The availability of the redshifts opened the way to the 3-dimensional studies of the distribution of galaxies.

Chincarini and Rood (1975) in the attempt to determine the size of the Coma cluster, following also discussion and suggestions given by Zwicky, realized that galaxies at the redshift of Coma were still present at very large angular distances from the cluster itself. Chincarini and Martins (1975) in the attempt to explain the discrepant redshift in the Seyfert Sextet realized that the distribution of galaxies in the redshift space was not homogeneous. Redshifts were indeed segregate. In other words regions populated by galaxies were separated by regions empty of galaxies.

The large scale clustering found in the redshift surveys was found to be in very good agreement with the picture that Peebles and collaborators were deriving by measuring the autocorrelation function of galaxies using two dimensional distribution of galaxies as listed in various catalogues: Shane and Wirtanen, Zwicky and the Jagellonian field (see Peebles 1980).

It is in this period and following also the work of Gregory et al (1981) that it became clear that in the Universe we had large scale structures and large regions void of galaxies. Independently this line of research was persued by Einasto and his collaborators (e.g. Einasto et al, 1980)). Fundamental in this context has been the theoretical work developed by Zeldovich and his collaborators (see e.g. Zeldovich, 1978).

In the early eighties enough observational and theoretical work had been done to allow an extensive meeting on this and related topics, IAU Symposium 104. The coupling between the very early Universe, related high energy physics and the observed distribution of matter was already in full development

The distribution of galaxies is not an isolated subject. It is related to the study of perturbations in the density and velocity field, to the formation and evolution of structures and galaxies, to the understanding of cluster membership, to phenomena of interaction between galaxies and the intergalactic gas, and deeply connected to the problematic of the missing mass (and its distribution).

In this review we will limit ourselves to the disribution of galaxies and refer for the rest to the various reviews and specialized papers presented at the present symposium. While M. Geller will discuss the recent results of the Cfa survey, we remind that in addition to the other large survey carried out at Arecibo, work is progressing in Brasil and South Africa. The work done in Soviet Union will be illustrated by Karachentsev.

2. SUPERCLUSTERING

The basic assumption here is that, in first approximation, the redshift is a measure of distance. We also refer to a large scale structure or a supercluster meaning an agglomerate of galaxies or systems which show a certain connection, and detected either by a percolation algorithm or visual inspection.

These assumptions are reasonable, however they also reflect the limits of our findings. Large scale motions would distort the topology (large peculiar velocities in clusters are easily recognizable) and the size and the form of a structure will depend on the length of the percolation vector or equivalently the density contrast over the mean. The volume of the sample itself will limit the size of the structures which can be detected.

Finally, as it has been often mentioned, the word "supercluster" means different things for different people and the word "filament" is, at times, improperly used.

The region of Coma has always been one of the most studied regions of
the sky. It is of some interest, therefore, to compare the distribution
as published by Chincarini and Rood (1975) about 10 years ago, Figure 1,
with a compilation made by Gavazzi (1986), Figure 2. In addition to the
published redshifts, in the plot are included about 110 new redshifts
obtained at Arecibo. The richness of details, structures and clustering
and the improvement over the years is remarkable.

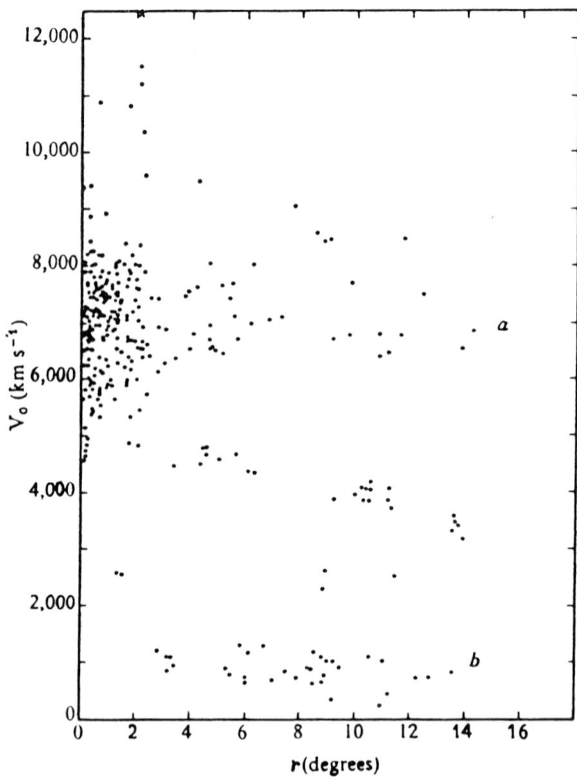

Figure 1. The distribution of galaxies in the Coma region,
Chincarini and Rood (1975).

In this contest we show in Figure 3 the velocity histogram prepared
about ten years ago according to the compilation of redshifts in Coma by
Gregory and Thompson (1978). Statistically we cannot distinguish
between 1 or 2 gaussian fittings. Gavazzi (1986) independently finds
the same thing , Figure 4. The two gaussian interpretation has now an
higher significance, however not so much because of statistics, but
mainly because of the fact that part of a filament (or pseudo-shell) may

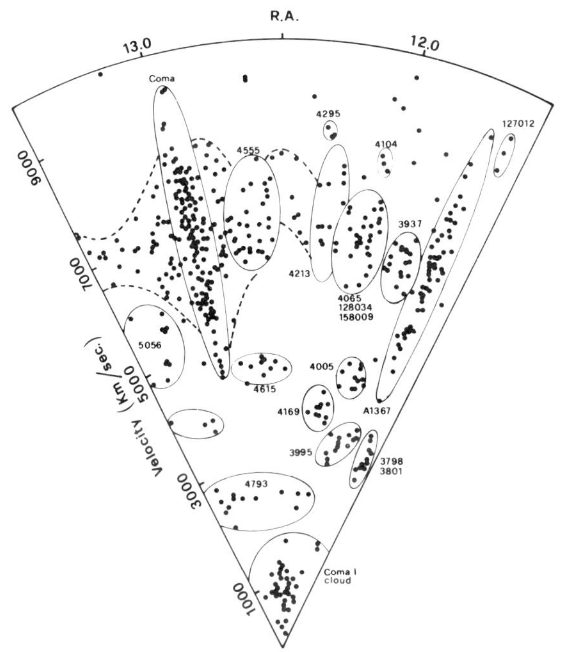

Figure 2.   The distribution of   galaxies   in   Coma   as   prepared   by
Gavazzi (1986).   In   the   plot we have 570 redshifts most of which are
from the literature.   110 redshifts from Arecibo are unpublished.

be   located in front of the cluster proper, Figure 5 left.   The filament
is visible, with various extension,   also   in   the   deep   survey   by   de
Lapparent et al (1986), Figure 6.   Such an interpretation would decrease
the velocity dispersion of the Coma cluster of about a factor   1.43   and
the mass of about a factor 2.
Comparison of Figure 5 left with Figure 5 center and right shows that
the   void   boundaries so sharply defined in the first are poorly visible
in the other two figures.   The western wall   is   well   defined   only   in
Figure   5.   left   because of   the compression in declination and it is
largely due to the effect of the   velocity   dispersion   in   the   cluster
A1367.   It is a flag of caution.

The well observed structures in the Perseus-Pisces region (Giovanelli
et   al,   1986)   and   Hydra-Centaurus   (da   Costa et al 1986a, 1986b) are
displayed in Figures 7, 8 and 9.
Figure 8 has been obtained using a percolation algorithm on   the   two
dimensional   distribution   of   galaxies   of   the   ESO-Uppsala   Catalogue
(Lauberts 1982).   The Hydra structure seems, unless we go   to   very   low

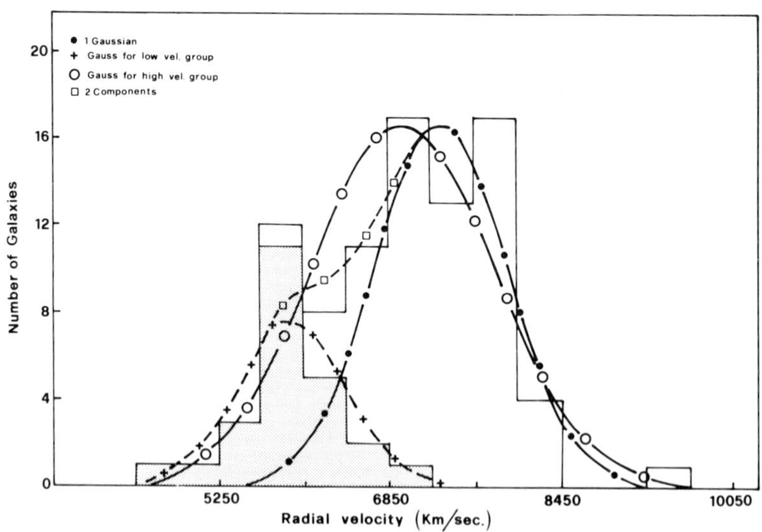

Figure 3.   Histogram of the redshifts for the Coma cluster as plotted in 1979 (according to the data listed in Gregory and Thompson 1978)

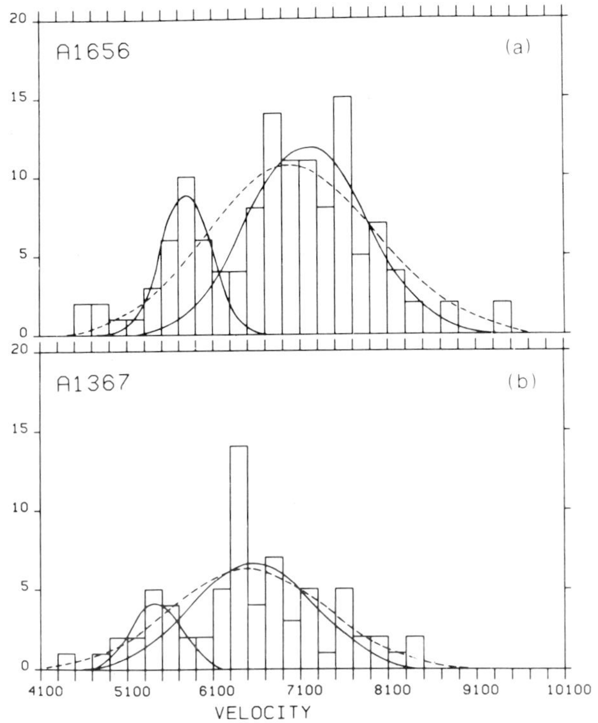

Figure 4. a) Redshifts histogram (128 galaxies) of a 4 square
digrees region centered on the Coma cluster.   b)   the same for A1367.

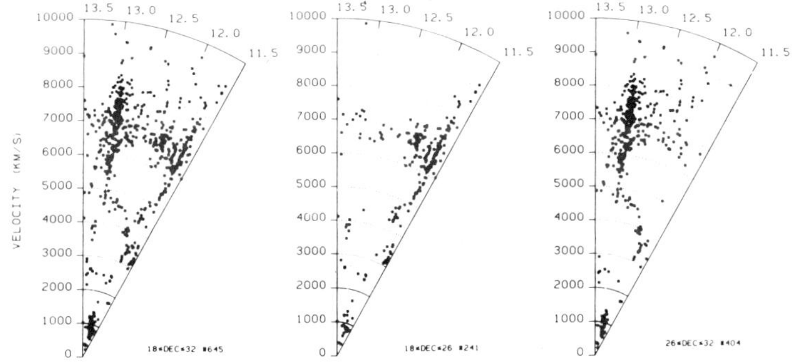

Figure 5. Left: wedge diagram of 465 galaxies in the region of
Coma.   Note   the   filament   at   about   V = 5800   Km/sec forming the
boundary of a void and located in front of the cluster.   Center   and
right: the two different cuts in declination show that the structure of
the void is irregular and/or that the cluster velocity dispersion
(A1367) may simulate in part the void boundary.

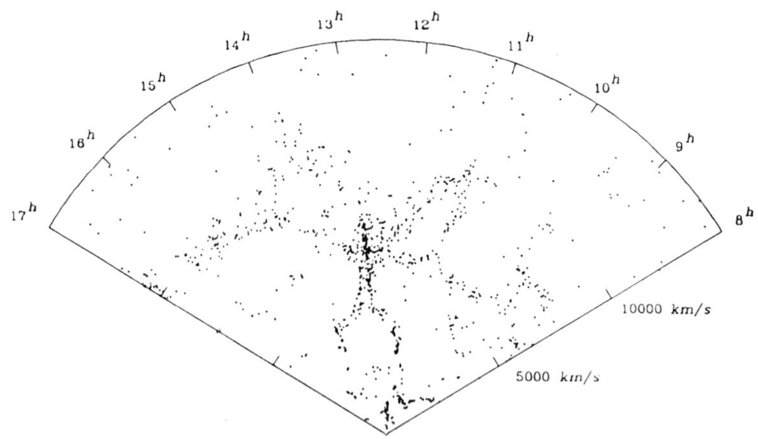

Figure 6. The distribution of galaxies in the region of Coma between 26:30 and 32:30 degrees in declination, according to the data of de Lapparent et al (1986).

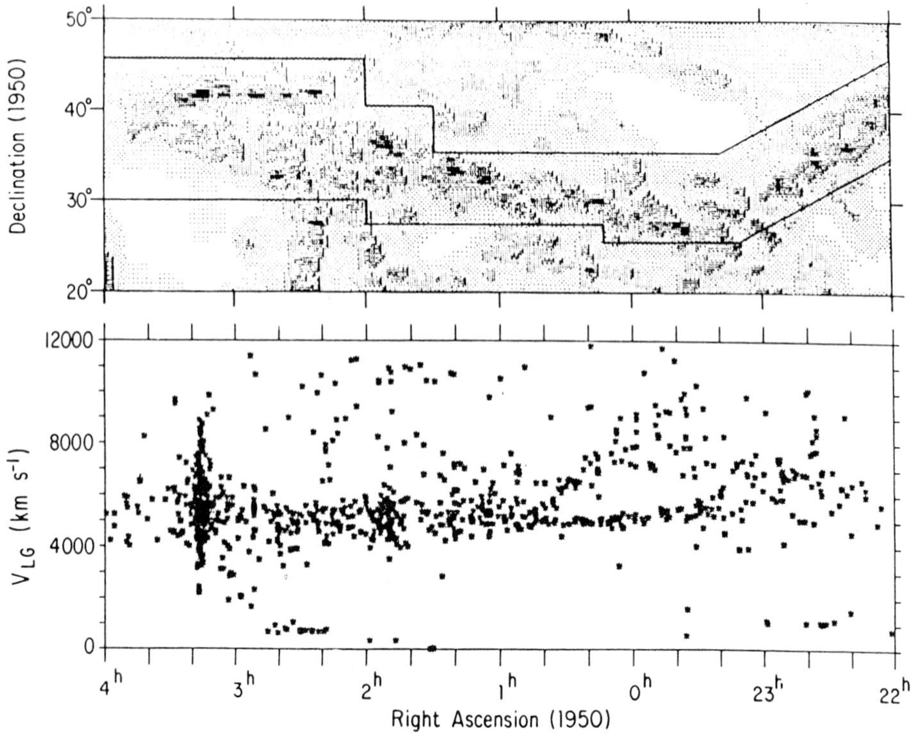

Figure 7. The Perseus-Pisces supercluster. Top: the density distribution of galaxies as seen projected on the celestial sphere. Bottom: The distribution of galaxies in depth (redshift).

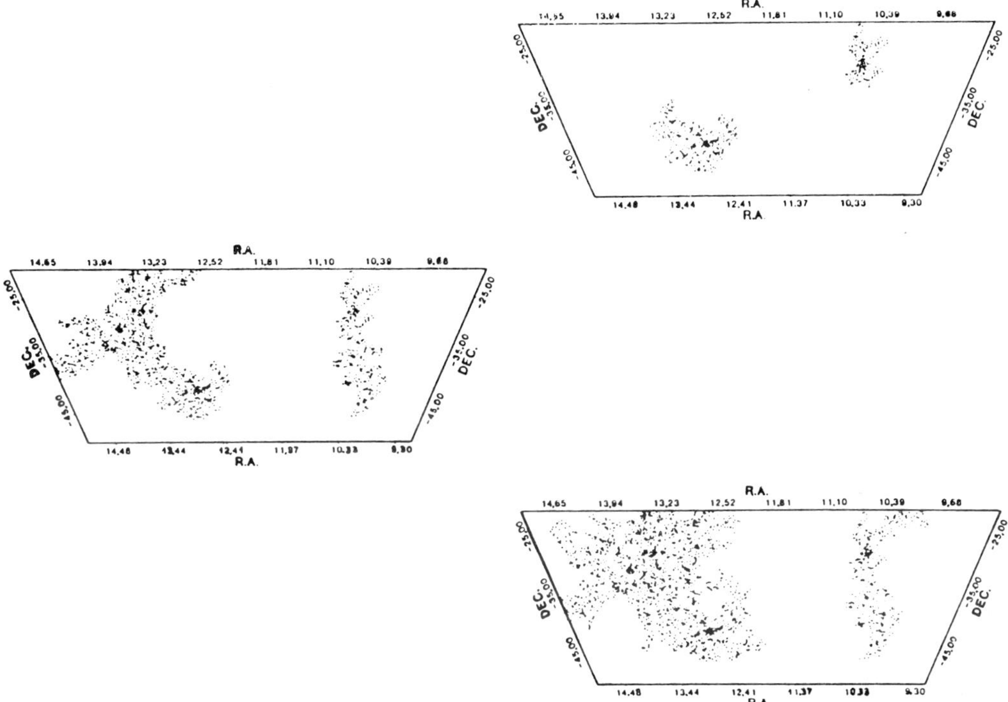

Figure 8. The Hydra-Centaurus structure as a function of the percolation vector. The percolation vector increases from the figure at the top to the figure at the bottom right (0.77, 0.80, 0.89 degrees). Note the lack of galaxies in the region between 11 and 12 hours in R.A.

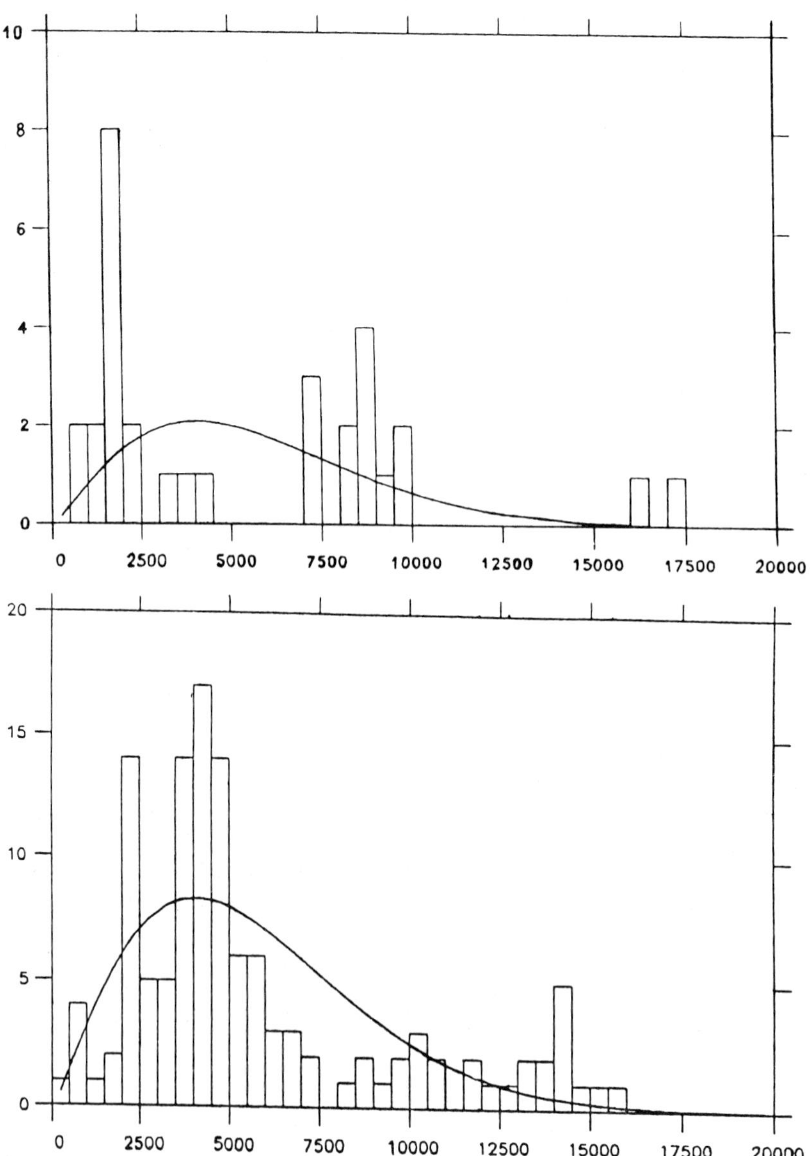

Figure 9. The region Hydrae-Centauri. Top: redshift distribution in the two dimensional gap between Hydra and Centaurus (see Figure 8) from 11 to 12 hours in R.A. and -27 -33 degrees in declination. Bottom: redshift distribution in Centaurus: $13^h \leqslant$ R.A. $\leqslant 14^h$ and $-33^\circ \leqslant$ DEC $\leqslant -25^\circ$.

density levels, disconnected from the Centaurus supercluster. To test for this effect redshifts were obtained in the region between 13 and 14 hours in R.A. and between -25 and -33 degrees in declination. The comparison between Figure 9 top and Figure 9 bottom clearly shows that there is a lack of galaxies with the redshift of Hydra-Centaurus (Da Costa et al 1986 b).

Since the mechanism of formation may have determined the present shape of superclusters it became important to determine some structural parameters to ease the comparison between observations and models. Doubts on disk shaped structures have been expressed by Oort (1983) and Chincarini (1982) and it seems that at least in the Perseus-Pisces supercluster there is some evidence of a filamentary structure.
The uncertainty on this matter is clearly apparent from a statement by Oort (1983) in his review article, when he discusses the central structure of Virgo:
" ..... it may either be an oblate distribution seen roughly edge-on, or possibly a prolate feature"
An analysis of shape parameters has been carried out on a complete sample with m ⩽ 14.5 by Chincarini et al (1986), see Vettolani et al (this conference).

A very puzzling possibility is the existence of agglomerates which exceed by a rather large factor the superclustering scale so far proposed. That we have continuity or connection among clusters and superclusters at very low galaxian density is a rather old concept and indications were available since some time. What is new, and very important if the reality can be confirmed, is that such connection and related agglomerates form a very large scale structure having some physical significance of its own. In other words it may be a larger unit cell in the Universe.
We refer in particular to the large structure embedding the Local Supercluster indicated by Tully (1986). He suggests that the larger stratified and flattened structure extends in layers of about 400 Mpc. The main and richer layer would be in the plane of the supergalactic equator. The geometry is shown in Figure 10.
Is this a real fact or could it be due to an observational bias?
Some possible biases are the following:
a) The supergalactic plane is favored by galactic extinction because it is located toward the north galactic pole. We notice however that long ago de Vaucouleurs (1976) has given clear evidence that the plane of the Local Supercluster is not due to an effect of obscuration. It seems therefore that a bias due to galactic obscuration is unlikely also in this case.
b) The percolation algorithm, or any clustering algorithm, is favored in connecting structures across the zone of avoidance because of 1) galactic exctintion, 2) the presence of the boundaries defined by the zone of avoidance 3) the incompleteness of the sample.
That the cluster catalogs may be affected by not well known observational biases it seems suggested by the recent work of Shectman

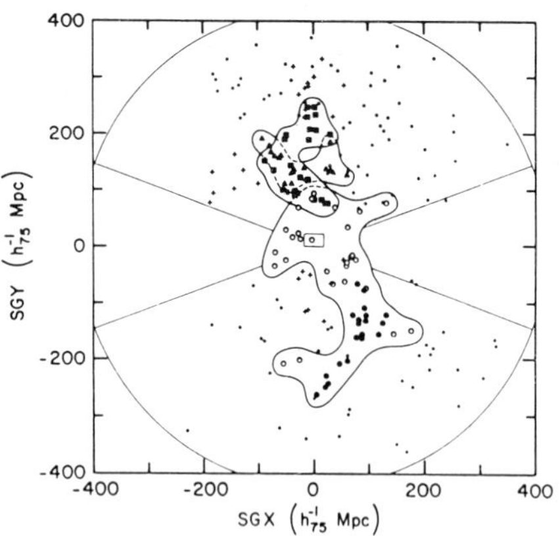

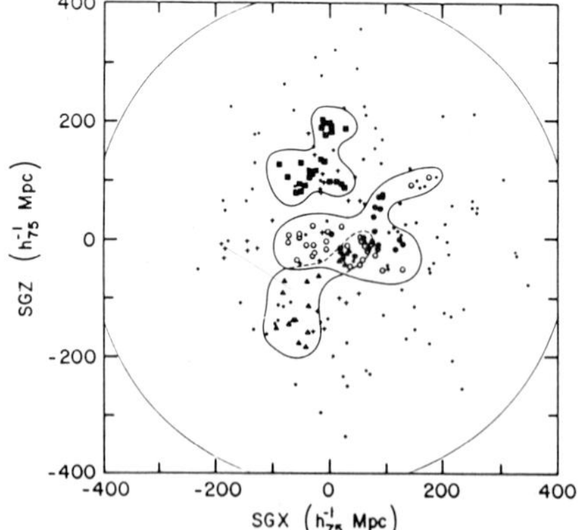

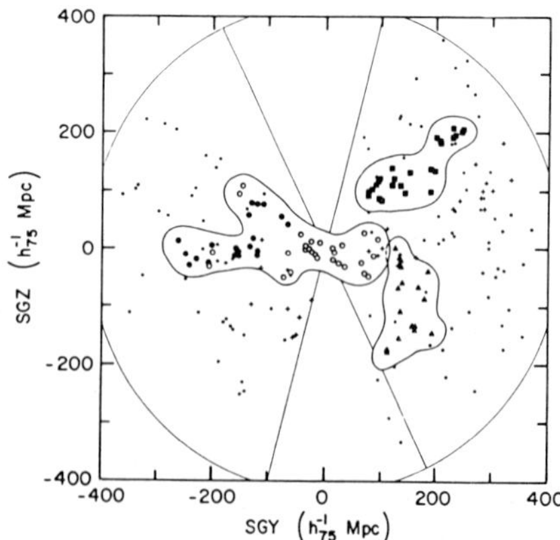

Figure 10. Distribution of clusters of galaxies according to Tully (1986). The supergalactic plane is Π(SGX,SGY).

(1985). The sample of clusters selected by Shectman from the Shane-Wirtanen galaxy counts shows that only 40% of these clusters are members of the Abell catalog. Shectman, furthermore, calls attention to an unexplained absence of clusters north of declination 57 degrees and to a lack of clusters in the range of galactic latitude between 40 and 50 degrees.

The evidence can be enforced or weakened only by future work, simulations and observations, and by complete and deeper samples. Some of the uncertainties will be probably resolved with the completion of the southern catalogue of clusters of galaxies, Abell, Corwin and Olowin (1987).

Assuming however that these structures are real and not due to an unknown observational bias, the important fact is not so much the size (we have other detections of large clusters of clusters) but the fact that the main plane of the structure is about parallel to the plane of the Local Supercluster. This calls for physical phenomena at the time of formation. Large scale clustering, up to 300 h Mpc has been evidenced by Batusky and Burns (1985a and 1985b), Bahcall and Soneira (1984), Bahcall and Burgett (1986). Indeed Abell (1961) in his analysis on the clustering of clusters gives an upper limit of about 1000 Mpc. He clearly stated that the distribution of clusters of galaxies is non-random.

Bahcall and Soneira (1983) put this fact on a very solid basis showing that the spatial correlation function of rich clusters of galaxies (Abell's catalogue) is approximately 18 times stronger than the correlation function of galaxies.

What is also attractive is the simple explanation given by Bahcall (1986). This is the quantized version of the simple minded observational fact which has been often stated: clusters are embebbed in larger structures.

Perturbations in density are necessarly related to perturbations in the Hubble flow so that large scale motions must be considered. These must ultimately be in agreement with the dipole anisotropy observed in the microwave background.

Fair understanding has been gained on the perturbations caused by the Virgo supercluster (for a comprehensive discussion on this matter see Tammann and Sandage 1985) while large uncertainties remain on larger scale motion.

The latest attempt in clarifiyng this matter is the work by Burstein et al (1986). Using the magnitude related parameter D_Σ , the diameter within which an elliptical galaxy reaches a mean surface brightness of 20.75 B mag arcsec^{-2} , and the relation $D_\Sigma - \sigma$ (σ is a measure of central velocity dispersion) the authors are able to predict distances to individual galaxies with accuracies of ± 23 %. With this new distance indicator and the accurate redshifts they interpret the observed systematic deviations from the Hubble flow as due to a bulk motion of approximately 700 km/s towards $l = 299$ and $b = 1$ degrees for galaxies within 60 h^{-1} Mpc of the Local Group, see Fig 11.

The detected motion is in agreement with the motion derived using

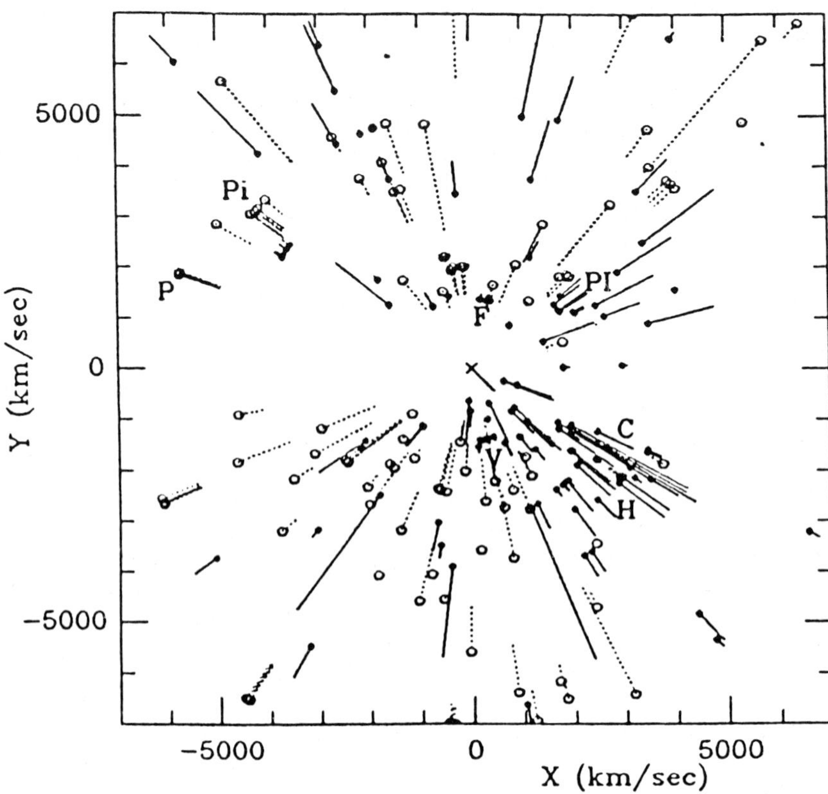

Figure 11. Motions of elliptical galaxies (Burstein et al 1986) with respect to the microwave background reference frame. The peculiar motion is represented as a solid line for a velocity vector pointing away from the position of the Local Group and as a dashed line for a velocity vector pointing towards the Local Group.

other samples, that is the sample of spirals by Aarson et al (1982), via the infrared Tully Fisher relation, and the Rubin-Ford sample (Rubin et al 1976)

Bahcall et al (1986 and these proceedings) using a complete sample of clusters of galaxies (subsample selected from the Abell Catalogue with $D \leqslant 4$ ($z \leqslant 0.1$) and $R \geqslant 1$) detect a strong asymmetry in the space distribution of cluster pairs. The strong elongation observed in the redshift direction corresponds to a velocity broadening of about 2000 km/s among clusters pairs.

The phenomenon, which appears to be real, is interpreted as partly due to the geometrical elongation of the superclusters along the line of sight. Peculiar velocities, on the other hand, could play a large role.

With the present data it is impossible to disentangle or distinguish between the two effects. Only in one case, the Hercules supercluster is a clean magnitude redshift relation which shows that cluster members satisfy the Hubble relation and are endeed spread along the line of sight.

It is unlikely, however, that the superclusters considered could be in equilibrium and that the velocity could be interpreted as due to a balance between kinetic energy and the gravitational potential energy. It seems more likely that such motions reflect perturbations in the Hubble flow due to overdensities or mass motions involving large volumes of space and reflecting primordial perturbations.

Infact the large scale motions discussed by Burstein et al (1986), and previously by Rubin et al (1976), the infall to Virgo and the dispersion detected by Bahcall et al (1986) are probably all due the same mechanism. A dispersion of about 2000 km/s on a scale of about 25 h^{-1} Mpc seems, however, rather large.

3. VOIDS

Recently attention has been focussed particularly on the regions of space which are void of galaxies, the "voids".

After Chincarini and Martins (1975), Chincarini and Rood (1975), Gregory and Thompson (1978), Tifft and Gregory (1978), Tarenghi et al (1979) clearly showed that on large scale galaxian redshifts were segregated in rather well defined intervals, a distribution structure-void, it became clear that such empty regions could help in distinguishing among clustering models.

The detection by Kirshner et al (1981) of a rather large void in Bootis convinced many of us that a pancake scenario, and related galaxy formation and evolution, could be the most likely explanation of the observed distribution of galaxies.

Vettolani et al (1985) approached the problem statistically using a sample complete to $m \leqslant 14.5$. In this study they came to the conclusion that 1) the voids observed in the sample used do not contradict a hierarchical distribution of matter in the Universe and 2) the presence of large voids as the one detected in Bootes could be explained in a hierarchical realization.

Soltan (1985) analyzed a similar sample, in this case a subsample of the Cfa redshift survey (Huchra et al 1982) with $M \leqslant -20.5$. As stated in Vettolani et al (1985) it is a matter of whether it is more reliable a sample with larger statistics and somewhat more complicated analysis or a sample with small statistics and a rather well defined volume. In practice Soltan (1985) reaches the same conclusions reached by Vettolani et al (1985) . If this is the case, voids could be a result of the gravitational clustering process and there is no need to invoke a special mechanism producing voids in the galaxy distribution: these form as a consequence of the clustering process.

However, using the wording by Occhionero et al (1984) it is a matter to decide whether we are dealing with condensations surrounded by cavities or with cavities surrounded by condensations.

Ikeuchi et al (1983) and Ostriker and Cowie (1981) propose that galaxies, and related structures, form in an intergalactic medium dominated by explosions. In this case the structures, rather than the voids, came second.

De Lapparent et al (1986), Fig. 6, claim that the distribution of galaxies observed in their sample ($m \leqslant 15.5$, a 6 times 117 degrees strip (0.2138 steradians) going through the Coma cluster, appears to have a bubble like structure with the galaxies distributed on the surface of the bubble. The bubbles have, according to the authors, a typical diameter of 25 h^{-1} Mpc, R 12.5 h^{-1} Mpc. Another striking feature emphasized by the authors is the sharpness of the boundaries of the high density regions which surround the voids (see Geller, these proceedings).

Statistics on voids, as we said, is rather scarce since a good understanding can be obtained only with deep samples, m 14.5, over large regions of the sky.

Using the zero order approximation given in Vettolani et al (1985) the number of voids of a given volume expressed in units of V_o , the volume of the sweeping vector ($V_o = 7^3$), is

$$N_{14.5} (V/V_o) = 3/2 (V/V_o) \qquad N_T (14.5)$$

and the corresponding probability is
$$P(\geqslant V/V_o) = 3/2 \int N (V/V_o) =$$

$$3/2N_T (14.5) (V/V_o)$$

Therefore, for the de Lapparent et al (1986) sample we find

$$N_{15.5} = N_{14.5} * V_{15.5}/V_{14.5} = 3.98 * N_{14.5}$$

where the volume of the sample has been increased by the amount

$$V_{15.5}/V_{14.5} = 10^{3/5(15.5 - 14.5)} = 3.98$$

with $N_T (\leqslant 14.5) \simeq 100$ over 4 sterad (Vettolani et al 1985) we derive
$$N = 43.9$$

The number of fields equivalent to the de Lapparent et al (1986) field in 4 steradians is 4 / 0.2138 = 18.71 and the number of expected 25 h^{-1} Mpc diameter voids is 43.9 / 18.7 = 2.3, a sizable number even if computed in a zero order approximation.

The number of voids observed in the de Lapparent et al (1986) sample is about 4. The approximate dimensions and volume as derived from their Figure 1 is shown in Table I.

Table I

void	V	V	Volume
	km/s	km/s	10 Mpc
1	3000	1590	31.6
2	8000	2400	19.3
3	7900	2170	160.0
4	9200	2000	23.7

If the topology of these large structures is confirmed and the number of rather large voids increases when compared to the simple hierarchical distribution of galaxies it may be hard to avoid the conclusion that standard gravitational clustering models do not match the observations.

While a bubble like structure may be somewhat unlikely, it seems that a sponge like structure is in agreement with most of the observations (see Gott, these proceedings). The data will show which one the Universe preferred. See however Ruffini (these proceedings) for another interpretation.

At the voids boundary we expect, both in the case of a hierarchy generated by gravitational phenomena and in the presence of cavities surrounded by superclusters (bubbles and shells) a galaxian density enhancement, Hoffmann et al (1983) (1983), Occhionero et al (1983), Peebles (1982) and Ostriker and Cowie (1981).

While de Lapparent et al (1986) find some indication of overdensity on the ridges, Soltan (1985), in a sample limited however at 14.5, but with a much larger volume, states that his analysis shows that galaxies do not create high-density regions around the voids. Any evidence on this is rather scanty also in view of the fact that one has to be carefull about the subtle effect of the velocity dispersion in groups and clusters, compare Fig. 2 with Fig. 12, where we have marked the presence of two groups. The alignment of the void wall, in the direction of the line of sight, is partly due to the velocity dispersion of the member galaxies. To better understand this and other problems related to the detailed structure of superclusters, via the Tully-Fisher relation we started years ago observations in the 21 cm line and infrared (H band, Gornengrat and KPNO).

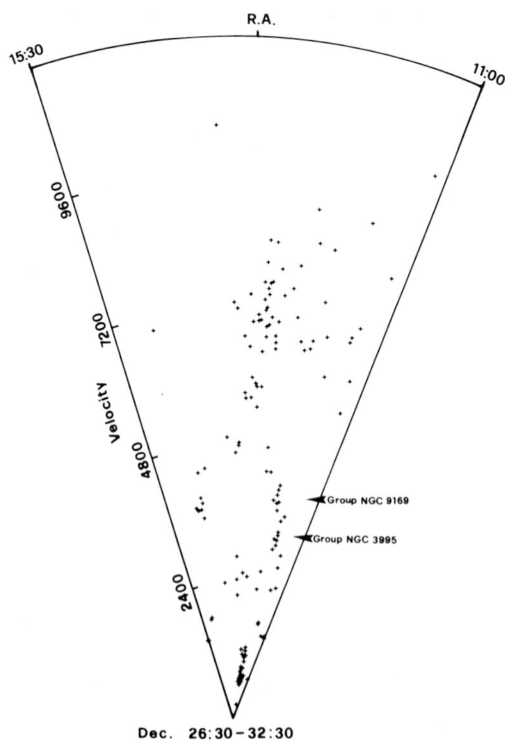

Figure 12.   Galaxies in the Coma region with $m \leqslant 14.5$ and declination between 26:30 and 32:30 degrees.   The figure shows how velocity dispersion in galaxies which are members of a group may simulate a void boundary and complicate the understanding of the boundary density.

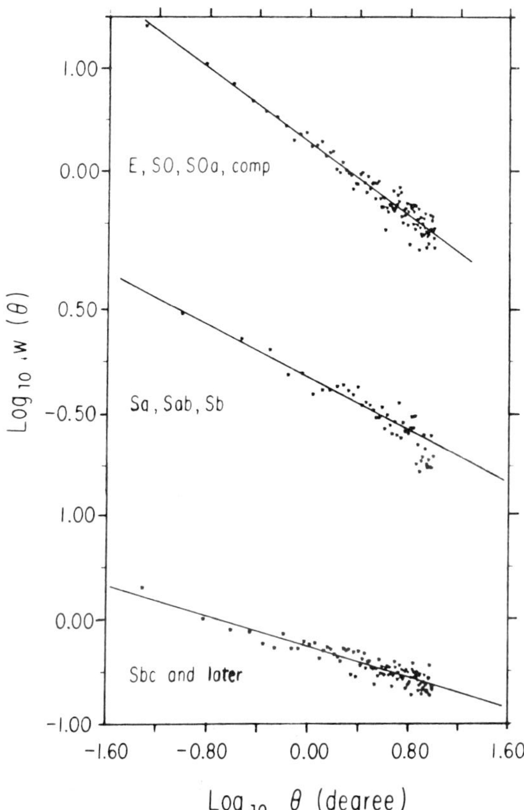

Figure 13. The autocorrelation function for various morphological types in the Perseus Pisces supercluster (Giovanelli et al 1986).

3. MORPHOLOGY OF GALAXIES AND STRUCTURES

Giovanelli, Haynes and Chincarini (1986), and references therein, have shown that the two points angular correlation function depends on the morphological type, Figure 13. That is late type galaxies are less clustered. Abell (1977) showed that it would be difficult to detect the Coma cluster by using solely spiral galaxies. In other words the more we go toward late types galaxies, the closer we approach a poissonian distribution.

The distribution of dwarf galaxies is known accurately only for the region of the Virgo cluster, Sandage et al (1984). The distribution of dE galaxies is somewhat clustered, Figure 14 left, while the irregular galaxies almost do not cluster at all. Indeed Sandage et al (1985) observe a lack of dwarf irregulars in the region of maximum concentration of dE galaxies. The result is confirmed, clustering of dE, by autocorrelating the sample of galaxies in Virgo.

The suggestion is that, unless we must distinguish between local and universal effects about the distribution of galaxies, the relation type-clustering strength holds down to the dwarf irregulars where any sign of clustering seems to be lost (note, however, that the sample is very small). Since dwarf galaxies are of low luminosity some correlation should be present also between clustering and magnitude.

Searching the literature we found 4 galaxies with magnitude $m > 14.5$ in the void of Figure 12. We did not, at present, a statistical search to check for the characteristics of faint galaxies in voids. In fact we do not have a good set of data, on the other hand we know that in other cases, Bootis void is one, going to fainter magnitudes we find a few faint galaxies in voids. The possibility is that dwarf galaxies may, to some extent, populate these regions of space and, perhaps, observationally support the idea of biassed galaxy formation.

However the question is: granted that we find some dwarf galaxies in the voids, are these galaxies located there because their formation is favoured by the low density region, whether or not dark matter is the solution to various cosmological problems, or because the distribution of dwarf galaxies does not show the clumpiness measured for bright galaxies. Virgo shows that regions of fairly high density of galaxies do not prevent the formation of dwarf galaxies. As for the type expected and number we must await for good data and for surveys similar to the one Binggeli (1985) is carrying out in non cluster fields.

To this end, however, it is of some interest to look at the distribution of types, diameters and surface brightness of dwarf galaxies in Virgo. Table II has been prepared using the data of Binggeli et al (1985). The surface brighness has been computed using the simple relation $SB = B + 5 \log D$ and at the distance of Virgo $(m-M = 31.7)$ 20 arcsec correspond to about 2.1 Kpc. The sample incompleteness begins to be very strong at $SB = 25.5$ (Binggeli et al 1986). Note however that their definition of surface brightness is not simple minded as the one above. The low surface brightness and small

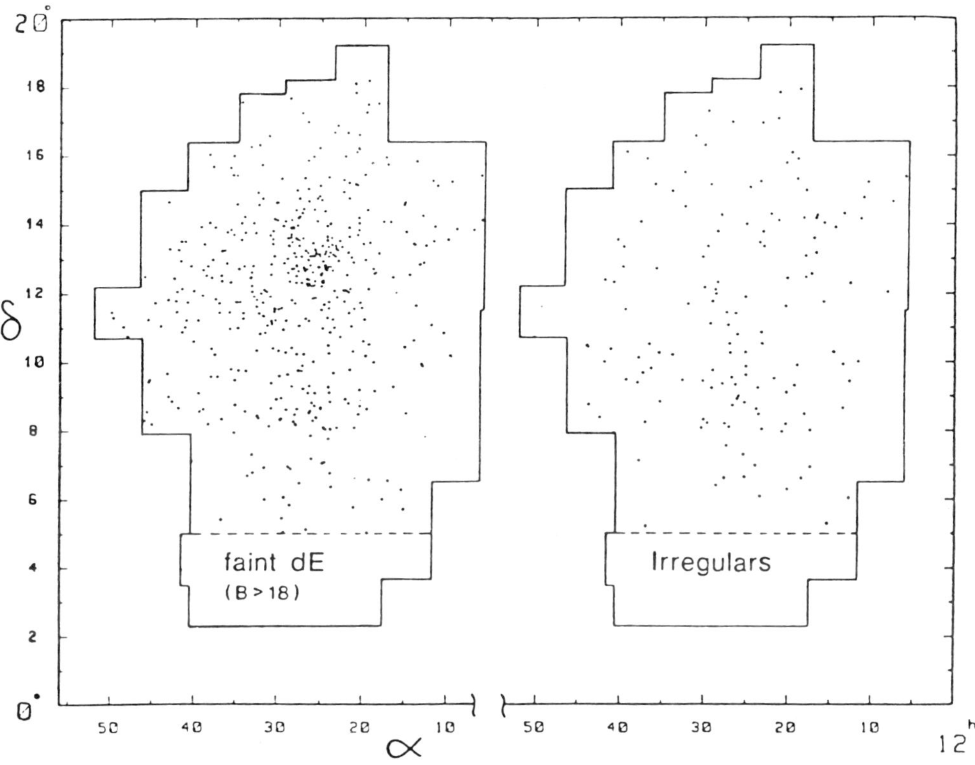

Figure 14. Distribution of dwarf elliptical and irregular galaxies in Virgo according to Sandage et al (1985).

diameter make difficult the detection of the peak of the distribution.
     Finally to estimate the expectation in the voids we have to account
also for the correlation between dwarf type and density since dwarf
ellipticals are more clustered than irregulars.

4.   A CURIOSITY: MICRO-SUPERCLUSTERS

     We find of some interest that structures similar to the one we
observe on the large scale in the Universe are also seen on much smaller
scale. We are familiar with colloids and areosols.
     If we have particles of gold or nickel, with a diameter of the order
of a few nanometers, in solution, the particles become ionized and repel
each other. Such force dominates over the Van der Waals forces and the
solution is rather stable against any form of aggregation.   The
repulsive electrostatic forces, however, can be shielded (or
neutralized) so that the Van der Waals attractive force acts when the
particles are close to each other and form aggregates. In other words
the solution is now unstable toward aggregation. As it is well known
when large aggregates form, the solution becomes opaque and the
aggregates deposit to the bottom.   In Figure 15 is shown the

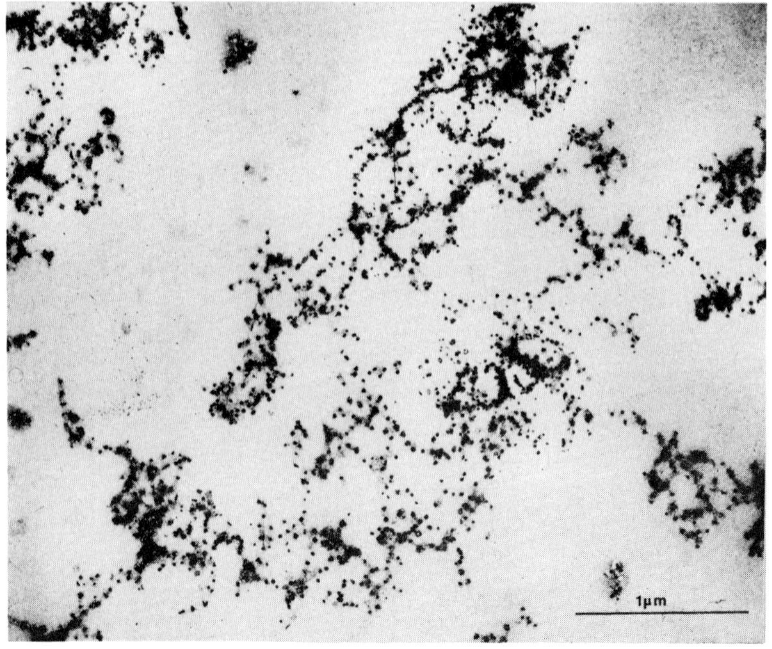

     Figure 15. A microsupercluster: agglomeration generated in a
nickel colloid. From Jullien et al (1985).

TABLE II

<SB>	dE	dS0	BCD	Im
19.5	0	0	0	0
20.5	0	1	0	0
21.5	2	8	0	0
22.5	12	20	1	4
23.5	127	18	16	14
24.5	395	9	17	60
25.5	453	6	6	51
26.5	147	0	3	18
27.5	11	0	1	0

<D> arcsec	dE	dS0	BCD	Im
20	411	1	21	21
25	282	0	16	24
35	204	2	12	23
45	87	8	6	18
55	71	3	4	16
65	31	1	2	13
75	21	6	1	8
85	15	7	0	13
95	8	3	0	3
110	12	2	0	3
130	5	12	0	5

reproduction of an aggregate of a colloid of Nickel prepared at Orsay by Mrs Belloni and J.L. Marignien and photographed with the electron microscope by J. P. Chevalier at Vitry (Jullien et al 1985).

The similarities with the distribution of galaxies on large scale is evident in spite of the fact that in this case the structures are somewhat more compact and large spaces exist among agglomerates. Are we learning something?

We do not want to push the similarity too far, but stress that also in this case we are dealing with fractals whose characteristics can be studied using the developments of Mandelbrot and the structures may be interpreted in a hierarchical model (see figure on page 1337 of Julien et al 1985).

This is an example of similarities between the micro and macrocosmos where similar structures are formed by forces of different nature.

REFERENCES

Aarson, M., Huchra, J., Mould, J., Shechter, P., Tully, R.B.
    1982 Astroph. J. 258, 64
Abell, G.O.: 1956 Astroph. J. Suppl. 3, 211
Abell, G.O.: 1961 Astron. J. 66, 607
Abell, G.O.: 1977 Astroph. J. 213, 327
Abell, G.O., Corwin, H.G., Olowin, R. 1987 in preparation
Bahcall, N. 1986 Astroph. J. 302, L41
Bahcall, N., Burgett, W.S. 1986 Astroph. J. 300, L35
Bahcall,N., Soneira,S.N. 1983 Astroph. J. 270, 20
Bahcall,N., Soneira,S.N. 1984 Astroph. J. 277, 27
Bahcall,N., Soneira,S.N., Burgett, W.S. 1986 Astroph. J. in press
Batusky, D.J., Burns,J.O. 1985a Astroph. J. 299, 5
Batusky, D.J., Burns,J.O. 1985b Astron. J. 90, 1413
Binggeli, B. 1986 private communication
Binggeli, B., Sandage, A., Tammann, G.A. 1985 Astron. J. 90, 1681
Burstein, D., Davies, R.L., Dressler, A., Faber, S.M.,
    Lynden-Bell, D.,Terlevich,R., Wegner, G. 1986 in
    "Galaxies Distances and Deviations from the Hubble
    Flow" B.F. Madore and R.B. Tully edts, Dordrecht,
    Reidel, page 123
Chincarini, G. 1982 in "The Large Scale Structure of the Universe
    lectures at the 3rd Escola de Cosmologia e Gravitacao
    at Rio de Janeiro, University of Oklahoma, Norman
Chincarini, G., Martins, D. 1975 Astroph. J. 196, 335
Chincarini, G., Rood, H.J.: 1975 Nature 257, 294
Chincarini, G., Rood, H.J.: 1979 Astroph. J. 230, 648
Chincarini, G., Vettolani, G., de Souza, R. 1986 in preparation
da Costa,L.N., Nunes, M.A., Pellegrini, P.S., Willmer, C.I.,
    Chincarini, L., Cowan, J. 1986a Astron. J. 91, 6
da Costa, L.N., Willmer, C.J., Pellegrini, P.S., Chincarini, G.
    1986b preprint
de Lapparent, V., Geller, M.J., Huchra, J.P. 1986 Astroph.
    J. 302, L1

de Vaucouleurs, G. 1958 Astron. J. 58, 30
de Vaucouleurs, G. 1966 Astroph. J. 203, 33
Dicke, R.H., Peebles, P.J.E., Roll,P.G., Wilkinson, D.T.
      1965 Astroph J. 142, 414
Einasto, J., Joeveer, M., Saar, 1980 Mont. Notices Royal Astr.
    Soc 193, 353
Gavazzi, G. 1986 private communication
Giovanelli, R., Haynes, M.P., Chincarini, G. 1986 Astroph.
    J. 300, 77
Gregory, S.A., Thompson, L.A. 1978 Astroph. J. 222, 784
Gregory, S.A., Thompson, L.A., Tifft, W.G. 1981 Astroph. J.
    243, 411
Hoffmann, G.L., Salpeter, E.E., Wassermann, 1983 Astroph. J.
    268, 527
Hubble, E. 1936 "The Realm of Nebulae, Yale
Huchra, J., Davis, M., Latham, D., Tonry, J. 1982 Astroph.
    J. Suppl. 52, 89
Ikeuchi, S., Tomisaka, K., Ostriker, J. 1983 Astroph. J. 265, 538
Jullien, R., Botet, R., Kolb., M. 1985 La Recherche 16, 1334
Kirshner, R.P., Oemler, A., Shecter, P.L., Shectman, S.A. 1981
      Astroph J. 248, L57
Lauberts, A. 1982 "The ESO/Uppsala Survey of the ESO(B)
    Atlas, ESO Garching
Neyman, J., Scott, E.L. 1959 in Handbuch der Physik 53, 416
Occhionero, F., Santangelo, P., Vittorio, N. 1983 Astron.
    Astroph. 117, 365
Occhionero, F., Santangelo, P., Vittorio, N. 1984 in IAU
    Symposium 104, G.O. Abell and G. Chincarini edts, Dordrecht,
    Reidel ,page 217
Oort, J.: 1983 Ann. Rev. Astron. Astroph. 21, 373
Ostriker, J.P., Cowie L.L. 1981 Astroph. J. 243, L127
Peebles, P.J.E. 1980 "The Large scale structure of the Universe",
    Princeton University Press, Princeton
Peebles,P.J.E. 1982 Astroph. J. 258, 415
Penzias, A,A., Wilson, R.W. 1965, Astroph. J. 142, 419
Rubin, V.C., Ford, W.K., Thonnard, N., Roberts, M.
    Astron. J. 81, 719
Sandage, A., Binggeli, B., Tamman, G.A. 1985 in "The
    Virgo Cluster" ,O.-G. Richter and B. Binggeli edts,
    ESO, Garching, page 239
Shane, C.D., Wirtanen, C.A. 1967 Publ. Lick Obs. 22, part 1
Shapley, H. 1935 Harvard Annals 88, N. 5
Shapley, H., Ames, A. 1932 Harvard Annals 88, N. 2
Shectman, S.A. 1985 Astroph. J. Suppl. 57, 77
Soltan, A. 1985 Mont. Notices Royal Astr. Soc 216, 537
Tammann, G.A, Sandage, A. 1985 Astroph.J. 294, 81
Tarenghi, M., Tifft, W.G., Chincarini,G., Rood,H.J.,
    Thompson, L.A. 1979 Astroph.J. 235, 724
Tifft, W.G., Gregory, S.A. 1978 in IAU N. 79, edited
    by M.S. Longair and J. Einasto, Dordrecht, Reidel,
    page 267

Totsuji, H., Kihara,T. 1969 Publ. Astron. Soc. Japan 21, 221
Tully, R.B. 1986 Astroph. J. 303, 25
Vettolani, G., de Souza, R., Marano, B., Chincarini, G. 1985
    Astron. Astroph. 1985, 144, 506
Zeldovich, Ya.B. 1978 in IAU N. 79, edited by M.S. Longair and
     J. Einasto,Dordrecht, Reidel, page 409
Zwicky, F. 1957 "Morphological Astronomy", Berlin, Springer

DISCUSSION

BURNS: Is there anything unusual about the galaxies at ~3600 km/sec that fall within one of the Coma voids? In particular, how do they compare with the dwarf emission-line galaxies found in Bootes?

CHINCARINI: The four galaxies I mentioned have been found in the literature and the published data are as follows:

$133624 + 2635$	$cz = 3994$	Type $= sp$	$m_j = 14.99$
$125718 + 2303$	3682	E	$m_V = 16.81$ (3.5 x 1.2 Kpc)
$125706 + 2822$	3651	SO	$m_V = 17.20$
$122648 + 2723$	3819	--	$m_V = 15.10$

DENG: It seems that too many filaments of clusters and super-clusters are along the line of sight in the diagrams you showed. It seems to be unreasonable. Do you think there are any effects existing in the observation which could cause this kind of distribution?

CHINCARINI: In the sample of Coma-A1367 I showed there are various groups, and the velocity dispersion of their members give the elongation, along the line of sight - I mentioned in the talk that in other cases distortions of the topology, at least in some regions, may be partly due to the large scale motion which various authors detected and similar to the one you referred to in the Hawaii meeting (see however R. Davies - this conference). However, I believe that the topological characteristics as found globally, probably sponge like, are by now demonstrated to be real.

TULLY: I can provide an update of the work on extremely large-scale structure that was mentioned. The original published work was based on an analysis of 214 rich clusters with redshifts less than 0.1 c, whereas 375 such clusters are now known. The concentration to the supergalactic plane is even stronger in the most recent sample. Of order 100 rich clusters participate in a structure that extends across 500 Mpc and has a FWHM thickness of 60 Mpc ($H_0 = 75$).

LARGE-SCALE STRUCTURE:

THE CENTER FOR ASTROPHYSICS REDSHIFT SURVEY

Margaret J. Geller, John P. Huchra, and Valérie de Lapparent
Center for Astrophysics
60 Garden Street
Cambridge, MA 02138 USA

ABSTRACT. Two slices of the Center for Astrophysics (CfA) redshift survey extension are now complete. The survey indicates that galaxies are distributed on the thin surfaces of "bubble-like" structures. The voids in the survey have diameters as large as 5,000 km s^{-1}. These structures challenge theories for the formation of large-scale structure in the universe and suggest new approaches to several problems in the field.

1. INTRODUCTION

Each of the large redshift surveys completed during the last ten years has caused significant evolution in our understanding of the nature of the distribution of galaxies. A perusal of the volumes from the the IAU Symposia in Tallinn and Kolymbari clearly demonstrates the continuing change in perspective (Longair and Einasto 1978; Abell and Chincarini 1982). In Tallinn, Joêveer and Einasto (1978) suggested that the large-scale distribution of galaxies has a "cellular" pattern in which rich clusters are connected by "filamentary" structures. The data at that time were incomplete and only adequate to hint at such structure. By 1982, the year of the meeting in Crete, several large redshift surveys were under way and some were complete (Davis, Huchra and Latham 1983; Kirshner *et al.* 1983 (KOSS); Giovanelli 1983). Voids, particularly ones as large as that in Boötes, and filaments like the one in the Pisces-Perseus region were the apparent features of the distribution in redshift space which commanded the attention of both theorists and observers. The ubiquity of such large-scale features was not clear.

Since the meeting in Kolymbari, the number of measured redshifts has more

A. Hewitt et al. (eds.), Observational Cosmology, 301–313.
© *1987 by the IAU.*

than doubled. There are now approximately 20,000 galaxies with measured redshifts in the catalog maintained at the CfA (Center for Astrophysics; Huchra *et al.* 1987a). Recently completed surveys continue to modify our picture of the large-scale distribution of galaxies. The deep surveys of Koo, Kron, Munn, and Szalay (1987) indicate that large voids are common at high redshift. The AAT surveys also reveal voids and thin structures perpendicular to the line-of-sight (Peterson *et al.* 1986). The continuing Arecibo survey delineates nearby voids and supports the interpretation of the structure in Pisces-Perseus as a "one-dimensional" filament (Haynes and Giovanelli 1986; Giovanelli *et al.* 1986). In this volume, Dr. Chincarini (1987) reviews these and a host of other observations. The extension of the CfA redshift survey, the subject of this talk, indicates that bright galaxies are distributed on thin sheets — two-dimensional structures — which surround (or nearly surround) vast voids. Large structures appear to be a common feature of all surveys large enough to contain them.

The continually changing picture reflects the attention which has been paid to the design of redshift surveys. Each of the surveys mentioned so far explores a new regime in a sort of "phase space" for observations of large-scale structure. A convenient set of parameters for comparing surveys are effective depth, maximum angular scale covered (solid angle is a less telling measure — the shape of the survey is important), and signal-to-noise (the number of galaxies available to define the structures). Surveys like the KOSS survey (Kirshner *et al.* 1986) of Boötes which consist of widely separated small probes are an efficient way of finding large voids. However, they have low "signal-to-noise" because they they cover only a small fraction of the volume spanned by the probes. Surveys like the CfA survey extension (Huchra *et al.* 1987b) which are complete over a region of large angular scale are less efficient for identifying large voids, but they are necessary for quantitative characterization of the distribution of galaxies over a range of scales.

2. DESCRIPTION OF THE SURVEY

The goal of the CfA redshift survey extension is to measure redshifts for all galaxies in a merge of the Zwicky *et al.* (1961 – 1968) and Nilson (1973) catalogs which have $m_{B(0)} \leq 15.5$ and $|b_{II}| \gtrsim 40°$. There will be $\sim$ 12,000 galaxies in the complete survey; 5,500 redshifts have already been measured. About 1,800 of these galaxies with measured redshifts lie in the "slices" for which the survey is now complete: (1) a slice with $8^h \leq \alpha \leq 17^h$ and $26.5° \leq \delta < 32.5°$ (de Lapparent, Geller, and Huchra 1986) and (2) a slice with $8^h \leq \alpha \leq 17^h$ and $32.5° \leq \delta < 38.5°$. More than 60% of the redshifts were measured with the Mount Hopkins 1.5-meter and the MMT. The mean external error in these measurements is $\sim$ 30 km s^{-1}.

Figure 1 shows the positions of the galaxies from the Zwicky-Nilson merge which have $m_{B(0)} \leq 15.5$, $8^h \leq \alpha \leq 17^h$ and $8.5° \leq \delta \leq 50.5°$. The grid is Cartesian in α

and δ. The deficiency of galaxies west of 9^h and east of 16^h is caused by Galactic obscuration. The bold ticks indicate show the location of the two complete survey strips. The Coma cluster is the dense region at 13^h in the 6° strip.

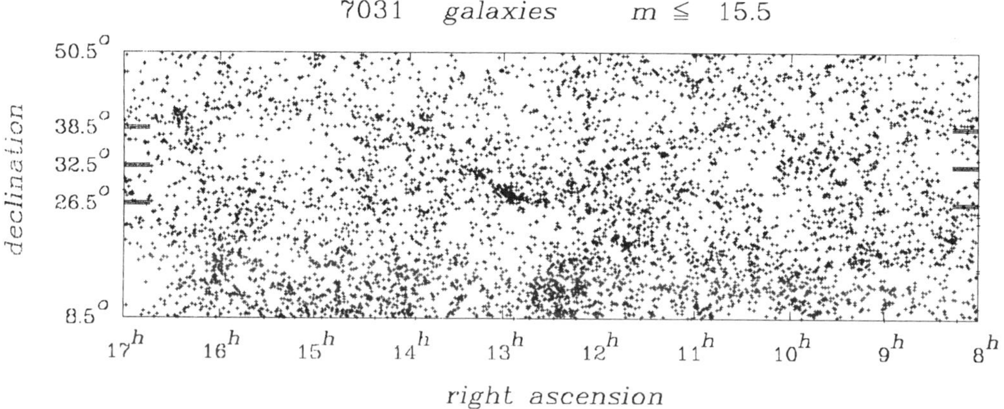

Figure 1. Positions of galaxies in the Zwicky–Nilson merge with $m_{B(0)} \leq 15.5$, $8^h \leq \alpha \leq 17^h$ and $8.5° \leq \delta \leq 50.5°$. The bold ticks indicate the declination limits of the complete redshift survey strips.

3. TOPOLOGY OF THE GALAXY DISTRIBUTION

Observations for the first strip of the survey were completed during the spring of 1986. Figure 2a is a plot of the observed velocity versus right ascension: the strip is 6° thick in declination. The plot includes only the 1067 galaxies with velocities less than 15,000 km s^{-1}. A galaxy with the characteristic luminosity $M^* = -19.4$ ($H_0 = 100$ km s^{-1} Mpc^{-1}; Davis and Huchra 1982) is at 10,000 km s^{-1} in this survey.

In Figure 2a, nearly every galaxy with a velocity less than 10,000 km s^{-1} is in an extended thin structure. The boundaries of the empty regions are remarkably sharp. Several of the empty regions are surrounded by thin structures in which the separation of galaxies is small compared with the extent of the enclosed void. The edges of some of the largest structures may be outside the right ascension limits of the survey. The only pronounced velocity finger in the distribution is the Coma cluster at $\sim 13^h$.

The thin structures in the distribution of galaxies are cuts through two-dimensional sheets; in this slice the structures are *not* one-dimensional filaments. If the

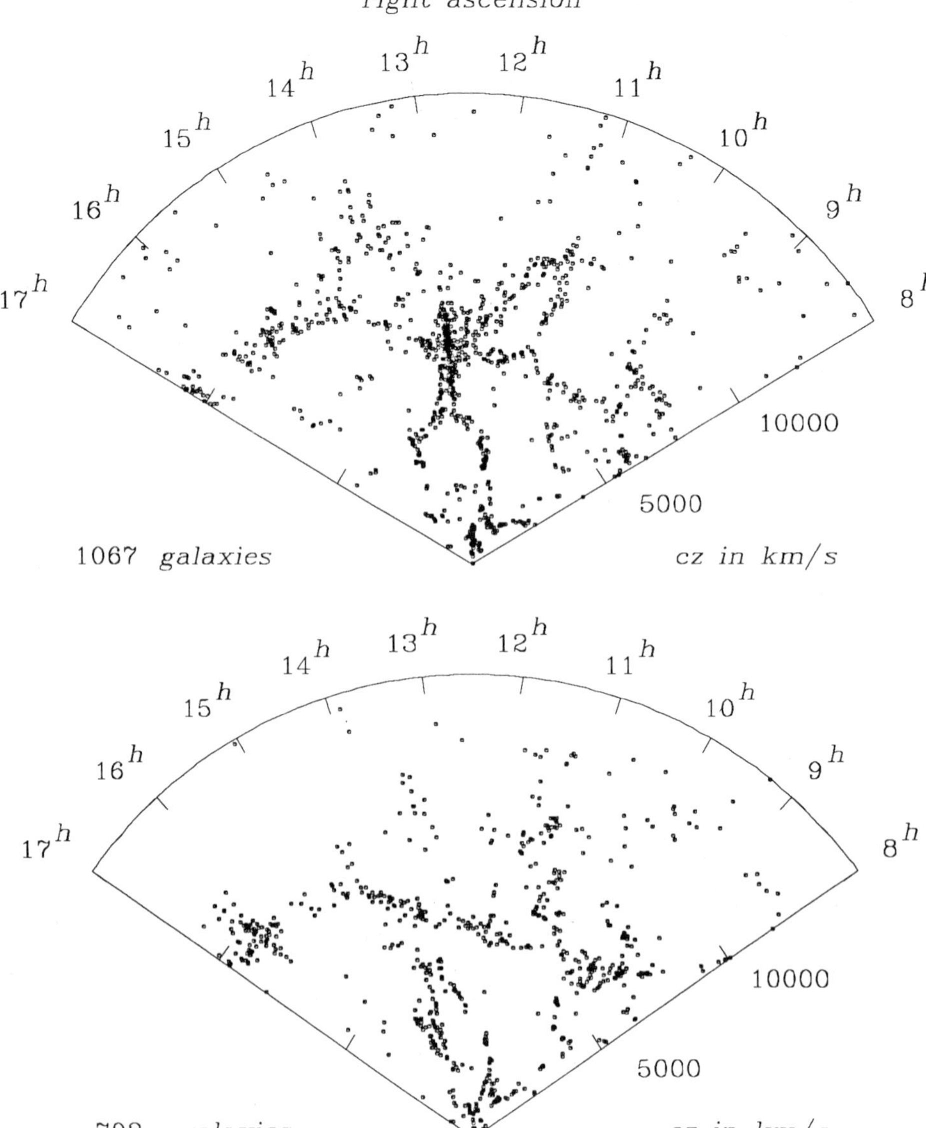

Figure 2a. (top) Observed velocity versus right ascension for the complete survey strip centered at $\delta = 29.5°$. The strip extends for 6° in declination. Only the galaxies with velocities $\leq$ 15,000 km s^{-1} are shown. **Figure 2b.** (bottom) Same as **a**) but for galaxies in the 6° declination strip centered at 35.5°.

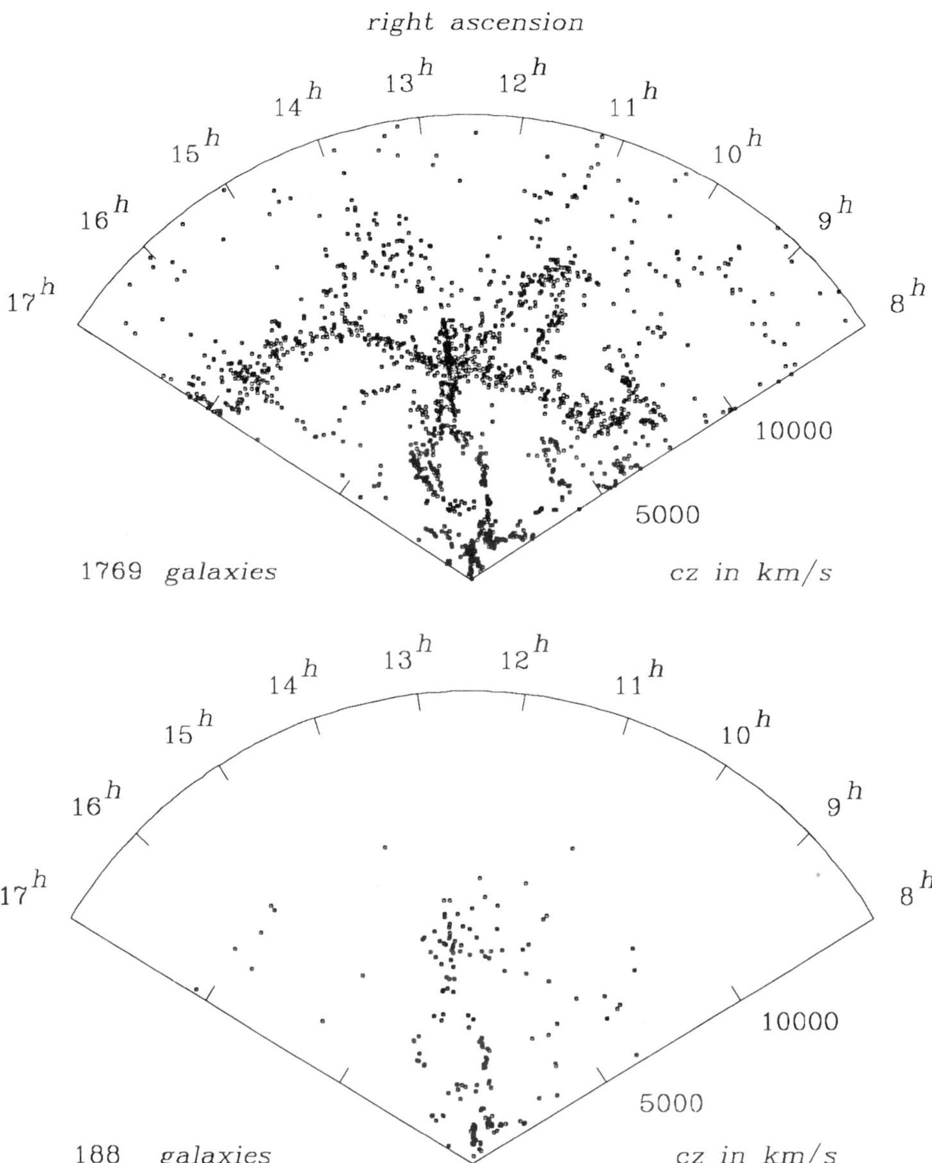

Figure 3. (top) Observed velocity versus right ascension for the two complete strips shown separately in Figures 2a and 2b. **Figure 4.** (bottom) Observed velocity versus right ascension for the survey strip centered at $\delta = 29.5°$ (see Figure 2a) but with a magnitude limit $m_{B(0)} = 14.5$.

$\sim$ 150 Mpc long structure which extends across the entire survey (from 9^h to 16^h between 7,000 km s^{-1} and 10,000 km s^{-1}) is a filament, a thin linear structure should be visible in projection on the sky. There is no such structure in Figure 1 (see de Lapparent, Geller and Huchra 1986). Because structure on the sky can be caused by patchy obscuration and/or by inhomogeneities in the galaxy catalog, structure in the distribution on the sky cannot provide complete proof (or disproof) of the filamentary nature of a structure in redshift space. A second argument against the filamentary nature of the structures in Figure 2a is that several thin, elongated structures lie in the survey slice: the intersection of a thin slice with a three-dimensional network of filaments is *a priori* unlikely to be a two-dimensional network of filaments. Of course, we could have been lucky (or unlucky).

A "bubble-like" structure in which the galaxies lie on thin surfaces surrounding voids accounts for the data. In such a structure, almost all slices will resemble Figure 2a. In this model, the 150 Mpc "filament" is made up of portions of the surfaces of adjacent "bubbles" and clusters like Coma lie in the interstitial regions (where several "bubbles" come together). We observe few, if any, pronounced velocity fingers poking through the shells.

Maps of adjacent slices support this geometric picture of the structure. During the spring of 1987, we completed the survey in the slice centered at $\delta = 35.5°$, just to the north of the strip in Figure 2a. Figure 2b shows the velocity distribution in the second slice. Once again most of the galaxies are in thin structures. Furthermore, these structures are a natural extension of the structures in Figure 2a; the structures are highly correlated in the two slices. The two closed structures at $\sim 11^h$ (9,000 km s^{-1} $\lesssim$ v $\lesssim$ 11,000 km s^{-1}) and at 14^h (7,000 km s^{-1} $\lesssim$ v $\lesssim$ 11,000 km s^{-1}) in Figure 2a are not clearly delineated in Figure 2b. Sampling of these structures may be affected by variations in the magnitude limit of the catalog.

Figure 3 shows the velocity distribution for the two complete strips taken together. The distribution remains remarkably inhomogeneous with empty voids outlined by thin structures. Because the surfaces are curved or inclined, the structures are thicker here than in Figures 2a and 2b. The curvature or inclination is most noticeable for the small void which is centered at 13^h20^m and $\sim$ 3,500 km s^{-1} (in front of the Coma cluster). The diameter of this structure is $\sim$ 2,000 km s^{-1}. The largest low density region in the survey is located between 13^h20^m and 17^h with 4,000 $\lesssim$ v $\lesssim$ 9,000 km s^{-1}. The diameter is $\sim$ 5,000 km s^{-1} (50 Mpc in the absence of large-scale flows). The underdensity in this region ($\lesssim$ 20% of the mean) is comparable with the recent estimates for the void in Boötes. The galaxies inside the structure which are at similar velocities in Figures 2a and 2b may form a tenuous structure.

It is instructive to compare these new surveys with the information available from the CfA survey (Davis *et al.* 1982; Huchra *et al.* 1983) to a limiting magnitude

$m_{B(0)} = 14.5$. Figure 4 shows the slice centered at 29.5° to this limit. Here an M^* galaxy has a velocity of ~ 6000 km s^{-1}. Because the "effective depth" of the earlier survey is comparable with the size of the largest structures in Figures 2a – 3, these structures could not be detected.

The nearby small void centered at $13^h 20^m$ and 3500 km s^{-1} can be seen in both Figures 2a and 4. The fainter galaxies in Figure 2a populate the same thin structure which is just barely detectable in Figure 4. Note that the fainter galaxies fill in gaps along the perimeter of the voids. This comparison and other deep probes through the 29.5° slice (Postman, Huchra and Geller 1986) indicate that the distribution of galaxies is independent of absolute luminosity for $M_{B(0)} \lesssim$ -17.4. The change in the fractional coverage of the perimeter of the void as a function of luminosity is a problem for attempts to distinguish between "sponge-like" (Gott 1987; Gott, Melott, and Dickinson 1986; Hamilton, Gott, and Weinberg 1987) and "bubble-like" structures. Because the fractional coverage of the surface area of any particular void is poorly defined, it is difficult to identify the borderline between the two topologies.

The 21-cm data (Haynes and Giovanelli 1986) also reveal sheet-like structures in the Perseus-Pisces region. The effective "signal-to-noise" of the data is somewhat lower for these surveys — only about 50% of the galaxies in a sample limited to $m_{B(0)} \lesssim 15.5$ are readily detectable at 21-cm. The smaller area AAT surveys (Peterson *et al.* 1986) provide further evidence of similar stuctures.

4. IMPLICATIONS

One of the important implications of the survey is obvious from inspection of Figure 3: the inhomogeneities in the distribution of galaxies are large compared even with the two slices taken together, a sample comparable in volume with the whole CfA survey to $m_{B(0)} = 14.5$. The new data are a clear demonstration that these samples are not large enough to be "fair". Figure 5 provides a more quantitative demonstration of the significance of the large-scale departures from homogeneity in the sample. The velocity histogram for the two slices differs markedly from the distribution (solid curve) predicted with luminosity function parameters from the $m_{B(0)} \leq 14.5$ sample ($\phi^* = 0.014$ Mpc^{-3}, $M^* = -19.4$, $\alpha = -1.3$). The departures are dominated by the large nearby voids and by the structures which run across the survey between 7,000 and 10,000 km s^{-1}, not by the core of the Coma cluster which contributes only ~ 100 galaxies to the histogram.

It is sobering that the largest structures we observe in the CfA survey extension are the largest we can detect within the constraints placed by the depth of the survey. The size of the inhomogeneities relative to the volume of surveys may underlie unexplained variations in traditional statistics of the galaxy distribution like the luminosity function (Schechter 1976; KOSS 1983; Bean *et al.* 1983; Davis and Huchra 1982) and the two-point correlation function at large scale (cf. Groth and Peebles 1977; Davis and Peebles 1983; Kirshner, Oemler, and Schechter 1979; Shanks *et al.* 1983). When the inhomogeneities are large compared with the sample volume, mean quantities are not well-defined.

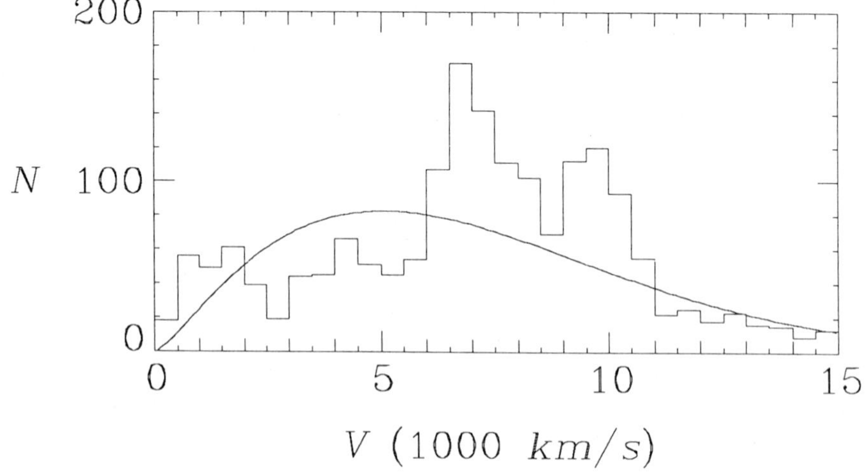

Figure 5. The velocity distribution for the entire complete sample displayed in Figure 3. The smooth curve is the distibution expected for a uniform galaxy distribution with the luminosity function parameters derived from the CfA survey to $m_{B(0)} = 14.5$.

In the absence of a "fair" sample, there are some statistical measures of the properties of individual structures which may be useful for comparing the data with simulations (see de Lapparent 1986; de Lapparent, Geller, and Huchra 1987). Both the voids and the surfaces may be characterized quantitatively. The frequent mention of the "size" of the void in Boötes is a demonstration of the power of a measure of the scale of the largest observed structures. The distribution of sizes of voids is an important test of models (the theoretical situation is reviewed by Dr. Dekel (1987) in this volume); the small-scale end is a constraint on hot dark matter models (Zel'dovich 1970; Doroshkevich *et al.* 1980; Centrella and Melott 1983) and the large-scale end is the most demanding for cold dark matter models (Davis *et al.* 1985) and for the explosive models (Ostriker and Cowie 1981; Ikeuchi 1981; Saarinen, Dekel, and Carr 1986). Determination of the distribution of sizes of voids requires samples much larger than those currently available.

If rich clusters lie in interstices between large voids, the cluster correlation function (Hauser and Peebles 1973; Bahcall and Soneira 1983; Postman, Geller, and Huchra 1986) reflects the spectrum of voids. The puzzling contrast between individual galaxies and clusters of galaxies as tracers of the large-scale matter distribution (cf. Kaiser 1984) will probably only be clearly resolved when galaxy and cluster redshift surveys overlap sufficiently.

Another interesting and possibly measurable characteristic of the voids is the elongation along the line-of-sight. If the density inside is low compared with the average surroundings, the voids should be expanding relative to the average cosmological flow and the structures should appear alongated in redshift space. The magnitude of the flow depends upon the details of the underlying physics for the formation of the structures (Peebles 1982; Hoffman, Salpeter, and Wasserman 1983; Bertschinger 1985). In a large enough sample containing many voids the intrinsic spatial geometry of the voids will average out and any net elongation along the line-of-sight may be interpreted as the result of residual expansion. The measurement of distances to galaxies in the structures offers a more direct probe for large-scale flows associated with the existence of voids. Many spiral galaxies lie in the extended sheets offering the possibility of using the infra-red Tully-Fisher technique (Aaronson, Huchra, and Mould 1979) to obtain limits at the few hundred kilometer per second level over scales of fifty megaparsecs.

The thinness ($\lesssim 500$ km s^{-1}) and coherence of the surfaces provide other constraints. The thickness as a function of orientation with respect to the line-of-sight may constrain the internal velocity dispersion and the spatial thickness. The "uniformity" of the surfaces may provide a constraint on Ω (see Peebles 1986). If $\Omega = 1$ and the distribution of galaxies marks the distribution of matter, it is unlikely that a smooth shell can persist over a Hubble time; gravity should cause the matter to clump up and "fingers" poking through the "shells" should be a more common feature in redshift space. If the actual density contrast in shells is small and the voids are filled with a nearly uniformly distributed dark matter (with Ω close to 1), the structures could still be in the linear regime. If, on the other hand, Ω is low (say $\lesssim 0.2$ as indicated by dynamical analyses of groups and clusters), the structure could set in early on and then just stretch with the universal expansion.

The study of the dynamics of individual rich clusters is a final example of a problem where the new survey data suggest a revised approach. The Coma cluster is an illustration of the difficulty. In Figure 2a the edge of the largest shell is projected nearly along the line-of-sight adjacent to the velocity finger of Coma. The structures are separated somewhat in declination, but they are sufficiently closely associated that the determination of the velocity dispersion of the cluster could be compromised by the spatial extent of the shell. Contributions from these shells would generally bias the velocity dispersions upward. The relationship between the topology of large-scale structure and the properties of individual clusters is also important for exploration of the infall pattern at large distances from the cluster center (see, for example Shectman 1982). It is not yet clear to what extent the gravitational field of a cluster drives the surrounding structure.

5. THE FUTURE

During the next year detailed quantitative comparison of the new data with simulations will further refine our understanding of the nature and origin of large-scale structure. The development of statistical measures to characterize the large-scale features in the data is an important first step in this process.

It is well to keep in mind that the maps of Figures 2–3 are maps in redshift space. Distance estimates for the galaxies in these structures are crucial to see how well the distribution in redshift space reflects the structure in three-dimensional position space. We have begun a program to study one of the largest structures. These measurements may indicate how the stuctures we observe fit together with the large-scale flows discussed by Dr. Davies (1987) at this symposium.

The discussion of Section 4 indicates several of the drivers for more extensive redshift surveys. In addition to completion of the CfA survey extension to the limiting $m_{B(0)} = 15.5$, we have begun (in collaboration with J. Thorstensen and G. Wegner at Dartmouth) to survey a $1° \times 100°$ strip to a limit of $m_{B(0)} = 17.5$. Both of these surveys should be complete within the next five years — perhaps they will bring further surprises as we sample more of the available "phase space."

REFERENCES

Aaronson, M., Huchra, J.P., and Mould, J. 1979, *Ap. J.*, **229**, 1.

Abell, G. O. and Chincarini, G. 1983, *Early Evolution of the Universe and Its Present Structure*, IAU Symposium 104, (Dordrecht: D. Reidel).

Bahcall, N. and Soneira, R. 1983, *Ap. J.*, **270**, 20.

Bean, A.J., Efstathiou, G., Ellis R.S., Peterson, B.A., and Shanks, T. 1983, *M.N.R.A.S.*, **205**, 605.

Bertschinger, E. 1985, *Ap.J.*, **295**, 1.

Centrella, J. and Melott, A. 1983, *Nature*, **305**, 196.

Chincarini, G. 1987, in *Observational Cosmology*, IAU Symposium 124, (Dordrecht: D. Reidel), this volume.

Ostriker, J. and Cowie, L. 1981, *Ap. J. (Letters)*, **243**, L127.

Davies, R. 1987, in *Observational Cosmology*, IAU Symposium 124, (Dordrecht: D.

Reidel), this volume.

Davis, M., Efstathiou, G., Frenk, C., and White, S. 1985, *Ap. J.*, **292**, 371.

Davis, M. and Huchra J.P. 1982, *Ap. J.*, **254**, 437.

Davis, M., Huchra, J.P., and Latham, D.W. 1983, in *Early Evolution of the Universe and Its Present Structure*, IAU Symposium 104, (Dordrecht: D. Reidel), p. 167

Davis, M. and Peebles, P.J.E. 1983, *Ap. J.*, **267**, 465.

Dekel, A. 1987, in *Observational Cosmology*, IAU Symposium 124, (Dordrecht: D. Reidel), this volume.

Doroshkevich, A., Kotok, E., Novikov, I., Polyudin, A., Shandarin, S., and Sigov, Yu. 1980, *M.N.R.A.S.*, **192**, 321.

Giovanelli, R. 1983 in *Early Evolution of the Universe and Its Present Structure*, IAU Symposium 104, (Dordrecht: D. Reidel), p. 273.

Giovanelli, R., Haynes, M., Myers, S.T., and Roth, J. 1986, *A.J.*, **92**, 250.

Gott, J.R. 1987, in *Observational Cosmology*, IAU Symposium 124, (Dordrecht: D. Reidel), this volume.

Gott, J.R., Melott, A. and Dickinson, M. 1986, *Ap.J.*, **306**, 341.

Groth, E.J. and Peebles, P.J.E. 1977, *Ap. J.*, **217**, 385.

Hamilton, A.J.S., Gott, J.R., and Weinberg, D. 1986, preprint.

Hauser, M. and Peebles, P.J.E. 1973, *Ap. J.*, **185**, 757.

Haynes, M. and Giovanelli, R. 1986, *Ap. J. (Letters)*, **306**, L55.

Hoffman, G.L., Salpeter, E.E., and Wasserman, I. 1983, *Ap. J.*, **268**, 527.

Huchra, J., Davis, M., Latham, D., and Tonry, J. 1983, *Ap. J. Suppl.*, **52**, 89.

Huchra, J.P. *et al.* 1987a, private communication.

Huchra, J.P., Geller, M.J., de Lapparent, V. *et al.* 1987b, in preparation.

Ikeuchi, S. 1981, *Publ. Astr. Soc. Japan*, **33**, 211.

Joêveer, M. and Einasto, J. 1978, in *The Large Scale Structure of the Universe*, IAU Symposium 79, (Dordrecht: D. Reidel) p. 241

Kaiser, N. 1984, *Ap. J. (Letters)*, **284**, L9.

Kirshner, R.P., Oemler, A. and Schechter, P 1979, *A.J.*, **84**, 951.

Kirshner, R.P., Oemler, A., Schechter, P.L., and Shectman, S 1983, in *Early Evolution of the Universe and Its Present Structure*, IAU Symposium 104, (Dordrecht: D. Reidel), p. 197.

Kirshner, R.P., Oemler, A., Schechter, P., and Shectman, S. 1983, *A. J.*, **88**, 1285.

Kirshner, R., Oemler, A., Schechter, P., and Shectman, S. 1986, preprint.

Koo, D., Kron, R., Munn, and Szalay, A. 1987, private communication.

de Lapparent, V. 1986, Ph.D. Thesis, Université de Paris VII.

de Lapparent, V., Geller, M.J., and Huchra, J.P. 1986, *Ap.J. (Letters)*, **302**, L1.

de Lapparent, V., Geller, M.J., and Huchra, J.P. 1987, in preparation.

Longair, M.S. and Einasto, J. 1978,*The Large Scale Structure of the Universe*, IAU Symposium 79, (Dordrecht: D. Reidel).

Nilson, P. 1973, *Uppsala General Catalogue of Galaxies (Uppsala Astr. Obs. Ann. V,)* **Vol. 1.**

Peebles, P.J.E. 1982, *Ap. J.*, **257**, 438.

Peebles, P.J.E. 1986, *Nature*, **321**, 27.

Peterson, B.A., Ellis, R.S., Efstathiou, G., Shanks, T., Bean, A.J., Fong, R. and Zen-Long, Z. 1986, *M.N.R.A.S.*, **221**, 233.

Postman, M., Huchra, J.P., and Geller, M.J. 1986, *A.J.*, **92**, 1238.

Postman, M., Geller, M., and Huchra, J. 1986, *A.J.*, **91**, 1267.

Saarinen, S., Dekel, A., and Carr, B. 1986, preprint.

Shanks, T., Bean, A.J., Efstathiou, G., Ellis, R.S., Fong, R., and Peterson, B.A. 1983, *Ap. J.*, **274**, 529.

Shectman, S. 1982, *Ap. J.*, **262**, 9.

Schechter, P.L. 1976, *Ap. J.*, **203**, 297.

Zel'dovich, Ya. B. 1970, *Astron. Astrophys.*, **5**, 84.

Zwicky, F., Herzog, Wild, P., Karpowicz, M., and Kowal, C.T. 1961-1968, *Catalog of Galaxies and of Clusters of Galaxies*, **Vols. I - IV**, (Pasadena:California Institute of Technology).

DISCUSSION

FANG: Since the relationship between the richness of clusters and the sizes of clusters or voids is important in fractal theory, is such a relationship found in your survey?

GELLER: We speculate that there is such a relationship, but we are far from having enough data to measure it!

ROWAN-ROBINSON: Do you have types for these galaxies so that you could remove the ellipticals, and hence the cluster 'fingers', and then see the bubbles uncontaminated by 'fingers'?

GELLER: We do not yet have types for all the galaxies in the survey, but we will in the near future.

DEKEL: (1) Isn't the Perseus-Pisces supercluster a filament?

(2) You've mentioned the 'finger of god' effect that generates elongated structures along the line of sight from velocity dispersion. Infall velocities generate thin walls perpendicular to the line-of-sight so one should be careful not to over-interpret the result in terms of "thin bubbles".

GELLER: (1) Before being absolutely convinced that the Perseus-Pisces supercluster is a filament I would like to see a careful analysis of the effects of observation in the region.
(2) What you say is true, but I think it is difficult to produce such thin, extended structures on the observed scales. The case will become clearer when we have better quantitative characterization of both the data and the models. The "bubble-like" geometry does not rule out your suggestion.

SZALAY: Peebles' argument for the homogeneity of the universe is based upon the excellent scaling of the angular correlations with depth. How can one reconcile this with the appearance of larger and larger structures as the redshift catalogues reach deeper?

GELLER: It's difficult! The only clue I see is that the correlation function is measured on scales small compared with the structures.

THE GALAXIAN SURFACE DENSITY IN THE NEARBY UNIVERSE

C. Balkowski, P. Chamaraux, P. Fontanelli
Observatoire de Paris, Section de Meudon
92195 Meudon Principal Cedex
FRANCE

ABSTRACT : A simple study of the galaxian surface density maps of the nearby Universe, obtained from CGCG, MCG and ESO catalogues, has allowed to evidence 3 new large-scale structures. Analysis of the redshifts of the galaxies in the corresponding zones shows that two of them are probably real three-dimensional entities.

INTRODUCTION

A number of studies have been recently devoted to the structure of the nearby Universe ($V_r \leq 10\ 000$ km s^{-1}), thanks to various redshift surveys (for instance Huchra et al., 1983). However the distribution of the nearby galaxies is still poorly known at low galactic latitudes and in the Southern hemisphere, by lack of redshifts. In this communication, we propose a preliminary investigation of the nearby large structures on the whole sky by drawing galaxian density maps and by using the existing redshifts. As a result, three new large structures have emerged, two of them are likely to be real (South of Hydra cluster and North of Lynx supercluster).

CONSTRUCTION OF THE MAPS

In order to determine the galaxian surface density σ, the whole sky has been covered by a 0.5° x 0.5° grid, and σ has been computed at the centre of each cell. The density has been defined in a way similar to Dressler's (1980) one, i.e. it corresponds to the average galaxian density within the circle centred at the cell centre and having a radius equal to the distance to the tenth nearest galaxy of this point.

Three catalogues of galaxies have been used, which cover well the depth zone $V_r \leq 10\ 000$ km s^{-1} we propose to investigate. These are the CGCG (Zwicky et al., 1963-1968) for the Northern hemisphere, the MCG (Vorontsov-Velyaminov, 1962-1964, 1968, 1974) for $-20^\circ < \delta < 0^\circ$, and the ESO catalogue (Lauberts, 1982) for $\delta \leq -20^\circ$. These catalogues do not present the same degree of completeness ; this effect has been tentatively taken into

A. Hewitt et al. (eds.), Observational Cosmology, 315–318.

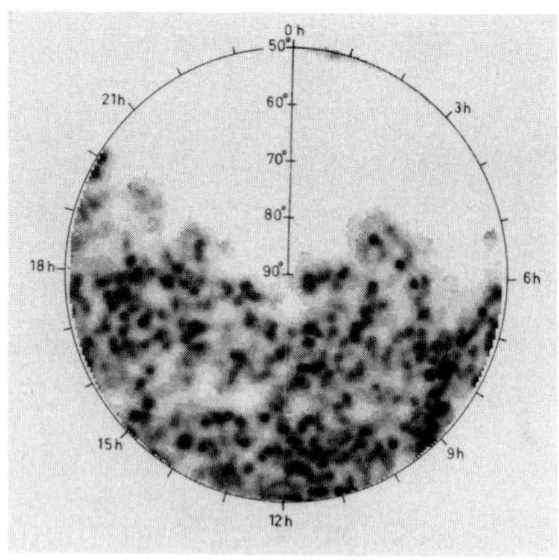

Fig.1 : Galaxian surface density
in the Northern polar cap
($50^{\circ} < \delta < 90^{\circ}$)

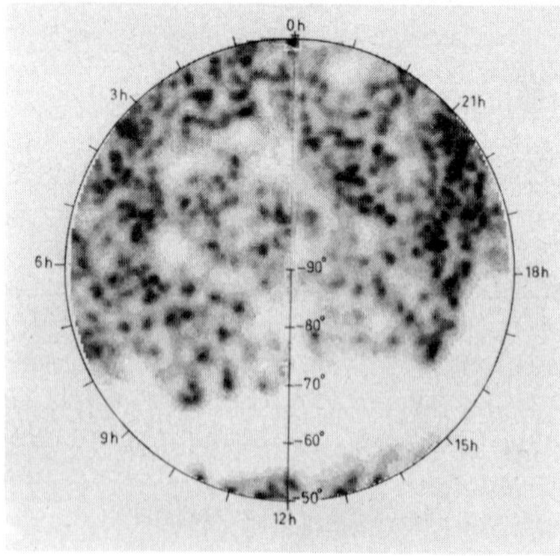

Fig.3 : Galaxian surface density
in the Southern polar cap
($-90^{\circ} < \delta < -50^{\circ}$)

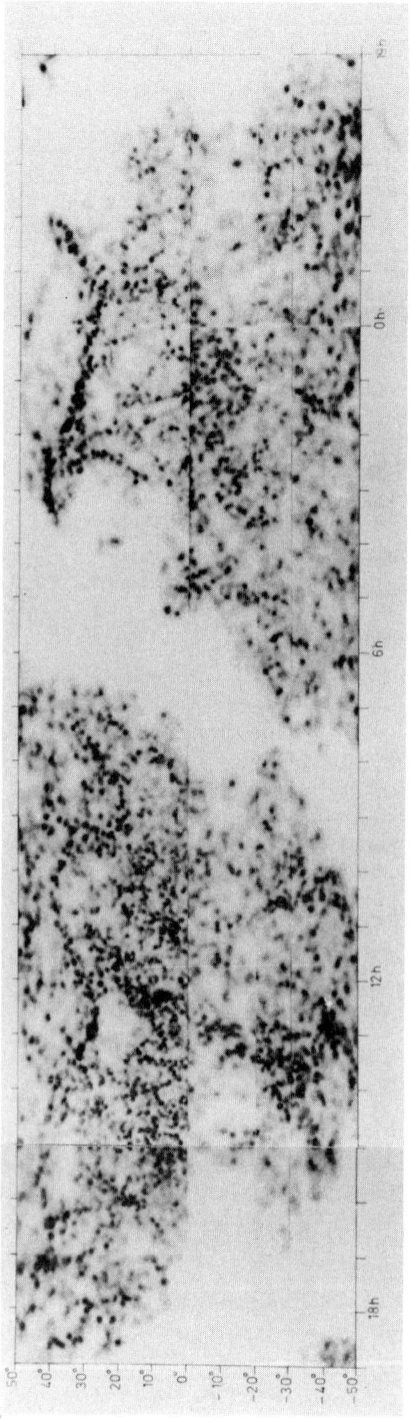

Fig.2 : Galaxian surface density in
the zone $-50^{\circ} < \delta < 50^{\circ}$.

account : in that way, zones having the same shading intensity on the maps and located in different catalogues correspond approximately to the same surface density in galaxies having a magnitude $m \leqslant 15.5$. Nevertheless, some difficulties still remain with the MGC, whose completeness varies with the right ascension !

The galaxian surface density for the whole sky is presented in the form of linear shade plots in 3 maps (Fig 1 to 3) which cover the zones $\delta \geqslant +50°$, $-50° \leqslant \delta \leqslant +50°$ and $\delta \leqslant -50°$, respectively. The darkest zones correspond to the highest densities ; the black pixels are for surface densities higher than 7 galaxies with $m \leqslant 15.5$ per square degree, i.e. 5 times the average density, and the white ones are for densities less than 0.35 galaxies per square degree.

ANALYSIS OF THE MAPS

On these maps, the irregularity and the complexity of the galaxy distribution is obvious (note also the zone of avoidance of the Milky Way, which goes across each map). A dozen of large-scale high density features appear, often as linear structures. Most of them are well known, in particular the conspicuous Perseus filament (α = 23h to 3h, $\delta \sim 35°$). But three seem to have escaped to previous investigations ; they are located North of the Lynx supercluster, South of the Hydra cluster and South of the Perseus supercluster, respectively. From now on, we will concentrate uniquely on these 3 structures, and we will test their tridimensional reality, by comparing the distributions of the available redshifts in the zone concerned to the distributions expected for galaxies uniformly distributed in depth and having a Schechter's luminosity function. In order to simplify this test, we consider only the galaxies having a magnitude $m \leqslant m_1$, where m_1 is the limit magnitude for which all the redshifts of the galaxies with $m \leqslant m_1$ are known in the zone, or a certain proportion of them which does not depend on m. Practically, m_1 = 14.5 and m_1 = 13.5 for the Northern hemisphere down to $\delta = -2°.5$ and the Southern hemisphere features, respectively. Of course, cluster galaxies are removed to perform the test.

THE NEW STRUCTURES

(1) In Figures 1 and 2, the main concentration seen between (α = 7h30, $\delta = 60°$) and (α = 10h, $\delta = 45°$) is the Lynx-Ursa Major supercluster, studied in details by Giovanelli and Haynes (1982).

For $\alpha < 7h30$, this structure seems to divide into two branches, one towards $\delta = 50°$ and the other one, cut off by the obscuration zone, towards high declinations. The first feature cannot be studied, by lack of redshifts. But the distribution of the redshifts of the galaxies in the second branch having $m \leqslant 14.5$ reveals a complete absence between 2500 and 3500 km s^{-1} and an excess between 3500 and 4500 km s^{-1} compared with what is expected for an homogeneous distribution. Although obtained from a small sample of 24 objects, this result suggests a real association of that branch with the Lynx Ursa Major supercluster, whose redshift range is similar.

2) In Figure 2, one can easily see the most prominent and best known

large-scale structure in the Northern hemisphere, namely the Perseus-Pisces supercluster, whose ridge extends from $\alpha = 22h$ to $\alpha = 4h$ and $\delta = 30°$ to $40°$. From the ridge four or five filaments seem to depart towards the South and reach a crowdy region around $\delta = 0°$ extending from $\alpha = 23h$ to $\alpha = 4h-6h$. To study the redshift distribution of the galaxies in this region, it is necessary to consider separately the zones North and South of $-2.5°$, respectively, because of different values of the limit magnitudes (cf. previous section). The redshift distribution in the zone 0h-4h, $-2.5°$ to $+10°$. shows a definite excess in the range 5000-7000 km s^{-1}, just as in the main ridge. No firm conclusion can be drawn concerning the Southern zone from $-10°$. to $-2.5°$ because the galaxies with measured redshifts are too bright, but there is however a hint of an excess in the same velocity range.

3) In the Southern hemisphere the Hydra cluster is well visible (Fig. 2) at 10.6h, $-27°$. A high density region located between $\alpha = 9.5h$ and 11h from $\delta = -30°$ to $-60°$ seems to be connected to the cluster ; in fact, it presents a velocity distribution with an excess between 2500 and 3500 km s^{-1}. compared to the case of an homogeneous spatial distribution, indicating a possible real association of the galaxies of this region to the Hydra cluster, which has also a mean velocity of 3500 km s^{-1}.

REFERENCES

Dressler, A. : 1980, Astrophys.J. <u>236</u>, 351.

Giovanelli, R., Haynes, M.P. : 1982, Astron.J. <u>87</u>, 1355.

Huchra, J., Davis, M., Latham, D., Tonry, J. : 1983, Astrophys.J.Suppl. <u>52</u>, 89.

Lauberts, A. : 1982, The ESO/Uppsala Survey of the ESO(B) Atlas, European Southern Observatory.

Vorontsov-Velyaminov, B.A., Krasnogorskaja, A. : 1962, Morphological Catalogue of Galaxies, Moscow University, Vol.1.

Vorontsov-Velyaminov, B.A., Arhipova, V.P. : 1963-1968, 1974, Morphological Catalogue of Galaxies, Moscow University, Vols. 2-5.

Zwicky, F., Herzog, E., Wild, P., Karpowicz, M., Kowal, C.T. : 1961-1968, Catalogue of Galaxies and of Clusters of Galaxies, Cal. Inst. Tech., Pasadena, Vols. 1-6.

TRACING SUPERCLUSTERS AND VOIDS WITH ABELL CLUSTERS

Jack O. Burns
Institute for Astrophysics, Univ. of New Mexico
Albuquerque, NM   87131   USA

David J. Batuski
Space Telescope Science Institute, Homewood Campus
Baltimore, MD   21218   USA

ABSTRACT.   The structure within the distribution of rich clusters is
described.   In particular, the Pisces-Cetus region is shown to contain
a dense supercluster, a 300 h_{75}^{-1} Mpc long filament of galaxies and
clusters, and a ring of 12 Abell clusters surrounding a $\approx$ 50 h_{75}^{-1} Mpc
diameter void.

Rich clusters of galaxies such as those found in Abell's catalog
may serve as important tracers of the large scale distribution of luminous matter in the Universe.   In principle, such clusters allow us to
efficiently and rapidly sample a much larger volume and, therefore,
larger scale sizes of structure than is currently possible in magnitude-
limited galaxy redshift surveys.   Recently, there have been suggestions
of some extremely large (>300 Mpc) supercluster structures composed of
Abell clusters linked by low density chains of galaxies (Batuski and
Burns, 1985a) and voids defined by the absence of rich clusters (Bahcall
and Soneira, 1982; Batuski and Burns, 1985b).   One general question
remains: how accurately do these clusters trace the distribution of
matter?

Two approaches to this problem have been taken.   First, one can
examine this question using the 2-point spatial correlation function
for rich clusters (Bahcall and Soneira, 1983).   Confusion becomes immediately apparent, however, in noting that the amplitude of this function depends upon richness (Bahcall, 1987).   The origin of this effect
is still unclear at present.   The relationship between clusters and
galaxies becomes clouded by this richness effect.   More important,
however, is to note that the 2-point function cannot provide a complete
description of the connectedness of Abell clusters.   Batuski and Burns
(1985b) have shown that there is much more large scale structure in the
multiplicity function of superclusters composed of Abell clusters than
in random distributions of pseudo-clusters constrained to follow the 2-
point function.   Thus, we conclude that the 2-point function is an
inadequate tool to address this question.

A. Hewitt et al. (eds.), Observational Cosmology, 319–322.
© 1987 by the IAU.
"""

Second, a more useful approach that we have undertaken is to identify regions that are overdense in Abell clusters using a percolation scheme, then follow-up on these candidate superclusters with redshift observations of individual galaxies within and between clusters. In this way, we directly address the physical connectedness of rich clusters. We have published the more extensive catalog to date of Abell cluster structures composed of 102 candidate superclusters and 29 candidate voids with $z \lesssim 0.13$ (Batuski and Burns, 1985b).

As an example of the structure seen within these superclusters, we show in Fig. 1 the Pisces-Cetus region containing a dense collection of rich and poor clusters from Batuski and Burns (1985a). This supercluster contains 36 Abell clusters with redshifts ranging between 0.03 to 0.08 (all of which are measured). This region is 5 times denser in rich clusters than the average found in Abell's catalog. The full range of large scale structures is represented here. There is a filament of clusters shown by the dashed contour in Fig. 1 that is 310 h_{75}^{-1} Mpc in length. It is the largest such structure presently known. The filament is traced well by both Abell and Zwicky clusters. In the nearest 90 Mpc (Perseus-Pisces), the clusters are very tightly linked in 3-D by CfA survey galaxies. Beyond this inner portion, our recent redshift observations show that galaxies continue to follow the clusters. Statistically, the filament is unlikely to be a chance 3-D alignment of clusters at the 0.3% probability level.

On either side of the filament are ~ 80 h_{75}^{-1} Mpc and ~ 120 h_{75}^{-1} Mpc regions that are devoid of clusters. They provide definition to the filament. Except for the volume nearest Perseus, the filament and void structures are not seriously affected by galactic obscuration.

At the heart of the densest portion of the Pisces-Cetus supercluster (near A189 and A195), we have also found a very interesting structure of clusters and galaxies. As shown in Fig. 2, a ring of 12 Abell clusters surrounds a region that is significantly underdense in galaxies. This structure is present in both the 2-D distribution of Zwicky galaxies and in the 3-D Cfa survey. The ring is tilted with respect to the plane of the sky oriented such that A400, A194, and A397 are closer to us than the remaining portion. There is also complex redshift structure within a "nest" of clusters to the south of the void. The structures seen here must still be viewed as tentative because of the limited galaxy redshifts presently available.

We conclude that in the Pisces-Cetus region and other candidate superclusters described by Batuski and Burns (1985b), there are highly interesting 50-300 h_{75}^{-1} Mpc sized structures illuminated by Abell clusters. Our work to date provides considerable evidence that rich clusters are good tracers of the galaxy distribution. Therefore, Abell clusters appear to provide an extremely useful tool for mapping the largest structures in the Universe.

We wish to thank J. Huchra and colleagues for providing us with some unpublished galaxy redshifts. J.O.B. was supported in this work by grants from the U.S. NSF (AST-8314558 and AST-8520115).

REFERENCES

Bahcall, N. 1987, this volume.
Bahcall, N. and Soneira, R. 1982, Ap.J.(Letters), 258, L17.
______ . 1983, Ap.J., 270, 20.
Batuski, D. and Burns, J. 1985a, Ap.J., 299, 5.
______ . 1985b, A.J., 90, 1413.

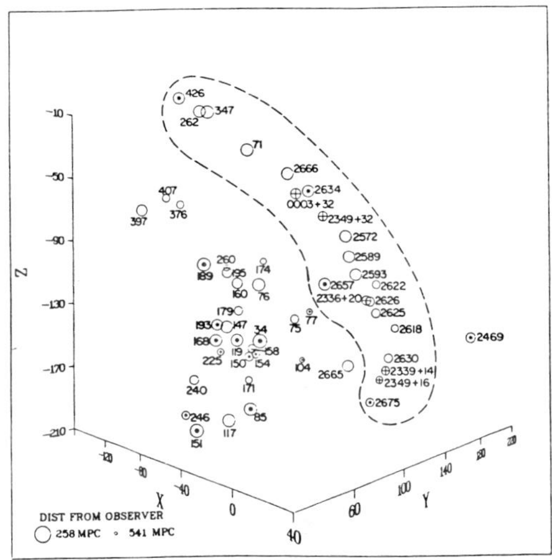

Fig. 1 - Plot of projection of the 3-D arrangement of clusters in the Pisecs-Cetus region. Units for the axes are Mpc. Earth is at position (0,0,0). Observer is 400 Mpc from the cube center. Numbers are Abell or Zwicky clusters. Note the filamentary arrangement within the dashed contour that is over 300 Mpc in length.

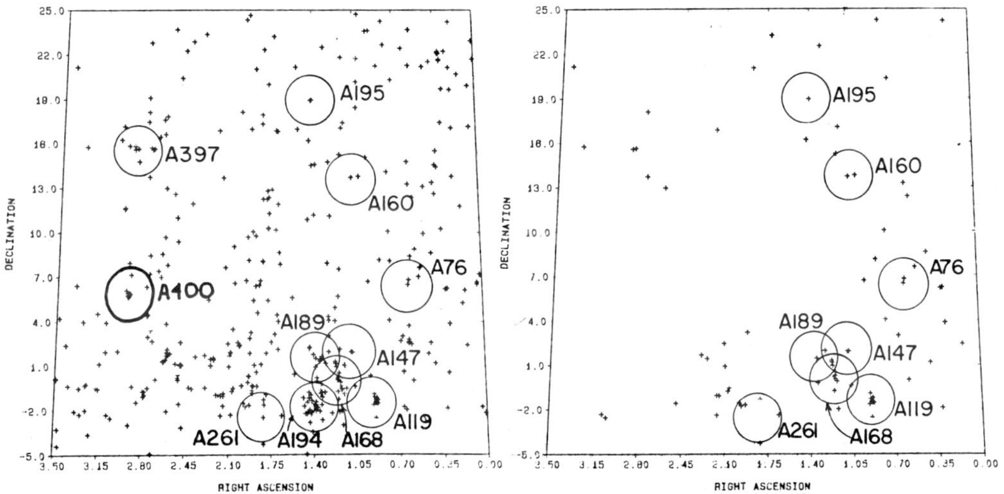

a) 5000-14,400 km/sec

b) 10,000-13,500 km/sec

Fig. 2 - Ring of 12 Abell clusters surrounding a 50 Mpc void of galaxies. Crosses are positions of CfA galaxies over a large (a) and smaller range (b) of velocities (distances). The ring is inclined at an angle with respect is the sky plane such that A189, A400, and A397 are closer to us. The void appears to be centered at about 12,000 km/sec. This region is near the center of Fig. 1.

DISCUSSION

KIANG: (1) I think I was the first one to suggest that the correlation
length for Abell clusters may be a function of sampling volume, through
a rather sophisticated model fitting method (Kiang 1966, MNRAS).   (2)
Saslaw and I produced a spatial correlation curve for Abell clusters,
using simply the magnitude of 3rd (10th?) brightest member as distance
indicator, about 20 years ago. This also showed the bump at the
large separation shown in the diagram given by Batuski et al. based
on extensive redshift data.

BURNS: (1) This possibility suggested again recently by Einasto et al.
(1986, MNRAS), is one of several possibilities to explain the richness
effect on the 2-point function. (2) This bump at ~200 Mpc (see
Batuski et al. these proceedings) appears to be a significant feature
in large volume samples of rich clusters. It appears consistent with
recent 6-m data described by Karachentsev (these proceedings) and the
appearance of galaxy sheets separated by ~200 Mpc in faint galaxy
surveys (Ellis, these proceedings). At present, this bump is not
reproduced in any models of large scale structure.

BAHCALL: Unless the volume studied is too small and therefore unrepre-
sentative of the sky average, the correlation functions do not depend
on the size of the sample volume. For example, several correlation
functions of galaxies were determined over quite different volumes
(Lick Counts, CFA Survey, Southern Surveys), all yielding essentially
the same results. Similarly, samples of different objects studied
over the same volume, such as the Lick galaxies, $R>1$ clusters, and
$R>2$ clusters (Bahcall and Soneira 1983, Ap.J.) yield intrinsically
different correlations.

SIMULATIONS OF LARGE-SCALE STRUCTURE COMPARED TO ABELL CLUSTER DISTRIBUTION

David J. Batuski
Space Telescope Science Institute, Homewood Campus
Baltimore, MD 21218 USA

Adrian Melott
Department of Physics and Astronomy, Univ. of Kansas
Lawrence, KS 66045 USA

Jack O. Burns
Institute for Astrophysics, Univ. of New Mexico
Albuquerque, NM 87131 USA

ABSTRACT. The amount of structure present among the Abell clusters out to redshift $z = 0.085$ has been compared with numerical supercomputer simulations (with 64^3 particles) of the isothermal, neutrino, and cold particle models for large-scale structure, assuming a flat universe and $H = 50$ km sec^{-1} Mpc^{-1}. High-density clusters of particles were identified in each simulation. Correlation and percolation tests were then used to compare the spatial distribution of these high-density points with the apparent superclustering among Abell clusters. While all of the models had some small superclusters (the neutrino model has too many), none came very close to possessing the extremely extensive structures found in the Abell clusters (generally, disagreement by 2σ or more).

A second set of simulations used the cold particle model with $\Omega = 0.2$ and 0.5. The structures found in these simulations were certainly larger than those of the $\Omega = 1.0$ cold particle case, but still $> 2\sigma$ too small in comparisons with the Abell clusters.

The spatial distribution of Abell clusters shows evidence of some very large-scale (~ 300 Mpc) structures in the Universe (*e.g.*, Oort 1983, Bahcall and Soneira 1984, Batuski and Burns 1985), and an important current question is how well models for large-scale structure can match the observed distribution of galaxies and clusters. To begin to answer this question on these very large scales, we created numerical simulations of three popular models (isothermal (IS), neutrino (NE), and cold particle (CP)) with a Cyber 205 supercomputer at Purdue University. We then looked for clusters within the simulations, and compared their spatial distributions with that of Abell clusters.

In the simulations, we used 64^3 particles, positioned on a 128^3 cloud-in-cell (CIC) grid to maximize spatial resolution. The density data were smoothed by another application of the CIC algorithm to a 64^3 grid, with each cell ~ 24 Mpc on a side. Thus, the volume for each simulation was $(\sim 1536$ Mpc$)^3$. The initial conditions for the simulations were the power spectra for the three models

A. Hewitt et al. (eds.), Observational Cosmology, 323–326.

considered, at $z = 2.5$, prior to which time collapse of density perturbations on scales of interest was linear. The models were then evolved gravitationally to the time $z = 0$, identified as the time that the slope of the mass two-point correlation function matched that observed for galaxies.

Each simulation volume was then searched for the grid points of highest mass density, which were considered analogous to Abell clusters in the Universe. Those points were selected that were above a mass density threshold which yielded the same number density of "pseudo-clusters" as that observed for the Abell clusters. For the percolation tests performed, each simulation was sampled with a volume of the same size and shape as that containing our Abell cluster sample.

This sample consists of 226 $R \geq 0$ clusters in the largely unobscured ($l > 30°$) portion of the sky, within $z < 0.085$. The sample is 85% complete in redshift measurements, and redshifts of unmeasured clusters were estimated by magnitudes of the tenth-brightest cluster galaxies. To make the Abell cluster data as directly comparable to the simulations as possible, these data were also smoothed with the CIC algorithm, to the same grid scale as the models. We used several tests on the different samples for $\Omega = 1.0$, $H = 50$ (and also $H = 75$, but the matches of models to observations were consistently worse than at $H = 50$). In Fig. 1, the two-point correlation functions of the four samples are shown. The Abell sample is quite strongly correlated for $r \sim 1.0$, much different from CP and IS, which have essentially $\xi = 0$ for $r > 0.8$, and even more different from NE, which shows significant anticorrelation near $r = 1$. The Abell cluster data has a $> 3\sigma$ "bump" in the $2.2 < r < 3.2$ range, also very different from the $\xi = 0$ of all the models in the same range.

Fig. 2 shows the results of one of our percolation tests, where the fraction of clusters identified as supercluster members is plotted as a function of the percolation parameter (b_p) used to define the superclusters. All the samples match well for b_p large, as nearly all the clusters are "connected" into superclusters, but the neutrino model has far too many clusters in superclusters at small b_p, and IS and CP have far too few in comparison to the Abell cluster sample.

Finally, Fig. 3 provides still another view of the structures present in the samples, through a multiplicity function analysis at $b_p = 0.7$. Almost half of the Abell clusters are found in superclusters of 10 or more members, while about 65% of clusters in each of the models are in smaller superclusters (2-10 member clusters). None of the models have more than 10^{-4} probability (χ^2 test) of being drawn from the same population as the Abell clusters, with these distributions.

We also looked at the CP model for $\Omega = 0.2$ (H had to be ≥ 100 to prevent conflict with the isotropy of the microwave background) and $\Omega = 0.5$ ($H \geq 50$). These simulations contained only slighty greater amounts of structure on larger scales, still disagreeing with the Abell case at the $> 3\sigma$ level in the correlation function and multiplicity function tests.

Thus, we conclude that the currently popular models for large-scale structure can not provide enough structure to match the observed distribution of Abell clusters. It is possible that incompleteness in the cluster sample contributes to the apparent large-scale structure. However, with the data currently available, some new model appears necessary, perhaps one employing cosmic strings to generate the required spectrum of very large-scale density perturbations (*e.g.*, Zel'dovich 1980).

REFERENCES

Bahcall, N. A., and Soneira, R. M. 1984, *Ap. J.*, **277**, 27.
Batuski, D. J., and Burns, J. O. 1985, *A. J.*, **90**, 1413.
Oort, J. H. 1983, *Ann. Rev. Astr. Ap.*. **21**, 373.
Zel'dovich, Ya. B. 1980, *M. N. R. A. S.*, **192**, 663.

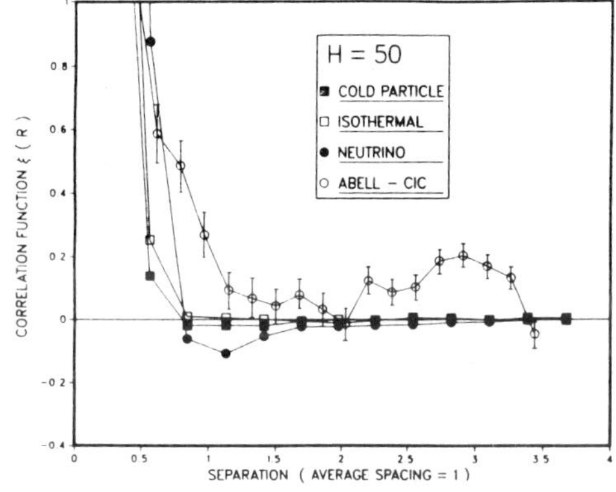

Figure 1 Two-point correlation functions for three simulations at $h_{100} = 0.50$, compared with $\xi(r)$ for the Abell clusters. Error bars on the Abell cluster curve reflect assumption of Poisson errors in numbers of pairs found in each separation bin. The scale for r is in terms of the average separation of nearest neighbors within the $R \geq 0$ Abell cluster sample, $r_{ave} \equiv \rho^{-1/3} = 42.5 h_{100}^{-1}$ Mpc, where ρ is the average number density of the Abell clusters.

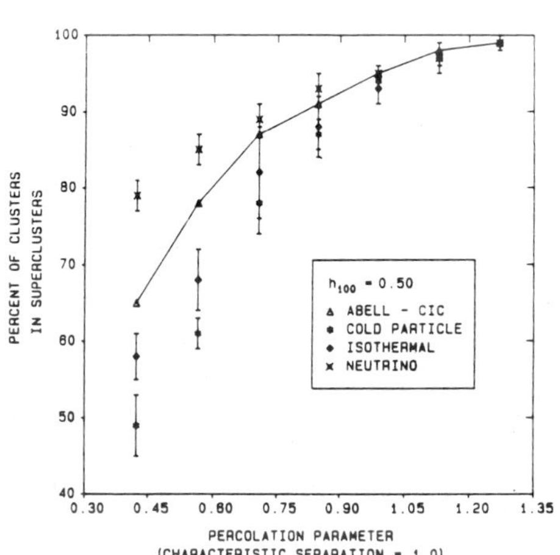

Figure 2 Percentage of pseudo-clusters in the models that were linked together into superclusters as a function of the maximum separation two pseudo-clusters could have and still be linked - the percolation parameter, b_p. Error bars represent the 1σ variation in four samples for each model. Observed function for the smoothed $R \geq 0$ Abell clusters sample is shown by the connected symbols.

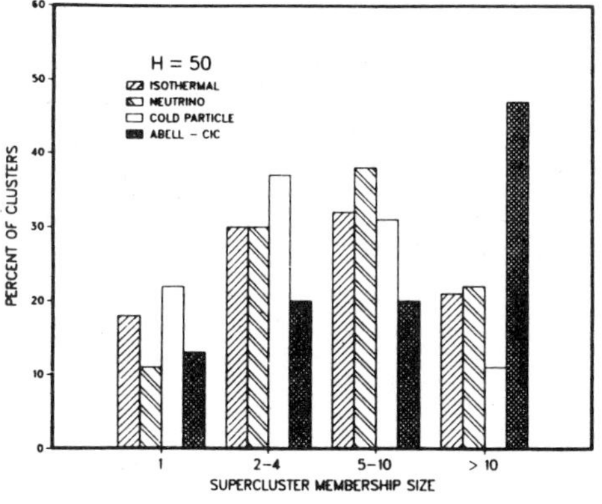

Figure 3 Average percentage of pseudo-clusters found in superclusters of various sizes in the models compared to superclusters found among the $R \geq 0$ Abell cluster sample population. Bins are number of clusters per supercluster, with a membership of "1" representing isolated clusters. The percolation parameter for the definition of a supercluster was $b_p = 71\%$ of the average nearest-neighbor separation of the Abell clusters, *i.e.*, for $h_{100} = 0.50$, $b_p = 60$ Mpc. Variations in the four samples of each of the models were roughly $\pm 9\%$ (one standard deviation) in the > 10 bin and $\pm 3\%$ in the other bins.

DISCUSSION

DEKEL: My feeling is that cosmic strings, which looked very promising at first as a way to explain the strong clustering of clusters, has become recently a theory which requires as much 'patching' as the other theories. For example, I don't think we understand why the cluster correlation function should be $\xi_{cc} \propto r^{-1.8}$ and of high amplitude. The loops about which clusters accrete don't seem to be distributed like 'beads along strings' as once thought, and the segments of infinite strings seem to be anticorrelated.

BATUSKI: I am disappointed to hear this. I had recently become quite excited about the possibility that cosmic strings could explain very large-scale structure.

ULMER: I doubt there is much effect, but I wonder how literally we should take the Abell Catalog given that some are superposition of line of sight clusters and that some clusters break up into unbound clumps when redshifts are measured.

BATUSKI: I also do not think that the effect is very large, in part because we have smoothed the Abell cluster data with the CIC algorithm, removing some of the details of the cluster distribution. Such superposition effects, as well as possible incompleteness of the $R \geq 0$ clusters with galactic latitude and redshift (even with the latitude and redshift limited sample we used to minimize the incompleteness), suggest that the Abell catalog does need observational refinement. These clusters are still the best available probes of the scales under consideration, however.

SHAPE OF SUPERCLUSTERS

G. Vettolani (1), G. Chincarini (2), R.E. de Souza (3)
(1) Istituto di Radioastronomia CNR, Bologna, Italy
(2) Osservatorio di Brera, Milano, Italy
(3) University of Sao Paulo, Sao Paulo, Brasil

ABSTRACT. Attempts to measure structural parameters of superclusters are described. Catalogues of superclusters have been prepared using a percolation algorithm applied to a complete sample of UGC galaxies brighter than the 14.5 magnitude. Both a two dimensinal (projected) and a three dimensional approach have been tried. The (conservative) result is that superclusters are certainly not spherical. Probably the departure from sphericity is towards elongated cigar-like structers. No evidence of relative alignements of superclusters has been found.

1. INTRODUCTION

In the last few years it has been established that visible matter in the Universe is distributed in large agglomerates of irregular form, the superclusters. Understanding their shape will strongly constraint the models of galaxy formation. The present contribution is an attempt towards the definition of some shape parameters of superclusters and the study of their statistical distribution.

2. THE PERCOLATION ALGORITHM

We have analyzed a sample which contains all the galaxies (3681) listed in the Uppsala Catalogue brighter than the 14.5 magnitude, selecting structures through a percolation algorithm.

This algorithm selects "friends of friends" and generates a system of sets. A set so selected is the ensamble of galaxies in which each member is separated from at least one other member by a distance smaller than a selected lenght R (percolation vector) and no other object of the sample is at such or smaller distance from any object of the set. Computing the percolation vector we take also into account the fact that in a magnitude limited sample the galaxy density is a decreasing function of distance, since, with increasing distance, only the more luminous galaxies enter the sample at encreasing distances. Furthermore we take into account the excess probability over a poissonian distribution of finding a galaxy at a distance R from another galaxy in agreement with the fact that the galaxy distribution is characterized by the two point correlation function.

Catalogues of structures have been obtained varying the percolation

A. Hewitt et al. (eds.), Observational Cosmology, 327–329.

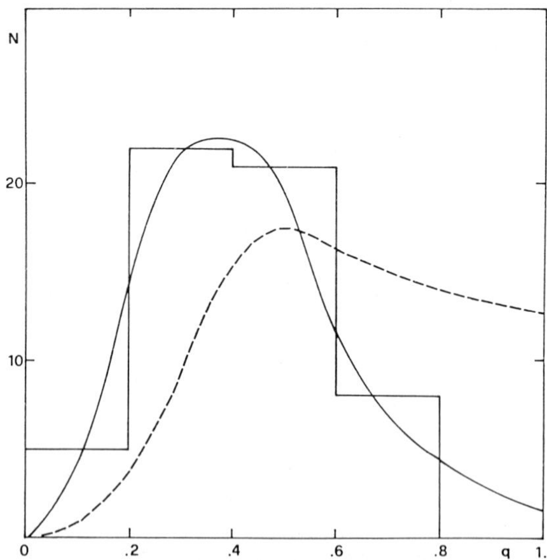

Figure 1. The flattening distribution of the structeres selected with R = 6 Mpc.   The dashed line and the solid line show the fitting with a distribution of oblate or prolate spheroids respectively.

vector R , and these structures have been analyzed in terms of shape
parameters. Note that increasing R is equivalent to select structures
at a lower density contrast above the mean.

2. SHAPE OF STRUCTURES

The underlaying assuption is that a supercluster can be idealized as
a symmetric distribution of points: an ellipse or a three axis
ellipsoid.

The flattening distribution of the structeres selected with a
percolation vector of 6 Mpc, as they apper projected upon the plane of
the sky, is given in Fig 1. The fittings by a distribution of oblate
(dashed line) or prolate (solid line) spheroids, whose true axial ratio
obeys a normal distribution, are also shown.   The best fitting is
clearly given by prolate ellipsoids (cigars), but simulations of a small
number of points drawn at random from a sphere indicate that there is a
tendency, when dealing with small statistics, to measure a prolate
spheroid. Some structures are very elongated indeed. An example is a
structure at RA 15 hours DEC +42 which has an axial ratio of 0.06.

In space structural parameters of superclusters have been defined by
a weighted tensor which is defined in a similar way to the tensor of
inertia of a set of points.   Each point represents a galaxy in a
structure (selected at a certain percolation vector) and the weights
account for the density variations as a function of depth in a magnitude
limited sample. The shape parameters are defined as the principal axes
of the correspondent three-axis ellpsoid. Here again we have a tendency
to have rather elongated structures , with a slight preference toward
oblateness, but the presence of velocity dispersion, or peculiar
velocities, tends to increase the depth of the structures and therefore
would tend to transform a structure which looks like a string as
projected on the sky into a disk or oblate form.

4. RELATIVE ORIENTATION OF CHAINS

Existence of relative alignements between large scale structures will
give us clues on the coherence length of the physical phenomena which
led to the supercluster formation.

We have searched for two possible kind of alignements: a)
paralleleism of the major axis of a chain and that of the nearest chain,
and b) orientation of the major axis of a structure in the direction of
its nearest neighbor. Both tests gave a negative results.

5. CONCLUSION

The analysis shows the need of deeper samples in order to improve
statistics, limitations of the approximation and above all the need of
an approach more sensitive to the rather complicate topology of the
Universe.   For example a sponge like topology (see Gott, these
proceedings) cannot be evidenced by our method.

The primary goal of our analysis, however, has been reached. That is
that there is no strong evidence, within such a sample, of pancake-like
structures.

AN ANALYSIS OF FIFTY-FIVE BRIGHT SOUTHERN CLUSTERS OF GALAXIES

R. P. OLOWIN
Department of Physics and Astronomy
The University of Oklahoma
440 West Brooks, Rm. 131
Norman, Oklahoma 73019   USA

ABSTRACT.   We present the results of an analysis of fifty-five bright clusters clusters of galaxies as a homogeneous sample taken from a new effort to catalogue galaxy clusters in the Southern Hemisphere. The positions of some 21,000 galaxies in clusters have been catalogued along with visual magnitudes, morphological types, position angles of extended objects and pertinent remarks. For all of the clusters, various cluster parameters have been determined and form the basis of comparative studies for these fundamental aggregates of matter in the universe.

1. The Cluster Catalogue

The aims of this study are to produce a homogeneous sample of galaxy clusters measured to a uniform limiting magnitude of mv = 19.0; catalogued with accurate positions relative to nearby astrometric standard stars; morphologically classified and population typed; and statistically analysed in a uniform fashion to deduce certain cluster parameters. The cluster parameters of interest include an estimate of cluster distance, center and richness; galaxy distributions as a function of morphological type, magnitude distribution and core radius. A succinct table (Table 1-a,b) gives the following information: Cluster Identification,Celestial Coordinates, ESO/SERC J Field Number, X-Y Coordinates measured from the Field center, Redshift as determined from the tenth brightest cluster member, Abell Richness, Abell Type and Bautz-Morgan Classification.

The Catalogue has been extracted from a new Survey of Southern Galaxy Clusters discovered on the ESO/SERC J Survey of the Southern Sky.

A. Hewitt et al. (eds.), Observational Cosmology, 331–333.

CATALOGUE OF BRIGHT SOUTHERN GALAXY CLUSTERS

N	RA (1950)	DEC (1950)	FLD	X	Y	Z	R	TY	B-M TYPE
01	00 00 45.6	-36 19 39.4	349	005.543	-070.735	0.042	1	R	I-II
02	00 03 13.9	-35 4 01.6	349	035.621	-003.767	0.072	2	R	II
03	00 06 49.1	-35 43 49.9	349	071.702	-039.543	0.057	2	IR	II-III
04	00 07 28.7	-57 14 52.8	149	051.069	-119.610	0.044	1	IR	III
05	00 18 08.3	-49 32 34.3	194	-107.031	024.617	0.050	1	R	II
06	00 22 49.7	-33 17 33.7	350	-015.272	094.597	0.038	2	RI	II
07	01 29 33.8	-51 32 55.7	196	-005.774	-082.259	0.040	0	RI	III
08	01 39 50.0	-42 23 49.0	297	-043.427	-127.677	0.050	1	RI	I-II
09	02 55 44.0	-52 56 09.1	154	087.306	109.669	0.049	2	I	II
10	03 44 04.7	-41 21 11.8	302	-103.679	-070.538	0.050	2	R	I
11	04 04 04.0	-39 00 28.8	302	101.219	054.375	0.042	1	IR	II
12	05 24 00.3	-45 01 46.1	253	-115.824	-001.271	0.061	0	I	III
13	06 21 39.4	-64 56 34.9	087	-083.921	004.016	0.038	0	R	II-III
14	06 25 02.5	-53 39 26.8	161	-088.653	070.926	0.049	2	R	I
15	06 26 25.1	-54 22 29.4	161	-076.713	032.793	0.048	2	R	II
16	12 51 41.2	-28 44 33.4	442	146.553	065.348	0.048	2	RI	II
17	13 03 25.3	-37 17 59.6	382	-095.233	-121.728	0.049	0	R	II
18	14 00 40.6	-33 44 13.9	384	005.223	069.735	0.020	0	RI	I
19	14 09 17.9	-32 50 02.6	384	102.148	117.241	0.044	1	I	III
20	14 09 28.3	-34 01 38.5	384	102.819	053.059	0.038	1	RI	II
21	14 30 26.3	-31 32 40.5	447	-043.965	-081.811	0.054	0	RI	II-III
22	19 56 35.4	-38 32 47.9	339	003.581	080.348	0.026	2	I	I-II
23	20 38 34.4	-35 23 58.1	401	-105.191	-022.456	0.065	1	R	III
24	20 48 40.7	-52 08 45.6	235	-095.831	-115.415	0.037	1	I	III
25	21 13 10.0	-59 36 24.2	145	-131.705	018.574	0.042	2	I	II

TABLE 1-a

CATALOGUE OF BRIGHT SOUTHERN GALAXY CLUSTERS

N	RA (1950)	DEC (1950)	FLD	X	Y	Z	R	TY	B-M TYPE
26	21 22 58.0	-35 00 38.8	403	-144.719	-001.784	0.052	2	RI	III
27	21 26 09.6	-51 04 18.4	236	-033.381	-057.179	0.048	0	I	III
28	21 29 13.5	-35 22 50.6	403	-075.807	-019.961	0.044	1	I	II
29	21 31 05.9	-53 50 08.2	188	028.755	064.519	0.050	1	I	II
30	21 31 14.2	-62 15 40.3	145	-008.471	-118.667	0.048	2	R	I
31	21 32 18.2	-52 44 02.1	188	039.266	123.490	0.044	1	I	III
32	21 41 46.5	-51 43 57.4	236	096.553	-093.917	0.043	2	I	I-II
33	21 42 50.8	-57 29 48.5	145	074.215	134.600	0.042	2	R	II
34	21 43 46.1	-44 06 15.9	288	-119.835	051.560	0.048	2	I	II
35	21 44 40.6	-46 13 32.3	288	-107.000	-061.752	0.043	2	R	II-III
36	21 50 32.0	-58 03 48.4	145	127.534	102.014	0.050	2	R	II-III
37	21 55 17.4	-60 35 13.4	146	-098.536	-032.630	0.048	2	R	III
38	21 58 08.7	-60 11 45.0	146	-080.549	-010.760	0.065	2	R	I
39	22 01 11.1	-58 18 36.8	146	-063.530	090.895	0.043	1	I	I-II
40	22 01 06.9	-50 18 11.9	237	008.125	-014.404	0.042	0	R	I-II
41	22 19 59.2	-50 23 08.1	190	-101.125	-022.117	0.040	2	R	II-III
42	22 21 25.9	-56 38 22.8	190	-087.461	-088.514	0.043	2	I	II-III
43	22 22 36.0	-56 05 57.1	190	-079.768	-059.436	0.043	1	I	III
44	22 24 46.3	-30 51 42.7	468	-143.511	-045.125	0.038	1	RI	II
45	23 16 35.3	-42 22 36.6	347	-075.547	-126.152	0.030	2	I	III
46	23 27 57.5	-39 37 06.8	347	038.794	021.784	0.049	1	I	I
47	23 34 24.3	-69 34 43.5	077	038.221	025.705	0.057	0	I	III
48	23 38 44.8	-30 30 04.2	471	-086.526	-026.689	0.044	1	I	I
49	23 44 55.4	-28 24 42.6	471	-015.479	085.995	0.023	1	R	I-II
50	23 59 05.8	-44 06 46.5	292	104.794	047.058	0.038	0	I	II

TABLE 1-b

Large-Scale Structure in the Universe: Spatial Distribution and Peculiar Velocities

Neta A. Bahcall
Space Telescope Science Institute
3700 San Martin Drive
Baltimore, MD, USA

Abstract. The evidence for the existence of very large scale structures, $\sim 100h^{-1}Mpc$ in size, as derived from the spatial distribution of clusters of galaxies is summarized. Detection of a $\sim 2000\ kms^{-1}$ elongation in the redshift direction in the distribution of the clusters is also described. Possible causes of the effect are peculiar velocities of clusters on scales of $10\text{-}100h^{-1}Mpc$ and geometrical elongation of superclusters. If the effect is entirely due to the peculiar velocities of clusters, then superclusters have masses of order $10^{16.5}M_{\odot}$ and may contain a larger amount of dark matter than previously anticipated.

I. Introduction

Rich clusters of galaxies are an efficient tracer of the large scale structure in the universe. Very large structures, of size $\sim 100h^{-1}Mpc$ or more, show up in the distribution of the rich clusters. ($h \equiv H_o/100\ kms^{-1}Mpc^{-1}$ is used throughout this paper). Clusters have mean separations of order $50h^{-1}Mpc$ and are therefore efficient in revealing these large scale structures that were not detected previously in the distribution of individual galaxies. The existence of large scale structures became evident by observing strong correlations among clusters, correlations that extend to separations as large as $100h^{-1}Mpc$ (Bahcall and Soneira 1983; Klypin and Kopylov 1983; Hauser and Peebles 1973), as well as by determining a specific catalog of superclusters in three dimensions (Bahcall and Soneira 1984). Other evidence, such as a giant void ($\sim 300h^{-1}Mpc$) of rich clusters (Bahcall and Soneira 1982b), and the extension of some rich superclusters to at least $100h^{-1}Mpc$ scale (Bahcall and Soneira 1982a), further support the existence of very large scale structures as a common feature in the universe.

The work was recently extended by Bahcall, Soneira, and Burgett (1986), who used the rich clusters to determine whether peculiar velocities exist on large scales. We find evidence for a large velocity broadening in the redshift distribution that corresponds to a cluster pair velocity of $\sim 2000\ kms^{-1}$. This suggests that dark matter may dominate even on these very large scales.

A. Hewitt et al. (eds.), Observational Cosmology, 335–348.

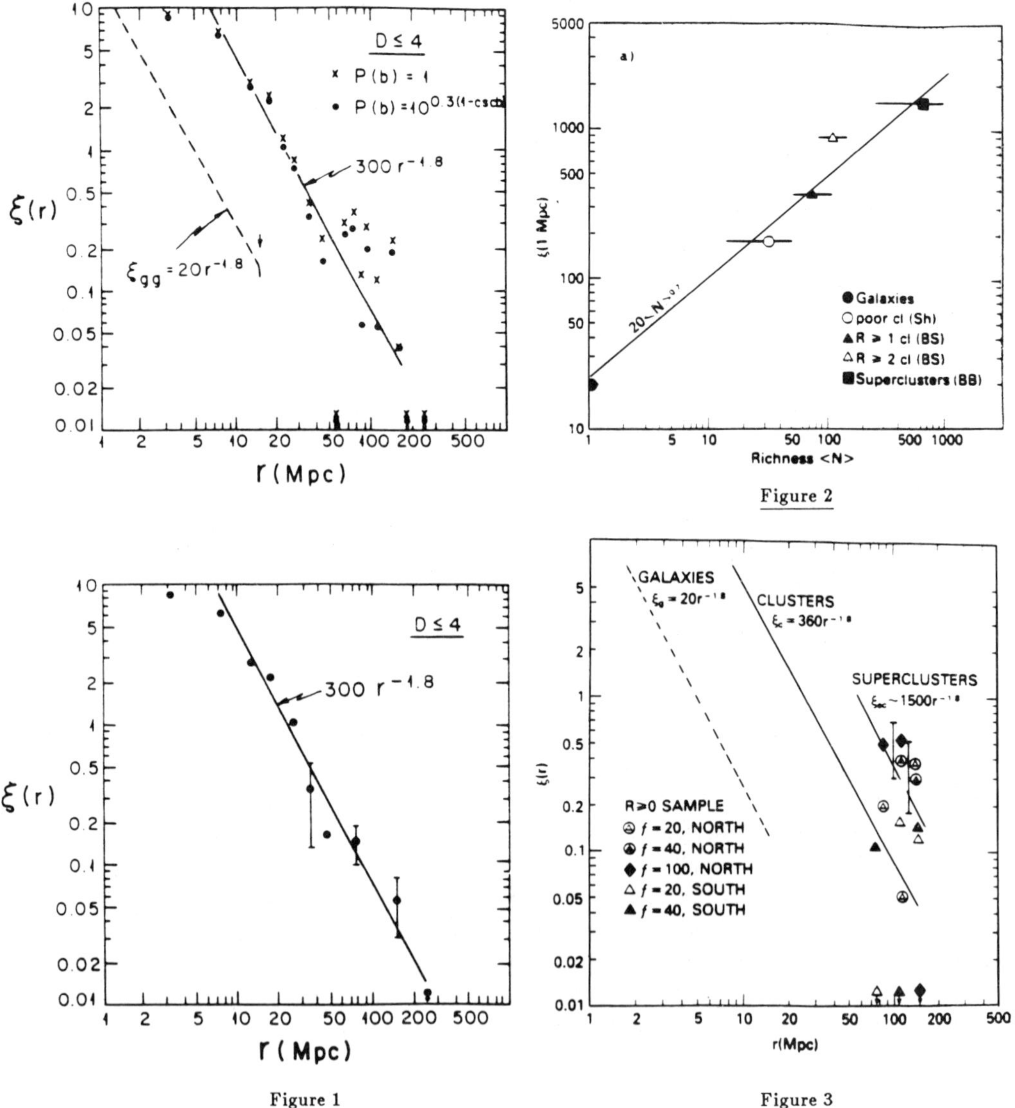

Figure 1: *(a)* The spatial correlation function of the $D \leq 4$ sample (Bahcall and Soneira 1983). The dashed line is the galaxy-galaxy correlation function determined by Peebles and co-workers. *(b)* Same as Fig. 1a but plotted in larger bins at large separations.

Figure 2: The dependence of the correlation function on the mean richness ($\propto$ luminosity) of the system (Bahcall and Burgett 1986). the points are for galaxies, poor clusters (Schectman 1985), $R \geq 1$ and $R \geq 2$ Abell clusters (Bahcall and Soneira 1983), and superclusters (Bahcall and Burgett 1986). The solid line indicates the approximate dependence of richness.

Figure 3: The spatial correlation of superclusters for the $r \geq 0$ sample (Bahcall and Burgett 1986). Different sub-samples are indicated by different symbols. No meaningful correlations are expected below $\sim 50h^{-1}\ Mpc$.

II. The Correlation Functions of Clusters and Superclusters

A. Clusters of Galaxies

The spatial correlation function of rich clusters of galaxies (Bahcall and Soneira 1983) reveals strong correlations among clusters and implies the existence of structures to scales of at least $\sim 100h^{-1}Mpc$. The above investigation is based on studying all rich clusters in the complete redshift sample of the Abell (1958) catalog to distance group $D \leq 4$ (i.e., $z \lesssim 0.08$). The study also includes, as a test, the determination of the angular correlation function of all 1651 clusters in the entire Abell catalog to $D \leq 6$ ($z \lesssim 0.2$).

The cluster correlation function is observed to be approximately 18 times stronger than the correlation function of galaxies (Figure 1). The cluster correlations, ξ_{cc}, for Abell richness groups $R \geq 1$ clusters can be expressed as (Bahcall and Soneira 1983):

$$\xi_{cc}(r) = 360r^{-1.8} \qquad r \lesssim 100h^{-1}Mpc \tag{1}$$

while the galaxy correlation function (e.g., Groth and Peebles 1977) is represented by:

$$\xi_{gg}(r) = 20r^{-1.8} \qquad r \lesssim 20h^{-1}Mpc \tag{2}$$

Bahcall and Soneira (1983) also found that the correlation function depends strongly on cluster richness, with rich clusters ($R \geq 2$) showing stronger correlations by a factor of ~ 3 as compared with the poorer ($R = 1$) clusters (both are consistent with an $r^{-1.8}$ power law). This result, combined with the lower correlation amplitude of individual galaxies lead the authors to the conclusion that progressively stronger correlations exist, at a given separation, for richer galaxy systems. This trend of increasing correlation strength with richness is shown in Figure 2. A recent study by Shectman (1985) of the correlations of still poorer clusters appears to be consistent with the trend suggested by Bahcall and Soneira. A phenomenological model that can explain the observed increase of correlation strength with richness was recently proposed by Bahcall (1986) and is summarized in section IV.

A representation of the observed correlations as a universal dimensionless correlation function, and its implication for a scale invariant clustering process is discussed in section III.

B. Superclusters

Bahcall and Burgett (1986) carried the study of rich galaxy clusters one step further by studying the spatial distribution of superclusters. The sample used was the Bahcall-Soneira (1984) complete catalog of superclusters to $z \leq 0.08$, where superclusters are defined as groups of rich clusters and identified by a spatial density enhancement of clusters. All volumes of space with a spatial density of clusters f times larger than the mean cluster density are identified in the above catalog as superclusters for a specified value of f. The supercluster selection process was repeated for various overdensity values f, from $f = 10$ to $f = 400$, yielding specific supercluster catalogs for each f value. A total of 16 superclusters are cataloged for $R \geq 1$ and $f = 20$, and 26 superclusters for $R \geq 0$ and $f = 20$. The spatial correlation among the superclusters was determined for the samples of different richness and overdensity values. Because of the large size of the superclusters themselves, no meaningful correlations are expected at small separations ($\lesssim 50h^{-1}Mpc$). In addition, no detectable correlations are expected at very large separations ($> 200h^{-1}Mpc$) since this scale is comparable to the limits of the sample. Any observable correlations are therefore expected only in a separation "window" around $\sim 100h^{-1}Mpc$.

The results, presented in Figure 3, reveal correlations among superclusters on a very large scale: $\sim 100\text{-}150h^{-1}Mpc$. Because of the small size of the supercluster sample, the statistical uncertainty is appreciable; the observed effect is at the 3σ level (as compared with numerical simulations of random catalogs). In addition, all the samples with different overdensities and cluster richnesses show a similar effect at a similar scale length. The results imply the existence of very large scale structures with scales of $\sim 100\text{-}150h^{-1}Mpc$.

Figure 3 also shows that the supercluster correlation strength is stronger than that of the rich cluster correlations by a factor of approximately 4. It is approximately two orders of magnitude stronger than the galaxy correlation amplitude. While this enhancement is observed in the $\sim 100\text{-}150h^{-1}Mpc$ range, it is possible that the super-cluster correlation function also follows an $r^{-1.8}$ law. If the correlations follow an $r^{-1.8}$ law, then the function would satisfy the relation

$$\xi_{sc,sc}(r) \simeq 1500r^{-1.8} \simeq (r/60)^{-1.8}. \tag{3}$$

The implied correlation scale of superclusters would be $60h^{-1}Mpc$, as compared with $5h^{-1}Mpc$ for the correlation scale of galaxies (Groth and Peebles 1977) and $25h^{-1}Mpc$ for rich ($R \geq 1$) clusters (Bahcall and Soneira 1983). This increase is correlation strength is consistent with the earlier prediction of Bahcall and Soneira (1983) of increased correlations with richness (luminosity) of the system. The supercluster correlation amplitude fits well the predicted correlation curve (Figure 2).

III. A Universal Correlation Function

The increase of correlation strength with richness (Figure 2) implies that rich, luminous systems are more strongly clustered, at a given separation, than poorer systems. The power-law of the correlation functions is also observed to be identical in the various systems studied. Either initial conditions, or superceding evolution, may be responsible for these observed clustering phenomena.

In Figure 4 we plot the amplitude of the correlation functions of the various systems (galaxies, poor and rich clusters, superclusters) as a function of the mean separation of objects in the sample, d (see Bahcall and Burgett 1986). The mean separation is related to the mean spatial density of objects in the sample, n, through $d = n^{-\frac{1}{3}}$. It is apparent from Figure 4 that the correlation strength increases with the sample's mean separation. Moreover, a dimensionless correlation function normalized to the sample's mean separation, d, appears to yield a constant, universal function for all systems studied

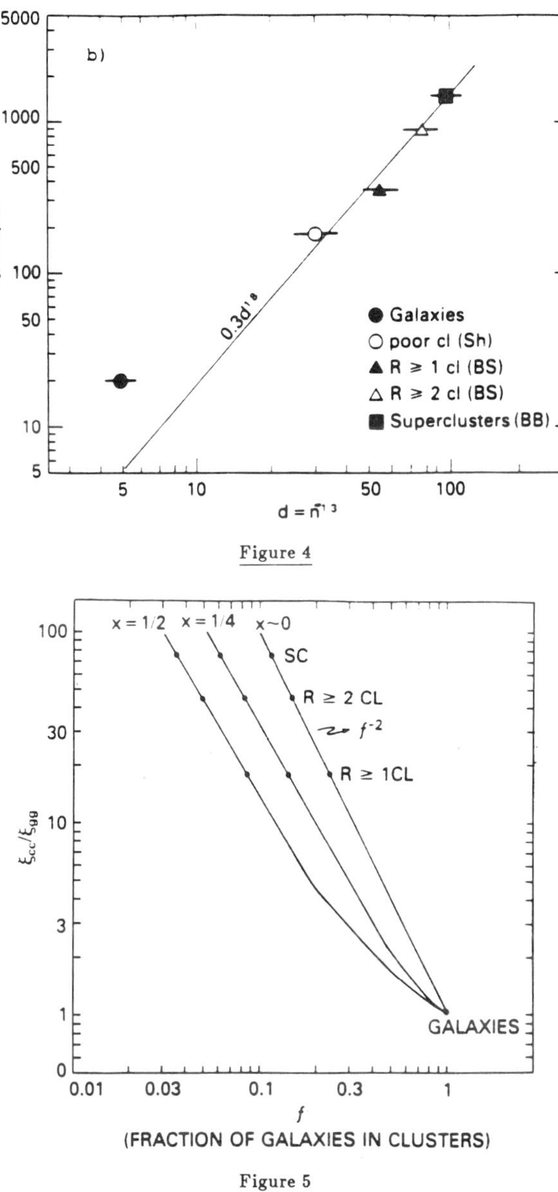

Figure 4

Figure 5

Figure 4: The dependence of the correlation function on the mean separation of objects in the system (Bahcall and Burgett 1986). The indicated ranges in the mean separation represent a $\pm 50\%$ variation in the mean density n (where $d = n^{-\frac{1}{3}}$). The solid line represents a $d^{1.8}$ dependence.

Figure 5: The ratio of the cluster to galaxy correlation functions predicted from Bahcall's (1986) model (eq. 5) is plotted as a function of the fraction f of galaxies associated with the clusters. The different curves represent different values of the "field" (non-cluster) correlation strength, ξ_{gg}^{f}. In terms of the ratio parameter $x = \xi_{gg}^{f}/\xi_{gg}$, the curves represent $x = 0, \frac{1}{4}$, and $\frac{1}{2}$, as indicated. The observed correlation strengths of $R \geq 1$ clusters, $R \geq 2$ clusters, and superclusters is indicated by the points.

(with some enhancement for galaxies, as described below). This universal dimensionless correlation function has the form

$$\xi_i(r) \simeq 0.3(r/d_i)^{-1.8}, \tag{4}$$

where the index i refers to the system being considered, and d_i is its mean separation. The correlation function of galaxies is stronger than given by equation (4) by a factor of about four (see also Figure 4).

The universality of the correlation function implies a scale-invariant clustering process (Szalay and Schramm 1985). The stronger dimensionless galaxy correlations may imply gravitational enhancement on smaller scales. If a non-linear process, other than gravity, participates in galaxy formation, and this process is scale-invariant, the created structure will have a single power-law correlation function, the slope of which is related to the geometry of the structure, i.e., its fractal dimension. Small scale gravitational clustering may break the scale invariance and increase the dimensionless correlation amplitude for galaxies.

Cosmic strings, serving as the primary agent in the formation of galaxies and clusters was recently suggested as one model for such a scale-invariant infrastructure (Turok 1986). The model yields a scale-invariant correlation function similar to that observed, with a power-law of -2.

IV. A Phenomenological Clustering Model

A phenomenological model that explains the observed difference between the galaxy and rich cluster correlation functions was recently proposed by Bahcall (1986). The model is based on the fraction of galaxies that participate in the clustering, and explains the observed trend of increased correlation strength with richness. A summary of the model is presented below.

The galaxy correlations depend, at least partially, on the rich cluster correlations since clusters contain galaxies. It can be shown that if all galaxies were members of rich clusters, the two correlation functions should be approximately the same on large scales. In this case they both trace the same large scale structure. The fraction of galaxies in clusters is clearly less than unity. The effect of the fraction of galaxies, f, that are associated with rich clusters, on the relation between the cluster and galaxy correlation functions, is the basis for the model. The parameter f represents the probability that a randomly chosen galaxy is correlated with a rich cluster. These associations may include large structures (separations of tens of Mpc), comparable to the separations observed in the cluster correlation function (and well above the standard Abell radius of $1.5h^{-1}Mpc$).

The galaxy correlation function contains contributions from three terms: galaxy pairs from the fraction f of galaxies that are cluster members; pairs from the fraction $1-f$ of galaxies that are non-cluster members ("field"); and cross-term pairs. Inserting the analytic expressions for each of these terms into the expression for the overall galaxy correlation function yields:

$$\left(\frac{\xi_{cc}}{\xi_{gg}}\right)^{\frac{1}{2}} = \frac{1 - (1 - f)(\xi_{gg}^{f}/\xi_{gg})^{\frac{1}{2}}}{f} \tag{5}$$

The above ratio of the cluster to galaxy correlation strength depends on two parameters: the fraction of galaxies in clusters, f, and the ratio of the "field" galaxy correlation strength, ξ_{gg}^{f} (i.e., the correlation of the 1-f fraction of galaxies outside the rich clusters) to the overall galaxy correlation ξ_{gg}. If all galaxies were associated with rich clusters, i.e., $f = 1$, then the galaxy and cluster correlations are identical, as expected. However, for any fraction $f < 1$, the galaxy correlations will be smaller than the parent cluster correlations due to the reducing effect of the less clustered "field" galaxies. Figure 5 represents graphically relation (5). The curves are the expected $\xi_{cc}/\xi_{gg}(f)$ relations for selected values of the parameter $x \equiv \xi_{gg}^{f}/\xi_{gg}$.

The observed correlation strengths discussed in sections II and III are represented by the data points. The observed ratio $\xi_{cc}/\xi_{gg} \simeq 18$ for $R \geq 1$ clusters yields a fraction of galaxies in clusters that ranges from $f \simeq 25\%$ for $\xi_{gg}^{f}/\xi_{gg} \simeq 0$ to $f \simeq 15\%$ for $\xi_{gg}^{f}/\xi_{gg} \simeq \frac{1}{4}$. Therefore, if approximately 20% of all galaxies are associated with rich $(R \geq 1)$ clusters, the galaxy correlation function will be, as observed, ~ 18 times weaker than the cluster correlations.

The model suggests that the fraction of galaxies associated with rich clusters is considerably larger than previously expected; most of these galaxies are distributed in the outer tails of the clusters, which may extend to at least $\sim 30h^{-1}Mpc$. Most clusters are therefore predicted to be embedded within much larger structures.

The model makes testable predictions that can be studied with complete redshift surveys. The model may also explain the negligible correlations observed among Ly-α clouds in QSO spectra (Sargent et.al. 1980). If these clouds cannot exist in rich clusters due to the high pressure in the latter, and therefore belong mostly to the non-cluster component, negligible correlations would be expected for the clouds.

V. Peculiar Velocities on Large Scales

The discussion in the previous sections summarizes evidence for the general existence of structures on the scale of $\sim 10\text{-}150h^{-1}Mpc$. A question of critical importance is what are the velocity fields in these structures. Peculiar velocities of clusters on these scales may indicate the existence of large amounts of (dark) matter. Recently, Bahcall et.al. (1986) used the complete redshift sample of rich clusters (§II) to study the possible existence of peculiar motion and/or structural anistropy on large scales. They find strong broadening in the redshift distribution that corresponds to a cluster pair velocity of $\sim 2000 \, kms^{-1}$. These findings are summarized below.

The distribution of clusters in space was studied by separating the three-dimensional distribution into its components along the line-of-sight (redshift) axis and the perpendicular axes projected on the sky. All clusters were assumed to be located at their Hubble distances as indicated by their redshifts. A cartesian coordinate system was set up in which cluster pair separations in Mpc were determined. The z-axis of the system was defined along the line-of-sight direction, and the x and y axes were in the plane projected

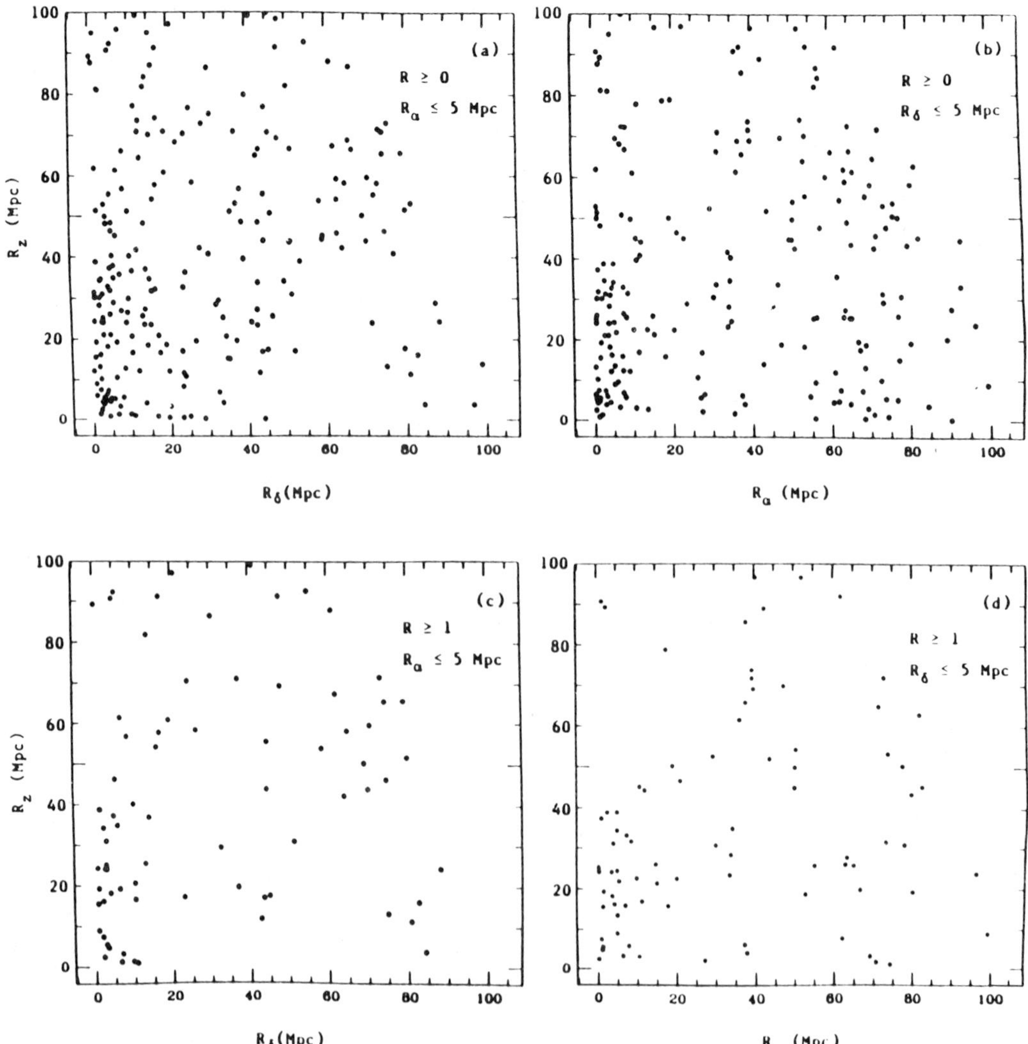

Figure 6: Scatter-diagrams of cluster pair separations in Mpc in the R_z-$R\alpha$ and R_z-$R\delta$ planes (Bahcall et.al. 1986). Figure 6a-6b and 6c-6d represent, respectively, the $R \geq 0$ and $R \geq 1$ richness samples. The elongation in the redshift direction is apparent in all cases.

on the sky. For simplicity, we used the α and δ coordinate system as the x and y axes, respectively. A scatter-diagram of the cluster pair separations in the z direction (R_z) versus their separations in α or δ $(R_\alpha$ or $R_\delta)$ was then determined for each sample. For a meaningful comparison between any two components, the third component was restricted to a given small value, which was varied in order to check the dependence of the results on the choice of the limit of the third component. In addition, only pairs with a total spatial separation less than $100h^{-1}Mpc$ (i.e. $R_\alpha^2 + R_\delta^2 + R_z^2 \leq 100h^{-1}Mpc$) were considered. This restriction limits the outer bounds of the sample studied but has no effect on the relevant inner points.

If all clusters were located at their Hubble distances with negligible peculiar motion, and if the sample was not dominated by elongated structures in a given direction, a symmetric scatter-diagram should be observed. In this case, a similar distribution should be observed along each of the three axes. If a large peculiar velocity exists among clusters, it would manifest itself as an elongated distribution along the z-direction in the R_z-R_α and R_z-R_δ diagrams. Such an elongation is normally interpreted as peculiar motion. However, the effect may also be caused by geometrically elongated structures, if they dominate the sample (with elongation toward the z-direction; see below).

Our results are presented in Figures 6 to 8. The scatter-diagrams are plotted in Figure 6 in both the R_z-R_α and R_z-R_δ planes, for both the $R \geq 0$ and $R \geq 1$ samples. Frequency distribution histograms representing these diagrams are presented in Figure 8. *A strong and systematic elongation in the z-direction exists in all the real samples studied.* Scatter-diagrams for sets of random catalogs were also generated for comparison; two representative examples are shown in Figures 7a-7b. As expected, no conspicuous elongation appears in any of the random catalogs. As an additional test, we have also determined the scatter diagrams in the projected plane, R_α-R_δ, of the cluster samples; typical results are shown in Figures 7c-7d. Again, as expected, a symmetric distribution is observed in this plane. These tests strengthen the conclusion that the observed elongation is real. The effect of elongation is strong; statistically it corresponds to approximately 8σ in a single sample (assuming, for illustrative simplicity, Gaussian statistics). It is therefore unlikely that the observed redshift elongation is a chance fluctuation. The effect becomes more apparent in the larger $R \geq 0$ sample; *this is expected if the effect is real.*

To determine what velocity could cause the observed effect, we convolved the frequency distribution observed along the projected axis, which is unperturbed by peculiar motion, with a Gaussian velocity distribution. This convolved distribution should match the broadened distribution observed in the redshift direction. The best fit is obtained for a Gaussian velocity width of $\sqrt{2}\sigma \simeq 2000\ kms^{-1}$. The fits are shown as the smooth curves in Figure 8.

What causes the elongation? One (standard) interpretation is peculiar motion. Another possibility is a true geometrical elongation in the shape of the large-scale structures. I discuss below both possibilities.

A. Peculiar Velocity Among Clusters

To estimate a supercluster mass which may support this velocity, we use a typical

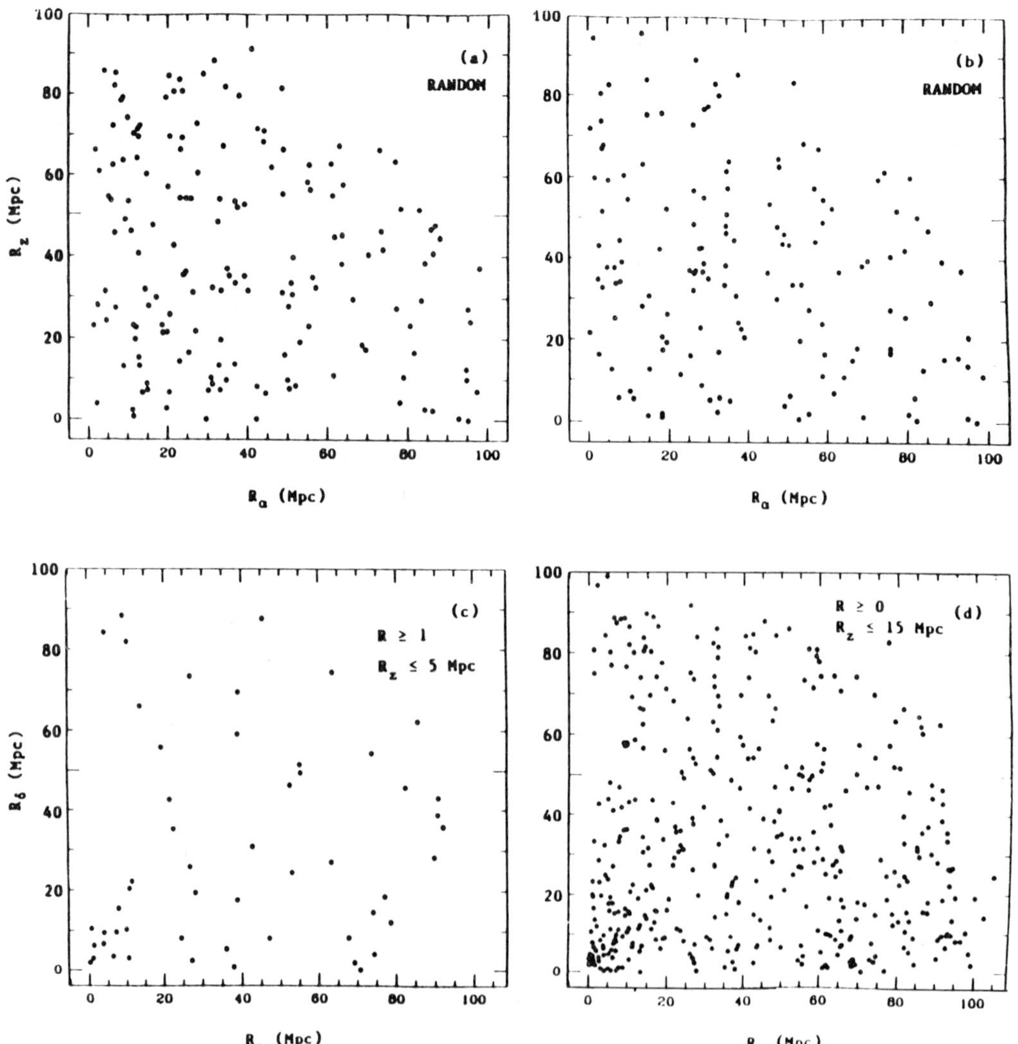

Figure 7: Same as Figure 6 but for typical random distribution of clusters (Fig. 7a-7b), and for the projected distribution (i.e. R_δ-R_α plane) of the actual cluster samples (Fig. 7c-7d). No elongation is expected in either case and none is observed. The clustering of clusters is apparent in the data sample of Fig. 7c-7d.

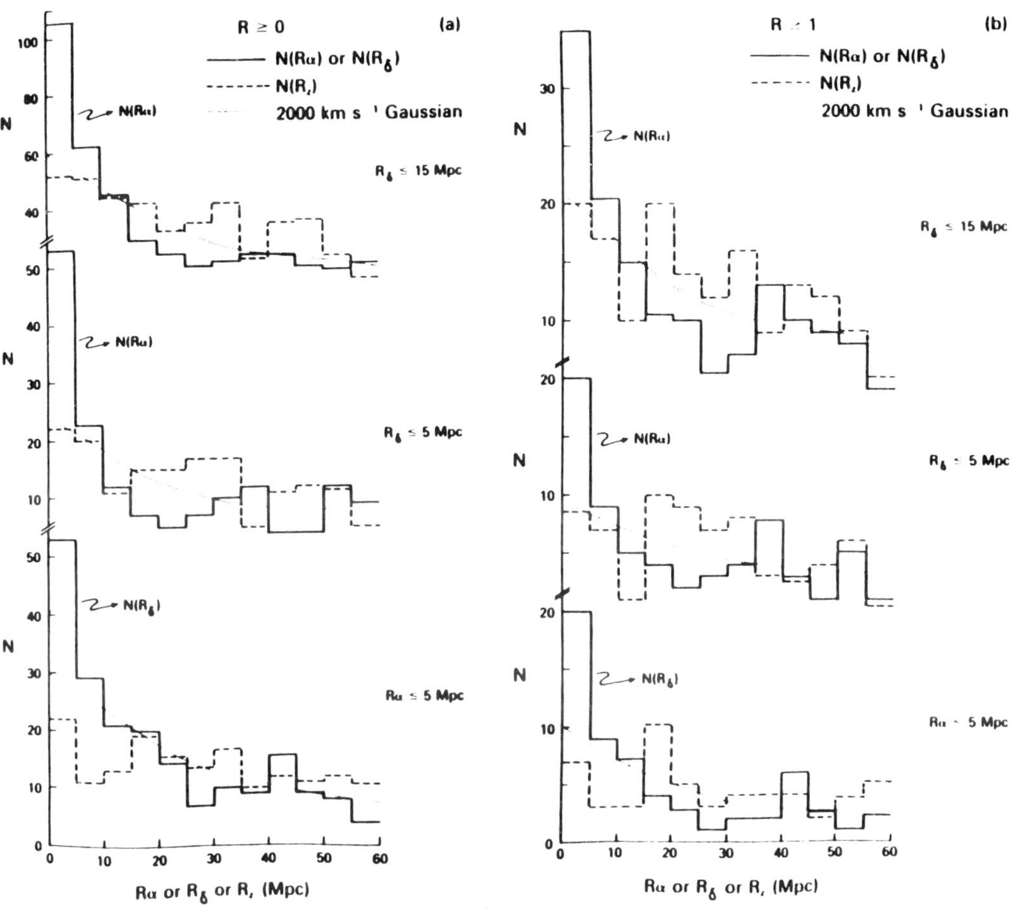

Figure 8: Histograms representing the distribution of pairs along the redshift, R_z, and projected, R_α and R_δ, directions, as determined from the scatter-diagrams, are shown for the $R \geq 0$ (Fig. 8a) and $R \geq 1$ (Fig. 8b) samples. The number of cluster pairs as a function of projected separation, $N(R_\alpha)$ or $N(R_\delta)$, is represented by the solid histogram. The number of clustered pairs in the redshift direction, $N(R_z)$, is represented by a dashed histogram. The dotted curve represents a convolution of the projected distribution $(N(R_\alpha)$ or $N(R_\delta))$ with a Gaussian of 2000 kms^{-1} width.

supercluster size of $\sim 25h^{-1}Mpc$ (=cluster correlation scale-length) and assume the virial relation $M \propto v^2 r$. This yields a typical supercluster mass of

$$M_{sc} \simeq 3 \times 10^{16} M_\odot \tag{6}$$

This mass is comparable to the mass of ~ 30 rich clusters while typically only ~ 3-5 rich clusters are members of a supercluster. Even when the luminous tails of clusters are accounted for the result may still imply an excess of dark matter in superclusters as compared with clusters. Using an observed luminosity and/or density profile of r^{-3} or $r^{-2.5}$ around a rich cluster, we estimate an M/L for superclusters that is typically two to three times that of rich clusters, i.e., $M/L \sim 500$.

B. Geometrical Elongation of Superclusters

The elongation observed in the scatter diagrams may also be caused, at least partially, by a geometrical elongation of superclusters. If the most prominent superclusters are elongated toward the line-of-sight direction, an apparent elongation in the distribution of pair separation along this axis may result. This preferred direction of elongation may occur either by chance, in a small sample of superclusters, or in cases of specific structures such as large cells or shells, when the most distant face-on components lie beyond the redshift limit of the sample. In this latter case, the sample may be dominated by large cells ($\sim 100h^{-1}Mpc$) that are cut off by the sample's redshift limit near their outermost edge. Accurate distance indicators such as Tully-Fisher type relations are needed in order to distinguish between peculiar velocity and geometrical elongation of large scale structures.

References

Abell, G.O., 1958, Ap.J.Suppl. **3**, 211
Bahcall, N.A., 1986, Ap.J.Letters, **302**, L41.
Bahcall, N.A., and Burgett, W., 1986, Ap.J.Letters **300**, L35.
Bahcall, N.A., and Soneira, R.M., 1982a, Ap.J.Letters **258**, L17.
Bahcall, N.A., and Soneira, R.M., 1982b, Ap.J. **262**, 419.
Bahcall, N.A., and Soneira, R.M., 1983, Ap.J. **270**, 20.
Bahcall, N.A., and Soneira, R.M., 1984, Ap.J. **277**, 27.
Bahcall, N.A., and Soneira, R.M., and Burgett, W.S., 1986, Ap.J. **311**, (December 1).
Groth, E., and Peebles, J.P.E., 1977, Ap.J. **217**, 385.
Hauser, M.G., and Peebles, P.J.E., 1973, Ap.J. **185**, 757.
Klypin, A.A., and Kopylov, A.I., 1983, Soviet Astronomy Letters, **9**, 41.
Sargent, W.L.W., Young, P.J., Boksenberg, A., Tytler, D., 1980, Ap.J.Suppl. **42**, 41.
Schectman, S., 1985, Ap.J.Suppl. **57**, 77.
Szalay, A.S., and Schramm, D.N., 1985, Nature **314**, 718.
Turok, N. 1986, preprint.

DISCUSSION

BURNS: (1) R=0 clusters can be very useful in tracing supercluster structure if selected carefully (in terms of distance) and restricted to certain uses. In particular R=0 clusters connect rich superclusters together as in Hercules. (2) Can these ~2000 km/sec peculiar velocities between clusters effectively "wash out" the claimed super-cluster and void structures in the distribution of Abell clusters?

BAHCALL: (1) One should be very careful when using any sample which is incomplete, such as R=0 clusters. Any project that requires completeness, or that may be sensitive to incompleteness effects, such as correlation functions, percolations, or the statistical determination of structure and shape, could be highly affected by such incompleteness, and meaningful interpretation may be difficult. They may be used only as an additional check, or trend, of the complete sample results.

(2) The 2000 km s^{-1} between pairs of clusters in super-clusters may be due to either velocity or actual geometrical elongation, as I described. If a large part of the effect is indeed due to motions, it would obviously distort any interpreted geometry and shape of the large structures that do not account for such velocities. The reverse, of course, is also true, i.e., geometry will distort velocities.

KARACHENTSEV: I think we must be very careful with the "universal" slope of correlation functions. For the centers of ~500 binary galaxies a correlation function has the power law parameter -1.0, but not -1.8, as for standard ones by Peebles. This difference is not caused only by selection effects.

BAHCALL: Yes. The "universal" correlation function applies only on large scales, from poor clusters, to richer clusters, superclusters etc. It does not apply on the small scale of galaxies and binaries, where gravitational forces are expected to yield additional clustering. Indeed, as I briefly noted, the galaxy correlation function is approximately three times stronger than the universal function on the large-scale.

DENG: How many clusters are in your sample?

BAHCALL: The complete statistical redshift survey used for the direct spatial correlation function determination contains 104 clusters - to z $\lesssim$ 0.1. However, it is tested and compared with the full sample of almost 2000 clusters to z $\lesssim$ 0.2, using the angular correlation function, where the scaling scales exactly as expected. Therefore, the effective number is probably close to 2000 clusters. For the velocity dispersion determination, we used both the 104 clusters in the complete $R \geq 1$ redshift sample, as well as the larger $R \geq 0$ sample of ~200 clusters with measured redshifts.

SPATIAL DISTRIBUTION OF GALAXIES: BIASED GALAXY FORMATION, SUPERCLUSTER-VOID TOPOLOGY, AND ISOLATED GALAXIES

Jaan Einasto and Enn Saar
Tartu Astrophysical Observatory
202444 Toravere
Estonia, U.S.S.R.

ABSTRACT. To study distribution of galaxies and voids, the rectangular box under study is divided into cubic cells, and mean density of particles in cells is derived. For any density level cells can be divided into 'filled' or 'empty' ones if their density is higher or lower than a threshold density. The length and volume of the largest connected system, as well as the number of systems of connected cells are derived for observed, model and random samples. The comparison of results demonstrates that galaxy formation is biased, supercluster-void topology is sponge-like in a wide threshold density interval, and that there are no isolated galaxies in voids.

1. INTRODUCTION

Most widely used quantitative methods to describe the spatial distribution of galaxies include correlation analysis and recently introduced cluster and percolation analyses. These methods complement each other and describe different aspects of the spatial distribution of test particles.

The study of the supercluster-void topology, the comparison of the distribution of isolated galaxies and models with biased galaxy formation with observations have raised the question: how to analyze these problems in quantitative terms. By experimentation we have found that a suitable method to study these problems is the division of a box under study into small cubic cells and to investigate the distribution, clustering etc. of filled and empty cells.

In the following we give a short description of the cell method, an overview of observational data and theoretical samples used and basic results obtained. This work is a result of collaborative efforts, more detailed reports are in preparation. We thank our collaborators Mirt Gramann, Maret Einasto and Adrian Melott for permission to use our joint results prior to detailed publication. Our special thank is due to Dr John Huchra, whose compilations of redshifts of galaxies have made this investigation possible.

A. Hewitt et al. (eds.), Observational Cosmology, 349–358.
© 1987 by the IAU.

2. METHOD

Consider a cubic box of side L, containing N test particles, galaxies or particles from numerical simulations. Split this volume into k^3 cubic cells of side l = L/k. Each cell contains in the mean

$$\rho_m = N/k^3 \tag{1}$$

particles. Calculate a continuous density field by smoothing the density of each particle using a hat function centered at the particle's actual position and having the hat side equal to the grid size l. Mark cells as 'empty' or 'filled' if the density of cells is lower or equal/higher than a given threshold density.

Now we can calculate the filling factors, study clustering properties of empty and filled cells separately etc. In contrast to classical cluster analysis where individual particles are used, in the cell method we consider cells of one class, either filled or empty. If two neighboring cells have a common sidewall, we count both cells as members of a system. The advantage of the cell method lies in the fact that identical procedures can be used to study the clustering of filled and empty regions whereas classical cluster analysis is not suited to study clustering of voids. We can vary the density threshold which divides cells into filled and empty ones. This corresponds to the variation of the neighborhood radius in the classical cluster analysis.

The cell method has been used previously to study large scale distribution of galaxies by Guberman et al. (1983) and Gott, Melott and Dickinson (1986) in two and three dimensional cases, respectively. In present paper we develope this method further by using a wide variety of statistical quantities over a broad threshold density interval. In the following we use dimensionless densities

$$\delta = \rho/\rho_m. \tag{2}$$

3. DATA

Observational data are based on the Huchra's (1983) compilation of redshifts which is complete in the Northern Galactic Hemisphere up to 14.5 apparent magnitude. We have used three samples from this compilation, located in a box of side L = 20 h^{-1} Mpc (h is the Hubble constant in units of 100 km/s/Mpc) and the center coordinate in supergalactic coordinates $x_0 = 0$, $y_0 = 15$ h^{-1} Mpc, $z_0 = 0$, which is close to the center of the Local Supercluster. The samples are absolute magnitude limited. Sample Virgo A contains galaxies brighter than $M_0 = -17.5$ (absolute magnitudes are calculated for H = 100 km/s/Mpc). This limit corresponds to the limiting magnitude of the CfA survey at the far side of the box, i.e. data in this sample are complete. Sample Virgo B contains dwarf galaxies in absolute magnitude interval $-15.0 > M > -17.5$ and includes numerous dwarfs from H radio surveys. This sample is not complete, but it is the best sample of dwarf galaxies available. Sample

Virgo C is the sum of samples Virgo A and B.

Observed distribution of galaxies is compared with theoretical samples found from N-body calculations. Three model samples were used. The first model sample, denoted Mel, is based on an axion dominated universe, containing 64^3 particles in a 64^3 mesh (Melott 1986). The second model, denoted GR-5, is based on a cold dark matter model with nonzero cosmological constant, containing also 64^3 particles in a 64^3 mesh. Parameters of the model were tuned to have for the present epoch (expansion factor a = 5.2) $\Omega_\lambda = 0.8$ and $\Omega_{matter} = 0.2$ (Gramann 1986). The third model, denoted M, corresponds to a neutrino dominated universe and was calculated for 32^3 particles in a 32^3 mesh (Melott et al. 1983). In all models several cases were considered, without and with biased galaxy formation. The first case includes all test particles. In other cases particles in low density environment were considered as primeval ones and rejected.

For comparison three samples with randomly distributed particles were used, containing 64^3, 32^3 and 16^3 particles and denoted by R-64, R-32 and R-16, respectively. Data on samples are given in Table 1.

TABLE 1. Data on samples used

Sample	L	Bias level	N	$\log \delta_{perc}$	C_{perc}
Virgo A	$20h^{-1}$		524	0.70	0.028
Virgo B			486	0.55	0.049
Virgo C			1010	0.65	0.036
M_o	$32h^{-2}$	2.7	4089	0.00	0.017
M_1		2.7	4139	0.00	0.017
M_2		2.7	4294	0.00	0.017
M_3		2.7	4449	0.00	0.017
Mel_{axion}	$32h^{-2}$	0	262144	0.25	0.056
		1	39944	0.35	0.039
GR-5	$40h^{-2}$	0	262144	0.55	0.031
		1.5	50823	0.35	0.039
R-64	32	0	262144	0.05	0.255
R-32		0	32768	0.12	0.254
R-16		0	4096	0.16	0.250

4. TESTS BASED ON CELL METHOD

Figs 1-3 present plots of three quantities, the relative length of the largest system, $\lambda_{max} = l_{max}/L$, the volume fraction of the largest system,

$C_{max} = V_{max}/V$, and the number of systems of multiplicity 4 and larger. The length of the system, l_{max}, was calculated as the maximum length of the system in coordinates x, y, and z. The total filled volume fraction, C, was also calculated, but respective curves are not plotted here since they are quite similar to curves for largest systems, i.e. most of the volume is occupied by the largest system.

At a threshold density higher than the density of the highest peak in a respective sample, there are no filled regions, and the total volume of the box is occupied by one large system of 'empty' cells. Respectively, for filled regions $\lambda_{max} = C_{max} = n_4 = 0$, and for empty regions $\lambda_{max} = C_{max} = n_4 = 1$. With decreasing density threshold there appear filled systems which at threshold density $\delta = \delta_{perc}$ form a percolating system: the length of the largest system is just equal to the length of the box, $\lambda_{max} = 1$. The percolating threshold density is related to the percolation parameter B (Shandarin 1983, Einasto et al. 1984) as follows

$$B = 3 \ / \ \delta_{perc}.$$

At a low threshold density the whole space is occupied by one large 'filled' system and the situation 'filled' versus 'empty' regions is reversed in comparison to high threshold density situation.

Some details of the tests depend on mean density of test particles in cells. The mean density ρ_m depends on the resolution parameter k and enters as an independent parameter of the method. Some quantities are, however, almost independent of mean density. Thus percolating density only little changes with ρ_m, and critical volume fraction of 'filled' cells (the volume fraction at the percolating density) of random samples is practically constant for all mean densities considered, as seen from Table 1.

Differences between random and other samples lie in the percolating density and the critical volume fraction. All random samples percolate at a threshold density $\delta_{perc} \sim 1$, whereas all observed and simulated samples percolate at much higher threshold densities. On the contrary, all random samples have the critical volume fraction $C_{perc} = 0.25$, but all observed and model samples have $C_{perc} = 0.02 - 0.05$. Both parameters reflect the presence of a connected network of filaments in observed and model samples which make the percolation easier than for random samples.

There exist also differences between various model samples, in particular, between model samples with and without biasing galaxy formation.

5. BIASED GALAXY FORMATION

Early comparisons of the observed distribution of galaxies (Joeveer, Einasto and Tago 1977, Einasto, Joeveer and Saar 1979) with numerical simulations (Zeldovich 1978) demonstrated a striking difference between theory and observations: in the real world the space between superclusters is empty whereas in simulations there exist a rarefied field population between densely populated structures. Quantitatively

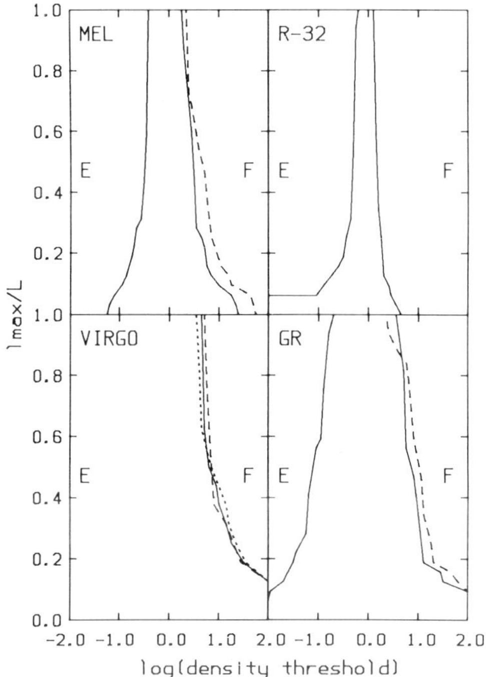

Fig. 1. The plot of the length of largest system in units of the box
size, $\lambda = l_{max}/L$, versus relative threshold density. Panel VIRGO
describes observed samples, panels MEL and GR axion-dominated models (GR
model includes non-zero cosmological constant), and panel R-32 – a
Poisson sample. Curves corresponding to filled and empty regions are
labeled by F and E, respectively. Bright and faint galaxy samples are
plotted by dashed and dotted curves, and the total sample by solid line
(panel VIRGO). In panels MEL and GR unbiased samples are plotted by
solid lines, biased samples by dashed lines.

this difference is seen in the relative fraction of populations of
isolated particles (Einasto, Klypin and Shandarin 1983, Einasto et
al. 1984). This difference was explained by Zeldovich, Einasto and
Shandarin (1982) by the absence of galaxy formation in low density
regions.
    The term 'biased galaxy formation' was suggested by Kaiser (1984)
who independently noticed that galaxy formation occurs basically in
high-density regions. Biased galaxy formation has been recently widely
discussed (Efstathiou et al. 1985, Bardeen et al. 1986, Kaiser 1986,
Melott and Fry 1986), however detailed physical mechanisms which lead to
the bias are not fully understood (Rees 1985, Silk 1985).
    Tests provided with the cell method give further quantitative
evidence for biased galaxy formation. As seen in Figs 2 and 3, unbiased
model samples have some similarity to random samples. This similarity

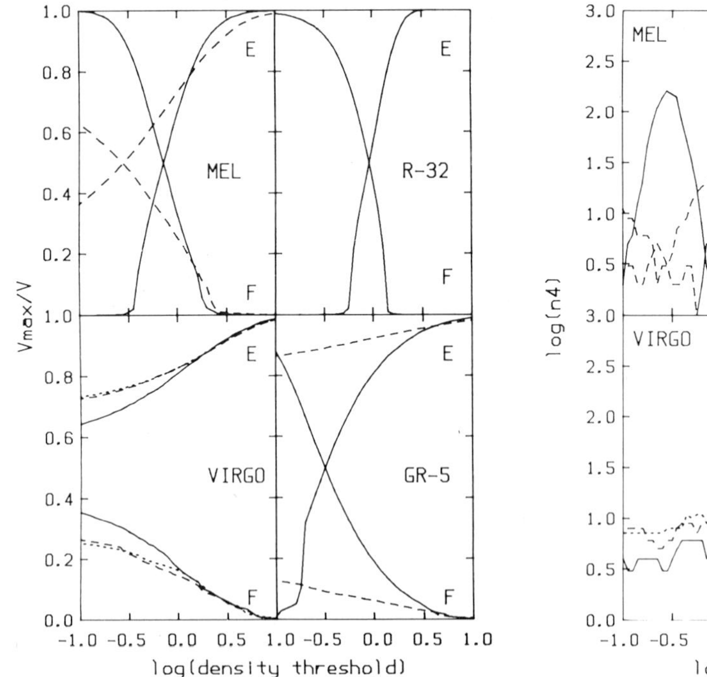
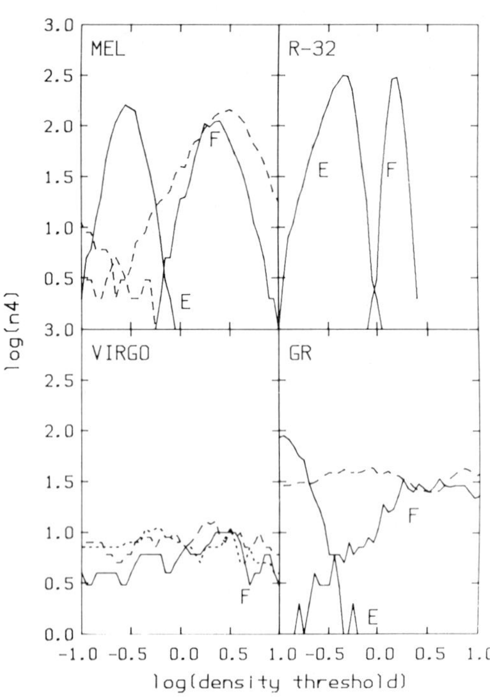

Fig. 2 (left). The plot of the volume fraction of the largest system, $C_{max} = V_{max}/V$, versus relative threshold density. Designations as in Fig. 1.

Fig. 3 (right). The plot of the number of systems of multiplicity 4 and larger, n_4, versus relative threshold density. Designations as in Fig. 1.

is, however, only a qualitative one. E.g. the unbiased model sample GR-5 has mean density $\rho_m = 8$, as in the random sample R-64, but the volume fraction curve of this sample is similar to that of random samples between R-32 and R-16, which have much smaller mean density. This similarity is due to the fact that unbiased model samples have a population of low density particles distributed more or less randomly.

Model samples can be brought into agreement with observations if particles from low density regions are removed, i.e. by introducing a bias in galaxy formation. Changing the biasing threshold it is possible to bring model curves plotted in Figs 2 and 3 (as well as the curve for total filling factor C) into agreement with observations. Filling factors C and C_{max} are especially sensitive to the biasing level, thus this test can be used to derive 'observational' biasing level of model samples.

Observed samples are centered on the core of the Virgo supercluster which has higher than an average filling factor. Model samples

correspond to much larger boxes which must have lower filling factors.
Thus we have used a biasing parameter which supplies a C_{max} curve by a
factor of two lower than the observed curve.

6. SUPERCLUSTER-VOID TOPOLOGY

There has been some discussion concerning the supercluster-void
topology. Zeldovich (1983) argued from theoretical considerations that
primeval matter at some low density must have cellular topology, i.e. low
density regions are from all sides surrounded by high density regions.
Also from theoretical considerations Gott, Melott and Dickinson (1986)
demonstrated that at median density the topology is of sponge-type,
i.e. both high and low density regions form penetrating percolating
systems. From observational point of view both possible topologies were
discussed by Joeveer, Einasto and Tago (1978) with no conclusive
results. More detailed observational data became available in early
80-ies. These data suggest that there are no isolating surfaces between
voids (Einasto and Miller 1983, Einasto et al. 1984).
      The cell method used here indicates that at low density unbiased
model samples have cellular topology which confirms theoretical
predictions by Zeldovich. At densities close to median density all
samples have sponge-topology conforming to the Gott, Melott and Dickinson
study. Observed and biased model samples have sponge-topology in a wide
range of threshold densities. At high threshold densities all samples
have the topology of isolated islands in a continuous ocean of voids.
      Recently interest to supercluster-void topology was renewed by the
discovery of large empty bubbles by de Lapparent, Geller and Huchra
(1986), surrounded apparently by continuous sheets of galaxies. Data
available on nearby voids do not support the view that surfaces
surrounding bubbles are actually continuous. If sheets of galaxies
around bubbles observed in the de Lapparent et al. survey do not have
structure radically different from nearby sheets observed in the Local
Supercluster then these distant sheets should also have holes which make
percolation of voids possible.

7. ISOLATED GALAXIES

Currently the most popular galaxy formation scenario is the biased cold
dark matter model (White et al. 1986). One particular problem with this
scenario is the presence or absence of isolated galaxies. As suggested
by Dekel and Silk (1986), giant galaxies should be formed only at highest
density peaks of the initial density field whereas dwarf galaxies should
be formed either everywhere or at lower density peaks. In both cases
there should exist a population of isolated dwarf galaxies.
      Correlation and conventional cluster analyses are not sensitive to
the presence or absence of a relatively small population of isolated
galaxies. The cell method used here is more appropriate for this
purpose. We have found that the most sensitive test is the number of
small systems with cell multiplicity 1 to 3.

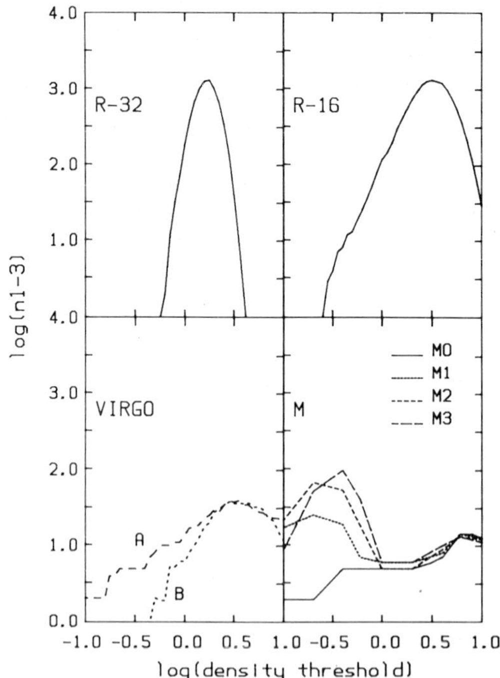

Fig. 4. Number of systems of multiplicity 1-3 versus relative threshold
density. Panels R-32 and R-16 describe Poisson samples, panel VIRGO -
observed samples, and panel M - a neutrino-dominated model. Bright and
faint galaxy samples in panel VIRGO are plotted by dashed and dotted
curves, respectively.

        Cluster analysis (Einasto et al. 1984) demonstrates that there
exists a small population of relatively isolated galaxies in observed as
well as in model samples. Traditional cluster analysis gives, however,
no answer to the question, are these galaxies outlying members of larger
systems or do they form a population of truly isolated galaxies, more or
less randomly spread over the whole space.
        If the spatial distribution of galaxies is studied by the cell
method, then isolated galaxies create a population of systems of low
multiplicity, 1, 2 or 3 (due to density smoothing one galaxy increases
the density in 1, 2 or 3 neighboring cells). If 'isolated' galaxies are
actually outlying members of larger systems then at low threshold
densities these systems of low multiplicity join up with larger systems.
On the other hand, if these galaxies are really isolated, then at these
same threshold densities they create an excess of systems of low
multiplicity. This effect is seen in panel d of Fig. 4. Curve labeled
M_o is based on the Melott neutrino dominated model (Melott et al. 1983)
with biased galaxy formation. All test particles from low density
regions are removed, and voids contain no test particles. Samples M_1,

M_2, and M_3 contain all particles of sample M_0 plus 50, 205 and 410 randomly located additional particles (1%, 5% and 10% of the original population). An excess of systems of low multiplicity in samples M_1, M_2 and M_3 is clearly visible. Sample M_0, as well as both observational samples have no excess of systems of low multiplicity in respective density interval.

This test demonstrates that galaxies at low density tail are related to other galaxy systems and that voids are really empty. This result is valid for both bright and faint galaxy samples.

8. FINAL REMARKS

The cell method used here complements other quantitative methods used earlier and is no substitute for them. Real world has a complex structure and there exist no single statistics which characterizes all aspects of the distribution of galaxies. Tests applied in this paper, suggest that

(i) there is strong observational evidence for biased galaxy formation,

(ii) at a broad relative density interval superclusters and voids have sponge-like topology, i.e. both filled and empty regions form infinite percolating and intertwined systems,

(iii) voids are really empty and contain neither bright nor faint galaxies.

REFERENCES

Bardeen, J., Bond, J. R., Kaiser, N. and Szalay, A. 1986,
     Astrophys. J. Suppl. (in press).
Dekel, A. and Silk, J. 1986, Astrophys. J. **303**, 39.
de Lapparent, V., Geller, M. and Huchra, J. 1986, Astrophys. J. **302**, L1.
Efstathiou, G., Davis, M., Frenk, C. S. and White, S. D. M. 1985,
     Astrophys. J. Suppl. **57**, 241.
Einasto, J., Joeveer, M. and Saar, E. 1979, Tartu Astr. Obs. Preprint
     A-2.
Einasto, J., Klypin, A. A., Saar, E. and Shandarin, S. F. 1984,
     Mon. Not. R. astr. Soc. **206**, 529.
Einasto, J., Klypin, A. and Shandarin, S. 1983, Early Evolution of the
     Universe and Its Present Structure, eds. G. D. Abell and
     G. Chincarini, Reidel, p. 265.
Einasto, J. and Miller, R. H. 1983, Early Evolution of the Universe and
     Its Present Structure, eds. G. O. Abell and G. Chincarini, Reidel,
     p. 465.
Gott, J. R. III, Melott, A., and Dickinson, M. 1986, Astrophys. J. (in
     press).
Gramann, M. 1986, Preprint.
Guberman, S. A., Doroshkevich, A. G., Kotok, E. V. and Shandarin, S. F.
     1983, Astrofizika, **19**, 97.
Huchra, J. 1983, Redshift compilation.

Joeveer, M., Einasto, J. and Tago, E. 1978, Mon. Not. R. Astr. Soc. **185,** 357.
Kaiser, N. 1984, Astrophys. J. **273,** L17.
Kaiser, N. 1986, Inner Space/Outer Space, eds. E. W. Kolb, M. S. Turner, D. Lindley, K. Olive and D. Seckel, Univ. Chicago Press, p. 258.
Melott, A. L. 1986, Preprint.
Melott, A. L., Einasto, J., Saar, E., Suisalu, I., Klypin, A. A., and Shandarin, S. F. 1983, Phys. Rev. Lett. **51,** 935.
Melott, A. L., and Fry, J. N. 1986, Astrophys. J. **305,** 1.
Rees, M. 1985, Mon. Not. R. astr. Soc. **213,** 75.
Schaeffer, R. and Silk, J. 1985, Astrophys. J. **242,** 319.
Shandarin, S. F. 1983, Pis'ma v Astr. Zh. **9,** 195.
Shandarin, S. F. and Zeldovich, Ya. B. 1983, Comments Ástrophys. **10,** 33.
Silk, J. I. 1985, Astrophys. J. **297,** 1.
White, S. D. M., Frenk, C. S., Davis, M., and Efstathiou, G. 1986, Astrophys. J. (in press).
Zeldovich, Ya. B. 1978, The Large Scale Structure of the Universe, eds. M. S. Longair and J. Einasto, Reidel, p. 409.
Zeldovich, Ya. B. 1982, Pis'ma v Astr. Zh. **8,** 195.
Zeldovich, Ya. B., Einasto, J. and Shandarin, S. F. 1982, Nature **300,** 407.

DISCUSSION

LACHIEZE-REY: A smooth variation of the filling factor with scale is an indication of scale invariance. So the fit between the observations and biased galaxy formation is the consequence of the fact that the real distribution of galaxies, and the one predicted by biased galaxy formation, both obey some scaling law. A fractal model exhibits such a characteristic.

THE VOID PROBABILITY FUNCTION AS A STATISTICAL INDICATOR

Marc Lachièze-Rey and Sophie Maurogordato
DPh/SAp
CEN Saclay
91191 Gif sur Yvette Cedex
France

Recent results concerning the galaxy distribution at scales $< 100\ h^{-1}$ Mpc ($H_o = 100\ h\ kms^{-1}\ Mpc^{-1}$) show a number of characteristics which cannot be described by conventional statistical models. Correlation functions, for instance, can in no way give account of the presence of voids of the cellular (or spongy) appearance of the local galaxy distribution (M. Geller, this conference). There is clearly a need for new kinds of statistical models and statistical indicators.

Among those we wish to emphasize the advantages of the void probability function (VPF), and its particular convenience for studying the galaxy distribution.

For a given sample of galaxies, the VPF is defined as the probability that a randomly selected volume, of specified size V and shape contains no galaxy. This quantity has many advantages. Firstly it can be calculated from dynamical or statistical models. It can be predicted, if necessary, from the set of all correlation functions, or from the set of count probabilities, for which it acts as a generator function (see White 1979)[12]. Moreover it can be used as a link between various approaches like BBGKY hierachy and correlation functions, presence of voids, and percolation.

The VPF is not difficult to calculate from any sample and is not very sensitive to the particular features of the galaxy distribution. It also presents the advantage that results in 1, 2 and 3 dimensions are easy to link. For these reasons, we developed a method for calculating the VPF (and associated indicators like low-order number counts) from any sample. Applications of this method to the 2-dimensional center for Astrophysics (CfA) catalog have already been published[1]. We present below some first results for the 3-dimensional CfA catalog.

SCALING

An important remark should be made, that the VPF is basically a function of 2 variables, the density of the sample and the size V of the volume. This means for instance that, for 2 different samples similarly clustered but with different densities, the VPF take different

A. Hewitt et al. (eds.), Observational Cosmology, 359–362.
© *1987 by the IAU.*

values for each volume. This contrasts with the 2 point correlation functions ξ which does not vary with the density of the sample.

For this reason, 2 approaches exist for the VPF. Either to consider it as a function of n at a fixed V, or as a function of V at fixed n.

The first approach, essentially used by Saslaw et al.[6], involves different samples (or subsamples of a same catalog), assumed to be similarly clustered. The VPF is then evaluated for each of them. A function of n (or of x = nV, but where V remains fixed) is obtained, which can be compared to the predictions of some models.

The second method is more natural and significant since it does not imply the analysis of a serie of different samples. Its results are well suited for comparison with models which, in general, concentrate on the V-dependence of the structure, for a given density. There is however one difficulty: being a function of the 2 variables n and V (and not only of V like $\xi(V)$), the VPF does not allow, a priori, to compare various samples with different densities ; or to analyse a catalog where the density varies from place to place, like any non volume-limited catalog, the CfA in particular.

There is however one way of surmounting such difficulties, if what we will call a scaling hypothesis is verified[10]. This latter can be expressed by the fact that the VPF (or more exactly its logarithm, normalized to the poissonian value – nV, i.e. $\chi \equiv \dfrac{-\log(VPF)}{nV}$) can be described as a function of some scaling variable q only, constructed from V and n. In other words, the V-dependence and the n-dependence of the VPF can be deduced each one from each other. This scaling hypothesis plays a very important role for the following reasons. Firstly it allows us to reconcile the two approaches, since the V and n dependence describe the same information. Secondly, it will allow us to compare the clustering properties of different samples having different densities and to check, for instance, if they are or not similarly clustered. Similarly it gives the possibility to explore different parts of a non homogeneous catalog (like for instance the CfA), and to combine them for a study of the whole catalog. The only condition is to plot χ as a function, not of n or V, but of the scaling variable q. Moreover the scaling hypothesis is predicted by a whole class of statistical models, called hierarchical models. These have been recently reviewed by Fry[5] as well as their predictions for the VPF. This makes very usefull to test if the scaling hypothesis is verified for the real galaxy distribution[10,11].

Before knowing the validity of the scaling hypothesis, and in order to check it, we must define complete volume-limited samples. We extracted three samples from the CfA catalog that we call:

     faint   –17   < M < -19, D < 20 Mpc, 366 galaxies
     medium   –17.5 < M   , D < 27 Mpc, 488 galaxies
     and bright –18.5 < M   , D < 40 Mpc, 396 galaxies.

For each of them we derived the VPF and tested the scaling hypothesis. For this task, we constructed catalogs with the same clustering properties but different densities, just throwing away randomly a given proportion of the galaxies present.

The result is that for each sample the scaling property is very well verified with the scaling variable $q = nV\, \bar{\xi}(v)$, as proposed in the Schaeffer models, $\bar{\xi}(v)$ being defined as $\frac{1}{V^2} \int_V d^3v_1 d^3v_2\, \xi(r_{12})$,[10] where the evaluation of ξ by Davis and Peebles[8] have been used. Independently of any model, this implies that the clustering properties of the matter in universe, at a given scale, are related to its properties at an other scale. A result that any proposed model has to take into account.

This scaling property allows to compare the different samples previously defined. We have been able to show that, for any value of the scaling variable q, χ decreases with the faintness of the samples.

In other words, bright galaxies appear much more clustered than faint ones. Note that we used the same value of ξ to calculate the scaling variable q for the 3 samples. This may be not correct if faint and bright galaxies have not the same correlation functions. However in such a case the conclusion that they are differently clustered would be unchanged.

Finally we have compared our measures with the available theoretical predictions. The result is a good agreement with the $\nu = 1$ model proposed by Schaeffer[10]. Additionally a similarity in shape appears with our calculations for biased galaxy formation (Maurogordato + Lachièze-Rey[7]), although the level is not the same. It should be remarked however that the comparison between theory and observations presently involves a normalization by the correlation functions which forbids any definitive conclusion.

In conclusion, the galaxy distribution appears to be scale invariant, in the sense previously defined. A result favouring the class of hierarchical models (including biased galaxy formation) and allowing more specific use of the VPF.

It appears also that bright galaxies are more clustered than faint galaxies in the CfA catalog.

Comparison with theoretical models is only tentative but predictions from the Schaeffer model ($\nu = 1$) and the biased galaxy formation both appear attractive.

REFERENCES

(1) F.R. Bouchet, M. Lachièze-Rey, 1986, Ap. J. (Letters), **302**, L37.
(2) J.N. Fry, 1983, Ap. J., **267**, 483.
(3) J.N. Fry, 1984a, Ap. J. (Letters), **277**, L5.
(4) J.N. Fry, 1984b, Ap. J., **279**, 499.
(5) J.N. Fry, 1986, Ap. J., **306**, 358.
(6) V. de Lapparent, M.J. Geller, J.P. Huchra, 1986, Ap. J. (Letters) **302**, L1.
(7) S. Maurogordato, M. Lachièze-Rey, 1986, in preparation.
(8) P.J.F. Peebles, "The large-scale structure of the Universe", (Princeton University Press).
(9) W.C. Saslaw, and A.J.S. Hamilton, 1984, Ap. J., **276**, 13.
(10) R. Schaeffer, 1984, Astr. Ap. (Letters), **134**, L15.

(11) R. Schaeffer, 1985, Astr. Ap. (Letters), **144**, L1.
(12) S.D.M. White, 1979, MNRAS, **186**, 145.

LARGE-SCALE DISTRIBUTION OF GALAXIES WITH DIFFERENT LUMINOSITIES

X. Y. Xia
Department of Physics, Tianjin Normal University,
Tianjin, China.
Z. G. Deng
Department of Physics, Graduate School,
Academia Sinica, Beijing, China.
Y. Y. Zhou
Center for Astrophysics, University of Science and
Technology of China, Hefei, China.

Analyses for complete samples of galaxies and clusters of galaxies showed that the two-point correlation function of galaxy-galaxy and cluster-
cluster have form of power law with power indices about -1.8 (Peebles,
1980; Bahcall and Soneira, 1983). Because the completeness of a sample
means that we have observed completely the objects brighter than a given
apparent magnitude in certain sky region. But, we know that a complete
sample will be lack of faint objects at distant region. If we attempt to
avoid the influences of any non-intrinsic properties on the analysis, we
have to use the samples which are complete in certain interval of absolute magnitude.

In this work, we present a tentative analysis based on the data
given by CfA survey (Huchra, et al., 1982). To determine the absolute
magnitude of galaxies we must know their distances from us. If we adopt
the observed redshifts as the indicator of distances, the peculiar motion
of galaxies will distort the positions along the line of sight and may
influence the reliability of the analysis results. So, we have to investigate if the influence of the peculiar motion of galaxies is severe for
the analysis. Geller and Huchra (Geller and Huchra, 1983) have presented
a catalog of galaxy groups included in the CfA survey with average velocity and the dispersion of velocity σ for each group. We can use the dispersion of velocity to estimate the peculiar motions of galaxies. Because
the peculiar motion will produce larger effect for nearer galaxies, we
would limit our sample to consist of galaxies with observed redshift larger than 1000 km/s. Under this restriction, there is no group with value
of $\sigma/\bar{v}$ larger than 25%, and more than 71% groups with values of $\sigma/\bar{v}$ less
than 10%. It means that if we take the redshift as the indicator of ga-
laxy distances for the restricted sample, the error caused by the peculiar motion will not be fatal to our analysis.

According to the method given by Geller and Huchra, we have corrected the velocity of each galaxy for a dipole Virgo-centric flow, then
use the finally obtained redshift value to calculate the distance of

A. Hewitt et al. (eds.), Observational Cosmology, 363–366.

every galaxy. Because faint galaxies can only be observed within a small redshift and the very bright galaxies are too few to give a significant statistical analysis, we restrict our sample to galaxies with absolute magnitude M between -19.0 and -22.0. We divide the sample into three sub-samples A, B, and C with different luminosities. There are 392, 533, and 387 galaxies within the absolute magnitude ranges -19.0- -20.0, -20.0- -21.0, and -21.0- -22.0 in samples A, B, and C, respectively.

We have calculated the two-point correlation functions for each sub-sample and compared them with the average two-point correlation function of Monte Carlo samplings in the same regions with the same numbers of objects as those of the subsamples. If the difference between the two-point correlation functions of our subsamples and the corresponding average correlation functions of Monte Carlo samplings is larger than 3σ, we say that our sample is significantly clustering (or anti-clustering) on the corresponding scales.

In the figures, we give the two-point correlation functions of samples A, B, and C (solid lines), respectively. We can see from these figures that at small scales the faint and bright galaxies are all clustering, but at larger scales they approach to random distribution. The regression analysis of the two-point correlation function of each group in the significantly clustering range of scales gives

$$\xi(r) = 55.76 \; r^{-1.54}, \qquad \text{for group A,}$$

$$\xi(r) = 30.39 \; r^{-1.06}, \qquad \text{for group B,}$$

$$\xi(r) = 41.47 \; r^{-0.87}, \qquad \text{for group C.}$$

The correlation coefficients are 0.98, 0.95, and 0.85, respectively. It shows that the power indices decrease monotonously as the luminosity of galaxies decreasing. The clustering scales of these three sub-samples are about 10 Mpc, 19 Mpc, and 30 Mpc, respectively.

From the analyses given above, we may draw following conclusions:

1. The galaxies with different luminosity may have quite different characters in their large-scale distribution. It seems that the fainter galaxies are clustering in smaller scale but the brighter galaxies are clustering in larger scale.

2. The systematic change of the power indices in the two-point correlation functions and in the clustering scales might indicate that the large-scale distribution of luminous matter in the universe may depend on the luminosity of galaxies. In order to know the real distribution of whole luminous matter we must determine the large-scale distribution of galaxies with various luminosities.

3. The regularities appeared in the results of analyses seem to show that they could not be caused by the approximation of neglecting the peculiar motions, because the peculiar motion does not like to be possible to give a systematic change as the luminosity changing.

4. CfA survey is not deep enough to give a sample for making analyses in a large enough region with faint enough galaxies. So that, the conclusions we obtained may still be influenced by the local conditions and the peculiar motions of galaxies. If one could get data from survey

which is complete to $15^m.5 - 16^m.0$, more dicisive conclusions can be obtained.

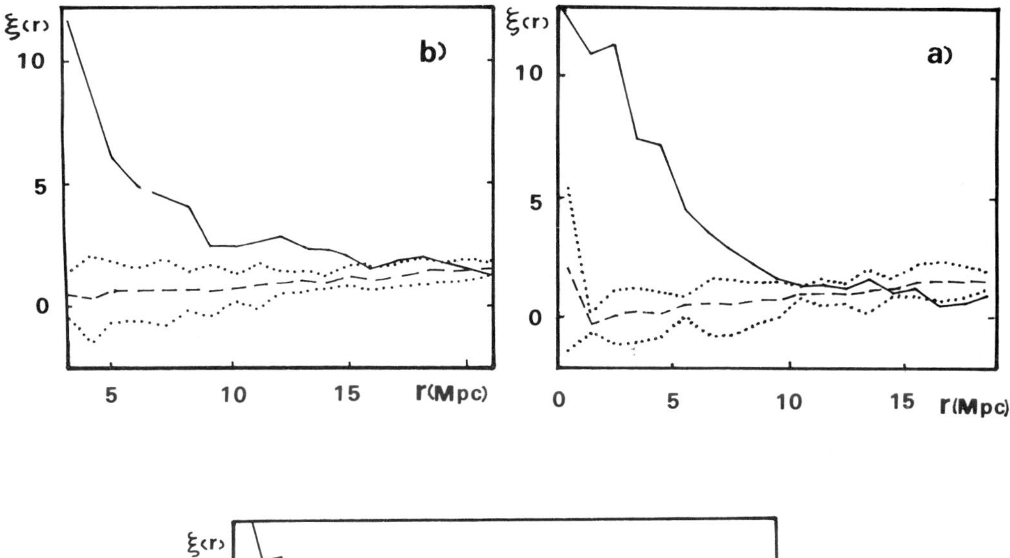

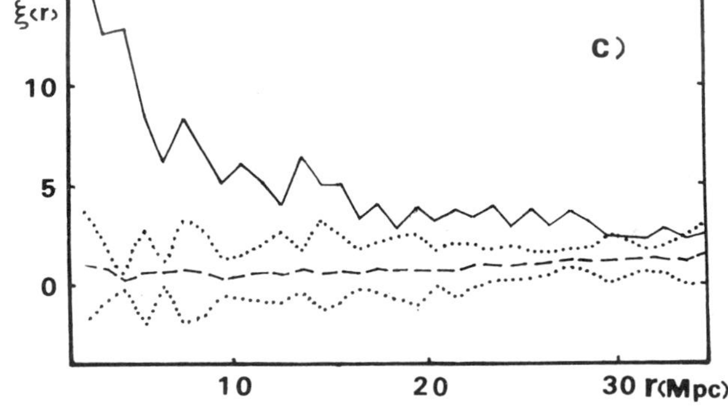

Figure. The two-point correlation functions for subsamples A (fig.a), B (fig.b), and C (fig.c). The dashed lines are the average two-point correlation function given by Monte Carlo samplings, and the dotted lines are the ±3σ lines.

REFERENCES:

Bahcall, N. A., and Soneira, R. M. 1983, Ap. J., 270, 20.
Geller, M. J., and Huchra, J. P. 1983, Ap. J. Suppl., 52, 61.
Huchra, J. P., et al. 1982, Ap. J., 257, 423.
Peebles, P. J. E. 1980, The Large-Scale Structure of the Universe,
     Princeton Univ. Press.

DISCUSSION

CHINCARINI: Did you determine the percent of morphological type for each luminosity bin you used?

XIA: There are about 24% elliptical and lenticular galaxies, 66.6% spiral galaxies and 9.3% irregular galaxies in group A. In groups B and C, the percentage of different morphological galaxies are respectively 26.5%, 68.9%, 4.6%, and 25.2%, 66.9%, 7.9%.

THE DISTRIBUTION OF FAINT GALAXIES

Richard Ellis
Physics Department
Durham University
England   DH1 3LE

ABSTRACT: Faint galaxy data are reviewed in the context of standard evolutionary models and our understanding of the statistical properties of galaxy populations. The differences in number magnitude counts from group to group can largely be understood via fluctuations induced by large scale clustering. However count slopes still present convincing arguments for an extra component of faint blue galaxies beyond B $\sim$ 21. Colours provide an useful tool in estimating redshift distributions but the uncertainties are large for the bluest galaxies of interest. However, new faint object redshift surveys are now underway and promise to determine definitively the nature of this extra component. We discuss preliminary results from one of these surveys. Neither distant luminous galaxies nor intrinsically faint nearby galaxies appear to be very numerous to B = 21.5. Many of the galaxies with 0.2 < z < 0.4 show spectral signatures of prominent star-formation. If such objects are somehow related to the excess counts, the traditional redshift-dependent evolutionary theory may require revision.

1. Introduction

A long-held view in observational cosmology is that the spectral evolution of galaxies and growth of large scale structure may be quantified from detailed studies of complete samples of faint galaxies. One disadvantage of this "all-object" approach is that the statistical properties of the various components of the galaxy population need to be well-understood before any evolution can be inferred. However, only with extensive and objectively-controlled surveys might one recognise selection effects otherwise hidden in probes of galaxy evolution based on studies of individual objects.

Historically the subject of faint galaxy studies is one of a gradual increase in detail. We begin with Olber's assessment of the finite integrated background and its implications for the effects of redshift and the evolutionary timescale, later come Hubble's counts of faint galaxies and his attempts to determine the curvature radius,

A. Hewitt et al. (eds.), Observational Cosmology, 367–381.

these led to the modern realisation of the vast information content of multi-passband Schmidt and 4 metre telescope photographic plates for the subject of galaxy evolution. Previous reviews (Ellis 1982, Kron 1982, Koo 1984) have been limited to analyses of these photographic data and generally concluded that there is some evidence from the number and colour distributions of faint (B > 21) galaxies for mild evolution over recent look-back times.

The evolution claimed is only statistical; it is not straightforward, for example, to identify any distant evolving component from photometric data alone. Koo (1985) and Loh and Spillar (1986) have investigated the use of multicolour data to determine approximate redshifts with interesting results. Their approach is to some extent being overtaken by a new development. Progress in multiple object instrumentation such as fibre optic couplers (Ellis et al 1984) and multislit aperture plates (Koo 1983) at last makes it practical to collect genuine spectra to sufficiently faint limits. Redshift distributions allow us direct cosmological and evolutionary tests hitherto not possible. Spectral details can also be examined independent of any search for luminosity or colour changes. Finally, we can look forward to another increase of detail in our knowledge of faint galaxies via ultraviolet and morphological data from the Hubble Space Telescope.

This review is the first to discuss the new faint object redshift surveys. At least two such surveys are in progress and here I briefly review preliminary results from one of them. There are implications for our understanding of how galaxies evolve with redshift and the distribution of broad populations of galaxies seen in large volumes of space.

2. Statistical Properties of Normal Galaxies

A number of local redshift surveys have been performed in recent years but few have a well-defined photometric scale and cover large enough volumes to be representative. The B $\sim$ 16 - 17 magnitude-limited pencil beam surveys of Kirschner and colleagues (Kirschner et al 1978,1979,1981) and the Durham/AAT group (Peterson et al 1986) have shown that galaxies are not a single parameter (luminosity) family. Analytical representations are usually fitted to local luminosity functions (LFs) classed by morphology (Ellis 1982) or colour (Koo 1981). A subset of the Durham/AAT redshift survey (hereafter DARS) has recently been extended to infrared passbands (Mobasher et al 1986). The combination of morphologies and restframe B - K colours provides a valuable check on different ways to estimate the population mix and LFs. B - K offers a wide baseline with which to discriminate stellar populations and a reasonable correlation with morphology has already been shown (Aaronson 1978).

Table 1 shows numbers and Schechter (1976) LFs classed by morphology and colour to B = 16.85 from the DARS. We adopt a fixed faint end slope of $\alpha = -1.25$ found by maximum likelihood fitting to the total sample (Bean 1983). The "B" photometric system here is the Kodak IIIa-J/Schott GG395 (= b_j) system at the 26.5 mags arcsec^{-2} isophote used in most faint studies (c.f. Peterson et al 1979, Shanks et al 1984). Table 1 shows a significant fading of the characteristic luminosity M_B for both later types and bluer colours. When spectral energy distributions are assigned to each type/colour class (from Pence (1976) in the optical and King and Ellis (1985) in the near ultraviolet), the no-evolution count-magnitude-redshift N(m,z) predictions for 20 < B < 23 are virtually identical whether one classifies by colour or morphology (c.f. Table 2).

Table 1

Type and Colour-dependent Luminosity Functions ($H_o = 50$)

Colour	No	M_B	Type	No	M_B
B - K > 3.65	74	-21.39	E/S0	97	-21.67
3.65 > B - K > 2.9	61	-21.24	Sa-Sbc	117	-21.20
2.9 > B - K	32	-20.52	Sc-Im	46	-21.31:
			Unclassed	39	-20.87:
			All(classed)	260	-21.28
All	167	-21.24	All	299	-21.48

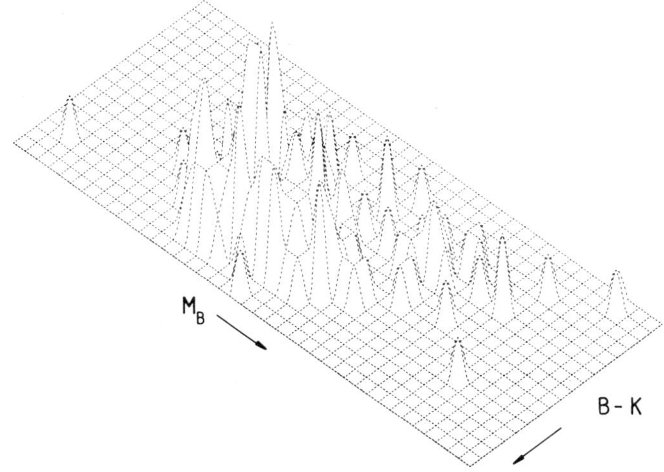

Fig. 1. Joint distribution of absolute B magnitude and restframe B-K colour for 170 DARS galaxies to B=17; note the tail of faint blue galaxies (H_o=50).

Although much is known about the LF of dwarf (M_B > -18.5) galaxies in *clusters*, little is known about their overall space density. Fig. 1 shows these systems in the DARS are mostly blue Scd-Im galaxies which do not fit a standard Schechter function very well. They cannot be accounted for in faint predictions in a straightforward way yet their fraction of the total population *increases* at fainter apparent magnitudes since they suffer almost no redshift dimming. To B $\sim$ 17 there are 17 (5%) DARS galaxies with M_B > -18.5 in excess of the Schechter LFs. With the exception of a few dEs all show strong Oxygen emission lines with excitation typical of star forming regions. As Kron (1982) pointed out, it is premature to attribute the observed excess of 20-40% of faint galaxies at B $\sim$ 21 - 22 to luminosity evolution over the last few Gyr. A 5% excess of intrinsically faint blue galaxies at B $\sim$ 17, if representative by volume, would contribute 30% at B = 21.5. Only deeper redshift surveys can clarify this possibility.

3. Counts and Colours

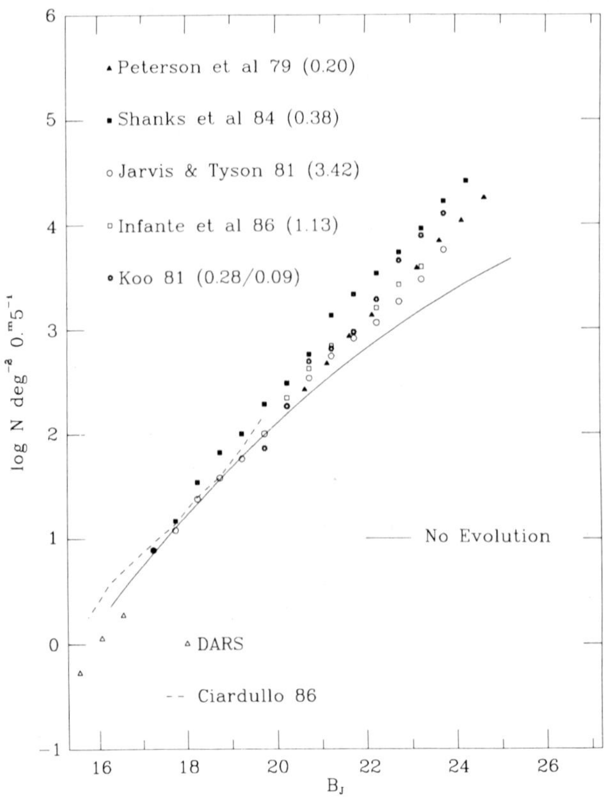

Fig. 2. Differential number magnitude counts in the photometric system defined in $2. The area (deg^2) appropriate to each published analysis is given. The model (solid line) assumes no evolution and properties of Table 1 with an absolute normalisation determined from Schmidt counts.

Standard evolutionary models (Tinsley 1978,1980, Bruzual 1983) predict a much greater sensitivity, at a fixed lookback-time, to changes in luminosity rather than colour (e.g. $0^m.4$ in observed B for each $0^m.1$ in B-V for models reproducing present day ellipticals with ages 16 Gyr). Since numbers increase steeply with distance, counts are still the most sensitive and objective probe. The models assume that galaxies of an individual type or colour class share a unique redshift-dependent evolutionary track irrespective of mass. Redshifts and colours for faint galaxies will allow us to examine this assumption.

The comparison of published counts (Fig. 2) shows substantial variations in the absolute number densities from group to group; it might be asked is the photometry reliable? The situation has now clarified considerably due to an increase in the number of 4m plates scanned and an international collaborative venture (Ellis and Koo 1986) to examine various software packages used for detection and photometry of faint galaxies. Standard intensity-calibrated photographic data was sent to each group who used their own algorithms for detection, photometry and image classification. Koo and I collated and compared the various catalogues. A preliminary analysis shows that for detection the groups agree remarkably well to about B = 23.5 - 24.0, close to our estimate of the completeness limit of the data. However, the galaxy photometry becomes seriously discrepant amongst some groups beyond B $\sim$ 23 (Fig. 3). Only part of this effect is the expected difference between isophotal and "total" magnitudes (Ellis 1982, Kron 1982, Shanks et al 1984); in those cases where the group catalogues listed the local sky estimate, it was apparent that its determination is a major uncertainty also. Deeper CCD data has now been procured to determine improved independent parameters for the objects. Meanwhile we conclude that the scatter in the galaxy counts arising from different measurement schemes should be small provided comparisons are made brighter than B $\sim$ 23.

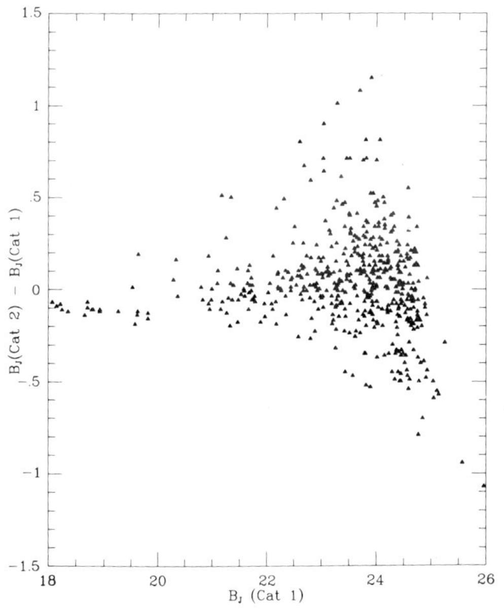

Fig. 3. A comparison of two different photometric reductions applied to a standard photographic data set supplied to a number of groups by Ellis and Koo (1986).

In collaboration with colleagues at Durham, the photographic counts have been extended to 5 more 4 metre fields; each has good CCD calibrations to better than $\pm 0^m.20$. In the range $20 < B < 22.5$ the fluctuations between 1 deg^2 areas are about 5 times those expected from Poisson noise with a peak-to-peak range of $\sim$ x 1.6, i.e. consistent with Fig. 2 using one reduction technique. Furthermore, comparisons of Schmidt plate counts with IRAS and H I data (Fong et al 1986) appear to rule out Galactic extinction as a source for the fluctuations and we conclude large scale clustering is most likely responsible (see $4). These fluctuations have serious repercussions for deeper magnitude count programmes using CCDs.

Fig. 2 also includes the no-evolution prediction derived from the LFs discussed in $2; the dependence on q_O is minimal. For reasons that are not well-understood, the absolute normalisation of the DARS is $\sim$ x 2 lower than that obtained from deeper Schmidt counts (Shanks et al 1984, Ciardullo 1986). Because of this and the fluctuations discussed above we will quantify the deviation of the observations from the model via the slope of the counts in the range $21 < B < 23$ (Table 2).

Table 2

Galaxy Count Slopes $21 < B < 23$

Observations:		No evolution models:	
Peterson et al 1979	0.46	King and Ellis (1985)	0.365
Koo 1981	0.47	Koo and Szalay (1985)	0.36
Jarvis and Tyson 1981	0.40	This work:	
Shanks et al 1984	0.48	morphological LFs	0.340
Infante and Pritchet 1986	0.40	B – K LFs	0.339

Adopting the Schmidt count normalisation, the excess at $B = 21.5$ is about 30% rising to a factor of 2 by $B = 22.5$. The most puzzling aspect of the observed slope is its constancy over a very wide range of apparent magnitude; unpublished CCD data analysed by Tyson (private communication) suggests the slope extends to $B = 26$ *with no features*. Although any slope steeper than 0.40 continued indefinitely would produce a divergent extragalactic background light (EBL), even to $B = 28$ the contribution is still an order of magnitude less than the most stringent lower limit on the direct measurement of the EBL (Dube et al 1979).

Koo (1981, 1986) has shown the slope of the counts depends on wavelength in a fairly systematic way - decreasing from a Euclidean 0.6 in the U to a sub-Euclidean 0.4 in F and N, i.e the extra objects are blue - a point originally made by Kron (1980) on the basis of two passbands. The colour of a galaxy depends on its restframe spectral energy distribution (SED) and redshift. It might thus be possible to infer approximate redshift distributions of faint galaxies from multicolour data alone. However if any spectral evolution is present, this will distort the form of the SED and the inferred N(z). Precise modelling may salvage the situation but the approach is rather restrictive when the evolutionary behaviour of galaxies is largely unknown.

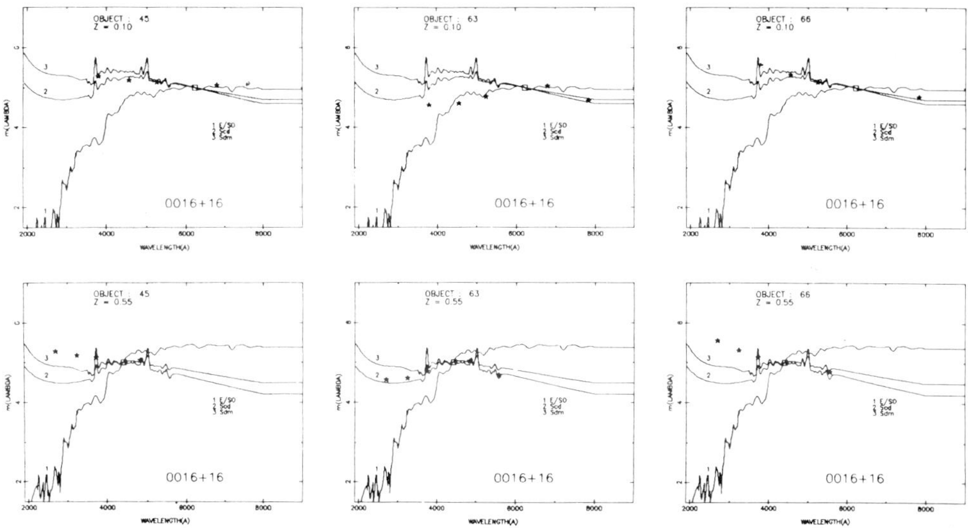

Fig. 4. Uncertainties in estimating redshifts from multi-colour data: 6 colour CCD data (*) for blue galaxies in the cluster 0016+16 (Ellis et al. 1985) compared with local SEDs (lines) assuming z=0.10 and z=0.55. Object 63 has a spectroscopic redshift of 0.54.

First we review the application of multicolour redshift estimation in distant clusters where there is often detailed CCD photometry of high precision *and* spectroscopic redshifts. Couch et al (1983) have developed a CCD imaging system based on 6 intermediate-band filters for monitoring evolution in early type galaxies. Simulations using present day SEDs suggest a greater sensitivity to fainter limits than that based on broad-band photographic UJFN. For early type galaxies the method is extremely successful and spectroscopically supported (MacLaren 1987). However, even with 5% precision over 4000-9000 A, the redshifts of later type galaxies are very difficult to determine to better than $\pm$ 0.2 in z (Fig. 4; c.f. Koo (1986) Fig 9). In addition to difficulties associated with featureless SEDs, there is evidence that "abnormal" systems unrecognised at present epochs will confuse matching algorithms (Ellis et al 1985). Colours alone may therefore be unable to identify and provide redshifts for hitherto unknown classes of objects. On the other hand the combination of colours *and* spectra within a given magnitude range may be a very powerful way of extending an otherwise restricted data set to include   many hundreds of faint galaxies (c.f Loh and Spillar 1986).

Koo (1986) claims from his photographic UJFN estimated redshifts and over a hundred spectroscopically determined, that his method is accurate to ± 0.05 in z. (A similar claim is made by Couch et al for their 6 colour CCD-based system but upheld only for the E/S0s and early/mid-range spirals).   Koo estimates the fraction of field galaxies with rest-frame colours bluer than B - V = 0.7 has doubled at z $\sim$ 0.4 and that, presumably, this contributes to the excess of field counts since the uv and blue luminosity functions will be shifted brightward.   However, the redshift estimates for these bluest galaxies will be the most uncertain and again only spectroscopic surveys can rule out the alternative hypothesis that these extra blue galaxies are, in fact, intrinsically faint galaxies at lower redshifts.

4. Faint Redshift Surveys

8 years after the original magnitude count papers of Kron(1978), Peterson et al (1979) and Tyson and Jarvis (1979) it is now possible to obtain redshifts at interesting limits. Faster spectrographs and detectors are partly responsible but the major factor is multiple object work that encourages the observational astronomer to make the necessary long exposures.   At the IAU Symposium 104 in Crete, the subject was just beginning with surveys reported by Koo (1983) and Ellis (1983).   These surveys deeper than B = 20 now have over 200 redshifts each (Table 3) and I will discuss preliminary results derived from the AAT survey done in collaboration with Broadhurst and Shanks.

Table 3

Deep Redshift Surveys

	Koo, Kron and Szalay	Ellis, Broadhurst and Shanks
Telescope:	KPNO 4.0m	AAT 3.9m
Instrument:	CryoCam + TI CCD	RGO + IPCS / FORS + GEC CCD
Multiplex gain:	20 holes/8 slits	50 + 50 fibres
Range (A)	4500 - 7500	3700 - 6100 / 5000 - 10,000
Resolution (A)	15	4 / 15
Exposures (hr)	1 - 4	4 - 6
Magnitude limit	R < 20	20 < B < 21.5

The only other complete survey known to the author is that of Tritton and Morton (1984) who catalogued $\sim$ 750 objects in a single field of 0.31 deg^2 area to B = 19.5 - 20 using low dispersion objective prism spectroscopy. Since *all* objects were scrutinised spectroscopically the ratio of stars/galaxies checks how many extragalactic objects are masquerading as stars e.g. because of their compact nature. The number of galaxies found by spectroscopic means was about 140 in close agreement with that expected from image classifications. To date only 45 of these have reliable redshifts from grating spectroscopy though further work is planned (Morton, private communication).

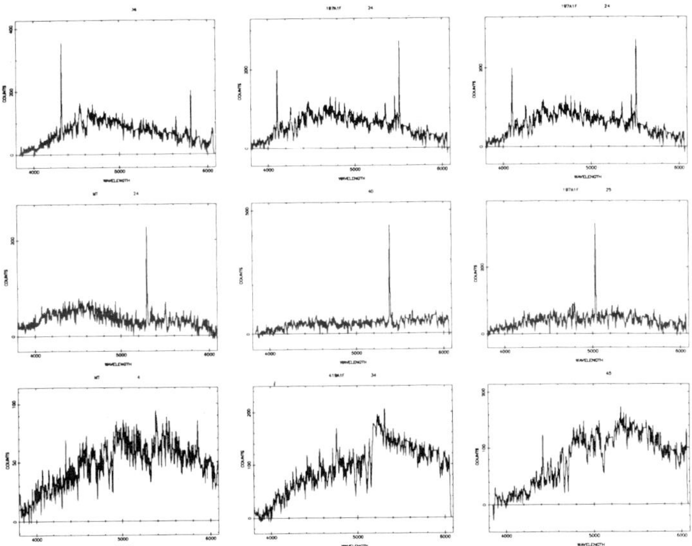

Fig. 5. Sample spectra for galaxies with <B>$\sim$21-21.5 from the AAT faint survey; (top) 0<z<0.22 showing [0 II] and [0III], (centre) 0.22 <z<0.45 showing [0 II] alone and (bottom) absorption line objects with various redshifts.

The AAT faint survey is complementary to that being pursued at KPNO because the galaxies are selected in B purposely to investigate the count excess. The multiplex advantage with fibres is larger with fewer geometrical restrictions on the target distribution. Longer exposures can thus be afforded and the high success rate on a first exposure ($\sim$ 75% in $\sim$ 4 hours) is obtained with intermediate (rather than low) dispersion. Slits are doubtless better for the faintest targets with B > 22 (we have recently commissioned a wide field multislit spectrograph on the AAT (Ellis and Taylor 1986) with which deeper surveys can continue) but fibres offer many advantages in the window 20 < B < 21.

The AAT survey is currently 85% complete to B = 21 and 75% complete to B = 21.5, with $\sim$ 200 redshifts secured. Since the AAT galaxies are confined to a narrow magnitude slice, the z distribution

is the most informative probe of any evolution particularly at the low
(z < 0.1) and high (z > 0.4) ends. Completeness is thus an important
issue. Fortunately we can check the observed redshift distribution is
representative in a number of ways. A large proportion of spectra (
50%) to B = 21.5 show both [O II] 3727 and [O III] 5007 yielding
unambiguous redshifts for 0 < z < 0.22 (Fig. 5).   A smaller
proportion (35%) show a single emission feature taken to be [O II]
with 0.22 < z < 0.6 since we can often readily identify the Balmer
absorption lines, also no single feature is ever seen in the low z
range where 5007 could also be detected.

We observe a marked decline in N(z) for z > 0.3 well within the
range discussed above, i.e.   there is no detectable high z tail
corresponding to luminous evolving galaxies. The Faint Object Red
Spectrograph (FORS) covers the window 0.5 - 1 microns at sufficient
dispersion for [O II] to be visible if present thus we would probably
detect any galaxies with emission lines with z < 1. For absorption
lines, only the low redshift range z < 0.5 is well covered, the
principal features being Ca II and the Balmer sequence; the FORS
resolution is mostly inadequate. Failures turn out to be mostly
objects whose continuum signal is too faint for redshift work - a few
are extended low surface brightness objects. Very few objects are
featureless at this dispersion.

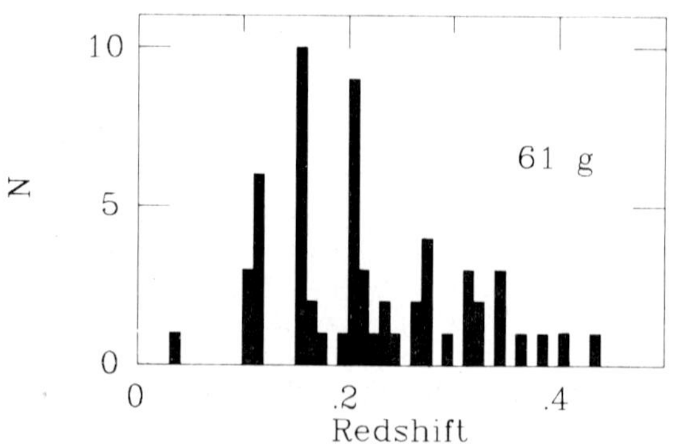

Fig. 6. Preliminary
redshift distribution
for one of 5 fields
in the AAT faint galaxy
survey.

The complete catalogue and analysis will be presented elsewhere
but a preliminary N(z) for one of the 5 fields is reproduced in Fig.
6. Analysis of the distribution for all fields indicates a reasonable
agreement between the data and the no evolution prediction. The mean
redshift for 200 galaxies with 20 < B < 21.5 is <z> = 0.222 ± 0.010,
compared with 0.231 for no evolution . Were the early type galaxies
to evolve in luminosity along the lines of Bruzual's μ =0.5 model, <z>
would rise to 0.25. Actually this would not be sufficient to explain
the steep count slope to B = 23; stronger evolution is required
pushing <z> above 0.30. Although more complete data is desirable, the
absence of any such high z tail in the observed distribution suggests

there is not yet any evidence for luminosity evolution over the last 5 Gyr.

The faint count excess cannot reasonably be due to large numbers of low luminosity galaxies either. We can match the steep count slope by incorporating a sizeable population of $M_B > -18.5$ galaxies. 12% to B = 16.85 are needed to raise the no evolution slope to 0.41 (c.f. 5% observed in DARS - $2). This would produce a low z hump in N(z) reducing <z> to 0.18. No such hump is observed. Incompleteness can be tested further by restricting the analysis to our 4 best fields with 20 < B < 21 where we have 85 galaxies representing a 90% complete sample. Again the no evolution model is a good fit to the observed distribution.

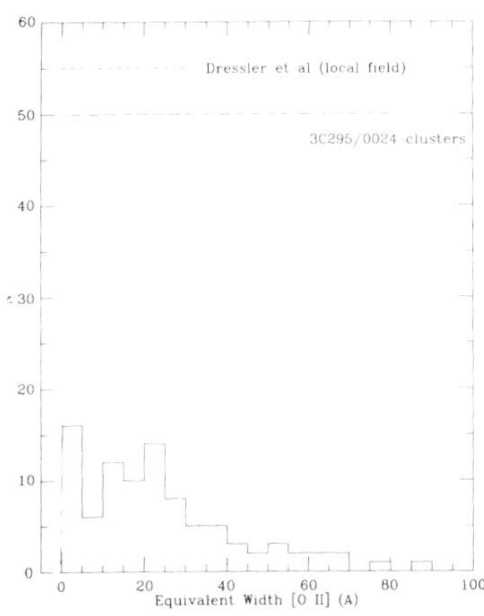

Fig. 7. Distribution of restframe equivalent widths for [O II] for the AAT faint galaxy survey (solid) and the nearb y DARS (dashed) with ranges shown for nearby field and distant cluster samples observed by Dressler et al.

One possible clue as to the nature of any excess comes from the surprising proportion of spectra showing strong [O II] emission, many have restframe equivalent widths > 20 A. The luminosities of these galaxies are not unusual but they seem to be undergoing proportionally much larger bursts of star formation than their counterparts in the nearby DARS surveys (Fig. 7). Interestingly, only the low luminosity objects nearby show such large [O II] widths, whereas at <z> ∿ 0.25 the effect is widespread amongst luminous galaxies.

There are two selection effects that might make distant galaxies show stronger emission lines - the aperture effect and the K-correction. The AAT fibre aperture at <z> = 0.2 is 13 kpc diameter whereas that in the DARS used for the comparison in Fig. 7 at <z> = 0.05 is a rectangle of 5 x 20 kpc. Dressler et al (1982,1985) have also observed a small sample of nearby spirals through a large aperture and find a range of [O II] widths similar to that in the DARS. Thus the aperture effect does not seem to be responsible. The K-correction effect whereby blue-selected galaxies at high z are more likely to be uv strong can be checked also since the relation between restframe colour and [O II] strength has been calibrated by Dressler et al. Colours are available for some of the faint AAT sample and too few uv galaxies are present to account for the emission line distribution.

To summarise, therefore, the redshift survey discussed has so far failed to find any obvious component responsible for the excess galaxy counts. There seems to be little evidence for any *systematic* evolution in any significant subset of the population over the last 5 Gyr. It may be that the count excess is an artefact of an incorrect no evolution slope and that, to B = 21.5, none would be expected. This seems unlikely given our reasonable knowledge of the field population, its mix and LF. The absolute normalisation of the galaxy counts does, however, play a vital role in assessing the excess (Fig. 2) and more well-calibrated counts in the 17 < B < 20 region are needed to check the effects of fluctuations.

On the other hand, it is interesting to speculate whether the strong emission line objects are related to the excess counts. Statistically complete large aperture spectroscopy of nearby galaxies is urgently needed to verify the trends suggested in Fig. 7. However, we note that a burst of star formation could, for a few Gyr, raise the luminosity of a low mass spiral (such as those in Fig. 1) to that required for appearance in the faint survey. Because only a subset of the LF might be prone to such bursts at any epoch (such as implied by the biased theory of galaxy formation), we would need to study the change in *shape* of the LF with epoch rather than to follow a systematic brightening of *all* galaxies with lookback-time.

5. Large Scale Structure and Galaxy Correlations

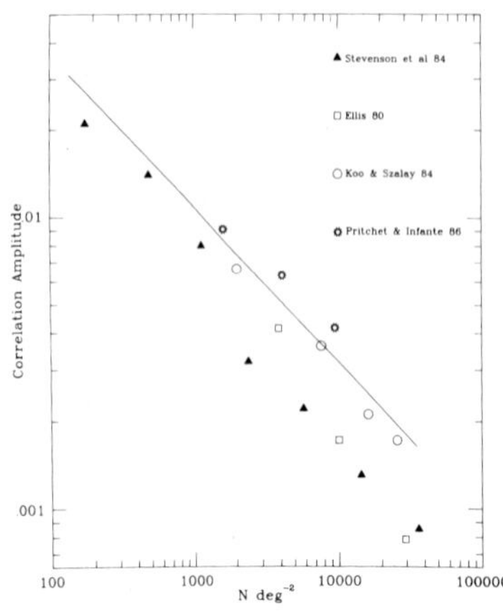

Fig. 8. Angular correlation amplitudes at 1 degree separation versus galaxy surface density for deep samples surveyed in B. The model shows the expected behaviour for stable clustering and no evolution.

Various groups have analysed the angular correlations at faint magnitudes (Ellis 1980, Koo and Szalay 1984, Stevenson et al 1985, Pritchet and Infante 1986). Their results are inconsistent (Fig. 8).

Ellis   and   Stevenson et al fitted a -0.8 power law to their two point function w(θ) but found an amplitude lower than expected for either virialised   or comoving structures to B = 24.   At B = 24 a significant fraction of galaxies would have 0.5 < z < 1 and they concluded   there was some evidence for growth in clustering over the last 10 Gyr.

Koo   and Szalay found a flatter slope of -0.5 in their faintest samples but found room for some luminosity evolution. It is not possible to distinguish between the two forms of evolution via w(θ) alone, one needs the counts as well. Since the redshift distributions so   far observed cast some   doubts   on   the   traditional   N(m) interpretation, it is interesting to reconsider these conclusions based on the correlation analyses.

Interpreting   correlation amplitudes for faint data is complex beyond the   need for an adequate description of evolution in the   LFs. For example, Davis and Geller (1978) showed ellipticals are more strongly clustered than spirals. In the DARS sample the -1.8 power law amplitude ratio for the spatial correlation function is about x 1.8. K-corrections increase the proportion of spirals in faint samples (from 60% at B = 16.85 to 80% at B = 24 for no evolution) and this affects the expected amplitudes at the 20% level. Contamination of data from spurious uncorrelated images (noise, stars) lower the amplitude according to the square of the contamination factor. Considering the software algorithm tests discussed earlier, these might cast doubt on the deepest points in Fig. 8 but not those at < 10^4 galaxies/deg^2 densities where discrepancies are still present.

Any correlation evolution might be more apparent in 3-D via   the new redshift surveys since the major problem with the angular function is the projection effects which reduce the amplitude observed to ∿ 5% on scales of 1 arcmin to B ∿ 23. The spatial correlation function in the faint AAT survey does show an amplitude tantalisingly lower than expected in the virialised case but about twice the number of galaxies is needed before any significance can be attached to this result. It will be particularly interesting to examine whether the star-forming galaxies discussed in $4 are distributed any differently from the remainder.

Finally a comment on the large scale distribution of galaxies as exemplified in Fig. 6. In common with other talks at this symposium we can identify large sheet-like structures in the 3-D distribution for some but not all of our narrow angle (∿20 arcmin) fields. Our sample covers perhaps the largest volume so far surveyed and thus might be expected to be the most representative of the universe on large scales - the average depth is 1000 Mpc (H_O = 50). The sheets explain in a natural way the fluctuations in the galaxy counts that puzzled the observers for so many years. There is a good correlation between the number of sheets "skewered" by each survey cone and the absolute number of galaxies counted to deeper limits in that area.

Hubble's goal of using the faint counts to confirm fairly preciselythe homogeneity of the Universe on large scales seems that bit further away.

6. Acknowledgements

The deep AAT redshift survey would not have been possible without the dedication and energy of Peter Gray who built the fibre device FOCAP. Thanks are also due to Warrick Couch and Ray Sharples for vital support at the telescope. I thank Tom Broadhurst, Tom Shanks and Laurence Jones for allowing me to use preliminary analyses of data gathered with their collaboration. Any errors or unorthodoxy in its interpretation is the author's responsibility. Useful discussions with Dick Fong, David Koo, Iain MacLaren, Gus Oemler and Jo Silk are acknowledged. Finally I thank Chinese colleagues Chen Jian-Sheng, Zhou Zhen-Long and Wang Shun-de for their warm hospitality during my stay in Beijing.

References

Aaronson, M 1978 Ap. J. 221, L103.
Bean, AJ 1983, Ph. D. thesis, University of Durham.
Bruzual, GB 1983 Ap. J., 273, 105.
Couch, WJ, Ellis, RS, Carter, D and Godwin, J 1983, MNRAS 205,1287.
Ciardullo, R 1986, Ph. D. thesis, Yale University.
Davis, M and Geller, MJ, 1976, Ap. J. 208, 13.
Dressler, A and Gunn, J 1982 Ap. J. 263, 533.
Dressler, A, Gunn, J and Schneider, DP 1985 Ap. J. 294,70.
Dube, RR, Wilkinson, DT and Wickes, WC 1979 Ap. J. 232, L51.
Ellis, RS 1980 Phil Trans R Soc London, 296, 355.
Ellis, RS 1982 in *Origin and Evolution of Galaxies*, eds Jones, BJT
        and Jones, JE, (Dordrecht : Reidel), p255
Ellis, RS 1982 in *Early evolution of the Universe and its Present
        Structure*, eds. Abell, GO and Chincarini, G, P87
Ellis, RS and Koo, DC 1986, in preparation.
Ellis, RS and Taylor, K 1986, AAT Newsletter 38.
Ellis, RS, Gray, PM, Carter, D and Godwin, J 1984, MNRAS 206, 285.
Ellis, RS, Couch, WJ, MacLaren, I and Koo, DC 1985 MNRAS 217, 239.
Fong, R, Jones, LR, Shanks, T, Stevenson, PRF, Strong, AW, Dawe, JA,
        Murray, JD 1986, MNRAS in press.
Infante, L , Pritchet, C and Quintana, H. 1986, A. J. 91, 217.
King C and Ellis, RS 1985 Ap. J. 288, 456.
Kirschner, RP, Oemler, A and Schechter, PL 1978 A. J. 83, 1549.
Kirschner, RP, Oemler, A and Schechter, PL 1979 A. J. 84, 951.
Kirschner, RP, Oemler, A, Schechter, PL and Schectman, S 1981 Ap. J.
        248, L57.
Koo, DC 1981 Ph. D. thesis, University of California at Berkeley.

Koo, DC 1982 in *Early Evolution of the Universe and its Present
    Structure*, eds. Abell, GO and Chincarini, G, P105
Koo, DC 1984 in *Astronomy from measuring Machines*, ed Reid, IN and
    Hewitt, PC. RGO publications, P112.
Koo, DC 1985 A J 90, 418
Koo, DC 1986 STScI Preprint 116
Koo, DC and Szalay, AS 1984 Ap. J. 282, 390.
Kron, R 1978 Ph.D. thesis, University of California at Berkeley
Kron, R 1980 Ap. J. Suppl., 43,305.
Kron, R 1982 Vistas in Astronomy, 26, 37.
Loh, E D and Spillar, E J 1986 Ap. J. 303, 154.
MacLaren, I 1987 Ph. D. thesis, University of Durham
Mobasher, B, Ellis, RS and Sharples, RM 1986 MNRAS 223,11.
Pence, WD 1976 Ap. J. 203, 39
Peterson, BA, Ellis, RS, Kibblewhite, EJ, Bridgeland, M, Hooley, A,
    and Horne, D. 1979 Ap. J. 233, L109.
Peterson, BA, Ellis, RS, Efstathiou, G, Bean, AJ, Shanks, T.
    Fong, R and Zhen-Long, Z. 1986 MNRAS 221, 233.
Pritchet, C and Infante, L 1986 A. J. 91,1
Schechter, PL 1976 Ap. J. 203, 297.
Shanks, T, Stevenson, PRF, Fong, R and MacGillivray, HT
    1984 MNRAS 206, 767
Stevenson, PRF, Shanks, T, Fong, R and MacGillivray, HT
    1985 MNRAS 213, 953.
Tinsley, BM 1978 Ap. J. 220, 816.
Tinsley, BM 1980 Ap. J. 241, 41.
Tritton KP and Morton, DC 1984 MNRAS 209, 429.
Tyson, JA and Jarvis JF 1979 Ap. J. 230, L153.

DISCUSSION

BAHCALL: Are the star-burst galaxies in the field fainter than
those found in the z ~ 0.3-0.4 Butcher-Oemler clusters?

ELLIS: Not significantly in luminosity. A similar limiting
magnitude was reached by Dressler et al. for clusters at redshifts
slightly deeper than our average field redshift.

EVOLUTION OF VERY FAINT FIELD GALAXIES AND QUASARS

David C. Koo
Space Telescope Science Institute
and
Richard G. Kron
Yerkes Observatory, University of Chicago

This paper reports preliminary results of two long-term redshift
surveys that are near completion. One consists of nearly 400
redshifts of field galaxies; the sample is faint enough (B $\lesssim$ 22) to
test models of galaxy luminosity and color evolution (and possibly
cosmology) and to search for the presence of very large-scale
structures among distant field galaxies. The other consists of over
60 spectra of quasar candidates of similar faintness; the
identifications and redshifts of bona-fide quasars provide strong
constraints on the evolution and shape of the luminosity function of
distant quasars. Almost all of the observations have been made with
the 4m telescope at Kitt Peak National Observatory with the Cryogenic
Camera in multiaperture mode. The spectral range covered 4500Å to
7500Å with 15Å FWHM resolution and 4Å per pixel; simultaneous
exposures of an hour or two were made for about 10 objects within the
5 arcmin field of view.

The quasar candidate sample is concentrated over an area of 0.3
deg^2 in Selected Area 57 at the North Galactic Pole and consists of
all stellar-like objects which lie away from the positions of expected
Galactic stars in the UBV two-color diagram (Koo, Kron, and Cudworth
1986). We emphasize that our selection does NOT exclude quasars of
high redshift (z > 2.2) as in other photometric UV-excess surveys.
Among the 77 candidates complete to B $\gtrsim$ 22.5, we now have spectra for
64. The 32 genuine quasars have redshifts ranging from 0.9 to 3.1,
with a median of 1.5. We also found 11 stars, generally at B < 21, as
well as 16 narrow emission-line galaxies with redshifts between 0.2
and 0.8. Five remain unidentified.

Details of our spectroscopic work will be published elsewhere.
In summary, we find:
1) To B ~ 22.5, we have spectroscopically confirmed a quasar surface
    density of over 100 deg^2, the highest among any existing survey.
2) The redshift distribution among our quasars with B < 21 is almost
    identical to the fainter ones. As already hinted by the
    predominance of UV-excess candidates, relatively few (7) quasars
    with z > 2.2 were found, supporting the view that high redshift
    quasars are rare over a large range in luminosities.

383

A. Hewitt et al. (eds.), Observational Cosmology, 383–388.
© 1987 by the IAU.

3) The turnover in the number counts of faint quasars is also seen in the shape of the luminosity function of quasars at redshifts $z < 3$, derived by combining our data with those of others from brighter surveys.

4) Using the break from the steeply rising bright portion of the luminosity function as a fiducial point, we find one interpretation to be that luminosity evolution in the form $(1+z)^4$ is dominant over density evolution, at least up to $z \sim 2.5$. In fact, the data suggest that the overall number of quasars may even have been fewer in the past. If so, quasars of $z < 4$ may NOT have ionized the intergalactic medium. If constant, the more rapid drop beyond $z \sim 2.5$ may signal the major epoch of quasar formation.

The galaxies of the field survey were selected randomly within several 5 arcmin areas, with brighter (in the red band) objects given higher weights. These areas were located in three fields of high galactic latitude (SA 57 at 1305+30; SA 68 at 0015+15; and Hercules No.1 at 1720+50) for which we have UBVI photometry from 4m prime focus plates that go deeper than our sample for spectroscopy. Our field galaxy survey represents a six year effort, largely because of our attempt to define and achieve a sample that is "complete", not necessarily in the sense of finishing all objects to a single magnitude limit over a given area of sky, but rather in the sense of having representative (i.e. unbiased by color, surface brightness, or strength of spectral features) subsamples useable for statistical analysis. To date, we have secured over 400 redshifts, with over 300 now constituting a statistically complete sample.

Although the reduction of the redshift survey is still underway because of our desire to improve the accuracy, reliability, and completeness, we are already able to report some interesting, though preliminary, results. One of the most striking is the observed clumpiness of the redshift distributions, especially in SA 57 (see Figure 1a). This field shows both very strong overdensities (clustering) and underdensities ("voids") on scales of ~ 100 Mpc. Our survey depth is not easily defined, but the faintest galaxies have $B > 22$ and the median redshift is 0.24. A detailed analysis of the clustering (and its evolution) of field galaxies and of its consistency with various cosmological scenarios is underway in a collaborative project with A. Szalay. In the meantime, our data can be compared to cold dark matter N-body simulations (White et al. 1986), which also show large fluctuations in the redshift distribution.

Another result comes from comparing the colors and redshifts of our data for three fields to that predicted from galaxy spectral evolution models, similar to (but not exactly the same as) that of Bruzual and Kron (1980). Our data (Fig. 2) contain far fewer low luminosity galaxies than predicted by our improved models and suggest at most mild luminosity evolution of galaxies at redshifts $z > 0.4$.

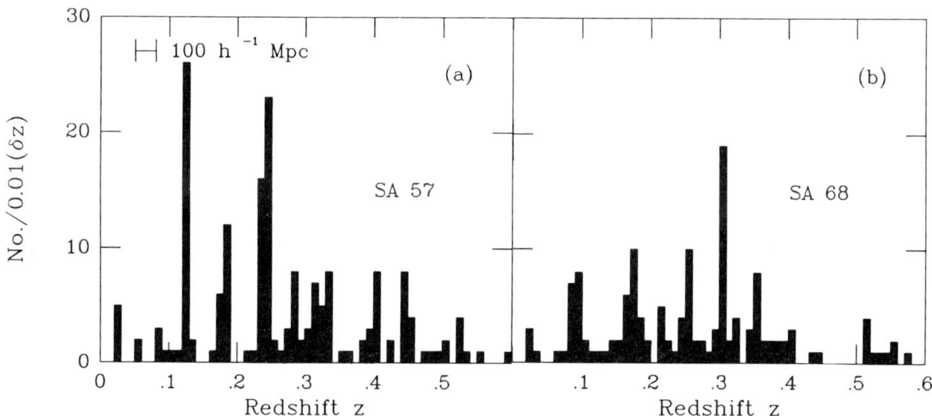

Fig. 1   a) Histogram in bins of 0.01 in z of redshifts for 148 galaxies within a 0.3 deg^2 field at the North Galactic Pole (SA 57). The magnitude and completeness limits are very complex, but roughly represent a sample brighter than B ~ 22.   The "void" between z of 0.13 and 0.17 is particularly noteworthy.   b) Histogram of 134 redshifts in another 0.3 deg^2 field (SA 68); depth is similar to that in SA 57.

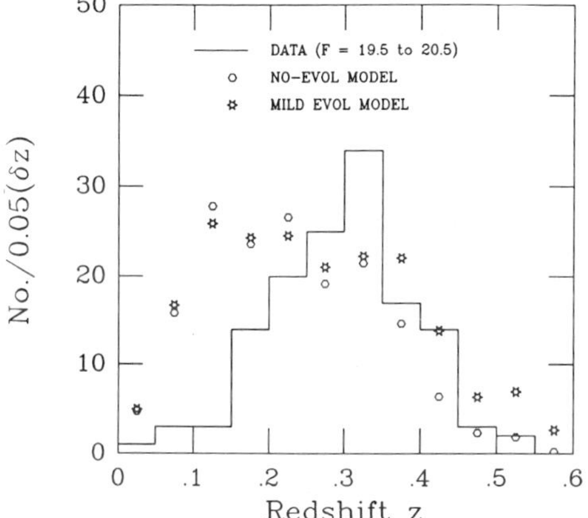

Fig. 2   Histogram of galaxies with red (F) mag between 19.5 and 20.5 in redshift bins of 0.05 compared to that of no-evolution (in color or luminosity) and evolution models that fully account for the completeness (equivalent areal coverage at different mag) and photometric errors of our survey.

These distant galaxies do, however, appear to be relatively brighter
in the rest frame ultraviolet (i.e. bluer) than the current model
predictions, an example of which is shown in Table I. Their spectra
often possess strong emission lines indicative of substantial star
formation (unfortunately, both the spectral energy distributions in
the UV and the integrated strengths of emission lines remain uncertain
for local galaxies). Not all galaxies have undergone similar color
changes, however; our reddest galaxies up to redshifts of ~0.55
possess intrinsic colors that do not differ from those of red galaxies
today (see review by Spinrad 1986). Finally, we note that neither
the deep QSO nor the deep field galaxy redshift survey has yielded any
good candidates for primeval galaxies.

We acknowledge the support of Kitt Peak National Observatory,
Department of Terrestrial Magnetism (Carnegie Institution of
Washington), and NSF Grants AST 81-21653 and AST 83-14232.

REFERENCES

Bruzual, G. A. and Kron, R. G. 1980, Ap. J., **241**, 25.
Koo, D. C., Kron, R. G., and Cudworth 1986, PASP, **98**, 285.
Spinrad, H. 1986, PASP, **98**, 269.
White, S. D., Frenk, C. S., Davis, M., and Efstathiou, G. 1986,
     Ap. J., in press

Table I
Color of Distant Galaxies
(z = 0.35 to 0.45)

| | DATA | | No-Evol | | Evol | |
Red Mag	Blue	Red	Blue	Red	Blue	Red
<19	1	0	1	1	3	9
19 - 20	7	3	3	5	8	11
20 - 20.5	4	5	4	7	9	12
Total	12	8	8	13	20	32

Blue and red correspond to observed U - F colors less and greater than
1.5. U(3620 Å) is about 2500 Å rest and F(6100 Å) is about 4400 Å
rest. This division is expected to divide galaxies with rest frame
B-V colors less and greater than 0.7.

DISCUSSION

WINDHORST: For what fraction of the field galaxies are the spectra
featureless, meaning, that no redshifts have been measured. Is this
a serious problem for the determination of the redshift distribution?
Do you have any clue from your multicolor analysis what their redshift
range might be? (Question directed to both D. Koo and R. Ellis.)

KOO: Our ability to secure redshifts depends on both S/N and the
strength of emission and/or absorption features - "featureless" is
difficult to define but we do have spectra without strong features.
Our redshift distribution is meaningful only for the statistically
complete sample. For galaxies without redshifts, multicolors can
provide estimates of distances, but these may be unreliable, especially
for redshifts greater than about 0.6.

ELLIS: We are statistically complete but only to a brighter limiting
magnitude than that adopted in our survey (see previous paper). The
most successful redshift feature in our wavelength range is [OII]3727
which we could detect to z ~ 0.6 yet our redshift distributions fall
well before this limit is reached. Thus I think it unlikely that
incompleteness will affect significantly the high z region of the
distribution, though we are continually striving to finish off the
remaining faint galaxies without redshifts both with the fibre system
and our new multislit spectrograph.

FILIPPENKO: Have you checked whether or not known QSOs with $3 \lesssim z \lesssim 4$
have colors which could be confused with stars in your color-color
diagrams? If they do, then you may have missed a substantial number
of high-z objects.

KOO: Yes, we have looked at all the QSO's with $z \gtrsim 2.5$ with UBV
colors in the Veron and Veron catalog, and estimate an incompleteness
of ~20%-30% for $z \gtrsim 3$. Marano et al. have checked their multicolor
sample against a slitless survey in the same field and conclude that
the incompleteness is at this level. Compared to model predictions
we have probably not missed substantial numbers, though independent
checks are needed to confirm this. An astrometric and variability
survey is already underway for this purpose.

SILK: There are many uncertainties in primeval galaxy models which
must be evaluated before they can be ruled out even at moderately low
redshift by means of optical searches. One of the most serious
problems may be that IRAS observations of extreme starbursts suggest
that primeval galaxies may be extremely dusty objects.

KOO: I agree totally. Our faint galaxy and quasar survey places a strong constraint on, for example, Meier's model (1976, Ap. J. <u>207</u>, 343) of primeval galaxies, but it certainly does not rule out others (see my review in 1986 <u>Spectral Evolution of Galaxies</u>, ed. C. Chiosi and A. Renzini, p. 419).

CHAPTER V

THE THEORY OF GALAXY FORMATION AND LARGE-SCALE STRUCTURE

GALAXY FORMATION: CONFRONTATION WITH OBSERVATIONS

Joseph Silk
Astronomy Department
University of California
Berkeley, CA   94720
USA

ABSTRACT. The implications for galaxy formation of inflationary cosmology are reviewed. In particular, I explore some implications of the hypothesis that galaxies form from adiabatic, gaussian density fluctuations in a cold dark matter-dominated universe. Topics discussed include protogalaxies and the epoch of galaxy formation, Lyman alpha clouds, dark halos and dwarf galaxies. Finally I describe how environmental biasing may arise as a consequence of tidally induced star formation in protoclusters.

I. INTRODUCTION

A notable development in cosmology, and in particular, in galaxy formation theory, over the past few years has been the gradual emergence of a standard model. This is due in large part to the contribution from particle physics, which has provided a theoretical framework for describing the evolution of the very early universe. Without this input, the initial value problem, namely the unknown initial conditions that characterized the density fluctuations at early epochs, presented an insuperable obstacle. I am going to tell you a modern myth. It has one advantage over the myths of ancient times – it is falsifiable. This brings it into the realm of science. The myth begins a long, long time ago with the creation of the universe. Once upon a time there was nothing, nothing at all. Not space, nor even time. This was the realm of quantum gravity. Our understanding of quantum gravity is inadequate, at present, to describe the initial singularity predicted by the classical regime of the Big Bang theory, although recent developments in the theory of higher dimensional space–time descriptions of physics, such as superstrings, have persuaded many physicists around the world that development of a unified theory of quantum gravity is imminent. From the Planck time, $t_{pl} \sim 10^{-43}$ s, onward, however, classical gravity suffices, and the very early universe provides

391

A. Hewitt et al. (eds.), Observational Cosmology, 391–413.

a natural laboratory for studying the grand unification of the electro–weak and strong nuclear interactions.

As the universe expands and adiabatically cools, from an energy or temperature of about 10^{19} GeV at the Planck time to about 10^{15} GeV, the symmetry is spontaneously broken, and the more familiar interaction strengths of the present universe emerge. The symmetry breaking is characterized by a first–order phase transition which triggers a brief period of exponential expansion or inflation (see e.g. Turner 1985 for a review of inflation). Preexisting inhomogeneities and anisotropies, of more or less generic type, are erased, while quantum fluctuations are amplified up to the horizon scale. Because of the nature of the free quantum fields, the fluctuations are gaussian: this is effectively guaranteed by the central limit theorem. Since the horizon inflates to a scale far larger than our present horizon, this means that we are left with a density fluctuation spectrum that can eventually power growth of large–scale structure on all observable scales. No preferred scale is selected for these fluctuations: one characterizes them by saying that the associated curvature fluctuations are of constant amplitude on all scales. Curvature is measured by the density fluctuation amplitude at horizon crossing, and one has the scale–invariant Zel'dovich–Harrison fluctuation spectrum $(\delta\rho/\rho)_H \approx$ constant. At the same time, the erasure of any preexisting global curvature by inflation means that the universe must be flat to a high degree of precision: specifically

$$\Omega = 1 \pm O(|\,\delta\rho/\rho\,|). \tag{2}$$

The density parameter Ω is defined to be the density expressed relative to the critical value for closure, $3H_o^2/8\pi G$. Inflation thereby accounts for a paradox implicit in the Friedmann equation, which can be expressed in the form

$$\Omega \approx 1 + Kt, \tag{3}$$

where K is a constant of order unity that depends on the curvature of the universe. If today, at $t \approx 6 \times 10^{17}$ s, Ω were equal, for example to 0.1, one would then have great difficulty is understanding why, at t_{pl}, Ω must have been equal to unity to within about 60 decimal places. Inflation naturally accounts for this coincidence. While inflation was by no means inevitable, nor did it necessarily lead to an acceptable amplitude for the density fluctuations, it has resolved enough problems in our understanding of the very early universe for us to regard it as a yardstick by which any rival theory has to be compared. I shall therefore adopt it as the standard model in the ensuing discussion.

This standard model also incorporates two other successes that result from the application of particle physics to cosmology. These are baryon genesis and light element nucleosynthesis. The universe today consists predominantly of baryons, not antibaryons, and the baryon number, a conserved quantity, can be expressed as $n_B/n_\gamma = 10^{-8}\Omega h^2$, where $h \equiv H_o/100\text{km}^{-1}\,\text{s}^{-1}\,\text{Mpc}^{-1}$. This manifests itself at early epochs as a slight baryon assymmetry, and can be understood as arising from weakly CP and baryon number violating processes that occurred following the symmetry breaking phase when inflation ended and the universe reheated to very high temperature, in the range 10^{10} GeV $- 10^{16}$ GeV. Since a universal baryon number results, this also means that one is unlikely to have had any entropy fluctuations: apart from one possible exception (in an axion–dominated universe), the surviving density fluctuations are adiabatic.

Primordial nucleosynthesis occurred much later, when weak interactions froze out at $T \sim 1$ MeV, leaving a residual neutron abundance of about 10 percent. Most of these neutrons were subsequently synthesized into ^4He at $T \sim 0.1$ MeV, and traces of other light elements, most notably $^2H, ^3He$, and ^7Li, were also produced. Remarkedly, these abundances agree with the observed abundances, to within uncertainties of corrections for galactic evolution contributions to light element synthesis (Yang et $al.$ 1984). One critical parameter has to be adjusted, namely the baryon density, and

$$\Omega_b h^2 \approx 0.03 \tag{4}$$

is required for consistency. There are no other plausible sources for ^4He and ^2H, both of which must be of pregalactic origin, and explaining their abundances to within one or two percent (in the case of ^4He) or to within a factor of 2 (in the case of ^2H) by adjustment of one plausible parameter is a powerful argument for a big bang origin.

One may now summarize the standard model of the early universe: $\Omega = 1, \Omega_b \approx 0.1$, and adiabatic, gaussian density fluctuations with a scale–invariant spectrum. I add one more ingredient, namely cold dark matter. Dark matter is needed in order for $\Omega >> \Omega_b$, and cold dark matter, by which is meant any species of weakly interacting massive particle, provides a generic description. Such particles are plausible stable relics from supersymmetry: candidates include photinos, higgsinos, or scalar neutrinos. These heavy particles generally have mass of order 1 GeV, and so have negligible velocity dispersion when the universe first becomes matter dominated at $T \sim 10$ eV and effective fluctuation growth commences. Axions, while light, are non–relativistic when first produced at $T \gtrsim 10^{10}$ GeV, and so are also a form of cold dark matter.

I regard the principal alternative, hot dark matter, characterized either by a universe dominated by a light, ~ 30 eV, neutrino or by adiabatic fluctuations in a baryon–dominated universe, as being less likely, both because of the recent experimental limits ($m_{\nu_e} < 18$ eV) (Fritschi et $al.$ 1986) on neutrino mass and because of theoretical objections to a hot dark matter–dominated universe that arise in simulations of the large–scale galaxy distribution (Frenk *et al.* 1983). Free streaming of hot dark matter suppresses fluctuation growth on mass scales below

$$1.8m_{pl}^3/m_\nu^2 \sim 10^{15}(30eV/m_\nu)^2 M_\odot$$

and requires a top–down scenario for forming structure in the universe today in which cluster–mass structures anisotropically collapsed, pancaked, and fragmented into galaxies. The large–scale structure constraints from galaxy correlations mean that such theories fail to account for galaxy formation at the epoch when quasars are seen (to a redshift that now exceeds 4), unless some additional ingredient is added that allows galaxies to form prior to pancaking.

Cold dark matter provides a hierarchical or bottom–up scenario for the development of structure. Fluctuations grow first on small scales, and eventually develop on very large scales. The only free parameter is the initial amplitude of the Zel'dovich Harrison scale–invariant spectrum; the value required in order to fabricate large–scale structure by today is

$$(\delta\rho/\rho)_H \sim 10^{-4}. \tag{5}$$

Any smaller value would not generate sufficient structure; any larger value would generate excessive structure. For the moment, there is no natural explanation for the origin of the amplitude (5), although anthropic arguments have been given in the context of inflationary scenarios. The constraint (5) can be reinterpreted as a constraint on the potential that drives inflation, and presumably an explanation will eventually emerge from particle physics (Steinhardt and Turner 1984).

A unique fluctuation spectrum therefore arises in the standard model, and its amplitude is adjusted to yield the observed correlation length in the galaxy distribution. This in itself is a non–trivial adjustment, since there is a natural scale that is imprinted on the residual, post–recombination, fluctuation spectrum, namely the comoving horizon scale at equality of matter and radiation densities,

$$L_{eq} \approx 10(\Omega h^2)^{-1} \text{Mpc}. \tag{6}$$

Small scales enter the horizon during radiation domination and undergo only logarithmic growth (Mészáros 1974); consequently the residual fluctuation spectrum is nearly flat at $L << L_{eq}$, as expected for sub–horizon growth of $\delta\rho/\rho$ proportional to $t^{2/3}$ during the matter–dominated era.

Numerical simulations have provided a quantitative test of this theory. The slope of the simulated correlation function steepens with time: it only agrees with the observed function

$$\xi(\nu) = (r_o/r)^{-1.8}, \quad r_o = 5h^{-1}\text{Mpc} \tag{7}$$

at a unique epoch (or equivalently by today, for a specified Hubble constant). The standard model is found to give a good fit to the slope and amplitude of the observed galaxy correlations if galaxy formation occurred recently and $H_o \approx$ 50 km s^{-1}Mpc^{-1}, provided that an additional parameter is added in order to allow mass to be more uniformly distributed than the luminous galaxies (Davis *et al.* 1985).

This additional parameter is specified as follows. In addition to galaxy correlations, one also has dynamical information on galaxy peculiar motions. Regarding galaxies as test particles, these allow a determination of Ω_{dyn}, which is measured on scales of up to ~ 10 Mpc to be

$$\Omega_{dyn} \approx 0.2(\pm 0.1). \tag{7}$$

To reconcile this with our standard model ($\Omega = 1$), the dominant mass contribution must be smoothly distributed on scales up to $\sim 10h^{-1}$ Mpc. One can quantity this: the correlation length for the mass distribution must be between one–half and one–third of that observed for the galaxies. Then the measured Ω_{dyn}, which for a given velocity measurement is inversely proportional to $\int_o^r \xi(r)rdr$, or the potential energy, can be reduced relative to $\Omega = 1$ by about a factor of 3–10.

This renormalization of the underlying mass distribution relative to that of the galaxies is parametrized by the introduction of a biasing threshold (Bardeen *et al.* 1986). For initially gaussian fluctuations, the rare peaks are more clustered than the common peaks. Denoting a peak by $\nu\sigma$, where σ is the rms density fluctuation amplitude, then imposition of a threshold for galaxy formation of $\nu\sigma$, such that galaxies only form in peaks greater than or equal to $\nu\sigma$, biases the correlation function: it is enhanced (Kaiser 1984; Politzer and Wise 1984) by a factor $\sim \{\exp[\nu^2\xi(r)] - 1\}/\xi \approx \nu^2$. Galaxy formation at 2.5 σ peaks, with galaxies defined to be fluctuations filtered over a comoving scale of 0.5 h^{-1} Mpc, is consistent with the observed correlations and the number density of galaxies for the standard model. Filtering is necessary in order to smooth over the smaller peaks which, while containing less mass, would dominate number counts. One of the topics to be discussed below will be the physical origin of this biasing towards rare peaks.

There are additional hurdles in the realm of large–scale structure that the standard model of biased cold dark matter satisfactorily overcomes. The predicted cosmic microwave background fluctuations are low, with $\delta T/T \sim (3-5) \times 10^{-6}$, and below current upper limits of about 3×10^{-5} (on a wide range of angular scales, from 1$'$ up to 90°) (Kaiser and Silk 1986). Large voids are produced in the simulated galaxy distribution as a consequence of the biasing threshold, and filaments and sheets of galaxies are present that resemble those observed (White *et al.* 1987). The frequency of rich clusters is comparable to that of the Abell clusters, when summed over richness class.

These are the principal successes arising from confrontation with observation on large scales. There are also difficulties, or potential difficulties, lurking ahead. The most serious of these concerns the cluster–cluster correlations. Clusters are themselves clustered, and the strength of this clustering is such that the correlation length is about five times larger than the galaxy correlation length (Bahcall and Soneira 1983). The standard model simulations produce some enhancements of cluster correlations, but the cluster correlation length is only enhanced by a factor of two relative to the galaxy correlation length (Barnes *et al.* 1985). However systematic errors in the cluster–cluster correlation function are very large, especially on large scales (Ling *et al.* 1986), and it is probably premature to reject the standard model on this account. A potentially even more serious problem for biased cold dark matter is the recent report of cosmic drift velocities of galaxy clusters over scales of up to $\sim 60h^{-1}$ Mpc that amount to ~ 600 km s^{-1} (Dressler *et al.* 1987). The model predicts velocities of ~ 100 km s^{-1} on such large scales: however, the reality of the cosmic drift velocities have yet to be definitively confirmed.

My philosophy is the following: the standard model of biased cold dark matter is aesthetically appealing, it has fewer free parameters than any other model of large–scale structure, and its predictions have been quantified in more detail than for any other models. The theory is a falsifiable theory. At present, the possible problems are minor compared to the advantages. I will take the attitude that the cluster–cluster correlations, for example, are a possibly major hurdle looming ahead for the standard model, but do not yet provide sufficient motivation to abandon cold dark matter. Of course, there is the likelihood that an improved model of large–scale structure would represent a variant on the standard model rather than being something radically different. In the absence of any alternative, we are surely justified in believing that the germ of the ultimate truth is already present in a model that is so well–motivated and has already met many challenges.

This review therefore considers galaxy formation in the context of the standard model, although many of the conclusions can readily be generalized to other models. I will emphasize observable implications of the standard model, including the epoch of protogalaxy formation and the role of Lyman alpha clouds and dark halos. In a final section, I will describe a biasing scheme and how environment plays an important role in determining galaxy morphology.

II. PROTOGALAXIES

Normalization of the large–scale structure to yield the observed galaxy correlation function fixes the rms mass density fluctuations today on a luminous galaxy filtering scale of 0.5 h^{-1} Mpc such that collapse has occurred for $\nu\sigma$ density fluctuations, represented by a spherical "top–hat" model with density contrast 1.69 at collapse, at a redshift $z_g = 2.5\nu/1.69 - 1 = 2.7$ for $\nu = 2.5$. The number density of 2.5σ fluctuations then agrees with the observed frequency of luminous galaxies. This normalization yields rms mass fluctuations $(\delta M/M)_{\rm rms} \approx 1$ at 3 h^{-1} Mpc and corresponds to a correlation scale for the 2.5σ fluctuations today of about $8h^{-1}$ where the correlations have unit amplitude (Bardeen *et al.* 1986).

The obvious difficulty with this scheme is that primeval galaxies are few and far between. Searches in random fields have not found any out to a redshift of about 5, with Lyman alpha emission as the principal search criterion (Koo 1986). To compare properly theory with this observational constraint, one has to examine carefully the epoch of formation of the luminous cores of galaxies, rather than of their halos. In fact, in a hierarchical scheme such as that of cold dark matter, the cores of galaxies can form long before the associated halos (Silk and Wyse 1986).

To evaluate this quantitatively, one may evaluate the probability that within a specified filter scale R_g corresponding to a luminous galaxy, a core of filtering scale R_c or equivalently, mass M_c, has undergone nonlinear collapse by some redshift z_c. The result (Silk and Szalay 1987) is that, provided $M_c << M_g$,

$$\frac{1 + z_c}{1 + z_g} \approx \left(\frac{M_g}{M_c}\right)^{1/3}. \tag{8}$$

On scale $R_g \approx 1\ h^{-1}$ Mpc, the redshift of collapse is $z_g \approx 2$. From equation (8), one infers that core collapse of $\sim 10^9 M_\odot$ of cold dark matter would have occurred very early, by $z \sim 20$. Thus it should be possible, in principle, by baryonic dissipation and continued contraction to very high density to form the $\sim 10^8 M_\odot$ black holes that are believed to be the central engines in quasistellar objects and active galaxies long before these systems become visible at $z \lesssim 4$. Infalling gas clouds and stars trapped by encounters with a dense gaseous disk fuel the accretion disk, which is believed to extend for $\sim 10^{13} - 10^{14}$ cm around the hypothesized supermassive black hole. This fueling is likely to be associated with the continuing collapse and growth of the core.

One cannot distinguish within the limits of the statistics of gaussian fluctuations between the alternative possibilities that the core consists of multiple clumps, or that the substructure is destroyed as the core grows. However this does not really matter for the prediction of protogalactic luminosity. One may simply assume that a substantial fraction of the baryonic component of the non-linear mass at any given epoch is used in star formation. Galaxies in effect form from inside out. The protogalactic luminosity then grows according to equation (8)

$$\frac{dM_c}{dt} \propto t. \tag{9}$$

The initial core, containing $10^7 - 10^8 M_\odot$ of baryons, will efficiently recycle matter and eventually form stars of abundance approaching the nucleosynthetic yield, about one–third solar abundance for a solar neighborhood – type initial mass function (IMF). If the IMF is enhanced in massive stars relative to the local IMF, as seems necessary to account for the observed metallicity towards the centre of our galaxy, the yield will increase, but on the other hand, the increased energy input into the gas will tend to redistribute widely the enriched stellar ejecta. As the core grows, some baryonic infall into the central regions will occur because of gas dissipation after each generation of massive star formation, and the mean abundance of the surviving low mass stars will decrease with galactic radius. It does not seem possible, given the uncertainties in yield and in gaseous dissipation modelling, to provide any precise estimate of the resulting metallicity gradient in the old stellar population. The core builds up an enhanced metallicity, approaching and possibly surpassing the nucleosynthetic yield due to infall of debris from ejecta of massive stars.

This mode of inside–out formation neatly circumvents one of the principal objections to forming stars prior to the epoch z_g of collapse of a galactic mass. A dense, metal–rich stellar core forms without requiring that the protogalaxy be predominantly gaseous at this epoch. Since galaxy formation commences very early, at $z \sim 10$, most of the protogalactic luminosity is radiated prior to z_g. The peak luminosity at z_g (which is of order unity) is reduced relative to that in a model in which the protogalaxy was wholly gaseous at z_g by a factor of roughly $\exp(t_g/t_*)$, where t_* is the duration of the vigorous star formation period. Since $t_* \lesssim 10^9 yr$ in a single burst model of protoellipticals and $t_g \sim 10^{10} yr$, the effective reduction in visibility of multiple burst protogalaxies may be very considerable indeed. It may be necessary to partially suppress long–lived, low mass star formation in these bursts in order for ellipticals to be sufficiently red by the present epoch.

Moreover, the synthesis of heavy elements at large redshift in massive stars strongly suggests that dust grains will be produced early in the course of proto-galactic evolution. This means that the emission of a protogalaxy is likely to be strongly modulated by dust. Certainly, Lyman alpha emission will be weak, as indeed is the case for the apparently young metal–poor dwarf galaxies with HII region–like spectra (Hartmann *et al.* 1984). Moreover, there may well be sufficient dust produced to absorb and reradiate the bulk of the luminosity in the far in-frared spectral region near 100μ. IRAS extreme starburst galaxies, whose infrared luminosity may exceed the optical luminosity by a factor of up to ~ 100, may conceivably be prototypes of protogalactic starbursts; indeed this will be argued below for other reasons.

III. LYMAN ALPHA CLOUDS

Luminous galaxies are formed from rare (2.5σ) fluctuations according to the normalization of the biased cold dark matter fluctuation spectrum. The fate of many of the recently collapsing run-of-the-mill lower mass fluctuations will be, I shall argue below, to form dwarf galaxies. The least massive fluctuations that collapse early, at $z \gtrsim 4$, have a different fate, however. It is at about this epoch that the intergalactic medium must have become highly ionized. Quasars, now seen out to a redshift of 4.01, show no absorption trough blueward of the Lyman alpha emission, indicating that the intergalactic medium contains very little atomic hydrogen. Specifically, the most recent study (Steidel and Sargent 1986) of the Gunn–Peterson limit sets an upper limit of 0.05 on the relative continuum depression due to Lyman alpha absorption, and indicates a neutral density $n_{HI} \lesssim 3 \times 10^{-12} \mathrm{cm}^{-3}$ at $z = 3$. Collisional ionization requires a temperature of $\gtrsim 10^6$ °K, and a correspondingly larger energy input of ~ 1 keV per baryon is needed to counter radiative energy losses. Photoionization is far less extravagant energetically; since the recombination time is very long, only ~ 20 eV per baryon is needed at $z \sim 4$. However, the photoionization source is unknown. The recent realization that quasars undergo luminosity (rather than density) evolution limits the ionization efficiency of the ionizing photon flux from quasars, to at most 10 percent of the flux required for IGM ionization if $\Omega_b \approx 0.1$ (Shapiro 1986). The most likely source is the ionizing radiation produced by massive stars during the early phases of protogalactic evolution. If the heavy elements seen in extreme population II stars were processed by such stars at $z > 4$, then the resulting photon flux is ~ 100 eV per baryon. The typical photon energy would be ~ 10 eV, and there would clearly be a sufficient number of energetic photons to photoionize the intergalactic medium to the required degree.

This photoionization flux takes time to build up, requiring protogalactic activity, and also could not be effective at heating and maintaining the high ionization fraction while Compton cooling is significant at $z \gtrsim 10$. The luminosity evolution of the quasars most probably monitors and reflects protogalactic activity. Indeed the only possible example of a primeval galaxy was found by searching in the vicinity of, and at the same redshift as, one of the most distant quasars (Djorgovski et al. 1986). Thus a reasonable guess would be that the metagalactic photoionizing flux has developed to its peak value by $z \sim 3$. Later phases of protogalactic activity may well be shrouded by dust.

Consider now the fate of an isolated 1σ fluctuation chosen with a filtering scale of 50 kpc. This goes non–linear at $z \approx 5$. It will begin collapsing, but becomes photoionized as the metagalactic ionizing flux builds up. The probability of there being a nearby luminous protogalaxy increases rapidly with redshift since these objects are identified with rare fluctuations. Prior to photoionization, the gas is very cold. Trace H_2 cooling will maintain the gas temperature at about 1000

K, and the comoving Jeans length is about 10 kpc. Once the gas is photoionized, however, it is heated to about 30,000 K. The Jeans length increases to about 70 kpc, and the gas becomes unbound. These estimates are somewhat modified by allowance for the gravitational effect of the dark matter, the presence of which allows a stable bound state for an isothermal distribution of gas. A stabilized distribution of neutral gas may still become unbound when photoionized.

For filter scales between roughly 10 kpc and 70 kpc, the photoionized gas is freely expanding. These expanding clouds are found to have parameters very similar to the Lyman alpha forest clouds seen in absorption towards quasars at $z \gtrsim 2$ (Bond, Szalay and Silk 1987). A grid of one–dimensional hydrodynamic models has been run to obtain cloud density profiles, which are found to be approximately gaussian. For $\nu \approx 1 - 2$, the cloud HI column densities span the range $10^{13} - 10^{16}$ cm $^{-2}$ at the onset of photoionization, and the distribution of cloud column densities at a specified redshift is fixed by the range in ν on given filter scale: it is found to be approximately proportional to N_{HI}^{-2}, the high N_{HI} clouds corresponding to higher ν clouds. The cloud frequency is $\sim 20(30\text{kpc}/R)^3$ Mpc^{-3}, and there is rapid evolution of N_{HI} with redshift, this quantity decreasing as $(1 + z)^{6-7}$ due to the free expansion of the clouds and the fact that $N_{HI} \propto \delta^2$, where δ is the cloud overdensity. All of these properties are in good accord with observations of Lyman alpha clouds. These clouds may therefore be the missing link between primordial density fluctuations and galaxies.

IV. DARK HALOS

The cold dark matter scenario predicts that dark halos should be a universal phenomenon. Detection of a single isolated galaxy with a symmetrical, Keplerian fall–off in its rotation curve would be a serious blow against cold dark matter. In fact, all spiral galaxies studied to date have either flat rotation curves, characteristic of a dark halo, or else asymmetric rotation curves, suggesting of tidal interaction with a neighbor or warping (van Albada and Sancisi 1986). Dwarf irregular galaxies often have rising rotation curves. Asymmetries in the gas distribution can mask the effect of a dark halo.

Observations of elliptical galaxies are consistent with their possessing dark halos. Evidence for dark halos around ellipticals comes from the presence of extensive hot gas bound to ellipticals (Forman et al. 1985), and the modelling of the multiple shells observed at very faint isophote levels (Hernquist and Quinn 1986).

The dark halos that are inevitable in the presence of cold dark matter account in a natural way for the angular momentum of galactic disks, and the amplitudes and the shapes of galaxy rotation curves. In the absence of a dark halo, the specific angular momentum is about an order of magnitude too small to account for that observed in disk galaxies. However dissipative infalling gas torques up against the relatively inert cold dark matter as an isolated protogalaxy forms, and after a reasonable collapse factor of about 5–10 in radius results in formation of a rotationally supported disk (Efstathiou and Jones 1980; Fall and Efstathiou 1980).

The specific angular momentum measures the product of characteristic size and maximum rotation velocity. The size of the luminous galaxy is determined by baryonic dissipation, to be discussed in the following section. The maximum rotation velocity v_m is determined solely by the depth of the dark matter potential well that constitutes the halo. Biased cold dark matter results in the observed range (100–300 km s^{-1}) for v_m in disk galaxies (Frenk et $al.$ 1985). The biasing is crucial for this to work out correctly, for it is the weakened correlations of the dark matter (relative to the luminous matter) on Mpc scales that are critical. The dark matter correlation scale, appropriate to 1σ fluctuations is about $3h^{-1}$ Mpc, equivalent to a velocity of 300 km s^{-1}. It is precisely this velocity which is a measure of the potential energy of the dark halo at turn–around, and manifests itself as the velocity dispersion of the dark halo after collapse and virialization have occurred.

The most recent success in the continuing saga of cold dark matter concerns the shape of galactic rotation curves. The rotation curves are flat because secondary infall of cold dark matter accumulates in a quasi-spherical halo with a density profile $\rho \simeq r^{-2}$. However the rotation curve predicted for a system containing only cold dark matter would decline gradually to zero at the center of the galaxy. This is quite different from what is observed in disk galaxies where the rotation curve remains flat well within the presumably baryon–dominated half-light radius (typically several kpc), and then declines steeply only within a kpc or so of the centre. In our own galaxy, for example, the dark halo only dominates the baryonic disk and bulge components outside the solar circle, yet the rotation curve is flat into a galactocentric distance of about 2 kpc.

This apparently paradoxical situation for a non–baryonic halo (Bahcall and Casertano 1985) has now received a simple explanation in the cold dark matter model (Blumenthal et $al.$ 1986). Consider a protogalaxy containing cold dark matter together with a baryonic component amounting to about ten percent of the mass of the system, as expected if $\Omega_b \approx 0.1$. The gas dissipates and contracts to form a dense bulge and disk. In doing so, it pulls in some of the dissipationless cold dark matter, which becomes trapped by the locally strong gravity field of the disk. As much as ten percent of the dark matter can be dragged in, and about fifty percent of the spherically averaged density may consist of cold dark matter at one baryonic scale length from the galactic centre. The resulting rotation curve accordingly remains flat well within a baryonic scale length.

V. DWARF GALAXIES

Even dwarf galaxies contain substantial amounts of dark matter. Mass measurements have been performed using HI observations for gas–rich dwarfs and from radial velocities of the brightest stars for nearby dwarf spheroidals. It is found that the ratio of mass to blue luminosity appears to rise systematically with decreasing luminosity, with (Aaronson 1987)

$$M/L_B \underset{\sim}{\propto} L_B^{-1/3}. \tag{10}$$

This is consistent with the observed surface brightness of dwarfs, for which $L_B \underset{\sim}{\propto} R^4$, provided that the halos all formed at nearly the same epoch, whence $M \underset{\sim}{\propto} R^3$. One might expect this to arise for low mass galaxies with the cold dark matter fluctuation spectrum.

A simple model which accounts for the decreasing surface brightness and increasing dominance of dark matter towards lower luminosities appeals to baryonic mass loss (Dekel and Silk 1986). Gas outflow is driven by supernovae when the dwarf galaxy is predominantly gaseous, and systematically strips the shallowest potential wells. The stripping occurs because of a galactic wind that arises when supernova remnants overlap before having been decelerated by ambient gas below the escape velocity from the dwarf galaxy potential well. As a consequence of the wind, which arises if the velocity dispersion (one–dimensional) of the dwarf galaxy is below ~ 60 km s^{-1}, the remaining baryons that have already formed stars remain bound but adiabatically expand. The cold dark matter halo helps trap these stars, systematically dominating the shallowest potential wells or least massive systems, which lose proportionally more gas and expand the most. Apart from providing a plausible explanation for the trends in surface brightness and in mass–to–light ratio, this model also accounts for the low metallicities observed for dwarf galaxies. Mass loss means that the process of recycling and progressive enrichment of the gas ejected by massive stars is interrupted, and a systematic decrease of metallicity with decreasing dwarf galaxy velocity dispersion is predicted.

Dwarf protogalaxy stripping also lends to a natural biasing mechanism. A potentially luminous galaxy must satisfy two conditions. One is that its mean dispersion σ must exceed ~ 60 km s^{-1}, in order to avoid protogalactic stripping. The second condition is that the mean surface density Σ (or equivalently, density) of the old stellar component must exceed a critical value $\Sigma_{cr}(\sigma)$ at any specified value of σ. This arises because a necessary condition for efficient star formation in a gaseous protogalaxy is that the characteristic energy dissipation time–scale, determined by atomic cooling processes, be shorter than the characteristic dynamical time–scale, typically a crossing time (or a free–fall time) (Silk 1977a; Rees and Ostriker 1977). Whether one considers a uniformly contracting gaseous sphere or, more realistically, an ensemble of clouds orbiting and colliding in the protogalactic potential well, one obtains a simple functional form for $\Sigma_{cr}(\sigma)$ that directly reflects the adopted cooling curve (Silk 1983, 1985). The simplest way to derive this result is to note that in order for a shock to be radiative, necessary for strong density enhancement, one has to satisfy the condition

$$v_s t_{cool} \lesssim R_{cloud}, \tag{11}$$

where $v_s(\propto \sigma)$ is the shock velocity and the postshock cooling time scale $t_{cool} \approx 3kT_s(\Lambda(T_s)n)^{-1}$ with the postshock temperature equal to T_s (proportional to v_s^2) and post–shock density equal to n. Equation (11) then reduces to a condition of the form $\Sigma_{cloud} \gtrsim \Sigma_{cr}(\sigma) \equiv v_s 3kT_s/\Lambda(T_s)$. Now a necessary condition for effective dissipation triggered by cloud–cloud collisions and ensuing shocks to have occurred during the epoch of formation of the old stars is that the surface density of the old stellar component Σ_g averaged over the galaxy exceeds $\Sigma_{cloud}(\sigma)$. This guarantees that a typical cloud collides with another cloud within one mean dynamical time-scale for the protogalaxy. Finally, we have $\Sigma_g > \Sigma_{cr}(\sigma)$ as a necessary condition for effective protogalactic energy dissipation.

It is straightforward to compare $\Sigma_{cr}(\sigma)$ for a gas of primordial composition with estimates of Σ_g as a function of σ for various galaxy types as well as for galaxy groups and clusters. Data on rotation curves, on the Faber–Jackson relation $(L \propto \sigma^4)$, and on the surface brightness–effective radius relation can be used for the galaxies, and luminosity is converted to mass using values appropriate for the observed stellar population. One immediately notes that galaxies lie in the dissipative regime $(\Sigma_g > \Sigma_{cr})$, whereas galaxy groups and clusters do not (Faber 1982; Gunn 1982; Silk 1983). This lends some credence to the underlying assumption about the importance of efficient dissipation in protogalactic star formation.

Now Σ_g represents a fossil memory of the galactic potential well at the termination of the protogalactic phase some 10^{10} yr ago. Apart from rare mergers, little dynamical evolution has subsequently occurred. One would ideally like to compare Σ_g with the predicted evolution tracks of density fluctuations. This is not presently feasible, but one at least can simply compute the locus of non–linear cold dark matter fluctuations, defined by setting $\mid \delta\rho/\rho \mid_k = 1$ and multiplying the resulting density by a factor appropriate to the collapse and virialization of a spherical shell of cold dark matter. The derived density (or surface density) represents the initial conditions for either 1σ or 3σ fluctuations that have dissipationlessly collapsed to form a halo of cold dark matter. Within this structure, the baryons can dissipate and contract further to form the dense stellar systems that are the galaxies. Cold dark matter fluctuations are inferred to provide a natural, even if not a unique, framework for the initial conditions that have given rise to galaxies and galaxy clusters (Blumenthal *et al.* 1984).

Natural biasing is now seen to arise as a consequence of the joint condition for luminous galaxy formation, namely that proto–galaxies not be stripped and undergo efficient star formation. The 1σ fluctuations of galactic mass fail this condition, but a window in the density, velocity dispersion plane opens up for 3σ fluctuations that allows luminous galaxies to form. This could explain why the luminous galaxies are more correlated than the underlying matter, the distribution of which is represented by the 1σ fluctuations. Dwarf galaxies should therefore be the true representative of the underlying dark matter correlations, and measurement of their correlation length and peculiar velocities would yield an unbiased measure of Ω.

Something is missing from this model, however, for it is well known that different morphological types are systematically correlated with mean density and velocity dispersion. To account for this, on has to examine a further aspect of the galaxy distribution, namely the environment.

VI. ENVIRONMENTAL IMPACT

The obvious explanation for morphological differences appeals to differences in dissipation rate. To achieve the higher density and binding energy of early type galaxies requires more dissipation. At the same time, there is a seemingly contradictory requirement, namely that the early type systems apparently underwent more efficient star formation than late–type systems. Stars must form early and rapidly to form a spheroid whose structure is largely due to anisotropic but random stellar motions, whereas formation of a thin disk, which is rotationally supported, requires strong gaseous dissipation. One way out of this dilemma is provided by dissipative collapse models (Larson 1975; Carlberg 1984 a,b), which successfully reproduce the luminosity profiles of ellipticals, and allow the possibility via continued infall and dissipation, of accounting for the high core densities. In the cold dark matter scenario of hierarchical formation, similar conditions arise between mergers of gaseous protogalaxies of comparable mass. These mergers largely destroy any preexisting disks, account for the low rotation of ellipticals (Negroponte and White 1983), and preferentially occur in denser regions where the early type galaxies consequently predominate. However mergers are only an effective solution if the galaxies involved are initially gas–rich protogalaxies. Mergers of mature disk galaxies cannot produce the observed frequency of early type galaxies with large bulge–to–disk ratios (Ostriker 1980). Moreover, the star formation rate must be enhanced by the mergers and be efficient, otherwise residual gas will dissipate and contract into a disk that eventually forms stars.

I shall argue that it is the environment that controls the efficiency of star formation. The key to efficiency is the initial mass function. Low mass stars ($\lesssim 3M_\odot$) mostly lock up mass in long–lived remnants whereas massive stars lose mass and recycle enriched matter. High efficiency is achieved if one can have

many cycles of star formation for a given initial mass of gas, and this requires an
IMF enriched in massive stars. The star formation rate can then be high, without
prematurely exhausting the gas supply. Precisely this situation is believed to
occur in starburst galaxies, where the star formation rate, measured by the far
infrared luminosity, is enhanced by up to a factor of 100 relative to the time–
averaged formation rate of the old star component, measured by the blue or visual
luminosity (Rieke *et al.* 1985). The IMF appears to be bimodal even in our own
galaxy, where massive star formation seems to be enhanced relative to low mass
star formation in the spiral arms (Güsten and Mezger 1983). A bimodal IMF leads
to reasonable estimates of the gas depletion time–scale in spiral disks (Sandage
1986), and to a simple explanation of the paucity of metal–poor dwarfs in the solar
neighborhood (Larson 1986).

A further clue comes from the observation that starbursts are apparently
triggered by tidal interactions between neighboring galaxies (Joseph and Wright
1985). On a milder level of enhancement of star formation rate, spiral arms are
of course associated with density waves that enhance the rate of cloud–cloud collisions. It is likely that tidal stirring and enhancement of cloud peculiar velocities
leads to cloud growth and that massive molecular clouds are especially favorable
sites for massive star formation. One reason may be that such clouds become
gravitationally unstable, and crude arguments about fragmentation favor a larger
critical mass for the minimum or characteristic fragmentation scale in the warm,
relatively massive molecular clouds (Silk 1977b; Larson 1985). Another possibility
is that non–linear interactions between HII regions and supernova remnants either
induce further massive star formation (Klein *et al.* 1986) or suppress low mass
star formation: such activity is only likely to occur in the more massive clouds
where the mass reservoir favors the simultaneous formation of several OB stars.

One can try to relate all of this to protogalaxies by the following speculative
reasoning. Star formation rates in protogalaxies can be inferred from population
synthesis modelling of the observed spectral energy distribution in elliptical and
spiral galaxies (Bruzual 1983). This constrains the low mass star formation rate,
that is to say, the stars which contribute to the observed light over time–scales
of $10^8 - 10^{10}$ yr. One finds that in a protoelliptical, the star formation rate was
about 300 $M_\odot \mathrm{yr}^{-1}$ over $\sim 10^9$ yr for an L_* galaxy ($L_* \approx 10^{10} h^{-2} L_\odot$), declining
abruptly thereafter, compared to a rate that is a few percent of this, and relatively
constant or even slowly increasing with time, for a spiral or irregular galaxy where
there is continuing star formation over $\sim 10^{10}$ yr (Gallagher, Hunter and Tutukov
1984). It is tempting to identify protoellipticals with the starburst phenomenon:
since with bimodal star formation, the integrated birth rates of massive and low
mass star formation are comparable, one finds a comparable enhancement of the
specific star formation rate in protoellipticals to that seen in starbursts (Silk 1985).
If this analogy holds, then protoelliptical star formation would also be bimodal,

and correspondingly efficient at recycling gas and rapidly forming the enriched stellar population of the spheroid before dissipation could lead to disk formation. To form a disk apparently requires a continuing gas reservoir over $\sim 10^{10}$ yr, as can occur for isolated galaxies. In galaxy protoclusters, however, the gas supply is likely to be strongly regulated by tidal interactions and collisions between galaxies, as well as by stripping due to hot, intergalactic gas.

However it is not so much the gas supply as the star formation rate that is the primary driver of galactic morphology. Rapid, efficient star formation exhausts the gas supply before a disk can form. Protogalactic mergers may also play a role, but again, the result of a merger will be to temporarily boost the star formation rate. More prolonged stimulation of star formation may arise from tidal heating between protogalaxies in a protocluster, and this could help account for the striking correlation between morphology and local density of galaxies. This effect can be significant only for protogalaxies, which were much larger in cross–section than galaxies are today.

For example, the rate of tidal heating between a galaxy and its neighbors in a cluster can be written

$$\dot{E}/E = n_g \sigma_E v_g \qquad (12)$$

where n_g and v_g are the galaxy number density and rms velocity dispersion in the cluster and $\sigma \approx 30 r_e^2 (v_*/v_g)^2$ is the cross–section for energy exchange in a tidal encounter between identical protogalaxies with de Vaucouleurs profiles, effective radii r_e, and velocity dispersions v_* (Aguilar and White 1985). This expression is appropriate if the protogalaxy consists of an ensemble of many small gas clouds. The tidal heating enhances their mutual collision rate and thereby presumably stimulates star formation, as apparently happens in environments as diverse as spiral arms and in starbursts. Active stirring is guaranteed if the characteristic heating time–scale $E/\dot{E}$ is comparable to the protogalaxy dynamical time–scale. Adopting values characteristic of a protocluster $(v_{100} \equiv v_o/100\text{km s}^{-1}, n_{100} \equiv n_g/100\text{Mpc}^{-3})$, one finds that the protogalaxy (mass $M_{11} \equiv M/10^{11} M_\odot$) binding energy scales with galaxy density as $v_* = 76 M_{11}^{3/5} (n_{100}/v_{100})^{1/5}$ km s^{-1}. Some further dissipation is required to enhance v_* to the value appropriate for a luminous galaxy. This suggests that if tidal heating indeed governs star formation efficiency in a protogalaxy of specified mass, the parameters of the resulting stellar distribution will be such that velocity dispersion increases with increasing density of galaxies, producing higher surface brightness spheroids in denser regions.

Consideration of the observed range of $n_g = 10^{-2} - 10^4$ Mpc^{-3} suggests that relatively cold, and consequently very extended, protogalaxies form slowly in the field, allowing disks to develop and the star formation rate to be low, whereas in dense regions, protogalaxies would be compact, form stars rapidly, and not have sufficient time for slow dissipative formation of a stellar disk. Dynamical relaxation should ensure that there is a large dispersion in galaxy parameters about the mean defined by this predicted correlation. By comparison, the observed dependence of morphological type on local density (Dressler 1980) is reasonably consistent with theoretical expectation in the very simple model that I have presented. SO galaxies form an intermediate class, and it may be that mergers in protocluster cores are responsible for converting proto–SO's to protoellipticals. Tidal heating therefore could lead to a natural biasing scheme when galaxies are studied in a volume–limited sample. This mechanism accounts for the dependence of galaxy morphology on luminous galaxy density, because spheroid formation is enhanced in regions where tidal heating stimulated more efficient and vigorous star formation.

VII. CONCLUSIONS

Cold dark matter provides a remarkably detailed prescription for galaxy formation and for the large–scale structure of the universe. Whether it provides a successful explanation for all of the data is not yet clear. Certainly, there are numerous successes, including on large scales, the uniformity of the microwave background, the correlation function of galaxies, the formation of galaxy groups and clusters, the peculiar velocities of galaxies, and the concentration of luminous matter relative to dark matter. On smaller scales, highlights include explanation of the range of galaxy masses, the density profiles of dark halos and of luminous cores, the shapes of galaxy rotation curves, and some of the characteristics of dwarf galaxies. We do not yet have a good understanding of the origin of the galaxy luminosity function or of the tight correlations found between characteristic parameters of luminous galaxies such as surface brightness, effective radius, and central velocity dispersion. However, refinements in modelling of galaxy formation are likely to soon improve this situation.

Of greater concern is the lack of any explanation in a cold dark matter–dominated universe of the cluster–cluster correlations or of the large–scale drift velocities of galaxy clusters. While definitive confirmation of these, still uncertain, observational results would be fatal for the standard model of biased cold dark matter with scale invariant gaussian density fluctuations, there seem to be several possible variations on this theme that may then arise, phoenix–like, in a resurrected theory of large–scale structure. As far as galaxy formation is concerned, however, the current theory appears to be on a more secure footing, in part because many of the observational characteristics of galaxies have been thoroughly studied, and the theoretical goals are well defined. One has the advantage of being able to treat galaxies as fossils, and develop cosmogonical theories to explain the vast array of rich phenomena that we observe in nearby systems. The next major step will be to test these theories by directly searching for young galaxies and for protogalaxies in remote regions of the universe. For the present, however, one may speculate that galaxy formation theory will undergo few radical changes despite the uncertainty of our understanding of large scale structure.

This research has been supported in part by grants from NASA and NSF. I am indebted to various colleagues for stimulating and enlightening discussions, including R. Wyse, A. Szalay and M. Davis.

REFERENCES

Aaronson, M. 1987, in *Nearly Normal Galaxies: From the Planck Time to the Present*, ed. S.M. Faber (New York: Springer Verlag) (in press).

Aguilar, L. and White, S.D.M. 1985, *Astrophys. J.*, **295**, 374.

Bahcall, J.M. and Casertano, S. 1985, *Astrophys. J. Lett*, **293**, L7.

Bahcall, N. and Soneira, R. 1983, *Astrophys. J.*, **270**, 20.

Barnes, J., Dekel, A., Efstathiou, G. and Frenk, C. 1985, *Astrophys. J.*, **295**, 368.

Bardeen, J., Bond, J.R., Kaiser, N. and Szalay, A. 1986, *Astrophys. J.*, **304**, 15.

Blumenthal, G., Faber, S., Primack, J., and Rees, M., 1984, *Nature*, **301**, 584.

Blumenthal, G., Faber, S.M. Flores, R.A. and Primack, J.R. 1986, *Astrophys. J.*, **301**, 27.

Bond, J.R., Szalay, A. and Silk, J. 1987, in preparation.

Bruzual, G. 1983, *Astrophys. J.*, **273**, 105.

Carlberg, R. 1984a, *Astrophys. J.*, **286**, 403.

Carlberg, R. 1984b, *Astrophys. J.*, **286**, 416.

Davis, M., Efstathiou, G., Frenk, C. and White, S.D.M. 1985, *Astrophys. J.*, **292**, 371.

Dekel, A. and Silk, J. 1986, *Astrophys. J.*, **303**, 39.

Djorgovski, S., Spinrad, H., McCarthy, P., and Strauss, M. 1986, *Astrophys. J. Lett.* (in press).

Dressler, A., 1980, *Astrophys. J.*, **236**, 351.

Dressler, A., Faber, S.M., Burstein, D., Davies, R.L., Lynden–Bell, D., Terlevich, R.J. and Wegner, G. 1987, *Astrophys. J.* (in press).

Efstathiou, G. and Jones, B.J.T. 1980, *Comm. Astrophys.*, **8**, 169.

Faber, S.M. 1982, in *Astrophysical Cosmology*, ed. H.A. Brück, G.V. Coyne and M.S. Longair (Vatican: Pontificia Academia Scientarum), p. 219.

Fall, S.D.M. and Efstathiou, G. 1980, *Mon. Not. Roy. Astr. Soc.*, **193**, 1895.

Forman, W., Jones, C. and Tucker, W. 1985, *Astrophys. J.*, **243**, 102.

Frenk, C., White, S.D.M., Efstathiou, G. and Davis, M. 1985, *Nature*, **317**, 595.

Frenk, C., White, S.D.M. and Davis, M. 1983, *Astrophys. J.*, **221**, 417.

Fritschi, M. *et al.* 1985, *Physic Letters B*, **173**, 485.

Gallagher, J.S., Hunter, D.A. and Tutukov, A.V. 1984, *Astrophys. J.*, **284**, 544.

Gunn, J.E. 1982, in *Astrophysical Cosmology*, ed. H.A. Brück, G.V. Coyne and M.S. Longair (Vatican: Pontificia Academia Scientarum), p. 233.

Güsten, R. and Mezger, P.G. 1983, *Vistas in Astronomy*, **26**, 159.

Hartmann, L.W., Huchra, J.P. and Geller, M.J. 1984, *Astrophys. J.*, **187**, 487.

Hernquist L. and Quinn, P. 1986, *Astrophys. J.* (in press).

Joseph, R.D. and Wright, G.S. 1985, *Mon. Not. Roy. Astr. Soc.*, **214**, 87.

Kaiser, N. 1984, *Astrophys. J. Letters*, **284**, L9.

Kaiser, N. and Silk J. 1986, *Nature* (in press).

Klein, R., Sandford, M.T. and Whitaker, R.W. 1986, *Astrophys. J.* (in press).

Koo, D. 1986, in *Spectral Evolution of Galaxies*, ed. C. Chiosi and R. Renzini (D. Reidel: Dordrecht), 419.

Larson, R.B. 1975, *Mon. Not. Roy. Astr. Soc.*, **173**, 671.

Larson, R.B. 1985, *Mon. Not. Roy. Astr. Soc.*, **214**, 379.

Larson, R.B. 1986, *Mon. Not. Roy. Astr. Soc.*, **218**, 409.

Ling, E., Frenk, C. and Barrow, J.D. 1986, preprint (Univ. of Sussex).

Mészáros, P. 1974, *Astron. Astrophys.*, **37**, 225.

Negroponte, J. and White, S.D.M. 1983, *Mon. Not. Roy. Astr. Soc.*, **705**, 1009.

Ostriker, J. 1980, *Comm. Astrophys.*, **8**, 177.

Politzer, H.D. and Wise, M. 1984, *Astrophys. J. Letters*, **285**, L1.

Rees, M.J. and Ostriker, J.P., 1977, *Mon. Not. Roy. Astr. Soc.*, **179**, 541.

Rieke, G.H., Catri, R. M., Black, J.H., Kailey, W.F., McAlary, C.W., Lebofsky, M.J. and Elston, R., 1985, *Astrophys. J.*, **290**, 116.

Sandage, A., 1986, *Astron. Astrophys.*, **161**, 89.

Shapiro, P. 1986, preprint.

Silk, J., 1977a, *Astrophys. J.*, **211**, 638.

Silk, J., 1977b, *Astrophys. J.*, **214**, 157.

Silk, J., 1977, *Astrophys. J.*, **214**, 784.

Silk, J., 1983, *Nature*, **301**, 574.

Silk, J. 1985, *Astrophys. J*, **297**, 1.

Silk, J. 1987, in *Star-Forming Regions*, ed. J. Jugaku and M. Peimbert (D. Reidel: Dordrecht).

Silk, J. and Wyse, R.F.G. 1986, in *Structure and Evolution of Active Galactic Nuclei*, ed. G. Giuricin, F. Mardirossian, M. Mezzetti, and M. Rumella (D. Reidel: Dordrecht).

Silk, J. and Szalay, A. 1987, in preparation.

Steidel, C.C. and Sargent, W.L.W. 1986, *Astrophys. J.* (in press).

Steinhardt, P. and Turner, M. 1984, *Phys. Rev.*, **D29**, 2162.

Turner, M. 1985, in *Fundamental Interactions and Cosmology*, ed. J. Audouze and J. Tran Thanh Van (Editions Frontieres: Gif–sur–Yvette), 267.

Van Albada, T.S. and Sancisi, R. 1986, *Phil. Trans. R. Soc. Lond.*, **A000**, 000.

White, S.D.M., Frenk, C., Davis, M. and Efstathiou, G. 1987, *Astrophys. J.* (in press).

Yang, J. *et al.* 1984, *Astrophys. J.*, **281**, 493.

DISCUSSION

BURNS: Your model would seem to predict that dwarf galaxies in voids are gas-poor. Is this inconsistent with recent reports by Moody and Kirshner of emission-line galaxies, that would suggest a gas-rich environment in the heart of the Bootes void?

SILK: I would expect the incidence of gas-rich dwarfs to be much less in voids than elsewhere. Also the emission line galaxies found in the Bootes void are relatively luminous compared to the dwarfs that are unable to retain their gas.

TURNER: You must identify a fluctuation of a particular number of σ with an L* galaxy. This must give you three independent quantities: the correct galaxy number density, the appropriate fraction of the Universe's total mass associated with galaxies, and the appropriate mass per galaxy (i.e., internal velocities). How well does this three constraints and one parameter game work?

SILK: Amazingly, it all works out! The procedure is the following. Fix the correlation length to normalize the fluctuation spectrum and the galaxy number density to specify the number of σ. Then cosmic coincidence number one is that the cosmic virial theorem yields Ω (luminous) equal to about 0.1.
  Cosmic coincidence number two is that the biased cold dark matter spectrum of fluctuations simultaneously specifies the correct maximum

rotational or internal velocities for the dark halos. Specifically,
the mass correlation length is unity at a Hubble velocity today of
300 km s-1, and this is precisely what is needed to give the binding
energy of dark halos that are forming recently.

WEBB: Could you give any further details on the confining mechanism
for your Lyα cloud model. Also, are there observational predictions
such as a correlation between N_{HI} and b for example? Can any quantitative statements be made concerning the amplitude and scale of any
clustering.

SILK: In the Lyα cloud model I described, the absorbing clouds are
freely expanding and unconfined. They are more weakly clustered than
galaxies.
     The correlation length predicted is about one-half of that expected
for luminous galaxies at the same redshift. For the low column density
clouds ($\lesssim 10^{14}$ cm^{-2}) the internal velocity dispersion of the cloud
is determined mostly by the sound speed (~20 km s^{-1}), and should be
nearly independent of HI column density. At higher column densities,
the differential Hubble expansion velocity across the cloud is
progressively more important and v should increase by a small factor.

ROWAN-ROBINSON: You are predicting that dwarf ellipticals, formed
from 1-σ fluctuations, are more uniformly distributed than the 2.5-σ
giant ellipticals. What are you predicting about spirals? If they
form from the rare 2.5-σ fluctuations, would they not show the same
degree of clustering as ellipticals?

SILK: Spirals and ellipticals alike form from 2.5 σ fluctuations.
The cluster environment helps convert spirals to elliptical by
tidally induced star formation or mergers. Hence the spirals avoid
dense regions and are more weakly clustered than ellipticals.

ELLIS: Searches for spectacularly luminous primeval galaxies
have so far failed and already suggest the formation redshift of such
objects would be $\gtrsim$ 5. What is the advantage of invoking a more
gradual onset of luminosity in young galaxies?

SILK: If most of the heavy elements seen in the old star components
are formed early in the protogalaxy's history when it was still
largely gaseous and just an agglomeration of interacting clouds, then
its peak luminosity will be substantially reduced. This means that
protogalaxies could be present at low redshift and not be easily
detectable for two reasons: first, the predicted fluxes are perhaps
an order of magnitude less than expected in standard models, and

second, the prior phase of heavy element synthesis makes it very
likely that they are dusty objects. Perhaps UV or optical wavelengths
are not the optimum spectral regions to search for primeval galaxies.

CHEN: (1) What is the life time of the free expansion model of Lyα
clouds?   (2) What is the behavior of the Lyα clouds before the birth
of the QSO which provides the energy for the re-ionization?

SILK: The radius doubling time for a Lyα cloud in our model is about
$\sim 2 \times 10^4$ yr. Prior to the ionization flash, the potential Lyα clouds
are density fluctuations that are just entering the non-linear regime.
Smaller scales ($\lesssim 10^7$ $M_\odot$) may already have collapsed.

GIRAUD: What do you do with those S0 galaxies which are not in
dense environments?

SILK: I presume that protogalaxies are still predominantly gaseous
at the onset of galaxy clustering. Hence much of the early star
formation can be enhanced or induced by tidal interactions and by
mergers. This means that the spheroidal components of galaxies will
be largest where star formation is most efficient and most rapid. If
this is correct, one can then understand why ellipticals form in the
densest regions. The star formation enhancement criterion, however,
depends in a systematic way on local density and galaxy peculiar
velocity, so that bulge sizes should systematically decrease as one
goes from rich clusters to poor clusters and to the field. S0 galaxies
are therefore an intermediate population that dominate the intermediate
range of environment between rich clusters and the field. This
distinction is not so rigid however that one would not expect to find
some S0's in the extreme environments as well, simply because the
local star forming history is presumed to be dominated by stochastic
events.

HARTWICK: What are the observational parameters of a primeval galaxy
in the cold dark matter scenario?

SILK: The final collapse of the halo of a typical primeval galaxy is
very recent (redshift $z \sim 1$), but pieces of the primeval galaxy form
much earlier. For example, the core collapses as early as $z \sim 10$.
This means that the peak luminosity is considerably less (by perhaps
a factor of 10) than if the onset of formation was instantaneous.
Moreover, the early star formation should produce a considerable amount
of dust. A primeval galaxy should be dusty, have a luminosity perhaps
100 times larger than a mature galaxy, and be visible at $z \gtrsim 1-2$.

TOWARDS UNDERSTANDING THE LARGE-SCALE STRUCTURE ?

Avishai Dekel
Racah Institute of Physics
The Hebrew University
Jerusalem 91904, Israel

ABSTRACT. Although some theories, such as that of cold dark matter, are quite successful in explaining certain aspects of the formation of structure, we seem not to approach a satisfactory theory which can easily account for all the observational constraints on all scales. Most difficult to explain are the indicated clustering of clusters and bulk velocities on very large scales, when considered together with the structure on galactic scales and the isotropy of the microwave background. If these observations are correct, the only scenarios that can work are hybrids of certain sorts, which involve somewhat *ad hoc* choices of parameters; they are not the theories that would have emerged naturally from first principles, and they do not satisfy the criteria of simplicity and elegancy. I will discuss the currently popular scenarios and the apparent difficulties they face.

1. INTRODUCTION

The current situation in the two front lines of physics is curiously very different; in particle physics there is one 'theory of everything' – superstrings – but no experimental constraints in the foreseen future, while in cosmology, to the contrary, we have a growing body of interesting observations of the large-scale structure, most of which are discussed in this symposium, but we also have *many* theoretical scenarios to account for them. It is not that there is an observation which we cannot explain; it is that each basic theory seems to explain a different subset of the observations, or it has to be modified or patched whenever a new observational constraint is reported. At each step we drift away from the simple, elegant theory we should wish for. Thus, although the field is very fruitful observationally, and despite the great effort and many ingenious ideas from the theoretical side, we do not seem to get closer to a 'theory of everything' for the formation of large-scale structure in the universe.

I would list the most relevant observational constraints as follows:
(a) Systematic properties of galaxies – relevant to galaxy formation on scales $\leq 1\ h^{-1}Mpc$ (for reviews see Silk 1987; Faber 1987; Efstathiou and Silk 1983).

415

A. Hewitt et al. (eds.), Observational Cosmology, 415–432.
© 1987 by the IAU.

(b) Quantities that measure the clustering of galaxies on scales $1 - 10\ h^{-1}Mpc$; the galaxy-galaxy correlation function, $\xi_{gg}(r)$, with the associated galaxy pair velocity dispersion, $v_{gg}(r)$, and the galaxy number density contrast within the Local Supercluster (LSC), δ_g, with the associated peculiar infall velocity, v_p (for reviews see Davis and Peebles 1983; Dekel, Einasto and Rees 1987).

(c) The structure of superclusters of galaxies on scales $10 - 100\ h^{-1}Mpc$, described in different contexts as 'filamentary structure', 'pancakes', 'bubbles', etc. (see a review by Oort 1983; Tully 1986; de Lapparent, Geller and Huchra 1986; Giovanelli, Haynes and Chincarini 1986).

(d) The presence of 'voids' – regions of very low number density of galaxies with typical dimensions $10-50\ h^{-1}Mpc$ (perhaps limited only by the size of the available samples) (See Oemler 1987 for a review).

(e) The clustering of rich clusters on scales up to $\sim 100\ h^{-1}Mpc$, expressed in terms of superclusters of clusters or by the cluster-cluster correlation function, $\xi_{cc}(r)$, which, for Abell clusters of richness ≥ 1, becomes linear only beyond $25\ h^{-1}Mpc$ and stays positive out to $100\ h^{-1}Mpc$ (e.g. Bahcall and Soneira 1983) and the associated cluster pair velocities of $\sim 1000\ km\ s^{-1}$ (Bahcall 1986).

(f) On a similar scale, the reported 'bulk motion' of $\sim 600\ km\ s^{-1}$ relative to the microwave background of the whole body of galaxies, groups and clusters around us in a sphere of diameter $\sim 100\ h^{-1}Mpc$ (Rubin et al. 1976; Burstein et al. 1986; Collins, Joseph and Robertson 1986).

(g) The upper limits on temperature fluctuations in the microwave background, $\delta T/T$, on the various angular scales (1° corresponds to $100\ \Omega^{-1}h^{-1}Mpc$) (e.g. Uson and Wilkinson 1984; Melchiorri et al. 1981).

(h) The indications for evolution of clustering in the distribution of high redshift objects such as quasars (Shaver 1987), Lyman-α clouds (Sargent 1987), and galaxies in very deep surveys (Koo 1987).

When constructing a theoretical scenario to account for these observational constraints, there is a large multi-dimensional parameter space available. Let me list just the options which are conservatively regarded as plausible:

(a) The values for the cosmological parameters: the density parameter Ω can vary roughly in the range $0.1 - 2$, the cosmological constant Λ may be zero or non-zero ($< 3H_0^2$), and the Hubble constant H_0 is 'either' 50 or $100\ km\ s^{-1}\ Mpc^{-1}$ (which is less important in the present context).

(b) The nature of the dark matter (DM): Is it *baryonic*? Is it made of *'hot'* particles like $10 - 100\ eV$ neutrinos? Is it *'cold'* particles like axions, or $> keV$ photinos? Is it some sort of *unstable* DM which decays on a time scale slightly smaller than the Hubble time? Is it some mixture of the above?

(c) The origin and nature of the initial density fluctuations: Have they originated in an *inflation* phase or were they induced by something like cosmic *strings*? Is the local distribution function, $p(\delta)$, *Gaussian*, as would be the result of the former, or *non-Gaussian*, as predicted by the latter? Is the initial spectrum a power-law, $\langle|\delta_k|^2\rangle \propto k^n$, and if it is, what is the power n? Are the fluctuations *adiabatic* or

isothermal (isentropic fluctuations or iso-curvature fluctuations)?
(d) Are physical processes other than gravity important in the formation of structure? In particular, do nuclear *explosions* dominate the cosmogonic process?
(e) What is the relationship between the spatial distribution of galaxies and the underlying mass distribution? If galaxy formation is *biased* – how is it biased?

If I had to sketch the simplest, most elegant scenario – a theorist's dream – I would probably choose the following:
(a) A flat Einstein-deSitter universe, with $\Omega = 1$ to a great accuracy and $\Lambda = 0$, as would naturally have emerged from inflation, without much fine-tuning. Otherwise, in the Friedmann equation

$$1 - \Omega - \Lambda/(3H^2) = -k/(aH)^2, \tag{1}$$

the initial conditions would have to be fine-tuned such that the present curvature radius a is 'just' of the order of the horizon radius H^{-1}, or the cosmological constant has a very specific non-zero value.
(b) The DM is made of the only kind of massive particles we know – baryons. Obvious astrophysical candidates are 'Jupiters', such as this planet we live on, or some kind of dead stellar remnants.
(c) The structure have grown from inflation-generated fluctuations: their local distribution function is Gaussian, $p(\delta) \propto exp[-\delta^2/(2\sigma^2)]$, based on the central limit theorem, their power-spectrum is scale-invariant (Harrison-Zeldovich) with $n = 1$, and they are adiabatic (curvature) fluctuations, preserving the baryon to photon ratio as a constant of nature which is determined by microphysics.
(d) Gravitational processes are dominant and explosions do not complicate matters.
(e) Galaxies trace mass without a bias – this would allow a straight-forward comparison between theory and observation.

I will try to describe in this talk how the observational constraints force us to drift away from this ideal scenario.

2. THE NATURE OF THE DARK MATTER

Can the universe be purely *baryonic*? The observed abundances of deuterium and He^3 imply, in the standard big-bang nucleosynthesis scenario, that the contribution of baryons to the mean density is $\Omega_b < 0.2$ (e.g. Audouze 1987). Thus, unless we are willing to consider non-standad nucleosynthesis theories (e.g. Rees 1985; Hogan 1987) or some segregation mechanisms for certain elements (e.g. Braun, Dekel and Elitzur 1987, in progress), we have to assume that baryons cannot account for $\Omega = 1$. The other serious problem of a pure baryonic universe is that the required density fluctuations, especially if adiabatic, are inconsistent with the observation that $\delta T/T < 10^{-4}$ on scales of 4.5' (Uson and Wilkinson 1984); the latter indicates small amplitudes for the density fluctuations at decoupling which cannot make the observed nonlinear structure by the present time (see Kaiser and Silk 1986 for a review). The small-angle temperature fluctuations had to be smeared out by reionization of the intergalactic medium no later than $z \sim 30 \ (\Omega_b h/0.1)^{-2/3}$ to provide an optical depth > 1 for Compton scattering. This would require early formation of stellar objects that could have ejected the required energy into the IGM. The simple adiabatic fluctuations cannot give rise to such objects because they damp out on the relevant scales due to photon diffusion and viscosity (Silk 1968); A special kind of ('isothermal'?) fluctuations that could survive damping on small scales is required. So, a model which is dominated by the simplest type of DM requires a special open universe, and non-trivial initial fluctuations.

The above difficulties can be eased if the DM is *non-baryonic* and only weakly interacting. It is then not subject to nucleosynthesis constraints so it can amount to $\Omega = 1$, and because it was never coupled to the radiation like the baryons, the fluctuations could start growing before decoupling ($z \sim 10^3$), when the universe turned matter dominated at $z_{eq} \simeq 2.5 \times 10^4 \Omega h^2$, and thus yield non-linear structure today with only small temperature fluctuations at the last scattering surface. The major types of non-baryonic particles are classified as 'hot' and 'cold', corresponding to whether they became non-relativistic at or much before z_{eq}. The fluctuations were damped out by free-streaming on all scales smaller than a critical coherence length, λ_{fs}, which is the horizon scale at the time when the particles became non-relativistic. Given the initial spectrum of fluctuations, the resultant spectrum at z_{eq}, after which gravitational growth occurs, can be calculated for each type of DM. The shape of this spectrum, and in particular the presence or absence of a critical coherence length, determines whether the subsequent formation of structure would be 'top-down' or 'bottom-up' respectively. The scenarios predicted for each type of DM can then be confronted with the structure as observed today and as evolving from $z \sim 3$. Before we proceed it should be born in mind that while particle theories can easily predict numerous candidates of either type, there is no actual detection of such particles; the scenarios of non-baryonic DM are therefore *a priori* based on a speculative assumption. The experimental effort directed at detecting such particles is therefore such a crucial development.

3. CAN Ω BE 1? DO GALAXIES TRACE MASS?

The mass estimates on scales of $\sim 10 \ h^{-1} Mpc$ are based on on methods such as modeling our infall into the LSC, or applying a cosmic virial analysis to pairs of galaxies. In a linear, spherical model for the LSC, for a given observed infall velocity of the shell containing the Local Group towards Virgo, Ω is related to the mean density enhancement δ interior to this shell by $\Omega \propto \delta^{-1.7}$. For given *rms* pair velocities the cosmic virial theorem gives $\Omega \propto \xi^{-1}$, where $\xi(r)$ is the two-point correlation function. The observed results suggest $\Omega \simeq 0.1 - 0.3$. How can we then have $\Omega = 1$? The low estimates are obtained using the quantities corresponding to *galaxies*: their number overdensity δ_g or the galaxy correlation function $\xi_{gg}(r)$. If, however, the galaxies cluster more than the matter, such that (in the linear approximation)

$$\delta_g = f\delta \quad \text{and} \quad \xi_{gg}(r) = f^2 \xi(r) \tag{2}$$

(the latter is the obtained value of Ω is larger and compatible with $\Omega = 1$ if $f \simeq 2 - 3$. This requires *biased galaxy formation.*

Perhaps the strongest argument for the need of bias is provided by the common existence of *voids* (a review by Oemler 1987; Geller 1987). The number density of galaxies in these voids seems to be typically less than 10% of the mean. (This is only a crude estimate; in the Bootes void, for example, no bright galaxy was found in a volume which contains on the average 32 galaxies.) Based on spherical models (e.g. Hoffman, Salpeter and Wasserman 1982), a similar underdensity in the mass corresponds at decoupling to $|\delta| \geq 10^{-2}$ if $\Omega \simeq 1$, and $\geq 5 \times 10^{-2}$ if $\Omega \simeq 0.1$. This is incompatible with the microwave isotropy on angles $10' - 1°$ in any of the cosmogonic scenarios, unless reionization has washed out the fluctuations. The large-scale N-body simulations demonstrate the difficulty: even in pancake scenarios (e.g. White, Frenk and Davis 1983; Centrella and Melott 1983; Dekel and Aarseth 1984) such large regions are not found with a density below 25% of the mean; they cannot be substantially evacuated dynamically by the present epoch, which is defined by matching the correlation functions of the simulated mass and the observed galaxies. The situation is worse if $\Omega < 1$, where the voids are evacuated even less efficiently. To estimate the real mass density in voids consider a 'toy' universe which consists of superclusters (sc) and voids of uniform densities both in the matter and in the galaxies. The relation

$$(\delta_g/\delta)_{void} = (\delta_g/\delta)_{sc} = f \tag{3}$$

results trivially from the definition of the density contrasts. Assuming that the LSC and the Bootes 'void' are typical, we can adopt the corresponding observed values for the *galaxies*: $\delta_g \simeq 2.5$ and -0.9 respectively. If $\Omega = 1$, the real mass overdensity in the LSC must be $\delta \simeq 0.85$ $(f \sim 3)$, so we get for the *mass* in the voids $\delta \simeq -0.32$. (The fractional volume in the voids is then 73% and the mass fraction is $\sim 50\%$.) These mass densities in superclusters and in voids are both compatible with $|\delta| \simeq 9 \times 10^{-4}$ at decoupling. If most of the mass is non-baryonic, this corresponds to $\delta T/T \simeq 3.5 \times 10^{-5}(\Omega h^2)^{-1}$, which is compatible with the isotropy constraints if Ωh^2 is not much smaller than unity. An open universe with $\Omega \simeq 0.2$ would be in trouble: it would require no bias in the LSC, and therefore a real deep mass underdensity of 10% in the voids. The corresponding $\delta T/T \sim 5 \times 10^{-3}$ would be hard to reconcile with observations.

Further support for bias is provided by the N-body simulations which show that neither the scenario of cold DM (CDM) nor the neutrino scenario can reproduce the observed distribution of galaxies unless it is *biased.* In both cases the matter correlation function steepens in time, and the stage of the simulation to be regarded as the present epoch is determined by matching its logarithmic slope to the observed $\gamma \simeq 1.8$ of galaxies. The CDM correlation length at this time turns out to be only $r_0 \simeq 1 \ (\Omega h^2)^{-1}$ (Davis *et al.* 1985). Hence, for the galaxies to match the observed

$r_0 \simeq 5\ h^{-1} Mpc$, with $\Omega = 1$, the galaxies must be biased by $\xi_{gg}(r) = (5 - 20)\ \xi(r)$ (for $h = 0.5 - 1$), in agreement with the required value of f. In the case of $\sim 30\ eV$ neutrinos, at the time when the slope is 1.8 the structure is still young (e.g. Dekel and Aarseth 1984); collapse to pancakes must have occurred at $z \sim 1$. This poses a timing difficulty for any 'dissipative pancake' scenario which assumes that galaxies are 'daughters' of pancakes, since various lines of evidence suggest that galaxies began to form at $z \geq 3$ (e.g. based on high-redshift quasars and galaxy candidates). The difficulty is also one of scaling: the neutrino correlation length, by the time when its slope is right, has already grown to be $r_0 \simeq 8\ (\Omega h^2)^{-1}$, which is too large in comparison with galaxies unless $\Omega h > 1$. If the galaxies form only in the collapsed regions the constraints become even tighter. But note that the bias required here is therefore of an opposite sense: the galaxies should somehow be *less* clustered than the neutrinos.

4. MECHANISMS OF BIASED GALAXY FORMATION

Biasing is motivated by theoretical considerations as well. Based on the observed correlation between galaxy type and environment (Dressler 1980), it would be astonishing if galaxy formation were not significantly affected by environmental effects which could segregate the galaxies from the underlying mass. After considering the physical processes that might be involved in galaxy formation, one can be easily led to the conclusion that a bias of one sort or another is expected in almost every cosmogony.

Counting the general possibilities, the bias could be determined in each protogalaxy *autonomously*, e.g. by its background density, or it may be a result of *feedback* from other galaxies. This feedback influence may propagate by gas transport to limited distances, or by radiation or fast particles to larger distances. The result might be *destructive*, suppressing galaxy formation *locally* (causing 'underclustering') or *far away* (causing 'over-clustering'), but it could also be *constructive*, enhancing galaxy formation in the neighborhood of other galaxies (e.g. explosions). I will elaborate on this using several of the 'standard' scenarios as examples.

4.1. A Uniform Component

The universe may be dominated by 'ultrahot' weakly interacting particles which do not cluster because they have velocities $> 10^3\ km\ s^{-1}$. But if the 'ultrahot' particles are relics of an early epoch, their mass would have always been dynamically dominant over the baryons, and would have inhibited gravitational clustering altogether (e.g. Hoffman and Bludman 1984). This would also yield an unacceptably fast expansion timescale during nucleosynthesis. A way around this difficulty involves supposing that these particles arise from nonradiative *decay of heavy particles* with lifetimes only slightly shorter than the age of the universe (Turner, Steigman and Krauss 1984; Flores 1987 for a review). Assuming that these decay products are substantially lighter than their unstable parents, they would be very 'hot'. Galaxies (halos) and clusters would have formed during the era of matter domination by the unstable particles, but they then expanded or even became unbound as a result of the decay. An elaboration of this idea suggests that the universe finally becomes dominated by a stable primordial CDM species (Olive, Seckel and Vishniac 1985) which helps explain the survival of structure on both small and large scales; but this requires certain *ad-hoc* fine-tuning among various DM components. Several astrophysical considerations constrain the allowable parameters in this scheme, and they may already eliminate it all together. For example, the decay epoch is bound to be $1 + z_D \leq 0.5$ based on the isotropy of the microwave background (Silk and Vittorio 1986) and the requirement that galaxies and the cores of rich clusters remain bound after the decay (Efstathiou 1985). The observed gravitational lenses, if due to typical galaxies, require $1 + z_D \leq 3$ (Dekel and Piran 1986). A lower limit of $1 + z_D \geq 5$ can be obtained from the dynamics of the Local Supercluster (Efstathiou 1985; Hoffman 1986) but this limit is very model dependent. On the other hand, assuming a quite general scenario for galaxy formation, we find that galactic rotation curves would not have remained flat if the universe were dominated by relativistic decay products (Flores *et al.* 1986).

The universe could be flat ($k = 0$) with $\Omega < 1$ if a non-zero *cosmological constant* contributed to the curvature such that $\Omega + \Lambda/(3H_0^2) = 1$. In some respects this idea resembles the alternative discussed above, but contrariwise, the Λ-term is unimportant at early epochs and so it would not have had such a serious inhibiting effect on galaxy formation. It is found in N-body simulations (Davis *et al.* 1985) that the large-scale structure in a flat CDM scenario with $\Lambda \neq 0$ is quite successful in reproducing the two and three point galaxy correlation functions and their peculiar velocities (with no further biasing in the galaxy formation). It is also compatible with the isotropy of the microwave background (Vittorio and Silk 1986). However, for the Λ contribution today to be comparable to the ordinary matter, the required fine-tuning is as *ad-hoc* as the one we intended to avoid by adopting $\Omega = 1$ (Peebles 1984).

4.2. Biasing In Hierarchical Clustering (e.g. Cold Dark Matter)

An enhanced clustering of galaxies over the background matter can arise in a 'bottom-up' scenario if galaxies formed only from exceptionally *high peaks* of the density distribution smoothed on galactic scales; peaks with an overdensity δ above a threshold $\nu\sigma$, where $\sigma^2 \equiv \langle \delta^2 \rangle$. If the local distribution function of δ has a steeply decreasing tail, like a *Gaussian*, and the power spectrum is not a white-noise, high peaks occur with enhanced probability in the crests rather than the troughs of a large-scale fluctuation mode, so they display enhanced clustering (Kaiser 1984). In a Gaussian process, in the region where $\xi(r) \ll 1$, the enhanced correlation function of high ν peaks is approximated by (Politzer and Wise 1984; Jensen and Szalay 1986)

$$\xi_{peaks}(r) \simeq exp[(\nu^2/\sigma^2)\,\xi(r)] - 1, \qquad (4)$$

which becomes $\xi_{peaks}(r) \simeq (\nu/\sigma)^2 \xi(r)$ where $\xi_{peaks} \ll 1$. The crucial question is what astrophysical mechanism prevents lower-amplitude peaks from also turning into galaxies, thereby neutralizing the effect. One has to come up with a mechanism that would produce a fairly sharp *cutoff* in the efficiency of (bright) galaxy formation at $\nu \sim 2.5$; the number density of such peaks in the case of CDM being comparable to that of bright galaxies.

N-body simulations (White 1987) suggest that the dissipationless dark halos which are the 'parents' of bright galaxies – those with velocity dispersions $> 200 \ km \ s^{-1}$ – are themselves more clustered than the overall mass distribution. This is essentially because the linear growth rate of galactic-scale perturbations is significantly affected by whether they are embedded in a peak or a trough of larger fluctuations. The growth rate is boosted up or suppressed if the background mimics an $\Omega > 1$ or an $\Omega < 1$ model respectively. A weak point of this scheme is that the resultant bias would show up in the distribution of bright galaxies more than in the distribution of galaxies of lower luminosities – an effect which is not supported by observations (e.g. Eder, Schombert, Oemler and Dekel, in preparation).

The dissipative gas contraction to the centers of the dark halos and the subsequent star formation would have an important role in the final bias. As a simple example, the high-ν peaks would collapse earlier, and have higher density at turnaround, than more typical fluctuations on a given mass scale. This could, in principle, in itself account for the bias if star formation were highly sensitive to (for instance) Compton cooling on the microwave background (Rees 1985) – an effect that depends on time like $t^{-8/3}$.

The bias may result from processes intrinsic to the protogalaxies, which depend only on the local background density. For example, Dekel and Silk (1986) have argued that in a bottom-up scenario the 'normal' bright galaxies *must* originate from high density peaks $(2\sigma - 3\sigma)$ in the initial fluctuation field, while typical $(\sim 1\sigma)$ peaks either cannot make a luminous galaxy at all because the gas is too hot and too dilute to cool in time, or, if their virial velocity is less than $\sim 100 \ km \ s^{-1}$, they make diffuse dwarf galaxies by losing a substantial fraction of their mass in supernova-driven winds out of the first burst of star formation. This would lead to a *selective bias*, in which the normal bright galaxies are biased towards the clusters and superclusters, while the dwarf galaxies do trace the mass, and should provide an observational clue for the real distribution of the DM. The evidence for such a segregation between the high and very low surface-brightness galaxies is still inconclusive (see Haynes 1986).

There are several ways whereby the first galaxies $(> \nu)$ could have influenced their environment so as to modify the formation of later galaxies $(< \nu)$, but many of the physical processes that have been considered would not seem to do the job very convincingly (e.g. Bardeen 1985; Peebles 1986). In order to unbind a protogalaxy one has to heat the intergalactic gas to temperatures above $\sim 100 \ eV$, corresponding to the potential well of a typical galaxy. Unfortunately, photoionization by available sources (like quasars) is capable of heating the gas only to a few eV, the binding energy of hydrogen. Furthermore, in order to be relevant, any feedback influence must propagate sufficiently fast over large distances, from proto-clusters to proto-voids, and maintain a continuous suppression of galaxy formation for a long time. It would be hard to expect that any mechanical heat source, such as explosive winds, would be capable of doing the job.

UV radiation is capable of carrying the influence and perhaps affecting the IMF in protogalaxies after the redshift $z \sim 3$ corresponding to the (apparent) peak of activity of quasars (Silk 1985). The first generation of protogalaxies might have fragmented efficiently via H_2 cooling into a 'normal' stellar population, but the radiation then photodissociated the H_2 molecules, making the fragmentation less efficient and thus leading to massive stars. The latter would be highly disruptive via supernova-driven winds, eliminating bright galaxies and leaving behind only diffuse 'failed galaxies'. However, an anti-bias might arise instead, if the fragmentation via H_2 were so efficient that it led to a population dominated by unseen 'Jupiters', while the later inefficient fragmentation ended up with a 'normal' visible population. Also, a similar suppression of H_2 may result from shock heating in the vicinity of luminous objects, so perhaps more likely is a local negative feedback effect, which would produce 'under-clustering'.

Alternatively, 'cosmic-ray' particles from first generation galaxies may raise the Jeans mass by heating the gas (if $< 0.1c$) or raising its pressure (if relativistic), provided that they can diffuse appropriately (Rees 1985). A constant pressure gradient may produce a constant drift of the baryons which, if larger than the escape velocity from the DM potential wells, would be sufficient to prevent further galaxy formation (see more in Dekel and Rees 1987).

To summarize, the 'standard' CDM scenario must be biased, and the origin of the bias can be understood. The biased CDM scenario is very successful in explaining observed properties of galaxies such as the $L - -\sigma$ type relation for 'normal' galaxies (Blumenthal $et\ al.$ 1984), the galactic angular momentum (White $et\ al.$ 1986), and the properties of dwarf galaxies (Dekel and Silk 1986). It can even marginally account for the observed filamentary structure and voids in the galaxy distribution on scales up to a few tens of megaparsecs (Frenk $et\ al.$ 1986; but see a reservation in Dekel 1984a based on alignmemnt of clusters).

4.3. Biasing in Pancake Scenarios (e.g. Neutrinos)

A bias is generated automatically in any 'top-down' scenario where the perturbations below a critical length of a few tens of megaparsecs have all damped out, as in the neutrino scenario or in the case of adiabatic perturbations in a baryonic universe. First, there are motions from 'proto-voids' to 'proto-pancakes' associated with the large coherence length; collapse into flat pancakes accompanied by streaming toward their lines of intersections ('filaments'), and toward the 'knots' where rich clusters form. The gas then contracts dissipatively into the high density regions, within which the conditions become ripe for cooling and galaxy formation; galaxies are thus limited to very specific regions.

However, if the efficiency of galaxy formation were similar in all the collapsed regions, this natural bias would make the timing-scaling difficulties described in §3 more severe; how can 'pancakes' collapse soon enough to form galaxies at $z > 3$ without producing large-scale clustering of excessive amplitudes? Galaxy formation must be *suppressed* in the high density regions (or, less likely, enhanced in the low density regions). for instance, galaxies might have formed preferentially in the sheet-like pancakes and not in the denser filaments and clusters, maybe because cooling was more efficient behind shocks of planar geometry, or because galaxy formation was, for some reason, more efficient at later times when most of the pancake galaxies form. N-body simulations in which the formation of a galaxy at a given position and time is determined taking into account suppressing feedback effects from nearby quasars (Braun, Dekel and Shapiro 1987) demonstrate that the required anti-bias could be easily obtained with a reasonable choice of values for the physical parameters such as the quasar output energy and lifetime and the cooling rate of the heated gas.

A complication arises because while an anti-bias can eliminate the timing problem, it has to be reconciled with the indications for a positive bias summarized in §1. The solution may be a complicated combination of anti-bias on scales of clusters and bias on scales of superclusters and voids. Another difficulty is the big clusters formed by the neutrinos; the gas must be prevented from concentrating in their cores to avoid producing excessive x-ray sources (White *et al.* 1984)

In Summary, the 'standard' neutrino scenario must be 'anti-biased'. Its great appeal is in explaining the distribution of galaxies in 'pancakes', 'filaments' and the 'voids' between them. This scenario is less specific as far as galaxy formation is concerned; although the timing problem might be solved by anti-bias, it may still be hard to explain certain facts such as the existence of galaxies away from pancakes, the failure to detect any alignment between the orientation (and angular momentum) of a galaxy and its parent pancake (Dekel 1985), and the possible presence of dark halos in dwarf galaxies (Tremeine and Gunn 1979).

What may help alternatively is the non-dissipative pancake scenario, such as would arise from a *hybrid* picture (Dekel 1981; 1983; 1984a; Dekel and Aarseth 1984): If galaxies form independently of pancakes, from another component of density fluctuations, the timing constraint becomes irrelevant: galaxies could have formed at $z > 3$ and large-scale pancakes at $z \sim 1$. Galaxies would not be limited to pancakes but rather be present everywhere, subject to the biasing mechanisms that are relevant in general 'bottom-up' scenarios. Such hybrids could involve two types of DM, baryonic and/or non-baryonic, or two types of initial fluctuations, adiabatic and isothermal. The hybrid scenarios can be successful where the single-DM models fail, e.g., in reproducing simultaneously the observed structure on galactic scales and on supergalactic scales, and in smearing the anisotropies in the microwave background (see also §5.1).

4.4. Useful Observations and Conclusions Rgarding Biasing

A few key observations may be helpful in distinguishing between the possible biasing mechanisms. In relation with the voids one would like to answer questions like:

(a) How big are the voids and how empty are they? We need to quantify the data using a meaningful statistic, to confirm (or disprove) our suspicion that no theory can account for the voids without biasing.
(b) Are voids empty of galaxies of all types, or only those types that are most conspicuous? Any evidence that galaxies of different types display unequal degrees of large-scale clustering would be relevant here, and most interesting would be the spatial distribution of very low surface brightness dwarfs.
(c) How much gas is there in the voids? Absorption systems along the lines of sight to quasars passing through voids may be detectable. If the gas is at $10^5 K$, too hot for 21-cm and too cold for x-ray, perhaps some features characteristic of neutral He may reveal its presence.

The relationship between 'parent' halos and 'daughter' galaxies could also have interesting implications on the biasing scheme (Rees 1985):
(d) Are there any 'barren' galactic-mass dark halos with no luminous galaxy within them? Such objects may be associated with either small halos of a shallow potential wells or with halos too big to let the gas cool in a Hubble time. (They may, perhaps, be candidates for invisible gravitational lenses.)
(e) Are there any galaxies which lack dark halos? Such galaxies may form from regions where the baryons had been compressed to ~ 10 times the DM density, indicating a certain type of biasing mechanism.

I have tried to argue here that the idea of biasing is not just an *ad-hoc* idea introduced by theorists to save the attractive $\Omega = 1$ model when confronted with apparently conflicting evidence. A bias is essential in order to understand the large-scale structure, in particular the big voids and the superclustering of galaxies, and to reconcile any of the above cosmogonic scenarios with the observed universe. What might have looked at first as a frustrating idea for astronomers became an interesting observational search which requires non-trivial interpretation. On the theoretical side, the biasing mechanism is intimately related to the cosmogonic scenario and the nature of the DM. Although some of the proposed biasing mechanisms may seem *ad-hoc*, others are very plausible. In some cases the bias improves the consistency of the cosmogonic scenario and the observations, and in others it introduces new problems. The moral is, in any case, that the default assumption to be made is *not* necessarily that galaxies trace the mass. Instead, a physical 'biasing' scheme should be considered and the possible scenarios are numerous (Dekel 1986; Dekel and Rees 1987 for more details).

5. DIFFICULTIES ON VERY LARGE SCALES

In contrast to the successes on galactic scales and up to $\sim 10\ h^{-1}Mpc$, the recent indications for significant structure on scales $\sim 100\ h^{-1}Mpc$ introduce a non-trivial difficulty for any of the 'standard' scenarios discussed so far. In the case of CDM, the cluster-cluster correlation function is expected to be proportional, in the linear regime, to the matter two-point correlation function, ξ (see eq. 4). But, with a Zeldovich spectrum, ξ becomes negative at $\simeq 20\ h^{-1}Mpc$, while ξ_{cc} is observed to be positive out to $\sim 100\ h^{-1}Mpc$. In the case of neutrinos ξ_{cc} was found numerically (Barnes *et al.* 1985) not to be much larger than ξ, and to be independent of the cluster richness – in disagreement with the observations.

The observed large scale bulk velocity, if real, introduces a similar difficulty. The mean-square mass fluctuation and bulk velocity in a spheres of radius R are both related to the power spectrum via

$$\left(\frac{\delta M}{M}\right)_R^2 \propto \int_0^\infty dk\ k^2\ \langle|\delta_k|^2\rangle\ W_R(k), \tag{5}$$

and

$$v_R^2 \propto (a_o H_o)^2\ \Omega^{1.2} \int_0^\infty dk\ \langle|\delta_k|^2\rangle\ W_R(k), \tag{6}$$

where the window function $W_R(k)$ can be approximated by the step function: $W_R(k) \simeq 1$ for $0 < k < 1/R$ and it vanishes elsewhere. The *rms* fluctuation of the number of galaxies is observed to be $\delta N/N = 1$ in spheres of radius $8\ h^{-1}Mpc$. So for a given spectrum, assuming $\delta M/M \leq \delta N/N$ as is appropriate for CDM, one can predict an upper limit for the *rms* bulk velocity on any given large scale. For spheres of $100\ h^{-1}Mpc$ in diameter in the 'standard' CDM the predicted velocity is $< 150\ km\ s^{-1}$ – way below the observed value of $\sim 600\ km\ s^{-1}$.

These observations indicate that we need more power on very large scales. But on the other hand, the observed upper limits on $\delta T/T$ constrain the amplitude of the spectrum from above on various scales. Finding a scenario that would satisfy simultaneously the opposite constraints is a non-trivial task. I consider below two possible solutions.

5.1. A Hybrid Open Universe

Lowering Ω may be helpful; the spectrum shifts to larger scales roughly in proportion to $(\Omega h^2)^{-1}$, like the scale corresponding to the horizon at z_{eq}. In particular, there are two possible characteristic scales of relevance. Baryonic fluctuations, if $\Omega h^2 > 0.05$, develop a secondary peak on a very large scale corresponding to the baryon-photon Jeans scale just prior to recombination,

$$\lambda_J \simeq 25\ (\Omega h^2)^{-1} Mpc. \tag{7}$$

Neutrinos develop a critical coherence length due to free streaming within the horizon until z_{eq}, at a comoving length

$$\lambda_\nu \simeq 14\ (\Omega h^2)^{-1} Mpc. \tag{8}$$

If $\Omega h^2 \sim 0.1$, the resultant 'feature' is on a very large scale; with the normalization $\delta M/M \leq 1$ at $8\ h^{-1}Mpc$ there is more power on large scales, as required.

Consider, for example, an *open* CDM model where $\Omega_{cdm} \simeq 0.1$. Baryons, based on nucleosynthesis arguments, are likely to contribute a comparable density, $\Omega_b \simeq$

0.1. This is, therefore, a natural *hybrid*, where CDM fluctuations are responsible for the formation of galaxies while baryonic fluctuations, because of λ_J, give rise to the structure on very large scales (Dekel 1984a; 1984b). The properties of galaxies and their distribution on scales $\sim 10\ h^{-1}Mpc$ are reproduced very well in such a model (Blumenthal *et al.* 1984; Davis *et al.* 1985). No bias in the formation of galaxies is required, which would be consistent with the formation of big voids only if we have overestimated the emptiness of the voids in §2. We have recently looked at the formation of large scale structure in this model in some detail (Dekel, Blumenthal and Primack 1987). The cluster correlation function comes out right (confirming Dekel 1984); the clusters are 'super-biased' into forming in 'superpancakes'. The predicted *rms* bulk velocity is $\simeq 600\ km\ s^{-1}$ (calculated independently by Bond 1987). There is no difficulty with the $\delta T/T$ isotropy on large angular scales (also Silk and Vittorio 1987), but there is a marginal difficulty on small angles. This difficulty can be removed by reionization, which could naturally occur in this hybrid model due to the early formation of subgalactic objects from the CDM component of the fluctuations. Another way out would be invoking a non-zero cosmological constant (Vittorio and Silk 1986).

Other options for hybrids that may work in a similar way are a mixture of CDM and $\sim 10\ eV$ neutrinos each contributing $\Omega \simeq 0.1$, or an open baryonic universe with a mixture of adiabatic and isothermal fluctuations.

Thus, here are scenarios which seem to work, but they were patched up to do so. The choice of parameters is somewhat *ad hoc*; it is not the choice which arise naturally from first principles, or based on simplicity and aesthetics arguments.

5.2. Non-Random Phases – Cosmic Strings

An alternative way to produce large scale density fluctuations without producing large thermal fluctuations involves non-Gaussian statistics. If the density fluctuations began as quantum fluctuations of a free scalar field during the era of inflation they are indeed expected to be Gaussian (Bardeen *et al.* 1985), but it is also possible that the fluctuations arose from a different mechanism, in which they would not in general be Gaussian, and have non-random phases. A specific model that incorporates this feature is the scenario in which the density fluctuations were induced by *cosmic strings* (see a review by Vilenkin 1985). The strings are generic objects which form in a phase transition in many potentially plausible theories of the microphysics of the early universe. They are curvature singularities which are born with a topology of random-walks. They turn into closed smoother 'parent' loops on entering the horizon and then chop themselves into (possibly) stable 'daughter' loops, all in a scale-free self-similar fashion. The spectrum of fluctuations represented by the loops is scale-invariant ($n = 1$), which is quite appealing. It was argued, based on pioneering low-resolution string simulations (Albrecht and Turok 1985), that the loop-loop correlation function has a general shape close to that of galaxies or clusters (Turok 1986), $\xi(r) \propto r^{-2}$, as expected from 'beads' along locally-linear 'strings'. Density fluctuations (whose spectrum is determined by the nature of the DM) are induced in the DM by accretion onto the loops, so the galaxies and clusters that form are expected to be aligned in space along the same 'parent' linear structures. Studying a 'toy' string model that incorporates DM gravity (Primack, Blumenthal and Dekel 1986), we found the *phase-correlations* to have a very pronounced effect on the correlation functions of galaxies and clusters that are defined as peaks above a density threshold, while the matter correlation function (and the fluctuation spectrum), and therefore the temperature fluctuations, are of low amplitude. It seemed to provide a natural galaxy biasing mechanism, as well as an appropriate excess of cluster clustering on very large scales.

I do not think, though, that the clustering of loops is well understood yet. First, contrary to previous claims, the correlations on scales larger than the horizon are found to be negligible because the strings' random walk is self-avoiding (E. Vishniac, private comm.; Blumenthal, Dekel and Primack 1987, in preparation). Second, it is not obvious that the notion of 'beads along strings' is at all relevant; newer simulations (Albrecht 1987) show no such effect. Also, high loop velocities tend to smear out their correlations on scales slightly smaller than the original parent loops ($\sim$ the horizon). It is therefore crucial to study this fragmentation process and the associated velocities in more detail before a serious attempt is made to understand the formation of large scale structure from cosmic strings. We are currently running high resolution string simulations for this purpose.

It turns out that the string theory yet suffers from further difficulties. Peebles (unpublished 'screed') have listed a number of problems concerning the properties of galaxies such as the origin of their angular momentum and their luminosity function. There are ideas of how to overcome these difficulties (e.g. Turok 1987), but they involve *ad hoc* 'patching' of the theory. Another problem is that the most appealing string model where the DM is 'cold' cannot reproduce the large bulk velocity. Only if the DM is 'hot' can string-induced fluctuations be associated with high velocities on the order of 500 $km\ s^{-1}$ (R. Brandenberger, private comm.). Thus, the cosmic-strings picture seems to follow the familiar route: after it emerged as a very appealing elegant theory which can 'naturally' explain a certain set of observations that are in conflict with the other scenarios (e.g. the cluster correlations), it has reached a stage where a quantitative confrontation with the various aspects of the observed structure forces *ad hoc* 'patching', which is not very satisfactory.

6. THE EXPLOSIONS SCENARIO

The theories discussed above all assume that the present structure arose from small amplitude density fluctuations which originated in the early universe, and that it is determined by the spectrum and statistics of these fluctuations. An alternative approach, based on the concept that the present structure is determined by physical processes in late cosmological epochs and is not sensitive to the exact initial conditions, is represented by the picture of explosions (Ostriker and Cowie 1981; Ikeuchi 1981). Here, nuclear energy from first generation objects helps gravity in forming further galaxies and enhancing their clustering. The exploding galaxies produce spherical blast waves that push the gas out of their interiors. The shells expand, cool and fragment into a new generation of galaxies, the last generation forming at $z \simeq 7$, after which the shells cannot cool efficiently anymore. Based on several astrophysical constraints it has been argued that the individual 'bubbles' of galaxies cannot be bigger than $\sim 10\ h^{-1}Mpc$ in radius (e.g. Carr and Ikeuchi 1985). But, the subsequent interaction of the bubbles with each other generates clusters and superclusters and makes the empty interiors of the shells grow significantly. Contrary to previous worries, it turns out that the resultant structure on scales $1 - 30\ h^{-1}Mpc$ resembles the observed structure quite well (Saarinen, Dekel and Carr 1986; Weinberg, Dekel and Ostriker 1987); it can reproduce the galaxy correlation function, the appearance of sharp edges in the distribution of galaxies and the occurrence of big voids. It can even account for the required 'bias' of the galaxy distribution (which arise from gas that was swept out into shell surfaces) relative to the DM (which partly still fills the shell interiors).

However, it is not clear how the explosions can be responsible for the clustering of clusters and the high velocities on scales as large as $\sim 100\ h^{-1}Mpc$. The only plausible way out is again 'patching' the theory with a component of primordial fluctuations on large-scales – a hybrid. For example, 'wakes' behind cosmic strings can give rise to the 'seeds' required for triggering the explosion scenario, which would be correlated appropriately on very large scales (Rees 1986). Also, superconducting strings combined with primordial magnetic fields can give rise to explosion-like phenomenon (Ostriker, Thompson and Witten 1986).

7. CONCLUSION

The field of the formation of large scale structure is, to my mind, in a dissatisfactory phase. Our 'standard' scenarios, which are sometimes very successful in explaining some of the observations, need 'patching' and *ad hoc* fine-tuning when confronted with the whole set of observations on all scales. This, by no means, indicates a breakdown of conventional physics; just that we should look for better ways of applying it. Quoting two of the participants in this symposium, Dr. Norman's conjecture is that: "Some observations must be wrong!", while Dr. Yahil says: "Theoreticians should think harder!". My feeling is that either, or both, are correct! This situation is not necessarily frustrating, considering the fact that the observational constraints accumulate rapidly and continuously improve qualitatively. It implies that, on the contrary, there is much more (and better) work to be done in this field, by all of us, and that there is a hope for significant improvement in our understanding of the large scale structure in the near future.

REFERENCES

Albrecht, A. 1987, in Nearly Normal Galaxies, ed. S. Faber (New York: Springer-Verlag) , in press.
Albrecht, A. and Turok, N. 1985, Phys. Rev. Lett. **54**, 1868.
Audouze, J. 1987, this volume.

Bahcall, N. 1986, in Galaxy Distances and Deviations from Universal Expansion, eds. B.F. Madore and R.B. Tully (Dordrecht: Reidel).

Bahcall, N. and Soneira, R. 1983, Astrophys. J. **270**, 20.

Bardeen, J., Bond, J.R., Kaiser, N. and Szalay, A. 1986, Astrophys. J. , in press.

Bardeen, J. 1985, in Inner Space/Outer Space, eds. E.W. Kolb and M.S. Turner (University of Chicago Press) , p. 212.

Barnes, J., Dekel, A., Efstathiou, G. and Frenk, C. 1985, Astrophys. J. **295**, 368.

Blumenthal, G.R., Faber, S.M., Primack, J.R. and Rees, M.J. 1984, Nature **311**, 517.

Bond, J.R. 1987, in Nearly Normal Galaxies, ed. S. Faber (New York: Springer-Verlag) , in press.

Braun, E., Dekel, A. and Shapiro, P. 1987, in preparation.

Burstein, D., Davies, R., Dressler, A., Faber, S.M., Lynden-Bell, D., Terlevich, R. and Wegner, G. 1986, in Galaxy Distances and Deviations from Universal Expansion, eds. B.F. Madore and R.B. Tully (Dordrecht: Reidel).; this volume.

Carr, B.J. and Ikeuchi, S. 1985, M.N.R.A.S. **213**, 497.

Centrella, J. and Melott, A. 1983, Nature **305**, 196.

Collins, A., Joseph, R.D. and Robertson, N.A. 1986, Nature **320**, 506.

Davis, M., Efstathiou, G., Frenk, C.S. and White, S.D.M. 1985, Astrophys. J. **292**, 371.

Davis, M. and Peebles, J.P.E. 1983, Ann. Rev. Astron. Astrophys. **21**, 109.

de Lapparent, V., Geller, M.J. and Huchra, J.P. 1986, Astrophys. J. (Lett.) **302**, 1.

Dekel, A. 1981, Astron. Astrophys. **101**, 79.

Dekel, A. 1983, Astrophys. J. **264**, 373.

Dekel, A. 1984a, in the Eighth Johns Hopkins Workshop on Current Problems in Particle Theory, eds. G. Domokos and S. Koveski-Domokos, (Singapur: World Scientific) p. 191.

Dekel, A. 1984b, Astrophys. J. **284**, 445.

Dekel, A. 1985, Astrophys. J. **298**, 461.

Dekel, A. 1986, Comments on Astrophysics, in press.

Dekel, A. and Aarseth, S.J. 1984, Astrophys. J. **283**, 1.

Dekel, A., Blumenthal, G.R. and Primack, J.R. 1987, in preparation.

Dekel, A., Einasto, J. and Rees, J. 1987, Rev. Mod. Phys., in preparation.

Dekel, A. and Piran, T. 1986, Astrophys. J. (Lett.) , in press.

Dekel, A. and Rees, M.J. 1987, Nature , submitted.

Dekel, A. and Silk, J. 1986, Astrophys. J. **303**, 39.

Dressler, A. 1980, Astrophys. J. **236**, 351.

Efstathiou, G. 1985, M.N.R.A.S. **213**, 29.

Efstathiou, G. and Silk, J. 1983, Fundamentals of Cosmic Physics, **9**, 1.

Faber, S. 1987, in Nearly Normal Galaxies, ed. S. Faber (New York: Springer-Verlag) , in press

Flores, R. 1987, in Nearly Normal Galaxies, ed. S. Faber (New York: Springer-Verlag) , in press.

Flores, R., Blumenthal, G.R., Dekel, A. and Primack, J.R. 1986, Nature , in press.

Frenk, C.S., White, S.D.M., Efstathiou, G. and Davis, M. 1985, Nature **317**, 595.

Geller, M. 1987, this volume.

Giovanelli, R., Haynes, M.P. and Chincarini, G. 1986, preprint.

Haynes, M. 1987, in Nearly Normal Galaxies, ed. S. Faber (New York: Springer-Verlag) , in press.

Hoffman, Y. 1986, M.N.R.A.S. , in press.

Hoffman, Y. and Bludman, S. 1984, Phys. Rev. Lett. **52**, 2087.

Hoffman, G.L., Salpeter, E.E. and Wasserman, I. 1982, Astrophys. J. **263**, 485.

Hogan, C. 1987, in Nearly Normal Galaxies, ed. S. Faber (New York: Springer-Verlag) , in press.

Ikeuchi, S. 1981, Pub. Astron. Soc. Japan **33**, 211.

Jensen, L.G. and Szalay, A.S. 1986, preprint.

Kaiser, N. 1984, Astrophys. J. (Lett.) **284**, L9.

Kaiser, N. and Silk, J. 1986, preprint.

Koo, D. 1987, this volume.

Melchiorri, F., Melchiorri, B., Ceccarelli, C. and Pietranera, L. 1981, Astrophys. J. (Lett.) **250**, L1.

Oemler, A. Jr. 1987, in Nearly Normal Galaxies, ed. S. Faber (New York: Springer-Verlag) , in press.

Olive, K., Seckel, D. and Vishniac, E. 1985, Astrophys. J. **292**, 1.

Oort, J. 1983, Ann. Rev. Astron. Astrophys. **21**, 373.

Ostriker, J.P. and Cowie, L. 1981, Astrophys. J. (Lett.) **243**, L127.

Ostriker, J.P., Thompson, C. and Witten, E. 1986, preprint (Princeton).

Peebles, P.J.E. 1984, Astrophys. J. **284**, 439.

Peebles, P.J.E. 1986, Nature **321**, 27.

Politzer, D.H. and Wise, M.B. 1984, Astrophys. J. (Lett.) **285**, L1.

Primack, J.R., Blumenthal, G.R. and Dekel, A. 1986, in Galaxy Distances and Deviations from Universal Expansion, eds. B.F. Madore and R.B. Tully (Dordrecht: Reidel).

Rees, M.J. 1985, M.N.R.A.S. **213**, 75P.

Rees, M.J. 1986, preprint.

Rubin, V.C, Ford, W.K., Thonnard, N., Roberts, M.S. and Graham, J.A. 1976, Astron. J. **81**, 687.

Saarinen, S., Dekel, A. and Carr, B. 1986, Nature , in press.

Sargent, W. 1987, this volume.

Shaver, P. 1987, this volume.

Silk, J. 1968, Astrophys. J. **151**, 459.

Silk, J. 1985, Astrophys. J. **297**, 1.

Silk, J. 1987, this volume.

Silk, J. and Vittorio, N. 1986, preprint (Berkeley).

Silk, J. and Vittorio, N. 1987, Astrophys. J. , in press.

Tremeine, S. and Gunn, J.E. 1979, Phys. Rev. Lett. **42**, 407.

Tully, B. 1986, in Galaxy Distances and Deviations from Universal Expansion, eds. B.F. Madore and R.B. Tully (Dordrecht: Reidel).

Turner, M.S., Steigman, G. and Krauss, L.M. 1984, Phys. Rev. Lett. **52**, 2090.

Turok, N. 1986, preprint.

Turok, N. 1987, in Nearly Normal Galaxies, ed. S. Faber (New York: Springer-Verlag) , in press.

Uson, J.M. and Wilkinson, D.T. 1984, Nature **312**, 427. **146**, 235.

Vilenkin, A. 1985, Phys. Reports **121**, 264.

Vittorio, N. and Silk, J. 1986, preprint (Berkeley).

Weinberg, D., Dekel, A. and Ostriker, J.P. 1987, in preparation.

White, S.D.M. 1987, in Nearly Normal Galaxies, ed. S. Faber (New York: Springer-Verlag) , in press.

White, S.D.M., Davis, M. and Frenk, C.S. 1984, M.N.R.A.S. **209**, 27p.

White, S.D.M., Frenk, C.S. and Davis, M. 1983, Astrophys. J. (Lett.) **274**, L1.

White, S.D.M., Frenk, C.S., Davis, M. and Efstathiou, G. 1986, preprint.

DISCUSSION

BAHCALL: In the hybrid model of cold dark matter plus baryons, are there problems with the observed isotropy of the microwave background radiation?

DEKEL: There are no difficulties on scales of a few degrees and up - those which should directly reflect initial density fluctuations. There is a marginal discrepancy on the scale of a few arc-minutes, which can be avoided if either 1) reionization by the first objects that emerged from the CDM fluctuations has smeared out temperature fluctuations on scales of a few degrees and less, or 2) there is a non-zero cosmological constant.

SILK: You used an upper limit of 10 percent on the void density contrast in reaching an important conclusion about justifying the need for biasing. However the observed 2σ limit on luminous galaxies in one of the largest voids in Bootes is 25 percent; moreover the void contains at least six emission line galaxies.

DEKEL: The statistics certainly need to be done better, with better data. Nevertheless, in the Bootes void, 32 galaxies were expected and none found! The situation is similar in the other voids that are found very frequently in every redshift survey. Ten percent mean number density is, I believe, a reasonable estimate for what we observe in voids. But this will become clearer in the future.

NORMAN: For biasing and anti-biasing theories, to make these truly scientific they need to be falsifiable! What are the best tests for these concepts that observers here should go out and measure?

DEKEL: I agree that this is crucial, and have listed suggestive tests in my paper. But the point is that, based on the numerous possibilities for biasing processes, it would be astonishing if they did not affect the distribution of galaxies relative to the matter. I refer you to a Nature review by Rees and myself for a more comprehensive discussion.

THE SPONGE-LIKE TOPOLOGY OF LARGE SCALE STRUCTURE IN THE UNIVERSE

J. Richard Gott, III
Princeton University Observatory
Princeton, NJ 08544

ABSTRACT. We describe and apply a quantitative measure of the topology
of large scale structure: the genus of density contours in a smoothed
density distribution. For random phase (gaussian) density fields, the
mean genus per unit volume exhibits a universal dependence on threshold
density, with a normalizing factor that can be calculated from the
power spectrum. The topology of the observational sample is consistent
with the random phase, cold dark matter model.

Recent deep redshift surveys by de Lapparent, Geller and Huchra (1986)
and Haynes and Giovanelli (1986) have renewed interest in the topology
of large scale structure in the universe.

If we look at a distribution of galaxies we see a series of points
distributed in 3-dimensional space. We would like to know the smooth
underlying distribution from which the points could have been obtained
by a sampling process. To begin, therefore, we smooth the data with a
gaussian smoothing window $W = e^{-r^2/\lambda^2}$ where r is the distance and λ is
a smoothing length picked to be larger than the mean galaxy-galaxy
separation. Information about the topology is then carried in the
density contour surfaces of this smoothed density distribution.

The topology of an object is mathematically specified by its
genus. We may define the genus of a contour surface as

$$g_s = \text{(no. of holes)} - \text{(no. of isolated regions)} \qquad (1)$$

where 'hole' means 'hole' like a donut has (Gott, Melott, Dickinson
1986). Suppose we have a density contour that shows 50 isolated
spherical clusters, then $g_s = -50$. A contour may also have a multiply
connected, sponge-like topology, in which case its genus is positive.
In what follows we study g_s as a function of threshold density (Gott,
Weinberg, Melott 1986; Weinberg, Gott, Melott 1986).

An important advantage of this method of looking at topology is
that in the standard big bang - inflationary model we can relate the
topology seen today to that present in the initial conditions. Imagine
looking at the small-amplitude density fluctuations present at a red-

A. Hewitt et al. (eds.), Observational Cosmology, 433–436.

shift of z = 25. These fluctuations arise from random quantum fluctuations that are gaussian with random phases with a power spectrum P(k)
which can be calculated directly from the primordial Zel'dovich inflationary spectrum. We now smooth the data with a smoothing length λ,
and construct density contours. Hamilton, Gott, and Weinberg (1986)
and Bardeen et al. (1986) have shown that the mean genus per unit
volume of these surfaces will be given by

$$g_s = N(1 - \nu^2)e^{-\nu^2/2}. \tag{2}$$

Here $\nu = \delta_c/\xi(0)^{1/2}$ is the number of standard deviations by which the
contour threshold density δ_c departs from the mean density. And
$N = [<k^2>/3]^{3/2}/4\pi^2$ is an appropriate moment of the smoothed power
spectrum. The amplitude of the $g_s(\nu)$ curve depends on N and therefore
on P(k), but the form $g_s(\nu) \propto (1 - \nu^2)e^{-\nu^2/2}$ is completely independent of
the initial power spectrum. N is positive definite so that g_s is positive (sponge-like topology) for $\nu=0$, f=0.5 (the median density contour) regardless of P(k). If we make a cut where the fraction of the
volume in the high density region is less than f = 16% ($\nu > 1$) then
$g_s < 0$ and we encounter isolated clusters. If we choose a contour such
that the fraction of volume in the high density region is greater than
f = 85% ($\nu < -1$) then $g_s < 0$ and we find isolated voids.
     Now we evolve the model to the present epoch, and generate biased
subsets of the mass distribution in the usual way (Melott and Fry 1986).
     Figure 1 plots the $g_s(\nu)$ curves for the initial, final, and biased
conditions (averaged results of four simulations) together with the
theoretical curve $g_s(\nu) = N(1 - \nu^2)e^{-\nu^2/2}$ expected for the CDM initial
conditions. Since Ñ is determined from P(k), which is known for the
CMD model, there are no free fitting parameters for this at all. The
agreement is remarkable.
     Why is this so? As long as the fluctuations stay in the linear
regime they just grow in place increasing in amplitude, and the
topology changes not at all. In the cold dark matter model the scale
at which the mass covariance function goes to unity, r_m = 3.6 Mpc, is
considerbly smaller than our smoothing length λ = 10 Mpc, so we are
mainly looking at fluctuations that are just now beginning to come out
of the linear regime and whose topology remains unchanged from the
initial conditions. Biasing increases the 'contrast' of the picture,
but it basically leaves the luminosity density a monotonic function
of the underlying mass density. Thus contours of constant luminosity
density are contours of constant mass density -- just the values are
shifted. If we draw contours as a function of the volume enclosed we
find that the biased and unbiased data sets are essentially identical.
     Figure 2 shows results from a volume limited sample of the
northern CfA redshift survey. It is a cube with a side length of 58
Mpc and includes galaxies brighter than 0.72 L_* out to a maximum redshift of 5000 km/s (see GMD for details). As in the analysis of the
N-body simulations, we use a periodic boundary condition for smoothing.
The smoothing length λ = 10.8 Mpc. Given the small volume of the
region surveyed, we would like to see results from larger observational samples before drawing any firm conclusions. Nonetheless, the

data as they stand are remarkably consistent with the random phase model and with the amplitude expected for a CDM power spectrum.

This work was partially supported by NASA Grant NAGW-765.

References

Bardeen, J. M., Bond, J. R., Kaiser, N., and Szalay, A. S. 1986, Ap. J., 304, 15.
Gott, J. R., Melott, A. L., and Dickinson, M. 1986, Ap. J., 306, 341.
Gott, J. R., Weinberg, D., and Melott, A. L. 1986, in preparation.
Hamilton, A. J. S., Gott, J. R., and Weinberg, D. 1986, Ap. J., in press.
Haynes, M. P., and Giovanelli, R. 1986, Ap. J., 306, L55.
de Lapparent, V., Geller, M. J., and Huchra, J. P. 1986, Ap.J., 302, L1.
Melott, A. L. and Fry, J. N. 1986, Ap. J., 306, 1.
Melott, A. L., Weiberg, D. H., and Gott, J. R. 1986, in preparation.
Weinberg, D. H., Gott, J. R., and Melott, A. L. 1986, in preparation.

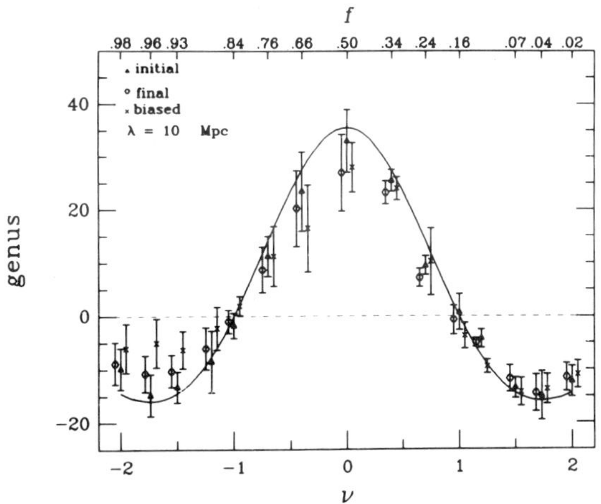

Figure 1. Cold Dark Matter Simulation Results

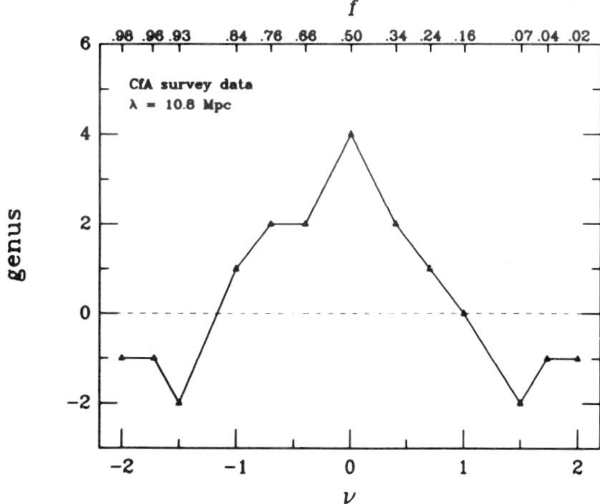

Figure 2. CfA Survey Results

DISCUSSION

PECKER: In what sense does your technique differ from that of
Mandelbrot in building either cheese and bubbles, - or sponges, using
various 3-dimensional fractal distributions?

GOTT: Mandelbrot has modeled galaxy distributions by starting with a
uniform high density region and excising random spherical voids from
it. With just a few voids excised this gives a swiss cheese topology,
but if enough voids are excised so that they percolate and the voids
form one connected low density network then this will make a sponge.
Mandelbrot has also done models where galaxies are laid down on a
random walk. In the sponge-like topology produced by random phase
perturbations clusters are linked to each other by a network of
filaments and voids are linked to each other by a network of tunnels.
It is important to note that the topology is a three dimensional issue.
A thin slice of swiss cheese and a thin slice of a sponge can look
identical, both showing holes. The question is how the low density
regions are connected up in three dimensions. In this regard it is
interesting to note that the $\Omega = 1$, cold dark matter, biased models
(by our group and also by White et al.) which are known to have a
sponge-like topology in three dimensions, do reproduce remarkably well
in thin slices the cellular appearance seen in the thin slices of the
de Lapparent, Geller and Huchra survey.

SYSTEMATIC PROPERTIES OF GALAXIES: IMPLICATIONS FOR GALAXY FORMATION

S.Okamura, K.Kodaira, and M.Watanabe
Tokyo Astronomical Observatory
Mitaka, Tokyo, 181 Japan

1. INTRODUCTION

We have compiled a data base of V-band two-dimensional luminosity distribution for 261 galaxies in the Virgo and the Ursa Major regions
(Watanabe 1983) using the 105-cm Schmidt telescope at the Kiso Observatory, Tokyo Astronomical Observatory. This is one of the largest
collections of homogeneous surface photometry available to date, though
the sample is biased for bright and large galaxies with $m(V) \lesssim 14$mag
and/or $D_{26} \gtrsim 2'$. Among the sample galaxies we have selected some 200
certain members($V_0 < 3500$km/s) of the Virgo cluster and the Ursa Major
clouds and performed various analyses on them to investigate systematic
properties of galaxies. The two clusterings lie at nearly equal distance of $(m-M) \sim 31.1$ (Aaronson and Mould 1983). In the present paper
we discuss the result of spheroid(bulge)/disk decomposition and
velocity-luminosity relation for galaxies.

2. SPHEROID(BULGE)/DISK DECOMPOSITION

We use an empirical model consisting of a $R^{1/4}$-law spheroid and an exponential disk whose radial luminosity distribution is given by,

$$I(R) = I_{0,S} \, \text{dex}[-3.33\{(R/R_{0,S})^{1/4} - 1\}] + I_{0,D} \, \exp(R/R_{0,D}). \qquad (1)$$

Main results obtained from decomposition for 167 galaxies are summarized
in the following:
(1) We find a difference in the parameter correlation, $\log R_0$ versus
$\mu_0 = -2.5\log I_0$, between elliptical galaxies and bulges of disk galaxies.
Bulges are, on the average, less luminous in absolute magnitude and
have both fainter μ_0 and larger R_0 even at the same absolute magnitude
than ellipticals. This result, together with the kinematical difference (e.g. Davies et al. 1983), may suggest different formation history
for ellipticals and bulges.
(2) Bulge parameters cover very wide ranges (1.9dex in $\log R_0$ and 10mag
in μ_0) while disk parameters are confined within (relatively) narrow
ranges (0.7dex and 4mag, respectively). It seems that some self-

437

A. Hewitt et al. (eds.), Observational Cosmology, 437–439.
© *1987 by the IAU.*

regulating mechanism is working in the process of disk formation under varying influences of bulges.
Details of the analysis is published elsewhere (Kodaira et al. 1986).

3. VELOCITY-LUMINOSITY RELATION AND LOG D_{26} VERSUS SB DIAGRAM (DSBD)

If $M(R_{dyn})$ is the dynamical mass of a galaxy within a radius R_{dyn}, the dynamical velocity at that radius is expressed by definition as,

$$V_{dyn}^2 \propto G\, M(R_{dyn})/R_{dyn}. \tag{2}$$

We introduce a 'photometric velocity parameter' defined by,

$$V_{ph}^2 = a' \cdot L(R_{ph})/R_{ph} = a \cdot B \cdot D, \tag{3}$$

where $L(R_{ph})$ is the luminosity integrated within the photometric radius R_{ph}, B the mean surface brightness within R_{ph}, D the photometric diameter $2R_{ph}$, and a' and a are dimensional scaling factors to yield the relationship, $V_{dyn}^2 = (M/L)V_{ph}^2$, with the same dimension for V_{dyn} and V_{ph}. Here (M/L) is a measure of mass-to-luminosity ratio of the galaxy defined by,

$$\left(\frac{M}{L}\right) = \frac{M(R_{dyn})}{L(R_{dyn})}\ \frac{L(R_{dyn})}{R_{dyn}}\ \left(\frac{L(R_{ph})}{R_{ph}}\right)^{-1} \tag{4}$$

This (M/L) can be considered to be the physical mass-to-luminosity ratio within R_{dyn} multiplied by a correction factor due to the difference between R_{dyn} and R_{ph}.

We compute V_{ph} using our photometric data. As V_{dyn} we use the 21-cm linewidth of neutral hydrogen compiled by Richter and Huchtmeier (1984) and the central velocity dispersion σ compiled by Whitmore et al. (1985) for spirals and ellipticals, respectively. Thus the meaning of V_{dyn} is quite different between spirals and ellipticals. However, *differential* behavior of (M/L) within spirals and within ellipticals can be derived. Fig. 1 shows (M/L) plotted against luminosity. We find no obvious dependence of (M/L) on L for spirals while there is a well-defined systematic dependence of $(M/L) \propto L^{0.4}$ for ellipticals.

Kodaira et al. (1983) introduced the diameter versus surface brightness diagram (DSBD) as a diagnostic tool to investigate the nature of galaxies and found

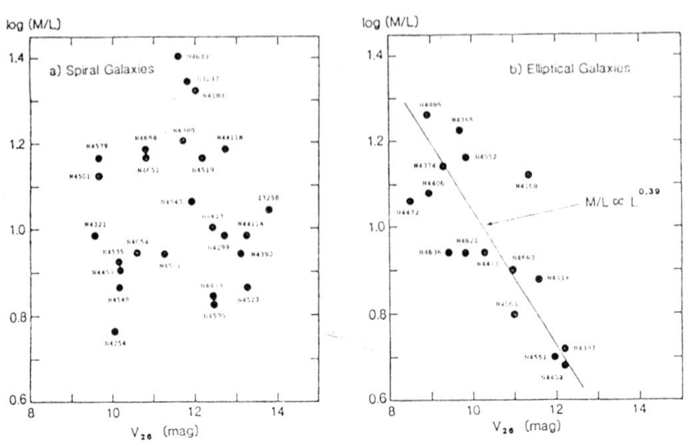

Fig.1 log(M/L) plotted against apparent magnitude.

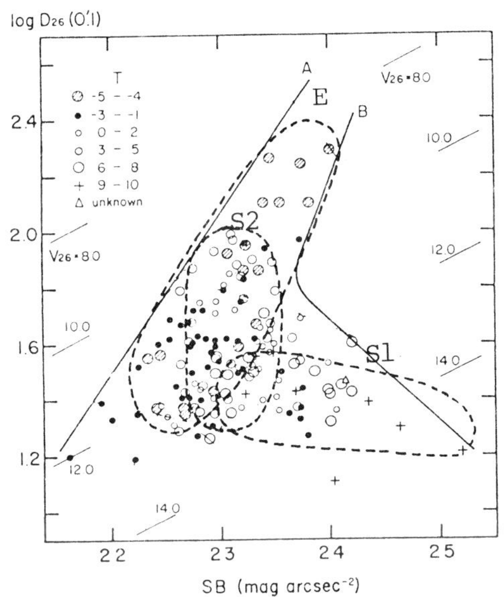

Fig.2 Sequential distributions of galaxies in DSBD.

that ellipticals and spirals are sequentially distributed in DSBD. Fig.2 illustrates the distributions. If we characterize the distributions by the gradient, $\beta=\Delta(\log D_{26})/\Delta(SB)$, we obtain the velocity-luminosity relation of,

$$L \propto V_{dyn}^{\gamma}, \tag{5}$$

with

$$\gamma = 2(1-5\beta)[\alpha(1-5\beta)+(1-2.5\beta)]^{-1}, \tag{6}$$

using the exponent of $(M/L) \propto L^{\alpha}$ relation derived above. For the elliptical branch denoted as E in Fig.2, we have $\beta \sim 0.6$ and $\alpha \sim 0.4$. These values lead to $L \propto \sigma^3$, which agrees well with the observation (e.g. Tonry 1981; Dressler 1984). Spirals appear to form two branches denoted as S1 ($\beta \sim 0$) and S2 ($\beta \sim \infty$) in Fig.2. With $\alpha \sim 0$ for spirals we obtain $L \propto V_H^2$ for S1 and $L \propto V_H^4$ for S2, respectively. It should be noted that most of gradients of Tully-Fisher relation reported so far lie between -10 ($\gamma=4$) and -5 ($\gamma=2$).

The present analysis suggests that the functional form of velocity-luminosity relation is determined by both the shape of the sequential distribution in DSBD, i.e., the diameter-mean surface brightness relation, and the systematic dependence of (M/L) on L. Although DSBD is a purely photometric diagram, it may contain essential information on kinematical properties of galaxies as well.

REFERENCES

Aaronson,M., and Mould,J.R. 1983, Ap. J., 265, 1.
Davies,R.L., Efstathiou,G., Fall,S.M., Illingworth,G., and
    Schechter,P.L. 1983, Ap. J., 266, 41.
Dressler,A. 1984, Ap. J., 281, 512.
Kodaira,K., Okamura,S., and Watanabe,M. 1983, Ap. J. (Letters), 274, L49.
Kodaira,K., Watanabe,M., and Okamura,S. 1986, Ap. J. Suppl. (Dec. 1986).
Tonry,J.L. 1981, Ap. J. (Letters), 251, L1.
Richter,O.-G., and Huchtmeier,W.K. 1984, Astr. Ap., 132, 253.
Watanabe,M. 1983, Annals Tokyo Astron. Obs., 19, 121.
Whitmore,B.C., McElroy,D.B., and Tonry,J.L. 1985, Ap. J. Suppl., 59, 1.

SEARCH FOR PROTO-CLUSTERS AT METER WAVELENGTHS

G.Swarup and R.Subrahmanyan
Tata Institute of Fundamental Research
Post Box 1234, Bangalore 560012
India

Although current models of galaxy formation differ widely, it seems likely that just prior to their formation, a significant fraction of the matter in the Universe was in the form of neutral hydrogen (HI) clouds at redshifts z greater than about 3, beyond which the quasar density decreases rapidly. In this paper we have taken a semi-empirical approach for estimating values of measurable parameters for a variety of scale sizes and masses expected in the epoch prior to galaxy formation and have compared these with the results of searches made so far.

Searches for the redshifted 21-cm line radiation from HI clouds have been carried out by the Jodrell Bank group at z = 3.28 and 4.92 (Davies et al. 1978) and by the Cambridge group at z = 8.4 (Bebbington 1986). The observations indicate upper limits for the masses of the proto-clusters of about $1 - 10 \times 10^{15} M_\odot$, or that the number of such objects in the early Universe is $\lesssim 10^6$.

According to the adiabatic perturbation model considered by Sunyaev & Zeldovich (1972;1974), super-clusters of mass $\sim 10^{14} - 10^{16} M_\odot$ condense first and galaxies are formed later by fragmentation at $z \sim 5$ to 10. In their model, only a few thousand proto-clusters are observable over the entire sky in a bandwidth of a few MHz in the frequency range of about 150 (z = 8.5) to 327 MHz (z = 3.3). On the other hand, in the isothermal perturbation scenario or in the recent cold dark matter models, galaxies form first and aggregate hierarchically to form clusters at later epochs (Peebles 1980; Blumenthal et al. 1984). The observational strategies for the detection of HI clouds differ considerably for the two cases (Hogan and Rees 1979; Oort 1984). Since the HI condensations expected in the adiabatic model are relatively rare, a large region of the sky needs to be searched. However, for the hierarchical clustering scenario, where large scale clustering of relatively low mass condensates are expected, highly sensitive observations are desirable, but searches over only a few square degrees may suffice. In both cases we may have to search over a large redshift range. An extensive search for proto-clusters is one of the major objectives of a Giant Meterwave Radio Telescope being set up in India (Swarup 1984).

In our model a fraction β_z, of the total mass M contained in a cell of characteristic size d_z, condenses into a cloud of size a_z. At any epoch only a fraction f_z of these cells are observable in their 21-cm emission, depending upon their formation rates and lifetimes. It is assumed that 75% of the baryonic mass is hydrogen, of which half is neutral at the epoch of condensation. Hence,

441

A. Hewitt et al. (eds.), Observational Cosmology, 441–444.
© 1987 by the IAU.

the neutral hydrogen mass $M_{HI} = 0.375 \beta_z (\Omega_b / \Omega_o) M$, where (Ω_b / Ω_o) is the ratio of baryon to total density parameter. We take $(d_z / a_z) \sim 2$ at the epoch of formation. The relationship between d_z and M is given by

$$d_z = [M / \rho_c \Omega_o]^{1/3} (1+z)^{-1}$$

where $\rho_c = (3H_o^2 / 8\pi G)$ is the critical density and $H_o = 100h$ km s^{-1} Mpc^{-1}. Hence

$$d_z = 7.13 [M (10^{14} M_\odot)]^{1/3} h^{-2/3} (1+z)^{-1} \Omega_o^{-1/3} \text{ Mpc}.$$

The clouds being optically thin, the peak line flux density is given by

$$S_p = \frac{g_2}{g_1 + g_2} \cdot A_{21} \cdot \frac{h\nu_e}{4\pi} \cdot \frac{M_{HI}}{m_p} \cdot \frac{1}{D^2(1+z)^2} \cdot \frac{1}{\Delta\nu_o}$$

where $\quad D = (2c/H_o \Omega_o^2) [\Omega_o z + (\Omega_o - 2)\{(\Omega_o z + 1)^{1/2} - 1\}] / (1+z)$.

Thus, $\quad S_p = 56.17 \dfrac{M_{HI}(10^{14} M_\odot) h^2 \Omega_o^4 [\Delta\nu_o^W (\text{MHz})]^{-1}}{[\Omega_o z + (\Omega_o - 2)\{(\Omega_o z + 1)^{1/2} - 1\}]^2}$ mJy.

The observed angular size of the condensate is

$$\theta_a = (a_z / D)(1+z)$$

$$= 0.573 \frac{h \Omega_o^2 a_z(\text{Mpc})(1+z)^2}{[\Omega_o z + (\Omega_o - 2)\{(\Omega_o z + 1)^{1/2} - 1\}]} \quad \text{arcmin.}$$

The observed width, $\Delta\nu_o^W$, of the red-shifted 21-cm line ($\nu_e = 1420$ MHz) due to the virial motion of HI gas is

$$\Delta\nu_o^W = \alpha_z (2\beta_z GM/3a_z)^{1/2} \nu_e (1+z)^{-1} c^{-1}$$

$$= 2.54 [\alpha_z/(1+z)] [\beta_z M(10^{14} M_\odot)/a_z(\text{Mpc})]^{1/2} \text{ MHz}$$

where $\alpha_z = \Delta\nu_o^W / \Delta\nu_o^V$ is the degree of virialization in the condensate.

Table I gives the calculated values of the observed peak line flux density S_p, virial line-width $\Delta\nu_o^W$ and angular size θ of the HI cloud of size $a_z = 0.5 \cdot d_z$ for $z = 3.34$ and 8.47. We have assumed $\Omega = 1$ and $H_o = 75$ km s^{-1} Mpc^{-1}. Taking $\beta_z = 0.5$ and $\Omega_b = 0.1$, the mass of neutral hydrogen in the condensate is $M_{HI} \simeq M/50$. The Observed line width may be smaller by a factor $\alpha_z \lesssim 1$, as the clouds may not be virialized at the epoch of formation. If the scale size of the clouds is smaller than $0.5 d_z$ at the epoch of condensation, the observed angular size θ would be correspondingly smaller.

It is seen from Table I that the line flux densities for super-cluster size clouds with $M \gtrsim 10^{16}$ are expected to be $\gtrsim 5$ mJy, which should be observable using some of the existing large radio telescopes, provided searches are made over a sufficiently large area of the sky and at several redshifts. However, for

Table 1. Calculated values of measurable parameters from Proto-clusters

Mass($M_\odot$)		$z = 3.34$ (327 MHz)				$z = 8.47$ (150 MHz)			
M	M_{HI}	d_z (Mpc)	S_p (mJy)	$\Delta\nu_o^v$ (MHz)	θ (arcmin)	d_z (Mpc)	S_p (mJy)	$\Delta\nu_o^v$ (MHz)	θ (arcmin)
5×10^{16}	10^{15}	15.8	18.86	3.29	28.34	7.24	3.47	2.23	21.82
5×10^{15}	10^{14}	7.33	4.06	1.53	13.15	3.36	0.75	1.03	10.13
5×10^{14}	10^{13}	3.40	0.87	0.71	6.10	1.56	0.16	0.48	4.70
5×10^{13}	10^{12}	1.58	0.19	0.33	2.83	0.72	0.035	0.22	2.17

clumpiness on smaller scales, higher sensitivity and higher angular resolution is required as obtainable by a synthesis radio telescope. Even in the case of hierarchical clustering of small clouds, the HI lines are likely to be sufficiently separated in frequency as discussed below.

The observed separation between lines, $\Delta\nu_o^s$, depends on the redshift distribution of clouds :

$$\Delta\nu_o^s = H_o d_z f_z^{-1/3} (1 + \Omega_o z)^{1/2} \nu_e c^{-1}$$

$$= 0.473 \, d_z(\text{Mpc})h \, f_z^{-1/3} (1 + \Omega_o z)^{1/2} \quad \text{MHz.}$$

Hence, the ratio of virial line width to the line separation is

$$\frac{\Delta\nu_o^w}{\Delta\nu_o^s} = 0.282 \, \alpha_z \beta_z^{1/2} f_z^{1/3} \Omega_o^{1/2} \left(\frac{d_z}{a_z} \right)^{1/2} \left(\frac{1 + z}{1 + \Omega_o z} \right)^{1/2}$$

The maximum value of $\Delta\nu_o^w / \Delta\nu_o^s = 0.28 \, (d_z/a_z)$ is < 1, if $(d_z/a_z) \lesssim 3$ at the epoch of formation of condensates. According to our estimates, the values of the above parameters are likely to be : $f_z \sim 0.2$, $\alpha_z = 0.5$, $\beta_z = 0.5$, $\Omega_o = 0.3 - 1.0$, $d_z/a_z \sim 2$ to 3. Hence $\Delta\nu_o^w / \Delta\nu_o^s \sim 0.07$ to 0.10. Searches for the $\sim 10^{15-16}$ $M_\odot$ super-clusters which are expected to form by the growth of adiabatic perturbations are perhaps best done at larger redshifts ($\sim 6 - 10$), when these structures may start fragmenting. On the other hand hierarchical clustering of relatively low mass condensates are more likely to be detected at lower redshifts (~ 3 to 6) when large scale clustering of the smaller condensates might have occured.

Recently, the Westerbork group has put upper limits to the line flux density of about 10 mJy at 327 MHz ($z = 3.34$) at a resolution of about 1 arcmin (private communication). This corresponds to $M_{HI} \sim 10^{14}$ $M_\odot$ over a scale size of $\lesssim 0.06$ Mpc. Since many such clouds may exist in the field of view of a few square degrees of the Westerbork synthesis radio telescope, it may be possible to increase the sensitivity for the detection of a hierarchical distribution of low mass clouds by appropriate power spectrum analysis.

The detection of primordial HI clouds is a challenging observational problem. The searches made at the Jodrell Bank and the Cambridge Observatories have constrained clumping only on the largest scales of epochs corresponding to $z \sim 3.3$, 4.9 and 8.5. The Westerbork observations have put a much lower limit of $\sim 10^{13}$ $M_\odot$ for $z \sim 3.3$ but only over a few square degrees of the sky. For

constraining various theoretical models and in order to have a detailed understanding of the evolution of the primordial gas and its condensation, it would
be desirable to reach an order of magnitude higher sensitivity, say, a fraction
of a mJy at several discrete frequencies corresponding to z in the range of
about 3 to 10.

REFERENCES

Bebbington, D.H.O., 1986, M.N.R.A.S., **218**, 577.
Blumenthal, G.R., Faber, S.M., Primack, J.R., Rees, M.J., 1984, Nature, **311**,517.
Davies, R.D., Pedlar, A., Mirabel, I.F., 1978, M.N.R.A.S., **182**, 727.
Hogan, C.J., Rees, M.J., 1979, M.N.R.A.S., **188**, 791.
Oort, J.H., 1984, Astron. Astrophys. **139**, 211.
Peebles, P.J.E., 1980, in The Large Scale Structure of the Universe (Princeton:
    University Press).
Sunyaev, R.A., Zeldovich, Ya.B., 1972, Astron. Astrophys. **20**, 189.
Sunyaev, R.A., Zeldovich, Ya.B., 1974, M.N.R.A.S., **171**, 375.
Swarup, G., 1984, GT-ITR-001, pp. 1-60, Radio Astronomy Centre, P.B. 8,
    Ootacamund 643001, India.

DISCUSSION

DAVIES: Our recent Jodrell Bank measurements at 151 MHz (z = 8.4)
were most sensitive to protoclusters with velocity widths in the range
200 to 800 km s^{-1}. Protoclusters with masses greater than 3 x 10^{14}
solar masses would have been detected if they had a substantial
fraction of their mass in this velocity range. No such protoclusters
were detected, although we would have expected a few in the volume of
space sampled. I would comment that ideally a velocity width of more
than 2000 km s-1 should be sampled in this type of observation. Could
I ask the theorists what they consider would be the best redshift
interval for the protocluster search?

SWARUP: We are undertaking observations over bandwidths of ~10 MHz
corresponding to $\gtrsim$ 10,000 km s^{-1}. Perhaps Dr. Silk may have a comment
regarding your query.

SILK: We would expect to find protogalactic or protocluster hydrogen
clouds at the redshifts of the most remote quasars. This is where
we might expect to see galaxy formation. Indeed the only example
of a primeval galaxy is an object at a redshift of 3.2 found by
optical search adjacent to a quasar at the same redshift.

CHAPTER VI

THEORIES OF COSMOLOGY

ALTERNATIVE COSMOLOGIES

Jayant V. Narlikar
Tata Institute of Fundamental Research
Homi Bhabha Road
Bombay 400 005
India

ABSTRACT. This review highlights some of the cosmological theories proposed as alternatives to the standard hot big bang model. Specific ideas discussed here are the matter - antimatter symmetric cosmologies, the empirical two-component model, the G-varying cosmologies, the chronometric cosmology and a simplified quantum cosmology. It is argued that many alternative cosmologies have contributed useful concepts and offered observational tests that have enriched the field of cosmology as a science.

1. INTRODUCTION

Speaking at the Vatican Conference of 1970, Fred Hoyle [1] made the following comment on the state of physics and cosmology:
"I think it is very unlikely that a creature evolving on this planet, the human being, is likely to possess a brain that is fully capable of understanding physics in its totality. I think this is inherently improbable in the first place, but, even if it should be so, it is surely wildly improbable that this situation should just have been reached in the year 1970."
Hoyle's reposte was meant as a caution against the categorical claims made at the time about the state of the universe and the knowledge of physics they were based on. The details of the 1970 - argument are not important; but even in the relatively short span of a decade and a half Hoyle's caution has been vindicated. The attempts at unification of fundamental forces and their implications for the structure of the universe, the discovery of superclusters, strings and voids, the possible implication of dark matter for galaxy formation, etc. are fresh inputs into cosmology that did not exist in 1970. By a similar token one may question the finality of many of the statements that are being made today, at this conference and elsewhere.
Alternative cosmologies describe attempts to understand the universe through theories and models other than the standard one. The motivation for exploring alternatives comes from Hoyle's cautionary remark just mentioned, against putting all one's eggs in one basket. Even amongst the believers in the standard picture all except the hard

447

A. Hewitt et al. (eds.), Observational Cosmology, 447–459.

core would be willing to admit that the picture has still a number of
difficulties and enigmas to resolve, although they may attach varying
degrees of significance to them in a purely subjective judgment. To
name a few examples: the problem of singular origin, flatness and
horizon, the photon to baryon ratio, the absence of a workable theory
of galaxy formation, the apparent mismatch of the age of the universe
vis-a-vis the ages of its constituents, the large dimensionless ratios
of physical constants, etc. So the need for exploring alternatives
is justified.

There is another motivation for exploring alternatives to a
standard theory. The conflicting predictions of two rival theories act
as a greater stimulus to the observer than the predictions of just the
standard theory. Cosmology as a science stands to gain considerably
from such an exercise, whatever its outcome.

Here I shall review a few alternative cosmologies. To fix ideas
I have taken standard cosmology to mean relativistic models with a hot
big bang beginning, followed at some early stage by an inflationary stage
of transient duration, followed by a near $\Omega = 1$ Friedmann model. Is all this
really well established? Judging by the output of literature on the
early universe it seems to me that workers in the field have made up
their minds that whatever its drawbacks inflation did occur; it only
remains to decide when, how and for how long. Likewise, the proponents
of hidden mass subconsciously assume that the unseen component must be
so much as to make $\Omega = 1$.

Limitations of time and space necessarily make this presentation
sketchy. For details see references [2, 3]. Although my main brief
is to highlight theoretical ideas, I cannot do so without referring to
some relevant observations. I shall proceed in what I consider to be
an increasingly radical order of departure from standard cosmology.

2.   MATTER-ANTIMATTER SYMMETRIC COSMOLOGY

During the 1950s and 1960s Alfven and Klein produced cosmological models
with perfect symmetry between matter and antimatter [4, 5]. The baryon
symmetric plasma was supposed to have been separated by hydromagnetic
processes in the early universe. In reference [2] a detailed
discussion of these models is given.

The idea of a baryon symmetric universe was resurrected in the
modern framework of grand unified theories. In standard cosmology CP
violation in GUTs is used to generate a net baryon number. The intention
behind the gymnastics is to produce a photon to baryon number ratio
in the range $10^8 - 10^{12}$ where it remained essentially frozen to this
day.

However, as argued by Stecker and his colleagues [see for example,
Ref. 6] spontaneous soft CP violation within the context of specific
GUTs based on SU(5) and SO(10) gauge groups, leads to the universe
breaking into domains of predominantly matter separating domains
predominantly made of antimatter. In other words, on a sufficiently
large scale containing several domains the universe has no net baryon
number.

For details of these theories see [7]. Since photons on which
most astronomy is based treat matter and antimatter alike it is not
easy (except for 'local' observations within the Galaxy or possibly
the Local group) to say what the overall composition of the universe
is like. Steckar [7] reviews indirect evidence in the form of the
gamma ray background spectrum over $\sim 0.5 - 200\text{MeV}$ as well as the
antimatter composition of cosmic rays to make a case for the baryon symmetric universe. Future more direct tests of the theory may come from
observations of the cosmic neutrino background by underwater detectors.

3.   THE STEADY STATE THEORY

From its inception in 1948 till the mid-sixties the steady state
cosmology had provided a competing alternative to the big bang. Here I
shall be concerned more with Hoyle's approach using field theory [8]
than with the deductive approach of Bondi and Gold [9] using the
perfect cosmological principle. Contrary to the generally held view
(even by professional physicists and astronomers), the steady state
theory in Hoyle's version does not violate the law of conservation of
matter and energy. This can be seen almost straight away in the
C-field version outlined by Maurice Pryce [private communication, but
see Ref. 10] where the entire theory is derived from an action
principle and therefore, by Noether's theorem, obeys the conservation laws.
    The C-field appears as an extra scalar field on the right hand
side of Einstein's equations. Thus while matter may be created
continuously, the total energy and momentum of all physical quantities
stays conserved. The C-field itself has to modes. In the creative mode
there is an exchange of energy between it and other fields leading to
matter creation. In the noncreative mode, the universal expansion
simply dilutes the intensity of the C-field and there is no matter
creation.
    The unusual feature of this version of the steady state theory lay
in the fact that the C-field has negative energy and negative stresses.
In a somewhat different theoretical framework proposed by McCrea [11]
in 1951, the steady state expansion was achieved by intrinsic negative
stresses in the cosmological medium.
    Although these concepts were considered of doubtful validity by
the theoreticians of the 1950s and the 1960s, they have become respectable
in the 1980s. In the typical inflationary scenario the same de Sitter
expansion of the steady state theory is achieved by negative stresses
in the vacuum, produced by a phase transition. Moreover, the idea of
'cosmic baldness' in which the de Sitter expansion wipes away relic
inhomogeneities in the inflationary model is strongly reminiscent of the
same concept in the steady state theory [10].
    Although the perfect cosmological principle strait-jacketed
the universe in a steady format, the C-field version permitted radical
departures from it. In the bubble universe version [12] the presently
observable universe was seen as a tiny fraction of a large super-dense
steady state universe. While the creation of matter was supposed to go
on steadily in the latter, the former arose as a phase transition in

which the C-field switched to the non-creative mode. Again, this concept
of a Friedmann-like bubble in a de Sitter spacetime bears a striking
similarity to the inflationary models of today. [for a detailed comparison see Ref. 13].

The particle physicists were unhappy with the steady state model
because it implied nonconservation of baryons. That objection need
not hold today since the baryon number is no longer believed to be a
conserved quantity in the wider framework of grand unified theories.
It is worth mentioning that probably the first application of a result
from particle physics to astrophysics and cosmology is found in the
'hot universe' model of Gold and Hoyle [14]. Here the created matter
was in the form of neutrons. The creation of a neutron and its subsequent
decay generating high kinetic temperature led naturally to the idea of
superclusters of galaxies. Superclustering on a scale of 30-100 Mpc
was not viewed with favour by astronomers three decades ago: it is now
taken as an accepted feature of the observable universe.

It is evident therefore that the steady state model generated a
number of ideas that were considered too radical at the time they were
proposed but which have subsequently found their way in the standard
scenario in different contexts. The main observational difficulty
that the steady state model faced came, however, from the microwave
background radiation. The simplest and historically anticipated [15,
16] interpretation of this radiation is as the relic of a hot dense
primordial epoch in standard cosmology. This interpretation could not
be invoked in the steady state model except in its bubble version.

Is an alternative interpretation possible for this radiation? On
this question hangs the fate of not only the steady state cosmology but
also of any other cosmology that does away with a hot dense epoch.
Attempts in the past [17 - 20] based on thermalization of excess stellar
radiation by dust grains in intergalactic space have only partially
succeeded in providing an alternative to the relic picture. In
favour of these attempts is the result that the starlight generated by
making all the observed helium in stars or very massive objects would
have energy density comparable to that found in the microwave background
[17]. The detailed theory producing the correct spectrum and isotropy
still eludes us, however. This may be partly due to our imperfect
understanding of the intergalactic medium. We should also remember
that even in the standard picture, isotropy of the microwave background
presents a problem.

4.   AN EMPIRICAL APPROACH

The standard picture implies an evolving universe. Evolution is of two
kinds: (i) the change in the large scale geometry and the overall physical
characteristics of the universe and (ii) the change in the physical
properties of populations of discrete objects with the cosmic time.
What evidence do we have for evolution of either kind?

Except for the relic interpretation of the microwave background,
our studies of the universe take us to redshifts of $z \lesssim 4$, if the QSOs
are at cosmological distances and to $z \lesssim 2$, if we consider only galaxies.

Attempts to detect non-Euclidean geometrical effects of the first
kind by observations of discrete objects have always been foiled by their
mixing up with effects of the second kind. Moreover, the apparently
Euclidean effects seen in the log N - log S test and the θ - z test
for radio sources have been interpreted as a combination of both kinds
of evolution.

Burbidge and this author [21] therefore considered in 1980 an
empirical approach to cosmology wherein the universe was considered as
a two-tiered system. The larger structure (length scale $\sim$ 2 x 10^{30} cm,
time scale $\sim$ 3 x 10^{19}s) was provided by a radiation dominated universe
with general relativity deciding the dynamics while the shorter
structure (length scale $\sim$2 x 10^{28} cm, time scale $\sim$ 3 x 10^{17}s) was
provided by discrete objects whose distribution and dynamics were
essentially described with (i) Euclidean geometry and (ii) no
evolution. The shorter structure was interpreted as 'super supercluster (SSC)', and it was argued that the larger structure contains
numerous SSCs.

This model was consistent with all observations of discrete
objects as well as with the intensities of background radiations of
various kinds. The only objection to it, so far as I am aware, was
that the relatively small peculiar velocity of the Sun with respect to
the microwave background rest frame forces the Galaxy to be close to the
centre of the SSC.

Regardless of the validity of this approach, the question of the
evolution of discrete source populations needs to be examined ab-initio.
DasGupta et al [22] for example, have recently argued that the redshift-
flux density distributions in complete samples of radio galaxies like
the 3CR can be explained entirely in terms of non-evolving radio
luminosity functions.

5.   G-VARYING COSMOLOGIES

In spite of the successes and internal consistency of general
relativity, gravity is still on enigma at the microscopic level. For
this reason, many physicists have tried to venture into a field that
Isaac Newton himself was reluctant to tread (vide his famous remark
'Hypotheses non-fingo'.). In particular, the validity of the inverse
square law and the constancy of the gravitational constant G have been
questioned. Let us consider the spatial dependence first.

The Solar-System tests of Newtonian gravity and general relativity
inspire confidence in the validity of these theories at distances of
the order of a few astronomical units. The success of stellar structure
and evolution theory (solar neutrinos not-withstanding) provides
indirect test of validity over distances of 10^6 - 10^{12} cm. But what about
smaller and larger scales?

For example, the dynamical arguments deducing missing mass
(e.g. flat rotation curves, intra-cluster velocity dispersion, etc.)
could also be turned round to question the validity of the inverse
square law at galactic or clustering scale. So far as I am aware, no
alternative cosmology has emerged from such a radical interpretation.

More recently Fischbach et al [23] in a reanalysis of the classic
Eötvos experiment, have questioned the validity of the inverse square
law at the scale of a few metres. In particular, they suggest a
modification of the Newtonian potential energy of two masses m_1, m_2 at
a separation of r, from the value Gm_1m_2/r to

$$V(r) = -G \frac{m_1m_2}{r} (1 + \alpha e^{-r/\lambda}) \qquad (1)$$

where $\alpha = -(7.2 \pm 3.6) \times 10^{-3}$, $\lambda = (200 \pm 50)$m. These numbers have
been obtained empirically but can be explained, according to these
authors by a fifth force of repulsive nature coupling baryon number to
hypercharge.

Since the distance scale of this force, even if it is confirmed, is
only ~ 200 metres, it cannot be relevant to extragalactic astronomy
or cosmology

So far as alternative cosmologies are concerned the possible
temporal variation of G has generated considerable discussion.
Pioneering work in this field came from Dirac [24] in 1937. Dirac was
concerned with the largeness of large dimensionless ratios of physical
constants:

$$\frac{e^2}{Gm_p m_e} \sim 10^{40} , \qquad \frac{m_e c^3}{e^2 H} \sim 10^{40}. \qquad (2)$$

Here G and H (Hubble's constant) are macroscopic while e (electron
charge), m_e (electron mass), m_p (proton mass) are microscopic.
Arguing that it cannot be coincidental that two large dimensionless numbers
are comparable in magnitude, Dirac suggested that they are related.
Known as the 'large numbers hypothesis' it implies that since in an
evolving universe H is epoch dependent, so is G. More than three decades
later Dirac returned to this conjecture and gave quantitative formulation
[25, 26]. In particular, he considered two different time coordinates,
t_a for atomic processes and t_g for macroscopic gravitational processes
with

$$\beta = \frac{dt_g}{dt_a} \qquad (3)$$

an epoch-dependent quantity. Dirac offered two versions of cosmological
models, one in which there is additive creation and another with
multiplicative creation.

Later Canuto and others [27] offered a field theory, the so called
Scale Covariant Theory of Gravity (SCTG) for describing Dirac cosmologies, and have discussed extensively the observable consequences for
celestial mechanics of the Solar System, stellar and galactic evolution
and cosmology. The variation of G with epoch in these model is
usually of the form $G \propto \beta^n$ where n is a constant positive or negative
index.

The SCTG has been criticized by A.K. Kembhavai [28] on the
grounds that its scale invariance makes it indistinguishable from general
relativity and that the claimed observational differences are spurious.
This has been hotly debated by Canuto [29]. In any case it is clear
that the present formalism of SCTG does not clarify where and how the

scale-invariance symmetry is broken; nor does a unique G-β relation
emerge from the formalism of the theory.

The second path to G-varying cosmologies is via Mach's Principle.
Brans and Dicke in 1961 offered an interesting alternative to general
relativity [31] by starting from Mach's principle in the following way.
If we consider a typical Friedmann model we find the following relation
between the G, the characteristic cosmological distance R and M the
mass contained within it:

$$\frac{1}{G} \sim \frac{2M}{Rc^2} \quad \Omega^{-1} \sim \frac{M}{Rc^2} \; . \tag{4}$$

Brans and Dicke took this relation as determining G^{-1} through inertial
contributions of all the masses m_1 , m_2 ,... at various distances
r_1 , r_2 ,... in the universe:

$$\frac{1}{G} \sim \sum_i \frac{m_i}{r_i c^2} \; . \tag{5}$$

Thus G is not a constant but its reciprocal behaves as a scalar field.
The Brans-Dicke theory therefore starts with an action in which the
standard Hilbert term of general relativity is modified thus:

$$\frac{c^3}{16\pi G} \int R \sqrt{-g}\, d^4x \;\to\; \frac{c^3}{16\pi} \int \phi R \sqrt{-g}\, d^4x \tag{6}$$

where ϕ is a scalar field. To determine its dynamical behaviour
typical (the standard massless) scalar field Lagrangian is used
with an effective coupling constant ω. The field equations tend to
those of general relativity as $\omega \to \infty$.

The accuracy of the Solar System tests have already forced ω to
be larger than 500, thus making the Brans Dicke theory practically
indistinguishable from general relativity in the weak gravity limit
[2, 3]. However, cosmologically the theory can still offer differences
from relativity. In particular it can lead to $|\dot{G}/G| \sim \omega^{-1} H$ at the
present epoch. The present accuracy of radar/laser ranging of the
Moon and the nearby planets does not rule out G-variation of this order.
The sign of $\dot{G}$ depends on the model chosen. The theory also differs from
standard cosmology in the 'early universe' phase, e.g. in the primordial
production of light nuclei [2, 3]. Its consequences in even earlier
phases of the universe have not been explored so far.

Another Machian theory of gravity was proposed by Hoyle and the
author [31] in 1964 and further explored later [32 - 35]. It is
based on a conformally invariant action principle and its form can
be uniquely deduced from considerations of symmetry. The theory is
naturally expressed in action at a distance format but can also be recast
as a field theory. On the Solar System scale it does not generate any
new results vis-a-vis general relativity.

However, cosmologically it produces new results under two scenarios.
It has models in which G varies with epoch. Canuto and the author
[36] and canuto et al [37] have shown that the G-carying cosmology is
consistent with whatever cosmological observations presently available ,
such as the Hubble relation, source counts, angular sizes, gamma ray

background, etc.

Earlier Rana and the author [38, 39] had fitted a G-varying cosmology to the microwave background spectrum including the data of Woody and Richards [40]. The fit could be obtained within 1σ at both short and long wavelengths thus apparently doing better than standard cosmology. However, since then the Woody-Richards data have been revised [41] and the new curve appears to be consistent with standard cosmology.

What is the observational status of $\dot{G}/G$? Laboratory measurements are not yet capable of measuring the small effect predicted by most theories. The best-bet seems to be to study the motions of the Moon and the planets for apparent secular effects. Reference [3] summarizes the state of affairs till $\sim$ 1982. The lunar motion is complicated by tidal interactions and although there have been claims [see 42 for example] that a clear nonzero value of $|\dot{G}/G|$ is indicated by the data, they are contested. However, the range measurements between tracking stations of Deep Space Network and the Viking landers on Mars as analyzed by Hellings et al [43] seem to rule out most G-varying cosmologies. Again, the accuracy of planetary ephemerides on which this analysis is based may be questioned [44].

A somewhat more radical application of HN theory was pointed out by the author [45]; to explain the apparent anomaly in the redshifts of quasars. For, in certain cosmological models the theory predicts a steady growth in the inertial mass m of a typical particle. Furthermore, light propagation in these models leads to a relationship

$$1 + z \propto m^{-1} \tag{7}$$

between redshift z and the typical particle mass m in the object. If a quasar is made out of material recently fired from a galaxy, it will exhibit larger redshift than the parent galaxy although both are physical neighbours. Das and the author have shown how the theory can be applied to the observations of anomalous redshifts of quasar galaxy associations [46]. Of course, it must be stated that in spite of numerous claims of observed cases of noncosmological redshft [for a review see Ref. 47] the issue is still considered controversial.

6.   CHRONOMETRIC COSMOLOGY

Introduced in 1976 by Segal [48] this cosmology also involves two time systems, one 'local' and the other 'global'. Globally the cosmos is a spacetime with the topology R x S. That is, the time coordinate, t, is given by the real number line R while the space is the three dimensional surface of a 4-sphere. Locally, however, the time coordinate x_0 is Minkowskian and different from t. A cosmological observation like the measurement of the redshift of a distant galaxy can distinguish between x_0 and t. The mathematical part of the theory describes how to connect the local Minkowski spacetime with the global R x S type of spacetime. Physically, one has to consider the operation of physics in the global spacetime and then deduce its observable consequences in the local Minkowski spacetime.

For example, the wave equation describing photon propagation in this spacetime gives a redshift-distance formula that is quadratic rather than linear. Nicoll and Segal have claimed that the redshift magnitude relation for QSOs satisfies the quadratic law rather than Hubble's law [49]. Even for galaxies, Segal claims that quadratic law gives a better fit than the linear law [50]. This inference is debated on the grounds as to whether the effect is real or attributable to selection effects.

Segal [51] has also considered background radiations, in particular, the microwave background which is usually claimed as the best proof of a hot big bang. In Segal's model the universe being closed and nonsingular, radiation would circulate round and round and an equilibrium Planckian spectrum reached in which emission from sources is balanced by absorbers. The actual temperature would depend on the total energy available to thermalize. Segal gives qualitative argument to relate the ratio of starlight to thermalized radiation, to the number of circuits of the closed universe that the radiation has made. By contrast the X-ray background in Segal's theory is not Planckian because it has not yet been fully randomized. The argument here is that the X-ray photons being more energetic, cannot be so easily absorbed and hence more circuits of the universe are needed for them. This, however, raises the question of time scales. What is the age of the universe? If it is infinite, then should not all processes have reached equilibrium?

7. QUANTUM COSMOLOGY

This discussion so far has centred around classical theories of gravity. Would any difference be made in the outcome of this discussion if gravity were quantized? A simple argument can be given to show that quantum gravity becomes relevant at the so called Planck scales of length and time

$$L_p = \sqrt{\frac{G\hbar}{c^3}} \sim 1.6 \times 10^{-33} \text{cm}, \quad T_p = \sqrt{\frac{G\hbar}{c^5}} \sim 5.4 \times 10^{-44} \text{s}. \tag{8}$$

That is, if we were to follow the standard model back to the big bang epoch $t = 0$ then we can no longer trust its classical foundations for $t \lesssim T_p$. Hence the important question of cosmology 'Did the universe originate in a big bang?' remains shrouded in quantum gravity.

There are several attempts underway to quantize gravity [see Refs. 52 - 54 for sample reviews]. The nonlinearity of general relativity and the intimate relationship of gravity with spacetime geometry make the task of quantization immensely difficult. It would be out of place to spend time on formal approaches in a conference devoted to observational cosmology. In any case the approaches described in the above references have little bearing on the cosmological question posed above.

There is, however, a more simple-minded approach that leads to interesting cosmological conclusions. In relativity the geometry is specified by a symmetric metric tensor g_{ik} which has ten components. These are subject to four coordinate conditions, thus leaving essentially six continuum degrees of freedom. In a fully quantized theory all 6 $\times \infty$

degrees of freedom must be quantized. The most important of these, especially for cosmology are the conformal degrees of freedom that set the scale of the expanding universe. For example, if the metric of the Minkowski spacetime of special relativity is multiplied by a conformal factor Ω^2 where Ω is a function of time only, we get the Robertson-Walker spacetime with k = 0. Robertson-Walker spacetimes with k = ± 1 are also obtained this way provided Ω is a suitable function of cosmic time and the radial coordinate. The spacetime singularity of the big bang comes from Ω = 0.

We may generalize this notion to metrics of the form $\bar{g}_{ik}\,\Omega^2$ where $\bar{g}_{ik}$ is the solution of classical Einstein equations, and explore the consequences of Ω treated as a quantum operator. Then $(\Omega-1)$ is the quantum conformal fluctuation from the classical metric and we have a range of 'nonclassical' cosmologies in the era $t \lesssim t_p$. A wavefunctional $\psi(\Omega)$ describes the quantum evolution of the universe. In particular, if the universe came out of a singularity, it would be characterized by functions Ω that tend to zero. The probability measure $|\psi|^2$ of all such functions in a suitably defined function space gives us the probability that the universe came from a singular state.

Recent work by the author [55] has shown that this probability is vanishingly small. In other words, given the quantum regime of Ω , it can be asserted that almost certainly the universe did not originate in a big bang singularity. It may have passed through a stage where its characteristic linear size was as small as L_p; but it was unlikely to have been zero as implied by standard (classical) cosmology.

Padmanabhan has constructed cosmological models which explicitly take account of the dynamics of quantum conformal fluctuations and [56] which are nonsingular. For t $\gg$ T_p , These models merge into the standard models. It is also possible to construct stationary state models in which the conformal fluctuations keep the universe unchanging at a finite size $\sim L_p$.

The elimination of the singularity removes the horizon problem since the past light cone in a quantum universe is not terminated at t = 0. The flatness problem is also resolved [57] if we assume that the universe evolved from quantum conformal fluctuations of an empty Minkowski spacetime. For, it can then be shown that with almost unit probability the universe would go into the k = 0 Robertson-Walker model.

Thus, incredible though it may seem, some of the observable features of the present day universe can be traced to its seeds in the quantum era. The details of this approach are given in a recent review [58; see also 59].

8.   CONCLUDING REMARKS

This brief survey of alternative cosmologies by no means exhausts the full range of ideas in the literature. Nor does it offer any unique alternative to the standard cosmology. Rather, it is meant to highlight

the important possibilities, both theoretical and observational, that
exist in the present literature once one ventures beyond the child's
garden of standard cosmological models. As mentioned here, we have
instances of ideas once considered nonstandard, gaining respectability
later. Of course, one can cite older references of this happening right
from the days of Copernicus through successive stages in which the
establishment view on the Sun's position in the Galaxy, the extragalactic
nature of the nebulae, and the value of Hubble's constant had to be
revised.

Alternative cosmologies stimulate observational checks on the
universe whenever they come up with alternatives to the standard
predictions. If cosmology is to be treated as science and not the
religious dogma that it once was, alternatives to the establishment
views have to be encouraged.

While I am willing to admit the possibility of all alternative
cosmologies being proved wrong, I find it hard to take the complacent
view that in standard cosmology we are so close to solving the
profound problem of the universe that deviating from it is a waste
of time. After all, to quote J.B.S. Haldane "The universe is not only
queerer than what we suppose, it is queerer than what we can suppose."

References

1.   Hoyle, F., 1970, in Study Week on Nuclei of Galaxies, p. 757
     ed. O'Connell, D.J.K. (North Holland, Amsterdam).
2.   Narlikar, J.V. and Kembhavi, A.K., 1980, Fundamentals of Cosmic
     Physics, 6, 1.
3.   Narlikar, J.V., 1983, Introduction to Cosmology (Jones and
     Bartlett, Boston).
4.   Alfven, H. and Klein, O., 1963, Ark.Fys. 23, 187.
5.   Alfven, H., 1965, Rev. Mod. Phys. 37, 652.
6.   Serjanovic, G. and Stecker, F.W., 1986, Phys. Lett. 96B, 285.
7.   Stecker, F.W., 1982 in Progress in Cosmology, p.1, ed. A.W.
     Wolfendale (Reidel, Dordrecht).
8.   Hoyle, F., 1948, M.N.R.A.S., 108, 372.
9.   Bondi, H. and Gold, T. 1948, M.N.R.A.S., 108, 252.
10. Hoyle, F. and Narlikar, J.V., 1963, Proc. Roy. Soc. A, 273, 1.
11. McCrea, W.H., 1951, Proc. Roy. Soc. A, 206, 562.
12. Hoyle, F. and Narlikar, J.V., 1966, Proc. Roy. Soc. A, 290, 143.
13. Narlikar, J.V., 1984, J. Astrophys. Astron., 5, 495.
14. Gold, T. and Hoyle, F., 1958, in Paris Symposium on Radio
     Astronomy, p. 583, ed. R.N. Bracewell (Stanford University Press).
15. Gamow, G., 1953, Kong. Dansk. Ved. Sels. 27 No. 10.
16. Alpher, R.A. and Herman, R., 1949, Phys. Rev. 75, 1093.
17. Hoyle, F., Wickramasinghe, N.C. and Reddish, V.C. 1968, Nature,
     218, 1124.
18. Narlikar, J.V., Edmunds, M.G. and Wickramasinghe, N.C., 1976, in
     Far Infrared Astronomy, p. 131, ed. M. Rowan-Robinson (Pergamon
     Press, New York).
19. Rees, M.J., 1978, Nature 275, 35.

20. Rana, N.C., 1981, M.N.R.A.S., **197**, 1125.
21. Narlikar, J.V. and Burbidge, G.R., 1981, Astrophys. and Sp. Sc., **74**, 111.
22. DasGupta, P., Narlikar, J.V. and Burbidge, G.R., 1986, Preprint.
23. Fischbach, E., Sudersky, D., Szafer, A., Talmadge, G. and Aaronson, S.H., 1986, Phys. Rev. Lett., **56**, 3.
24. Dirac, P.A.M., 1937, Nature, **139**, 323.
25. Dirac, P.A.M., 1973, Proc. Roy. Soc. A, **333**, 403.
26. Dirac, P.A.M., 1974, Proc. Roy. Soc. A, **338**, 439.
27. Canuto, V.M., Adams, P.J., Hsieh, S.H. and Tsiang, E., 1977, Phys. Rev. D, **16**, 1643.
28. Kembhavi, A.K. and Pollock, M.D., 1981, M.N.R.A.S., **197**, 1087.
29. Canuto, V.M., 1981, M.N.R.A.S., **197**, 1093.
30. Brans, C. and Dicke, R.H., 1961, Phys. Rev., **124**, 125.
31. Hoyle, F. and Narlikar, J.V., 1964, Proc. Roy. Soc. A, **282**, 191.
32. Hoyle, F. and Narlikar, J.V., 1966, Proc. Roy. Soc. A, **294**, 138.
33. Hoyle, F. and Narlikar, J.V., 1972, M.N.R.A.S., **155**, 305.
34. Hoyle, F. and Narlikar, J.V., 1972, M.N.R.A.S., **155**, 323.
35. Hoyle, F. and Narlikar, J.V., 1974, Action at a Distance in Physics and Cosmology (W.H. Freeman, New York).
36. Canuto, V.M. and Narlikar, J.V., 1980, Astrophys. J., **236**, 6.
37. Canuto, V.M., Narlikar, J.V. and Owen, J.R., Astron. Astrophys., **92**, 26.
38. Narlikar, J.V. and Rana, N.C., 1980, Phys. Lett.A, **77**, 219.
39. Narlikar, J.V. and Rana, N.C., 1983, Phys. Lett. A, **99**, 75.
40. Woody, D.P. and Richards, P.L., 1981, Astrophys. J., **248**, 18.
41. Melchiorri, F. ed., 1985, The Cosmic Background Radiation and Fundamental Physics (Italian Physical Society, Bologna).
42. Van Flandern, T.C., 1981, Astrophys. J., **248**, 813.
43. Hellings, R.W., Adams, P.J., Anderson, J.D., Keesey, M.S., Lau, E.L. and Standish, E.M., 1983, Phys. Rev. Lett. **51**, 1609.
44. Rana, N.C., 1986, Preprint.
45. Narlikar, J.V., 1977, Ann. Phys. **107**, 325.
46. Narlikar, J.V. and Das, P.K., 1980, Astrophys. J. **240**, 401.
47. Narlikar, J.V., 1986, in Quasars: Proceedings of the IAU Symposium No. 119 eds. G. Swarup and V.K. Kapahi (Reidel, Dordrecht).
48. Segal, I.E., 1976, Proc. Nat. Acad. Sci. **73**, 669.
49. Nicholl, J.F. and Segal, I.E., 1978, Proc. Nat. Acad. Sci., **75**, 535.
50. Segal, I.E., 1980, M.N.R.A.S., **192**, 755.
51. Segal, I.E., 1985, in The Cosmic Background Radiation and Fundamental Physics, p. 209, ed. F. Melchiorri (Italian Physical Society, Bologna).
52. Isham, C.J., Penrose, R. and Sciama, D.W. eds., 1975, Quantum Gravity - an Oxford Symposium (Clarendon Press, Oxford).
53. Isham, C.J., Penrose, R. and Sciama, D.W. eds., 1981, Quantum Gravity - an Oxford Symposium II (Clarendon Press, Oxford).
54. Christenson, S.M., ed., 1984, Quantum Theory of Gravity (Adam Hilger, Bristol).
55. Narlikar, J.V., 1984, Foundations of Physics **14**, 443.
56. Padmanabhan, T., 1983, Phy. Rev. D, **28**, 745 and 756.
57. Narlikar, J.V. and Padmanabhan, T., 1983, Ann. Phys., **150**, 289.

58. Narlikar, J.V. and Padmanabhan, T., 1983, <u>Phys</u>. <u>Rep</u>. <u>C</u>, **100**,
     151.

59. Narlikar, J.V. and Padmanabhan, T., 1986, <u>Gravity, Gauge Theories
     and Quantum Cosmology</u> (Reidel, Dordrecht).

DISCUSSION

YU XIN: (1) Is there any observational evidence that matter is being
continuously created in the universe? (2) Do you think that there
might be intrinsic spin in spacetime? - If so then spacetime singularities might be avoided.

NARLIKAR: One could interpret the explosive outpouring of matter in
quasars and active galactic nuclei as evidence for matter creation.
Regarding intrinsic spin, it is hard to cite evidence. However, the
Einstein-Cartan type theories do describe torsion in spacetime by
having a non-symmetric affine connection. There are cosmological
models in such theories (e.g. those discussed by Trautmann) which
are nonsingular.

TOPOLOGY OF THE UNIVERSE

L. Z. Fang and H. J. Mo
Center for Astrophysics
University of Science and Technology of China
Hefei, Anhui
People's Republic of China

ABSTRACT.   Topology of the universe is the remains of quantum cosmology.
The theoretical and observational aspects for the topology of the universe are discussed to show that the significances of topology of the
universe in present observations can shed some light on the properties
of the universe in the quantum cosmological era.

1. INTRODUCTION

"Is the Earth flat or spherical?" was a vexed question for ancient
scholars. Today cosmologists face a similar vexation: what is the
shape of the universe?
      The difficulty of determining the shape of the Earth in the ancient
days was due to that the observable area was much smaller than the
whole surface of the Earth. Even in that time, several ancient scholars
justfied the spherical shape of the Earth. This conclusion was obtained
from the following two reasons:
      a) The curvature is locally observable. It was found that the
curvature radius of the Earth is finite.
      b)The surface of the Earth is homogeneous, namely, the curvature
is the same everywhere.
      Similarly two points can also be found in modern cosmology:
a) spacetime curvature of the universe is locally observable;
b) cosmological principle, i.e. the space time curvature is the
same everyzwhere in the universe. However, from the above-
mentioned two points we would not be able to find definite conclusion
on the shape of the universe, because the metric do not determine
the topology of the spacetime as a whole.
      For instance, the metric of a flat universe(k=0) is given by
$ds^2 = -c^2 dt^2 + R^2(t)(dx^2+dy^2+dz^2)$. The spacelike section of t=const.
is an infinite 3-dimensional Euclidean geometry, $-\infty < x,y,z < \infty$.
If we do the identification of points (x,y,z,t) with $(x+la_x, y+ma_y, z+na_z)$
for all integers l,m,n, the spacial section becomes then 3-dimensional
torus T^3, in which the spacial volume is finite. Different topologies
can be formed from a metric by different identifications. T^3 is one of
18 topologically different types of identifications in the case of flat

461

A. Hewitt et al. (eds.), Observational Cosmology, 461–475.
© 1987 by the IAU.

universe

The following statement can ofen be found in text books: the spacial
volume of open(k=-1) and flat universe are infinite, close universe(k=1)
is finite. In fact, this statement implies that the topology of the
universe is simply connected. But there is no reason, both observatioal
and theoretical, to show that the topology of the universe must be
simply connected. Therefore, a fundamental question in cosmology is:
what is the topology of the universe?

The topology of the universe is not important in many cosmological
topics, such as the origin of microwave background radiation, nuclear
synthesis, formation of galaxies, etc.. In discussing these problems, we
only need the cosmological principle as boundary conditions, while the
topology is neglegible.

The development of quantum cosmology has made that the topology of
the universe would not be neglectable. The aim of quantum cosmology is
to study the universe in the Planck era, in which the main process was
the formation of spacetime itself. i.e. spacetime as a whole becomes
dynamical variable in quantum cosmology.

On the other hand, the theory of the topology of the universe is
testable Since Einstein's equations of general relativity determine
only the metric of spacetime, but not the topology, it means that the
topology of spacetime is invariant in the period described by classical
gravitation. Therefore, after the Planck time, i.e., the era of the
formation of spacetime topology, the topology never changes; the topology
observed in present universe is the same as that in Planck era, i.e.,
the era of cosmological age $\sim 10^{-43}$ sec.. In other words, cosmological
topology is the remains of Planck era.

The significance of cosmological topology can be expressed by
Table 1, in which we list all the important remains and eras of their
formation. Cosmology studies the evolution of the universe from the

TABLE I. Cosmic Remains of Various Eras

Remains	Age of the universe
Topology of spacetime	10^{-44} seconds
Matter-antimatter asymmtry	10^{-34} seconds
Abundances of elements	3 minutes
Microwave background radiation	10^{5} years
Objects with large red-shift	$\sim 10^{10}$ years ago

remains left over various cosmic eras. Therefore, the topology problem
may be the key of studying the universe at the earliest era, i.e. the era
of creation of the universe. From Table 1 one can also find a systematic
property: the earlier the remains, the stronger(or harder) the associated
interaction; the chronological order of the remains is particle-nuclear-
radiation-galaxy, which corresponds to the order of interaction intensity,

strong-electroweak-gravitation. Spacetime is the hardest. In most
problems of physics, spacetime is considered as a fixed plateform on
which all physical processes take place. This is the same as to assume
that spacetime is the hardest among all physical entities. We should
therefore search for the remains of Genesis in the properties of spacetime. This is just the topology of the universe.

2. POSSIBLE TOPOLOGIES OF THE UNIVERSE

Immediately after Einstein gave the cosmological model which has S^3 as
the configuration of the space, Klein noticed that it is equally possible
for the space to have P^3 , the 3-dimensional projective space as the
configuration. But the fact is that models with S^3 topology have
received far more considerations than those with P^3 topology. The reason
for this comes more from human's prejudice than from physical considerations. This preference eliminates from our interests many models that
are as possible to be a cosmological model as the ones we prefered.

The number of the candidates is very large, as matter of fact,
mathematicians have given an almost complete classification of the
topologies for 3-dimensional spaces(Scott 1983), before physical considerations, they are all possible candidates for the 3-dimensional spacelike hypersurfaces of spacetime, so we in fact have an infinite number
of candidates.

In cosmology, model chooing must rely on physical considerations
and observations. There are many physical considerations, such as the
orientability and the differentiability of the spacetime, etc., we will
give here physical considerations related to the operations of identifications to give various topologies. In static spacetime, the identifications can be made at any time, because the geometry as well as the
topology does not change with time. But for a model with expanding as
our universe is, in order that the identification at one time be the
same as that at another time, the universe must be expanding conformally
or the identification is made in some special way. Three classes of
models of this kind is of special interest they are the ones with
constant curvature. The one with positive curvature k=1(class I) is
locally diffeomorphic to S^3 , the one with null curvature k=0(class II)
is locally difeomorphic to E^3 and the one with negative curvature k=-1
(class III) is locally difeomorphic to H^3 . They correspond to the
density parameters of $\Omega > 1, =1, <1$ respectively. the metrices for these
three classes are given by

$$dl^2 = dr^2 + f^2(r)(d\theta^2 + \sin^2\theta \, d\varphi^2) \tag{1}$$

where

$$f^2(r) = \begin{cases} \sin^2 r & \text{(class I)} \\ r^2 & \text{(class II)} \\ sh^2 r & \text{(class III)} \end{cases}$$

As is the mathematical result, all spaces of the three classes can

be classified(not completely) topologically according to the discrete
subgroups of the isometry groups of S^3, E^3, H^3 respectively; they are
the quotient spaces of the kinds S^3/Γ , E^3/Γ , H^3/Γ respectively
with Γ the discrete subgroups acting without fixed points. It is equally
to say that they can be obtained from properly identifying points in S^3
E^3 and H^3 .

Class I. This class is obtained from properly identifying points
in S^3 , so one can write compactly the class as $\Sigma_1 \approx S^3/\Gamma_1$ where Γ_1
is the discrete subgroup of the isometry group SO(4) acting without
fixed points. Noticing that SO(4) is diffeomorphic to $SO(3) \times SO(3)/Z_2$.
So if we have the discrete subgroups for SO(3) , we can find Σ_1 by
binary identifications in both SO(3) group manifolds. It is well known
that there are 7 kinds of such discrete subgroups for SO(3) , they
are,Z_2, the 2-cyclic group; Z_n (n>2), the n-cyclic groups; D_m (m>2), the
dihedral groups; T, the regular tetrahedron groups; O, the regular
octahedron group; I, the regular icosahedron group. So there are infinite
number topologies for Σ_1. To see the identifications more clearly, one can
represent SO(3) manifold diagrammatically as follows: choosing two
balls S_1 , S_2 of radius π with antipodal points on each surface identified.
Rotation in the direction (θ, φ) by $\alpha < 2\pi$ is given by a point with radical
coordinate α and azimuthal coordinates (θ, φ) in S_1 , while rotations in
the direction of (θ, φ) by $2\pi \leq \alpha < 4\pi$ is given by a same point in S_2 .
Rotation with $\alpha = 4\pi$ will return to the original point. If one identifies
tow points on S_1 and S_2 with α-coordinates differ by 2π , one thus
obtains P^3 which is diffeomorphic to SO(3) manifold. So the two balls
thus constructed is diffeomorphic to S^3 , the identifications in S^3 can
be achieved(topologically) by binary identification of points in the two
balls.

Because of the topology complexity, it is very difficult to find
for all cases the coordinates of the identified points in terms of
(r, θ, φ) in equ.(1). But for the simplest case with P^3 configuration,
which is the only one in this class that has the same symmetry as S^3 ,
the identification can be found. If one considers S^3 as the 3-sphere in
4-dimensional space with equation $x^2 + y^2 + z^2 + w^2 = 1$, then P^3 is the identification $(x,y,z,w) \equiv (-x,-y,-z,-w)$ on S^3 , or, in terms of intrinsic
coordinates, the identification of $(r, \theta, \varphi) \equiv (\pi-r, \pi-\theta, \pi+\varphi \pmod{2\pi})$ on
S^3 .

Because of the compactness of the space in this class, multiple
images can even occur in S^3 case. As matter of fact, because the geodesic
for light is the great cycle on S^3 , it would be possible to find a
conter image in the opposite direction to the objects. But it has been
proved that for an expanding universe, is is impossible for such a conter-
image to occur in S^3 case, because in this case the horizon size will be
smaller than the length of the semicycles. But it is possible for P^3
case, if the deceleration parameter q_0 at present time were greater
than 1, there would be one or more images for one object(Narlikar and
Seshadri 1985).

Class II. This class is obtained by properly identifying points
in E^3 . The spaces in this class can be written as $\Sigma_2 \approx E^3/\Gamma_2$ where Γ_2
is the discrete subgroups of the isometry group for E^3 , that is $R^3 \times SO(3)$
acting without fixed points. So if E^3 is considered to be an infinite

crystal constituted of lattice cells, then class II can be obtained as identifications of points on the opposite faces of each cell(sometimes with a screw). The classification of this class is complete, there are 18 kinds of Γ_2, so we can have 18 different kinds of topologies. The first group consists of six compact, orientable members. They are the ones obtained by proper identifications of points on the opposite faces of rectangular lattice that is (i) identification of opposite surfaces that gives T^3, the 3 torus. (ii) identification of opposite faces, with one pair screwed by an angle π, (iii) identification opposite faces with one pair scewed by an angle $\pi/2$, (iv) identification opposite faces with all pairs screwed by an angle π and those obtained from the identi-fications of points on the lattice made by translating a hexagonal lattice plane a certain distance perpendicular to the plane, that is (v) iden-tification of the two hexagonal faces with one screwed by an angle $2\pi/3$ with respect to the other (vi) identification the two hexagonal faces with one screwed $\pi/3$ with respect to the other The second group consists of four compact, nonorientale members they can be obtained in the same way as in cases (i)-(iv) but with some reflections. The third group consists of four orientable but noncompact members, it is obtained from identifications(with screws) in less than three pairs of lattice faces. They are (i) type $\mathcal{E} = E^3$, with no identification at all, (ii) type $\mathcal{S}_i^\theta$, with Γ_2 generated by a translation and a screw motion of an angle θ, (iii) type $\mathcal{J}_1$, with Γ_2 generated by two independent translations and (iv) type $\mathcal{K}$, with Γ_2 generated by a translation and a screw motion of an angle π in the direction perpendicular to the translation. The four non-orientable, noncompact manifolds can be obtained in the same way. It should be pointed out that the only one which preserves the maximal symmetry $R^3 \times SO(3)$ is E^3 itself. others are homogeneous but anisotropy because all identifications can not commute with the rotational group $SO(3)$.

The coordinates of the identified points in E^3 for some cases can be found without great difficulty. For example, in order to get T^3-topology, one can identify point (x,y,z) in E^3 with points $(x+nL_1, y+mL_2, z+lL_3)$ with (n,m,l) the lattice numbers and (L_1, L_2, L_3) the lengthes of the rectangular lattice.

One reason for considering models in this class is that the infla-tion in the very early universe gave a flat spacetime, and, as will be discussed in the next section, compactness of the 3-space is prefered by quantum cosmology. If such is the case, we will inevitably have one of the six manifolds in the first group as our candidates. But they are anisotropic, so one must be very careful about the effects of such an anisotropy, when one considers an inhomogeneous model.

Whether or not one or more images will be observed for an object depends on whether or not the present horizon is larger than the smallest length of the lattice. But theoretically, one can choose so small a length scale for the universe as to have one or more images for one object, we do not face the problem as in class I.

Class III. This class can be obtained by properly identifying points in H^3. But the classification has not been completed in mathe-matics. There are three ways to proceed. The first is given by Löbell, Löbell(1931) found that H^3 can be fitted just once by the fundamental

region formed by 14-sided figure with two sides are regular rectangular
hexagons and the rest twelve rectangular pentagons. So we can find non-
trivial topologies by properly identifying points in these fundamental
regions. The second is given by noticing that the metric of 3-space
with constant negative curvature can locally be written as(Ellis 1967)

$$d\sigma^2 = dr^2 + ch^2 r d\tau^2 \tag{2}$$

where $d\tau^2$ is the metric on the 2-surface H^2 with constant negative
curvature. So the total 3-space can be written as $M^3 \approx R \times H^2$. Now the
identification on H^2 can be obtained, for example, by attaching hundles
on the surface, we thus have a group of 3-space with constant negative
curvature in the form of direct product. The final way is given by the
"translation" or "translation plus screw motion" on H^3 . As matter of
fact, H^3 can be globally imbedded in Minkowski space F^4 ; it is invariant under a simply transitive group of Bianchi type V and a 1-
parameter family of simply transitive group of Bianchi type VII_h . So one
can find the discrete subgroups of these isometry groups and obtain the
quotient space by these discrete subgroups. There are infinite number of
3-space in this class.

     Astronomical observations made so far gives an open universe($\Omega \lesssim 1$),
this plus the compactness of the three space required by quantum cosmology
(as is mentioned before) lead inevitably to a space with nontrivial
topology, this is one reason to study models in this class. The determinations of the coordinates for the identified points are extremely
difficult. As a simple example, Fagundes(1985a) considered an unrealistic
model with topology $H^2 \times E$, and found the images in the 2-dimensional
hyperbolic geometry H^2. As for the identification in H^3 , so far as we
know, researches are remaining to be done.

     We have just given a relatively detail list of 3-spaces which are
made from properly identifying points in the maximally symmetrical
spaces S^3 , E^3 , H^3 . The list is far from complete. There are many other
alternatives which can not be excluded from our candidates(for a more
complete list, see Fagundes(1985b)). The list given above only provides
 an illustration of how the constructions of models with different
topologies proceeds and of how many alternatives we can have for the
models of the universe.

3. CONSTRAINTS ON TOPOLOGY BY QUANTUM COSMOLOGY

We have seen in the last section how large a number of variants can be
for the candidates of the models of the universe. But we believe that
our universe exists with a certain specific topology. Now one question
is why our universe has such a topology not the other. To answer this
question, we therefore first turn to the observational results. That is
the candidates chosen must be consistent with the observations we have
for the universe today. The observational significances of topology of
the universe, or reversely speaking, the constraints on the possible
topologies of the universe by observations will be our main concern for
our latter sections. In this section we will consider the problem theo-

retically, that is the constraints on topology by quantum cosmology.

If we speak quantum cosmology in the sense of quantum events on a background spacetime, the constraints come from the instability of some background spacetimes. For example, some authors have considered the instability of de Sitter space, they found that quantum corrections of the energy-momentum tensor causes an instable de Sitter space; only anti-de Sitter space is stable under the quantum correction. Although the constraint on the topology(especially the multiply connectedness) of the space is very loose, it indeed sheds light on the problem. Remember that in the nontrivial topology, there is a nontrivial constraint to the energy-momentum tensor of the fields from the gravitational casemir effect. The instability is thus related to the global properties of the spacetime.

If one considers quantum cosmology in the sense of quantum creation of the universe, interesting constraints can be found.

First we discuss the problem in a broad sense, that is, we admit that the state of the universe is given by a wavefunction(cosmic wave-function) which gives the probability amplitude for the universe to distribute with respect to some characteristic parameters. Topology being a very important property of the universe, so quantum cosmology, intuitively speaking, will give a choice of the topology. The dynamics of the wavefunction is given by Wheeler-DeWitt equation, the quantum cosmological analog of the Schrodinger equation. So one can find the possibility for the universe to transite from the quantum euclidean era (motions with imaginary time) to the classical lorentzian era. The amplitude for such transition is given by the WKB wavefunction $\exp(-|I_{cl}|)$ where I_{cl} is the euclidean action on the classical trajectories. Then if one wants to find the possibility of a universe with a certain topo-logy, one must know how the topological effects can enter the action I_{cl}. Since I_{cl} is the integral of lagrangian density over the spacetime manifold, global properties of the spacetime that is implied in the action is the volume of the universe. For pure gravitational field, $I_{cl} \propto V^{2/3}$ where V is the volume of the universe(t=const.). One immediately find that the wavefunction is zero for $V \to \infty$, that means that the universe with infinite large volume can not occur from the quantum tuneling. One result of this is that, universe with null curvature(k=0) and negative curvature(k=-1) can occur with nonzero possibility only if it exists with nontrivial topology so that the space is compact. As mentioned before, models with k=0,-1 are preferred by both observations and infla-tionary theory, therefore nontrivial topology seems to be inevitable in this sense.

In the consideration of the transition, one must have in the field configuration a spacelike hypersurface which separates the field confi-guration into euclidean and lorentzian sections. This gives strigent constraints on the form of classical solutions of the model, this in turn gives some constraints on the topology. For example, if one con-siders the isotropic expanding universe E^3, if the effective cosmological constant Λ is positive, the expanding factor will be proportional to $\exp(Ht)$ with $H^2 = \Lambda/3$. Although $t \to -\infty$ is a singularity, there is no point in the scale factor that can separate the euclidean region and lorentzian region, the universe could then create from the classical singularity at $t \to -\infty$. If one considers such a model with T^3-topology, the space will be geodesically complete, so there is no singularity at $t \to -\infty$. In these cases, there is no sense for the quantum creation. But if one takes into account the gravitational casemir effect of the matter fields due to the nontrivial topology, the situation will change. For example, in the T^3-topology, the contribution of the casemir effect to the energy-momentum tensor will lead to a classical trajectory with the form $a(t) \propto (ch2Ht)^{1/2}$ (Zeldovich, et al. 1984), there indeed exists a minimum a_{min} for a, in this case the region $0 < a < a_{min}$ corresponds to the euclidean era and the region $a > a_{min}$ corresponds to the lorentzian region. There is no problem for the interpretation of the quantum creation of the universe. It concludes that if one insists on the quantum creation of the universe, there are some constraints on the topology.

Now we turn to a specific quantum cosmology theory which is invented by Hawking and his colleagues. The corner stone in this theory lies in the ground state proposal of Hartle and Hawking(1983) which states that the ground state wavefunction for the universe (Σ, h_{ij}, ϕ) is given by the path integral over all compact manifolds M(compact euclidean space after continuation) which has Σ as its boundary with metric h_{ij} and matter field ϕ on it. One can see that the compactness of M poses very strigent constraints on the forms of the classical solution for the models chosen, these constraints are the same as that posed by the existences of the barrier separating the euclidean and lorentzian regions, so the early discussion still holes in this case. There are other two questions in this proposal:

(i) for a given Σ, does M always exist with Σ as its boundary?

(ii) For a field on Σ : $\phi : \Sigma \to \bar{\Phi}$, can the field be extended to all of M such that there exists an induced mapping $\bar{\phi} : M \to \bar{\Phi}$?

The answer to the first question is positive if one considers a 4-dimensional spacetime, since for any 3-dimensional closed manifold Σ, there exists a 4-dimensional manifold M with its boundary diffeomorphic to Σ. But if one considers high-dimensional models, one can find manifolds which are not boundaries of any manifold. These cases should be excluded by the qantum cosmology.

The answer to the second question is not certain. For example, the conjectures made by Mkrtchyan(1986)(proved for some cases by Li(1986)) shows that if the topology of Σ is S^3 and the matter field assumes values on the group G, then the mapping $\bar{\phi} : M \to G$ which satisfies the conditions implied in Hartle-Hawking's proposal exists if and only if the solitonic charge of the mapping $S^3 \to G$ is zero. But this solitonic charge corresponds to the baryonic number in the model. So being zero is a contradiction. One way out of this may be the considering of spacetime manifolds with complicated topologies, but whether or not they can give nonzero baryonic number remains to be proved. The discussion here only serves to illustrate how quantum cosmology can give constraints on the topology of the universe.

We have seen that quantum cosmology indeed gives some interesting constraints(although very loosely) on the topology of the universe. The purpose of writing this section is not so ambitious as to give an affirmative answer to the question, but rather to show that, as to the choice

of topology of the universe, we can no longer rely on our costums or
prejudice, there are theories(the quantum cosmology) by our hands.

4.   OBSERVATIONS OF THE COSMOLOGICAL TOPOLOGY

As mentioned in the introduction, the spacetime topology is not a
dynamical variable in classical gravity. Topology is an invariant after
the Planck era. Therefore, the topology of the present universe is the
same as that formed by the event of the birth of the universe. The
prediction on the topology given by quantum cosmology may then be testable by observations.

Up to now quantum cosmology would not be able to predict the detail
of the topological type, but only on whether or not the universe is
multiply connected. So we limit the discussion on the observable properties which can be used to distinguish simple and multiple connecticity
of the cosmological space. In a simply connected space, the geodesic
between any two points is unique, namely, each object can be observed
only in one direction. While in a multiply connected space, the geodesics
between two points are sometimes multiple, namely, each object can be
observed simutaneously in several different directions. Let uz consider
a two-dimensional torus which can be constructed from a plane by identifying point (x,y) with points (x+la,y+mb) with all integers l and m.
An observer in such a torus will find that the observed picture is the
same as a plane, but the distribution of matter is periodic with "wavelength" a in the x-direction and b in the y-direction, and the picture
is like a plane lattice.

Is the above-mentioned periodicity in the distribution of matter
observable? It depends on the length scale of the horizon and of the
size of the universe. In a compactified space, the size of the universe
is given by the distance L between the most widely separated points. In
T^3 universe, $L \sim R(t)a_x, R(t)a_y, R(t)a_z$. Therefore, global properties which
are due to the effect of topology should have a length scale larger
than L. On the other hand, the scales of all observable properties
should be smaller than the horizon L_H. So the necessary condition for
the topological effects to be observed is

$$L_H > L \tag{3}$$

Since $L_H \sim cH_0^{-1}$, we have

$$L < cH_0^{-1} \tag{4}$$

This condition can also be obtained in the field theoretic regime.
Let $\mathcal{E}_t$ be subspace of synchonous space Σ_t. Let Φ_1 and Φ_2 be two fields
defined on Σ_t and identified on $\mathcal{E}_t$:

$$\Phi_1(\vec{x}) = \Phi_2(\vec{x}) , \qquad for \ \vec{x} \in \mathcal{E}_t \tag{5}$$

But they can differ with each other on $\Sigma_t - \mathcal{E}_t$. If the observer can
determine the properties of Φ_1 and Φ_2 only through interactions on $\mathcal{E}_t$.

Then there is no way to distinguish $\overline{\Phi}_1$ and $\overline{\Phi}_2$. The question is whether $\overline{\Phi}_1$ and $\overline{\Phi}_2$ can be defined on spaces with different topologies. If it is the case, the observer can not rely on the $\overline{\Phi}_1$, $\overline{\Phi}_2$ interactions on $\mathcal{E}_t$ to determine the topology. It has been shown that, in general, Φ_1, Φ_2 can be defined with different topologies(Unwin 1982), hence for one to determine the topology through interactions of Φ_1, Φ_2, the interacting region $\mathcal{E}_t$ must be equal or larger than Σ_t. $\mathcal{E}_t$ is nothing but the horizon, so the condition is the same as eq .(3).

In the case of eq .(4), the light rays cross the whole volume of the universe more than one time, an object can be seen in more than one directions.

An important result in observational cosmology is that eq .(4) can not be ruled out by observations. The lower limit of L can be obtained from the fact of no image twins in some surveys of galaxies,such as Shane-Wirtanen sample(Shane and Wirtanen 1967) which gives $L \gtrsim 200 h_0^{-1} Mpc$, where h_0 is the Hubble constant in unit of 100 km/s.Mpc. This lower limit is smaller than the present horizon L_H, which is about $3000 h_0^{-1} Mpc$. It concludes that present observations do not exclude the possibility of the multiply connecticity of the cosmological topology.

5. POSSIBLE EVIDENCES FOR MULTIPLY CONNECTED TOPOLOGY

The lower limit of the cosmic size given in last section is even smaller than the distance of objects with high redshifts, such as quasars. Therefore, from the distribution of quasars one can already find some evidences which seems to show the multiple connecticity of the cosmic space.

5.1. Periodicity in the Distribution of Quasar Redshifts

Resently, it has been demonstrated from the statistical analysis of the observational data that the distribution of the emission line red shifts of quasars have a periodic feature with respect to the argument $x \equiv F(z, q_0)$ defined by(Fang 1982)

$$F(z, q_0) = \int_0^z \frac{dz}{(1+z)(1+2q_0 z)^{1/2}} \tag{6}$$

where q_0 denotes the deceleration parameter. In the distribution of the red shifts there exists a set of peaks at z_n given by

$$F(z_n, q_0) = An + B \tag{7}$$

where n is zero or positive integer and A and B are constants.

This above periodicity might be interpreted as the existence of a large scale periodic perturbation in the density distribution of cosmic matter. According to this interpretation, A is related to the "wavelength" of such periodic perturbation as(Fang 1982)

$$A = H_0 \lambda_0 / c \tag{8}$$

the subscript 0 denoting the present value. From the statistical analysis, λ_0 has been estimated as

$$200h_0^{-1}\,\mathrm{Mpc} \lesssim \lambda_0 \lesssim 600h_0^{-1}\,\mathrm{Mpc} \qquad (9)$$

The constant B in eq .(7) is related to our position relative to the perturbation waves. However, it is very difficult to explain why B is not random depending on direction. If we were located in a preferred position near center of the large-scale spherical wave perturbation, the above relation would be explained but such interpretation would be not acceptable from the point of view of the cosmological principle.

If we insist on the simply connected universe, it will be very difficult to overcome this difficulty. But such a type of result can be obtained natually if we assume the multiply connected universe.

Let us consider the flat universe with topology T^3 for a simple example(Fang and Sato 1983). The redshifts of the multiple images of the source located at x_S in the x-direction is given from

$$x_S + na_x = \int_{t_n}^{t_0} \frac{c\,dt}{R(t)} \qquad \text{and} \qquad z_n + 1 = \frac{R(t_0)}{R(t_n)}$$

as

$$z_n + 1 = \left[1 - \frac{R(t_0)x_S}{2(c/H_0)} - \frac{R(t_0)a_x}{2(c/H)}n \right]^{-2} \qquad (10)$$

From eq .(7) for $q = 1/2$, $A = R(t_0)a_x/(c/H_0)$ and $B = R(t_0)x_S/(c/H_0)$. Considering the $R(t_0)a_x/2 \sim 400\,\mathrm{Mpc}$ and $R(t_0)x_S < a_x/2$, the quasars with $z < 3$ are the original image($n=0$) or the ghosts of $n=1$ or $n=2$.

In the case of the whole-sky correlation, however, the situation is more complicated. For example, in the case of $a_x = a_y = a_z = a$, the units of space periodicities are Ra, $\sqrt{2}Ra$, $\sqrt{3}Ra, \ldots$, depending on the observed directions. Then, the superposition of the two periodicities of Ra and $\sqrt{2}Ra$ produces an approximate periodicity of $0.5Ra$. In the actual situation, the observed periodicity may be the superposition of a few fundamental space periodicities.

In spite of the above ambiguity, we can predict as a general tendency that the "wavelength" of periodicity for the quasars in the given direction should be larger than the "wavelength" estimated from the whole-sky data. This result might be used to interpret the following observational evidence: the wavelength of the whole-sky data is about a half of the wavelength estimated from the quasars listed by Savage and Bolton(1979), which are the quasars in the two given directions of the south Galactic Polar region of $02^h00^m, -50°00'$ and $22^h04^m, -18°55'$.

5.2. Periodicity in the Distribution of Absorption Line Redshifts of Quasars

If the absorption lines in the quasar spectrum are due to absorptions of intervening objects, the distribution of absorption line redshifts should also show periodicity in a multiply connected universe.

It has been found(Chu, Fang and Liu 1984) that there are several peaks in the number distribution of absorption lines with respect to β given by

$$\beta = \frac{(1+z_{em})^2 - (1+z_{ab})^2}{(1+z_{em})^2 + (1+z_{ab})^2} \qquad (11)$$

where z_{em} and z_{ab} are the redshifts of emission and absorption, respectively. The peaks in β distribution were explained by so called "line-locking" mechanism. But to us, the meaning of such peaks is to provide more possibility of finding evidence of multiple connecticity. Indeed, the distribution of the peaks seems to be periodic. From the locations of the peaks, one can find the size of the universe L by the same way as that in the emission line case. An interesting result is that the size of the universe determined by β-distribution is about the same as that determined by emission line distribution.

5.3. Close Pairs of Quasars

It has been shown(Burbidge, Narlikar and Hewitt 1985) that the possibility of finding so many close pairs of quasars with different redshifts by chance projection is as small as $\lesssim 10^{-4}$. It seems too low to support the belief of cosmological origin of the redshifts. However, in a multiply connected universe, original images and its ghost images can locate in about the same direction. Considering this mechanism, one can show, close pairs of quasars do not contradict with cosmological hypothesis, it may be an evidence for multiply connected topology of the cosmological space.

In a simply connected universe, the probability of finding pairs with angular separation $< \theta$ by chance projection is given by

$$<s>_c =2.4 \times 10^{-7} \; \Gamma(<m) N f^2 \theta^2 \qquad (12)$$

where $<s>_c$ is the expected number of close pairs with angular separation less than θ, $\Gamma(<m)$ is the sky density of quasars brighter than magnitude m, expressed in units of $(arcdeg)^{-2}$, N is the total number of quasars listed in catalogues which are brighter than m, f indicates the fraction of such quasars whose fields have been searched for close companions out to θ.

In a multiply connected universe, the probability $<s>_c$ should be enhanced by a factor

$$\mathcal{J} = (1+\delta)^{l_m} \qquad (13)$$

where δ is the amplitude of inhomogeneity in quasar distribution, l_m is the the maximum order of ghost images. For weak clustering of quasars, $1+\delta \sim 1.3-1.5$. For the λ_0 given in eq .(9), $l_m \sim 3-5$. Therefore, the enhancement factor $\mathcal{J}$ is about four, which relax strongly the contradiction of the observed number of close quasar pairs with the expect number of eq .(12).

5.4. Association between Quasars and Galaxies

Since 1971, after the first finding of bright galaxies associated with
3CQSO, many statitical analyses have been done to check a significance
level of association between quasars and galaxies. Some analyses show
a positive correlation(Chu and Zhu 1983; Seldner and Peebles 1979). It
is very difficult to explain why the objects with such different redshifts are correlated in their positions. A possible way out of this
problem also lies in the assumption of multiple connecticity. In the
multiply connected universe, we can sometimes see the original image
and its ghost images in the same direction. If we see some clump where
the formation of quasar and galaxy were active, the association in
appearance between the small redshifts galaxies of the original image
and the large redshift quasars of the ghost images will result naturally.
Since the quasar is thought to be a short-lived phenomena lasting only
a few million years, we will see only one image of the quasar if $R_H \sim$
400Mpc and $z < 3$. According to this interpretation, the associated
regions may provide us a good sample for studying the evolutionary
relationship between galaxy and quasar: the remnents of the quasars
should belong to the same clump of the associated galaxies.

5.5. Length Scale of Clustering

In a simply connected topology, the length scale of clustering is admis-
sible,in principle, up to t_0 , the size of the horizon $cH_0^{-1} \sim 3000h_0^{-1}$Mpc.
However, in a compactified universe, the upper limit of length scale of
clustering is given by the size of the universe. Namely, if the result
of eq .(9) is correct, we can then predict that no clustering has scale
larger than $\sim 400h_0^{-1}$Mpc.

The largest clustering found today is superclusters which have the
length scale of about $100h_0^{-1}$Mpc. The scale of voids is also less than
$100h_0^{-1}$Mpc. It has been found from the distribution of quasars that the
clustering of quasars is much weaker than galaxies. All these seem to
show that the prediction on the evidence of a cut-off in the clustering
scale can be accepted by present observations. More conclusive result
could be obtained in the near future from the Space Telescope.

5.6. Isotropy in a Multiply Connected Universe

The identification of finding multiply connected topology from Robertson-
Walker metric will, in general, destroy the isotropy of the universe as
a whole. For instance, a T^3 universe of $a_x = a_y = a_z$ is anisotropic, but
with a symmetry of a cubic lattice. Since, even in this case, the
expansion of the universe can be isotropic, the small scale isotropy
of microwave background radiation is still ensured, as in the simply
connected case.

Multipole component of the background radiation is sensitive to the
global anisotropy. Therefore, we should discuss whether or not the
present observations on quadrupole component of background radiation is
inconsistent with the multiply connected model. It has been shown
(Fang and Mo 1986) that, at least in the case of $L_H > L$, the anisotropi-

cally multiply connected universe can not be ruled out by the observation of quadrupole component.

The large scale fluctuation in the background radiation temperature $T(\vartheta,\varphi)$ can be described by multipole expansion, i.e., spherical harmonic expansion:

$$\frac{T(\vartheta,\varphi)}{T_b} = 1 + \sum_{l,m} a_l^m Y_l^m(\vartheta,\varphi) \tag{14}$$

In a simply connected universe, the multipole component a_l is given by(Peebles 1982)

$$\langle a_l^2 \rangle = \frac{3\pi(2l+1)}{2l(l+1)} \delta^2 \tag{15}$$

where δ is the density fluctuation at small k. From eq .(15) one finds $a_l \sim \delta$.

In a T^3 universe of $a_x = a_y = a_z = a$, we have(Fang and Mo 1986)

$$\langle a_l^2 \rangle \cong \frac{4\pi(2l+1)}{(4\pi c/H_o \lambda_o)^6} \delta^2 \tag{16}$$

Therefore, in the case of $\lambda_o < cH_o^{-1}$ (i.e., $L_H > L$) one have $a_l < \delta$, namely, in multiply connected universe, the multipole component of the background temperature will even be smaller than that in simply connected case.

6. CONCLUSIONS

All the evidences mentioned in section 5 are only very tentative. Anyhow the further statistical analyses of quasar redshift will be important to check the "positive" evidence of the compactified universe or to improve the lower limit on its size. All these results will be important for quantum cosmology regargingless whether or not there are positive or negative evidences for multiple connecticity.

The significance of this research is, more importantly, in the aspect of methodology. It teaches us that the birth of the universe can also be studied by the following methods:

(i) in observational cosmology, to determine the spacetime topology of the universe as a whole by means of the distributions of objects, such as quasars and galaxies;

(ii) in theoretical cosmology, to find the model of the birth of the universe, which can explain why the spacetime topology is just as it is.

REFERENCES

Burbidge G.R. Narlikar,J.V.,and Hewitt,A.,1985,Nature,317,413.
Chu,Y.Q.,Fang,L.Z.,and Liu,Y.Z.,1984,Astrophys.Lett.,24,95.

Chu,Y.Q.,and Zhu,X.F.,1983,Astrophys.J.,271,507.
Ellis,G.,1967,J.Math.Phys.,8,1171.
Fagundes,H.,1985a,Astrophys.J.,291,450.
Fagundes,H.,1985b,Phys.Rev.Lett.,54,1200.
Fang,L.Z.,1982,in Astrophysical Cosmology, eds. H.A.Bruck,G.V.Coyne,
    and M.S.Longair, Vatican Press, Vatican City.
Fang,L.Z., and Mo,H.J.,1986,USTC preprint.
Fang,L.Z., and Sato,H.,1983,Communi.Theor.Phys.,2,1055.
Hartle,J., and Hawking, S.,1983,Phys.Rev.,D28,2960.
Li,M.,1986,Phys.Lett.,173B,
Lobell,F.,1931,Ber.Verhand Sachs,Akad.Wiss,Leipzig Math.Phys.KL,83,167.
Mkrtchyan,R.,1986,Phys.Lett.,172B,313.
Narlikar,J.V.,and Seshadri,T.,1985,Astrophys.J.,288,43.
Peebles,P.J.E.,1982,Astrophys.J.(Letters),263,L1.
Savage,A.,and Bolton,J.G.,1979,Mon.Not.R.Astron.Soc.,188,599.
Scott,P.,1983,Bull.London Math.Soc.,15,401.
Seldner,M.,and Peebles,P.J.E.,1979,Astrophys.J.,227,30.
Shane,C.D.,and Wirtanen,C.A.,1967,Pub.Lick Obs.,22,Part 1.
Unwin,S.,1982,Gen.Relati.Grav.,14,509.
Zeldovich, Ya.B.,and Starobinski,A.,1984,Sov.Astron.Lett.,10,135.

DISCUSSION

NARLIKAR: In 1984 Seshadri and I discussed elliptical topology of closed Friedmann models. We found that a QSO could have two images in diametrically opposite directions. However, the effect could operate only for $q_0 > 1$ and for redshifts under certain limits. Do you have any such theoretical constraints in your models?

FANG: For T^3 topology q_0 must be equal to 0.5, namely only that universe admits torus topology. In this case, there is no constraint on redshift for the formation of images with opposite directions. However different images are related in general, to different times of the sources.

LOH: In a torus universe, does volume increase approximately as z^3 even if $cz/H_0 > L$, where L is the "period" of the universe?

FANG: Yes, the geometrical properties of a torus universe are the same as those of the simply connected K = 0 universe.

CHAPTER VII

THE NON-STANDARD APPROACH

OBSERVATIONS REQUIRING A NON-STANDARD APPROACH

Halton Arp
Max-Planck-Institut für Astrophysik
Karl-Schwarzschild-Straße 1
D-8046 Garching bei München

ABSTRACT. Extragalactic astronomy rests on the assumption that red-shifts are always equal to velocities and that large velocities are unambiguous distance indicators. There are now many observations which disprove this fundamental assumption. We here review the evidence which has been accumulating for 20 years with particular emphasis on the most recent developments.

I. QUASARS

Since quasars are stellar appearing, high energy-density objects their large redshifts cannot automatically be interpreted as measures of their distances as is customarily done for galaxies. In fact since quasars do not exhibit a redshift-apparent magnitude relation there is no direct evidence that they obey a redshift-distance relation. We now cite evidence of six different kinds that quasars are much closer than their redshift distances. Each kind of evidence is conclusive in its own category but is supported by other evidence for non-velocity red-shifts in both quasars and galaxies.

1. Multiple Quasars Associated With Nearby Galaxies

Figure 1 shows the four galaxies now known which appear to have more than one quasar closely associated. Table 1 lists the data for these associations. The chance for accidental proximity in each of these systems is:

a) $\sim 10^{-5}$ for NGC 622. This was discovered in a UV excess search which included approximately 30 such galaxies. Note straight filament leading from galaxy to HII region adjacent to quasar.

b) $\sim 10^{-5}$ for NGC 470. The quasars were discovered in a UV excess survey which included about 5 galaxies this bright. In the disk of the spiral galaxy are seen irregularly shaped nebulosities which appear to be interacting with the quasar images.

A. Hewitt et al. (eds.), Observational Cosmology, 479–498.

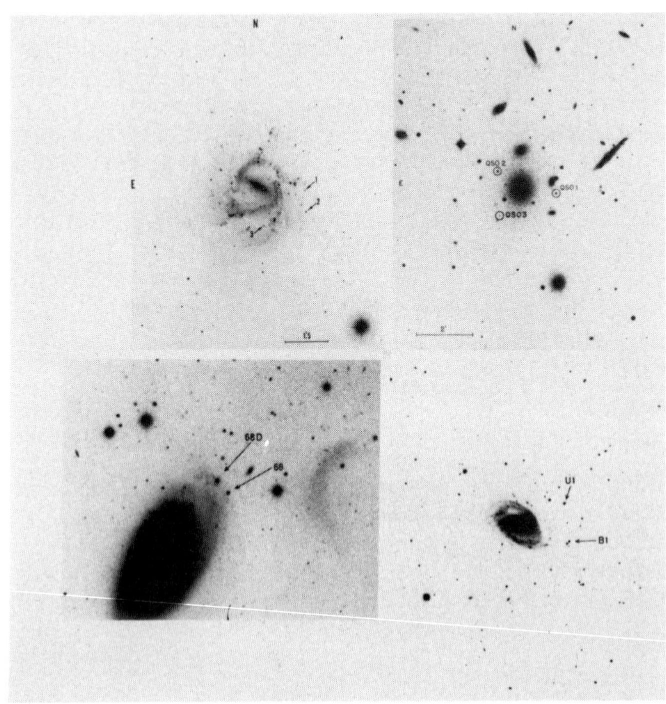

Figure 1. Multiple quasars close to galaxies. Clockwise from upper left: NGC 1073, NGC 3842, NGC 622, NGC 470

Table 1

Galaxy		Quasar				
Name	redshift	name	dist. (arc sec)	mag.	redshift	probability
NGC 622	0.018	UB1	71	18.5	0.91	0.001
		BSO1	73	20.2	1.46	0.02
NGC 470	0.009	68	95	19.9	1.88	0.015
		68D	85	18.2	1.53	0.002
NGC 1073	0.004	BSO1	104	19.8	1.94	0.01
		BSO2	117	18.9	0.60	0.006
		RSO	84	20.0	1.40	0.02
NGC 3842	0.020	QSO1	73	19.0	0.34	0.003
		QSO2	59	19.0	0.95	0.002
		QSO3 (radio)	73	21.0	2.20	0.01

c) $\sim 10^{-6}$ for NGC 1073. Quasars were searched for in this system because of the discovery of a compact radio source (RSO). The galaxy is so bright that it is one of the only 176 in the Hubble Atlas and one of the 1246 in the Shapley Ames Catalog. Three clumps of hydrogen in the disk, if rotated forward by ~ 20 degrees, correspond to the positions of the three quasars.

d) $\sim 10^{-7}$ for NGC 3842. The two brightest quasars form a pair of X-ray sources across the galaxy. The galaxy is one of the two brightest in an unusual cluster of galaxies which, like Virgo, has individual galaxies as X-ray sources.

It is usually not appreciated how close these quasars fall to low redshift galaxies. The vast majority of bright galaxies have not been inspected but the compound probability of finding two or three quasars in those that have is so small that only a few such cases are required to give a certain proof of the association of high redshift quasars with low redshift galaxies.

A more detailed review and source references can be found in [1].

2. Quantization of Quasar Redshifts

When the first quasar redshifts began to accumulate, Geoffrey Burbidge pointed out that there was a significant excess around the value z=1.95. In 1968 [2] he also identified peaks at z=0.06, 0.30 and 0.60. In 1971 and again in 1977 [3], K.G. Karlsson showed the quasar redshift distribution had broad peaks given by: $\Delta\log(1+z)=0.089$. This has been most lately confirmed for all quasars known through 1984 by Depaquit, Vigier and Pecker [3]. The positions of the peaks described by this formula (using z=0.30 as a zero point) are listed in Table 2.

Table 2 - Peaks in Quasar Redshift Distribution

z (all quasars from $\Delta\log(1+z)=0.089$)	z (NGC 1073 observed)
0.30	
0.60	0.60
0.96	
1.40	1.40
1.95	1.94

It is clear that the redshifts of the NGC 1073 quasars fit these predicted peaks much too closely to be accidental. The remaining redshifts in Table 1 fit less well but still confirm this average periodicity. It is important to note, however, that each group of quasars fits better with a slightly different constant in the equation $\Delta\log(1+z)=$ const.

For example, Fig. 2 shows that quasars selected by objective prism

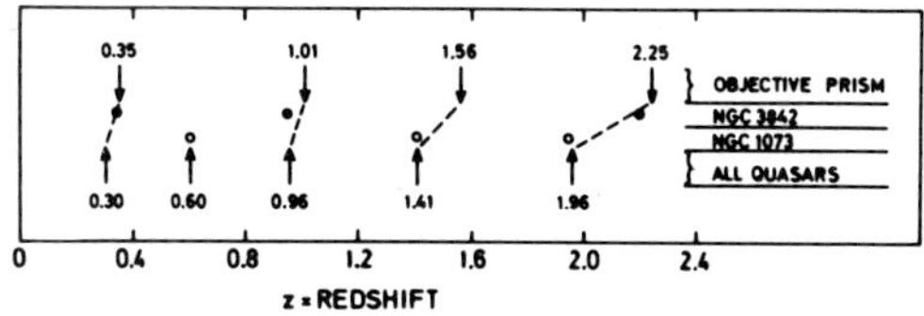

Figure 2. Redshift peaks observed for different groups of quasars

searches (dominated by the Sculptor region) show peaks z=0.35, 1.56 and

2.25 (with an uncertain peak at 1.01) [5]. These latter peaks, on average, fit better the formula with const. ∿0.096. The redshifts in the triple association NGC 3842 fit better with this slightly larger constant as Fig. 2 shows. A group of bright apparent magnitude quasars in the core of the Virgo Cluster fit an equation with const.=0.098 [6].

The adding together of different groups of quasars gives broad peaks as observed in the overall redshift distribution. But the fact that an individual group of quasars such as the NGC 1073 group can fit so exactly a particular value of redshift multiple, confirms that it is one physical group at the same distance despite the huge range in redshift of its quasars.

3. Single Quasars Associated With Nearby Galaxies

A single quasar falling improbably close to a low redshift galaxy is more common than the multiple quasars discussed earlier. In 1972 Burbidge, O'Dell and Strittmatter [7] showed significant associations of such quasar/galaxy pairs. Moreover they showed that the higher the redshift of the galaxy, the closer the quasar appeared in angular separation as if the whole association were viewed a greater distance. Fig. 3 shows this relation with the data available up to 1983 [8].

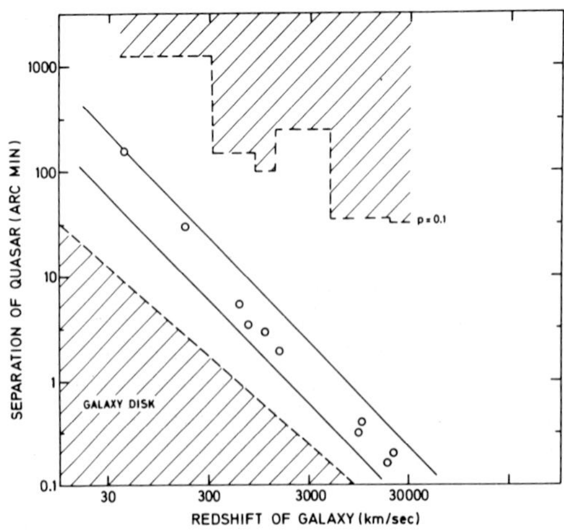

Figure 3. Apparent separation of quasar varies as distance of galaxy

In 1972 Arp began to show that quasars fell particularly close to a class of galaxies which were companions to large spirals [9]. These investigations culminated in 1981 and 1983 when it was shown that quasars were about 20 times more dense around companions to large, low redshift spirals than they were in the general field.[10] This result is shown in Fig. 4 and has a probability of about 7×10^{-16} of ocurring by accident.[1]

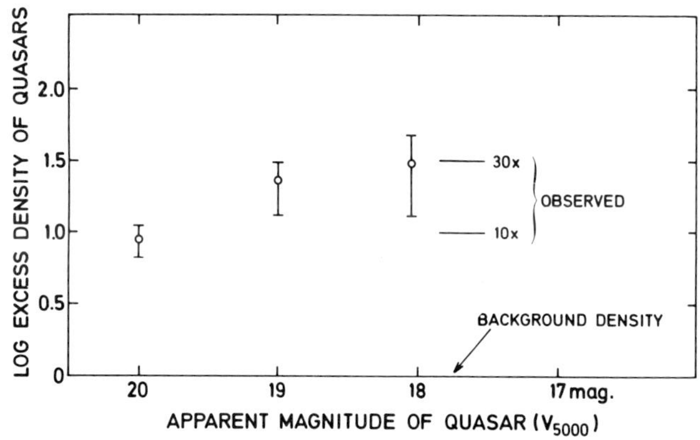

Figure 4. Density of quasars around companion galaxies of large, low redshift spirals

It is generally believed that subsequent criticisms of these probability calculations established that the association were not significant. That is not true, since the last and most elaborate analysis of
the statistics obtained only a 0.01 probability for chance association.[11] Though still significant, the reason this improbability is
not nearly so small as the one Arp calculated is because it ignores the
fact that the galaxies with which the quasars are associated have varying distances from us as shown in Fig. 3. Ignoring the differing galaxy
distances still gives a signficant association but not the overwhelming
significance which is obtained if this elementary fact of astronomy is
taken into account.[1]

4. Quasars Visibly Connected to Galaxies

Arguments about degrees of improbability of chance association
seem unnecessary, however, when it has been demonstrated that some
quasars are visibly connected to low redshift galaxies. The most famous
of these is NGC 4319/Mark 205 in which a luminous filament was detected
leading from the quasar (z=21,000 km/sec) back to the galaxy (z=1,700
km/sec). Long debate about the reality of this connection has finally
led to the acknowledgement of its reality.[12] The only remaining
escape is to have the luminous filament arise from a quasar in the background and to have it only accidentally project toward the galaxy,
mimicking a true connection. This possibility is ruled out, however, by
the image processing done by J.W. Sulentic on CCD, KPNO 4 meter images
of the center of NGC 4319. As shown in Fig. 5 the maximum intensity in
the center of the galaxy proceeds at an angle to the major axis directly into the luminous connection which leads to Mark 205.

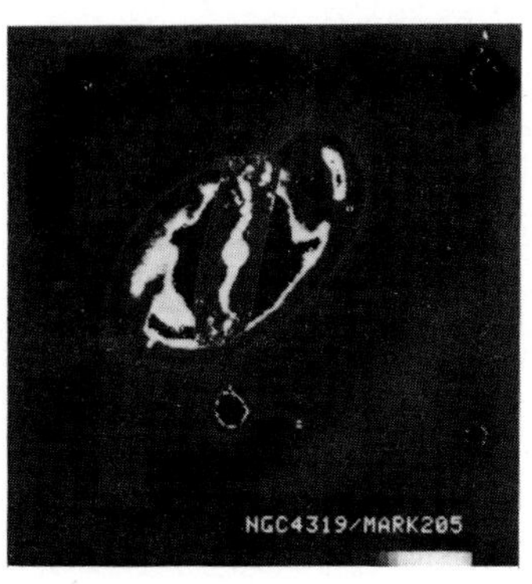

Figure 5. Image processing of NGC 4319 showing central spine leading to Mark 205

Another remarkable development in the analysis of NGC 4319 is shown in Fig. 6. Then the VLA radio map by J.W. Sulentic shows radio

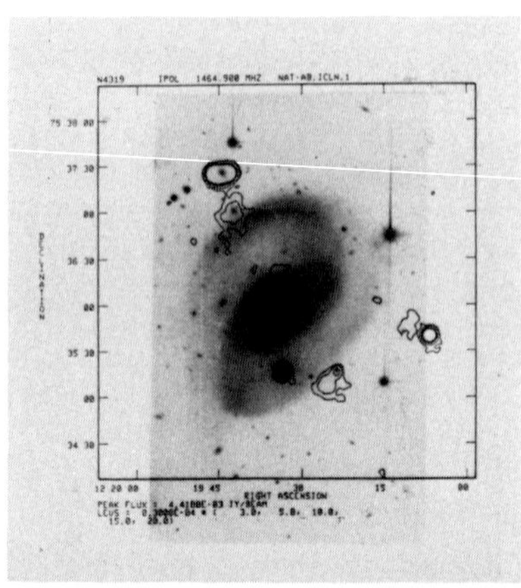

Figure 6. Map at 20 cm with VLA by J.W. Sulentic showing radio lobes across NGC 4319

lobes ejected in either direction from the nucleus of NGC 4319. Radio lobes across a spiral galaxy are unprecedented. These could only have come from the explosion that disintegrated the spiral arms at their roots and cleared out the Hα from the disk of NGC 4319 leaving only shock excited [NII] emission.[13] It was originally concluded that Mark 205 had been ejected from NGC 4319. We now see that the radio source Mark 205 is aligned approximately along the direction of the ejected radio lobes. Both the unusual optical and radio properties support the physical association of the quasar with the galaxy.

An even more straightforward example of a quasar connected to a galaxy is shown in Fig. 7. The galaxy, MCG 03-34-085, is chaotic, possesses a jet and exhibits an unusual spectrum.[14] The quasar has a luminous filament which emerges most strongly on one side of its stellar image and leads back to the base of the jet. The connection was discovered in 1984 on two independent, IIIa-J, U.K.schmidt, survey plates. It is being published for the first time only now.[1]

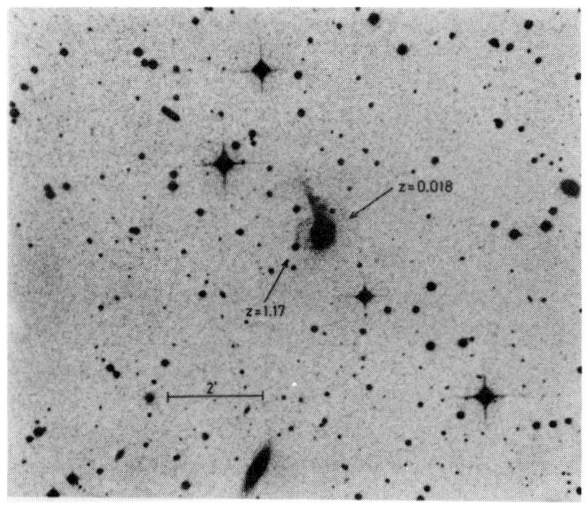

Figure 7. The quasar PKS 1327-206 connected to a low redshift, disturbed galaxy by a luminous filament

As an example of a quasar connected to a galaxy by a radio filament we show in Fig. 8. In the northern lobe of the radio galaxy 0844+31 there is a hot spot only 5 arc sec distant from a high redshift, bright apparent magnitude quasar. The chance of a quasar this

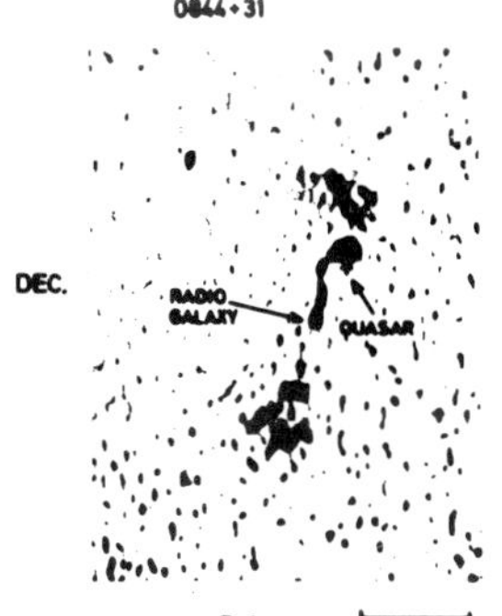

Figure 8. Radio map of ±2402 and quasar with z=1.83

bright falling this close to the hot spot is only 3×10^{-6}. Even if we take the significant distance to be from the quasar to the center of the lobe, a distance of 19 arc sec, the chance is only 4×10^{-5}.

Another well-known example, 3C 303, is listed in Table 3. The observations of P. Kronberg, E.M. Burbidge, H.E. Smith and R.G. Strom [15] show the radio jet from the galaxy terminates near three ultraviolet objects, two quasars and one peculiar extended object. The improbability of finding just the one quasar with measured redshift is $<10^{-4}$.

Table 3 - Quasars Connected to Galaxies or Close to Radio Lobes

Galaxy		Quasar				
Name	redshift	Name	distance (arc sec)	mag	z	probability
N6C 4319	1,700 km/sec	Mark 205	40	14.5	0.07	~ 0
MCG 03-34-085	5,400	PKS 1327-206	38	17.0	1.17	~ 0
N6C 5296	2,500	BSO #1	55	19.3	0.96	$\ll 10^{-3}$
3C 303	42,000	UV #C	20[+]	20	1.57	$< 10^{-4}$
I 2402	20,000	0844+31	70[+]	18.0	1.83	$\sim 10^{-5}$
0924+30	8,000	Compact source	497[+]	21.5	2.02	$\sim 10^{-5}$

[+] Distance from galaxy; quasar is much closer to hot spot or radio lobe.

How many chances do we have to find such close associations of quasars with radio jets? The most recent compilation of jets in galaxies with z<0.2 by Bridle and Perley [16] list only 75 examples. In addition to the two associations listed next to last in Table 3, my casual inspection of list of radio jets reveals as many as 11 associations at about the improbability level of 0.01 with already known active objects of higher redshift in the vicinity, some with good alignments to the radio jets.

5. Galaxies With Many Associated Quasars

In addition to galaxies with one, two or three very close there are a few galaxies with larger numbers of associated quasars. Almost always these quasars are distributed on a line or cone originating in the nucleus of the galaxy. The galaxies of origin are usually the most disturbed and disrupted of any known.

The galaxy with the longest, straightest optical jets known is the hot-spot nucleus spiral, NGC 1097. Search of the central 8.1 sq.deg. of an objective prism schmidt plate by X.T. He produced 43 candidate quasars around NGC 1097. Slit spectra of 33 showed these candidates were 94% true quasars. The concentration of quasars to the position of this ejecting galaxy is obvious.[17] The concentration builds up from expected background density at the edge of the field and reaches a maximum of greater than 20 times normal just north of NGC 1097 between the two strongest optical jets.

Fig. 9 shows that the radio and X-ray material extends from the galaxy nucleus out along the line of the jets. The six quasars near the strong northern jet are associated with individual patches of X-ray emission. Since the X-ray material is continuous with material in the active nucleus of the galaxy and has obviously been ejected, the X-ray quasars must also have been ejected.

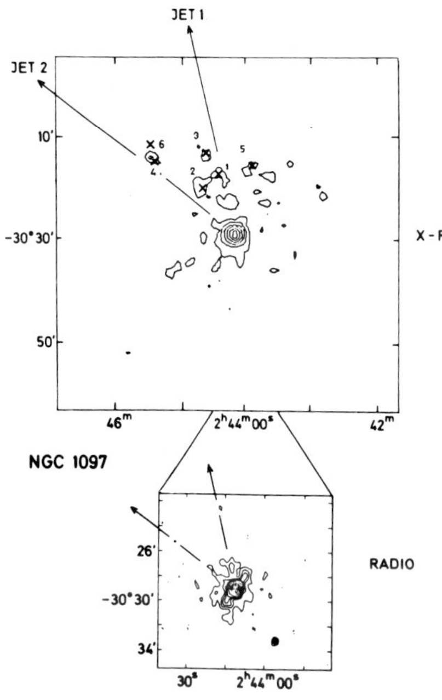

Figure 9. Relation of X-ray and
radio material to jets in NGC 1097.
Quasars in densest region numbered
1-6

In the Shapley-Ames Catalog of the brightest apparent magnitude
galaxies the two most disrupted are M82 and NGC 520. Fig. 10 shows that
in 1967 a line of radio sources was identified emerging NE and a cone
SW from NGC 520. By 1970 a line of radio bright quasar along one edge
of the cone had been discovered. In 1983 a double independent search of

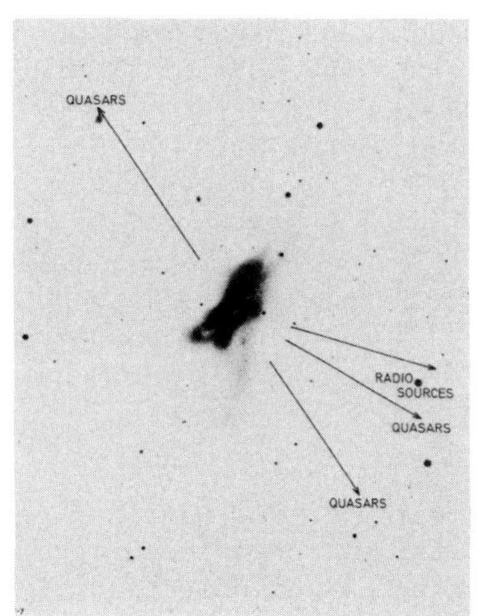

Figure 10. NGC 520 with lines of quasars
and radio sources indicated

the area around NGC 520 revealed a line of six quasars with $V_{5000} < 19.0$
mag. emerging from the galaxy.[18] The outer X-ray material around

NGC 520 is also roughly aligned along the direction of the quasars and
the radio material. The density, concentration and alignment of these
various objects around NGC 520 has a negligible chance of being accidental.

The brightest, very disturbed galaxy in the sky is M82. Fig. 11
shows that a unique group of quasars fall very close to it. Three of

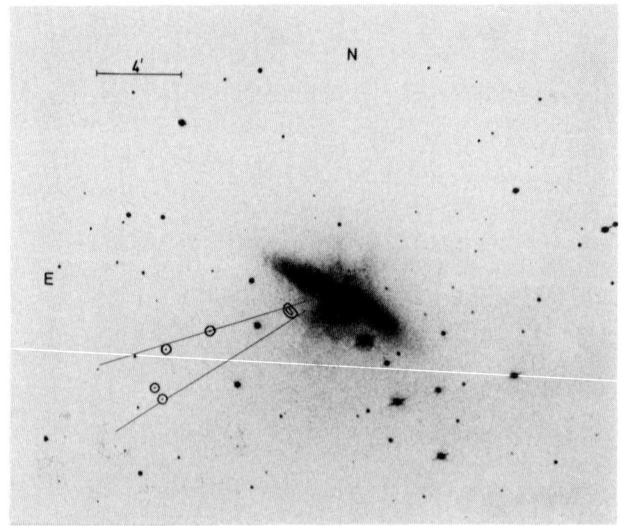

Figure 11. M82 with
four quasars and a
slightly extended radio
source emerging in a
cone from a notch in
its disturbed outline

them have redshifts of z=2.048, 2.054 and 2.040.[19] Together with a
fourth quasar [10] these objects define an ejection cone leading back
to the disturbed center of M82. Recently J.J. Condon [20] discovered a
radio source on the edge of M82. It fits exactly into the base of the
previously defined ejection cone as Fig. 11 shows.
    In Fig. 12 the X-ray map of the interior [21] is superposed on a

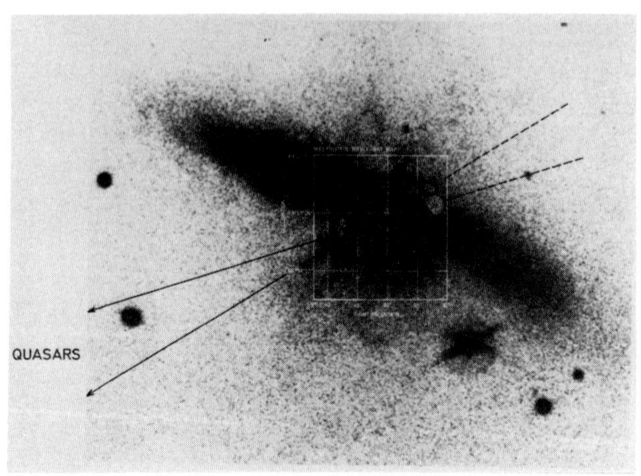

Figure 12. M82 with
X-ray map superposed in
interior and direction
of quasars marked

photograph of M82. It is seen that some of the X-ray emission coincides
with the optical emission filament coming out along the SE minor axis
of M82. The main body of the X-ray material, however, is extended out
along the line of the ejection cone to the quasars.

In summary, three of the most disturbed bright galaxies we know
all have the most numerous, close quasar associations we know. In each
case the quasars are aligned in ejection cones originating from the
galaxies. In each case radio and X-ray material confirm that these
quasars have been ejected out from the disrupted galaxy.

6. Where Do Most Quasars Belong?

The most conspicuous small groupings of quasars on the sky all
involve quasars with redshifts predominantly in the range 0.8<z<1.4.[22]
The relatively small scale of their separation indicates they are the
most distant and hence the most luminous at a given apparent magnitude.
Quasars of different redshift, particularly in the range 1.4<z<2.4
should be less luminous. Of these, the brightest in apparent magnitude
should be closest to us. Plotting all bright, high redshift quasars in
the sky shows that in the direction of the Local Group of galaxies they
are 3.5 times more numerous than in the opposite direction of the sky
towards the Virgo Cluster.[6] Fig. 13 shows the most conspicuous feature

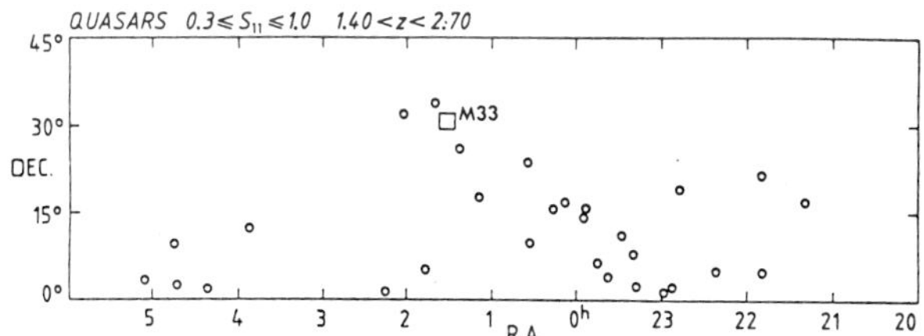

Figure 13. Radio quasars
in the direction of the
Local Group of galaxies

in the Local Group direction, a line of quasars from M33, the companion spiral to M31.

If we investigate the next nearest group of galaxies to us, the
Sculptor Group, we find it contains the bright Sc, NGC 300. From
NGC 300 there is a similar line of high redshift quasars.[6] Most
amazingly, there are also lines of hydrogen (HI) printing to both M33
and NGC 300 which are rotated only 20 and 25° from the previously discovered lines of quasars. The lines of HI have scales proportional to
the fact that M33 is twice as close to us as NGC 300.

Finally if we look at the apparent magnitudes of the quasars in
the M33 and NGC 300 lines we find that the latter one is just about
1.5 magnitudes fainter. Again this reflects the fact that NGC 300 is

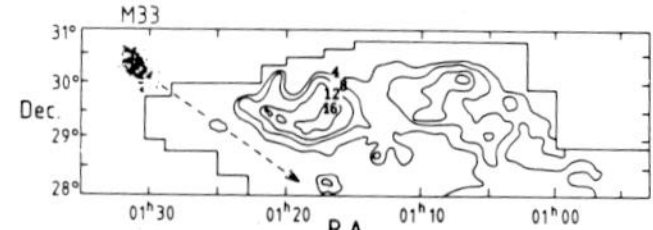

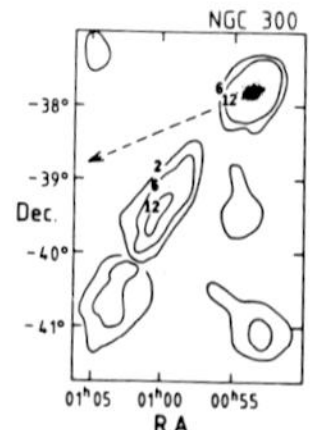

Figure 14. Lines of quasars (dashed) and hydrogen coming from nearest galaxies

just about twice as far away. Fig. 15 also shows that the quasars associated with M82 are fainter than 20th magnitude and hence we should

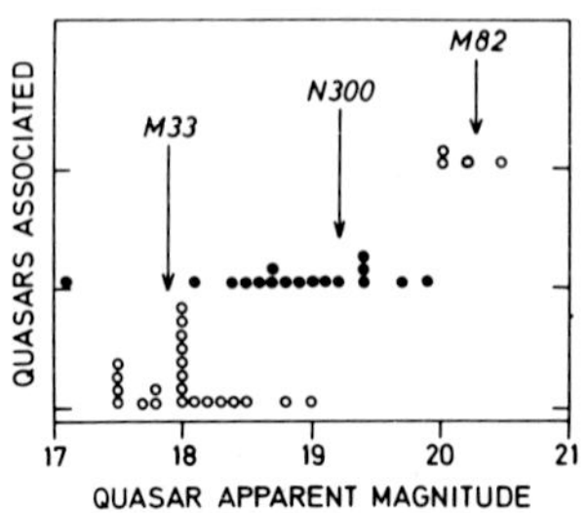

Figure 15. Apparent magnitudes of quasars ($z \sim 2$) become fainter as galaxies become more distant

not generally see, with present surveys, high redshifts quasars at much greater distances than M82.

The answer to the question of where most quasars belong is: Most high redshift quasars belong to very nearby galaxies and are projected at large angular distances around them on the sky. Quasars near $z \sim 1$ are more luminous, are seen to greater distances and form separate, smaller groupings on the sky. The overall distribution of quasars which is observed is a projection on the plane of the sky of nearby, large diameter groups and smaller, more distant groups, all of which contain a range in redshifts.

One closing note is that there is a subgrouping of quasars in Fig. 13 at the position of the anomalous quasar/galaxy, 3C 120. In 1973

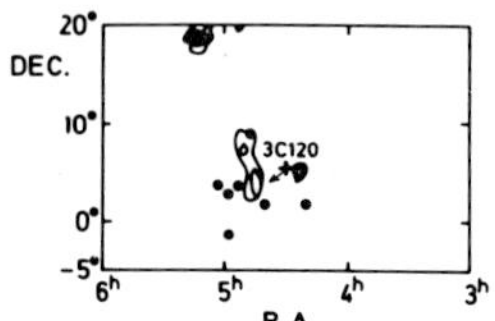

Figure 16. Quasars around 3C 120

it was concluded that the group of galaxies to which this unusual object belonged was the Local Group. Fig. 16 shows the same kind of high
redshift quasars that were associated with M33 are also closely
associated with 3C 120. Local Group HI clouds and other anomalous objects are also associated with 3C 120.[23] One consequence of placing
3C 120 in the Local Group rather than at its redshift distance
(z=0.033) is that its apparent superluminal expansion is reduced from
6 times the speed of light to about 0.04 the speed of light.

II. GALAXIES WITH EXCESS REDSHIFT

     The most common argument rejecting the foregoing evidence that
quasars are closer than their redshift distance is: Quasars are
physically similar to galaxies and galaxies can only have Doppler,
velocity redshifts. We briefly review here the evidence that many
galaxies have small, non Doppler components of redshift and that the
more a galaxy resembles a quasar, the greater the non-velocity redshift
it usually has.

1. Galaxies with Discordant Redshift Companions

     The prototype galaxy in this class is shown in Fig. 17. The

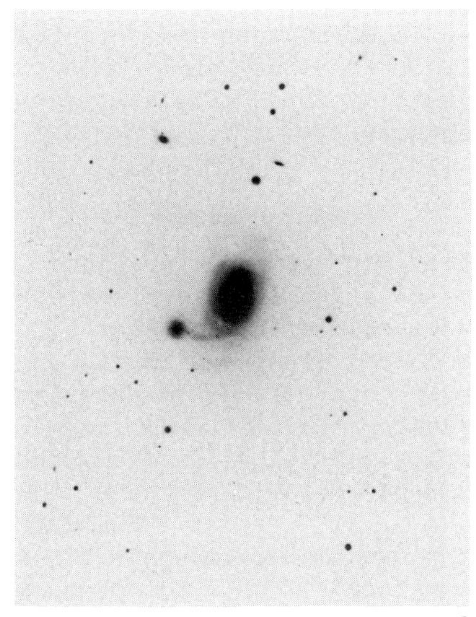

Figure 17. NGC 7603 with companion
discrepant by +8,300 km/sec

Seyfert galaxy has one luminous filament connecting it to the companion. The photograph shown here is an image processed composite by
N. Sharp of plates taken in 1973 by R. Lynds. Available now, the image
best shows the interaction of the companion with the filament from
NGC 7603.[1] It has always been clear that a companion running by the
main galaxy with a differential velocity of 8,300 km/sec could not
gravitationally pull out such a filament of stars. If they are connected the companion must have a large component of intrinsic redshift.

There are now 38 examples of discordant redshift companions interacting with 24 main galaxies.[24] A sample of these associations is
shown in Table 4. It is important to note that the companions with the

Table 4 - Sample of Interacting Galaxies with Large Discordant Redshifts

Main Galaxy	Comparison	Excess Redshift	Type of Spectrum
N6C 7603	comp SE	+ 8,300 km/sec	late absorption
AM 0059-402	comp S	+ 9,695	late absorption
AM 0213-283	comp N	+14,021	strong em. early abs.
AM 0328-222	comp S	+17,925	emission, early abs.
AM 2006-295	KN SW	+22,350	weak em., pec. abs.
NGC 1232	Gal B	+26,210	emission, early abs.
NGC 53	comp N	+32,774	emission, early abs.
AM 2054-221	comp E	+36,460	emission, late abs.

highest discordant redshift are in the same redshift range as the
smaller redshift quasars. (The last entry in Table 4 has z=45,000
km/sec or z=0.15. This is comparable to z=0.16 for the famous quasar
3C 273.) But as the redshifts of the discordant companion galaxies
approach that of the quasars, their spectra and general morphology do
also. (Note 3C 48, a prototypical quasar has strong emission and early
type absorption and is in the position of a companion to M33.) This
demonstrates that the physical characteristics of quasars and compact,
emission line galaxies are continuous and that therefore any proofs of
non-velocity redshifts in one class of objects supports proofs of non-
velocity redshifts in the other class of objects.

2. Groups of Galaxies

In general we should emphasize that the only empirical evidence
for the distance of an object comes from associating it with other objects of known distance. The galaxies we know best group together
around dominant S_b galaxies like M31 in the Local Group and M81 in the
M81 group. Every investigation made has shown that fainter galaxies in
groups like these have varying amounts of systematic, intrinsic redshifts.[25] In fact, close analysis of Shapley-Ames galaxies shows that
the only galaxy types which obey a linear Hubble relation are the
dominant S_b, non-companion galaxies.[26]
In order to give a vivid example of the systematic redshift of
fainter members of groups we turn to our own Local Group. Fig. 18 shows
M31 and the companions which classically belong to the group. However,
they only represent a spread in redshift from $cz_\circ$ (M31) = -86 to
$cz_\circ$ (NGC 404) = +142 km/sec. Groups characteristically contain a spread
in redshift of about 800 km/sec. Plotting galaxies up to $cz_\circ$ <700 km/sec
in Fig. 18 demonstrates that these less luminous, higher redshift
galaxies are members of the Local Group. They even define the line of
group members distributed along the M31 minor axis, the M31-M33 line,
which extends to Local group members in the region of 3C 120 (Discovered in connection with Figs. 13 and 16).

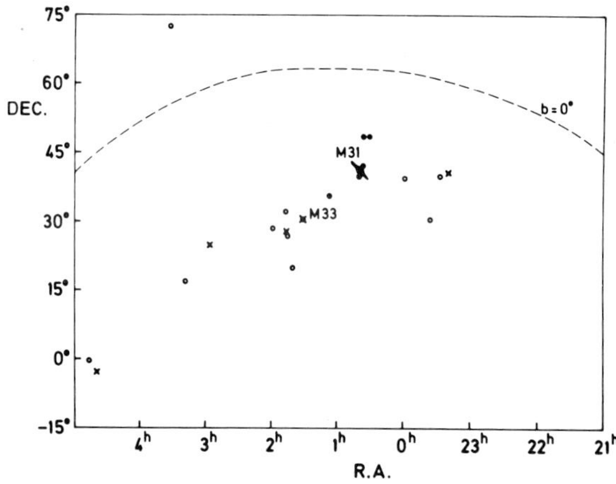

Figure 18. Filled circles are Local Group members. Crosses and open circles are spirals and dwarfs with $300 < cz_{\odot} < 700$ km/sec

Even if we restrict ourselves to just the traditionally accepted members of the Local Group and M81 groups we see from Fig. 19 that 21

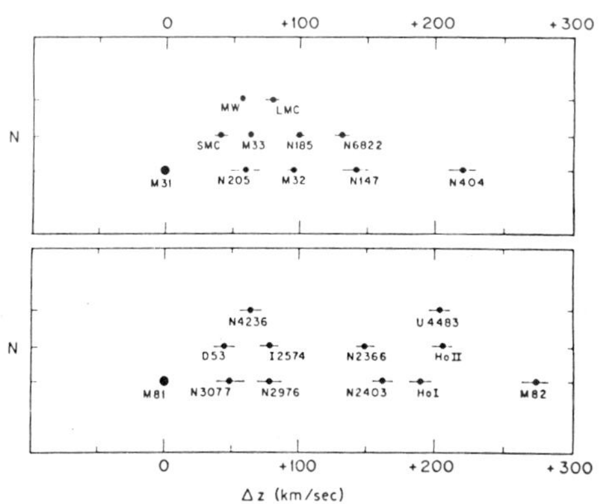

Figure 19. Differential redshifts in M31 and M81 groups

out of 21 major companions are redshifted with respect to the dominant galaxy. This has one chance out of two million of ocurring accidentally.[25] All redshift values and membership are so accurate that this result can never be changed. It is therefore a proof of non-velocity redshifts which must be accepted now.

3. Quantization of Galaxy Redshifts

The existence of systematic, intrinsic redshifts of galaxies has been established from large through small redshift discrepancies by the preceding analyses. If this evidence were not sufficient, there exists the further evidence of quantization of redshifts which is an independent demonstration of their non-velocity nature. Starting in 1976,

W. Tifft has shown in a number of careful analyses of different pairs
and groups of galaxies that their redshifts are quantized, principally
in steps of 72 km/sec. This was most recently confirmed by a maximally
accurate analysis of an independent sample of galaxies in groups by Arp
and Sulentic.[25]

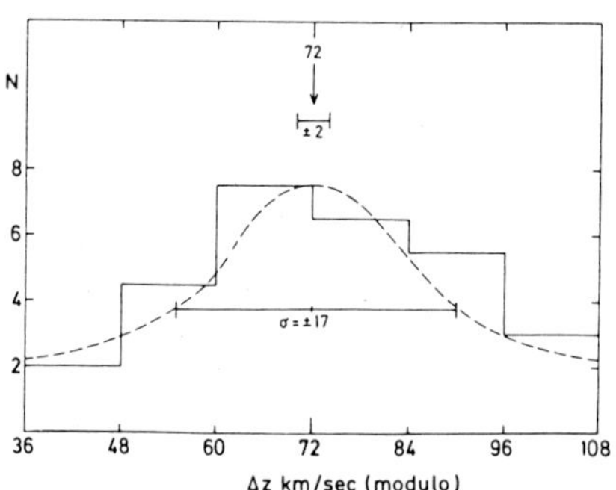

Figure 20. Distribution
around multiples of
72 km/sec for M31 and M81
companions

Can we test for quantization in the most nearby galaxies? Yes, if
we use a corrected value of solar motion near 250 km/sec instead of the
300 km/sec in current use. The corrected value comes from allowances
for non-velocity redshifts in our Local Group of galaxies, an analysis
which solves some current enigmas in conventional Local Group
motions.[28] Fig. 20 shows that the systematically redshifted companion
galaxies in the M31 and M81 groups (as shown in Fig. 19) are quantized
around 72 km/sec with an average deviation of only ±17 km/sec.

If we examine members of the closeby M31 and Sculptor Groups,
companions that range up to 13 multiples of 72.4 km/sec higher than the
dominant galaxies, we see in Fig. 21 that the residuals from modulo

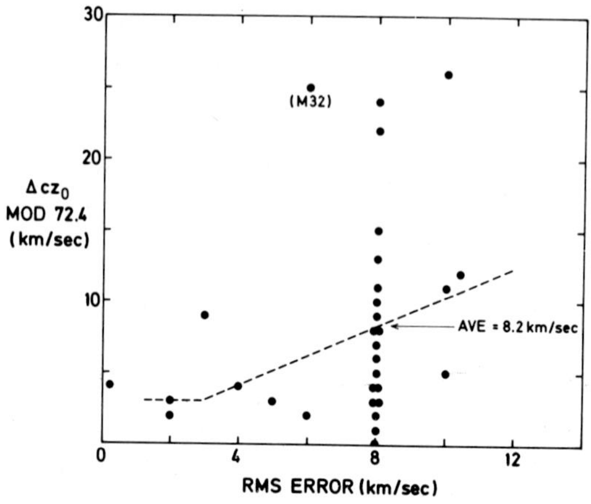

Figure 21. Residual from 72.4
km/sec multiple as a function
of accuracy of measures.[27]

72.4 km/sec are only the measuring errors of the redshifts themselves.
(An extra decimal place in 72 km/sec is needed because of the high

order multiples, up to 13. The value 72.46 km/sec was derived by Tifft
in 1977 in the Coma Cluster of galaxies - and 72.4 by Arp in the M31
and M81 groups shown in Fig. 21). A few galaxies are indicated in
Fig. 21 as having measured redshifts more accurate than ±8 km/sec. They
show residuals from modulo 72.4 km/sec of less than ±5 km/sec! This
introduces the challenging paradox that there is very little true
velocity motion left for orbital or peculiar motion around the central
galaxy.

Altogether, however, these results unify the findings from the
most extreme quasars through apparently quite normal galaxies. In the
quasars all possible empirical tests showed large to intermediate intrinsic redshifts that were quantized. In the physical continuity
connecting them to galaxies again we saw all possible tests demonstrat-
ing the presence of intermediate to small non-velocity redshifts. In a
continuation of the trend from the quasars, these smaller non-velocity
redshifts for the galaxies were also quantized but on a smaller scale.

4. Is There a Hubble Relation for Galaxies?

Among bright (Shapley-Ames) galaxies only the Sb's define a linear
Hubble relation. The Sc's (which are characteristically companions to
Sb's in groups) deviate above the Hubble relation as their apparent
magnitudes become fainter.[26] Malmquist bias is an inadequate ration-
alization for this deviation.

Fig. 22 shows the distance-redshift relation for the Sc's listed
by Sandage and Tammann in the Revised Shapley-Ames Catalog. The

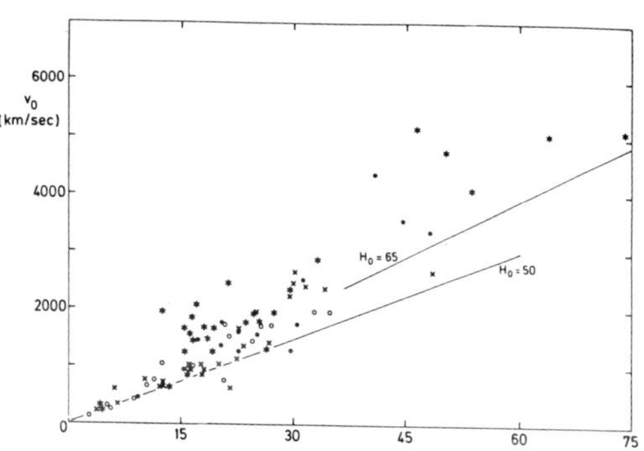

Figure 22. Redshift-distance
relation for Sc galaxies.
Distances in Mpc from Tully-
Fisher relation

distances are derived from the redshift independent, Tully-Fisher
criterion. Deviations above the line for H_o=50 km/sec/Mpc cannot be due
to streaming motions because they would have to be very large, increase
with distance and always be in the receding sense.

The paradoxical result is that the expansion of the universe would
have to change from H_o=50 within 15 Mpc to something over H_o=100 at
greater distance. This change in H_o as a function of distance was first
pointed out by G. de Vaucouleurs in 1972 [29] and reemphasized in a
series of papers by E. Giraud [30] (and this symposium).

The way out of this paradox is the same as for all the contradictory and discrepant results presented in this review. Namely, we must no
longer ignore observational evidence which demonstrates that redshifts,
from quasars to companion galaxies, contain appreciable components of
non-velocity redshifts.

References

[1] Arp, H. 1986 "Quasars, Redshifts and Controversies" published by
      Interstellar Media, 2153 Russell Street Berkeley 94705
[2] Burbidge, G. 1968, Ap.J. 154, L41
[3] Karlsson, K.G. 1977, Astron. Astrophys. 58, 237
[4] Depaquit, S., Pecker, J.-C., and Vigier, J.-P. 1984, Astronomische
      Nachrichten 305, 339
[5] Box, T.C. and Roeder, R.C. 1984, Astron. Astrophys. 134, 234
[6] Arp, H. 1986, J. Astrophys. Astron. (India) 7, 77
[7] Burbidge, G.R., O'Dell, S.L. and Strittmatter, P.A. 1972, Ap. J.
      175, 601
[8] Arp, H. 1983, Ap. J. 271, 479
[9] Arp, H. 1980, Ann. N.Y. Academy of Sciences 336, 94
[10] Arp, H. 1981, Ap.J. 250, 31
      Arp, H. 1983, Ap. J. 271, 479
[11] Zuiderwijk, E.J. and de Ruiter, H.R. 1983, N.N.R.A.S. 204, 675
[12] Cecil, G. and Stockton, A. 1985, Ap. J. 288, 201
[13] Sulentic, J.W. and Arp, H. 1986, "the Galaxy-Quasar Connection
      NGC 4319/Mark 205". Ap.J. submitted
[14] Kunth, D. and Bergeron, J. 1984, M.N.R.A.S. 210, 873
[15] Kronberg, P., Burbidge, E.M., Smith, H.E. and Strom, R.G. 1977,
      Ap.J. 218, 8
[16] Bridle, A.H. and Perley, R.A., 1984, Ann. Rev. Astron. Astrophys.
      Vol. 22
[17] Arp, H., Wolstencroft, R.D. and He, X.T. 1984, Ap. J. 285, 44
[18] Arp, H. and Duhalde, O. 1985, P.A.S.P. 94, 1149
[19] Burbidge, E.M. Junkkarinen, V.T., Koski, A.T., Smith, H.E., and
      Hoag, A.A., 1980, Ap. J. 242, L55
[20] Condon, J.J. 1983, Ap. J. Supp. 53, 459
[21] Kronberg, P., Biermann, P. and Schwab, F.R. 1985, Ap. J. 291, 693
[22] Arp, H. 1986, Nature 322, 316
[23] Arp, H. 1986, "3C 120 and the Surrounding Region of Sky" J. Astro-
      phys. Astr. (India) submitted
[24] Arp, H. 1982, Ap. J. 263, 70
[25] Arp, H. and Sulentic, J.W. 1985, Ap. J. 291, 88
[26] Arp, H. 1986 "The Hubble Relation: Differences Between Galaxy Typs
      S_b and S_c". Max-Planck-Institut für Astrophysik No.236; A.J. sub-
      mitted.
[27] Arp, H. 1986, "Quantization and Systematic Redshifts in the Two
      Nearest Groups of Galaxies". J. Astrophys. Astr. (India) submitted
[28] Arp, H. 1986, Astron. Astrophys. 156, 207
[29] de Vaucouleurs, G. 1972, IAU Symposium No.44, 353
      de Vaucouleurs, G. and Peters, W.L. 1986, Ap. J. 303, 19
[30] Giraud, E. 1986, Ap. J. 301, 7

DISCUSSION

BERGERON: We have obtained deep CCD imaging of the field around the
QSO-galaxy pair 1327-206. The existence of a faint east arm from the
galaxy is confirmed and further this feature is bending back toward
the galaxy and appears more as a loop. The NaI doublet seen in
absorption in the QSO spectrum at the redshift of the galaxy (z = 0.018)
exhibits a complex velocity structure which indeed agrees with the
disturbed morphological structure of the interstellar material.

Finally on the line of sight to the QSO there are also
absorbing clouds at much higher redshift (z = 0.853) than that of the
galaxy. This second absorbing system is very strong and shows multiple
velocity structure.

There is a poster on these observations.

ARP: It is typical of these systems where a high redshift object is
associated with a low redshift object, that there are often other
systems of radically different redshift also associated. 3C 120 is
an example of this. In the case we are discussing here, the alignment
of the objects of different redshift is so exact that the probability
of them being unrelated objects accidentally in the line of sight is
vanishingly small. In addition we have the connection by the luminous
filament of the quasar to the low redshift galaxy.

RUBIN: For many years Dr. Arp has been showing us that galaxies can be
very peculiar, based on skillful picture processing of beautiful
photographic images. Recent CCD images now show that many galaxies for
formerly thought to be "normal" are instead embedded in low light level
envelopes containing streamers, jets, rings, and shells. Such galaxies
must not be very rare, but their interpretation is still open to
question, and does not necessarily require a non-Hubble interpretation
of redshifts.

ARP: The objects which have supplied critical evidence for the physical association of objects of different redshift have all emerged from
normal surveys of the sky; Palomar Atlas, radio surveys, etc. It is
natural to study these objects with greater resolution and greater
sginal-to-noise. In many cases we then find additional, detailed
confirmation of the physical association of these objects. But these
additional confirmations are always judged in the context of what
galaxies normally show under higher resolution and lower surface
brightness levels.

It would be completely incorrect to take objects which have
gross peculiarities, for example, those in the Atlas of Peculiar

Galaxies, and imply that they are not peculiar because of anything that happens at very low light levels.

FILIPPENKO: You have shown us some very interesting results concerning NGC 4319 and Mark 205. I should point out, however, that Wallace Sargent and I are conducting a detailed spectroscopic survey of nearby galaxies, and we have found at least a few spirals in which [NII]$\lambda\lambda$6583, 6548 are the only emission lines (other than [OII]λ3727) visible throughout the nuclear region. Thus, NGC 4319 is not an exceptional case. I am not convinced that the [NII] emission is indicative of shocks produced by the ejection of Mark 205.

ARP: Starting with the observational work of the Burbidges, it is well known that strong [NII]/Hα ratios are characteristic of spiral nuclear regions. The point about NGC 4319 is that the [NII] is visible and Hα invisible in regions far from the nucleus. In fact, in the regions of the arms which spectra by V. Rubin and others show are characteristically rich in Hα and weak in NII.

I would doubt that you would argue that in the "few" spirals in which you find only [NII] in the nuclear regions, these nuclear regions were not excited by shocks.

Is there an alignment of quasars near NGC 520 ?

E. Gosset, J. Surdej, and J.P. Swings
Institut d'Astrophysique, University of Liège, Belgium

Abstract :

Three statistical tests applied to 65 UV excess objects in the
field of NGC 520 lead to no detection of any significant alignment.

Introduction

In 1985 we (Gosset *et al*, 1986) reported on the results of several statistical analyses of a 25 square degree field around NGC 450, i.e.
essentially on the detection of a clustering on a scale of order 10 arc
minutes. The present contribution deals no longer with clusterings per
se, but with searches for possible alignment(s) in another field containing an irregular galaxy, NGC 520.

The field of NGC 520 was chosen following the claim by Arp and
Duhalde (1985) that "six of the seven nearest quasars to NGC 520 are
aligned so accurately across the galaxy as to demonstrate physical association".

The three statistical tests briefly described below were applied
to a sample of 65 ultraviolet excess objects (including 48 bona fide
quasars) detected in a 4.5º x 4.5º field around NGC 520, on the basis of
U/B dual image Schmidt plates (see Swings *et al*, 1985; Gosset *et al*,
1986). As explained on several occasions (see e.g. Gosset *et al*, 1986),
the "plates have been scanned at least four times by two different per-
sons in an unbiased homogeneous and objective manner".

A. Hewitt et al. (eds.), Observational Cosmology, 499–502.
© *1987 by the IAU.*

Statistical tests

The three analyses performed on the 65 objects and the respective results are the following :

i) <u>the blunt angle technique</u>, described by Kendall and Kendall (1980) and Webster (1982). The objects are taken 3 by 3 to form triangles : for each of these, the minimum angle (bluntness) among the supplements of the angles at the vertices is selected; the statistic is a function of the distribution of those minimum angles. Even if the tolerance on the alignment angle is relaxed to 1 degree, no alignment is seen in the field of NGC 520 to better than 1 standard deviation;

ii) <u>the spherical blackboard technique</u>, developed by Kendall (1984). The shape of a triangle belonging to a plane can be represented by one point in a shape-space whose property is to be isometric with a two-dimensional sphere. The distribution of shapes can therefore be investigated through the distribution of the representative points on the above-mentioned sphere. In particular, the locus of the triangles that degenerate into a line is well defined. For the 43680 points (65 objects taken 3 by 3) of our sample, only a minor, almost insignificant, deviation from randomness occurs but its location on the spherical blackboard (L vertex of figure 2 of Kendall (1984)) corresponds to a degeneracy by coincidence of two vertices of the triangles. This clearly indicates a tendency for the objects to cluster and not to line up. Consequently, no statistically significant alignment is found by a three points'statistic in our sample of quasars or quasar-candidates in the field of NGC 520;

iii) <u>a third method</u> derived from a pattern recognition method developed for the automated detection of particle tracks in bubble chambers, has been adapted by one of us (E G) to serve as a statistical test for alignments. The basic idea is simple; each "point" (the object) is transformed into a "line" such that several aligned "points" will give rise to "lines" going through the same intersection point. In the space of the intersection points, an approximate alignment will correspond to a cluster : such a space can be investigated using a hierarchical clustering method like, for example, the single linkage cluster analysis based on the minimum spanning tree (MST, Gower and Ross, 1969). A statistic based on the MST edge-length distribution is actually very powerful, but this test leads to the detection of no significant alignment in the present case.

Conclusion

On the basis of the three tests described above, it is concluded that, as far as our sample is concerned, no significant alignment is detected in a field of 4.5° x 4.5° containing NGC 520.

References

ARP, H., DUHALDE, O. : 1985, Publ. Astron. Soc. Pac., **97**, 1149.

KENDALL, D.G., KENDALL, W.S. : 1980, Adv. Appl. Prob., **12**, 380.

KENDALL, D.G. : 1984, Bull. London Math. Soc., **16**, 81.

GOSSET, E., SURDEJ, J., SWINGS, J.P. : 1986, in Quasars, IAU Symp. 119, eds. : G. SWARUP, V.K. KAPAHI, 45.

GOWER, J.C., ROSS, G.J.S. : 1969, Appl. Statistics, **18**, 54.

SWINGS, J.P., SURDEJ, J., HENRY, A., GOSSET, E., ARP, H. : 1985, Rev. Mexicana Astron. Astrofis., **10**, 91.

WEBSTER, A. : 1982, Monthly Not. Roy. Astron. Soc., **201**, 179.

DISCUSSION

WAMPLER: If you created a _perfect_ line of 5-10 quasars and added it to your observed field would your 3-point statistical tests find or notice it?

SWINGS: The first method is not sensitive to the perfection of the line and, in our particular case, will require approximately eight to nine quasars to reach the 95% confidence limit. The second one will detect the perfection of the line (presence of the shape-points on the LM edge of the spherical blackboard); however, concerning the number of aligned objects, the computation of the probability will suffer from the same difficulties as the first method. An increase of the sample may add overwhelming noise to the possible signal.

The third method is _not_ a three-point statistical test; it is sensitive to the number of aligned objects but also reacts continuously to the perfection of the line. Six truly aligned quasars would be detected to approximately the 3 standard deviations' level.

ARP: This elaborate statistical test has been performed on a set of

quasars and candidates which are not relevant to the association
with NGC 520. NGC 520 is the second brightest very disturbed galaxy
in the sky. The quasars which are associated with it are conspicuously
bright. If you refer to the picture I show here (from P.A.S.P. 1986)
you see that just the confirmed quasars brighter than 19.0 magnitude
form the line through NGC 520.

Five of the quasars on this line were found from the original
searches of Swings and Surdej. Subsequent searches of the region on
plates taken with a different telescope by Arp and Duhalde revealed
one more quasar on the line and 5 off the line. This is one of the
most completely searched regions anywhere in the sky that I know of.
One can judge for oneself by looking at the picture whether or not the
existence of the line needs to be tested.

SWINGS: I can only restate the conclusion of the work described here,
i.e. that as far as our homogeneous sample is concerned, no significant
alignment is detected in a 4.5° x 4.5° field containing NGC 520.

REDSHIFT ASYMMETRIES AND THE MISSING MASS

Gene G. Byrd[1] and Mauri J. Valtonen[2]
[1]Department of Physics and Astronomy
University of Alabama, U.S.A.
[2]Department of Physical Sciences
University of Turku, Finland

ABSTRACT. We study the existence of missing mass in the outermost regions of galaxies not accessible to study by rotation curve methods. We
consider binary galaxies, groups and clusters of galaxies. Arp has previously explained redshift asymmetries in pairs or groups with "non-
Doppler redshifts". Instead, we propose the asymmetries indicate
contamination by optical pairs or by members which are not gravitationally bound to the group or pair. The group samples which are commonly
used to justify very high missing mass values in spiral galaxies (>>
the mass detected by rotation curves) also exhibit significant redshift
asymmetries. From this and other information, we conclude that spiral
galaxies do not possess very massive halos. Only the rare giant elliptical galaxies, such as the binary pair in the center of the Coma Cluster
of galaxies, apparently possess extremely massive halos. Dynamical
effects of such giants lead to overestimates of the mass of
clusters. The evidence indicates that missing mass sufficient to close
the universe is not concentrated in individual galaxies, groups or rich
clusters.

1. SPIRAL GALAXIES

Spiral galaxies are often found in binary systems and also many groups
are composed of spiral galaxies (Huchra and Geller 1982, HG hereafter).
Strong redshift asymmetries have been demonstrated to exist in spiral
dominated HG groups (Sulentic 1984). A total of 77 HG companions are
blueshifted relative to the primary versus 119 redshifted. To explain
the asymmetries, Sulentic favors the non-Doppler redshifts originally
suggested by Arp (1970). In contrast, if the groups are expanding populations of galaxies either due to the Hubble flow or by dynamical
ejection, the asymmetry is more conservatively explained (Byrd and
Valtonen 1985). Whatever the explanation for the asymmetries, clearly
groups of galaxies as defined by HG cannot be used for missing mass
estimates.

For the nearby groups Sculptor and M81, we explain the asymmetry as
a consequence of the group's expansion and their large angular sizes.
Using the HG membership and picking the edge-on spiral NGC253 as the

A. Hewitt et al. (eds.), Observational Cosmology, 503–506.

true Sculptor primary by mass, we predict 6 blueshifted, 9 redshifted
for M81 and NGC253. The observed numbers, 5 and 10, agree well with the
prediction (Byrd and Valtonen 1985, Valtonen and Byrd 1986). Members of
the Sculptor group in the front and back of the expanding population can
be identified (Richter and Huchtmeir 1984, Graham 1982). Using this information to identify reasonable subgroups in Sculptor lowers the virial
mass-to-light ratio by a factor of 20 from ~500 estimated by HG. Because
we are within the M31 (i.e. Local) group, the situation is more extreme
with a predicted asymmetry ratio of 0.4 members blueshifted to 5.6 redshifted. The observed HG members, 0 and 6 respectively, agree with
prediction.

     Binary pair data is more promising. Some small samples (e.g. Turner
1976, Peterson 1979, White et al. 1983) show redshift asymmetries which
may be due to contamination by optical companions (Valtonen and Byrd
1986). But the much larger sample of Karachentsev (1972) with redshift
information from Karachentsev (1980a) and Tifft (1982) appears relatively free of asymmetries. Thus, the conclusion of Karachentsev
(1980b), that spiral galaxies do not contain much more mass than that
detected in rotation curve studies, is strengthened.

     This smaller mass is supported by many independent studies.
Gottesman and Hunter (1982) obtain a low total mass for NGC3992 by
using its companions. Sandage (1986) concludes from the Hubble flow deviations of the outer members of the Local Group that the total mass of
the Group is so small it leaves no room for extremely massive halos in
our galaxy or the Andromeda Galaxy. Our studies of the dynamics of
spiral galaxies with satellite systems have led to the same conclusion
(Valtonen et al. 1984, Byrd et al. 1986, Thomasson et al. 1986). Although spiral galaxies are the common type of galaxy in the universe,
they cannot contribute to the closing density of the universe more than
one percent at most.

2. CLUSTERS OF GALAXIES

The Coma cluster of galaxies is composed largely of E or SO galaxies.
Its center is dominated by two supergiant E galaxies NGC4874 and
NGC4889. As one of the nearest rich clusters of galaxies, it has been
well studied and provides a good test case for determining missing mass.

     In traditional models (e.g. Kent and Gunn 1982), the center of the
cluster is placed at NGC4874. However, more members of Coma are blueshifted than redshifted relative to NGC4874 (204 versus 145). The difference is found in both bright and dim cluster members. Also, the
opposite asymmetry occurs relative to NGC4889. Doubt is thus cast on
tradition (Valtonen and Byrd 1986). The cluster apparently has two
massive centers associated with the two supergiant galaxies which are
orbiting one another. This conclusion is also supported by galaxy number
densities and velocities of galaxies near NGC4874 and 4889 (Bahcall 1973,
Quintana 1979, Sarazin 1980). Clusters with massive binaries, or even
with one dominant central galaxy, are known by theoretical studies to
eject smaller galaxies from the cluster and thus invalidate virial
theorem determinations of the cluster masses (Valtonen and Byrd 1979,
Valtonen et al. 1985, Saarinen and Valtonen 1985). Consequently, the

mass usually quoted for the Coma cluster ($\sim$2x10^{15}M$_\odot$ for H$_o$=75 km/s/Mpc) is definitely an upper limit and may exceed the true value by a large factor. We estimate that the mass of the Coma cluster is $\sim$10^{15}M$_\odot$, with $\sim$half in the central binary and the rest divided $\sim$equally between the intergalactic medium and cluster galaxies (Valtonen and Byrd 1986). Also, the analysis of x-ray spectral and imaging data suggest that the models based on a straightforward application of the virial theorem give excessive mass estimates by a factor of 2-3 (Cowie, Henriksen and Mushotzky 1986).

3. IMPLICATIONS FOR DISTRIBUTION OF MISSING MASS

We have seen evidence that spirals and ordinary elliptical galaxies do not have very massive halos and also that supergiant elliptical galaxies are exceptional objects made up almost entirely of dark matter. This E galaxy dark matter does not seem to exist in significant amounts outside the cluster core. Although supergiant elliptical galaxies are very massive, they are also rare. Therefore, their contribution to the overall mean mass density of the universe is probably no greater than the contribution of ordinary galaxies. Altogether, we estimate that only matter to generate a cosmological density parameter $\Omega\approx0.02$ exists in galaxies, groups and clusters. A less concentrated component is necessary for $\Omega=1$.

REFERENCES
Arp, H. 1970 Nature 225, 1033.
Bahcall, N.A. 1973, Ap.J. 183, 783.
Byrd, G.G., Saarinen, S. and Valtonen, M.J. 1986, M.N.R.A.S. 220, 619.
Byrd, G.G. and Valtonen, M.J. 1985, Ap.J. 287, 535.
Cowie, L.L., Henriksen, M.J. and Mushotzky, R. 1986, Ap.J., in press.
Gottesman, S.T. and Hunter, J.H. 1982, Ap.J. 260, 65.
Graham, J.A. 1982, Ap.J. 252, 474.
Huchra, J.P. and Geller, M.J. 1982, Ap.J. 257, 423.
Karachentsev, I.D. 1972, Commun. Special Astr. Obs. USSR 7, 3.
Karachentsev, I.D. 1980a, Ap.J.Suppl. 44, 137.
Karachentsev, I.D. 1980b, Astrofizika 16, 217.
Kent, S.M. and Gunn, J.E. 1982, A.J. 87, 945.
Peterson, S.D. 1979, Ap.J.Suppl. 40, 527.
Quintana, H. 1979, Ap.J. 84, 15.
Richter, O.-G. and Huchtmeier, W.K. 1984, Astr. Ap. 132, 253.
Saarinen, S. and Valtonen, M.J. 1985, Astr. Ap. 153, 130.
Sandage, A. 1986, Ap.J. 307, 1.
Sarazin, C.L. 1980, Ap.J. 236, 75.
Sulentic, J.W. 1984, Ap.J. 286, 441.
Thomasson, M., Sundelius, B., Byrd, G.G. and Valtonen, M.J. 1986 preprint.
Tifft, W.G. 1982, Ap.J.Suppl. 50, 319.
Valtonen, M.J. and Byrd, G.G. 1979, Ap.J. 230, 655.
Valtonen, M.J. and Byrd, G.G. 1986, Ap.J. 303, 523.
Valtonen, M.J., Innanen, K.A., Huang, T.-Y. and Saarinen, S. 1985
   Astr.Ap. 143, 182.
Valtonen, M.J., Innanen, K.A. and Tähtinen, L. 1984, Ap.Space Sci. 107, 209.
Valtonen, M.J., Valtaoja, L., Sundelius, B., Donner, K. and Byrd, G. 1986, preprint
White, S.D.M., Huchra, J., Latham, D. and Davis, M. 1983, M.N.R.A.S. 203, 701.

DISCUSSION

ARP: You are mistaken when you say I have included the Sculptor group in my list of 21 major companions all redshifted with respect to their central galaxy. Those 21 companions are only from the best known M31 and M81 groups where there is no ambiguity as to the dominant galaxy. That is the sample of companion galaxies which cannot be explained by an expanding group model because there are no relative blue shifts whatsoever.

BYRD: The M31, M81, and Sculptor groups were all included in the original discussion of this problem. There is no reason to exclude Sculptor which is about the same angular size as M81's group. Clearly NGC 253 is the most massive and luminous Sculptor spiral. These three groups form a nearby large angular size subsample of the Huchra-Geller list where the "geometric-expanding population" effect can produce redshift asymmetry. Exclusion of Sculptor increases the importance of the M31 group particularly if the many faint M31 group members are now included. The large predicted blue/red = 1/13 asymmetry ratio for the M31 group agrees with Arp's finding of all 21 companions redshifted within the expected random variations (±5).

A POSSIBLE TIRED-LIGHT MECHANISM

Jean-Claude PECKER
Collège de France & Institut d'Astrophysique du
CNRS, Paris

Jean-Pierre VIGIER
Institut Henri Poincaré, Paris.

Recent developments in physics and astrophysics
lead us to reintroduce a new tired-light
mechanism, implying an interaction between a
massive photon and the particles of Dirac's
vacuum.

1. The tired-light mechanisms have appeared in literature as
early as in the late twenties, when Zwicky (1929), worried by
the large values of the apparent recession velocities of
galaxies , conceived a mechanism according which photons
could lose energy through some interaction with the medium
located between the source and the observer. Without entering
in a detailed historical review of these attempts, we would
like to recall that we have suggested similar mechanisms
since 1971. The first idea we developed was (as did earlier
Finlay-Freundlich, 1953-1954, with the approval of Max Born,
1954 a et b) a photon-photon interaction. This mechanism
(Pecker, 1974) was difficult to accept and we looked instead
for possible interactions between the photon and some massive
pseudo-scalar boson, a particle to be specified. Again, as
noted by Schatzman (1979), this mechanism met a large
difficulty : in principle, a notable redshift should be
accompanied by a clear blurring of the images; in spite of
the fact that such an effect was strongly linked to the
unknown characteristics of the alleged particle, the
criticism was essentially valid. Our new suggestion replies
to Schatzman's criticism; it has the same qualities as other
mechanisms of the same kind, namely a qualitative prediction
of both redshift and cosmological background radiation, in
the reference-frame of any cosmology, either assuming a
static or an expanding universe.

A. Hewitt et al. (eds.), Observational Cosmology, 507–511.

2. One knows that the redshift - distance relation

$$z = f(d)$$

depends upon the kind of cosmological theory one adopts. In classical Friedmann-like universes, the law is :

$$z = H_o \frac{d}{c} + O(q_o, d^2)$$

the term of degree 2 being noticeable only for very large and being a possible way to determine the deceleration factor q_o . In Segal's chronogeometry, the tangentiality of the local space with the local minkowskian leads to a square law :

$$z = K_o d^2$$

Finally, the tired-light mechanisms (whatever the detailed description) lead to an exponential law :

$$1 + z = ln \left(H_o \frac{d}{c} \right)$$

or, in first approximation, to a linear law :

$$z = H_o \frac{d}{c} + H_o^2 \frac{d^2}{2c^2}$$

the term of degree 2 is noticeable only for large values of z, of the order 0,5 perhaps.
One should note, therefore, that recent determinations of the law $z = f(d)$ give a strong argument to revitalize the tired-light mechanism, or the Segal's chronogeometry. At small z , Giraud has shown (during this meeting) that biases of the Malmqvist's type cannot explain the non-linearity of relation $z(d)$; on the other side, at large z, LaViolette has shown the tired-light prediction to be very adequate to observed data (1986).

3. On the other side, physics gives new reasons to look at the nature of vacuum with new eyes. Physicists feel more and more that one should not consider as obsolete the de Broglie-Bohm-Vigier point of view according to which the deterministic pilot-wave description could be just as adequate as the Copenhagen point of view to describe the microphysics of interactions (Bell 1986). If this is true, if "empty" waves can travel without associated particles, then a material structure of the "vacuum" is needed. Recently, de Martini (1986) has provided an experimental evidence favoring such a material nature, and Badurek et al., through neutron interferometry, have given a strong, if not decisive, argument for the pilot-wave description (1986). And the Dirac

proposal (1951), revisited by Sinha et al. (1986), may be quite appropriate : The vacuum is a covariant superfluid medium, made of fermions and antifermions - massive of course. Our contribution, inspired in part by the famous Tolman, Ehrenfest, Podolski (1931) paper, is to suggest a description of the effects of an interaction between a massive photon and this Dirac's vacuum.

4. Let us assume that photons, of rest mass m_γ , interact with the vacuum's particle of mass m_o . There is, along the interaction path ℓ , a transfer of energy and momentum from the travelling photon to the vacuum's particles. It gives to the vacuum's particles a motion towards the trajectory (a sort of pinch effect). The loss of photon energy and of photon momentum can be computed. The acceleration towards the track resulting of this acceleration is

$$\int \frac{d^2 y}{dt^2}\, dt = - \frac{2\lambda p \ell}{y[((\ell/2)^2 + y^2]^{1/2}}$$

and the momentum transfer per vacuum particle :

$$\int m_o \frac{d^2 y}{dt^2}\, dt = - \frac{2 m_o m_\gamma \ell}{y[(\ell/2)^2 + y^2]^{1/2}}$$

where λ is the length of the tube of linear density p $(\lambda p = m_\gamma)$, y being the coordinate across the path, t the time.
The effect has a perfect geometrical symmetry, being in essence the result of an interaction between a photon along its trajectory with a strictly symmetrical potential. Hence, Schatzman's criticism does not apply to this mechanism.
The redshift-distance law is obviously of the "tired-light" type :

$$\frac{\Delta \nu}{\nu} = e^{kd} - 1$$

where the masses m_o and m_γ, which fix k, are still unknown.

5. A much more detailed publication is forthcoming. For the time being, our sole purpose is to attract the attention of cosmologists on a possible relevant phenomenon; we expect new developments to occur in laboratory physics before decisive progress on the astrophysical side. It seems to us that one should look at these now "exotic" theories with an open mind. The recent discussions have shown that the classical standard face of the Big Bang cosmologies has suffered many

complications; it has many scars indeed; but, if one has been
able to repair the standard theory, it is undoubtedly at the
expense of its beautiful simplicity which was, some one or
two decades ago, its best asset : it was, truly enough, a
purely aesthetic argument, of which the value, in our eyes,
has never been definitely convincing.

<u>References</u>

Badurek, G., Rauch, H., Tuppinger, D., sept. 1986, Physical
          Review A, sous presse.
Bell. J., 1986 (in press) Nobel Symposium "Possible worlds in
          arts and sciences", Ac. Roy. Sweden
Born, M., 1954a, Nachr. Ak. Wiss Göttingen, $\underline{7}$, 102.
De Martini, 1986, Physics Letters A, Vol. 115, p.421. Dirac,
          P. M., 1951, Nature, $\underline{168}$, 906.
Finlay-Freundlich, E., 1953, Nachr. Ak. Wiss. Göttingen Mat
          Phys. Kl. 7 , 95.
               1954a, Phil. Mag. $\underline{45}$, 303.
               1954b, Proc. Phil. Soc $\underline{A\ 67}$.
               1954c, Phys. Rev. $\underline{95}$, 654.
LaViolette, P., 1986, Ap. J., $\underline{301}$, 544-553.
Pecker, J.-C., Roberts, A.P., Vigier, J.-P., 1972, Nature,
          $\underline{237}$, 227-229.
Schatzman, E., 1979, Astr. & Astrophys., $\underline{74}$, 12.
Segal, I., 1976, Mathematical Cosmology and Extragalactic
          Astronomy, New York, Acad. Phys. Rev.
Sinha, K. P., Sudarshan, E. C. G., Vigier, J.-P., 1986,
          Physics Letters, $\underline{114\ A}$, 298-300.
Tolman, R. C., Ehrenfest, P., Podolski, B., 1931, Phys. Rev.,
          $\underline{37}$, 602.
Zwicky, F., 1929, Proc. Nat. Ac. Sc. , Washington, $\underline{15}$, 773.

DISCUSSION

DEKEL: Let me just clarify your quotation from my talk "no theory
really works." All I meant was that if <u>all</u> the observations discussed
here, and their interpretations at this stage are correct, (which is
not at all evident) then the scenarios we can come up with are hybrids
which are perhaps not as straightforward and elegant as we would wish
them to be. But they are still based on conventional physics within the
standard cosmology.

PECKER: I did not mean to imply that <u>you</u> expressed the need for "new physics." But, so far as I am concerned, comments such as yours encourage me (and should encourage others) not to close the possibility of new avenues. I feel that the standard big bang has become too much of a dogma for not always scientific reasons. And its weaknesses are now obvious.

PECKER: I did not mean to imply that <u>you</u> expressed the need for "new physics." But, so far as I am concerned, comments such as yours encourage me (and should encourage others) not to close the possibility of new avenues. I feel that the standard big bang has become too much of a dogma for not always scientific reasons. And its weaknesses are now obvious.

CHAPTER VIII

GALAXIES AND CLUSTERS

THE LUMINOSITIES AND SIZES OF DISK GALAXIES IN CLUSTERS

G. Giuricin[1], F. Mardirossian[2], M.Mezzetti[2]

1) Osservatorio Astronomico
   via G.B. Tiepolo 11, 34131 – Trieste, Italy
2) Dipartimento di Astronomia
   via G.B. Tiepolo 11, 34131 – Trieste, Italy

ABSTRACT.   Our examination of the absolute luminosities and sizes
of ∼ 500 lenticular and spiral members of several clusters has revealed
appreciable differences in the galaxy luminosity-diameter relationships
for different clusters.

1. INTRODUCTION

The study of the properties of galaxies located in different
regions can allow us to clarify the extent of the influences of
environment in the formation and evolution of galaxies. In this
study we shall focus our attention on the galaxy luminosity-size
relationship, considering the luminosities and radii of about 500
lenticular and spiral galaxies which are very likely to belong to
the clusters Virgo, Hercules, Pegasus I, A262, Zwicky 74-23, A1367,
and A1656.

2. SIZES AND LUMINOSITIES OF CLUSTER GALAXIES

For our sample of galaxies we have gathered together the isophotal
diameters a (reduced to the galactic pole and zero inclination)
in arc min at the 25 B- mag. arc sec^{-2} brightness level from the
RC2 catalogue (de Vaucouleurs et al., 1976) or (when not available
therein) from the UGC catalogue (Nilson, 1973). We have converted
the original UGC diameters according to the relations given in RC2.
We have also taken the corrected blue total magnitudes B_T^0 from RC2;
when they are not available therein, the values of B_T^0 come from
the Zwicky magnitudes converted by us into the B_T^0 system according

A. Hewitt et al. (eds.), Observational Cosmology, 515–517.
© 1987 by the IAU.

to Auman et al.'s (1982) precepts.

The distance of the galaxies of a given cluster was evaluated by means of the Hubble law with an assumed constant $H_o = 100 \text{Km s}^{-1} \text{ Mpc}^{-1}$ and the cluster galactocentric recession velocity.

The distance makes it possible to determine the absolute linear diameters D at the 25 B- mag. arc sec^{-2} brightness level and the absolute blue total luminosity L in the RC2 system.

3. ANALYSIS

A standard linear regression analysis of the fit of the diameters D (in Kpc) and luminosities L (in solar units) to the function

$$\text{Log } L - < \text{Log } L > = b (\text{Log } D - < \text{Log } D >) \tag{1}$$

– for lenticulars, early-type spirals, late-type spirals and irregulars and all spirals – evidences appreciable differences in the apparent and true slopes of the regression lines for different cluster.. We have evaluated the true slope s on the reasonable assumption that the r.m.s. error of Log D is on average 0.04 (this is the average r.m.s. error of the values of Log a listed in RC2). For example in the case of lenticulars, the slopes s of Virgo (s=1.97$\pm$0.12 for N=59) and Pegasus I (s=2.2$\pm$0.2 for N=5) are significantly higher than those of Coma cluster (s=1.25$\pm$0.17 for N=30); in this respect, little can be said about Hercules (s=1.48$\pm$0.33 for N=5) and Z74 – 23 (s=1.61$\pm$0.36 for N=8), for which the errors associated to the estimates of s are rather large. In particular, the smallest (D $\lesssim$ 15 Kpc) lenticulars of Virgo and Pegasus I are appreciably less luminous than the corresponding Coma members.

We have verified that the character of the luminosity–diameter relations does not appear to depend on the presently observed main properties of clusters, such as richness, Rood and Sastry's and Bautz and Morgan's classifications, velocity dispersion, viral radii, crossing time, fraction of spiral members, extent of the HI deficiency of cluster spirals.

Initial conditions occurring at the time of galaxy formation rather than secular processes are thus probably responsible for the observed differences in the power–law relationship between luminosity and diameter.

REFERENCES

Auman, J.R., Hickson, P., Fahlman, G.G., 1982, Publ. Astron. Soc. Pacific 94, 19.
de Vaucouleurs, G., de Vaucouleurs, A., Corwin, H., 1976, Second Reference Catalogue of Bright Galaxies (Univ. of Texas Press, Austin).
Nilson, P., 1973, Uppsala General Catalogue of Galaxies, Ann. Uppsala Astron. Obs. 6.

DISCUSSION

BIRKINSHAW: Might there be a cluster-dependent classification problem which could affect your results (as, for example, for anaemic spirals)?

GIURICIN: Cluster membership assignments and galaxy classifications are reliable for the relatively bright galaxies that we consider.

LUMINOSITY SEGREGATION IN DISTANT GALAXY CLUSTERS

Guy Mathez[1], Olivier Le fèvre[2], Yannick Mellier[1]
Geneviève Soucail[1] and Alain Mazure[3]

[1] : Observatoire
14, Av. E. Belin
F31400 Toulouse
France

[2] : CFHT Corp.
P.O. Box 1597
Kamuela
Hawaii 96743, USA

[3] : Observatoire
Place J. Janssen
92195 Meudon Cedex
France

ABSTRACT : Angular separation-magnitude plots drawn from deep photometry of galaxies in distant clusters show evidence for luminosity segregation which cannot be accounted for by field contamination. This segregation, interpreted in terms of dynamical friction, allows one to determine $(M/L)_g$, the mass-to-light ratio of galaxies **from their velocity dispersion. In A370 (z=0.37) more than 30 velocities have been measured by multi-aperture spectroscopy, leading to $(M/L)_g \sim 70$. Values greater than 100 are found in 5 other distant clusters by deriving the velocity dispersion from the richness. The luminosity segregation, observed even inside each of the clumps of 3C299, could result either from dynamical friction inside the clumps or from some early environmental influence .**

A few years ago, we started a program of deep photometry and multi-aperture spectroscopy of distant clusters involving many collaborators : A. Bijaoui (Nice), B. Fort and J.P. Picat (Toulouse), D. Gerbal (Meudon), R. Pello and E. Salvador-Solé (Barcelona). For the clusters A992, 1305+2952, 3C299 and 0016+16 electronographic BV photometry was first obtained at Canada-France-Hawaii 3.60 m Telescope prime focus and then VRI CCD photometry was made at the focal reducer of the Pic du Midi 2m Telescope (Le fèvre et al. 1986 a,b). For A370, CCD BVR photometry and then multi-aperture spectroscopy with the PUMA system was made at the CFHT Cassegrain focal reducer (Soucail et al. 1986). We would like here to focus on the somewhat surprising **luminosity segregation** found in each of these distant clusters (z=0.25 to 0.55) and on its implications on cluster dynamics.

Observational evidences for luminosity segregation, obtained in nearby clusters by various technics, were first very controversial. In distant clusters, we used the method of Capelato et al. (1980) of comparing the mean angular separations $\langle R_{ij} \rangle$, a characteristic of the point distribution in various magnitude bins. This method does not need any hypothesis on the cluster symmetry, but the price to pay is a major difficulty encountered in the field correction.

A. Hewitt et al. (eds.), Observational Cosmology, 519–521.
© *1987 by the IAU.*

In all distant clusters for which we obtained photometric data, the brighter the galaxies, the smaller are their $\langle R_{ij} \rangle$ hence the more concentrated their distribution (Fig. 1). Verified for the brightest galaxies only, such a property indicates some luminosity segregation above a critical luminosity L_c. This result, obtained as well in the photographic data of Mathieu and Spinrad for 3C295 as in the electronographic and CCD data on 5 distant clusters, is independent of the telescope and of the receptor.

Fig. 1 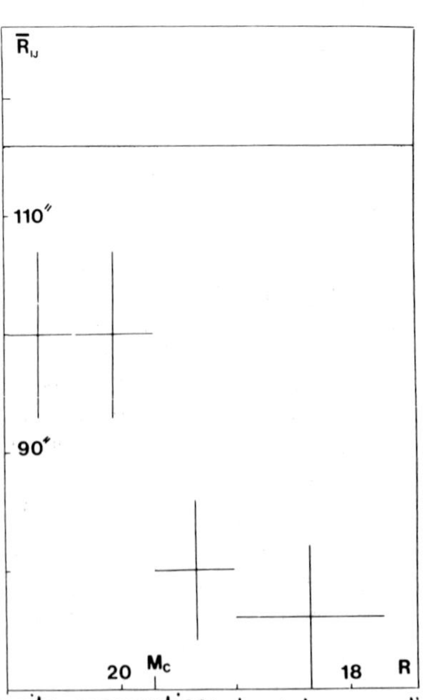

luminosity segregation: galaxy/galaxy mean distance in bins by R mag. The more luminous are the galaxies the more concentrated is their distribution. The horizontal line corresponds to an uniform distribution.

Similar results were (at least partially) attributed to contamination in nearby clusters. Since the mean angular separation is greater between field galaxies than between cluster galaxies, field contamination, supposed to increase with increasing magnitude, will mimic luminosity segregation. This is in general the case in nearby clusters, but **not** in distant ones. Two other arguments are in order : i) the luminosity segregation found in the Coma cluster from photometric data only is still observed among galaxies with Coma redshift (Mellier et al. 1986) and ii) the luminosity segregation found in the distant cluster A370 (Mellier et al. in preparation) from the photographic data of Butcher, Oemler and Wells (1983) **vanishes** in a composite frame consisting of data on the cluster plus its comparison field.

To look for dynamical implications, consider T_{df}, the characteristic time for dynamical friction (Chandrasekhar 1943), inversely proportional to the mass density ρ and to the particle mass m. It is then natural to interpret our finding of a critical luminosity L_c by assuming that the cluster age $T(z)$ is around the time necessary to achieve dynamical friction for particles of mass M_c (corresponding to the luminosity L_c), which writes :

$$T(z) \; \alpha \; \frac{v^3}{\rho \; M_c} \; \alpha \; \frac{v^3}{(M/L)_g^2 \; \rho_L \; L_c} \tag{1}$$

denoting by v the velocity dispersion, $(M/L)_g$ the mean galactic mass-to-light ratio and ρ_L the luminosity density. Once a $T(z)$ relation is determined from the choice of a cosmological model, eq. (1) expresses $(M/L)_g$ in terms of the observable quantities v, L_c and z.

Let us illustrate the method on the **distant cluster** A370 ($z=0.374$) by using our photometric and spectroscopic data, namely the total luminosity $L_v = 1.7 \; 10^{13} \; h_{50}^{-2} \; L_0$, the virial radius $r_{vir} = 0.5 \; h_{50}^{-1}$ Mpc and the velocity dispersion $v = 1380$ km/s. It leads to a **cluster** mass-to-light ratio of 120 h_{50} solar units (without evolution correction to the luminosity). The plot in Fig. 1 gives $L_c = 4 \; 10^9 \; h_{50}^{-2} \; L_0$ then a **galactic** $(M/L)_g = 70 \; h_{50}$, about half the cluster M/L, which seems to indicate that **half of the dark matter lies outside galaxies**. However, the things are less simple. For the other distant clusters, the photometric data, combined with a value of the velocity dispersion guessed from the richness $N_{0.5}$, lead to unacceptably high values $(M/L)_g > 100$ (Le fèvre et al. 1986c). The segregation observed inside the clumps of the cluster 3C299 rather indicates that either the dynamical friction occurs within the clumps before their merging or that the luminosity segregation results from some environmental influence undergone at galaxy formation.

REFERENCES

- Butcher, H., Oemler, A., Wells, D.C. 1983 ApJ. Sup. **52** 183
- Capelato, H., Gerbal, D., Mathez, G., Mazure, A., Salvador-Solé, E., Sol, H. 1980 ApJ. **241** 521
- Chandrasekhar, S., 1943, ApJ. **97** 263
- Le fèvre, O., Bijaoui, A., Mathez, G., Picat, J.P., Lelièvre, G. 1986a A&A **154** 92
- Le fèvre, O., Mathez, G., Picat, J.P., Mauron, N., Mellier, Y. 1986b, A&A Sup., in press
- Le fèvre, O., Mathez, G., Mazure, A., Salvador-Solé, E., Gerbal, D. 1986c, submitted
- Mellier, Y., Mazure, A., Mathez, G., Proust, D. 1986 submitted
- Soucail, G., Mellier, Y., Fort, B., Picat, J.P., Cailloux, M. 1986, submitted

ELEMENTAL ABUNDANCES AND TEMPERATURE DISTRIBUTIONS IN
THE INTRA-CLUSTER MEDIUM OF THE PERSEUS CLUSTER

M. P. Ulmer[1], R. G. Cruddace[2], E. Fenimore[3], W. A. Snyder[2], and G. Fritz[2]

1 Northwestern University, Evanston, Illinois, 60201
2 Naval Research Laboratory, Washington, DC 20375
3 Los Alamos Scientific Laboratory, Los Alamos, New Mexico, 87545

ABSTRACT. Temperatures and iron abundances in two distinct regions of
the Perseus Cluster, based on X-ray observations made with SPARTAN 1,
are reported. Iron abundance values relative to solar are $0.81^{+0.26}_{-0.16}$
and $0.41^{+0.41}_{-0.25}$, and the temperatures are $4.2^{+0.2}_{-0.1} \times 10^7$ K and
$7.1^{+0.8}_{-0.6} \times 10^7$ K, for the regions 0' to 5' and 6' to 20' from the cluster
center, respectively. The uncertainties are 90% confidence.

1. INTRODUCTION

Determination of the elemental abundances and temperatures of the intra-
cluster medium (ICM) of rich clusters of galaxies provides important
cosmological information relating to the formation and evolution of
large-scale structure, dark matter, and scenarios of the generation of
the heavy elements in clusters. The Perseus cluster has long been
recognized as an X-ray source, the primary emission mechanism being
thermal bremsstrahlung from an optically thin plasma. Iron line
emission was detected from the cluster, but, until now, these
measurements could not distinguish between two hypotheses: (1) the iron
line emission is spread throughout the cluster, and much of the intra-
cluster medium has been processed through stars; or, (2) the iron line
emission in concentrated in the cluster core, and much of the ICM is
mainly made up of primordial gas. We report here the first definitive
evidence for spatially varying temperature and abundance in the ICM of
the Perseus cluster. Details of the analysis and interpretation of our
data can be found in Ulmer et al. (1986).

OBSERVATIONS AND RESULTS

The observations were made with two nearly identical systems, each
consisting of a proportional counter and a passive collimator, on board
SPARTAN 1. Figure 1 shows the SPARTAN 1 instrument prior to deployment
with the remote manipulator arm of the space shuttle. The field of
view defined by the collimators was about 5'x 3°. After deployment, the
SPARTAN 1 scanned the sky in the narrow field of view direction.

A. Hewitt et al. (eds.), Observational Cosmology, 523–526.

Figure 1. The SPARTAN 1
instrument just prior to
deployment.

Figure 2 shows the counts versus time for a typical scan across
the Perseus Cluster. Due to limited statistics, we binned the data
into two regions, inner and outer. The inner region includes data
obtained when the field of view of the detectors was with ~5' of the
peak emission; the outer-region data was accumulated when the
instrument was ~6' to 20' from the peak emission.

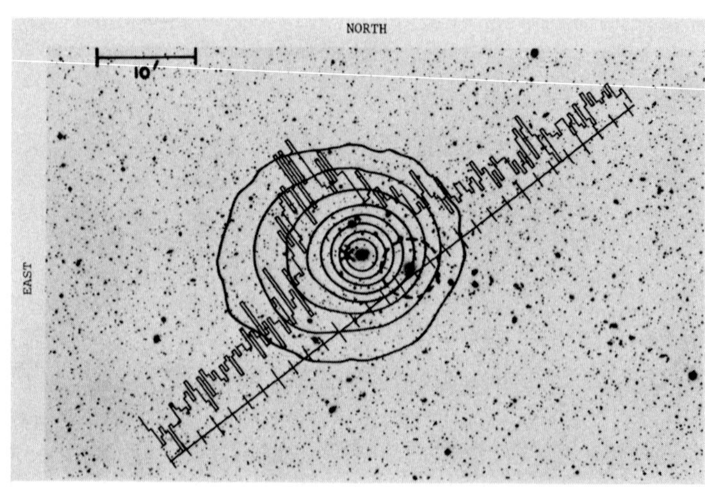

Figure 2. Counting
rate as a function
of time (the histo-
gram) for a typical
SPARTAN 1 scan super
posed on an optical
print of the
cluster. The con-
tours are from the
Einstein Observatory
X-ray image of the
cluster (Branduardi-
Raymont et al. 1978).

The complex spectrum for the inner region is shown in Figure 3. A power
law is needed in addition to a thermal spectrum to produce a good fit
to the data. We note that the best-fit spectral index is 2.05 $\pm$ .10
(photons cm $^{-2}$ s^{-1}) and that the absorption derived from the best fit
is consistent with the interpretation that the X-ray source is the
nucleus of NGC 1275. Details of the fitting procedure and spectra for
the outer region are given in Ulmer et al. 1986.

For the inner and outer regions, our derived best-fit iron abundance
values relative to solar are, respectively, $0.81^{+0.26}_{-0.16}$, and $0.41^{+0.41}_{-0.25}$,
and the temperatures are $4.2^{+0.2}_{-0.1}$ x 10^7 K and $7.1^{+0.8}_{-0.6}$ x 10^7 K. The
uncertainties are 90% confidence values. The abundance values being
nearly solar, we conclude that a substantial fraction of the ICM has
been processed through stars.

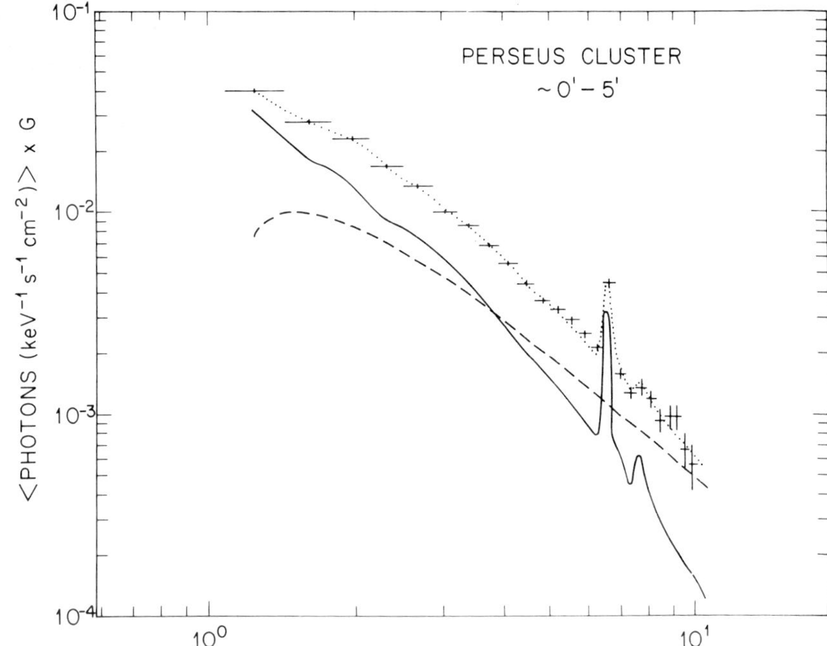

Figure 3. The best fit for spectra for the inner region.

It is difficult to determine whether or not there is an abundance
gradient. Our derived abundances for the inner region being higher than
that derived previously for the entire cluster (e.g. Mushotzky et al.
1978) may be taken as evidence for an abundance gradient in the
cluster. We caution, however, that there are several uncertainties
both in our derived abundances as well as those published elsewhere.
Some uncertainties introduced by the models can be more important than
the statistical errors. For example, the effect of the power-law
continuum on the computed elemental abundances has been ignored by
previous workers. Also, iron line emission being temperature-
dependant, some uncertainty is introduced by assuming an average
temperature for the entire cluster, as has been done prrviously, or for
the selected regions here. We conclude that any evidence from our
results for an abundance gradient in the Perseus cluster is marginal,
at best.

The lower gas temperature we derive in the central region of Perseus is
consistent with the high density and resultant short cooling time (less
than a Hubble time) in the cluster core (Fabian et al. 1978 ; Sarazin
1986 and references therein).

REFERENCES

Branduardi-Raymont, G. D., et al. 1981, Ap. J. 248, 55.
Fabian, A. C., et al. 1981. Ap. J. 248 55.
Mushotzky, R. F., et al. Ap. J. 225, 21.
Sarazin, C. L. 1986, Rev. of Modern Physics, 58, 1.
Ulmer, M. P., et. al. submitted to Ap. J. Letters, October 1986.

DISCUSSION

AUDOUZE: What are the other chemical elements that you expect to see with your observational device?

ULMER: The instrumental energy resolution did not permit us to directly detect other species. The other most prominent lines in the theoretical spectrum are from Mg, Si and S.

TIDAL TRIGGERING OF SEYFERT GALAXIES AND QUASARS: PHYSICAL SUFFICIENCY

G.G. Byrd[1], M.J. Valtonen[2], B. Sundelius[3], L. Valtaoja[2]

[1]Department of Physics and Astronomy, University
 of Alabama, Tuscaloosa, AL 35487 U.S.A.
[2]Department of Physical Sciences, University of Turku, Finland
[3]Onsala Space Observatory and Department of Astronomy,
 Chalmers University of Technology, Sweden

ABSTRACT. Excess optical and radio emission in nuclei of Seyfert
galaxies and quasars is observationally related to nearby companions.
Dahari found Seyferts are more strongly perturbed tidally by companions
than normal spirals are. QSO data suggests a similar relationship.
Quasar and Seyfert activity are thought to be fueled by gas inflow to
the nucleus. To study whether gravitational tides from companions
cause inflow from the spiral's disk, we use a self-gravitating 60,000
particle disk in Miller's polar n-body program. The perturbation to
produce levels of inflow required for activity matches Dahari's Seyfert
triggering level. QSO's are also perturbed by observed companions more
than necessary for high inflow. We reanalyze Dahari's data to find
that at least the majority of Seyferts are in multiple systems. Tidal
triggering thus could be the main cause of Seyfert activity.

1. INTRODUCTION

Seyfert galaxies and quasars both appear to occur in spiral
galaxies characterized by excess optical and radio emission in their
nuclei. In both, inflow of gaseous matter to an "engine" in the nucleus
is thought to fuel the activity (e.g. Simkin et al. 1980, Dahari 1984,
Hutchings and Campbell 1983).

Dahari (1984) studied ~100 Seyferts and a matched angular size
sample of about 300 control spirals. He searched for companions of
the Seyferts measuring the sizes of both and their separations from one
another. Dahari found that more Seyferts had companions than controls
did (37% to 21%). Using the empirical relation between mass and galaxy
size, Dahari's data gives (companion mass/Seyfert mass)/(separation/
Seyfert radius)^3 = the tidal perturbation of the Seyfert by the
companion. Dahari found that Seyferts are perturbed more strongly by
their companions than the controls are. Only 1% of controls are
perturbed at a level > 0.1 compared to 7% of the Seyferts. At levels
from 0.006 to 0.1, he found 9% controls versus 18% Seyferts. At lower
levels, the percentages were about equal. We see that companions may
trigger Seyfert activity.

A. Hewitt et al. (eds.), Observational Cosmology, 527–530.
© *1987 by the IAU.*

One might question the importance of tidal triggering of activity since Dahari found only 37% of Seyferts had companions. We examined Dahari's data further and found this to be a result of Dahari prudently not searching for small companions that might be confused with stars. Also, Dahari did not search for companions beyond three times the diameter of the Seyfert or control. The former selection effect is strongly redshift dependent. By examining the low redshift portion of the sample to remove the above selection effects, we find that 75 to 90% of Seyferts have companions. We used measurements by Dahari (1985) and other sources of data to insure that these percentages are for physical companions with redshifts close to the Seyferts' (Byrd et al. 1987).

The high companion percentage implies that tidal triggering may be the main cause of Seyfert activity. However, observational correlations can be fallacious. An example would be the rancher who thinks vultures kill the dead cows he finds because of the observational correlation between the two. Here, and in our study, a test of physical sufficiency would help validate the hypothesis.

2. METHOD

We test physical sufficiency by using simple computer models of tidally perturbed disk galaxies. The required levels of mass inflow for Seyfert activity are commonly cited between 0.1 and 1.0 $M_\odot$/yr. We ask whether levels of tidal perturbations in the model that produce the required inflow match Dahari's observed perturbations. We perturb our galaxies two ways: with a perturber in a direct parabolic orbit and with a distant perturber fixed relative to the disk spin (Byrd et al. 1986b).

Our computer model uses Miller's two dimensional polar n-body program to calculate the behavior of the self-gravitating galaxy disk using 60,000 particles. Besides this component, we insert an inert "halo" component whose mass relative to the disk is designated, Q_o. We define as "active" an encounter with a companion in which about 400 disk particles are thrown into orbits crossing the inner 5% of the disk radius. If we consider a 2×10^{11} $M_\odot$ galaxy with 1/3 the mass in a 10 kpc disk and 10% of the disk mass as gas, the minimum required inflow is about 10^8 $M_\odot$ of gas during the ~two disk revolutions of the encounter. This is about 0.2 $M_\odot$/yr, within the required range. Though beyond the scope of our simulations, these gas clouds should inelastically collide with other gas to fuel the nucleus.

3. RESULTS VERSUS OBSERVATIONS

We find that strong perturbations cause strong inflows and weak perturbations none. For disks with no halo and a simulated stellar velocity dispersion matching that of our galaxy, the transition occurs at perturbations of 0.006. The disk velocity dispersion determines the transition level. For disks with high mass halos, the perturbation for required inflow is ~0.1. Recall that there are about three times as many Seyferts perturbed at levels above 0.006 than normal galaxies are.

Seven times as many Seyferts are perturbed above 0.1 than normal galaxies. Within all the uncertainties, the agreement of the levels seems reasonable. The few apparently solitary Seyferts in the low redshift portion of Dahari's sample are mainly E or SO galaxies, expected for merger remnants. Simulations indicate disk galaxies quickly merge with close companions (Byrd et al. 1986a).

For QSO's, Stockton (1982) examined low redshift quasars with compact companions nearby and concluded that: the objects were only a few kpc from the QSO's, at least 20% of the QSO's may have such objects nearby, and that the objects are probably cores of galaxies that have interacted with the QSO galaxies. Companions of comparable masses several galaxy diameters (much less a few kpc) generate a tidal perturbation greater than 0.1, the triggering level we found. Thus observations of quasars are consistent with tidal triggering.

Bar distortions in the nuclear bulge have been suggested to feed the nucleus with disk material (e.g. Balick and Heckman 1982, Simkin et al. 1980). Strong tidal perturbations cause bars in our model so the two hypotheses do not really conflict. Feeding of the nucleus by gas _from_ the companion (Toomre and Toomre 1972, Keel et al. 1985), requires the two galaxies to be interacting. For feeding from the companion, both galaxies probably must have gas for the "fuel" to lose kinetic energy via inelastic collisions. This is supported by Dahari's (1985) observational study of strongly interacting pairs where many spirals were Seyferts but no ellipticals were active.

The authors thank R. H. Miller for use of his program.

REFERENCES
Balick, B., Heckman, T.M.: 1982, Ann. Rev. Astron. Astrophys. 20, 431.
Byrd, G.G., Saarinen, S., Valtonen, M.J.: 1986a, M.N.R.A.S., 220, 619.
Byrd, G.G., Valtonen, M.J., Sundelius, B., and Valtaoja, L.: 1986b,
    A. and A. 166, 75.
Byrd, G.G., Sundelius, B., and Valtonen, M.: 1987, A. and A. (in press).
Dahari, O.: 1984, A.J. 89, 966.
Dahari, O.: 1985, Ap. J. Supp. 57, 643.
Dahari, O.: 1985, A. J. 90, 1772.
Hutchings, J.B., Campbell, B.: 1983, Nature 303, 584.
Keel, W.C., Kennicutt, R.C., Hummel, E., van der Hulst, J.M.: 1985,
    A.J. 90, 708.
Miller, R.H.: 1976, J. Comput. Phys. 21, 400.
Miller, R.H.: 1978, A.J. 224, 32.
Simkin, S.M., Su, H.J., Schwarz, M.P.: 1980, Ap. J. 237, 404.
Stockton, A.: 1982, Ap. J. 257, 33.
Toomre, A. and Toomre, J.: 1972, Ap. J., 178, 623.

DISCUSSION

ULMER: How does tidal disruption relate to the production of the engine in the center?

BYRD: The tidal disturbance throws some disk gas into orbits
crossing the nuclear region. This gas then collides inelastically
with gas in other such orbits (or with gas already in the nucleus).
After a series of such collisions the gas could then be part of the
accretion disk of the black-hole.

We are modest in our aims, in that we simply tabulate the
amount of gas thrown into such "accretable" orbits. We do not
concern ourselves with processes near the black hole. We plan future
studies to model the inelastic collision process.

BOYLE: How many Seyferts in the Dahari sample show the typical
perturbations that your modelling predicts?

BYRD: Dahari's 1984 work on comparative perturbations of normal
spirals and Seyferts was not concerned with morphology, only
separations and relative masses. However, Dahari (1985) later
studied interacting pairs (which show distortions similar to those
seen in our models) and found there was a higher percentage of
Seyferts in the interacting than the non-interacting sample of
spirals.

AN INTERESTING PAIR OF ACTIVE GALAXIES

X.-T.He[1], X.H.Xiao[1], S.Okamura[2], T.Noguchi[2], H.Maehara[2]
and R.D.Cannon[3]
1. Department of astronomy, Peking Normal University, China
2. Tokyo Astronomical Observatory, University of Tokyo
   Mitaka, Tokyo 181, Japan
3. Royal Observatory, Blackford Hill, Edinburgh, EH9 3HJ
   Scotland

ABSTRACT. A pair of emission-line galaxies with separation 7.1
arcseconds was found from the survey of the UK Schmidt objective-prizm
plates. The spectroscopic observations by the 1.88m reflector of the
Okayama Observatory show the same emission-line features with Z=0.0441.

1. INTRODUCTION

Using objective-prizm plates taken with the UK 1.2m Schmidt telescope on
the Siding Spring Mountain in Australia, a systematic survey for
emission-line objects has been made in a 25 square degree field centred
on 10^h40^m and 5°00'. The survey work was carried out in the Royal
Observatory, Edinburgh and the Peking Observatory, and checked in the
Kiso Observatory. A pair of emission-line galaxy candidates numbered as
No.34 and No.35 was found in this survey with positions as following,

No.34 $10^h30^m51^s.61$ 7°23'38".1
No.35 $10^h30^m51^s.64$ 7°23'31".1   (1950.0)

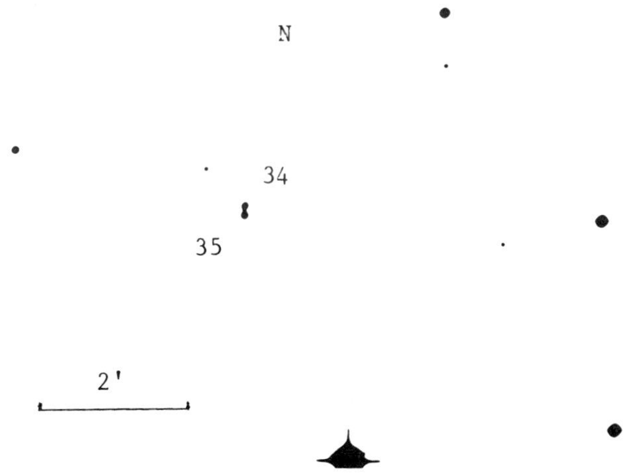

Figure 1. Finding chart
of the two galaxies No.34
and No.35

531

A. Hewitt et al. (eds.), Observational Cosmology, 531–534.

2.   SPECTROSCOPIC OBSERVATION

Both galaxies were observed on April 13, 16 of 1985 and April 12 of 1986
using a Cassegrain spectrograph with an image intensifier attached to
the Okayama 1.88m reflector. The exposure time were 111 min, 60 min and
120 min respectively with IIa-O plates. The wavelength range is $\lambda\lambda$ 3500-
7500 A with dispersion 237 A/mm at H_β. Both objects show the same
emission-line features.

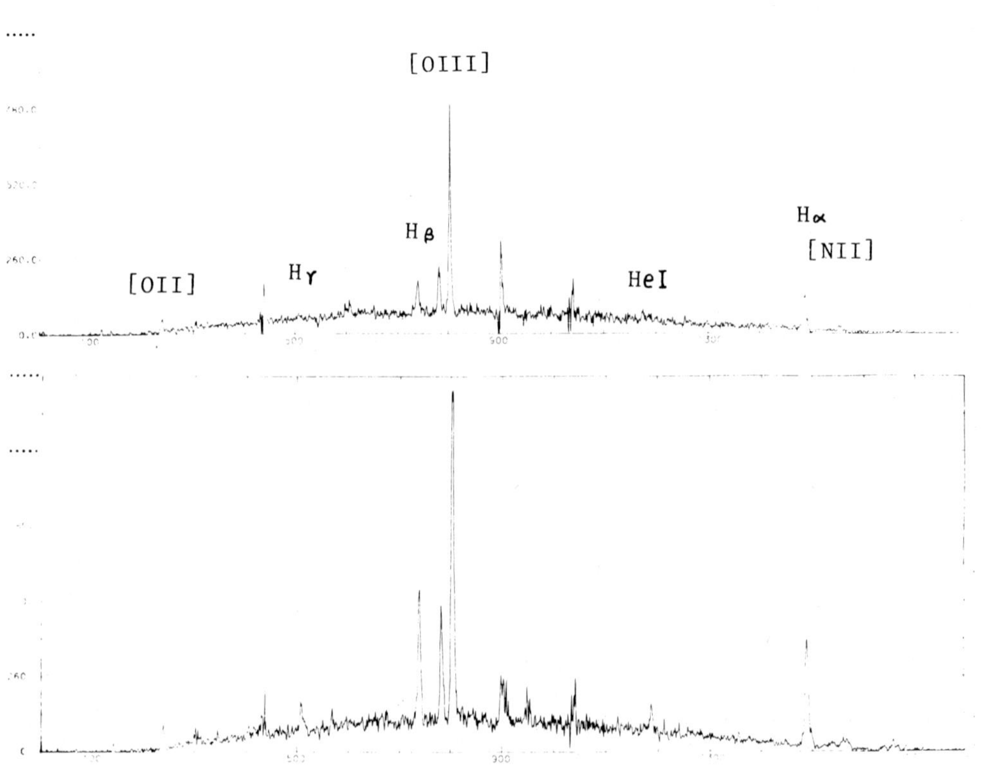

Figure 2. The tracing of the spectra for No.34 (above) and No.35 (down)
without spectrophotometric correction taken on April 12 of 1986.

    The emission lines and the relative strengths calibrated
spectrophotometriclly are listed in Table 1. There are emission line
(1), observed wavelength (2), redshift Z (3), equivalent width (4),
relative strength (5) and ratio with H_β (6) in the table. Each column
has two figures, the left one represents galaxy No.34 and right one
No.35. The figures with bracket are uncertain.
    The spectral classification is mainly based on the ratios of the
line intensities (Baldwin et al 1981). The line [NII] 6584 is unresolved
which is as a weak red wing on H_α. The [NII]/H_α is only about 0.1 which
likes HII region, also the absence of [NeV] and [OI] would argue against
as a nonthermal power source. Both objects may be classified as

extragalactic HII regions except [OIII] 5007/H$_\beta$ is smaller than typical HII regions (Veron and Veron 1985). This pair is quite similar with 1300+361 A,B (Halpern et al 1984). In the Virgo cluster region we have found ten similar compact, metal-poor, extragalactic HII regions (He and Impey 1986). We will refer to members of this class as blue compact dwarf galaxies (BCG).

The redshifts of both galaxies are exactly same as Z=0.0441$\pm$0.0002.

TABLE I Emission-lines, relative strengths and Redshifts

Emission line (1)	Wavelength of Obs. (2)		Z (3)		E.W. (4)		I (5)		I/I (6)	
	(34)	(35)	(34)	(35)	(34)	(35)	(34)	(35)	(34)	(35)
[OII] 3227	3891	3891	0.044	0.044	(24)	(30)	(631)	(948)	1.58	1.01
Hγ 4340	4529	4529	0.044	0.044	5	10	100	211	0.25	0.26
Hβ 4861	5076	5076	0.044	0.044	21	39	399	858	1.00	1.00
[OIII]4959	5178	5178	0.044	0.044	20	28	400	644	1.00	0.75
[OIII]5007	5227	5227	0.044	0.044	83	126	1884	3276	4.72	3.83
HeI 5876		6136		0.044		14		385		0.45
Hα 6563	6853	6853	0.044	0.044	111	192	2830	7200	7.10	8.39
[NII] 6584	6867	6872	0.043	0.044	13	16	331	600	0.82	0.70
[SII] 6717	7014	7018	0.044	0.045	(7)	(16)	(191)	(611)	0.48	0.71

3. Photometric Observation

The photometric calibration was made using the plates taken with the Kiso 105cm Schmidt telescope. The magnitudes in B-band are 16.89+0.43 (No.34) and 16.09+0.43 (No.35). The absolute magnitudes (H=50, q_0=1) are -20.23+0.43 and -21.03+0.43

The brightness does not show variable to compare the photographic plates of April 2 1986 with March 28 1984.

The separation between the two galaxies is only 7.1 arcseconds and about 3.0x10^{4}ly due to the distance of 8.7x10^{8}ly

4. Summary

A pair of blue compact dwarf galaxies
Z=0.0441+0.0002 and separation = 7".1
m$_b$=16.89+0.43 (No.34)   M$_b$=-20.23+0.43 (No.34)
    16.09+0.43 (No.35)   -21.03+0.43 (No.35)

References

Baldwin,J., Phillips,M. and Terlevich,R. 1981, P.A.S.P., **93**, 5.
Halpern,J., Marshall,H. and Oke,J. 1984, A.J., **89**, 1802.
He,X.-T. and Impey, C. 1986, M.N.R.A.S., **221**, 727.
Veron, M.-P. and Veron, P. 1985, ESO Scientific Report No.4.

DISCUSSION

VERON: Can you say anything about the [NII]/Hα ratio which would tell
if these objects are extragalactic HII regions or Seyfert 2?

HE: Yes, according to the ratios of line intensities both objects
may be classified as extragalactic HII regions.

A COMMENT ON THE SEYFERT ENVIRONMENT

Zou Zhen-long and Chen Jian-shen
Beijing Astronomical Observatory
Beijing, China

The study on the Seyfert environment has been of considerable interest
lately, since it is believed that the tidal force of a close companian
may drive gas into the inner region and either fuel nuclear activity
directly or lead to starbursts (Toomre et al., 1972, Byrd, 1987).
Many authors noticed an excess of Seyfert among close pairs of
galaxies (Adams, 1977, Kennicutt et al., 1984, Keel et al., 1984,
Dahari, 1984, 1985 a, b). In a recent paper of ours (Chen and Zou,
1986), a neighborhood statistical study has been done for the galaxies
in CfA sample (Huchra et al., 1983), and it is found that the fraction
of Seyferts is 3 times higher than the average one, if the distances
between them and its nearest neighbors are less than 20 kpc. The
result would strengthen the opinion that the existance of close
companian is a favourable environment to trigger Seyfert activities.

REFERENCES

Adams, T.F., 1977, Astrophys. J. Suppl., 33, 19.
Byrd, G.G., 1987, this Symposium.
Chen, J.S. and Zou, Z.L., 1986, Acta Astrophysica Sinica, 6, 169.
Dahari, O., 1984, Astron. J., 89, 966.
Dahari, O., 1985, Astrophys. J. Suppl., 57, 643.
Dahari, O., 1985, Astron. J., 90, 1772.
Huchra, J. et al., 1983, Astrophys. J. Suppl., 52, 89.
Keel, W.C. et al., 1985, Astron. J., 90, 708.
Kennicutt, R.C. et al., 1984, Astrophys. J., 279, L5.
Toomre, A. et al., 1972, Astrophys. J., 178, 623.

DISCUSSION

R.L. DAVIES: Seyferts tend to occur in earlier type spirals; we know
from the morphology-density relation, found by Dressler, that these

A. Hewitt et al. (eds.), Observational Cosmology, 535–536.
© 1987 by the IAU.

galaxies are more clustered than average. Have you taken this into
account in your analysis? Are there other selection effects that
might change your result?

ZOU: In our sample, the number of early type spirals ($0 < T < 5$)
with $d < 20$ kpc is 38, and the fraction of Seyferts in all of this
type of galaxies is 4.2‰. The observed Seyferts (6) are much more
than expected (1.6), so the selection effect could be ignored.

SEARCH FOR STRUCTURAL CHANGES IN THE CORES OF "NEARBY" RADIO GALAXIES

E. Preuss and W. Alef
Max-Planck-Institut für Radioastronomie
Auf dem Hügel 69
D-5300 Bonn 1
Fed. Rep. of Germany

Recent VLBI observations of the central radio components in the lobe-dominated radio galaxies 3C111 (Götz et al. 1987) and 3C390.3 (Alef et al. 1987) show drastic changes in the structure of the compact emission regions. In 3C111 a jet-like feature with two knots and a flux density of 600 mJy which was visible in 1980.4 had disappeared in 1985.4. This implies "superluminal behaviour" with a characteristic speed of $\gtrsim 3c$. In 3C390.3, there is also evidence for superluminal motion: radio knots appear to be ejected from the core at a rate of about one every 4 years and move out with an apparent velocity of 4.1 c ($H_0 = 50$ km/s/Mpc throughout this paper) in the direction of the north-west extended lobe.

These observations are clearly relevant to the question: are radio galaxies energized by relativistic jets, and if so, is the apparent structure and temporal behaviour of their central radio components strongly influenced by relativistic beaming effects. The beaming hypothesis in its simplest form (Scheuer and Readhead 1979, Blandford and Königl 1979, Orr and Browne 1982) leads to "unifying schemes" which predict that in a sample of randomly oriented radio sources with similar intrinsic parameters, observable properties such as core prominence, asymmetry, and the occurrence of superluminal motion are mainly determined by the viewing angle of the observer. The best test of this hypothesis would be the VLBI observation of a statistically complete sample of radio galaxies without bias in orientation. This has so far been impossible due to sensitivity limits and the lack of dedicated VLBI networks. Only about a dozen of the central components of radio galaxies have been mapped with milliarcsec resolution.

The observation of strong structural variability in 3C111 and 3C390.3 prompted us to review the status of VLBI observations of a complete sample of "nearby" radio galaxies, which are strong at the 3C survey frequency of 178 MHz. We may ask whether the VLBI results available for this sample, collected in an unsystematic fashion by various observers bears some statistical significance with regard to the relativistic beaming hypothesis.

Table 1 lists 39 radio galaxies along with their milliarcsec core characteristics, if available, and other relevant parameters. These objects form a distance limited (z < z(3C273)=0.158) subset of the

537

A. Hewitt et al. (eds.), Observational Cosmology, 537–541.

Table 1: Object (1)	P_{178} $(\frac{erg}{s.Hz})$ (2)	Radio type (3)	Size (kpc) (4)	$R_{5\ GHz}$ (5)	S_5(VLBI core) (Jy) (6)	$V_{app.}$ (7)	Refs. (VLBI) (8)
3C31	2.1E32	I	50	.06	<.12		1
3C33	8.6 33	II	380	.003			
3C35	2.1 33	II	980	<.02			
3C66B	5.0 32	I	150	.06	.12		2
3C76.1	5.6 32	I	40	<.01			
3C83.1B	7.6 32	I	140	.006			
3C84	7.9 32	I	180	.99	60. M	.4c	3
3C98	1.9 33	II	250	<.01	<.08		2
3C184.1	8.3 33	II	470	.005			
DA240	1.2 33	II	1960	.07	<.08		1, 2
3C192	3.4 33	II	300	<.004			
3C223	1.3 34	II	800	.007			
3C236	6.4 33	II	5860	.54	.32 M		4
3C264	4.9 32	I	30	.17			
3C272.1	8.1 30	I	11	.11	∿.13		2
3C274	8.4 32	I	6	.05	2.2 M	<.3c	5
3C285	2.2 33	II	230	<.01			
3C293	1.2 33	I	100	.80	<.08		2
3C296	3.2 32	I	180	.04			
3C303	1.0 34	II	60	.21	<.09		2
3C305	1.2 33	I	13	.09	<.08		2
3C310	7.1 33	I	360	.07			
3C314.1	6.9 33	I	520	<.03			
3C315	9.5 33	I	530	.13			
3C321	5.6 33	II	690	.03			
3C326	7.4 33	II	2630	.03			
N6109	4.1 32	I	690	.04			
3C338	1.2 32	I	34	.04	∿.09		2
N6251	2.5 32	I	2900	.58	.85 M	<.6c	6
3C382	3.0 33	II	280	.11	∿.08		2
3C386	3.0 32	I	120	.005			
3C388	9.2 33	II	70	.05	<.10		1
3C390.3	6.8 33	II	330	.09	.35 M	4.1c (?)	7
3C433	2.6 34	I	90	<.004			
3C442A	4.9 32	I	200	<.01			
3C449	1.5 32	I	150	.03			
3C452	1.6 34	II	730	.04	<.08		2
N7385	2.8 32	I	80	.09			
3C465	1.4 33	I	430	.10	.23		2

Meaning of Columns:

(2) Spectral power at 178 MHz; H_0 = 50 km/s/Mpc and q_0 = 0.5 throughout this table; e.g., 2.1E32 $\equiv$ 2.1·10^{32}; Note: P_{178} (CYGNUS A) = 1.2E36 erg/(s.Hz)

(3) Classification of the overall radio structure of double sources

(Fanaroff and Riley 1974): the brightness distribution either peaks near the centre (type I) or at the ends of the radio lobes (type II).

(4) Projected overall size

(5) Fractional core flux density S(arcsec core)/S(total) at 5 GHz

(6) Flux density of milliarcsec core at 5 GHz, i.e. maximum correlated flux measured in VLBI pilot observation or flux contained in VLBI map marked by "M", if any. Note: the milliarcsec core fluxes of the remaining sources (without entry in column (6)) are all <.09 Jy (<.02 for most of them), the upper limit to their arcsec core fluxes.

(7) Apparent transverse velocity of any component motion

(8) References to VLBI observations. The numbers in columns (2), (4) and (5) are based on data from Laing et al. (1983), Kühr et al. (1979), Bridle and Perley (1984), and references therein.

References to Table 1:
[1]Graham et al. (1981)   [4]Barthel et al. (1985)
[2]Preuss et al. (1977)   [5]Schmitt and Reid (1986)
[3]Romney et al. (1984)   [6]Jones (1986)
                                        [7]Alef et al. (1987)

catalogue by Laing, Riley and Longair (1983), which lists all radio sources with $S(178\ MHz) \geq 10$ Jy, $\delta \geq 10°$ and $|b| \geq 10°$. M82 has been omitted as it is both qualitatively and quantitatively very different from the other objects. The sample includes 16 classical double sources of type II. Note that some well known radio galaxies with relatively strong cores are not in this sample either because their galactic latitude is too low (Cyg A, 3C111) or because their 178 MHz fluxes are too weak (NGC315, 3C120). Of the 39 galaxies, 8 (2 of type II) have milliarcsec cores stronger than 0.1 Jy at 6 cm; of these, 5 have been mapped with milliarcsec resolution and for 4, there is some kinematic information available (Tab. 1). Note that 3C390.3 has the fourth strongest milliarcsec core in the sample, and the strongest in the subset of type II sources.

   3C390.3 is the first lobe-dominated radio galaxy for which the evidence available so far suggests that superluminal motion is present. Its core contains only 9% of the total flux density at 6 cm and ranks only as number 12 in the whole sample. The observation of superluminal motion, the relative core strength, and overall size of 3C390.3 are not incompatible with the relativistic beaming hypothesis if one assumes, e.g., a Lorentz factor of 5 and an angle of $22°$ between the source axis and our line of sight. However, if we interpret the observation in this way, only one or two more objects in the sample should show superluminal motion with $v \gtrsim 4$ c (e.g. Scheuer 1984). If the sample is restricted to the 23 powerful sources with spectral power $> 10^{-3}$ times that of Cyg A at 178 MHz, statistically no further superluminal sources should be present. Furthermore, from the viewpoint of the relativistic beaming hypothesis, it is disturbing that the objects with the largest overall sizes in the sample (Tab. 1), the giant radio galaxies 3C236, NGC6251, 3C326 and DA240, for which one does not expect substantial Doppler boosting, have some of the most prominent cores. This has also been noted by Saripalli et al. (1986) as a general feature of giant radio galaxies.

We conclude that to the extent that apparent superluminal behaviour reflects relativistic bulk motion, the observations of 3C111 and 3C390.3 (independent of specific models or "unified schemes") support the idea that relativistic bulk motion is present in the central components of radio galaxies. However, the relativistic beaming hypothesis in its simplest form (as basis of a unified scheme) seems, even on the grounds of the limited evidence available so far, hardly applicable to the sample of nearby radio galaxies in Table 1.

REFERENCES

Alef, W., Götz, M.M.A., Preuss, E., Kellermann, K.I.: 1987, (submitted)

Barthel, P.D., Schilizzi, R.T., Miley, G.K., Jägers, W.J., Strom, R.G.: 1985, Astron. Astrophys. 148, 243

Blandford, R.D., Königl, A.: 1979, Astrophys. J. 232, 34

Bridle, A.H., Perley, R.A.: 1984, Ann. Rev. Astron. Astrophys. 22, 319

Fanaroff, B.L., Riley, J.M.: 1974, Monthly Notices Roy. Astron. Soc. 167, 31P

Götz, M.M.A., Alef, W., Preuss, E., Kellermann, K.I.: 1987, Astron. Astrophys. (in press)

Graham, D.A., Weiler, K.W., Wielebinski, R.: 1981, Astron. Astrophys. 97, 388

Jones, D.J.: 1986, JPL Astrophys. preprint No. 122

Laing, R.A., Riley, J.M., Longair, M.S.: 1983, Monthly Notices Roy. Astron. Soc. 204, 151

Kühr, H., Nauber, U., Pauliny-Toth, I.I.K., Witzel, A.: 1979, MPIfR Preprint No. 55

Orr, M.J.L., Browne, I.W.A.: 1982, Monthly Notices Roy. Astron. Soc. 200, 1067

Preuss, E., Pauliny-Toth, I.I.K., Witzel, A., Kellermann, K.I., Shaffer, D.B.: 1977, Astron. Astrophys. 54, 297

Romney, J.D., Alef, W., Pauliny-Toth, I.I.K., Preuss, E., Kellermann, K.I.: 1984, IAU Symp. No. 110, 137

Saripalli, L., Gopal-Krishna, Reich, W., Kühr, H.: 1986, Astron. Astrophys. (in press)

Scheuer, P.A.G., Readhead, A.C.S.: 1979, Nature 277, 182

Scheuer, P.A.G.: 1984, ESA-SP 213, 149

Schmitt, J.H.M.M., Reid, M.J.: 1985, Astrophys. J. 289, 120

DISCUSSION

BIRKINSHAW: I'd like to comment on the importance of tracing the high-speed motions inferred from VLBI out to larger scales by high-precision astrometry (for example, using the VLA). I await the results of such studies (which I understand are presently in progress) with great interest.

PREUSS: Attempts to measure the core distance of brightness peaks in extended radio sources on larger scales (1" - 100") with high precision ($<0.''01$) are underway. This is tried with VLBI by phase referencing methods.

LONGAIR: I was intrigued by the fact that, of the four extended radio galaxies studied with VLBI, only 3C 390.3 showed evidence of super-luminal motion. In fact, it is the only one of your 4 sources which Julia Riley and I classified as "good" doubles in the sense that the extended structure of the double source contains hot spots (as edge-brightening) in both source components. Thus it may be that the probability of observing relativistic motion in "good" double sources is much higher than 1 in 4. The importance of extending the VLBI observations to a large sample of "good" doubles is very important.
    I agree with Dr. Miley that the parent populations from which one can select superluminal sources seems to be becoming smaller and smaller and I believe this poses significant problems for the simplest forms of relativistic beaming models.

CHAPTER IX

THE DISTRIBUTION OF RADIO SOURCES

RADIO SOURCE COUNTS AND THEIR INTERPRETATION

K. I. Kellermann
National Radio Astronomy Observatory
Charlottesville, Va., USA

J. V. Wall
Royal Greenwich Observatory
Herstmonceux, U.K.

ABSTRACT.   Radio source counts at several wavelengths are shown and discussed in terms of evolving populations.   The deepest counts now reach a surface density close to a million sources per steradian.   At this level essentially all of the luminous radio galaxies and quasars appear to be included, and the weaker sources apparently reflect a relatively nearby population of less luminous radio sources.

1. HISTORICAL BACKGROUND

Since the early years of radio astronomy, counts of radio sources as a function of flux density have been used together with optical identifications and measured redshifts to explore the distribution of radio galaxies and quasars as a function of luminosity and cosmic epoch.   At optical wavelengths complete surveys of quasars and galaxies are difficult, due to a variety of selection effects which can lead to incompleteness or unreliability.   Flux density-limited samples are affected by varying degrees of obscuration and sky brightness, as well as by the presence of bright spectral features.   Color-selected samples are also influenced by bright emission lines within the filter bandpass as well as by the large and uncertain K correction.   Moreover, morphological distinctions between galaxies and quasars are distance dependent, and for faint objects, quasar surveys may be contaminated by galactic stars.

At radio wavelengths, the sky is cold, absorbtion is negligible, and even for very distant sources, measurements of flux density and position are now routinely made to an accuracy of a few percent and a second of arc, respectively.

A. Hewitt et al. (eds.), Observational Cosmology, 545–564.

The extended radio sources which dominate the counts at low frequencies are optically thin, so that unlike many galaxies at optical wavelengths, the apparent radio luminosity is not orientation dependent. Except very close to the galactic plane, essentially all discrete radio sources appear to be extragalactic and there is a negligible galactic contribution.

The radio surveys, on the other hand, are affected by noise, confusion, and resolution which lead to systematic errors in determining the source count. However, these biases are well understood, and corrections to the observed count are readily determined (e.g., Murdoch et al. 1973, Windhorst et al. 1984). A more serious problem is that it is not possible from radio measurements alone to determine redshifts or even to distinguish between quasars and qalaxies. Despite dramatic improvements in the sensitivity of optical telescopes, the identification and spectroscopic analysis, especially for optically faint counterparts of radio sources is, however, still time consuming. Even for the 3CR catalogue of the strongest sources, the identification and redshift measurements only now approach completeness (Laing et al. 1983, Spinrad et al. 1985).

With complete optical identifications of selected radio samples, the radio luminosity function and its dependence on redshift may be derived in a straightforward way, provided that the geometry of the universe is assumed. However, since there are only a limited number of optical identifications, the determination of the radio luminosity function and its evolution has centered on the analysis of radio source counts, which may include thousands of objects distributed over a wide range of luminosity and redshift and supplemented by optical identifications and redshifts for restricted samples, primarily selected from strong source surveys. For the strong radio sources selected from low frequency surveys, about 80 percent are identified with bright elliptical or N galaxies of absolute magnitude near -23 (H_0 = 50 km/sec/Mpc) and about 20 percent with quasars and BL Lacs (Spinrad et al. 1985). At high frequencies, the strong sources are approximately equally divided between quasars and galaxies. At low flux densities, essentially all quasars have already been counted and the counts at both high and low frequencies are dominated by galaxies (e.g., Windhorst et al. 1986).

Although the radio measurements are now routinely made with high precision and cover a wide range of flux density, this was not always the case. Until the development, in the 1960's, of synthesis arrays and large steerable dishes operating at short wavelengths, the poor sensitivity and angular resolution of radio surveys led to large errors due to noise and especially confusion between nearby unresolved sources. In particular, the first extensive radio source surveys made at Cambridge gave a slope of the log N-log S relation apparently well in excess of the value of -3/2 expected from a uniformly filled static Euclidean universe. At the same time, however, surveys made in Australia indicated a slope much closer to the Euclidean value. Moreover, in the limited region of the sky where the two surveys overlapped, except for the brightest sources, there was little agreement on individual sources. It is now recognized, however, that the effect of large experimental errors in these early surveys were exaggerated by the use of integral counts, which suppresses features in the source count as well as underestimates the uncertainty due to non independent data points (Jauncey 1967, Crawford <u>et al.</u> 1970). Radio surveys now have sufficient resolution to avoid these difficulties. Even more important, however, due to the very large range in flux density (over 10^6 sources per ster) over which counts are now compiled, the slope of the very strongest sources is no longer a crucial part of the general interpretation.

In this review we summarize in Section 2 the modern radio source counts, with particular emphasis on the 5 GHz (6 cm) count which now reaches sources as faint as 25 micro Janskies. In Section 3 we discuss the interpretation in terms of cosmological evolution, and in Section 4 the analysis is extended to distinguish between the compact flat spectrum sources and the extended steep spectrum sources. In Section 5, we discuss alternate interpretations of the source counts, and in Section 6, we speculate on what might be learned from future measurements made with radio telescopes of improved sensitivity and resolution.

2. THE RADIO SOURCE COUNTS.

When presented in differential form, the steep slope of radio source counts makes them inconvenient for graphical presentation. It has become customary to show the differential source count normalized to the static Euclidean slope of -5/2 corresponding to a slope of -3/2 for the integral count. Alternately, some authors choose to show the unnormalized differential count multiplied by $S^{5/2}$. This has the advantage of not requiring the introduction of an arbitrary normalization factor which is usually taken as the number density of sources with flux density above 1 Jy.

In Figures 1 through 4 we show the normalized radio source counts at 408 MHz (74 cm), 1.4 GHz (21 cm), 2.7 GHz

TABLE I

5 GHz Source Surveys

Survey	Instr.	S_{lim}	n	Ω	References
S1[1]	140	800	23	0.27	Kellermann et al. 1968, Astrophys. Lett. 2, 10.
S2[1]	140	600	136	0.97	Pauliny-Toth et al. 1972, A. J. 77, 265.
S3[1]	140	600	123	1.06	Pauliny-Toth and Kellermann 1972, A. J. 77, 797.
S4[1]	300/100	500	269	1.71	Pauliny-Toth et al. 1978a, A. J. 83, 451.
NPS[1]	100	14	82	0.0046	Pauliny-Toth et al. 1978b, A & A Suppl. 34, 253.
S5[1]	100	250	185	0.37	Kuhr et al. 1981, A. J. 86, 854.
	100	10–18	118	0.0071	Pauliny-Toth et al. 1980, A & A 85, 329.
5C6[1]	100	20	96	0.0066	Maslowski et al. 1980, A & A 95, 285.
	300	15	186	0.0096	Leddon et al. 1980, A. J. 85, 780.
	210	32	53	0.0082	Wall et al. 1982, MNRAS 200, 1123.
	VLA	0.54	14	0.00016	Bennett et al. 1983, Nature 301, 686.
	300	35	480	0.069	Owen et al. 1983, A. J. 88, 1.
	VLA	0.35	25	5.9×10^{-5}	Fomalont et al. 1984, Science 225, 23.
	VLA	0.06	9	7.3×10^{-6}	Fomalont et al. 1984, Science 225, 23.
5C2[1]	100	17	106	0.0066	Maslowski et al. 1984, A & A 139, 85.
GB[1]	100	30	99	0.0092	Maslowski et al. 1984, A & A 141, 376.
MG	300	53–106	5974	1.87	Bennett et al. 1986, Ap. J. Suppl. 61, 1.
	300	20	625	0.035	Altschuler 1986, A & A Suppl. 65, 267.

[1]Included in compilation of Maslowski et al. 1984, A & A 141, 376.

(11 cm), and 5 GHz (6 cm), based on the best available data. Exploratory observations have also been made at frequencies as high as 10.7 GHz (2.8 cm) (Aizu 1987). The data at 408 MHz are discussed by Wall et al. (1980); the 1.4 GHz data by Windhorst et al. (1984), Windhorst et al. (1985), and Windhorst (1986); the 2.7 GHz data by Wall et al. (1981), Peacock and Wall (1981), Wall and Peacock (1985) and Wall et al. (1986). The 5 GHz data are summarized in Table I and discussed further in this paper.

Because of the wide range of flux density and source density involved, no single instrument can provide a complete count, even at a single frequency. Pencil beam instruments, large steerable dishes, and phased arrays are typically used to survey large regions of the sky to obtain statistically significant counts for the stronger sources with a relatively low surface density. Separate surveys made from the northern and southern hemispheres are necessary to cover the whole sky, and in Table II we list the major all sky catalogues which have been compiled from large scale radio surveys.

TABLE II

All Sky Catalogues

Freq. (MHz)	S_{min} (Jy)	n	Ω	References
178	10	173	4.05	Laing et al. 1983, MNRAS 204, 151.
408	10	160	10.1	Robertson 1973, Astr. J. Phys. 26, 403.
2700	2	233	9.81	Wall and Peacock 1985, MNRAS 216, 173.
1400	2	234	4.30	Bridle et al. 1972, A. J. 77, 405.
5000	1	518	9.81	Kuhr et al. 1981, A & A Suppl. 45, 367.

Synthesis instruments provide the most sensitive surveys, but only over very small regions of the sky typically covering 10^{-5} to 10^{-4} ster. Counts of very faint sources based on only a few such small fields may be subject to error if there is significant clustering. Nevertheless, in contrast to the early radio source surveys, modern data obtained by different authors using very different kinds of radio telescopes are in good agreement with respect to individual source position and flux density as well as in surface density.

Examination of Figures 1 through 4 shows five regions of the source count as follows:

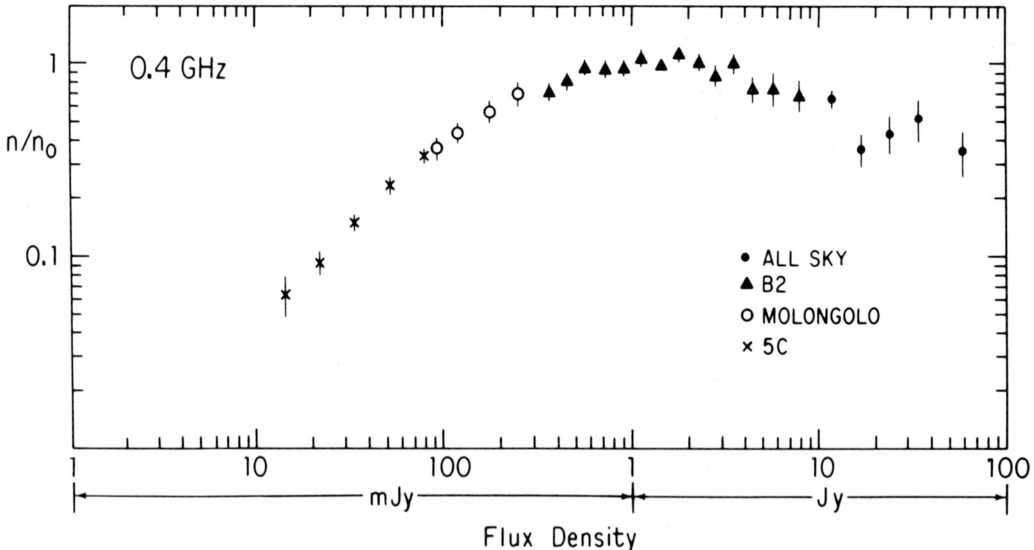

Fig. 1. Source count at 0.4 GHz normalized to a uniformly filled static Euclidean uviverse with a differential number count $n_0 = 1125 S^{-5/2}$. Data are taken from the all sky catalogue of Robertson (1973), from the Bologna B2 Survey (Colla et al. 1973, A&A Suppl. 11, 291), from the Molongolo Deep Survey (Robertson 1977, Austr. J. Phys. 1977, 30, 241), and from Pearson and Kus (1978, MNRAS 182, 273), all as compliled by Condon (1984).

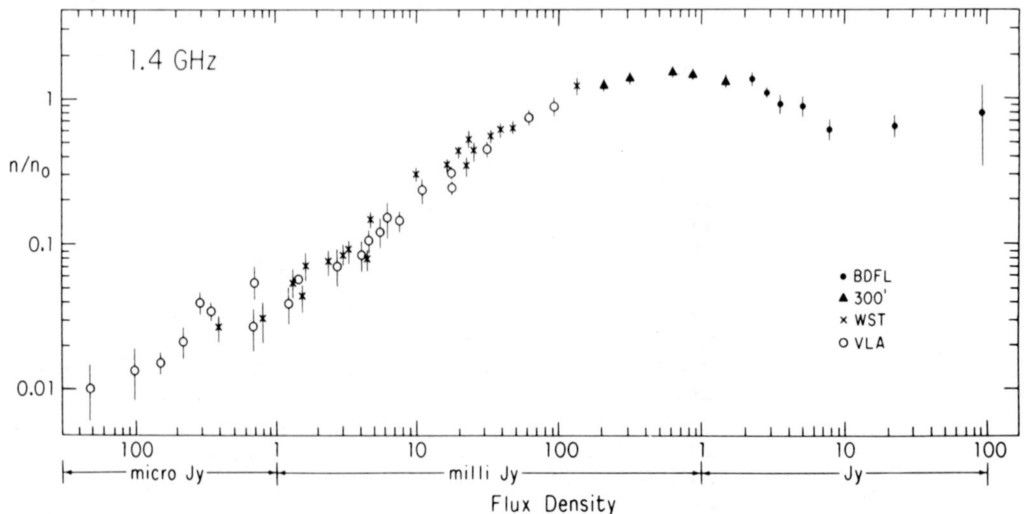

Fig. 2. Source count at 1.4 GHz normalized to a uniformly filled static Euclidean universe with a differential number count $n_0 = 225 S^{-5/2}$. Data are from the BDFL compilation (Bridle et al. 1972), the 300 foot GB and GB2 surveys (Machalski 1978, A&A Suppl. 65, 157), various Westerbork surveys as compiled by Windhorst et al. (1984, A&A Suppl. 58, 1), Oort and Windhorst (1985, A&A 145, 405) and Condon (1984).

1) Euclidean Region. At the highest flux densities corresponding to source densities of only a few tens of sources per ster, the source count is close to the Euclidean value reflecting the fact that the very bright sources contain many nearby objects (Wall and Peacock 1985).

2) Steep Rise. Between source densities of about 20 to 50 sources per ster, the observed number of sources appears to increase rapidly, with a slope which is in excess of the Euclidean value. But over the whole sky there are only about 500 sources that contribute to this part of the count, and so the statistics are limited. It is this rapid rise in source density which occurs only over a limited range of flux density, that led to the earlier claims of a "huge excess of weak sources." The best fitting slope in this region of the source count is significantly steeper than the Euclidean value, but only over a limited range of perhaps three to one in flux density.

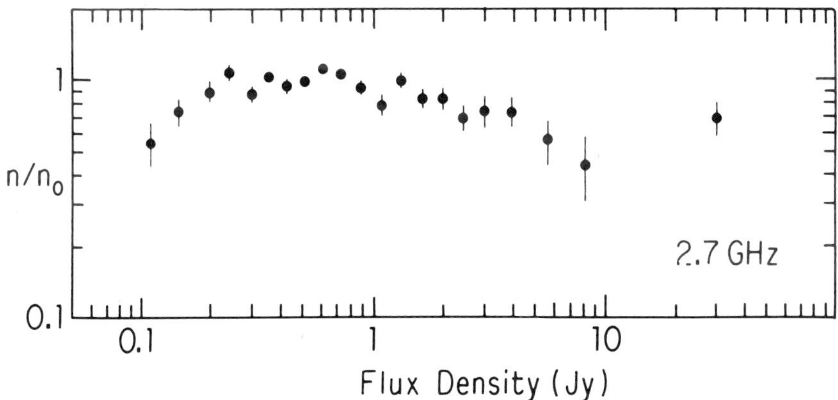

Fig. 3. Source count at 2.7 GHz normalized to a uniformly filled static Euclidean universe with a differential number count $n_0 = 150S^{-5/2}$. Data are from surveys made with the Parkes 210 ft radio telescope as compiled by Wall and Peacock (1985).

3. Euclidean Region. Particularly at the shorter wavelengths there is a plateau region that has a Euclidean slope over a range of flux density of up to 100 to 1, which extends up to a surface density of 2000 sources per ster.

4. Convergence. Below the plateau region the counts at all wavelengths drop relative to the Euclidean value for another two orders of magnitude in flux density. In this region the differential slope is about -3/2.

5. Euclidean Region. At the very lowest flux densities of less than about one mJy the count again steepens (flattens in the normalized presentation) to approximate the Euclidean slope.

3. INTERPRETATION OF THE COUNTS

There are too few very luminous sources at low redshifts to define the high luminosity end of the local
luminosity function and so it must be derived from the data
at intermediate or high redshift along with the evolution
function. The broad range of the radio luminosity function

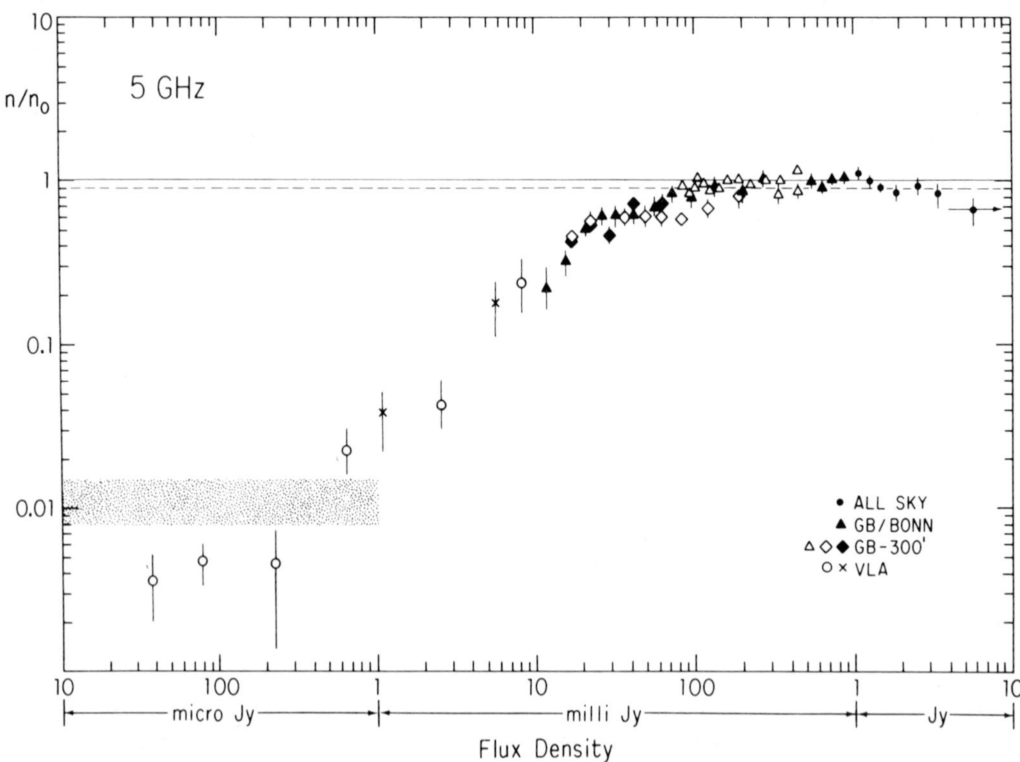

Fig 4. Source count at 5 GHz normalized to a uniformly
filled static Euclidean universe with a differential number
count $n_0 = 90 S^{-5/2}$ as indicated by a solid horizontal line.
The dashed horizontal line is drawn at $n_0 = 80 S^{-5/2}$ and
represents a somewhat better fit to the data above 100 mJy.
Data are taken from the All Sky compilation of Kuhr (1981),
the Green Bank-Bonn surveys made with the 140 foot, 300
foot, and 100 meter telescopes as compiled by Maslowski
(1984), the MG Survey (Bennett et al. 1986), the 300 foot
surveys of Owen et al. (1983) and Leddon et al. (1980), and
Altschuler et al. (1986), as well as the VLA surveys of
Fomalont et al. (1984) and Bennett et al. 1983. The points
at the three lowest flux density intervals are preliminary
values from a new deep VLA survey (Kellermann et al. 1986,
Highlights of Astronomy, pg 367). The horizontal stipled
area at the lower left represents the predicted counts based
on an extrapolation of IRAS observations and measured ratios
of 6cm to 60 micron flux density (Biermann et al. 1985).

compared with the range of flux density covered by the counts makes it difficult to untangle the geometry, the form of the luminosity function, and the way it changes with redshift. If the luminosity function is flat, i.e., dominated by strong sources, then flux density-limited samples are dominated by the most luminous sources and there is a well defined Hubble type relation between flux density and redshift. On the other hand, if the luminosity function is steep, then there is an inverse Hubble law, and the weaker sources are relatively nearby low luminosity objects and not distant powerful sources. Clearly in this case the interpretation of the source counts is not obvious, particularly in the absence of independent distance information. Indeed, the so-called "local hole" interpretation of the strong source count becomes a "distant hole" interpretation. Finally, there is a "critical luminosity function" in which each luminosity contributes equally at all flux densities. In this case, there is no Hubble relation at all, and each flux density range contains the same distribution of redshifts.

In the uniformly filled static Euclidean universe, with a differential source count slope of $-5/2$, the critical luminosity function is a power law with differential slope also equal to $-5/2$ (von Hoerner 1973). Derivations of the radio source luminosity function show that it is more complex than a simple power law and has a marked change in slope near 10^{25} W/Hz (e.g., Auriemma et al. 1977, Windhorst 1984, Meurs and Wilson 1984, Peacock 1985). For the more luminous sources, the slope is steeper than the critical value, while at lower luminosity the slope is flatter. Although in a relativistic universe, geometric effects and the K correction make the critical slope less steep than $-5/2$, the actual luminosity function appears to be close to the critical slope over a wide range of luminosity. Thus radio source counts show little relation between observed flux density and redshift. This conspiracy of nature has made the interpretation of radio source counts difficult, and at least in the past, a source of great controversy.

Although there is no radio equivalent of Hubble diagram for radio galaxies and quasars, it has long been known that galaxies selected as the result of the identification of radio sources have a remarkable narrow range of absolute magnitude which makes them ideally suited for studying the optical Hubble diagram (e.g., Sandage 1972). Although there is a much bigger dispersion in the optical magnitude of quasars, as pointed out by Wall and Peacock (1985) and shown in Figure 5, there is some evidence that quasars selected from radio identifications do loosely follow a Hubble relation.

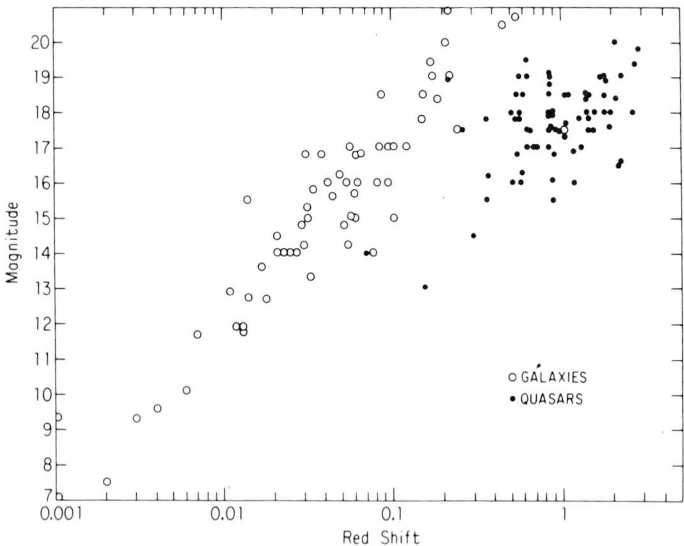

Fig. 5. Red shift magnitude diagram for all identified
sources with 6 cm flux density greater than 2 Jy.

3.1 Evolution

Many authors, (e.g., Robertson 1980, Wall et al. 1980,
Peacock and Gull 1981, Peacock 1985) have attempted to
interpret the observed source counts constrained by the
identification content and measured redshifts of flux
density limited samples, but the number of unknowns is still
too large to define uniquely the geometry, the local luminosity function, the evolution function, and possible
redshift cutoff. Because of the relatively narrow peak in
the count, particularly at long wavelengths, and in view of
the relatively broad radio luminosity function, essentially
all recent workers agree with the conclusions of Longair
(1966) that pure luminosity or pure density evolution,
corresponding to a horizontal and vertical translation of
the luminosity function respectively) does not work. It is
generally agreed that some sort of luminosity-dependent
density evolution, in the sense that only the most luminous
sources evolve, is necessary to reproduce the sharp peak in
the normalized source count.

For the strongest few hundred sources in the low
frequency 3CR catalogue, identifications and redshifts are
nearly complete (Spinrad et al. 1985). For these sources,
which constitute the steep part of the source count, the
luminosity-volume test has shown that the evolution is
confined to quasars and the most powerful radio galaxies,
and that the lower luminosity sources appear to be more
uniformly distributed (e.g., Laing et al. 1983).

It has been customary to describe the evolution function as a power law of the form $(1+z)^P$, an exponential law of the form $e^{t/\tau}$, or with a free form fit to the data, (e.g., Robertson 1980, Peacock and Gull 1981, Peacock 1985). The power law is mathematically convenient, but requires a cutoff at large redshift ($z \sim 2.5$) to obtain the observed convergence of the counts at low flux density. The exponential law has a straightforward interpretation in terms of cosmic look back time. The free form approach provides a tool for including all of the available data, but it is sometimes claimed that a physical understanding of the evolution is obscured. However, the **assumed** analytical forms, power-laws and exponential functions offer no real physics either.

The analysis of source counts, and of the free form approach in particular, uses the counts, identifications, and redshifts - incomplete as they are - to provide the best numerical description of space density and its epoch dependence. In view of the non physical basis, denoting such descriptions as "models" is misleading. Nevertheless, the results do have substantial significance:

i) the degree of evolution, and its dependence on epoch and radio luminosity are delineated, time scales are established, and the possibility of a redshift cutoff is explored;

ii) the derived luminosity functions may be compared with true physical models or used to predict such quantities as the angular size-flux density and the angular size-redshift relations, or the integrated radio background due to discrete sources:

iii) the overall statistical agreement between data sets and the ability of the conventional framework (e.g., geometry and Doppler redshifts) to account for the data is checked;

iv) the differential evolution of different radio source populations may be investigated (see Section 4);

v) the needed areas for further work are elucidated.

Condon (1984) has noted that because the local luminosity function has a rather well defined knee, simple luminosity evolution (horizontal translation of the luminosity function) will give an apparent luminosity dependent change in the space density in the required sense that the evolution appears to mostly affect the more luminous sources. Condon was able to fit the observed source counts at a number of wavelengths with a combination of simple luminosity and simple density evolution. He points out that more complex types of evolution, such as described by Wall, Pearson, and Longair (1980) or by Peacock and Gull (1981) are not excluded, but Condon's approach is attractive since it treats all sources in the same way and does not require that the strong sources behave differently from the weaker ones. A similar approach to Condon's has been used by Green (1986) to describe the evolution of the quasar optical luminosity function, without the need to postulate a different z dependence for the strong and weak sources.

Somewhat paradoxically, however, the counts of optically selected quasars and the results of the luminosity volume (V/V_{max}) test on optically selected quasars indicate an even more dramatic evolution than for radio sources (e.g., Green 1986). Taken literally, this leads to the surprising conclusion that there is a net negative evolution for radio quasars in the sense that a quasar at large redshift is less likely to become a strong radio source than a nearby quasar. This is in contrast to galaxies, where the probability of a bright elliptical galaxy becoming a strong radio source appears to increase dramatically with increasing redshift.

3.2 Faint Radio Sources

As shown in Figures 2 and 4, near a level of 1 mJy the slope of the source counts change, suggesting the emergence of a new population of sources. This change in the slope of the source count is seen both at 1.4 GHz (e.g., Windhorst et al. 1986) and 5 GHz (e.g., Kellermann et al. 1986) in a number of survey fields, and, is therefore, unlikely to be the result of an unusual clustering in one particular region of the sky. Van der Laan (1983) and Windhorst (1986) suggest that the observed increase in source density at sub mJy levels is due to a population of blue peculiar, interacting and merging galaxies. Wall et al. (1986) and Weistrop et al. (1987) find no evidence for any new population, and suggest that the sub mJy sources are relatively low luminosity radio galaxies at moderate distance. Normal spiral galaxies such as those found locally cannot account for the weak source population without evolution (Condon, 1984). However, radio observations of IRAS sources (e.g., Condon et al. 1982, de Jong et al. 1985) suggest that the sub mJy source count can be reproduced by a simple extrapolation of the -5/2 power law which describes the IRAS counts of moderately active (e.g., starburst) and relatively nearby (z < 0.3) galaxies (Biermann et al. 1985). Thus the faint radio sources which constitute the low end of the source count may not give us information on the evolution of quasars and radio galaxies, but may provide substantial information on the local radio luminosity function of certain types of active galaxies.

4. SOURCE POPULATIONS

The interpretation of radio source counts is complicated by the fact that extragalactic radio sources are an inhomogeneous collection of quasars, radio galaxies, compact, and extended sources, all of which may have a different local radio luminosity function which may have different individual lifetimes and which may show different cosmic evolution. For the most part, compact sources are identified with quasars and extended sources with radio galaxies, but the correlation is far from perfect. Some radio galaxies are dominated by a compact core while many quasars, such as most of those found in low frequency surveys like the 3CR survey, host extended radio sources.

The compact radio sources are most easily distinguished by their flat radio spectra, and are, therefore, most commonly detected at short wavelengths, where they represent about half of the strong source population. At long wavelengths, the source count is dominated by the steep-spectrum extended sources. Since there is only one population at low frequencies, the interpretation of the source counts are more straightforward than at high frequencies.

In Figure 6, we show the 6 cm source count separated into the two classes of flat spectrum (compact) and steep spectrum (extended) sources. The individual sub counts each peak more sharply than the combined count and at different flux densities, so that the near-Euclidean slope observed above 30 mJy for the combined 6 cm count results from combining the two sub counts. It is curious that the effect of evolution, the relativistic effect of the redshift on the observed flux density, the change in the volume element with redshift, and the combination of the two subcounts combine to mimic a uniformly filled static Euclidean universe over a wide range of flux density.

As shown in Figure 6, the strongest sources are roughly equally divided between the two classes. The steep-spectrum subcount peaks near a few hundred mJy while the flat spectrum sources peak near 1 Jy and drops rapidly at lower flux densities, where they contribute only about 1/3 of the total count. However, below 1 mJy, the flat spectrum population again contributes more than half of the total. Thus, independent of the indicated upturn in the count, the change in spectral distribution at sub mJy levels suggests a change in source population.

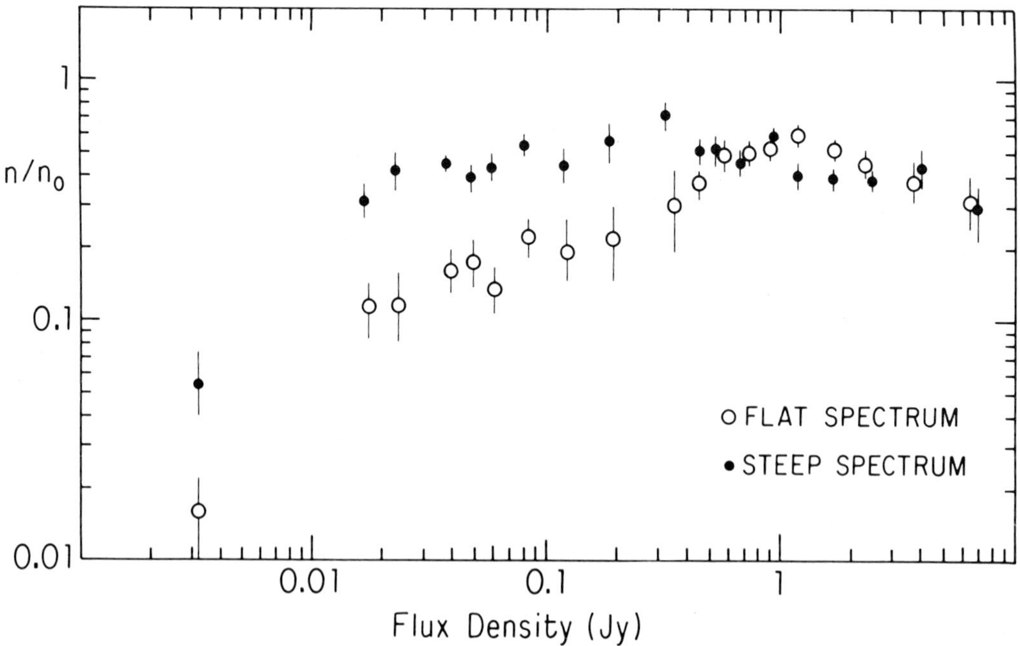

Fig. 6. Differential source counts for flat ($\alpha > -0.5$) and steep ($\alpha < -0.5$) spectrum souces normalized as in Fig. 4. Data are taken from the compilations of Pauliny-Toth et al. (1978a) and Condon (1984) and for the point at 3 mJy from Fomalont et al. (1984).

The early interpretations of the separate sub counts or of the corresponding luminosity-volume (V/V_m) test suggested that the flat spectrum population is distributed more uniformly than the steep spectrum sources (e.g., Schmidt 1976, Masson and Wall 1977). However, the data used in these investigations covered that part of the source count where the slope is not so steep and includes the relatively low power sources which have a more uniform space distribution, (e.g., Kellermann 1980). More recent analyses of the much more extensive material now available indicates that for both populations, the powerful sources evolve more strongly than the those of low luminosity, (e.g., Laing et al. 1983, Wall and Peacock 1985, Peacock 1985, Zawislak-Raczka and Kumor-Obyrk 1986).

5. ALTERNATIVE INTERPRETATIONS

For strong radio sources, modern surveys give source counts which are very close to that obtained 25 years ago by Australian workers and which have an initial slope not grossly different from that of a uniform distribution of sources in a static Euclidean universe. The earliest interpretation in terms of dramatic cosmic evolution was based on an inappropriate analysis of poorly understood data. Later, a more sophisticated analysis of the so-called P(D) statistics of the confusion limited surveys reinforced the need for a radio source population which has changed dramatically with cosmic epoch, an interpretation which in general terms is still widely accepted. However, the data and consequently the arguments, have changed, and although much more detailed and subtle, the general conclusions remain essentially unchanged.

Except for the strongest sources, the observed counts are characteristic of a homogeneous distribution of sources in a relativistic universe where the effect of the redshift on the volume element and flux density suppress the count at low flux densities below the -3/2 Euclidean value (e.g., Scheuer 1975). Therefore, it may be argued that the source counts can equally well be interpreted as a deficiency of a relatively small number of strong sources rather than a huge excess of weak sources (e.g., Kellermann 1972), and that it is inappropriate to base cosmological models on the distribution of these few strong sources. The number of "missing" sources in these "local hole" models is surprisingly small. Jauncey (1975) estimates that only 35 strong sources need to be added over the whole sky to reproduce a Euclidean slope for the 178 MHz 3CR count. As seen in Figure 4, at 6 cm the observed count is close to Euclidean over a wide range of flux density, except for about 20 "missing sources" over the whole sky.

In view of the apparent isotropy of the source counts, it is often argued that "local hole" models are untenable since it would imply that we are preferentially located near the center of such a hole in violation of Copernican principles. However, the isotropy measurements apply only to the weaker sources where the source count slope appears consistent with a uniform distribution of sources. For the strong sources, where the slope appears is steeper than the Euclidean value, there are too few sources to put firm limits on the isotropy or lack of isotropy, (e.g., Pauliny-Toth et al. 1978). Moreover, the apparent recent detections of superclusters, strings, and voids on scales of 10 to 100 Mpc suggests that the universe may be less smooth than previously suspected, and that the departure of a few dozen strong sources from the $-3/2$ law does not require large-scale cosmic evolution or departure from accepted Copernican concepts.

However, comparison of the counts with those expected from a realistic luminosity function based on identified samples and redshifts indicates that the counts at low flux density should fall off even faster than observed, and (independent of the strong source count) this is commonly accepted as the result of the evolution of the source population toward higher space density or greater luminosity at large redshift. Thus in its current form, the argument for evolution depends on the cosmological interpretation of redshifts. While we do not wish to raise the issue of non cosmological redshifts, we do want to emphasize that the source counts by themselves do not necessarily require departure of the radio source population from homogeneity on a cosmological scale.

The interpretation of radio source counts has thus "evolved" considerably over the years, although the basic conclusions have remained essentially unchanged, as it has proven possible to accommodate the observational refinements with a correspondingly refined interpretation. It is interesting to speculate how the field might have developed had the first source counts been compiled at 6 cm (e.g., Fig 1) instead of at long wavelengths, and had the early counts reached source densities of a few hundred thousand per ster thus not emphasizing the high flux density - steep count end.

6. The Future

At a source density of about 10^5 sources per ster essentially all of the evolving radio-powerful sources have been counted. Surveys to higher source density primarily contain less luminous sources which appear to be bright Seyfert or starburst galaxies with a luminosity of about 10^{22} Watts/Hz and only moderate redshift. While it would be of interest to investigate the cosmological evolution of these sources, there are two fundamental limits which are independent of any future advances in instrumentation, and which may limit the depth to which radio surveys can be extended.

The deepest radio surveys now reach a source density of almost 10^6 sources per ster, so that the mean separation between sources is only a few minutes of arc. Even a small further improvement in sensitivity will reach a source density where the separation between sources becomes comparable with the typical component spacing of individual sources;

the source count becomes a source component count, and
interpretation becomes unclear.

A further limit may be introduced by small scale fluctuations in the microwave background radiation.   The deepest
VLA survey reaches an rms noise level of only 5 micro Jy,
corresponding to an equivalent sky temperature less than one
milli degree at 18 arc second resolution, and they are sensitive to fluctuations of a few tenths a degree (Kellermann
et al. 1986).   This is approaching the level where many
models predict observable fluctuations due to the formation
of galaxies, and more sensitive surveys may be influenced by
this effect.

LITERATURE REFERENCES

Aizu, K 1987, this volume.
Auriemma, C., Perola, G.C., Ekers, R., Fanti, R., Lari, C.,
     Jaffe, W.J., Ulrich, M.H. 1977, Astron. and Astrophys.
     57, 41.
Biermann, P., Eckart, A, and Witzel, A. 1985, Astron.
     Astrophys. 142, L23.
Condon J.J, Condon, M.A., Gisler, G., Puschell, J.J. 1982,
     Astrophys. J. 252, 102.
Condon, J.J. 1984, Astrophys J. 287, 461.
Crawford, D.F., Jauncey, D.L., and Murdoch, H.S. 1970,
     Astrophys. J. 162, 405.
de Jong, T., Klein, U., Wielibinkski, R., and Wunderluch, E.
     1985, Astron. and Astrophys. 147, L6.
Green, R. 1986, I.A.U. Symposium No. 119, eds G. Swarup and
     V.K. Kapahi, Reidel, p 429.
Jauncey, D. 1967, Nature 216, 877.
Kellermann, K.I. 1972, Astron. J. 77, 531.
Kellermann, K.I. 1980, Physica Scripta 21, 664.
Kellermann. K.I, Fomalont, E., Weistrop, D., and Wall, J.V.
     1986, Highlights of Astronomy, Reidel 6, 367.
Laing, R.A., Riley, J.M., Longair, M.S. 1983, M.N.R.A.S. 204, 151
Longair, M.S. 1966 M.N.R.A.S. 133, 421.
Masson, C.R. and Wall, J.V. 1977, M.N.R.A.S. 180, 193.
Meurs, E.J.A. and Wilson, A.S. 1984, Astronomy and
     Astrophys. 136, 206.
Murdoch, H.S., Crawford, D.F.,and Jauncey, D. 1973,
     Astrophys. J. 183, 1.
Pauliny-Toth, I.I.K. et al. 1978, A.J. 83, 451.
Peacock, J.J. and Wall, J.V. 1981, M.N.R.A.S. 194, 331.
Peacock, J.J. 1985, M.N.R.A.S. 217, 601.
Peacock, J.A. and Gull, S.F. 1981, M.N.R.A.S. 196, 611.

Peacock, J. J., Perrymann, M.A.C., Longair, M.S., Gunn,
     J.E., and Westphal, J.A. 1981, M.N.R.A.S. 194, 601.
Robertson, J.G. 1980, M.N.R.A.S. 190, 143.
Sandage, A. 1972, Astrophys. J. 178, 25.
Scheuer, P. 1975, Galaxies and the Universe, Stars and
     Stellar Systems, vol 9, ed. Sandage, A., Sandage, M.,
     and Kristian, J., p 725, University of Chicago Press.
Schmidt, M. 1976. Astrophys. J. 209, L55.
Spinrad, H., Djorgovski, S., Marr, J., Aguilafr, L. 1985,
     Publ. Astron. Soc. Pac. 97, 932.
van der Laan, H. and Windhorst, R.A. 1982, Astrophysical
     Cosmology, p 349.
von Hoerner, S. 1973, Astrophys. J. 186, 741.
Wall, J.V. and Peacock, J.J. 1985, M.N.R.A.Ś. 216, 173.
Wall, J.V., Benn, C.R., Grueff, G., and Vigotti, M., 1986,
     Highlights of Astronomy 6, in press.
Wall, J.V., Pearson, T.J., and Longair, M.S. 1980,
     M.N.R.A.S. 193, 683.
Weistrop, D., Wall, J.V., Fomalont, E.B. and Kellermann,
     K.I., 1987, Astron. J. (in press).
Windhorst, R.A. Miley, G.K., Owen, F.N., Kron, R.G., and
     Koo, D.C. 1985, Astrohys. J. 289, 494.
Windhorst, R.A., van Heerde, G.M., and Katgert, P. 1984,
     Astron & Astrophys. Suppl. 58, 1.
Windhorst R. 1984, Ph. D. Thesis, University of Leiden.
Windhorst, R.A. 1986, Highlights of Astronomy, Reidel.
Zawislak-Raczka, J. and Kumor-Obyrk, B. 1986, M.N.R.A.S.
     222, 487.

DISCUSSION

SWARUP: Could you comment on why 6 and 21 cm source counts look so
similar, in contrast to that at 75 cm.

WALL: Not very constructively, you raise a point which has not received
enough attention from radio surveyors. The most striking similarity
between the 6 and 21 cm source counts is in the region of "convergence",
where the coincidence of the counts implies a median spectral index
near zero. However we know that sources selected at these flux density
levels do not all show flat high-frequency spectra, together with the
steep low-frequency spectra required to translate the 75-cm count to the
21-cm count. Some combination of more subtle effects must be responsible,
presumably including the count normalization used here, luminosity
functions and spectral indices. It all bears further investigation.

MILEY: Do you know of any attempts that have been made to study
luminosity evolution as a function of spectral index for the steep

spectrum sources? This is interesting since we know that the ultra-steep spectrum sources are much more luminous than the normal-spectrum sources.

WALL: I don't know of such attempts, and only now do we have data sets which permit us to reconsider it. The situation is not as simple as you suggest, because some cluster sources of relatively low radio luminosity also have very steep spectra. A criterion in addition to radio spectrum would be needed to select the sample.

WINDHORST: To what extent has the radio K-correction been taken into proper account by those who performed model descriptions of multi-frequency radio source counts? We know that the radio K-correction is negligible for low z, but the situation at $z \gtrsim 1$ might be more complicated due to spectral curvature. Donnelly, Partridge and I recently found in a deep 6-21-50 cm survey that radio sources with $V \gtrsim 23.0$ mag (probably $z \gtrsim 0.75$) have generally very significant concave or convex spectra, which illustrates this complication.

WALL: This complication is well known in the sense that no "flat-spectrum" radio source has a flat-spectrum; all are curved, either simply concave, simply convex, or in combinations. Single bulk K-corrections are typically used, which obviously do not consider individual spectra. However this assumption does not bias space densities and is at present certainly not the major source of uncertainty in these estimates.

SZALAY: You mentioned marginal evidence of large scale anisotropies in the bright sources. Could you give more details?

WALL: These 'anisotropies' were reported some 10 years ago, on the completion of cm-wavelength surveys of large areas of extragalactic sky (Pauliny-Toth, Wall and colleagues; see e.g. Proc. IAU Symp. 74). They are on scales of steradians, and are at a significance level of about 2σ. Because the very radio-brightest objects are involved, there is no way to improve the statistics.

PARTRIDGE: Jasper, you showed an overlay comparing two histograms, one of instrument noise alone, and one of instrument plus sky noise. The positive flux sides of the two histograms agree well except for a small tail, perhaps due to residual sources. But the negative sides don't agree well at all - there appears to be a lot of excess sky noise. If this additional variance is due to microwave background fluctuations, it should appear symmetrically on both sides of the histogram, but yours doesn't. Can you explain why? Is there a problem with the zero spacing flux?

WALL: It's possible, a shift of 1μ Jy/beam evens the balance, and
then provides excess signal both positive and negative. Wrong flux
in the zero spacing doesn't make this signal disappear. But
obviously we want to understand the limits on the zero spacing flux
(as well as other effects) before stronger statements of results
can be made.

BURKE: What cleaning procedure was used to derive your evidence for
background fluctuations?

WALL: We used standard NRAO 'cleans', from extremely light to heavy;
and we found that the excess signal does not depend primarily on the
degree, at least for moderate 'cleaning'. It is essential to carry
out numerical simulations to find the full and combined effects of
'cleaning' + noise + confusion by extremely faint sources. Even then,
residual instrumental errors may mean that yet another upper limit is
the best we can do.

BURKE: Since the CLEAN procedure is not well understood when applied to
very noisy data, I don't think your evidence for background fluctuations
is believable yet.

NARLIKAR: I have two comments. (1) You mentioned the shape of the
log N-log S curve at the bright end as arising from inhomogeneities in
the distribution of galaxies on the scale of 50-100 Mpc. In 1961, during
the Hoyle-Ryle controversy about the steep source counts, Hoyle and I
had suggested inhomogeneities of precisely this order to explain the
observed effect. However, at that time we were accused of proposing a
model that was too inhomogeneous to be realistic. The balance of opinion
has shifted in the last twenty five years. (2) In a recent work
P. DasGupta, G. Burbidge and I have constructed a non-evolving radio
luminosity function from the observed N-z plot for 3CR radio galaxies.
The S-z plot generated by this RLF does not differ from the observed S-z
plot in a statistically significant manner. We have repeated the exercise
for the Wall-Peacock catalogue of sources at high frequency, with the
same result. We would be interested in applying this technique to samples
at fainter flux densities once their redshift determinations are complete.
This exercise would tell us whether evolution is really necessary.

WALL: With regard to the second point, catalogues of the brightest sources
span a factor of 10 or so in apparent luminosity, while the absolute-
luminosity functions are known to cover at least 7 orders of magnitude.
This was a major criticism in the early days of source-count interpretation.
We now have counts for tens of thousands of radio sources at several
frequencies, spanning 7 orders of magnitude in intensity, and complete with
identifications and redshifts for many of the samples involved. Any
successful description of the space density of radio sources and its epoch
dependence must be able to account for these data.

THE 10 GHz LOG N - LOG S CURVE OBTAINED AT NRO

Ko Aizu
Physics Department, Rikkyo University
Nishi-Ikebukuro, Toshima-ku, Tokyo, 171, Japan
Makoto Inoue
Nobeyama Radio Observatory[†]
Nobeyama, Minamimaki, Nagano, 384-13, Japan
Hiroto Tabara and Tatsuji Kato
Faculty of Education, Utsunomiya University
Mine, Utsunomiya, Tochigi, 321, Japan

ABSTRACT. Using 45-m radio-telescope at NRO, we obtained log N - log S curves at 10 GHz for flat- and steep-spectrum sources separately. A simple model-fitting of evolution is given.

1. INTRODUCTION

The log N - log S curve of radio sources is an important method in observational cosmology. Here S means the flux density of radio sources and N = N(S) the differential number-spectrum per unit solid angle of the sky. One recent trend in these studies is to apply this method separately to QSO, radio galaxies and/or normal galaxies. This separation is made possible by a close correlation between radio spectral indices α (defined as $S \propto \nu^{-\alpha}$) and types of radio sources. At higher frequencies this correlation becomes closer and also in the larger flux-density regions QSO's dominate radio galaxies. So far the highest frequency of log N - log S curves was 5 GHz. To double this frequency we made observations at 10 GHz using 45-m telescope at NRO. Preliminary result was published in Aizu et al. (1986).

2. OBSERVATIONS

As the beam width 2.7' of the 45-m telescope at 10 GHz is too narrow for a survey work of moderate size under limited observation time (Seielstad 1983) we made flux-density measurements for complete samples obtained at 5 GHz. It is convenient to divide these samples into three

[†] Nobeyama Radio Observatory, a branch of the Tokyo Astronomical Observatory, University of Tokyo, is a facility open for general use by researchers in the field of astronomy and astrophysics.

565

A. Hewitt et al. (eds.), Observational Cosmology, 565–568.

parts A, B, and C:

A. Strong sources.   These are NRAO-Bonn strong survey (Kühr et al. 1981) of the northern sky and Parkes survey (Bolton et al. 1979) of the southern sky of declination range 0° ∿ -30°. The Galactic plane |b| < 10° is excluded. Different survey areas have different limiting flux densities (Aizu et al. 1986). For most sources flux densities at 10 GHz are available (Kühr 1981). We add newly fifteen sources. Altogether A sample has a solid angle of 5.02 sr and includes 758 sources, of which 121 are southern ones.

B. Intermediate sources.   Owen et al. (1983) made a fast survey for a selected area with a solid angle .0691 sr. In 1983 and 1984 we measured flux densities of 101 sources with $S_5 \geq 100$ mJy. A part of data of these measurements were already published (Tabara et al. 1984).

C. Weak sources.   Ledden et al. (1980) made a confusion-limited survey for a selected area with a solid angle 0.00956 sr, and Condon and Ledden (1982) made a detail study with VLA. In 1984 and 1985 we measured flux densities of 121 sources with $S_5 > 15$ mJy. Details of our new measurements will be published elsewhere.

3. LOG N - LOG S

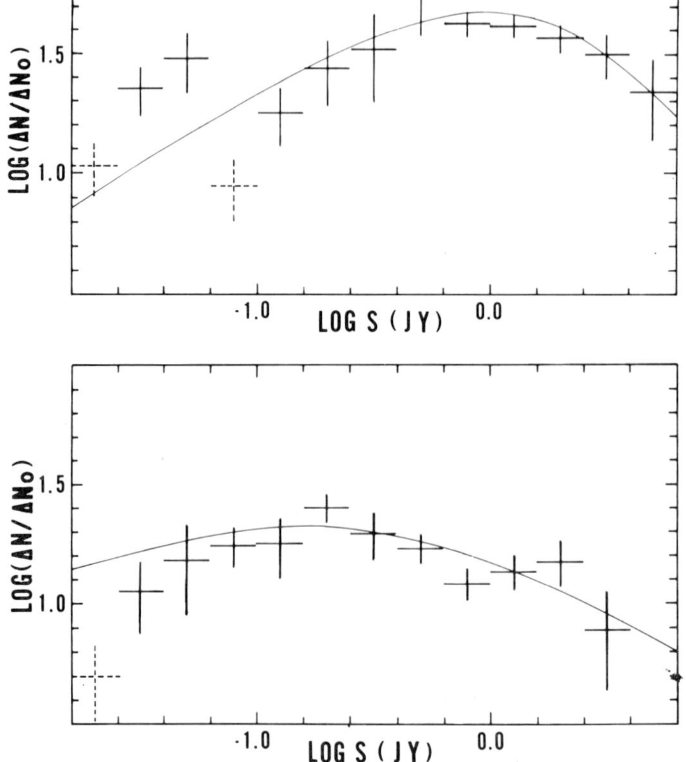

Fig.1. Log N-log S curve for flat sources at 10 GHz. Dashed cross means the statistical completeness of data is poor. A thin curve is given by a simple model calculation.

Fig.2. The same for steep sources. Symbols have the same meaning as in Fig.1.

The log N - log S curves are shown in Fig.1 for flat-spectrum sources and in Fig.2 for steep-spectrum sources. Here "flat" means the spectrum index between 5 and 10 GHz $\alpha \leq 0.5$ and "steep" means $\alpha > 0.5$.

In these figures we normalize N to the uniform Euclidean distribution $N_0 = \text{const} \times S^{-2.5}$ where S is in Jy. Here we assume const = 1 for convenience. Note that actually $\log(\Delta N(S)/\Delta N_0(S))$ are plotted in Fig.1, where $\Delta N(S) = \int_{S_1}^{S_2} N(S)dS$, and $\Delta N_0 = (S_1^{-1.5} - S_2^{-1.5})/1.5$. Here S_1 and S_2 are lower and upper bounds of a bin respectively. Dashed crosses are cases where the completeness is low.

Although the statistics is poor we could point out the followings:
(1) The overall features of these curves are similar to those at 5 GHz (Owen et al. 1983), but for $S > 1$ Jy the flat-spectrum sources dominate the steep-spectrum sources by a factor of three.
(2) The maximum for flat sources occurs at $S = 1$ Jy and for steep sources at 0.2 Jy. Comparing the two curves we find that in the region of $S < 0.2$ Jy the slope of the steep sources is flatter than that of the flat sources. These features (1) and (2) agree well with curves predicted by Condon (1984).
(3) For both kind of sources there are large dips around $S_{10} = .063$ ~ 0.1 Jy , which are transition regions from B sample to C sample. Although this may due to a new feature of C sample it is possible that this is simply statistical fluctuations.

4. MODEL

Theoretically the log N - log S curve can be expressed with a luminosity function (abbreviated to LF) $N(L, z)$ at the redshift z in the comoving system as $N(S)ds = \iint N(L, z)dLdV(z)$, where $dV(z)$ is the comoving volume element per unit solid angle at z, and the region of the double integral is limited by $S < 1/(4\pi R^2(1+z)^{\alpha-1}) < S + dS$.

The local LF $N(L, z=0)$ for QSO including BL Lac objects and for radio galaxies are constructed from A sample. Redshifts are restricted to a range 0 to 1.45. These LF are expressed in an analytical form $N(L) = A L^{-C} \exp(-BL)$, where parameters A, B, and C and luminosity ranges L_0 and L_1 are given in Table 1. The general LF are obtained as

Table 1. Parameters of analytical form of
luminosity functions

	QSO	Radio galaxies
A	14.88	4.421
B	0.0125	0.0063
C	1.886	2.297
Log L_0	31.8 $\sim$ 33.0	30.6
Log L_1	35.8	34.6

a solution of a continuity equation on the $L \sim t$ plane $\partial N/\partial t + \partial(pN)/\partial L = q$ (Cavaliere et al. 1970). Here p is the rate of luminosity change and q the birth rate function. The local LF is used as a final condition, and the formation redshift z_c is incorporated in our model under an assumption $N(L, z) = 0$ for $z > z_c$. As for p and q we adopt, as a first trial, simple forms $p = -L/t_1$ and $q = N/t_2$, where t_1 and t_2 are constants. Then the solutions have the form $N(L, z) = N(L)\exp(M\tau + BL(1-\exp(-M'\tau)))$, where M and M' are parameters and $\tau = 1 - (1+z)^{-1.5}$ is the look-back time in unit of $2/(3H_0)$ (Compare with Cavaliere and Szalay 1986). Here we assume the Hubble constant $H_0 = 75$ km $s^{-1}Mpc^{-1}$, and the deceleration parameter $q_0 = 0.5$. This form is similar to that currently used by many authors (for example, Schmidt and Green 1983). With suitable choice of parameters we can fit the forms of log N - log S curves as shown in Fig.1 and 2. It turns out that z_c is around 4 in both cases.

In conclusion, although improvement of statistics and more detailed theoretical studies are needed, the log N - log S curve method at higher frequencies is hopeful for us to obtain important parameters in the formation and evolution of QSO and radio galaxies.

We thank Dr. J. Machalski for a comment and Dr. A. Cavaliere for showing preprint before publication. Computations were carried out with Fujitu M360 at NRO and Rikkyo University.

REFERENCES

Aizu, K., Tabara, H., Kato, T., and Inoue, M. 1986, Astrophys. and Space Science, 119, 257.
Bolton, J.G., Savege, Ann, and Wright, Alan E. 1979, Aust. J. Phys. Astrophys. Suppl. Ser., 45, 367.
Cavaliere, A., Morrison, P., and Wood, K. 1970, Astrophys. J., 170, 223.
Cavaliere, A., and Szalay, A.S. 1986, Astrophys. J., in press.
Condon, J.J. 1984, Astrophys. J., 287, 461.
Condon, J.J., and Ledden, J.E. 1982, Astron. J., 87, 219.
Kühr, H. 1981, Astron. Astrophys. Suppl. Ser., 45, 367.
Kühr, H., Pauliny-Toth, I.I.K., Witzel, A., and Schmidt. J. 1981, Astron. J., 86, 854.
Ledden, J.E., Broderick, J.J., Condon, J.J., and Brown, R.L., 1980, Astron. J., 85, 780.
Owen, F.N., Condon, J.J., and Ledden, J.E. 1983, Astron. J., 88, 1.
Schmidt, M., and Green, R.F. 1983, Astrophys. J., 269, 352.
Seielstad, G.A. 1983, Publ. Astron. Soc. Pacific, 95, 32.
Tabara, H., Kato, T., Inoue, M., and Aizu, K. 1984, Publ. Astron. Soc. Japan, 36, 297.

MOLONGLO DEEP SURVEY AT 843 MHz

C.R. Subrahmanya* and B.Y. Mills
School of Physics
University of Sydney
N.S.W. 2006, Australia

ABSTRACT.   The Molonglo Observatory Synthesis Telescope has been used
for a deep survey of 25 southern fields of about 1° diameter. All
these fields have been observed on several occasions in 120 x 12-hour
sessions during 1983-85. Radio sources have thus been detected down to
a few millijansky at 843 MHz in 20 deg^2 of sky from these observations
and in a few selected fields, it has been possible to reach a detection
limit of 1 mJy. The first results on the source counts and optical
identifications are presented. By supplementing these data with some
existing observations of stronger radio sources, we have been able to
obtain source counts at 843 MHz for the entire range of flux densities
above 1 mJy.

1.   INTRODUCTION

The Molonglo Observatory Synthesis Telescope (MOST) is a multiple fan-
beam radio telescope operating at 843 MHz capable of synthesizing a
23 x 23 cosec δ arcmin^2 with a beam of 43 x 43 cosec δ arcsec^2
(Mills 1981; Durdin, Large & Little 1984; Crawford 1984). The diameter
of the synthesized field can be expanded to 46 or 70 arcmin by time-
multiplexing the hardware beams within each sampling interval. The large
collecting area of MOST (18000 m^2 total) makes it a very sensitive
telescope with an estimated thermal noise of about 0.2 mJy rms in a
12-hour observation of a very southern field.
   The deep survey with the MOST covers 25 well-separated fields
of diameter 70 arcmin. Most of these fields have been observed on
several occasions to reach a uniform detection limit of a few mJy while
confusion-limit has been reached in the case of central portions of the
best fields. In this paper, we present preliminary results from the
deep survey and for completeness, supplement the resulting source counts

---

* Present address :   Tata Institute of Fundamental Research,
                      P.O. Box 1234, Bangalore 560012, India.

A. Hewitt et al. (eds.), Observational Cosmology, 569–572.
© 1987 by the IAU.

with those at higher flux densities based on other surveys carried out
with the MOST.

2. DEEP SURVEY

2.1. Observations

The fields chosen for the deep survey are all in the far southern declinations where the telescope sensitivity is the best and they have been
chosen to be well away from catalogued sources. Unless the initial
large field map showed particularly bad grating responses or confusion,
each field has been observed on several occasions depending on the
quality of the maps. Out of the 25 fields, 19 have been mapped over the
large field of 70 x 70 cosecδ arcmin^2, 23 with the intermediate size
(46 arcmin diameter), and 9 with the normal field of 23 arcmin diameter.
The project involves a total of 120 x 12-hour observations carried out
between 1983 June and 1985 October, mostly during winter nights. Initial
reduction of data is essentially complete and the tabulation of sources
and work on optical identification are in progress. For the purpose of
this paper, the maps from different observations have been combined and
a preliminary compilation of sources based on a computer program has
been used to derive the source counts at 843 MHz. In the case of best
fields, the sources have been verified manually from contour plots before
estimating the counts.

2.2. Optical Identifications

The final radio positions are expected to be accurate to 1-3 arcsec in
most cases. Initial search for optical identifications are being made
on the film copies of the SRC-J and ESO-R sky surveys. Preliminary
results on some of the fields indicate 20-30 percent identifications to
the limit of SRC IIIaJ films ($\sim$23 mag). For one of the fields, a deep
plate has been specially obtained from the service photographic facility
of the Anglo Australian Telescope. Nearly half of the sources have been
identified on this plate. Further identifications and a follow up of
optical spectroscopy are being planned for all the fields.

2.3. Source Counts

The initial results on the source counts at 843 MHz are presented in
Figure 1, where the deep survey has been divided into 3 ranges of flux-
density. An area of 0.4 deg^2 of sky has been covered in the range
1-2 mJy, 2.3 deg^2 in the range 2-7 mJy and 20 deg^2 for 7-224 mJy. At
higher flux densities, the counts would be biased by the rejection of
strong sources in the initial selection of the fields and also the sky
coverage is not adequate to obtain good statistics. The Figure shows
the observed (uncorrected) source counts, normalised to a Euclidean
integral count of 100 $S_{Jy}^{-1.5}$ sr^{-1}.

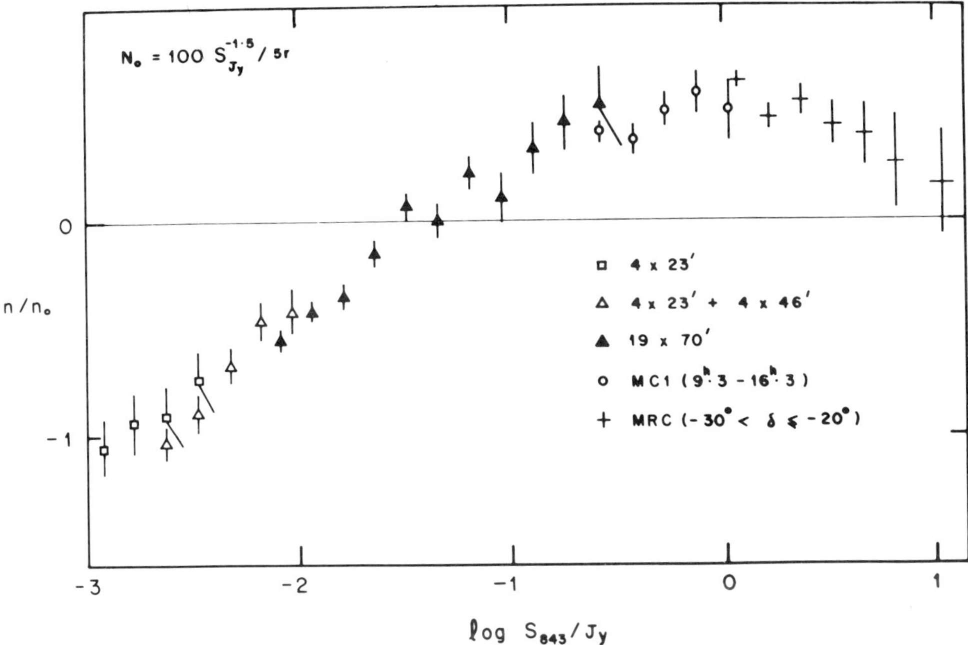

Figure 1. Normalised differential source counts at 843 MHz. All the
counts are derived from observations with the MOST. In the case of
deep survey, the number of independent fields and the individual field
diameters are indicated in the explanation of symbols.

The observed counts for $S_{843} > 1$ mJy correspond to a source density
of 150 deg^{-2} or about half million sources per steradian. This density
corresponds to about 50 beam areas per source, suggesting that we are
close to confusion limit. In the case of multiple observations of the
same field, the noise fluctuations of the sum and difference between
various maps confirm that the confusion is about 0.2 mJy rms.

3. COUNTS FOR STRONGER SOURCES

A large number of catalogued radio sources have been observed with the
MOST in the form of short scans at several hour angles. Sufficient
observations of complete samples now exist to enable us to derive the
counts reliably for the stronger sources not adequately represented in
our deep survey. Details of these observations will be published else-
where(Subrahmanya 1987, in preparation). For the sake of completeness,
we consider here the counts derived from two samples.

In the region $9^h < \alpha < 16^h$, $-20° > \delta > -22°$, flux-densities at 843 MHz
are available for all sources with listed (peak or integrated) $S_{408} \gtrsim 0.2$Jy
in the MC1 catalogue (Davies, Little & Mills 1973). Based on these, we
have derived the source counts between 0.2 and 1 Jy at 843 MHz, although

the derived counts may have been underestimated near 0.2 Jy due to some sources of inverted spectra below the flux density limit at 408 MHz sample. At the brighter end, about 700 sources in the Molonglo Reference Catalogue (MRC; Large et al 1981) with $S_{408} > 0.9$ Jy, $-20°>\delta> -30°, |b|>-20°$ have been observed with the MOST. In addition, the observations have also included inverted spectrum sources selected from Parkes catalogue which were likely to be stronger than 0.8 Jy at 843 MHz on the basis of the catalogued flux densities at 2.7 and 5 GHz. Using these observations, a complete sample has been defined with $S_{843} \geq 1$ Jy over 0.7 sr of sky, whose details will be published elsewhere(Subrahmanya 1987, in preparation). For the present purpose, we have included in Figure 1, the counts based on the integrated flux-densities of the above samples.

4. DISCUSSION

The radio source counts presented in Fig. 1 cover the widest range of flux-densities for which the counts have ever been derived from a single instrument. By including progressively larger areas of sky for stronger sources, we have been able to achieve reasonable statistics for the entire range of flux densities above 1 mJy. Our results at 843 MHz are in good statistical agreement with the source counts at 1.4 GHz reviewed by Windhorst et al 1985. However, this does not rule out field-to-field variations due to small scale statistical inhomogeneities in the radio source distribution. Since our deep survey covers a large number of independent fields, we are in a position to investigate such inhomogeneities in the counts and identification content of weak radio sources. The results of such studies will be published in due course.

ACKNOWLEDGEMENTS. The Molonglo Observatory Synthesis Telescope is financed by the Australian Research Grants Scheme (ARGS) and the University of Sydney. One of us (CRS) wishes to acknowledge the receipt of a post-doctoral research fellowship from ARGS during the course of this work.

REFERENCES

Crawford, D.F.: 1984, <u>Proc. URSI/IAU Symp. Indirect Imaging</u> (ed.) J.A. Roberts, Camb. Uni. Press, 373.
Davies, I.M., Little, A.G., Mills, B.Y.: 1973, <u>Australian J. Phys. Astrophys. Suppl.</u> **28**, 1.
Durdin, J.M., Large, M.I., Little, A.G.: 1984, <u>Proc. URSI/IAU Symp. Indirect Imaging</u> (ed.) J.A. Roberts, Camb. Uni. Press, 75.
Large, M.I., Mills, B.Y., Little, A.G., Crawford, D.F., Sutton, J.M.: 1981, <u>Mon. Not. R. Astron. Soc.</u> **194**, 693.
Mills, B.Y.: 1981, <u>Proc. Astron. Soc. Aust.</u>, **4**, 156.
Windhorst, R.A., Miley, G.K., Owen, F.N., Kron, R.G., Koo, D.C.: 1985, <u>Astrophys. J.</u>, **289**, 494.

ULTRADEEP OPTICAL IDENTIFICATIONS AND SPECTROSCOPY OF FAINT RADIO SOURCES

Rogier A. Windhorst and Alan Dressler
Mount Wilson and Las Campanas Observatories, Pasadena
Carnegie Institution of Washington

David C. Koo
Space Telescope Science Institute, Baltimore

1. ULTRADEEP CCD IMAGING OF MILLIJANSKY AND MICROJANSKY SAMPLES.

During the last two years we have completed an extensive project of direct CCD imaging with the Palomar 200 inch Four-shooter on several milliJansky and microJansky radio samples, now totaling 200 sources. The purpose is to study the nature and redshift distribution of the radio galaxies that cause the upturn in the milliJansky source counts (WMOKK) and to search for candidates of primeval radio galaxies.

For a complete sample of 70 mJy sources in Hercules accurate VLA positions (0.1", OKSW) are available, yielding about 60% identifications on Mayall 4 m plates (V<23 mag). In 10 minute Four-shooter exposures this fraction increased to 85% down to Gunn r=24.5, while 30 minutes integration in Gunn g and r yielded 97% identifications down to 25.5 mag. The first results are described in W86 and Windhorst and Koo (1986, in prep).

Still, some radio sources refused to show up at r=25.5 mag. These are believed to be the most distant objects, because their angular sizes are very small (expected at high z if $q_0 \ll 0.5$) and their radio spectra are very steep (due to the large radio K-correction at high z). For these sources we finally secured identifications after reaching a limit of r=26.5 mag, obtained in several hours integration, while taking spectra with the Four-shooter for brighter radio galaxies in the same field.

Figure 1 shows the resulting magnitude distribution for our now totally complete sample. The distribution rises strongly for 18<r<21, indicating the cosmological evolution of the mJy radio sources, and gradually declines for 21<r<27, perhaps heralding a redshift cut-off. This is in contrast with the field galaxy counts that continually rise down to r=27 and shows that the accurate VLA positions prevented significant background contamination. Because the identification sample is now 100% complete, optical selection effects are no longer important.

The models of C84 and P85 predict that 10-40% of the milliJansky population extends to z>4, which, if true, means that primeval galaxies must be present in our Four-shooter sample. The 100% complete identifications, and their in general very blue colors, leaves no room for sources that are optically invisible due to dust obscuration at z>3.5. Hence, either the Universe is not significantly obscured by dust at high redshifts, or faint radio sources do not extend to beyond z=3.5, or both.

573

A. Hewitt et al. (eds.), Observational Cosmology, 573–576.
© *1987 by the IAU.*

<u>Figure 1:</u> The r-magnitude distribution of the complete radio source sample in Hercules, which is **100% completely identified down to r=26.5.**

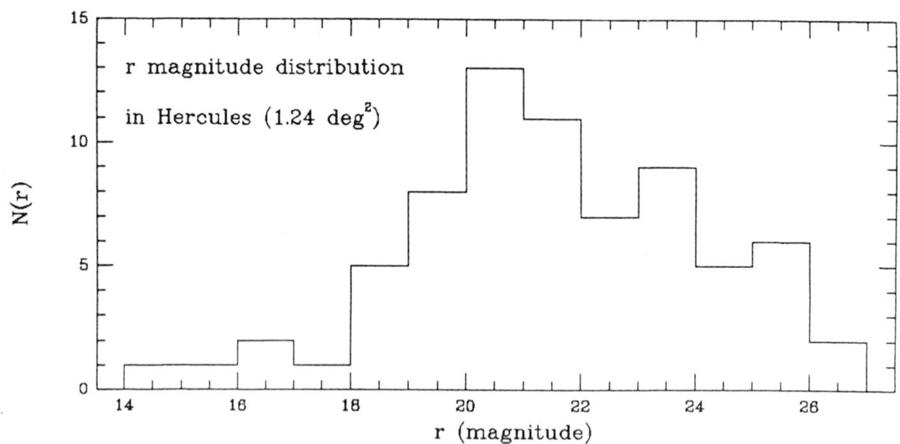

Most of our new Four-shooter identifications (r>22 mag) are clearly extended objects. Their colors are in general fairly blue (-0.5<g-r<1.0), although there exist a few redder identifications. The faint, blue radio objects are possibly actively star forming galaxies at z>0.5, probably the high redshift extension of the blue radio galaxy population discussed in section 2. The reddest radio galaxies are probably very high redshift (z>1) ellipticals with passively evolving spectral energy distributions, like their more nearby counterparts in faint radio samples (KKW, WKS).

It also became clear why the faintest radio sources were not detected before. Although their integrated magnitudes are in the range r=25.0-26.0 mag, they have all very low surface brightness. Some of them are 4-6" in diameter, and show several subclumps in the stacked CCD images. In general these objects showed up in both the deep g,r,i frames with colors in the range -0.5<g-r<0.7 and 0.0<r-i<0.5. (In each exposure the objects were offset and exposed by different pixels to minimize flat field errors). They were more easily recognized in images convolved to a FWHM of 2.5". An example is shown in Figure 2.a and b, which contains 4200 seconds in Gunn g and r added together. While hardly anything is visible at the radio source position in Figure 2.a, an object is clearly visible in Figure 2.b which contains the same data convolved to a FWHM of 2.5". We have detected several of these "fuzzballs". The faintest ones have surface brightness of the order of <10^{-3} x sky.

They have optical colors, morphology and radio structures similar to 3C326.1, although their radio fluxes are a factor 1000 fainter. The radio source 3C326.1 was recently identified by McCarthy et al. (this volume) with an actively starforming gascloud with V~24.0 mag and a size of 10". The object has a low ionization spectrum at a redshift of 1.82, and was tentatively identified as a primeval galaxy. If the "fuzzballs" in our milliJansky sample have the same <u>intrinsic</u> optical brightness and size as 3C326.1, their observed colors and surface brightness suggest that they could be similar galaxies, but at redshifts of 2.5-3.0, due to the $(1+z)^4$ dimming. Spectroscopy will be required to check whether these "fuzzballs" are high redshift radio galaxies in the process of formation.

2. SPECTROSCOPY OF MICROJANSKY RADIO SOURCES.

Since the discovery of the upturn in the 1.4 GHz source counts at milli-Jansky levels (W84, WMOKK), various explanations have been offered for this phenomenon. WMOKK suggested that a population of blue radio galaxies at intermediate redshifts might be responsible, with peculiar optical morphology, sometimes interacting or merging galaxies. C84's model held normal spiral galaxies responsible that underwent very strong cosmological evolution and implied that a large fraction is at very high redshifts. WBGV hypothesized that they are various kinds of unevolving low luminosity galaxies at low redshifts. Here we present the first results of a systematic spectroscopic study with the 100" du Pont CCD spectrograph at Las Campanas. We concentrated on a deep VLA field at 8h+17°, for which we also obtained deep Four-shooter coverage in Gunn g and r.

We obtained good quality spectra with a TI CCD at 20Å resolution for 18 radio sources with B<21.0. Four objects are Galactic stars, mostly of late type. They are usually variable radio sources and the radiation mechanism is possibly gyroresonance or gyrosynchrotron emission (LG). Radio stars do not dominate the sub-milliJansky population but their numbers are sufficiently large that they might induce field to field variations in sub-mJy source counts made at different galactic latitudes.

The sample contained only one QSO (at $z=1.92$), and four red galaxies with spectra and luminosities of passively evolving giant ellipticals at $z=0.2-0.3$. The remaining 9 objects were all blue galaxies, some having peculiar optical morphology on the CCD images, like the objects of KKW. Among the 9 blue galaxies, 8 have fairly narrow, and weak to moderately strong emission lines, usually Hα/S[II] and O[III]/Hβ with ~20-30Å equivalent widths. If the galaxies were at high enough redshift we also picked up O[II] 3727. Sometimes Mg b and Na were seen in absorption, as well as H and K and the G-band. These blue radio objects are like actively starforming galaxies, possibly similar to the ones found in spectroscopic studies of distant clusters (DG). For 15<B<21 mag their measured redshift range is $0.05<z<0.3$, indicating that sub-mJy radio sources are <u>not</u> local, low-luminosity dwarf galaxies. They have, on the contrary, L* like optical luminosities. The fact that the emission line fraction is significantly higher than the 30% in a field sample (DTS), combined with their sometimes peculiar optical morphology, makes it unlikely either that they are just normal spiral galaxies.

In conclusion, the blue, actively star forming galaxies constitute the majority group among the sub-mJy identifications and are most likely responsible for the upturn in the 1.4 GHz source counts at milliJansky levels. We expect that the fainter (r>20 mag), blue milliJansky and sub-milliJansky radio galaxies are of the same class, but at $z>0.3$.

REFERENCES

Condon, J.J.: 1984, <u>Astrophys. J.</u>, **284**, 44 (C84).
Dressler, A., Gunn, J.E.: 1983, <u>Astrophys. J.</u> **270**, 7 (DG).
Dressler, A., Thompson, I.B., Shectman, S.A.: 1985, <u>Ap.J.</u> **285**, 481 (DTS).
Kron, R.G., Koo, D.C., Windhorst, R.A.: 1985, <u>A. A.</u>, **146**, 38 (KKW).
Linsky, J.L., Gary, D.E.: 1983, <u>Astrophys. J.</u> **274**, 776 (LG).
Oort, M.J.A., et al.: 1986, <u>Astron. Astrophys.</u>, submitted (OKSW).

Peacock, J.A.: 1985, MNRAS, **217**, 601 (P85).
Wall, J.V., et al.: 1986, in "Highlights of Astronomy, Vol. VII", ed.
     J.-P. Swings (Reidel, Dordrecht), pg. 345 (WBGV).
Windhorst, R.A.: 1984, Ph.D. Thesis, University of Leiden (W84).
Windhorst, R.A., et al.: 1985, Astrophys. J., **289**, 494 (WMOKK).
Windhorst, R.A.: 1986, in "Highlights of Astronomy, Vol. VII", ed. J.-P.
     Swings (Reidel, Dordrecht), pg. 355 (W86).
Windhorst, R.A., Koo, D.C., Spinrad, H.: 1986, in "Galaxy Distances and
     Deviations from Universal Expansion", ed. B.F. Madore and R.B. Tully
     (Reidel, Dordrecht), pg. 197 (WKS).

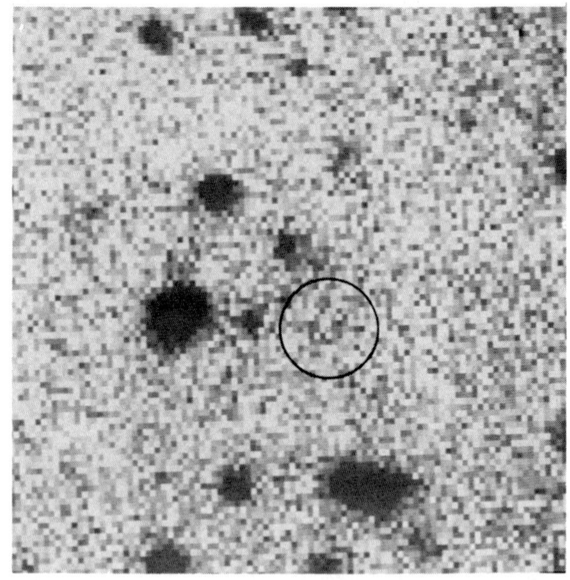

Figure 2.a: Stack of 4200 seconds of Four-shooter exposures in Gunn g and r on a radio source in Hercules. The image shown is 28x28" and has FWHM=1.2". The circle has 3" radius and is centered on the VLA position, which is known to 0.1". There is only marginal signal at the radio position.

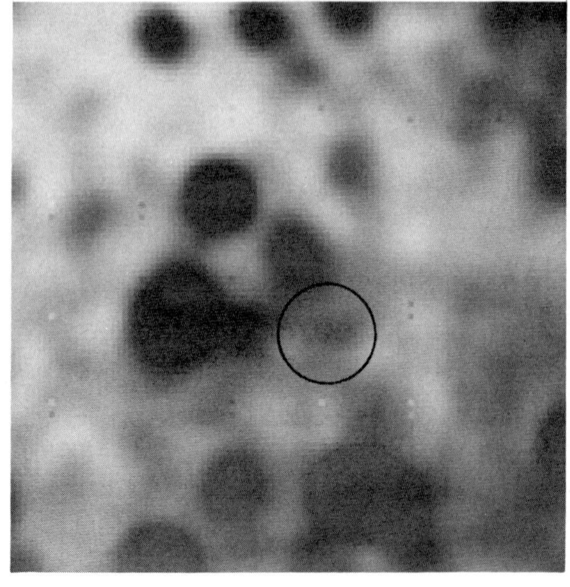

Figure 2.b: The same data convolved to a FWHM of 2.5". A very low surface brightness object now appears at the radio source position, with integrated V=25.7 mag. The grey scale has optimal contrast between -1 and +2σ. Apparently deviant pixel values are due to low level bit errors in the Grinnell.

CHAPTER X

THE DISTRIBUTION OF X-RAY SOURCES AND THE X-RAY BACKGROUND

THE DISTRIBUTION OF X-RAY EMITTING QUASARS IN SPACE

G. Setti*
European Southern Observatory
Karl-Schwarzschild-Str. 2, Garching bei München, FRG

ABSTRACT

This review is mainly concerned with the discussion of the
statistical properties of X-ray and optically selected samples of
quasars and Sy 1 nuclei and their intercomparison. The inconsistencies
which have been found are clearly exemplified by the discrepancy
between the observed and the predicted X-ray source counts. It is
shown that a satisfactory understanding of these problems has not yet
been reached. The last section deals with the long debated contribution of AGNs to the extragalactic X-ray background (XRB), which is
still uncertain within a factor of 2, although a lower bound of 30% at
the (now convenient) energy of 2 keV can be set.

INTRODUCTION

It is well known that a big step forward in the study of the X-
ray properties of large samples of objects has been provided by the
enhanced sensitivity of the Einstein Observatory. The optical identifications of several hundred sources detected by the Einstein
Observatory have led to the discovery of a substantial number of X-ray
selected AGNs. The published observational material can be convenient-
ly subdivided into two groups: on the one side the "surveys" such as
the High Sensitivity Survey (HSS) (Giacconi et al., 1979; Griffiths et
al., 1983) and the Medium Sensitivity Survey (MSS) (Maccacaro et al.,
1982; Stocke et al., 1983; Gioia et al., 1984), where the derived
samples are complete down to precisely defined X-ray flux limits, and
on the other side statistical samples of serendipitous sources found
in Einstein IPC and HRI fields (Chanan, Margon and Downes, 1981; Kriss
and Canizares, 1982; Reichert et al., 1982; Margon, Chanan and Downes,
1982; Margon, Downes and Chanan, 1985). In addition, a large number of
previously known quasars have been observed either in preselected

---

*) On leave of absence from the University of Bologna, Italy.

579

A. Hewitt et al. (eds.), Observational Cosmology, 579–592.
© *1987 by the IAU.*

heterogeneous samples (Tananbaum et al., 1979; Zamorani et al., 1981;
Ku, Helfand and Lucy, 1980) or as members of samples complete down to
well defined optical and/or radio limits (Tananbaum et al., 1983;
Marshall et al., 1984; Tananbaum et al., 1986). Observations of
Seyfert galaxies have also established the continuity of X-ray
properties between Sy 1 galaxies (nuclei) and quasars (Kriss,
Canizares and Ricker, 1980).

Studies of optically selected samples have led to some important
conclusions which are relevant to the subject of this review. These
can be summarized as follows:

a) The X-ray to optical luminosity ratios appear to be predominantly
   dependent on the optical luminosities of the objects, such that
   on average $L_x/L_{opt} \propto L_{opt}^{-0.2}$, although some dependence on the
   redshift cannot be excluded (Kriss and Canizares, 1985; Avni and
   Tananbaum, 1986).

b) Contrary to what happens in the radio domain, no more than a few
   percent of optically selected quasars (and Sy 1 nuclei) can be X-
   ray quiet - probably all are X-ray loud (Avni and Tananbaum, 1986).

Among other things, these findings tell us that it should be
possible, at least in principle, to make precise predictions about the
properties (counts, composition, evolution, etc.) of X-ray selected
samples based on the knowledge accumulated in optical samples. These
issues and the question of the contribution of quasars and Sy 1
nuclei* to the extragalactic XRB will be the subject of the following
discussion.

THE X-RAY NUMBER COUNTS RELATIONSHIP

Direct information is being obtained from the optical identification content of three surveys which are believed to be complete down
to different, but well defined X-ray flux limits. For the sake of
clarity, let us briefly summarize the results so far obtained.

---

*) The adoption of this terminology is only meant to provide a
convenient separation of the quasar phenomenology in two luminosity
domains, and it has been preferred to the term AGNs which comprises a
wider class of objects not discussed in this paper. The luminosity
separating the two classes falls in the range $M_B = -23 \div -24$
(depending on authors), but its precise definition is not critical for
the present discussion.

Whenever needed, cosmological parameters $H_o = 50$ km s^{-1} Mpc^{-1} and
$q_o = 0$ have been adopted throughout this paper.

The intermediate X-ray fluxes are covered by the Einstein Medium Sensitivity Survey (MSS). This survey is composed of serendipitous sources from a mosaic of Einstein IPC fields ($\sim$ 340) with different sensitivities ranging from 7×10^{-14} to 2×10^{-12} erg cm^{-2} s^{-1} in the energy band 0.3-3.5 keV. The area of the sky covered is $\sim$ 90 deg^2 of which only 0.3 deg^2 have been covered with the highest sensitivity mentioned above, so that from the sky coverage table provided by Gioia et al. (1984) it appears that only sources brighter than 1.6×10^{-13} ergs cm^{-2} s^{-1} may effectively be represented in the survey. By applying well defined selection criteria 112 sources have been detected above the 5σ level, all of which have been identified in the POSS with the result that almost exactly 50% are associated with quasars and Sy 1 nuclei, for which redshift and colours are also available. Thus, for the first time it has been possible to derive an observed X-ray number count relationship for this class of objects with the result that over a factor of about 30 in flux the integral counts can be represented by a power law of the form $N(>S) \propto S^n$ with n = -1.71 $\pm$ 0.15 (Fig. 1). Although a steep Log N - Log S was expected on the basis of the optical source counts and of the strong correlation

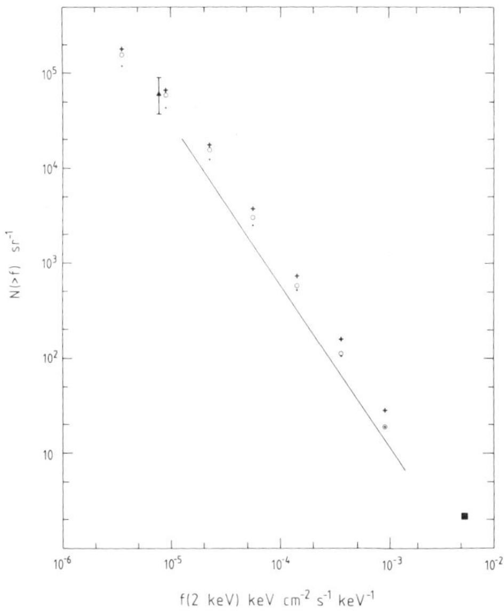

<u>Fig. 1</u> - The integral source counts versus the observed fluxes at 2 keV. The straight line corresponds to the MSS counts for AGNs, the filled square to the surface density of Sy 1 galaxies detected at the limit of the HEAO 1 survey, and the filled triangle to the HSS limit of Giacconi et al. (1979). The open circles correspond to the predicted integral counts from the optical Log N - B relationship given in the text, while the crosses refer to the same with the inclusion of object A in Fig. 2. The filled dots outline the modifications in the predicted counts after inclusion of a possible K-correction term as explained in the text.

between X-ray and optical luminosities, as emphasized by Gioia et al. (1984) this result is an important independent observational evidence of the existence of strong cosmological evolution for the combined sample of quasars and Sy type 1 nuclei free from the optical selection effects which could have biased the optical source counts (see Wampler and Ponz (1985) for a recent discussion of these effects). The slope of the X-ray counts appears to be significantly less steep than the slopes derived from optical counts for which values of the exponent $n = -2 \div -2.3$ are normally quoted, which seems to indicate that the cosmological evolution as measured at X-rays is less dramatic than at optical wavelengths. Maccacaro, Gioia and Stocke (1984) analyzed the MSS data by adopting a pure luminosity evolution model whereby the X-ray luminosity of the objects is assumed to increase exponentially as a function of the look-back time, and consistently derived a cosmological evolution coefficient much lower than those derived in the corresponding models applied to optical samples.

It should be noted that if luminosity evolution plays an important rôle in describing the spatial distribution of quasars, as it appears to be the case (e.g. Koo, 1986; Boyle et al., 1986; Schmidt, 1987), then the slope of the X-ray source counts should be less steep than that of the optical counts because of the inverse correlation between L_x/L_{opt} and L_{opt} (Avni and Tananbaum, 1982). However, care should be exerted when making a closer comparison between the source count slopes: since a substantial fraction of the objects in the MSS sample tend to have quite reddish colours ($\sim 30\%$, showing a $B-V \gtrsim 0.8$) and frequently, but not always, associated extended images (Gioia et al., 1984), they would be underrepresented or completely excluded in optically selected quasar samples based on colour criteria (UVX). Since on average these objects are intrinsically faint and of low redshift (Sy type 1 nuclei with $M_B > -22$), it is likely (but it should be proven observationally) that their effect is to flatten the overall X-ray Log N - Log S with respect to that which would be obtained by including only objects which meet the colour (and images) selection criteria of the optical quasar searches.

At the faintest fluxes the X-ray source counts are bound by the limits derived from the studies of the Einstein High Sensitivity Survey (HSS) fields. From the Draco and Eridanus fields Giacconi et al. (1979) derived an absolute limit of 19.2 ± 7.9 deg^{-2} to the total number of extragalactic sources ($> 5\sigma$) down to a flux limit of 2.6×10^{-14} ergs cm^{-2} s^{-1} in the energy range 1-3 keV. Primini et al. (1986) have analyzed available deep IPC exposures of 8 HSS fields (including a reanalysis of previously reported fields) to cover a sky area of $\sim 2,3$ deg^2. From the optical identification work on the sources detected above a 4.5σ level, 8 have been positively identified with quasars (with absolute magnitudes ranging around $M_B = - 24$) and 10 others are "probable extragalactic objects", leading to a surface density of X-ray selected quasars and Sy 1 nuclei $\lesssim 8^{+3}_{-2}$ deg^{-2} down to a flux limit approximately equal to that of Giacconi et al. reported above. Although this result is in agreement within the statistical errors with the Giacconi et al. limit, the central value is smaller by a factor 2.4 and it provides a much stronger constraint on the X-ray source counts.

At the very bright end of the flux scale information is provided by the HEAO 1 X-ray survey of Piccinotti et al. (1982), which covers $\sim 27,000$ deg^2 of the sky and is claimed to be complete down to a limiting sensitivity of 3.1×10^{-11} ergs cm^{-2} s^{-1} in the 2-10 keV energy band. It contains 19 Sy 1 galaxies and 1 quasar.

These results have been plotted in Fig. 1 as a function of the monochromatic 2 keV fluxes by converting the broad-band spectra with the corresponding power-law photon distributions assumed in the data analysis of the various surveys, i.e. photon spectral indices of 1.4, 1.5 and 1.65 for the MSS, HSS and HEAO 1 surveys, respectively. It should be noted that a downward extrapolation of the MSS source counts falls about a factor 3 below the HEAO 1 point. Since the MSS source counts are ill defined at the bright band, it is not clear to me whether the counts would tend to flatten to meet the HEAO 1 point or whether this discrepancy is due, at least in part, to a genuine scale difference between the Einstein Observatory and HEAO 1 fluxes, as recently suggested by Danese et al. (1986).

COMPARISON BETWEEN X-RAY AND OPTICALLY SELECTED SAMPLES

When the complete results of the MSS became available it was immediately realized (Setti, 1984; Zamorani, 1984) that there was a striking contradiction, an excess of a factor 2-3, between the number of quasars and Sy 1 nuclei predicted on the basis of the optical source counts and of the X-ray emission properties of optically (and radio) selected samples and the number actually found in the MSS. Similarly, one finds contradictory predictions with regard to the slope of the X-ray source counts, the redshift distributions, and so on. It should be stressed that these contradictions persist independently of the methods, evolution models, reference optical samples, luminosity and colour constraints which have been adopted in making the predictions (Setti and Woltjer, 1962; Maccacaro, 1984; Kriss and Canizares, 1985; Zamorani, 1985; Schmidt and Green, 1986; Avni and Tananbaum, 1986).

In particular, Schmidt and Green (1986) have made a study of a homogeneous set of data obtained for the Bright Quasar Survey (BQS), a sample of which has been observed with the Einstein Observatory (Tananbaum et al., 1986). X-ray fluxes have been derived by assuming a 1.5 power-law photon spectral index, and therefore are directly comparable to the MSS results; in order to minimize the intervention of optical selection effects their analysis has been confined to the quasar sample ($M_B < -23$). In the framework of the evolution models (luminosity-dependent density evolution) utilized to fit the BQS data, an overprediction of a factor 2, or so, in the number of quasars and a

higher average redshift with respect to the MSS results are again
found. By forcing agreement with the MSS quasar sample these authors
introduced a negative cosmological evolution of the X-ray luminosity
(at a given optical luminosity).

This has prompted Maccacaro and Gioia (1986) to analyze possible
causes of incompleteness of the MSS, with the conclusion that the MSS
AGNs sample is statistically complete. Franceschini, Gioia and
Maccacaro (1986) find that most of the discrepancies could be greatly
reduced, if not completely eliminated, if the intrinsic distribution
of the residuals in the L_x, L_{opt} correlation is narrower (and more
skewed at large L_x/L_{opt}) than the observed one, and discuss a number
of effects which could have artificially broadened the observed
distribution, among which the temporal variability represents a clear-
cut case since X-ray and optical data have usually been acquired at
very different epochs (see also Avni and Tananbaum, 1986, and
Zamorani, 1985).

In order to have a somewhat clearer and more qualitative percep-
tion of the various issues, without entering into the whole alchemy of
luminosity functions, evolution models, α_{ox}, etc., I have derived
X-ray source counts by making use only of the observed Log N - B
relationship(s) and of the observed ratios of X-ray to optical flux
for a well defined unbiased sample of optically selected objects, such
as the sub-sample of BQS observed with the Einstein Observatory
(Tananbaum et al., 1986). Of the 66 objects observed, 57 have been
detected and their 2 keV fluxes versus B magnitudes are plotted in
Fig. 2. A posteriori I have verified that the upper limits available

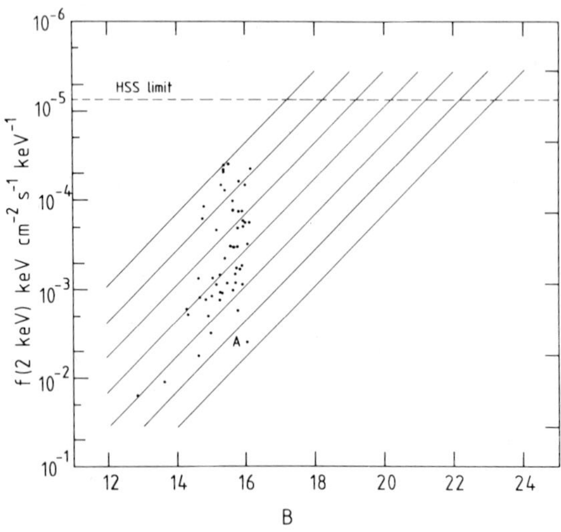

Fig. 2 - The observed fluxes at 2 keV from the BQS X-ray sub-sample
versus B magnitudes corrected for galactic absorption. The straight
lines correspond to loci of constant intrinsic X-ray to optical
ratios.

for the remaining objects are such that these objects can be neglected without substantially modifying the results of the present analysis. It is seen that most of the objects fall within one magnitude in B, while they are spread over more than two orders of magnitude along the X-ray flux axis. If for any given object the intrinsic ratio of X-ray to optical flux were preserved as a function of distance, then in the Log f_x - B diagram the objects would move along trajectories parallel to the diagonal lines shown in Fig. 2. It is seen that objects from a wide range of B magnitudes will contribute to any given X-ray flux interval and it is now a simple matter to derive the corresponding X-ray source counts by summation over the contributions obtained by drawing out objects from the optical Log N - B according to the percentage found in the reference sample per bin of B magnitudes. However, because of the inverse proportionality of L_x/L_{opt} with L_{opt}, and because the percentage of objects in any given absolute magnitude interval in the Log N - B relationship changes with B, the procedure described above must be qualified per luminosity bins. The percentages of objects per luminosity bin in the Log N - B relationship have been derived from the contents of the BQS (Schmidt and Green, 1983), BFG (Braccesi et al., 1970) and BF (Formiggini et al., 1980; Marshall et al., 1984) samples. Beyond B $\simeq$ 20 percentages have been calculated by assuming a quasar redshift cutoff z = 3, so that at B = 24 only nuclei with $M_B \gtrsim$ -24 would be present in the optical counts.

Several qualitative conclusions can be drawn immediately from the inspection of Fig. 2:

(a) According to the scheme outlined above, it is clear that the precise distribution of the objects at bright X-ray fluxes plays a key-rôle because of the coupling with the fainter B magnitudes where the surface density of quasars increases dramatically due to the very steep slope of the optical source counts. For instance, the contribution of the object marked A, found in the last bin, to the derived X-ray source counts is of the order of 25% (Fig. 1). This is precisely the effect discussed by Franceschini, Gioia and Maccacaro (1986). It is hard to believe that the estimate should depend so strongly on the inclusion (or exclusion) of just one object, and accordingly we shall ignore it in what follows. It should be noted, however, that to remove it from its position in the diagram of Fig. 2 would require that it be at least 0.6 magnitudes brighter at the time of the X-ray measurement. [Object A is the nucleus of the Sy 1 galaxy III Zw 2 and, since its absolute magnitude M_B = -22.7, it has been excluded from the quasar sample discussed by Schmidt and Green (1986). Evidence for cosmological evolution of Sy 1 nuclei of this luminosity has been discussed by Setti (1984).]

(b) A flattening of the X-ray source counts is predicted to take place well before the HSS limit when one begins to include objects with B $\gtrsim$ 20, the magnitude beyond which the optical source counts are known to flatten dramatically.

(c) Objects with magnitudes up to $B \sim 23$ contribute to the X-ray source counts down to the HSS sensitivity.

The adopted optical source count can be described as follows: $\text{Log } N(<B) = 0.53 + 0.85 \, (B-19)$ for $B < 19$, consistent with the BQS and BFG samples corrected for selection effects (Woltjer and Setti, 1982; Setti, 1984); $\text{Log } N(<B) = 1.57 + 0.38 \, (B-21)$ for $B > 21$, consistent with Koo (1986) and Marano, Zamorani and Zitelli (1986), while in the interval $19 < B < 21$ the counts have been smoothly interpolated so that at $B = 19.8$ an integral surface density of $\simeq 13$ obj. deg^{-2} is obtained, a factor ~ 1.5 below the surface density of the BF sample which appears to be in excess of the findings in other quasar surveys (Marano, Zamorani and Zitelli, 1986; Boyle et al., 1986). The integral X-ray source counts can now be derived and the results are plotted in Fig. 1. It is seen that the predicted counts (excluding object A) exceed by a factor from 1.5 to 2.1 the MSS counts and have a steeper slope. As can be immediately guessed from a quick glance at Fig. 2, the predicted X-ray counts are very sensitive to the precise value of the optical counts in the magnitude range $19 < B < 21$ (Woltjer and Setti, 1982; Anderson and Margon, 1986). A surface density of a full factor 2 below the value of the BF sample would flatten the predicted counts and the discrepancy with the MSS counts would not exceed a factor 1.8. In addition, if the average X-ray spectrum of quasars and Sy 1 nuclei in the BQS sample is as steep as indicated by the findings of Elvis et al. (1986), and if a spectral index $\simeq -0.5$ is representative of the average optical continuum (Richestone and Schmidt, 1988), then an important K-correction term must be taken into account (see also Schmidt and Green, 1986). In the diagram of Fig. 2 this is equivalent to saying that the (hypothetical) trajectories of the objects are curving upwards and a stretching of the source counts along the X-ray flux axis is inferred. Assuming that the difference between the average X-ray and optical continuum spectral indices is 0.5, the predicted source counts change as shown in Fig. 1 (filled dots)*. It is worth noting that the slope of the counts is now almost exactly equal (by coincidence?) to that of the MSS, but an excess of a factor 1.6 can still be registered**. A further flattening of the counts at the fainter fluxes would be introduced by another K-correction term (not considered here) because of the brightening of the objects at $z \gtrsim 2$ caused by the strong emission lines entering the B filter.

---

*) As discussed later, a much steeper X-ray spectrum implies a sizeable reduction in the 2 keV fluxes. This has not been taken into account since we are interested here only in the relative comparison with the MSS results.

**) The detailed analysis also shows that at the HSS limit most objects are comprised in the luminosity range $-22 > M_B > -26$, almost equally positioned around $M_B \simeq -24$.

Based on the above discussion, it should be noted that because of the steep slopes involved the difference between the predicted and observed number counts can be completely eliminated by a horizontal shift of only a factor 1.25-1.30 in the relative flux scale. This could be a genuine difference, but one may wonder if it could at least in part be explained by the presence of systematic effects such as, for instance, the different procedures (background subtraction) followed to derive the source fluxes for the MSS and for the BQS X-ray sub-sample as described in Maccacaro et al. (1982) and in Tananbaum et al. (1986), respectively. Moreover, if the steep X-ray spectra derived by Elvis et al. (1986) for 8 X-ray bright quasars of the BQS sample are representative of the optical as well as X-ray samples surveyed by the Einstein Observatory, then, by direct comparison with the data shown in Tananbaum et al. (1986), one finds that an average reduction of ~ 40% in the 2 keV fluxes displayed in Figs. 1 and 2 should be applied. While this would not modify the intercomparisons within the samples studied by the Einstein Observatory, it would further enhance by a large factor (~ 2.3) the difference between the HEAO 1 point and the downward extrapolated MSS counts. Equivalently, the number of Sy 1 nuclei predicted on the basis of the HEAO 1 data and a uniform space distribution would greatly exceed the one found in the MSS. Although this may certainly contribute to solving a selection problem pointed out by Schmidt and Green (1986), it seems to me that the general situation would be eased by a substantial upward scaling of the MSS fluxes. In addition, this may also contribute to the solution of another apparent discrepancy between the number and mean luminosity of clusters of galaxies predicted on the basis of the HEAO 1 survey and the findings of the MSS (Schmidt and Green, 1986).

In conclusion, I believe that at this stage the observational situation is so tight that it is important to make sure first that the X-ray fluxes have been calibrated exactly on the same scale.

THE CONTRIBUTION TO XRB

The question of the contribution of quasars and Sy 1 nuclei to the XRB has been the subject of many investigations over the past fifteen years. Despite the vast increase in the number of objects detected in the X-rays and the great advance in the understanding of their emission properties, this subject remains open to a great deal of incertitude as shown by the wide range of estimates arrived at by different authors (Cavaliere et al., 1981; Setti and Woltjer, 1982; Zamorani, 1982; Maccacaro, Gioia and Stocke, 1984; Anderson and Margon, 1986; Schmidt and Green, 1986). Obviously, in the discussion of this problem a key rôle is also played by the spectral shape of the sources. The differential energy spectra of a close-by sample of Sy 1 galaxies selected at hard X-ray energies can be fitted by power laws in the 2-20 keV interval, such that the sample shows a mean spectral index $\simeq$ -0.65 with a small dispersion around the mean (Mushotzky, 1984). Very little information is available on the hard X-ray spectra of quasars. Recently, results from the Einstein Observatory IPC (0.1-4 keV) have become available (Elvis, Wilkes and Tananbaum, 1985; Elvis et al., 1986) for a small sample of optically selected quasars and Sy 1 nuclei at relatively small redshifts ($z < 0.3$). Single power-law fits to the spectra are obtained with slopes on average much steeper (mean spectral index $\lesssim -1.00$) than the previously quoted value for the hard X-ray domain. The reason for this difference is not yet fully understood, although the hard and soft X-ray selection of the samples may play an important part (Elvis et al., 1986; also for references to previous work on the subject). The important point as far as the present discussion is concerned is that the spectra of the sources appear to be at variance with the spectrum of the XRB, which in the 2-20 keV interval can be represented by a power law with a spectral index -0.4 and an intensity at 2 keV of $\simeq 5.83$ keV cm^{-2} s^{-1} ster^{-1} keV^{-1}. Consequently, unless the X-ray spectra of quasars and Sy 1 nuclei undergo a strong cosmological evolution (become flatter at large z), these objects cannot explain the bulk of the XRB. In fact, the high precision measurements of the XRB in the 3-50 keV interval made with the HEAO 1-A2 (which can be accurately fitted by a 40 keV thermal bremsstrahlung) have been used to set a limit of $\lesssim 30\%$ in the 2-10 keV energy interval to the integrated contribution from sources having power-law spectra with mean spectral index $\simeq -0.7$ and a high energy cutoff (De Zotti et al., 1982). As a corollary, the same authors noted that any population of sources substantially contributing to the XRB in addition to AGNs must have a mean spectral index $\gtrsim -0.4$.

With the above considerations in mind, the contribution of quasars and Sy 1 nuclei to the XRB can be better discussed with reference to a fixed observed energy of 2 keV which represents a convenient compromise between the soft X-ray sensitivity of the IPC, where most of the statistical information on these objects has been obtained, and the hard X-ray where the extra-galactic XRB has been accurately measured. The most recent estimate of the quasar contribution to the XRB has been made by Schmidt and Green (1986). On the basis of the X-ray source counts derived from the BQS sample by forcing agreement with the number of quasars found in the MSS (as discussed in the preceding section), they estimate a contribution to the 2 keV background of 13% down to 8% if the steep spectral index found by Elvis et al. (1986) is considered typical of all quasars. Estimates, and in fact lower bounds, can be obtained by integration over the X-ray source counts derived from the optical counts independent of the particular choice of evolution models. With the specifications described in the previous sections with regard to the optical counts and their content, it is found that quasars down to B = 23 would contribute 19-21% of the 2 keV background depending on the adopted value of the counts around B = 20. If agreement with the MSS counts is forced and/or a steep slope for the X-ray spectra of a typical quasar is adopted, then contributions ranging from 7 to 11% are found, in good agreement with the findings of Schmidt and Green (1986).

A "minimal" contribution of 29% to the 2 keV background can be obtained from the HEAO 1 (local) sample of Sy 1, assuming a uniform spatial density and a mean spectral index -0.65 (Piccinotti et al., 1982; Schmidt and Green, 1986), with a statistical error of ± 6.5% due to the small size of the sample. Since two Sy 1 nuclei in the Elvis et al. (1986) sample show a very steep spectral index, a possible additional contribution from steep spectra Sy 1 nuclei should be considered. Moreover, a further contribution may result from evolution possibly present at the bright end of the Sy 1 luminosity function (Setti, 1984). A direct estimate of the Sy 1 contribution to the XRB via the counts cannot be done at present, not only because of the selection effects affecting both the X-rays and optical source counts at the bright end, but, even more importantly, because it would require the knowledge of the optical counts down to and deeper than B = 24, where the counts should be fully dominated by Sy 1 nuclei.

Therefore, while it appears ascertained that the combined contribution of Sy 1 nuclei and quasars to the 2 keV background must be $\gtrsim$ 30%, it may also exceed 50%. A more precise value for this contribution, together with an appropriate knowledge of the corresponding X-ray spectra not yet available and of the contributions from other classes of sources, are needed to properly derive the residual XRB above 3 keV, a prerequisite to the discussion, beyond the scope of the present review, of the nature of the source (or sources) which saturate the XRB. A recent discussion of these issues has been made by Giacconi and Zamorani (1986).

CONCLUSIONS

The apparent inconsistencies between the statistical properties of X-ray selected samples of quasars and Sy 1 nuclei and those predicted on the basis of X-ray observations of optical samples have not yet been satisfactorily understood. Although possible solutions have been proposed, further investigations are required including a verification that the X-ray flux measurements in different samples are calibrated on the same scale.

The problem has become even more complicated, if one may say so, after the discovery that the Einstein IPC spectra of optically selected X-ray bright quasars and Sy 1 nuclei are on average much steeper than the often quoted "canonical" value for AGNs, which indicates a widely spread distribution in the spectral indices. Because of the interplay between the spectral shapes and the corresponding X-ray fluxes, a proper intercomparison of different samples should be based on an adequate knowledge (not achievable at this moment) of the spectral characteristics of the corresponding populations of objects. To assume, for instance, that the IPC spectra of the optically selected X-ray bright quasars are representative of all optically selected quasars, or that the X-ray and optically selected samples have similar average spectra, may turn out to be a gross oversimplification leading to contradictory results.

These uncertainties are fully reflected in the estimates of the combined quasar and Sy 1 nuclei contribution to the XRB which span a factor of 2, although a lower bound of 30% at the monochromatic energy of 2 keV seems well established. It should be noted that the additional estimated contributions from different objects (such as galaxies and clusters of galaxies) make it likely that the integrated contribution of known classes of sources to the 2 keV XRB exceeds 50%. This, together with the poor knowledge of the X-ray spectra applicable to the various classes of objects (with the exception of clusters), renders very uncertain any extrapolation of the contribution to the background above 3 keV.

It has become customary to terminate papers dealing with this subject by deferring the solution of all these difficult problems to the launch of future X-ray observatories such as ROSAT, AXAF and XMM. While this will certainly be the case, I believe that in the meantime a tremendous amount of important work can be done with the data already acquired. In particular, the extension of the MSS now under way is of the greatest importance since it will provide a much larger sample of X-ray selected AGNs, thereby allowing a much closer comparison with the optically selected samples. The MSS probes the quasar distribution down to B ~ 21 (Fig. 2) and it is hoped that the increased size of the sample will permit to achieve statistically meaningful results at this faint end not affected by well known optical selection biases. Similar considerations apply to the HSS, for which it would obviously be of greatest importance to complete the optical identifications. This is clearly not an easy task because objects down to B ~ 23 may be involved. But it is worth trying since, at least in principle, the identification content of the HSS, together with the corresponding surface density of the objects, can provide extremely important constraints on the evolution of quasars.

REFERENCES

Anderson, S.F., Margon, B.: 1986, in Quasars, IAU Symp. 119, G. Swarup and V.K. Kapahi (eds.), Reidel, Dordrecht, p. 247.
Avni, Y., Tananbaum, H.: 1986, Astrophys. J. **305**, 83.
Avni, Y., Tananbaum, H.: 1982, Astrophys. J. **262**, L17.
Boyle, B.J., Fong, T., Shanks, T., Peterson, B.A.: 1986, in Structure and Evolution of Active Galactic Nuclei, G. Giuricin, F. Mardirossian, M. Mezzetti and M. Ramella (eds.), Reidel, Dordrecht, Astrophys. Spa. Sci. Library, Vol. 121, p. 491.
Braccesi, A., Formiggini, L., Gandolfi, E.: 1970, Astron. Astrophys. **5**, 264.
Cavaliere, A., Danese, L., De Zotti, G., Franceschini, A.: 1981, Astron. Astrophys. **97**, 269.
Chanan, G.A., Margon, B., Downes, R.A.: 1981, Astrophys. J. **243**, L5.
Danese, L., De Zotti, G., Fasano, G., Franceschini, A.: 1986, Astron. Astrophys. **161**, 1.

De Zotti, G., Boldt, E.A., Cavaliere,, A., Danese, L., Franceschini,
     A., Marshall, F.E., Swank, H.J., Szymkowiak, A.E.: 1982, Astrophys.
     J. 253, 47.
Elvis, M., Green, R.F., Bechtold, J., Schmidt, M., Neugebauer, G.,
     Soifer, B.T., Matthews, K., Fabbiano, G.: 1986, Astrophys. J. 310,
     in press.
Elvis, M., Wilkes, B.W., Tananbaum, H.: 1985, Astrophys. J. 292, 357.
Formiggini, L., Zitelli, V., Bonoli, F., Braccesi, A.: 1980, Astron.
     Astrophys. Suppl. 39, 129.
Franceschini, A., Gioia, I.M., Maccacaro, T.: 1986, Astrophys. J. 301,
     124.
Giacconi, R., et al.: 1979, Astrophys. J. 234, L1.
Gioia, I.M., Maccacaro, T., Schild, R.E., Stocke, J.T., Liebert, J.W.,
     Danziger, I.J., Kunth, D., Lub, J.: 1984, Astrophys. J. 283, 495.
Griffiths, R.E., et al.: 1983, Astrophys. J. 269, 375.
Koo, D.C.: 1986, in Structure and Evolution of Active Galactic Nuclei,
     G. Giuricin, F. Mardirossian, M. Mezzetti and M. Ramella (eds.),
     Reidel, Dordrecht, Astrophys. Spa. Sci. Library, Vol. 121, p. 317.
Kriss, G.A., Canizares, C.R.: 1985, Astrophys. J. 297, 177.
Kriss, G.A., Canizares, C.R.: 1982, Astrophys. J. 261, 51.
Kriss, G.A., Canizares, C.R., Ricker, G.R.: 1980, Astrophys. J. 242,
     492.
Ku, W.H.-M., Helfand, D.J., Lucy, L.B.: 1980, Nature 288, 323.
Maccacaro, T.: 1984, in X-Ray and UV Emission from Active Galactic
     Nuclei, W. Brinkmann and J. Trümper (eds.), MPE Report 184, 63.
Maccacaro, T., Gioia, I.M.: 1986, Astrophys. J. 303, 614.
Maccacaro, T., Gioia, I.M., Stocke, J.T.: 1984, Astrophys. J. 283,
     486.
Maccacaro, T., et al.: 1982, Astrophys. J. 253, 504.
Marano, B., Zamorani, G., Zitelli, V.: 1986, in Structure and
     Evolution of Active Galactic Nuclei, G. Giuricin, F. Mardirossian,
     M. Mezzetti and M. Ramella (eds.), Reidel, Dordrecht, Astrophys.
     Spa. Sci. Library, Vol. 121, p. 339.
Margon, B., Chanan, G.A., Downes, R.A.: 1982, Astrophys. J. 253, L7.
Margon, B., Downes, R.A., Chanan, G.A.: 1985, Astrophys. J. (Suppl.)
     59, 23.
Marshall, H.L., Avni, Y., Braccesi, A., Huchra, J.P., Tananbaum, H.,
     Zamorani, G., Zitelli, V.: 1984, Astrophys. J. 283, 50.
Mushotzky, R.F.: 1984, in High-Energy Astrophysics and Cosmology,
     COSPAR/IAU Symp., G.F. Bignami and R.A. Sunyaev (eds.), Advances in
     Space Research, Vol. 3, p. 157.
Piccinotti, R., Mushotzky, R.F., Boldt, E.A., Holt, S.S., Marshall,
     F.E., Serlemitsos, P.J., Shafer, R.A.: 1982, Astrophys. J. 253,
     485.
Primini, F.A., et al.: 1986, in preparation.
Reichert, G.A., Mason, K.O., Thorstensen, J.R., Bowyer, S.: 1982,
     Astrophys. J. 260, 437.
Richstone, D.O., Schmidt, M.: 1980, Astrophys. J. 235, 361.
Schmidt, M.: 1987, these proceedings.
Schmidt, M., Green, R.F.: 1986, Astrophys. J. 305, 68.
Schmidt, M., Green, R.F.: 1983, Astrophys. J. 269, 352.

Setti, G.: 1984, in X-Ray and UV Emission from Active Galactic Nuclei,
    W. Brinkmann and J. Trümper (eds.), MPE Report 184, 243.
Setti, G., Woltjer, L.: 1982, in Astrophysical Cosmology, H.A. Bruck,
    G.V. Coyne and M.S. Longair (eds.), Pontificia Academia
    Scientiarum, Vatican City, p. 315.
Stocke, J.T., Liebert, J., Gioia, I.M., Griffiths, R.E., Maccacaro,
    T., Danziger, I.J., Kunth, D., Lub, J.: 1983, Astrophys. J. 273,
    458.
Tananbaum, H., Avni, Y., Green, R.F., Schmidt, M., Zamorani, G.: 1986,
    Astrophys. J. 305, 57.
Tananbaum, H., Wardle, J.F.C., Zamorani, G., Avni, Y.: 1983,
    Astrophys. J. 268, 60.
Tananbaum, H., et al.: 1979, Astrophys. J. 234, L9.
Wampler, E.J., Ponz, D.: 1985, Astrophys. J. 298, 448.
Woltjer, L., Setti, G.: 1982, in Astrophysical Cosmology, H.A. Bruck,
    G.V. Coyne and M.S. Longair (eds.), Pontificia Academia
    Scientiarum, Vatican City, p. 293.
Zamorani, G.: 1985, Astrophys. J. 299, 814.
Zamorani, G.: 1984, in X-Ray Astronomy '84, M. Oda and R. Giacconi
    (eds.), Institute of Space and Astronautical Science, Tokyo,
    p. 419.
Zamorani, G., et al.: 1981, Astrophys. J. 245, 357.

DISCUSSION

PECKER: When evaluating the counts of X-ray sources, in relation to
optical counts, you use the relation $L_X = L_O^{\beta}$ ($\beta \sim 0.7$-0.8). But,
I understood that some data indicated that indeed β is a function of
the X-ray energy; this effect should definitely affect the z-dependence
of the predicted counts for sources observed in a well-defined energy
range. What could be said now about it?

SETTI: I am not aware of data proving the existence of such an effect.
If it existed, it would simply mean that the slope of the X-ray spectra
would depend on the optical luminosity, and this cannot be excluded
at present.

X-RAY SURVEYS AS TOOLS TO INVESTIGATE THE COSMOLOGICAL EVOLUTION OF QUASARS, BL LAC OBJECTS AND CLUSTERS OF GALAXIES

Isabella M. Gioia[1,2], Tommaso Maccacaro[1,2] and Anna Wolter[1]

1) Harvard-Smithsonian Center for Astrophysics, Cambridge, MA

2) CNR - Istituto di Radioastronomia, Bologna, Italy

ABSTRACT

We present a progress report on a major extension of the Einstein Observatory Medium Sensitivity Survey (MSS). The basic properties of the extragalactic sources identified with Active Galactic Nuclei (AGN) and clusters of galaxies are discussed. Results from previous work are briefly summarized.

1. INTRODUCTION

In the last two years a major extension of the MSS has been carried out at the Center for Astrophysics. The Medium Survey consists of X-ray sources serendipitously discovered with the Image Proportional Counter on board the Einstein Observatory, in the energy range 0.3-3.5 keV, at fluxes between 10^{-11} and 10^{-13} ergs cm^{-2} s^{-1}. A detailed description of the survey and of its selection criteria can be found in Maccacaro et al. 1982 and Gioia et al. 1984.

2. PUBLISHED MSS

While some investigations can be successfully carried out utilizing incompletely identified samples or samples not well defined, some types of studies are possible only with completely identified and homogeneously selected samples. The optical identification of all the sources (112) in the MSS allowed us to extract modestly-sized but complete samples of astronomical objects which have been used for the following purposes:

(a) to derive the LogN-LogS relationship for the extragalactic population as a whole and for AGN and clusters of galaxies separately (Gioia et al. 1984);

(b) to determine the AGN X-ray luminosity function and cosmological evolution and their contribution to the diffuse X-ray background (XRB) (see Maccacaro, Gioia and Stocke 1984);

(c) to study the relationship between X-ray selected and optically selected AGN (Franceschini, Gioia and Maccacaro 1986);

(d) to show, despite the small numbers of BL Lacs in the survey (5), that their number-count relationship differs significantly from that of AGN (Maccacaro et al. 1984).

We have been so far prevented from studying the cluster luminosity evolution by the small number (18) of clusters detected in the published survey.

For AGN, the X-ray selection favors the discovery of relatively low luminosity, low redshift objects (i.e. high luminosity Seyfert galaxies), a subset of AGN not easily found by other means. For clusters of galaxies, the X-ray selection is an effective method to discover high redshift clusters (up to z = 0.5). This should allow us to compare their dynamical and morphological properties with those of nearby clusters as well as distant clusters found through optical searches.

A. Hewitt et al. (eds.), Observational Cosmology, 593–595.

For BL Lac objects, X-ray selection offers the opportunity to obtain a well defined, flux limited
sample of objects, and to compare their properties with those of optically or radio selected
quasars. Moreover, an X-ray survey is perhaps the best technique to find, if they exist, radio
quiet BL Lacs. With these motivations in mind we have undertaken a large program of extending
the Survey.

3. EXPANDED MSS

Over the last two years we have been working at this project by analyzing new IPC fields
from the Einstein Data Bank. The gathering of the X-ray data has been now completed.
The identification process for the new sources found and the collection of information at other
wavelengths, such as radio or infrared, is still under way. The expanded MSS not only has a larger
number of sources but it is also qualitatively better than the published survey. The reprocessing
of the data uses improved software and calibrations which allow a better determination of the
IPC background and of the source flux density. For each field a background map has been
generated, which takes into account the non-uniformities of the detector. This and the results
from an extensive simulation analysis (Maccacaro, Romaine and Schmitt, this volume) have
allowed us to lower our detection threshold to include the 4 σ sources and to utilize for the
survey a much larger portion of the IPC field. Table 1 presents the relevant parameters of the
expanded MSS.

TABLE 1

Expanded Medium Survey

IPC images analyzed	1439	(345)
Area for IPC field (sq.deg.)	0.54	(0.2)
Total area of sky (sq.deg.)	780	(90)
Sources detected	836	(112)
Significance of detection	≥ 4	(≥ 5)
Sources identified*	434	(112)
Sources with radio data*	311	(102)
Active Galactic Nuclei:*	189	(56)
BL Lacs or candidates:*	16	(5)
Clusters of galaxies:*	56	(18)
Normal galaxies:*	14	(3)

* **As of August 1986**

Numbers in parenthesis refer to published MSS

4. BASIC PROPERTIES OF THE EXTRAGALACTIC SAMPLES

As of today only half of the sources have been identified. Their properties are thus somewhat biased in favor of the apparently brighter objects, or in favor of objects which are active also at other wavelengths.

The X-ray selected AGN are low-luminosity, low-redshift objects ($<m_v> = 17.6$; $<z> = 0.4$; $<M_v> = -23.7$). At the beginning, the identification of "serendipitous" sources with this type of object was considered somehow disappointing. However, it is now clear that it is important to study the evolution properties of these low-luminosity AGN and compare them with the properties of high luminosity quasars to better understand the AGN phenomenon. As expected from the fact that they are X-ray selected, the MSS AGN are strong X-ray emitters with luminosities in the range 10^{42} - 10^{47}. Most of them, however, do not have enough counts to allow a determination of their spectral parameters, either because detected off-axis, or just above the threshold of the detector. It would be very interesting to analyze their X-ray energy distribution since quasars and Seyfert galaxies are the major contributors to the diffuse XRB and the XRB spectrum is at variance with the spectrum of the very few AGN for which a spectral determination exists. To this end we have undertaken to extract the available information on the AGN energy distribution from their hardness ratio (HR), which is a measure based on two X-ray "colors".

X-ray selection favors the detection of high redshift clusters. Of the 56 identified clusters, 44 have already a redshift determination ($<z> = 0.18$). Thirteen clusters have a z > 0.2, seven have a z > 0.3. When the identification process will be completed the number of X-ray selected clusters will be large enough (about 100 expected) to allow the study of their evolutionary properties.

The MSS project would not have been possible without the continuous collaboration of John Stocke, Simon Morris and Rudolph Schild. Isabella M. Gioia acknowledges travel support from the Smithsonian Institution under the "Research Opportunities Fund" program. Partial financial support for this work comes from the Smithsonian Scholarly Studies Program Grant SS48-8-84 and from NASA contract NAS8-30751.

REFERENCES

Franceschini, A., Gioia, I.M, and Maccacaro, T., 1986, Ap.J., 301, 124.

Gioia, I.M., Maccacaro, T., Schild, R.E., Stocke, J.T., Liebert, J.W., Danziger, I.J., Kunth, D. and Lub, J., 1984, Ap.J., 283, 495.

Maccacaro, T., et al., 1982, Ap.J., 253, 504.

Maccacaro, T., Gioia, I.M., and Stocke, J.T., 1984, Ap.J., 283, 486.

Maccacaro, T., Gioia, I.M., Maccagni, D. and Stocke, J.T., 1984, Ap.J.Letters, 284, L23.

Maccacaro, T., Romaine, S., and Schmitt, J., this volume.

LOGN-LOGS SLOPE DETERMINATION IN IMAGING X-RAY ASTRONOMY

Tommaso Maccacaro[1,2], Suzanne Romaine[1] and Jurgen H.M.M. Schmitt[3]

1) Harvard-Smithsonian Center for Astrophysics, Cambridge, MA.

2) CNR - Istituto di Radioastronomia, Bologna, Italy.

3) Max Planck Institute fur Physik und Astrophysik, Institute fur
Extraterrestrische Physik, Garching bei Munchen, FRG.

ABSTRACT

We briefly discuss the problem of estimating the slope of the number-counts relations for the specific case of imaging X-ray surveys. Results have been obtained from extensive simulations of Einstein Observatory imaging X-ray data. We conclude that the bias which affects the X-ray number-counts slope determination is much smaller than that which affects the radio number-counts slope.

1. INTRODUCTION

The analysis of the number count relationship for a given class of astronomical objects is not merely an academic or strictly mathematical matter but carries important information on the nature of the sources under study and on the geometry of the Universe.

Objects which do not evolve with cosmic time and are uniformly distributed in the Universe, are characterized by a logN-logS of integral slope 1.5 as long as they are at small redshifts where the geometry of the Universe is well approximated by the Euclidean geometry. Under these hypotheses, any class of objects has the same logN-logS, regardless of its luminosity function. In this case, the amount of information derived from the study of the number-count relation is rather limited. It has allowed, however, the establishment of the extragalactic nature of the unidentified high galactic latitude sources discovered with early X-ray satellites. As soon as objects at high redshifts are sampled, a flattening of the logN-logS is expected. This flattening carries information on the geometry of the Universe and on the luminosity function of the sources. Furthermore, if the objects under consideration exhibit some form of evolution with cosmic time, then their number count relationship will again deviate from the 1.5 slope. Indeed, any departure from this slope is the signature, to be interpreted, of some effect.

2. LOGN-LOGS SLOPE DETERMINATION

It is customary to describe the flux distribution function of the underlying source population with a power law of slope α. The task is to obtain the best estimate for α. Techniques to estimate the number-counts slope have been developed mostly by radioastronomers. Crawford, Jauncey and Murdoch (1970) and Murdoch, Crawford and Jauncey (1973) (MCJ) have shown that the Maximum Likelihood Method (MLM) provides the minimum variance best estimate of the slope α and that in the presence of measurement errors, this estimate is biased and systematically overestimates the slope of the true distribution. The bias is due to the presence of a lower flux limit S_{th} in conjunction with the measured errors. Sources with fluxes just above the threshold

597

A. Hewitt et al. (eds.), Observational Cosmology, 597–600.
© 1987 by the IAU.

S_{th} may be lost because of "negative" error fluctuation, while sources with fluxes just below the threshold may enter the survey because of "positive" error fluctuations. In general the two effects do not cancel out. Their relative importance depends on the number of sources present above and below the adopted threshold (i.e. on the slope α), as well as on the detailed shape of the distribution of the measurement errors. MCJ have tabulated correction factors to remove the bias in the source counts slope determination and have shown that these corrections can be determined only if sources with flux density at least 5 times the rms error (i.e. SNR$\geq$5) are accepted in a survey.

3. THE CASE OF IMAGING X-RAY SURVEYS

X-ray surveys differ in their statistical properties from surveys at longer wavelengths in several respects. First the measurement errors are determined by the integrated flux, i.e. the total number of counts recorded, rather than by the flux itself. Next and more important, the measurement error distribution is an asymmetrical Poisson distribution rather than a symmetrical Gaussian distribution. Further, it is necessary to subtract background counts since they often constitute a significant portion of the observed signal. Finally, most of the Medium Sensitivity Survey (MSS) exposures as well as large parts of the planned ROSAT all-sky survey are or will be photon limited rather than background or confusion limited. For the case of the MSS the statistical error on α exceeds at the moment the applied bias correction. However, as discussed by Gioia, Maccacaro and Wolter at this Symposium, the MSS has been extended and now contains more than 800 sources. Furthermore, the planned ROSAT all-sky survey will yield a 100 fold increase in the number of X-ray sources. The statistical error on the determination of the next generation of LogN-logS curves will be significantly smaller. It is thus worthwhile to reconsider the problem of the logN-logS slope estimation in the light of the features intrinsic to an imaging X-ray survey. We have approached the problem both analytically, analogously to MCJ, and empirically through Monte Carlo simulations. The two methods are complementary in the sense that analytical calculations provide checks of the Monte Carlo simulations under simplifying assumptions, whereas instrumental effects (i.e., non uniformity of the detector response, background inhomogeneities) can be far more easily incorporated into a Monte Carlo computation. The analytical work is presented by Schmitt and Maccacaro (1986). Here we discuss the results of our Monte Carlo simulations.

A large number of IPC images were simulated, containing point-like sources with fluxes distributed according to a logN-logS of known slope α_t. These images were analyzed with the same software used for the real IPC images (see Harnden et al. 1984). We have simulated IPC images of about 5000 s exposure ($\sim$13 background counts per detection cell). The threshold for source detection has been set to SNR $= 3$ (SNR is defined as the ratio between the source net counts and the square root of the source plus background counts. For a 5000 s IPC image this threshold corresponds to a flux of $\sim 2 \times 10^{-13}$ ergs/cm^2/s in the 0.3 - 3.5 keV band. Sources were created with fluxes 5 times smaller than this value and according to 3 different flux distributions, simulating logN-logS with (integral) slopes of 1.5, 1.8, and 2.0. We are still in the process of creating simulated data to reduce the uncertainties due to the finite sampling statistics. The preliminary results we report here are based on the analysis of 20586 ($\alpha_t = 1.5$), 23592 ($\alpha_t = 1.8$), and 8948 ($\alpha_t = 2.0$) sources. For these three cases we have derived, using the MLM, the best estimate of α using all the detected sources (SNR ≥ 3) or only those sources detected above SNR $= 4$ and SNR $= 5$. The results are shown in Figure 1 where the bias (α/α_t) is plotted as a function of the threshold SNR. For comparison, the bias as derived by MCJ for the case of $\alpha_t = 1.5$ is also indicated. It is evident that in the case of imaging X-ray surveys the bias in the determination of the number-counts slope is much smaller than that which has been determined for radio surveys. This is mainly due to the fact that in the former case the measurement error distribution is an asymmetrical Poisson distribution while in the latter case it is a symmetrical Gaussian distribution.

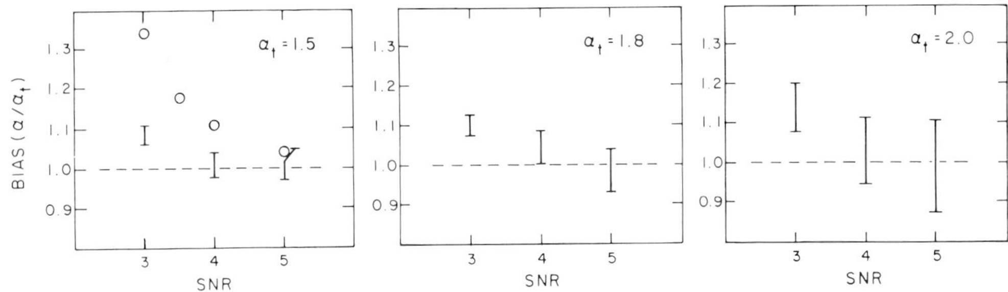

Figure 1. Bias in the slope determination versus the threshold SNR used. In the $\alpha_t = 1.5$ case the bias derived by MCJ for radio surveys is also indicated (open circles).

This leads to a very small number of spurious sources detected above the threshold SNR = 3. In fact, at least 17 net counts have to be recorded on top of the 13 average background counts in order to satisfy the requirement that SNR = 3. But the probability to observe 30 counts or more, when 13 are expected is 3.8×10^{-5}, much smaller than the probability associated to the 3σ level in the Gaussian statistics. Furthermore, the algorithm used to detect sources discriminates "real" sources from "spurious" sources on the basis of the shape as well as the amplitude of the candidate source. In other words, a fluctuation which satisfies the requirement of SNR $\geq$ 3 can be rejected because the spatial distribution of the counts may be too dissimilar from the expected distribution (point response function).

4. CONCLUSION

X-ray surveys differ in their statistical properties from surveys at longer wavelengths. This requires a different treatment of the data when determining the slope of the number-counts relationship. For X-ray surveys, the bias in the determination of the logN-logS slope is much smaller than that which has been determined for radio surveys. It is possible to lower the threshold for inclusion of sources in the sample to SNR = 4 and still be able to make valid corrections for the bias. This results in an increase of 50% or more in the sample size, with a significant reduction in the errors due to finite sampling statistics. We thank Donna Irwin for her care in preparing this manuscript for publication. This work has received partial financial support from NASA contract NAS8-30751.

REFERENCES

Crawford, D.G., Jauncey, D.L., and Murdoch, H.S., 1970, Ap. J., 162, 405.

Gioia, I.M., Maccacaro, T., Wolter, A., this volume.

Harnden, F.R., Jr., Fabricant, D.g., Harris, D.E., and Schwarz, J., 1984, SAO Report 393.

Murdoch, H.S., Crawford, D.G., and Jauncey, D.L., 1973, Ap. J., 183, 1.

Schmitt, J.H.M.M., and Maccacaro, T., 1986, Ap. J., 310, in press.

DISCUSSION

SCHMIDT: Dr. Setti stated that an extrapolation of the MSS towards higher fluxes with the MSS shape yields two or three times fewer sources than Piccinotti's HEAO1-A2 counts. Can you comment on the significance of this?

MACCACARO: I am not too concerned about this disagreement. First there is a difference in the energy range of the two surveys and this contributes to uncertainty in the comparison. Second, it may not be correct to extrapolate the MSS AGN $\log N - \log S$ (which has a slope steeper than the Euclidean slope) to the flux range of the Piccinotti survey. At these fluxes in fact, only very, very low redshifts are sampled and therefore the $\log N - \log S$ has to have the Euclidean slope.

WAMPLER: You said that in your model the probability of getting an extra 18 counts on top of the background was 10^{-5}. In the real data some 3σ sources were removed because their "shape" was wrong. Have the real data been shown to be governed by counting statistics? If the probability of getting a false event was really 10^{-5} I would not have expected many false events.

MACCACARO: The real data are governed by what I would call "modified Poisson" statistics - the underlying statistics is obviously Poisson statistics (we are dealing with a small number of integer counts) but is "modified" in the sense that after the amplitude of a fluctuation is checked, its shape is also checked. This is likely to remove   more "spurious" sources than "real" sources.

THE EXOSAT HIGH GALACTIC LATITUDE SURVEY

P. Giommi and G. Tagliaferri
EXOSAT Observatory, ESOC
Robert Bosch Str. 5,
6100 Darmstadt,
West Germany.

ABSTRACT. We have performed a survey of $\sim$570 square degrees of high galactic latitude sky using a large number of EXOSAT X-ray images. 130 serendipitous sources were detected and $\sim$60% of them have been identified with catalogued objects. The large majority of the remaining sources have faint optical counterparts on POSS or ESO plates and are expected to be extragalactic. The comparison of our results with those of the Einstein Medium Sensitivity Survey and an analysis of the correlation between the spatial distribution of EXOSAT serendipitous sources and the amount of galactic Hydrogen column density indicate that the average spectral index of extragalactic X-ray sources is steeper than the canonical AGN slope.

The EXOSAT High Galactic Latitude (HGL) Survey is derived from the search for serendipitous X-ray sources in a sample composed of a large number ($>$ 400) of EXOSAT CMA fields (see Taylor et al. 1981 and De Korte et al. 1981 for a description of the instrumentation). The EXOSAT CMA is characterised by a limiting sensitivity in the range 1.E-13 - 2.E-12 erg/cm^2/sec in the very soft energy band 0.05-2 keV. The instrument response is approximately constant in the central part of the field of view (a circular region of radius 12 arcminutes) and rapidly changes at larger off-axis angles until the sensitivity becomes one tenth of the central value at about 45 arcminutes from the centre. The conversion between CMA counts and X-ray flux is a strong function of the Hydrogen column density (N_H) along the line of sight. In order to properly calculate the limiting sensitivity of the Survey for each field we have estimated the appropriate amount of N_H from the 21 cm measurements of Stark et al. 1986. The area of sky covered by the EXOSAT Survey is plotted in fig. 1 as a function of limiting sensitivity for different assumptions on the source spectral slope. For comparison the sky coverage of the Einstein Observatory Medium Sensitivity Survey (MSS) (Gioia et al. 1984) is also plotted in fig. 1 as a dashed line.

601

A. Hewitt et al. (eds.), Observational Cosmology, 601–605.
© 1987 by the IAU.

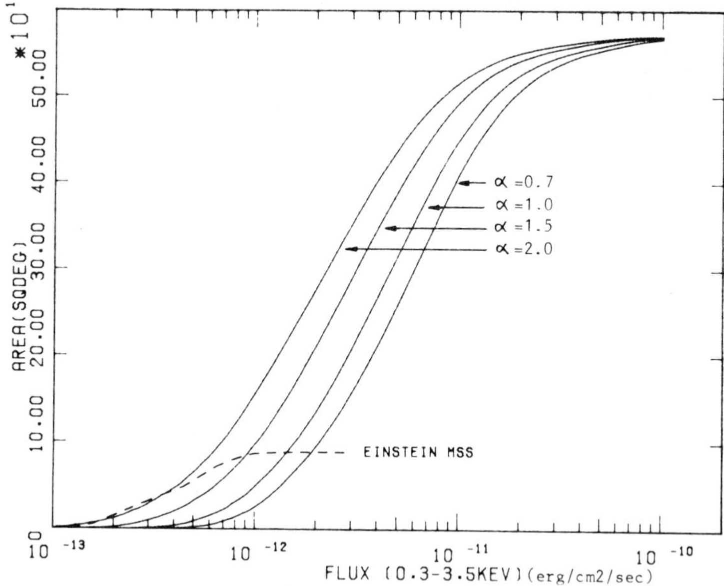

Fig. 1.

Sky coverage of the EXOSAT High Galactic Latitude Survey for different assumptions of the extragalactic sources energy spectral slope (solid lines). The sky coverage of the Einstein Medium Sensitivity Survey is shown as a dashed line.

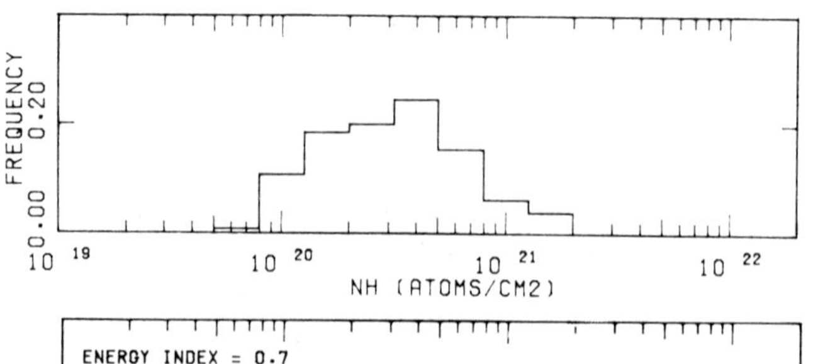

Fig. 2a.

NH distribution of all the fields in the EXOSAT High Galactic Latitude Survey.

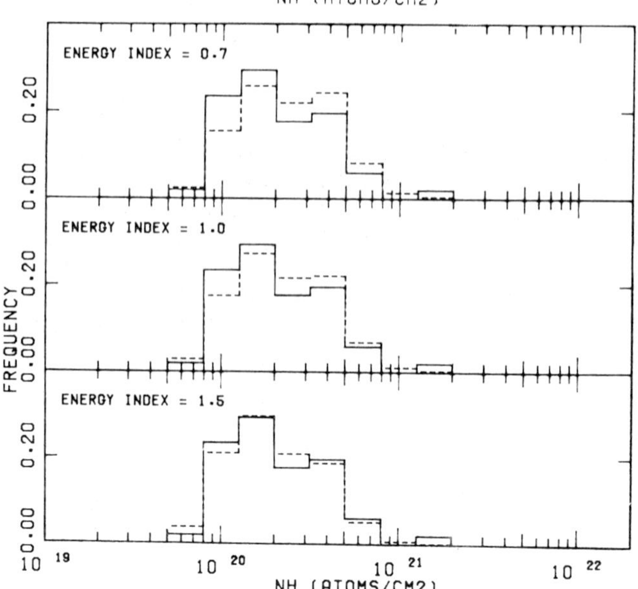

Fig. 2b.

NH distribution of extragalactic or unidentified sources with mv 15 (solid histogram) and predicted distributions for three different values of the spectral index (dashed histogram).

130 serendipitous sources were detected above a threshold that was chosen so that only a small fraction of detections could have been missed because of the non-optimal efficiency of the automatic source detection algorithm near the sensitivity limit. 74 sources have been identified with previously catalogued objects (56 Stars, 2 Catacylsmic Variables, 1 Galaxy, 11 AGN and 4 BL-Lacs), the remaining 56 are still unidentified.

Using the Log N-Log S derived by Gioia et al. 1984 we have estimated the number of extragalactic objects expected in the EXOSAT HGL Survey assuming that the source energy spectrum is well represented by a power law in the energy range 0.05-3.5 keV. The results are summarised in table 1.

TABLE 1

Energy Index	# of sources (extragalactic)
0.50	18.1
0.75	26.2
1.00	37.5
1.25	53.5
1.50	74.8
1.75	103.0

Since the expected number of sources is a strong function of the assumed spectral index, the comparison between expectations and actual results provides information on the average spectral slope of extragalactic serendipitous sources.

Only about 60% of the sources in our sample have been so far identified. However, we note that the large majority of the EXOSAT HGL serendipitous sources have optical counterparts with mV 15 or fainter on POSS or ESO plates. From the Einstein MSS we know that, with very few exceptions, X-ray serendipitous sources with optical magnitude in the range 15-21 are extragalactic (Stocke et al. 1983). If we assume that all unidentified sources with mV $\geq$ 15 in our sample are extragalactic, the total number of extragalactic objects is 57. This number, when compared to the expectations of table 1 indicates that the average energy spectral index of these sources is $>$ 1, steeper than the canonical slope of 0.7 found in samples of AGN in the 1-20 keV band (Mushotzky et al. 1980, Rothschild et al. 1983, Petre et al. 1984, Reichert et al. 1985).

Since the EXOSAT CMA is so sensitive to interstellar absorption extragalactic serendipitous sources are preferentially detected in fields where the amount of N_H is low. The fraction of extragalactic sources expected in a given interval of N_H is a function of the Log N-log S slope and of the source spectral shape. Starting from the Log N-log S of Gioia et al. 1984 we have calculated the predicted distribution shown in fig. 2b (dashed histograms) for three different assumptions on the source spectral index. The actually measured distribution of serendipitous sources is shown in fig. 2b (solid histogram). The agreement between predictions and measurement is acceptable only for very steep energy indices ($\sim$1.5) confirming the indications obtained before. The N_H distribution of all fields in the Survey is shown in fig. 2a.

The percentage of stars in the EXOSAT HGL Survey is $\geq$ 43% and is significantly higher than the value found in the Einstein MSS (25% Stocke et al. 1983). This result was however expected and it is mainly due to the very soft energy band pass of the CMA. A similar effect has to be expected also in the future ROSAT whole sky survey, which will be performed in the 0.1-2.0 energy range (Trümper 1984).

REFERENCES

De Korte, P.A.J. et al. 1981. Space Sci.Rev., 30, 495.
Gioia, I.M., Maccacaro, T., Schild, R.T., Stocke, J.T., Liebert, J.N.,
    Danziger, I.J., Knuth, D., and Lub, J. 1984. Ap.J., 283, 495.
Mushotzky, R.F., Marshall, F.E., Bolt, E.A., Holt, S.S., and Serlem-
    itsos, P.J. 1980. Ap.J., 235, 377.
Petre, R., Mushotzky, R.F., Krolik, J.H., and Holt, S.S. 1984. Ap.J.,
    280, 499.
Reichert, G.A., Mushotzky, R.F., Petre, R., and Holt, S.S. 1985.
    Ap.J., 296, 69.
Rothshild, R.E., Mushotzky, R.F., Baity, W.A., Gruber, D.E., Matteson,
    J.L., and Peterson, L.E. 1983. Ap.J., 269, 423.
Stark, A., Heiles, C., Baily, J., and Linke, K. 1986. In preparation.
Stocke, J.T., Liebert, J.W., Gioia, I.M., Griffiths, R.T., Maccacaro,
    T., Danziger, I.J., Knuth, D. and Lub, J. 1983. Ap.J., 273, 458.
Taylor, B.G., Andresen, R.D., Peacock, A., and Zobl, R. 1981.
    Space Sci.Rev., 30, 479.
Trümper, J. 1984. In X-ray and UV Emission from Active Galactic Nuclei
    MPE Report 184.

DISCUSSION

VERON: How have you assigned an optical magnitude to the unidentified X-ray sources?

GIOMMI: By measuring the diameter of the candidate optical counterpart on Palomar or ESO plates.

CANIZARES: Two comments. First, I presume that most of your objects ought to be at low z, given your flux limits and the results of the Medium Survey, so your objects are not the ones that make up the bulk of the X-ray background. Second, we have been using Einstein data to study the composite spectra of high z quasars and the preliminary result is that both radio loud and radio quiet quasars have reasonably flat spectra - closer to the canonical 0.7. These are objects with z > 1, so our spectra correspond to ~1-10 keV in the quasar rest frame.

GIOMMI: Yes, many of our sources are expected to be at low redshift. In addition, a non-negligible fraction of them is expected to be BL Lac objects, i.e. sources with steep spectra that evolve less than quasars.

CORRELATION OF X-RAY AND INFRARED EMISSION OF RADIO-QUIET QSOs

D. M. Worrall
Harvard-Smithsonian Center for Astrophysics, Cambridge, MA, U.S.A.

ABSTRACT. A correlation of the form $l_x \propto l_{ir}^\gamma$, with $\gamma = 0.83$, is found between 2 keV monochromatic X-ray luminosity and 1.65 μm monochromatic infrared luminosity for a sample of optically selected radio-quiet QSOs. This is equivalent to an average 1.65 μm to 2 keV spectral index of $\alpha_{ir/x} = 1.28 + 0.05 \log(l_{ir}/10^{31} \text{ ergs s}^{-1} \text{ Hz}^{-1})$. The ratio of X-ray to infrared luminosity decreases with increasing infrared luminosity in a manner similar to the relationship between X-ray and optical luminosity. The standard deviations of the distributions of data about the best-fit functions are comparable for the X-ray versus infrared and X-ray versus optical fits. Thus, contrary to a previous claim, the infrared luminosity is not better than the optical luminosity at predicting the X-ray emission from QSOs.

I. INTRODUCTION

There is general agreement that optically selected QSOs exhibit a dependence of monochromatic 2 keV X-ray luminosity, l_x, on monochromatic 2500 Å luminosity, l_o, of the form $l_x \propto l_o^\alpha$, with α in the range 0.6-0.9 (*e.g.*, Avni and Tananbaum 1982, 1986; Kriss and Canizares 1985). At a rest wavelength of 2500 Å, QSO spectra are generally flat and may be dominated by thermal emission from an accretion disk (*e.g.*, Shields 1978; Malkan and Sargent 1982). However, the near-infrared (1.65 μm) radiation is dominated by a steeper component, which generally fits a power law of energy index $\alpha_{ir} \gtrsim 1.0$ and often extrapolates well to X-ray emission at $\sim$ 2 keV (*e.g.*, Malkan and Sargent 1982; Elvis *et al.* 1986). Malkan (1984) reports that for Seyfert 1 galaxies and QSOs, X-ray and near-infrared luminosity exhibit a much tighter correlation than any other correlations between continuum bands. Here this issue is examined by comparing the X-ray and near-infrared correlation with the X-ray and 2500 Å correlation for a carefully selected sample of radio-quiet QSOs.

II. SAMPLE

The sample has been extracted from the optically selected sample of Avni and Tananbaum (1986). Objects with a spectral index between 5 GHz and 2500 Å greater than 0.35 have been deleted because their X-ray emission is enhanced by some process involving their radio emission (*e.g.*, Ku, Helfand and Lucy 1980; Worrall *et al.* 1986). Of the 134 remaining objects, 69 have near-infrared photometric measurements. Objects in the Palomar Bright Quasar Survey account for $\sim$ 3/4 of these, and their infrared measurements are given by Neugebauer *et al.* (1986). Data for the

607

A. Hewitt et al. (eds.), Observational Cosmology, 607–609.
© *1987 by the IAU.*

other QSOs are from a variety of published sources. Optical and *Einstein Observatory* Imaging Proportional Counter X-ray data are as given in Avni and Tananbaum (1986) and Tananbaum *et al.* (1986), except that the X-ray luminosities are corrected to values appropriate for an X-ray spectral index of 1.0 rather than 0.5 (see Wilkes and Elvis 1986). Worrall (1986) gives a full description of the data.

III. RESULTS

Figure 1 shows separate plots of monochromatic X-ray luminosity versus optical luminosity and infrared luminosity. From visual inspection, the standard deviation of the data around the regression line appears to be roughly the same for both plots. This has been confirmed by fitting functions of the form $l_x/10^{27} = A(l_o/10^{31})^\alpha$ and $l_x/10^{27} = C(l_{ir}/10^{31})^\gamma$ using the fitting method of Avni *et al.* (1980), which treats both detections and upper limits. In this method, the distribution of residuals around the best-fit function is assumed to be Gaussian, and the standard deviation of the distribution, σ_G is treated as a free parameter. Taking only the data shown as filled symbols in Figure 1, the values found for σ_G are 0.38 ± 0.06 for the upper graph and 0.41 ± 0.08 for the lower graph (90% confidence errors for one interesting parameter). The values increase to 0.50 ± 0.08 and 0.52 ± 0.08, respectively, if all the data are fitted. These results are contrary to those of Malkan (1984), and imply that the infrared luminosity is not significantly better than the optical luminosity as an estimator of the X-ray emission from these QSOs. The fit to the upper graph gives $\log A = -0.11 \pm 0.08$ and $\alpha = 0.79 \pm 0.08$, consistent with previous results. The lower graph gives $\log C = -0.40 \pm 0.11$ and $\gamma = 0.83 \pm 0.12$. (All errors are given as the projection of the 90% confidence error surface for one interesting parameter onto the axis for that parameter.)

The ratio of X-ray to infrared luminosity decreases with increasing infrared luminosity in a manner similar to the relationship between X-ray and optical luminosity. The result for the X-ray versus infrared correlation is equivalent to an average 1.65 μm to 2 keV spectral index of $\alpha_{ir/x} = 1.28 + 0.05 \log(l_{ir}/10^{31})$.

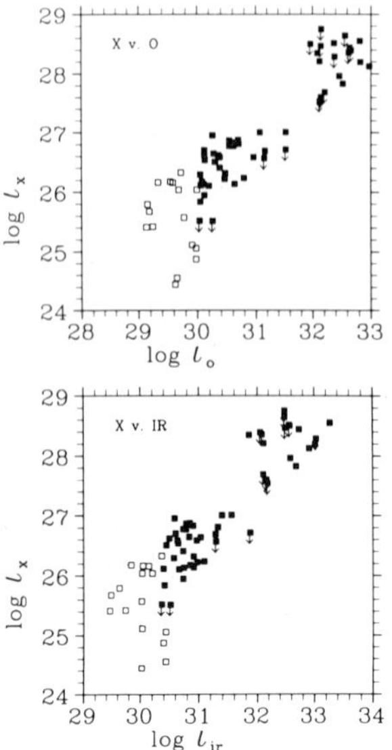

Figure 1. Logarithmic plots of monochromatic X-ray luminosity at 2 keV versus monochromatic optical luminosity at 2500 Å (above) and monochromatic infrared luminosity at 1.65 μm (below). Luminosities are in units of ergs s^{-1} Hz^{-1} for the source frame and have been calculated assuming a Friedmann cosmology with $H_o = 50$ km s^{-1} Mpc^{-1} and $q_o = 0$. Objects for which the monochromatic optical luminosity is $< 10^{30}$ ergs s^{-1} Hz^{-1} are shown by open symbols: these data show the greatest scatter, attributed to the larger and more uncertain X-ray absorptions at low luminosities (Zamorani 1986).

I am grateful to G. Neugebauer for providing an early preprint of Neugebauer *et al.* (1986). This work was supported by NASA Contract NAS8-30751 and Smithsonian Institutional funds.

REFERENCES

Avni, Y., Soltan, A., Tananbaum, H., and Zamorani, G. 1980, *Ap.J.*, **238**, 800.

Avni, Y., and Tananbaum, H. 1982, *Ap.J. (Letters)*, **262**, L17.

Avni, Y., and Tananbaum, H. 1986, *Ap.J.*, **305**, 83.

Elvis, M., Green, R.F., Bechtold, J., Schmidt, M., Neugebauer, G., Soifer, B.T., Matthews, K., and Fabbiano, G. 1986, *Ap.J.*, **310**, in press.

Kriss, G.A., and Canizares, C.R. 1985, *Ap.J.*, **297**, 177.

Ku, W.H-M., Helfand, D.J., and Lucy, L.B. 1980, *Nature*, **288**, 323.

Malkan, M.A. 1984, *X-ray and UV Emission from Active Galactic Nuclei*, ed. J. Trümper and W. Brinkmann (Garching: Max-Planck-Institut), p121.

Malkan, M.A., and Sargent, W.L.W. 1982, *Ap.J.*, **254**, 22.

Neugebauer, G., Green, R.F., Matthews, K., Schmidt, M., Soifer, B.T., and Bennett, J. 1986, *Ap.J. Suppl.*, in press.

Shields, G. 1978, *Nature*, **272**, 706.

Tananbaum, H., Avni, Y., Green, R.F., Schmidt, M., and Zamorani, G. 1986, *Ap.J.*, **305**, 57.

Wilkes, B.J., and Elvis, M. 1986, *Ap.J.*, submitted.

Worrall, D.M. 1986, *Ap.J.*, submitted.

Worrall, D.M., Giommi, P., Tananbaum, H., and Zamorani, G. 1986, *Ap.J.*, in press.

Zamorani, G. 1986, *Quasars*, eds. G. Swarup and V.K. Kapahi (Reidel), p223.

THE COSMIC X-RAY BACKGROUND

Elihu Boldt
NASA/Goddard Space Flight Center
Code 666
Greenbelt, Maryland 20771
USA

ABSTRACT. As the sky in the microwave band is dominated by a cosmic
background, so too is the X-ray sky. In this presentation the observational situation regarding the extragalactic X-ray background is
reviewed, emphasizing data obtained from HEAO-1 and the Einstein
Observatory (HEAO-2). Spectral characteristics and spatial variations
are described and discussed within the context of what is currently
known about individual extragalactic sources. It is concluded that the
bulk of the cosmic X-ray background is yet to be understood. Possibilities range from genuinely diffuse emission to an "as yet" unknown
large population of unresolved discrete sources. The role of cosmological effects is examined and could be important.

Most of our current knowledge about the cosmic X-ray background and the
sources contributing to it comes from observations carried out with
instruments aboard the HEAO (High Energy Astronomy Observatories) of
NASA's program in space astrophysics. With mechanically collimated
detectors HEAO-1 provided an all-sky survey (i.e., coverage corresponding to 41×10^3 deg^2) over a broad band of photon energies ranging from
0.1 keV to over an MeV. Two of the HEAO-1 experiments were developed
especially for the measurement of sky background. The HEAO-1 A2 experiment, a collaboration between Goddard Space Flight Center and Penn
State University, was based on multi-celled gas proportional counters
with various fields of view ranging from 4.5 deg^2 to 18 deg^2; these
detectors were extremely low noise devices that allowed for an absolute
measurement of sky surface brightness for energies up to about 60 keV.
The HEAO-1 A4 experiment, a collaboration between the University of
California at San Diego and M.I.T., employed solid scintillation
counters to detect harder X-rays collimated with fields of view corresponding to 34 deg^2 and 200 deg^2. These experiments established the
broadband spectrum (Marshall et al. 1980; Rothschild et al. 1983) and
large-scale spatial structure (low-level anisotropy) of the cosmic X-ray
background (Boldt 1981; Shafer 1983). The HEAO-1 all-sky survey
detected about 300 bright X-ray sources via the A2 experiment and about
70 hard X-ray sources (mostly galactic) via the A4 experiment.

A. Hewitt et al. (eds.), Observational Cosmology, 611–615.
© 1987 by the IAU.

Appropriately classified HEAO-1 objects have been used for estimating
the present-epoch luminosity function for extragalactic sources
(Piccinotti et al. 1982).

The HEAO-2 Einstein Observatory, a consortium led by SAO, was
based on a grazing incidence telescope for soft X-rays, mainly below
about 3 keV in energy. Having arc-second resolution it could be used
for detecting relatively faint sources with sufficient precision to
obtain optical identifications. The high sensitivity surveys (HSS)
covered 1.4 deg^2, yielding 19 sources down to the level 2.6 x 10^{-14}
ergs cm^{-2}s^{-1} (1-3 keV); see Giacconi et al. (1979) and Griffiths et al.
(1983). The medium sensitivity surveys (MSS) are still being investigated (Gioia, this symposium), but a subset of 112 sources corresponding to 89 deg^2 coverage has already been completely identified (Gioia
et al. 1984). Based on data from the imaging proportional counter in
the focal plane of the Einstein Observatory telescope, sky surface
brightness could be estimated with arc-minute resolution, yielding upper
limits on small scale fluctuations of the background (Hamilton and
Helfand 1986).

The cosmic X-ray background spectrum measured with HEAO-1 over the
band 3-100 keV may be characterized by an optically thin isothermal
model with kT = 40 keV (Marshall et al. 1980; Rothschild et al. 1983;
Gruber 1984). This energy spectrum may be well approximated by an
expression of the following form (Boldt and Leiter 1984):

$$dI/dE = A (E/3 \text{ keV})^{-\alpha} \exp(-E/B) \tag{1}$$

where E is photon energy (in keV), A = 5.6 keV/(keV cm^2 s sr), B = kT
= 40 keV and α = 0.29. Although data from the HEAO-2 Einstein Observatory are for soft X-rays (mostly below about 3 keV) HEAO-1 results on
the extragalactic background are clearly uncontaminated by galactic
effects only for energies greater than about 3 keV. Hence, comparisons
between HEAO-1 and HEAO-2 results are most reliably made in a relatively small band of spectral overlap at E $\sim$ 3 keV. Using X-ray data on
extragalactic objects obtained with HEAO-1 and HEAO-2, in association
with optical data, Schmidt and Green (1986) have estimated the contribution of discrete sources to the cosmic X-ray background, albeit in
the HEAO-2 band (< 3 keV). Extrapolating their estimate to an evaluation at E = 3 keV, the composite contribution of all galaxies
(including AGN and quasars) and clusters of galaxies amounts to 40%-
55% of the extragalactic background. With HEAO-2 HSS sources having
power-law spectra characterized by $\alpha \geq 0.7$ at 3 keV (see Setti, this
symposium, for a review), the faintest sources resolved with the
Einstein Observatory would account for $\leq$ (21±9)% of the background at
that energy. Furthermore, the observed energy spectrum of the extragalactic background is much flatter than that of known sources (Boldt
1981; DeZotti et al. 1982); at $\sim$ 3 keV the effective power-law index
is only 0.4 (Marshall et al. 1980).

Subtracting the estimated spectral contribution of discrete
sources to the cosmic X-ray background (normalized to 40% at 3 keV)
yields a residual spectrum (Leiter and Boldt 1982) that is well fitted
by an analytical expression of the form given by equation (1), under

the constraints:

$$0.2 > \alpha \geq 0 \tag{2}$$

$$30 \text{ keV} > B \geq 23 \text{ keV}. \tag{3}$$

Assuming that this residual background arises from optically thin thermal bremsstrahlung, the constraint provided by (2) implies that $kT > 200$ keV for the radiating plasma (Boldt and Leiter 1984). For an isothermal plasma at a redshift (z) to produce this residual background, the constraint provided by (3) then requires that $z > (kT/B)-1 = 6$ (i.e., well beyond any known quasar). Alternatively, Guilbert and Fabian (1986) have considered a hot IGM (intergalactic medium) that is heated to $kT > 200$ keV at $z > 3$ and then adiabatically cooled (for $z < 3$) to $kT = 14(1+z)^2$ keV due to the expansion of the universe. This hot IGM model provides a remarkably good broadband spectral fit to the residual cosmic X-ray background but requires that the baryonic mass involved be greater than 20% of the closure value. Since such a hot IGM would induce a spectral distortion of the 2.7 K blackbody background amounting to an effective temperature variation as much as 0.1 K, Danese (personal communication) has suggested that current microwave spectral measurements are already getting close enough to the precision needed for providing a meaningful diagnostic of the model.

Severe integral constraints on the possible discrete source contribution to the residual background (unresolved with the HEAO-2 Einstein Observatory) have been obtained by evaluating upper limits to the corresponding apparent surface brightness fluctuations observed on the scale of arc-minute pixels. In this way Hamilton and Helfand (1986) conclude that the number of point sources needed would have to exceed 5×10^3 degree^{-2}, much larger than the estimated total number of quasars (Schmidt and Green 1986). Extrapolating the known background spectrum (at > 3 keV) into the 1-3 keV band appropriate for HEAO-2 makes it possible for us to use this lower limit on the number of sources to estimate an upper limit on their average X-ray intensity, viz:

$$<S> \leq 10^{-15} \text{ ergs cm}^{-2} \text{ s}^{-1} \text{ (1-3 keV)}. \tag{4}$$

This upper limit on the average intensity is only about 4% of the threshold sensitivity used for the HEAO-2 HSS of faint point sources.

Based on the spectral constraints for the residual background provided by equations (1-3) we can approximate the spectrum characteristic of those discrete sources that would be needed, viz (the appropriately redshifted spectrum viewed by us):

$$dS/dE \sim \exp[-E/(23 \text{ keV})]. \tag{5}$$

Normalizing to the narrow band limit given by equation (4), we integrate this spectrum to obtain an upper limit on the average value for the apparent X-ray luminosity of such sources. Assuming a Hubble constant $H_0 = 50$ km s^{-1} Mpc^{-1} and $q_0 = 0.5$ the corresponding luminosity upper limit becomes:

$$\langle L \rangle \quad < \quad 10^{45.2} \text{ ergs s}^{-1} \qquad \text{for} \quad z \leq 4 \qquad (6)$$

$$\langle L \rangle \quad < \quad 10^{44.5} \text{ ergs s}^{-1} \qquad \text{for} \quad z \leq 2 \qquad (7)$$

A clue as to the possible nature of such sources is obtained by noting that these luminosities are comparable to the bolometric luminosity of a typical Seyfert galaxy; the median mass value of the central compact object inferred for such galaxies is about 10^8 $M_\odot$ (Wandel and Mushotzky 1986). Hence, if the sources of the residual background are early precursor AGN powered by accretion (Leiter and Boldt 1982) then the mass build-up in a Hubble time could well be comparable to that of the central object characteristic of present-epoch Seyfert galaxies. For such a situation these compact precursor Seyfert galaxies would initially have to be of relatively small mass, radiating at the Eddington luminosity limit or close to it. If the luminosity of an accretion powered AGN starts out at the Eddington limit and remains constant (i.e., constant accretion rate) then the central mass would increase about a hundred-fold after a Hubble time (Boldt and Leiter 1986) and the source would now be radiating at $\sim 1\%$ of the Eddington limit, as observed for Seyfert galaxies (Wandel and Mushotzky 1986). While these compact precursor AGN are still young and close to being Eddington limited, non-thermal electron emission mechanisms are suppressed (Cavaliere and Morrison 1980) and a significant electron-positron "pair photosphere" would be produced, thereby maintaining $kT < mc^2$ (Rees 1984) for the thermal source and yielding a comptonized thermal spectrum of the flattened form required for the residual X-ray background. Even though these relatively low luminosity precursor AGN are numerous (compared with quasars) they might not be easy to find in usual optical surveys (Boldt and Leiter 1984).

   If the residual X-ray background is found to be genuinely diffuse and a hot IGM origin becomes untenable, then we may have to consider a non-standard cosmology. For example, Segal (1987) has suggested that the residual background could be a cosmological X-ray albedo associated with discrete source radiation which has undergone multiple traversals of the universe.

REFERENCES

Boldt, E. 1981, Comments Ap., 9, 97.
Boldt, E., and Leiter, D. 1984, Ap. J., 276, 427.
Boldt, E., and Leiter, D. 1986, in Structure and Evolution of Active
       Galactic Nuclei, ed. G. Giuricin et al. (Dordrecht, Holland:
       D. Reidel Publishing Co.), p. 383.
Cavaliere, A., and Morrison, P. 1980, Ap. J., 238, L63.
De Zotti, G. et al. 1982, Ap. J., 253, 47.
Giacconi, R. et al. 1979, Ap. J., 234, L1.
Gioia, I.M. et al. 1984, Ap. J., 283, 495.
Griffiths, R.E. et al. 1983, Ap. J., 269, 375.

Gruber, D. 1984, in X-ray and UV Emission from Active Galactic Nuclei,
      eds: W. Brinkman and J. Trumper (Garching: Max Planck Inst.
      ISSN 0340-8922), p. 129.
Guilbert, P., and Fabian, A. 1986, MNRAS, 220, 439.
Hamilton, T., and Helfand, D. 1986, Ap. J., submitted.
Leiter, D., and Boldt, E. 1982, Ap. J., 260, 1.
Marshall, F. et al. 1980, Ap. J., 235, 4.
Piccinotti, G. et al. 1982, Ap. J., 253, 485.
Rees, M. 1984, Ann. Rev. Astron. Astrophys., 22, 471.
Rothschild, R. et al. 1983, Ap. J., 269, 423.
Schmidt, M., and Green, R. 1986, Ap. J., 305, 68.
Segal, I.E. 1987, Ap. J., in press.
Shafer, R. 1983, Ph. D. dissertation, University of Maryland, NASA TM
      85029.
Wandel, A., and Mushotzky, R. 1986, Ap. J., 306, L61.

DISCUSSION

ULMER: When creating the residual cosmic X-ray background by subtracting the spectra of Seyfert galaxies did you worry about K
corrections or spectral evolution?

BOLDT: In the spectral band of interest (3-100 keV) we assume pure
power-law spectra for Seyfert galaxies, consistent with HEAO-1
observations. After subtracting the expected contribution of such
unresolved foreground sources we conclude that the dominant sources
of the residual background have a substantially different spectrum
(i.e., an exponential form, implying severe spectral evolution).

CHAPTER XI

QUASI-STELLAR OBJECTS: DISTRIBUTION AND GENERAL PROPERTIES

COSMIC DISTRIBUTION OF OPTICALLY SELECTED QUASARS

Maarten Schmidt
Palomar Observatory
California Institute of Technology
Pasadena, CA 91125, USA

ABSTRACT. Counts of optically selected quasars as a function of magnitude and redshift show the effects of strong evolution. If quasars have relatively short life times, then the observed numbers at a given redshift are mostly determined by their birth rate and mean luminosity over their lifetime. In this case the evolution of the luminosity function can be described by density evolution, where the rate of evolution may depend on luminosity and other properties. On the other hand, if all quasars were formed at large redshift and have been decaying in luminosity since that time, then the evolution of the luminosity function is best described in terms of luminosity evolution. We discuss some of the consequences of luminosity evolution for the mass of quasars and for the X-ray background.

We explore the observational aspects of the redshift cutoff of quasars. The situation is complicated by the unavoidable bias in slitless surveys against weak-line objects. Since quasar emission lines show a wide range of equivalent widths, a spectral survey will be characterized by a distribution of limiting continuum magnitudes rather than by a single value. The decline in the space density of quasars at large redshift may depend on luminosity, and may also have structure, such as a steep drop, but not a total cutoff, in density at a redshift near 3.

1. INTRODUCTION

The cosmic distribution of quasars shows evidence for strong evolution of quasar properties with redshift. This finding is based on the statistics of quasars in both apparent magnitude and redshift. The evolution exhibits itself in a redshift dependence of the luminosity function, which mediates all quasar statistics.

We will discuss the observational evidence for evolution and in particular the concepts of density and luminosity evolution. We also review surveys of high redshift quasars in connection with the decline of quasar numbers around a redshift of 3.

A. Hewitt et al. (eds.), Observational Cosmology, 619–625.
© *1987 by the IAU.*

2. QUASARS SELECTED BY COLOR

For redshifts below 2.2, the ultraviolet excess is an effective way to
select quasar candidates. Spectroscopic observations are required to
confirm the quasar identification and to obtain the redshift. For a
recent review, the reader is referred to Green (1986), who collected
all available samples based on spectroscopically confirmed redshifts
and showed the luminosity function in seven intervals of redshift.
Since that time, Koo et al. have published a survey for quasars to
B = 22.5 selected on the basis of multiple colors, variability and
proper motions. Boyle et al. report in this volume on a survey of
about 170 quasars with B < 20.9, z < 2.2, which will add much weight to
the derivation of the luminosity function.

The observed variation of the luminosity function with redshift
may be alternatively described as an increase of density with redshift
at given luminosity, or as an increase in luminosity with redshift.

At this stage, it is useful to remember that the observed luminosity function represents the properties of the ensemble of quasars.
The number and luminosities of quasars at any given time are determined by the birth rate of quasars and their subsequent light curve
L(t). Disentangling the contributions of birth rate and light curve
will in general be difficult. If all quasars had individual light
curves showing decay at a 30 million year time scale, we would not be
able to derive that from a luminosity function that shows variation on
a time scale of around a billion years.

2.1. Density Evolution

If we assume that quasars have life times or time scales much shorter
than the age of the universe, then their observed numbers mostly reflect their birth rates integrated over their life time. If the birth
rate varies with cosmic time, density evolution will result.

The first evidence for evolution of optically selected quasars
(Schmidt 1970) was based on only 23 objects. With so few objects in
hand, little else could be done than to parametrize the co-moving space
density as a function of redshift only. This scenario is now called
pure density evolution (PDE).

In the early 1980s, sufficient survey material became available
that the density increase could be investigted in more detail. Assuming
that the co-moving density changed with cosmic time as exp(-kt),
Schmidt and Green (1983) discussed the available samples. They found that
k could be well determined for quasars of M(B) around -27. Applying
this same value of k to quasars regardless of their luminosity yielded
an overestimate of quasar numbers with B < 21 by a factor of 20. Since
most of these quasars are less luminous than M(B) = -27, k had to be
smaller for quasars of lower luminosity. This started the concept of
luminosity-dependent density evolution (LDDE). Schmidt and Green (1983)
opted for a simple linear dependence of k on M(B), since the available
material did not allow more detailed conclusions. On the basis of new
material, including that presented by Boyle et al, in this volume, the
dependence of k on M(B) may have to be reevaluated.

All recent authors in the field agree the PDE is inconsistent with the observations. Some authors have argued that LDDE does not fit the observations. If so, that may be a consequence of new observational material that was not available at the time of the Schmidt and Green (1983) paper. In principle, LDDE can fit any observations, since it simply describes the density increase at given luminosity--if there is a need, the density law can be made to depend on other properties beyond luminosity.

2.2 Luminosity Evolution

The alternative approach to the one sketched above is to assume quasars are long-lived and to interpret the change in the luminosity function as a consequence of evolution of luminosity. This follows if one assumes that all quasars were formed at an early epoch. In that case, the total quasar density does not change and the luminosity function will shift in the luminosity coordinate L, reflecting the light curve $L(t)$ of individual quasars. This is the case of pure luminosity evolution (PLE), introduced by Mathez (1978).

Boyle _et al_. (this volume) find that for redshifts below 2.2, PLE provides a good representation of the available statistics. The luminosity evolution required between z = 0 and z = 2.2 amounts to about 5 magnitudes. In other words, some 5 magnitudes of the bright end of the local luminosity function of Seyfert galaxies is shifted into the quasar luminosity range at redshift 2.2.

There are at least two problems with the concept of long-lived quasars and PLE that have to be addressed. The first one concerns the X-ray background. As shown by Schmidt and Green (1986), even if there is no evolution at all, Seyfert galaxies with luminosities $M(B) > -23$ contribute 29% of the X-ray background. If there is evolution of Seyfert galaxies as required in PLE, then the X-ray background contributed by Seyferts far exceeds the total measured background.

The second concern has to do with the total mass of quasars. In the case of PLE, a quasar like 3C 273 will have radiated around 3×10^{65} ergs over the past 10 billion years. The corresponding mass based on a 10% efficiency of accreted mass into radiated energy is around 10^{12} solar masses. The corresponding Schwarzschild radius of about a light month may be in conflict with the time scale of observed quasar variability.

Problems with either the X-ray background or the mass of quasars can be avoided if the life time of quasars is much less than that of the universe. To maintain PLE, dying quasars would have to be replaced in equal numbers by newly born ones to keep the number constant. Such an accidental situation is avoided in LDDE, where there is no imposed equality between the number of births and deaths.

3. QUASARS SELECTED BY EMISSION LINES

For redshifts larger than 2.2, Lyman-alpha emission enters the B filter and the ultraviolet excess usually disappears. Most surveys of quasars of larger redshift are based on slitless spectral surveys, in which selection of candidates is made on the basis of the presence of line emission.

Such surveys have yielded large numbers of quasars. The most remarkable result that has emerged is that quasars of very high redshift are rare and that there may be a redshift cutoff. The first convincing evidence for a decrease in quasar density (with increasing redshift) came from Osmer (1982) who failed to find any quasars with redshift in the range 3.7-4.7 in an area of 5 square degrees down to magnitude 20. No quasars with redshifts substantially larger than 4 have been found yet. Since the quasars with the largest observed redshifts have magnitudes as bright as 17 or 18, it is clear that such quasars could be easily detected at larger redshifts if they existed.

Veron (1986) has reviewed results of published emission-line surveys and concludes that the quasar density rises with redshift up to 2.45, drops by a factor of 5, then rises with redshift up to 3.8 to drop steeply once more. On the other hand, Hazard and McMahon (1985) state, on the basis of a UK Schmidt telescope objective prism plate, that the density of quasars peak at a redshift of 2.

The reliability of emission-lines surveys has sometimes been questioned. In particular, there is uncertainty about their completeness or their limiting magnitude (Clowes 1981; Smith 1983; Veron 1983). This is best illustrated by a comparison of the surface densities of quasars found in Schmidt surveys and in 4-meter telescope surveys, which differ by a factor of up to 10 at overlapping magnitude (Osmer 1980). As pointed out by Clowes (1981) and Smith (1983), faint objects in a survey will only be picked up if the emission line is strong.

Schmidt, Schneider, and Gunn (1986a,b) use a strict algorithm for the acceptance of an emission-line object as a candidate quasar, namely that the emission line be detected with a given minimum signal-to-noise ratio and that it have a minimum observed equivalent width. Since a given emission line exhibits a large range of equivalent widths in quasar spectra, a survey at any given redshift has a distribution of limiting magnitudes, corresponding to the equivalent width distribution. In addition, due to the response function of the detector, the limiting magnitude at given equivalent width is a function of wavelength and hence of redshift.

The complexities of a survey that has a distribution of limiting magnitudes, rather than a single value, makes it somewhat difficult to construct luminosity functions from the survey observations in a straightforward manner. It is, of course, possible to compare observed numbers of quasars at different redshifts with predictions based both on luminosity function models and on an adopted distribution of the (zero redshift) equivalent widths of the main emission lines.

Schmidt et al, (1986a,b) used such a procedure to compare results from two surveys for high redshift quasars with a smooth extrapolation of luminosity function models constructed for z < 2.2, discussed in the preceding section. No quasars with z > 3.5 were found in these two surveys, whereas the extrapolated models predicted that around 70 should have been observed. This is entirely consistent with Osmer's (1982) finding. Schmidt et al. also did not find any quasars with redshift over 3, where the extrapolated models would predict that 20 should have

been observed. This was rather surprising, since general quasar catalogues contain some 50 quasars with redshifts larger than 3, based on
slit spectroscopy.

There is no evidence at present that an abrupt shift in the distribution of equivalent widths of emission lines causes the shortfall
of redshifts larger than 3 in systematic surveys. Since almost all
high-redshift quasars in the quasar catalogues are quite bright, we
initially proposed (Schmidt et al. 1986a) that the redshift at which
quasar densities reach a maximum and turn around might be a function of
luminosity, in the sense that quasars of higher luminosity have a
larger redshift cutoff. When a survey that went less deep over a larger
area still produced no quasars with redshift over 3 (Schmidt et al.
1986b), we suggested the possibility that the density of quasars might
exhibit a steep drop, by a factor of 10 or 20, at a redshift near 3,
with a subsequent slow decline to zero at a redshift beyond 4. This
could explain why small samples yield no quasars with z > 3, while a
collection of several thousand objects such as the entire quasar catalogue yields a substantial number.

Much additional work will be needed to clarify the situation. It
is essential that further spectroscopic surveys of quasars address the
detectability of the emission lines in quantitative fashion. Improved
understanding of the evolution and the redshift cutoff of quasars will
depend critically on insight into the completeness of quasar surveys.

This research was supported in part by the National Science
Foundation under grant AST-8314134.

Note added October 15, 1986:
Schmidt, Schneider and Gunn have recently observed slit spectra of
emission-line candidates of a 4-Shooter spectroscopic transit survey
covering 14 square degrees. A first list of 75 candidates was selected
from a total of 120,000 spectra observed. A total of 16 quasars was
found, of which 9 have redshifts between 3.0 and 3.8. Schmidt et al.
conclude that there is no evidence for a sharp drop in quasar density
at a redshift just below 3: instead, the observations suggest that
there is a gradual decline of quasar density beyond a redshift of 3.

REFERENCES
Clowes, R.G. 1981, M.N.R.A.S., 197, 731.
Green, R.F. 1986, in IAU Symposium 119 on Quasars, held in Bangalore,
    December 1985.
Hazard, C., and McMahon, R. 1985, Nature, 314, 238.
Mathez, G. 1978, Astron. Atrophys., 68, 17.
Osmer, P.S. 1980, Astrophys. J. Suppl., 42, 523.
Osmer, P.S. 1982, Astrophys. J., 253, 28.
Schmidt, M. 1970, Astrophys. J., 162, 371.
Schmidt, M., and Green, R.F. 1983, Astrophys. J., 269, 352.
Schmidt, M., and Green, R.F. 1986, Astrophys. J., 305, 68.
Schmidt, M., Schneider. D.P., and Gunn, J.E. 1986a, Astrophys J.,
    306, 411.

Schmidt, M., Schneider, D.P., and Gunn, J.E. 1986b, Astrophys. J.,
    310, 000.
Smith, M.G. 1983, in 'Quasars and Gravitational Lenses', Proc. of
    the 24th Liège Colloquium (Université de Liège), p.4.
Veron, P. 1983, in 'Quasars and Gravitational Lenses', Proc. of the
    24th Liège Colloquium (Université de Liège), p. 210.
Veron, P. 1986, Astron. Astrophys. (submitted).

DISCUSSION

BOLDT: What is the value of q_o that you assumed in evaluating the
number of quasars expected at high redshifts?

SCHMIDT: I assumed q_o = 0.5; but the predictions differ little for
q_o = 0.1.

SETTI: Referring to your statement that Sy 1 should not evolve otherwise their integrated emission would exceed the XRB, it should be
remembered that in the case of luminosity evolution models, this
may not be the case because it appears that the X-ray luminosity does
not increase as fast as the optical luminosity $(L_x \propto L_o^{0.7-0.8})$.

SCHMIDT: Since there is a large observed scatter of the (L_x, L_o)
correlation, luminosity evolution may not follow the mean relation.
Even if the X-ray luminosity evolution is so reduced, the XRB predicted
for Seyferts on the basis of optical PLE still exceeds the observed
background.

LONGAIR: I believe important information on the cut-off at large
redshifts is provided by radio source surveys which are not subject to
the awkward selection effects which beset the emission line surveys.
Peacock, Dunlop and their colleagues have been carrying out a survey
of about 180 radio sources selected at 2.7 GHz with flux densities
$S_{2.7} \geq 0.1$ Jy. This sample is designed to be most sensitive to the
presence or absence of radio sources with flat radio spectra with
$z \gtrsim 4$. The identifications and spectra of the sources in this sample
are well advanced and already it is clear that there are few objects
with very large redshifts in the sample. A banded V/V_{max} test shows
that at redshifts greater than about 2, the comoving space density
of radio quasars decreases from a redshift of 2 to 4 by a factor of
about 5. This is direct evidence for the behavior of the flat-spectrum
source population predicted from Peacock's analyses of the source
counts. These show that the maximum comoving density of flat-spectrum
sources occurs at redshifts of about 2-2.5 and that the numbers are

falling off at large redshifts. It is important to emphasize that this sample is selected purely by radio criteria and seems to be producing the same type of distribution found by Dr. Schmidt in the emission-line surveys for radio quiet quasars.

SPATIAL DISTRIBUTION OF QUASARS

Yaoquan Chu and LiZhi Fang
Center for Astrophysics
University of Science and Technology of China
Hefei, Anhui, The People's Republic of China

1. INTRODUCTION

The distribution of quasars has become one of the most interesting problems in observational cosmology. This is due mainly to
the development of theory of the formation of large scale structure in the universe. In recent years, several scenarios of clu-
steringhave been proposed. In the adiabatic case, the clustering
process is from larger scales to smaller ones, i.e., the first
systems to form out would be on the scale of superclusters, then
these systems fragment to form smaller scale systems such as galaxies. In the isothermal case, the clustering is from smaller scales
to larger ones, namely, galaxies condense out at first and larger
scale systems, such as clusters and superclusters, then form later
via hierachical build-up processes. In the universe contain two
components, the scenario of clustering might be different from
both standard adiabatic and isothermal cases(1). According to this
new scenario, there should be two kinds of small scale objects,
one is formed due to fragment of larger scale systems, another is
formed before large scale systems form.
Therefore, one of the crucial problems in cosmology is to distinguish which scenario is the right one. For doing this test we
need information of the largescale structure at different cosmological time, i.e., different redshifts. However, up to now, most
of the information of the structure of the universe is limited to
a very tiny fraction of the observable universe, it comes from
the study of distributions of galaxies, all of which with redshifts $Z < 1$. From the $Z < 1$ information, it would not be able to
judge which one of scenarios agree with the actual situation
better. Hence the distribution of quasars become important. It can
tell us the distribution of matter (at least the luminous matter)
in the time $Z \sim 1 - 3$, just earlier than that of $Z < 1$ galaxies.
Work on the quasars distribution has often been hampered by the
lack of good complete samples. Recently, a large number of new
identifications of quasars have been done from the surveys using

A. Hewitt et al. (eds.), Observational Cosmology, 627–638.
© *1987 by the IAU.*

objective prism or grating prism techniques. The total number of
confirmed quasars is about 3000 and quasars' candidate about 10^4.
Several surveys are already available in studying the spatial distributions of quasars.

The quasars surveys have been investigated in Sect. 2 from the
view of point of clustering analysis; In Sect. 3 the statistical
methods employed in the analysis are discussed. The statistical
results and their implication in the clustering scenario are given
in Sect. 4 - 6.

2. QUASARS SURVEYS

The difficulty of to do the statistical analyses of quasars
clustering is the lack of available samples of quasars. Most of
homogeneous surveys for quasars listed only brighter quasars, or
covered only very small area. For example, the Schmidt-Green sample have only 46 confirmed quasars but spread 10714 square degree
on the sky; Braccesi et. al. found 8 confirmed quasars in their
37.2 square degree UVX survey; Kron and Chiu's sample cover only
0.1 square degree. These samples can not be used to do statistical
analyses.

In recent years, a large number of quasars surveys have been
carried out. A complete catalogue of main slitless-spectrum surveys can be found in "The Asiago Catalogue of Quasars Edition
1985" (2). About 10000 quasars canditates have been identified
using these methods. In complementary to this methods, the ultra-
violent-excess (UVX) search methods is best suited for quasars
with redshift $Z < 2$. Several wide field samples which are comparatively available for the tests are listed as follows:

a, Savage and Bolton's survey of two $5° \times 5°$ fields near the
South Galactic Pole; (3)

b, CTIO optical quasars sample given by Osmer; (4)

c, South Galactic Pole sample of Shanks et. al.; (5)

d, $01^h 12^m$, $-35°$ field sample given by Savage et. al.; (6)

e, Smith and He's survey of a 40 square degree field centred
at $01^h 44^m$ and $-40° 00'$; (7)

f, S.A. 94 field optical quasars sample given by Barbieri et.
al.

3. STATISTICAL METHODS

The advances and widespread availability of computers have made
the statistical analysis of large catalogue of data to be a tractable work. Now the distribution on one, two and three dimension
space of quasars and quasars' candidates have been investigated
using variously statistical methods, many of which have already
been used in investigations of galaxy clustering. The most popular
method is the correlation function (CF). Of considerable practical
importance has been the fact that the correlation function have
been widely used in study of galaxy and cluster of galaxies distribution. It makes fairly easy to compare the results between

quasars and galaxies distribution. Another popular statistic is
the nearest neighbor test (NNT). The basic idea of NNT is to compare the mean nearest neighbor distance between objects with that
expected for a random distribution sample. The NNt is a useful
complement to the CF approach, because the probability of not finding a objects in a given distance region depends on all order of
correlation function. Although the NNT do not give the information
about the type of clustering, but the test is powerful for detecting the presence of weak cluster.

Beside these, more statistical methods have been employed. For
example, the power spectrum (PS) is fourier or spherical harmonic
trasform of the correlation function; The binning analysis (BA) is
also known as the cluster cell methods; The percolation test,which
have beensuggested by Zeldovich and his colleagues, seems to be
sensitive to the clustering pattern.

4, CLUSTERING OF QUASARS

All the statistical analyses of quasars clustering are listed in
Table 1. It can be seen from Table 1 that the results from different researcher are not completely the same with each other. Osmer
(4) and Webster (9) showed no clustering in the quasars distribution from the CTIO sample, while Arp had claimed that there are
some quasars groups and other inhomogeneity structure in this sample. For the quasars of two $5° \times 5°$ survey done by Savage and
Bolton, Chu and Zhu (10) found weak clustering on about 100 Mpc
size in one $5° \times 5°$ sample and no clustering in the other one. But
Deng et. al. gave same traces of possible void structure.

The differences among the results is due mainly to the sample,
and the statistical methods. Namely, the number of quasars in all
survey are not large enough to do statistical analyses with high
confidence; The criteria of clustering in different statistical
methods may not be equivalent; The last but not least point is
selection effects.

In spite of the above-mentioned differences, a conclusion seems
to already be acceptable: The clustering of quasars is rather weak,

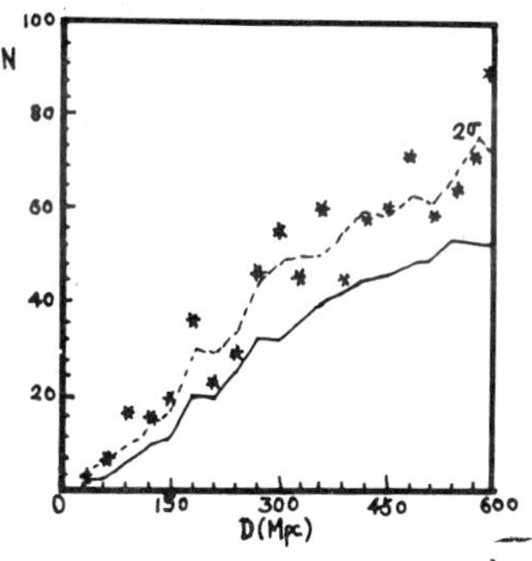

Fig.1 CF for QSOs in (02^h,-50)

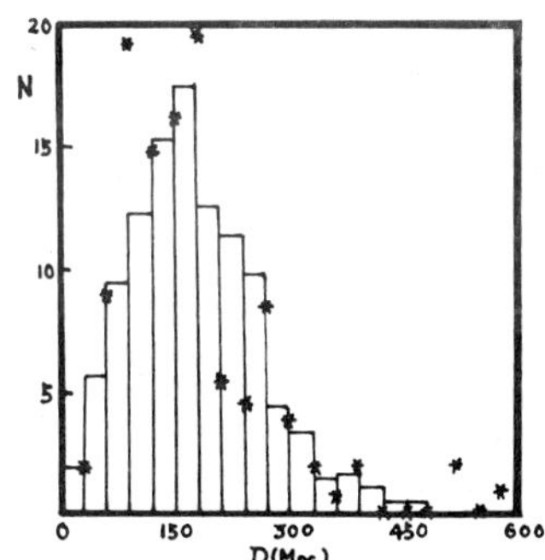

Fig.2, NNT for QSOs in (02^h,-50)

Table 1: Statistical Analysis for Quasars Clustering

Survey Name	Survey Tech.	Area	No. of Quasars	Analysis Author	Statistical Methods*	Results
CTIO 21^h-04^h $-40°$	Prism+ Grism	15x($5°$x$5°$)	174	Osmer(1981)	NNT CF BA	No
			108	Weberst (1982)	PS	No
SB 02^h--$50°$	Prism+ UVX	$5°$ x $5°$	116	Chu & Zhu (1983)	NNT PS CF	Weak
			43 $1.9 < Z < 2.5$	Fang,Zhou et. al. (1986)	CF PE	No
SB 22^h--$18°$	Prism+ UVX	$5°$ x $5°$	124	Chu & Zhu (1983)	NNT PS CF	No
			29 $1.9 < Z < 2.5$	Fang,Zhou et. al. (1986)	CF PE	No
SGP	UVX	$11.5deg^2$	293	Shanks et. al.(1983)	CF	Yes
01^h12^m $-35°$	Prism+ UVX	5.75x$5°.3$	325	Savage et. al.(1984)	CF	Weak
			106 $1.6 < Z < 2.5$	Fang,Zhou et.al. (1986)	CF PE	No
S.A. 94	Prism+ UVX	2^h43^m - 3^h04^m $-2°18'$ - $2°51'$	190 350	Barbieri & Chu (1986)	CF	Weak
NGC 450	UVX	$25deg^2$	140 94 60	Gosset, Surdet & Swings(1985)	MBA NNT CF PS EKS	Yes

* NNT: Nearest Neighbours Test
  CF: Correlation Function
  BA: Binning Analysis
  PS: Power Spectrum
  PE: Percolation
  EKS: Extended Kolmogorov Smirnov
  MBA: MUltiple Binning Analysis

at least, it is weaker than that of galaxies. For instance, Fig. 1 shows the correlation function for the quasars in the field (02^h, -50°) of Savage-Bolton survey. In Fig.1 the observed data are systematically deviated from the mean expected correlation function for a random distribution. This is an evidence for a clustering in this field. However, the deviation between the observed and the random data is only 2σ level. This means the clustering is rather weak. The nearest neighbor test also shows the same result(Fig.2).

Weak clustering in the quasars distribution is very significant. It implies that at the time $Z\sim2$ the visible objects, at least quasars, cluster weakly at the scale of superclusters. Therefore, the superclusters of galaxies should be formed after the formation of quasars, or the formation of galaxies and quasars are due to different mechanism, the former has strong clustering, the later weak. Both above-mentioned pictures are inconsistent with the adiabatic scenarios.

5, REDSHIFT DEPENDENCE OF QUASARS CLUSTERING

It is very interesting to see from Table 1 that all results of samples containing UVX quasars (or quasars' candidates) always show more strong clustering than others. We know that the redshift of quasars found by objective prism and grism are higher than that of quasars found by UVX methods. Therefore the stronger clustering in UVX samples may imply that the strength of quasars clustering depends on the redshift of quasar: more larger redshift more weaker clustering.

For showing the redshift dependence, we should do the clustering analyses for quasars in different redshift ranges. This analysis have be done using Savage-Bolton sample, which consists of two classes of quasars identified by both objective prism technique and the UVB two colour method, the redshifts in this sample spreads on a more broad region than that of other surveies. It is convenient to do the comparision between quasars with larger and smaller redshilft.

The results of the nearest neighbor test are given in Table 2, in which N is the number of quasars, $\langle D_i \rangle$ denote the sample's

Table 2: Nearest Neighbor Test for Savage-Bolton Sample

redshift	N	quasar data $\langle D_i \rangle$ Mpc	Monte Carlo data D_i^* Mpc	$\hat{\sigma}$	δ	$1-P(\delta)$
			$(02^h, -50^o)$ field			
$Z < 2$	62	141.7	159.0	79.6	-1.72	96%
$Z > 2$	48	201.0	205.9	83.2	-0.40	66%
			$(22^h, -18^o)$ field			
$Z < 2$	57	146.7	165.8	77.9	-1.84	97%
$Z > 2$	26	207.1	193.0	75.9	> 0	-

mean of nearest neighbor separation, $D_<^*$ and $\hat{\sigma}$ are the mean and standard deviation for Monte Carlo samples respectively. As a measure of statistical significance, we define the following function

$$\delta = N^{1/2} \frac{\langle D_< \rangle - D_<^*}{\hat{\sigma}} \tag{1}$$

The distribution of δ is asymptotically normal with mean 0 and variance being 1. Therefore $1 - P(\delta)$ in Table 2 is the probability of clustering to be found in the sample. The main results from Table 2 are that an apparent clustering at 95% significant level for the quasars of $Z < 2$ in both fields, and there is no evidence of clustering for $Z > 2$. These results can be more clearly seen in Figs. 3 and 4, in which the distributions of the nearest neighbor distance for each field are ploted. The observed $Z < 2$ distribution

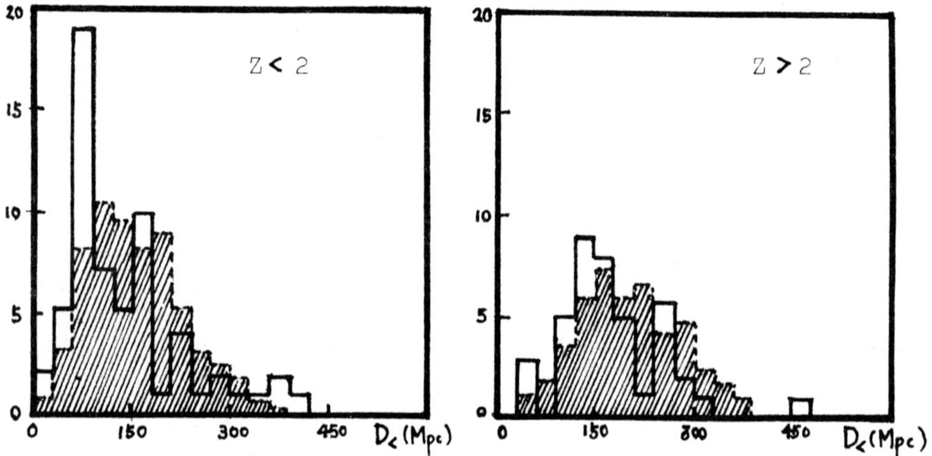

Fig.3 The nearest neighbor distribution for quasars with $Z < 2$ and $Z > 2$ for (02^h, $-50°$) of Savage-Bolton sample.

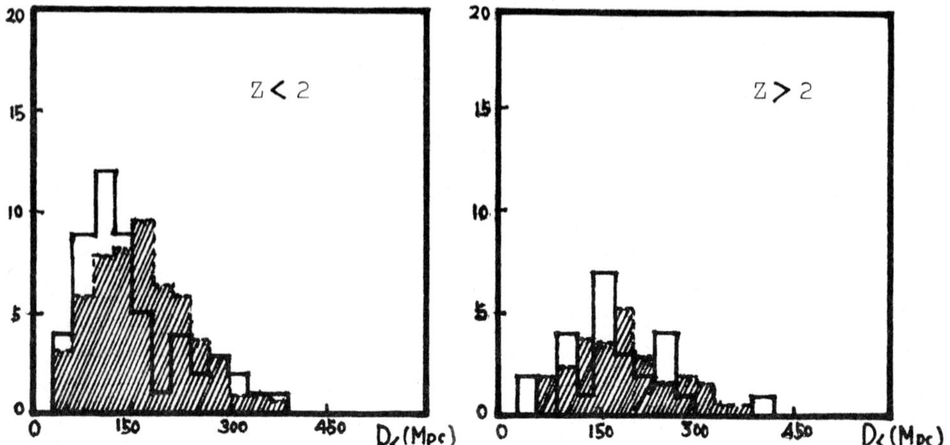

Fig. 4 The nearest neighbor distribution for quasars with $Z < 2$ and $Z > 2$ for (22^h, $-18°$) of Savage-Bolton sample.

(solid line) deviate obviously from the Monte Carlo samples (dashed line) on scale of 50 - 100 Mpc. It means that the distribution of $Z < 2$ quasars does have $\sim$100 Mpc clustering. The $Z > 2$ distribution does not show the difference from that of random sample, namely, there is not distinguishable inhomogeneity.

The clustering of quasars in the sample of Shanks et. al. (5) (see Table 1) has been analysed by 2-dimensional correlation function method. They found that the UVX stars are clustered, while 'probable' quasars identified by objective prism show no evidence of clustering. Therefore, the difference of clustering between the UVX stars and the objective prism quasars can also be seen as an evidence for the difference between $Z < 2$ and $Z > 2$ quasars.

Quasars listed in Smith-He catalog (7) are identified by the objective prism method. By using power spectrum analysis, it is found no evidence for clustering of quasars in this catalogy. CTIO sample is also given by the objective prism method. Most of redshifts of quasars in this sample is of $Z > 2$. Osmer already claimed no evidence for large clustering can be found from CTIO sample.(4,9)

In one word, a common result in these studies on large structure of quasars from different samples is that the inhomogeneities on the scales of about 100 Mpc exist in the distributions of $Z < 2$ quasars, but not of $Z > 2$ quasars.

The above mentioned conclusion on the evolution of quasar clustering imply that the dark matter in the universe consists, at least, of two kinds of noninteracting particles, one is dominant component with larger velocitydispersion, such as massive neutrinos; one is non-dominant component which is more weakly interacting, more massive particle with smaller velocity dispersion (11). Since more weakly interacting particles decoupled at higher temperature, its number density should be lower than neutrinos. Thus it is possible that the universe is still dominated by massive neutrinos, while the more massive particles are only to be a non-dominant component. The Jeans mass and length (or the free streaming length) of the more massive particle component are much smaller than that of neutrinos, for instence, the Jeans mass of m=1 kev noninteracting particles can be equal to about 10^{12} $M_\odot$. In such a two component dark matter universe the small scale $\sim 10^{12}$ $M_\odot$ perturbations can avoid to be erased by neutrino free streaming, and it can be kept in the non-dominant component until the recombination. Thus the small scale perturbations can also collapse before the formation of large scale structure. Therefore, the picture for small scale structure formation in two component dark matter model is quite different from the universe containing only one kind dark matter. In the latter the small scale structure formed only after the large scale structure collapses, namely, the clustering process is from larger scales to smaller ones. The first objects to be condensed out would be that of mass about 10^{15} $M_\odot$, then smaller scale objects such as galaxies form due to fragmentation. Thus, all the visible objects with small scales should have redshifts less than 2. However, in the two component dark matter model, it is possible to form small scale structure before the collapse of the large scale structure. After the recombination, small scale clustering can take place in both the non-dominant dark matter and the baryons. These clustering processes are independent of large scale clustering.

According to this clustering picture, there should be two kinds of visible objects with the scale less than superclusters. One is to be formed by fragmentation of the large scale structure; one is not to link strongly with the formation of large scale structure. The distribution of the first kind of objects should have marked large scale inhomogeneity, while the second kind should have not such structure. Thus, this scenario leads to the following conclusion: the small scale objects with $Z > 2$ should distribute more uniformly, especially on the scale of superclusters, than $Z < 2$ objects. This is just the results of statistical analyses to be discussed above.

6, OTHER EVIDENCE FOR THE EVOLUTION OF QUASAR CLUSTERING

Binggeli pointed out (12) that the major axis of a cluster tends to point to the nearest neighbor cluster. If the distance between a cluster and its nearest neighbor is less than 35 Mpc, the angle between the major axis of the cluster and the direction to the nearest neighbor is always smaller than $45°$. No such correlation exists if the nearest neighbor distance is large than 35 Mpc. Therefore, the orientation correlation is also an evidence of large scale inhomogeneity in the distribution of clusters.

A similar study for radio double sources has been done recently (13). Using a complete sample of radio double sources by Condon et. al.(14), one found some statistical evidences of the correlation between the orientation of the double sources and the direction to their nearest neighbor radio sources, the correlation scale is also on about tens Mpc, but the correlation is not so obvious as that of clusters of galaxies.(Fig. 5). Not all of radio sources are quasars. Since these sources are selected by the criterion of the separation between two components to be less than 1.5 arcmin, many of them should probably be quasar.

Recently, Zhu (15) did the analyses on the alignment of quasars. She also took $5° \times 5°$ survey of Savage-Bolton as sample. The result shows that the phenomenon of alignment in quasars is not obvious, i.e., no identifiable difference from sample given by Monte-Carlo method.

The Ly-α absorption lines in quasar spectrum might indicate the intergalactic clouds. The redshift distribution of Ly-α absorption

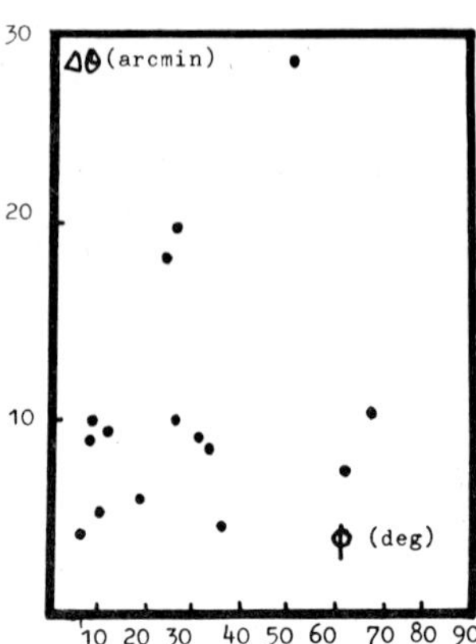

Fig. 5, Orientation distribution for radio double sources

lines should then show the clustering of intergalactic clouds. Sargent et. al. studied (16) six quasars with very rich absorption lines, the redshifts of which cover from 1.7 to 3.3. They did not find any evidence of the inhomgeneity in the absorption line distribution. By using different method, which is especially available to analyse the inhomogeneity with scale of a few hundreds Mpc, it is found that (17) the absorption lines distribution inhomogeneous, but the amplitude of the inhomogeneity is very small.

An evidence for the difference between $Z < 2$ and $Z > 2$ quasars has also been found from the research on the evolution of optically selected quasar. Veron (18) have built the luminosity function of quasars at various redshifts, using the number-magnitude relation for quasars and the redshift distributions at various apparent magnitudes. He found that the evolution law for small redshift is quite different from that for large one; the evolution is very fast for $1 < Z < 2.3$, while at some $Z > 2.3$ the evolution has to stop and even to reverse. This shows again that $Z \simeq 2$ is a crucial time of the large scale structure formation.

7, QUASAR CLUSTERING AND DARK MATTER CLUSTERING

The result of quasar clustering can be used in finding information of the clustering of dark matter in the universe. In a dark matter dominant universe, the total density inhomogeneity should be dominated by the density distribution of dark matter. Therefore the information of dark matter distribution is necessary in studying the large scale structure in the universe. Obviously, it is difficult to determine the length scale and the amplitude of the inhomogeneity of dark matter. Nevertheless, some information about such inhomogeneity has been found from the gravitational effects of the dark matter. For instance, the stochastic perturbation in the gravitational field due to the density inhomogeneity will lead to luminosity fluctuations for distant sources. Therefore, the amplitude of the inhomogeneity can be found from the observed differences of luminosities between sources which originally had the same luminosity.

As mentioned in Sects. 4,5, the spatial distributions of quasars from many catalogues and surveys show that the distributions of quasars are quite uniform or of very weak clustering, especially for the quasars with large redshifts. Since the lensing effect is one of the reasons for giving the observed differences or inhomogeneities of luminosities of quasars, hence an upper limit to the amplitude of inhomogeneity of dark matter can be derived, assuming that all the luminosity inhomogeneity of quasars comes from the lensing effect.

The results in Sect.4 and 5 are only on the number density inhomogeneity of quasars, it cannot be used as a measure of the luminosity of quasars. We can develop a modified power-spectrum analysis to find the required luminosity inhomogeneity. It is a modified power- spectrum analysis (M-PSA) with the basic statistics as follows:

$$I^l_{uv} = \frac{2}{\sum\limits_{j=1}^{\hat{n}} m_j} \left| \sum_{j=1}^{n} m_j \exp[i(ux_j + vy_j)] \right|^2 \qquad (2)$$

where m_j is the magnitude of j-th quasar. In order to increase the power of thetest for luminosity inhomogeneity, we introduce, as in the case of number inhomogeneity, the statistics Q:

$$Q = \Sigma_I / \Sigma_V \qquad (3)$$

with

$$\Sigma_I = \sum_{(uv)} I^l_{uv} \qquad (4)$$

$$\Sigma_V = \frac{1}{n} \left(\sum_{j=1}^{n} m_j \right) \sum_{(u,v)} 2, \qquad (5)$$

where the sums in Σ_I and Σ_V are summed over all the (u,v) with $\lambda_{uv} > \lambda_c$. If any inhomogeneity exists, Q will deviate from unity and Q-1 can be used as a measure of the inhomogeneity amplitude.

Table 3: The M-PSA Statistics Q for quasars of Savage-Bolton Sample.

$(02^h, -50°)$ field			$(22^h, -18°)$ field		
$1/\lambda_c$	all quasars	Z > 2 quasars	$1/\lambda_c$	all quasars	Z > 2 quasars
1	0.50	0.95	1	2.66	1.12
2	1.63	0.85	2	0.73	1.00
3	0.85	0.68	3	1.17	1.05
4	1.39	1.15	4	0.70	1.18
5	1.00	0.96	5	0.80	1.32
6	0.90	1.24	6	0.79	0.80
7	1.00	1.04	7	1.15	1.22
8	1.15	1.30	8	0.82	0.76
9	1.08	0.99	9	0.94	1.07
10	1.04	0.97	10	0.95	1.09

The results of M-PSA for quasars in the Savage-Bolton survey are shown in Table 3. We note immediately from the Table that the mean of (Q-1) is about 0.1. The luminosity distribution of quasars with Z > 2 are more uniform than that of all quasars, especially for longer length scale $1/\lambda$ =1,2,3.

It appears that the mean luminosities of quasars located at different directions are the same to within about 10%. From this we find that the upper limit to the total density inhomogeneity:

$$\delta < \lambda_{10}^{-1/2} h_0^{-1/2} \Omega^{-1} \qquad (6)$$

seems to be less than the inhomogeneity amplitudes of galactic distribution by a factor of three to five. This conclusion should be considered as a very rough one. Nevertheless, the link between the distribution of quasars with Z > 2 and the large scale inhomogeneity of dark matter merits our attention.

This upper limit may already be used to test models on the formation of large scale structure in the universe. For instance, the so-called 'biasing' model of the cold matter scenario predicted that light is not an unbiased trace of mass and that there are many (most) clumps of cold matter and baryons which are not shining. This hypothesis implies that galaxies are formed only in high density regions which are the peaks in the primordial fluctuation spectrum. If the peaks lie above some threshold τ (in units of rms density fluctuation), the amplitude of overall density inhomogeneity is lower, relative to galaxies, by a factor of τ roughly. In particular, $\tau = 2.5$ is considered by requiring the fit of the 2-point correlation function of galaxies on the scale of $5h_0^{-1}$ Mpc. Therefore, the biasing model agrees, marginally, with the upper limit. However, if the amplitude of luminosity inhomogeneity of $Z > 2$ quasars is confirmed, by further systematic search, to be smaller than 0.1, the biased dark matter scenario would be in difficulty.

References
(1),L.Z. Fang, Y. Chu & X. Zhu, Ap.S.S., 115(1985),99.
(2),C. Barbieri, Y. Chu, et. al., "The Asiago Catalogue of QSQs edition 1985", in press.
(3),A. Savage & J.G. Bolton, MNRAS, 188(1979), 599.
(4),P.S. Osmer, Ap.J., 247(1981), 761.
(5),T. Shanks, R. Fong, et. al., MNRAS, 203(1983), 181.
(6),A. Savage, et. al., MNRAS, ___(1984),
(7),X.T. He, Y. Chu, L.Z. Fang & M.G. Smith, Acta Astronomia Sinica, 27(1986), No.2,144.
(8),C. Barbieri & Y. Chu, preprint.
(9),A. Webster, MNRAS, 199(1982),683.
(10),Y. Chu & X. Zhu, Ap.J., 267(1983),4.
(11),L.Z. Fang, S.X. Li & S.P. Xiang, A.Ap., 140(1985),77.
(12),B. Binggeli, A.Ap., 107(1982), 338.
(13),L.Z. Fang, Y. Chu & X. Zhu, A.Ap., 143(1985), 241.
(14),J.J. Condon, M.A. Condon & C. Hazard, A.J., 87(1982),739.
(15),X. Zhu, Acta Astrophysica Sinica, 6(1986),no.1, 1.
(16),W.L.W. Sargent, P.J. Young, Ap.J.Suppl., 42(1980),41.
(17),Y. Chu, L.Z. Fang & Y.Z. Liu, Astrophys.Lett.,24(1984),95.
(18),P. Veron, in "Quasars and Gravitational Lensens", 24th Liege Astrophysical Colloquium, 1983.

DISCUSSION

BAHCALL: With the large number of QSOs available in your sample, can you separate the redshift subsamples into smaller bins (e.g., z = 0 to 1, 1 to 2, and 2 to 3) to see if the change of the clustering strength is gradual or rather sharp at z ~ 2?

CHU: I agree that this is a good idea, but so far the sample is not big enough for us to divide it into smaller bins.

BOYLE: On the fields in which you find clustering, what fraction of the QSOs have spectroscopic confirmation and/or redshifts?

CHU: One of the fields which we studied is a field worked on by Savage and Bolton (1979). Most of the quasars in this sample have been observed spectroscopically, either using slit spectra or the objective prism method.

CLOWES: We have recently completed an automated search for quasars in SA 94 using the same objective-prism plate as Prof. Barbieri did for his visual search. Prof. Barbieri appears to have missed quasars, particularly in the central regions of the plate. I suspect that any 2-D clustering analysis of his data is likely to be invalid.

CHU: We think it will be interesting to reanalysis the quasar distribution using your catalogue in SA 94. By comparing both results we can get some idea of the selection effects.

ARP: For the clustering of quasars at redshifts less than $z < 2$, what are the redshift ranges involved in the individual clusters?

CHU: At the present stage of our statistical analysis no individual cluster can be identified.

DEKEL: Any clustering evolution should be compared with the growth rate in a theoretical model. Can you comment on the significance of the detected evolution in this respect?

CHU: No, but our motivation for doing the clustering analyses of quasars with redshifts $z > 2$ and $z < 2$ separately was to compare with the model that predicts stronger clustering for objects at $z < 2$ than for objects with $z > 2$.

THE COSMIC EVOLUTION OF QUASARS AT HIGH REDSHIFTS

P. Véron
Observatoire de Haute-Provence
04870 Saint Michel l'Observatoire
France

ABSTRACT.   The analysis of all published slitless quasar surveys suggests that the space density of quasars increases quickly with redshift up to $z \sim 2.45$, then drops abruptly by about a factor of five, increases again up to $z \sim 3.8$ at least to drop once more below the present day limit of sensitivity.

UV excess quasars ($z < 2.3$) are known to evolve quickly with cosmic time, their space density (or luminosity) being much smaller now than at $z = 2.3$ (see for instance Véron 1983). Weedman (1985) has shown that the quasar space density still increases at $z > 2.3$; however surveys of high redshift quasars suggest that a redshift cut-off exists either near or below $z = 3$ (Schmidt et al. 1986), or in the range $3.7 < z < 4.7$ (Osmer 1982).

A number of objective-prism or grism surveys have been carried out with the IIIaJ Kodak emulsion. These surveys are most efficient at picking out quasars with visible Ly α ($1.8 < z < 3.4$). The lower limit is set by the ultraviolet transmission of the atmosphere and the telescope optics between 3300 and 3600 Å; the upper limit is set by the Lyα + NV emission passing beyond the red limit of the IIIaJ emulsion at $\lambda \sim 5380$ Å.

These surveys belong to two groups : the Schmidt telescope (the Curtis Schmidt and the UK Schmidt) surveys, reaching a magnitude of ~ 18.5, and the large telescope (Kitt Peak and Cerro Tololo 4 m and CFHT 3.6 m) surveys reaching about two magnitude fainter. The two groups contain 232 and 331 Ly α quasars respectively.

Fig. 1 shows the histograms of the redshift distribution of the Ly α quasars separately for the Schmidt surveys and for the large telescope surveys. One of the important features of this figure is that the two redshift distributions are almost identical over the whole range covered. This indicates that the luminosity function is a power law ($N (>L) \propto L^{-n}$) in the absolute magnitude range of interest, the index n being independent of redshift. The fact that the surface density of quasars increases by a factor of ~ 20 when their apparent (or absolute) magnitude increases by 2.0 magnitudes indicates that $n \sim 1.6$.

The redshift distributions shown in Fig. 1 raise slowly by a factor of 2 between $z = 1.8$ and 2; this is due to the combine decline of sensitivity in the UV of the IIIaJ emulsion, of the atmosphere and of the telescope optics. The

A. Hewitt et al. (eds.), Observational Cosmology, 639–641.

most important feature is the sharp drop (by a factor of ~ 5) of the quasar
surface density at z = 2.45, followed by a flat wing extending up to the
observational limit of z = 3.4. At z = 2.45, Ly α appears at 4200 Å; there is no
obvious reason why the various surveys, should have a drop in sensitivity at
this wavelength; the only feature common to all surveys is the use of the IIIaJ
emulsion which has a wavelength sensitivity extending almost flat to λ 5300 Å.
Therefore we believe that this drop is real rather than due to any
observational effects.

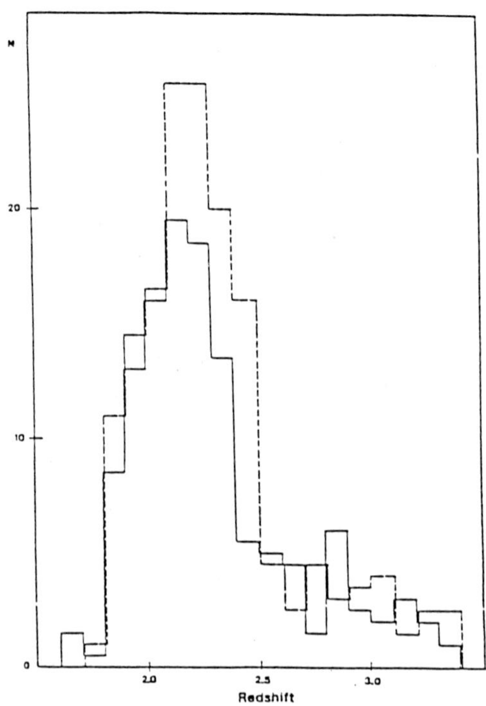

Fig. 1. The observed redshift distribution for Lyα
quasars found in IIIaJ slitless surveys.
(-) the Schmidt telescope surveys;
(--) the large telescope surveys.

  A number of surveys using other techniques sensitive to redshifts much
larger than 3.4 have been published; they suggest that the wing of high
redshift quasars may extend to z = 3.8. Until recently all efforts to find
quasars at redshifts larger than this value have been vain suggesting that there
was a cut-off at this limit. The announcement of the discovery of a quasar at
z = 4.00 (Hewett, this conference) shows that this is probably not the case.
  Knowing that the luminosity function is a power law, the redshift
distribution can be easily used to determine the cosmic evolution of the quasar
space density. It is found that the quasar space density increases quickly with
redshift up to z ~ 2.45, drops by about a factor of 5 at this value, and then
increases again at least up to z = 3.8.

References

Osmer, P.S. 1982, Astrophys. J. 253, 28.
Schmidt, M., Schneider, D.P. and Gunn, J.E. 1986, Astrophys. J. 306, 411.
Véron, P. 1983, in "Proceedings of the 24th Liege International Astrophysical
        Colloquium", Institut d'Astrophysique de Liège, p. 210.
Weedman, D.W. 1985, Astrophys. J. Suppl. Ser. 57, 523.

DISCUSSION

WINDHORST: This is a general question to all observers who have made
QSO surveys using slitless techniques. If everyone would cut-off his
sample more severely, e.g. at 1-1.5 mag brighter than they are currently
doing would the redshift distributions then agree better? I realize
that the numbers will be smaller, but the samples would be less biased.

VERON: There is no basic disagreement between the redshift distributions shown by various authors. Most of these distributions are
based on very small samples and suggests a cut-off in the redshift
range 2.5-3.5. My analysis, based on a large sample of more than
500 QSOs, shows clearly a drop at $z \sim 2.45$ which, if real, would
certainly be acceptable by other authors. However it may be due to
some as yet unrecognized observational bias.

The Clustering and Evolution of Optically-Selected QSOs

B.J. Boyle
University of Edinburgh, Blackford Hill,
Edinburgh EH9 3HJ, U.K.

T. Shanks and R. Fong
University of Durham, South Road,
Durham DH1 3LE, U.K.

B.A. Peterson
Mount Stromlo and Siding Spring Observatories
Woden, ACT 2606, Australia

Abstract
We are in the process of compiling a large catalogue of faint (B < 20.9 mag),
UVX selected QSOs with complete spectroscopic identification using the fibre optic
(FOCAP) system at the Anglo-Australian Telescope. From the 220 QSOs thus far
identified we find that QSO evolution is most simply parameterised by a uniform
increase in luminosity towards higher redshifts. We also find evidence for strong
QSO clustering at scales < $10h^{-1}$ Mpc.

Data
Candidate QSOs are first selected using the ultra-violet excess (UVX) technique
from COSMOS (MacGillivary and Stobie 1985) machine measurements of UK
Schmidt telescope U and J plates. Details of the application of the UVX
techinque to the COSMOS measurements may be found in Shanks et al. (1983) and
Boyle et al. (1985). At present, faint (B < 20.9 mag) UVX stellar objects (U-B <
-0.35) have been identified on seven, well separated 4°5 x 4°5 Schmidt plate areas
(see Boyle et al. 1987). Subsequent spectroscopic observations with the FOCAP
fibre optic system (Gray 1984), operated in conjunction with the IPCS and RGO
spectrograph, at the Anglo-Australian Telescope have enabled spectra to be
obtained for more than 600 of these UVX objects in 13 40 arcminute diameter
fields. At the magnitude limit of our survey, the surface density of UVX objects
(120 - 150 per square degree) means that the number of UVX objects in the 0.35
square degree FOCAP field is ideally matched to the number of fibres (45 - 50)
available. Of the 600 UVX objects surveyed 220 are QSOs (the 9000 second
integrations being sufficient to obtain unambiguous redshifts for 85% of the QSOs),
17 are white dwarfs, 25 are narrow emission line galaxies, the remainder being
Galactic subdwarfs.

A. Hewitt et al. (eds.), Observational Cosmology, 643–647.

Results

a. The Number Magnitude Counts
In figure 1 we present the differential number-magnitude relation for QSOs with
z<2.2 (the redshift range for which the UVX technique is complete) found in this
survey. Also presented are the surface densities of z<2.2 QSOs obtained from
other complete spectroscopic surveys of UVX objects. At bright magnitudes (B <
19.5 mag) we see good agreement between the different surveys, with the QSO
number-magnitude relation following a steep (dlogN/dB=0.86) slope. Beyond B =
19.5 mag, however, the counts obtained from our survey show a sharp break to a
much flatter slope (dlogN/dB=0.33), with the integral surface density at B = 20.9
mag being 36 ± 4 per square degree, a factor of five lower than that predicted
from a simple extrapolation of the steep slope seen at B < 19.5 mag. The error
quoted on the surface density is obtained from the r.m.s. field-to-field variation
and is consistent with a Poissonian distribution for the QSOs. We have established
(see Boyle 1986) that this turn-over is certainly real and not caused by any
magnitude dependent selection effect in our data. Indeed the surface densities for
faint (B > 20 mag) QSOs presented here are in good agreement with those
reported by other authors at this conference (Marano, Crampton, Koo) based on
smaller but similar spectroscopic surveys of UVX-selected objects.

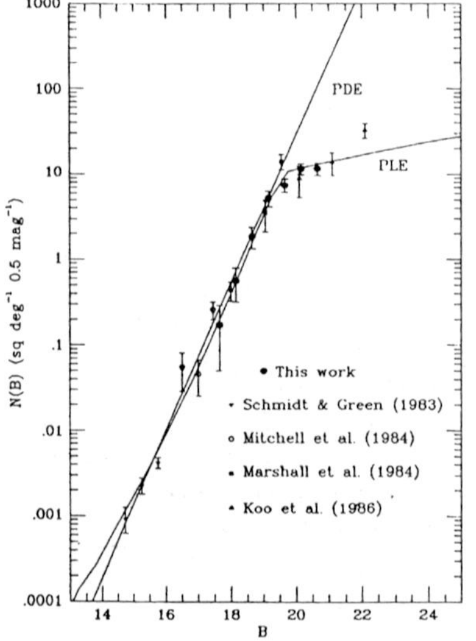

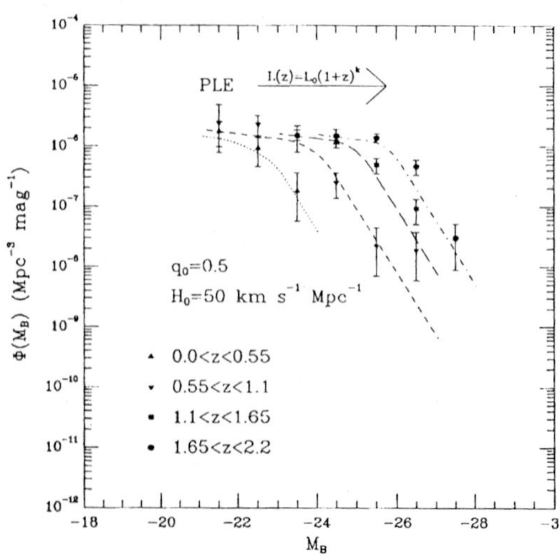

Figure 1 Number-magnitude counts   Figure 2 QSO luminosity functions
          for spectroscopically   derived from our survey.
          confirmed QSOs with
          z<2.2

b. QSO Evolution

It is well known (see e.g. Cavaliere et al. 1983), that a pure density evolution model for QSOs is strongly ruled out by the appearance of a break in the QSO number-magnitude relation; such a feature being reflected in the shape of the QSO luminosity function itself. By tracking the position of this feature in QSO luminosity functions determined in different redshift ranges we may directly establish the form that the evolution takes. In figure 2 we present the QSO luminosity function as determined from our data in four separate redshift intervals between $0<z<2.2$. From figure 2 we see that, over the absolute magnitude and redshift ranges sampled by our survey, the QSO luminosity function may simply be represented by two power laws, with a steep slope ($\Phi(>L) \propto L^{-3.6}$) beyond a 'break' feature and a much flatter slope ($\Phi(>L) \propto L^{-1.2}$) fainter than the break. Futhermore, the redshift dependence of the luminosity function in this range may be expressed as a uniform shift towards higher luminosities at higher redshifts. More detailed statistical analysis (Boyle et al. 1987) reveals that a power law luminosity evolution of the form $L \propto L_0(1+z)^{3.7}$ provides the best fit to the data, with the predicted $z=0$ QSO luminosity function being similar to the Seyfert luminosity function derived by Cheng et al. (1985). A physical interpretation of this 'pure luminosity evolution' is, however, far from straightforward. The observed evolution could represent either the actual evolution of individual, long-lived (few billion years) QSOs or the evolution in average properties of succesive generations of short-lived (few million years) QSOs. Although the long-lived model may be a more natural explanation of pure luminosity evolution the requirement that massive ($10^9 - 10^{10}$ $M_\odot$) exist in the centres of low redshift QSOs and Seyferts may severely constrain this model.

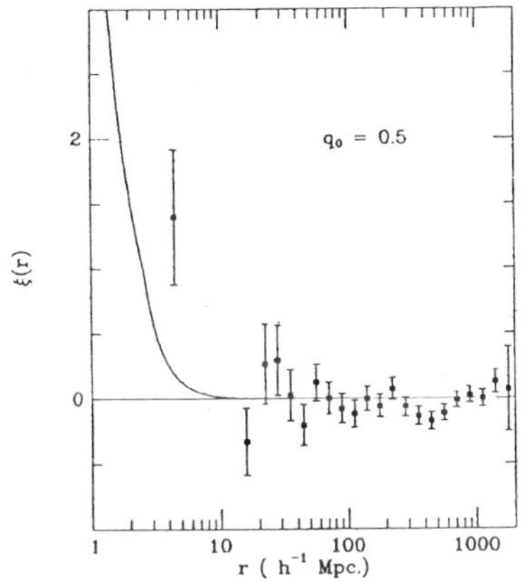

Figure 3   QSO 2-point correlation function. r is a comoving coordinate computed in an Einstein-de Sitter universe. Solid line is the predicted galaxy correlation function at z=1.5

c. QSO Clustering

We have also investigated the spatial clustering of QSOs in this survey using the two-point correlation function; the high surface density and accurate spectroscopic redshifts obtained for the large number of QSOs in this survey enabling the first determination of the QSO correlation function at scales $<30h^{-1}$ Mpc based on a homogeneous QSO catalogue. The correlation function is shown in figure 3. At large scales ($>50h^{-1}$ Mpc) the amplitude of the correlation function is consistent with a random QSO distribution. At small scales ($<10h^{-1}$ Mpc), however, we find tentative evidence (3σ) for clustering amongst QSOs. Although statistics here are poor, we find 12 QSO pairs with separations $<10h^{-1}$ Mpc, where only 4 are expected based on a random hypothesis. Indeed it appears that QSOs may cluster more stongly than galaxies at these epochs ($<z>_{QSO}= 1.5$) based on predictions of the galaxy correlation function at $z=1.5$ assuming stable clustering (Shanks et al. 1987), with only 5.5 pairs expected were QSOs to cluster like galaxies.

Conclusions

We have presented the results obtained from a large catalogue of UVX-selected QSOs with complete spectroscopic identification. We find that the number-magnitude relation for UVX QSOs exhibits a sharp break at B = 19.5 mag, consistent with a feature in the QSO luminosity function. This feature may be used to track QSO evoultion, which we find to be simply expressed as a uniform increase in luminosity with increasing redshift. Tentative evidence for strong QSO clustering at scales $<10h^{-1}$ Mpc has been found but more data will be needed to confirm this result and to enable the evolution with redshift of the correlation function to be determined.

References

Boyle, B.J. 1986 Ph.D. Thesis, University of Durham.
Boyle, B.J., Fong, R., Shanks, T. and Clowes, R.G. 1985, MNRAS, 216, 623.
Boyle, B.J., Fong, R., Shanks, T. and Peterson, B.A. 1987, MNRAS, in the press.
Cavaliere, A., Giallongo, E. and Vagnetti, F. 1983, Ap.J., 269, 57.
Cheng, F.Z., Danese, J., De Zotti, G. and Franceschini,A. 1985, MNRAS, 212, 857.
Koo, D.C., Kron, R.G. and Cudworth, K.M. 1986, PASP, 98, 285.
Gray, P.M. 1984, in "Instrumentation in Astronomy" SPIE proc., V445, 57.
MacGillivray, H.T. and Stobie, R.S. 1985, Vistas in Astronomy, 27, 433.
Marshall, H.L., Avni, Y., Braccesi, A., Huchra, J.P., Tanenbaum, H., Zamorani, G. and Zitelli, V. 1984, Ap.J., 283, 50.
Mitchell, K.J., Warnock, A. and Usher, P.D. 1984, Ap.J., 287, L3.
Schmidt, M. and Green, R.F. 1983, Ap.J., 269, 352.
Shanks, T., Fong, R., Green, M.R., Clowes, R.G. and Savage, A. 1983, MNRAS, 203, 181.
Shanks, T., Boyle, B.J., Fong, R. and Peterson, B.A. 1987, MNRAS, in the press.

Postscript. One week after this conference a further 3 clear nights were obtained for the spectroscopic survey, resulting in a further 130 QSOs being identified, bringing the total in the catalogue to 350.

DISCUSSION

BIRKINSHAW: What criteria in <u>angular separation</u> and <u>redshift separation</u> do you use to define a quasar "pair"?

BOYLE: We simply calculate the comoving separation between QSOs using the standard Osmer (1980) procedure. The pairs that I specifically referred to in the talk were those with a comoving separation of $<10h^{-1}$ Mpc, the scales at which we observe the correlation function to be positive.

SILK: Can you comment on the correlation length that you have measured for the quasar distribution and on the large-scale homogeneity limit that you can get?

BOYLE: The correlation length of the clustering is approximately $5h^{-1}$ Mpc (although it is still tentative) and the homogeneity limit at large scales is $\Delta\delta/\delta \lesssim 10\%$ although stronger limits will be placed on this figure as we increase our sample size.

CHEN: What do you mean the 10% unidentified objects? Does it mean by spectroscopic work?

BOYLE: Less than 10% of the objects for which we have spectra remain unidentified. They mainly comprise objects at the faint magnitude limit of our sample where insufficient signal-to-noise was obtained to reliably identify stellar features or weak emission lines. We expect most of these objects will be galactic stars, since, at low S/N stellar absorption features will be more difficult to detect than broad emission lines. However, none of our conclusions are affected even if we assume that <u>all</u> the unidentified objects are QSOs.

ON COSMOLOGICAL EVOLUTION OF QUASARS

Jerzy Machalski
Astronomical Observatory, Jagellonian University
ul. Orla 171,
PL-30244 Cracow, Poland

1. INTRODUCTION

Hereafter the term "quasar" is applied to both optically selected (and/or spectroscopically confirmed) UVX objects, and
to radio-detected objects of that kind. A recent attempt to
model number-counts of radio-selected QSOs (Condon, private
communication) has revealed that the counts cannot be modelled by simply translating the general radio luminosity function (RLF) leaving the cosmological evolution unchanged.
While it may be true that the radio sources in QSOs evolve
in the same way as do all radio sources (i.e. translation
function), the radio sources are probably not always in QSOs.
In particular, it may be that increasing the luminosity (especially the core luminosity) of a radio source increases
the probability that it is in a QSO. If that is true, a radio
detection rate of optical QSOs should be strongly dependent
on the optical (core) luminosity, i.e. on the absolute magnitude.

Recently published data on faint optically- and radio-
selected QSOs (Sec.2) seem to provide a straightforward observational evidence fot the above hypothesis. In Sec.3 the
radio detection rate of QSOs at constant redshift is discussed. This detection rate, in a function of apparent magnitude,
can be considered as a ratio of radio-loud QSOs to all QSOs
in a function of their absolute magnitude M_B. A question of
whether such detection rate is also dependent on redshift is
considered in Sec.4.

2. THE DATA USED

(1) Optical QSO samples:
Bright QSO sample (BQS) of Schmidt and Green (1983) complete to an effective mean limit of B=16.16 mag.
Medium-bright QSO sample (MBQS) of Mitchell et al.(1984)
complete to B=17.65 mag.

A. Hewitt et al. (eds.), Observational Cosmology, 649–653.
© 1987 by the IAU.

"AB" and "BF" samples selected by Marshall et al. (1984) complete to B=18.25 mag and B=19.80 mag, respectively.
Hoag and Smith's (1977) (HS) and Sramek and Weedman's (1978) (SW) samples were partly used at high redshifts.
These complete samples were supplemented with QSO candidates up to J=22.5 mag found in SA57 and SA68 (cf. Koo et al.1986).

(2) Radio QSO samples selected at 1.4 GHz:
"2-Jy" sample (BDFL) almost complete to $B \simeq 20$ mag, comprising majority of 3CR quasars.
"GB2" sample of QSOs stronger than 0.25 Jy (Machalski and Condon 1986) and with complete photometry up to B=18.0 mag.
"LBDS" sample (Kron et al. 1985) with $S \gtrsim 1$ mJy and complete to $J \simeq 23$ mag.
The Parkes (PKS) sample of Peacock et al. (1986) was partly used.

3. DEPENDENCE OF RADIO EMISSION OF QSOs ON M_B

This dependence can be determined for objects either with radio luminosity higher than a chosen limiting luminosity, P_{lim}, or with/above chosen radio-to-optical luminosity ratio R. A meaning of both characteristics is different; the first dependence specifies a probability of radio emission (at $P \geq P_{lim}$) as a function of M_B, the second one — a degree of correlation between optical and radio luminosities.
a) Determination of the detection rate (at $\log P(W/Hz) \geq 26.4$) on M_B (H_o=50; q_o=0.5) for $1<z<2.2$ (Table 1 below) suggests

ΔM_B B_{lim}	Opt. sample	N	$N(sr^{-1})$	Radio sample	N	$N(sr^{-1})$	Ratio (%)
-23,-25 22.52	SA57	18±3	205000	LBDS	2.5±0.5	800	0.4
-25,-27 20.52	SA57	2	22800	LBDS	1	325	1.1
	BF	18	34300	GB2	47	$\gtrsim$240	>0.8
-27,-29 18.52	MBQS	8	>270				
	AB	5	440	GB2	7	>35	>8.0

that the probability of radio emission increases about 20 times if M_B increases from -23 mag to -29 mag, reaching about 10% at the highest M_B. A straightforward confirmation of this fraction comes from the radio measurements of the BQS QSOs (BQS sample is large one though not deep enough to detect a QSO with M_B=-27 mag at z=2.2). There is one radio-loud QSO among 13 QSOs with $-29<M_B<-27$ and within $1<z<2.2$, i.e. $\sim$8%. The GB2 data give only a lower limit to the detection rates since the flux-density limit of the sample (250 mJy) is much

greater than 52 mJy necessary to detect a source with logP=
26.4 at z=2.2. The numbers in Table 1 emphasize how the
existing QSO statistics are poor yet.
    b) The data in Sec.2 enable one to calculate the radio de-
tection rate for const R (hereafter the R-parameter is determined between $1.2\,10^{15}$ Hz and $5\,10^{9}$ Hz). Sky densities of the
optical and radio (logR$\geq$3) QSOs in a function of apparent B
or J magnitude, for z<1 and 1<z<2.2, are shown in Fig.1. The

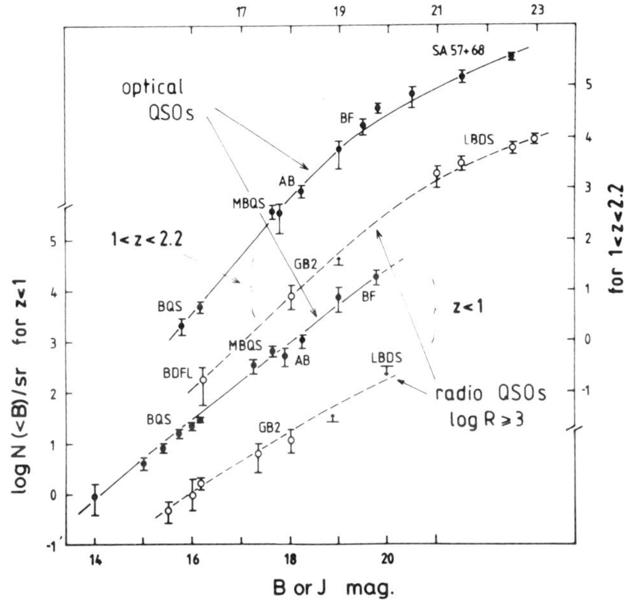

Fig.1. Detection rate of QSOs with
logR≥3 in two intervals of redshift

SA57+68 and LBDS survey's data indicate that there is a rapid
change of differential sky density of both optical and radio
QSOs around 20—21 mag. The detection rate decreases slowly
with decreasing magnitude, from about 10% for the brightest
objects to about 3% at the fainter end. A similar situation
is seen in z<1 range, though there are almost not radio-loud
QSOs fainter than B≈20 mag at z<1. 10:3 ratio of the detection rates at different redshifts suggests a large degree of
correlation between optical and radio luminosities. After
appoximation that the curves in Fig.1 represent detection
rates at <u>constant</u> z, one can transform the N(<B) vs. B relation into the volume density vs. M_B relation. Resultant optical luminos ty functions for both optical and radio QSOs
in two redshift intervals are shown in Fig.4. These functions
clearly show very similar cosmological evolution of the both
populations in the optical domain.

4. DEPENDENCE OF RADIO EMISSION ON REDSHIFT

Another aspect of the radio emission of QSOs is a problem of the variation of detection rate with redshift. Again the data in Sec.2 were used to look for this dependence. Observed sky densities of the optical and radio ($\log P \geq 26.4$) QSOs vs. magnitude for two intervals of M_B are shown in Fig.2. For lower

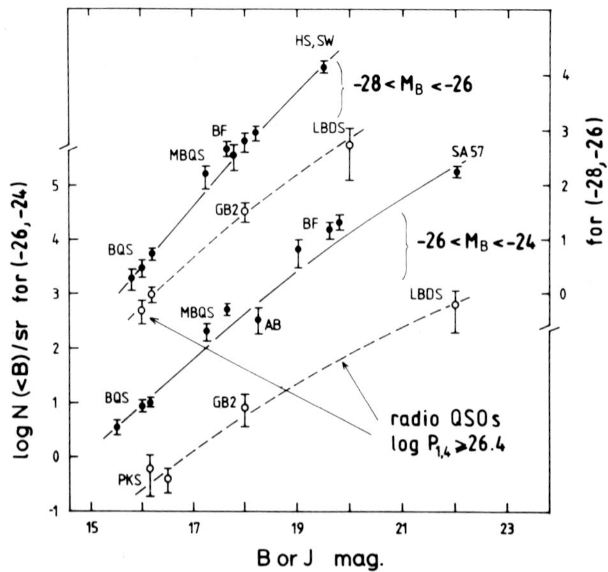

Fig.2. Detection rate of QSOs with $\log P(W/Hz) \geq 26.4$ in two intervals of M_B

luminosities ($-26 < M_B$) the detection rate decreases from about 3% to less than 0.5%. For $M_B < -26$ it changes from 25% to about 2.5%. The cosmological evolution (Fig.4) cannot explain these differences; there must be an intrinsic dependence on both optical luminosity and redshift. In fact, the dependence on M_B is proved in Table 1. It is also clearly seen in Fig.3 which shows $N(<B)$ vs. z relations for optical and radio QSOs with two different M_Bs. Since some samples, used here, are not complete in z, mean redshifts have been computed from B magnitudes assuming <u>constant</u> M_B in Fig.2. The sky density of radio-loud QSOs with $\overline{\log P(W/Hz) \geq 26.4}$, but with different M_B, are almost identical within statistical errors (bars in Fig. 3), though sky density of optical QSOs with the same optical luminosities differ significantly. Thus, the dependence of radio emission on M_B is unquestionable.

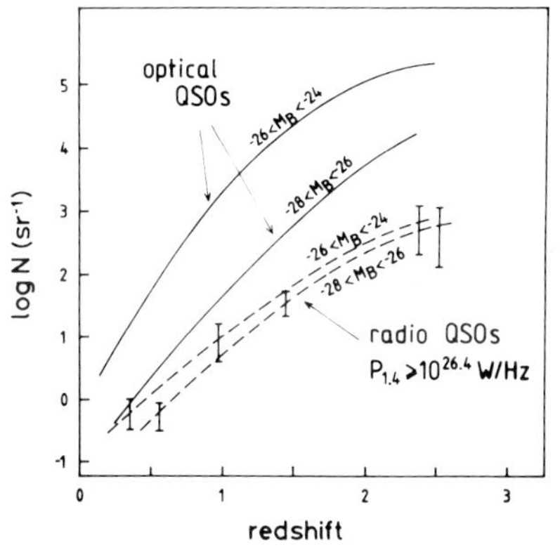

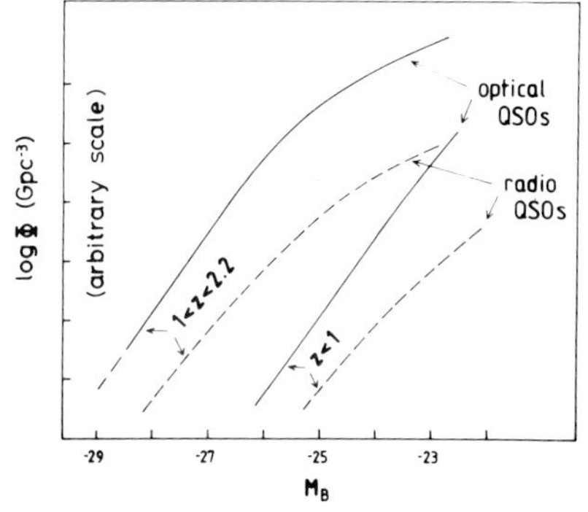

Fig.3. Sky density of QSOs vs. redshift

Fig.4. Luminosity functions for two ranges of redshift

A dependence on z can be interpreted (Peacock et al.1986)
either by different probability distribution of the radio
emission of QSOs at high redshift, or, having postulated two
distinct populations of quasars, by different proportion of
radio-loud to radio-quiet QSOs at high z. A paucity of data
on faint radio QSOs, especially uncompleteness of redshifts,
does not allow now to solve this ambiguity.

CONCLUSIONS

1. Probability of being radio-loud (const L_{radio}, const z)
is strongly dependent on M_B; the detection rate at const R
only slightly decreases with decreasing M_B, implying some
$L_{radio} - L_{opt}$ correlation
2. Probability of radio emission (at const M_B) decreases
with increasing redshift.
3. Evolution functions of radio-quiet and radio-loud QSOs
in the optical domain are almost identical, suggesting some
luminosity evolution also in the radio domain.

REFERENCES

Koo, D.C., Kron, R.G., Cudworth, K.M.: 1986, Publ. astr. Soc.
    Pac. 98, 285.
Kron, R.G., Koo, D.C., Windhorst, R.A.: 1985, Astron. Astrophys.
    146, 38.
Machalski, J., Condon, J.J.: 1986, Astron. J. 91, 998.
Marshall, H.L., Avni, Y., Braccesi, A., Huchra, J.P., Tananbaum,
    H., Zamorani, G., Zitelli, V.: 1984, Astrophys. J. 283, 50.
Mitchell, K.J., Warnock, A., Usher, P.D.: 1984, Astrophys. J.
    (Lett) 287, L3.
Peacock, J.A., Miller, L., Longair, M.S.: 1986, Monthly Not.
    R. astr. Soc. 218, 265.
Schmidt, M., Green, R.F.: 1983, Astrophys. J. 269, 352.

THE SPACE DISTRIBUTION OF FAINT CFHT QUASARS

D. Crampton
Dominion Astrophysical Observatory, Victoria, B.C. Canada
A.P. Cowley
Arizona State University, Tempe, AZ, USA
F.D.A. Hartwick
University of Victoria, Victoria, B.C. Canada

The space distribution of quasars discovered in our CFHT blue grens survey is discussed in detail. Redshifts for about 200 of the quasar candidates show that the sample is relatively complete for $0.2 < z < 3.4$ and $m < 20.5$. Two-thirds of the quasars have $z < 1.8$ and only 5% have $z > 2.5$, indicating that high redshift quasars are rare. The surface density of quasars brighter than $m = 20.5$ is 30 deg^{-2}. Seven quasars with $z = 1.165 \pm 0.007$ discovered in one of the fields have typical separations of ≈ 20 Mpc and may belong to a very large structure. Statistical tests on our data indicate that clustering among quasars is not common, however. The luminosity dependent density evolution models proposed by Schmidt and Green (1983) combined with a redshift cutoff at high redshift are consistent with our data and that of Schmidt and Green (1983), Marshall et al. (1984), Koo, Kron and Cudworth (1986), and Schmidt, Schneider and Gunn (1986). The model indicates that there was a broad maximum in the comoving density of quasars near $z = 1.7$. The results will be reported in detail in the March 1987 issue of The Astrophysical Journal.

A. Hewitt et al. (eds.), Observational Cosmology, 655.
© 1987 by the IAU.

Counts of optically selected quasars in the magnitude range
19<J<22

V.Zitelli(1), B.Marano(1), and G.Zamorani(2)

1- Dipartimento di Astronomia
   via Zamboni 33, 40126 Bologna (Italy)
2- Istituto di Radioastronomia
   via Irnerio 46, 40126 Bologna (Italy)

ABSTRACT. We present the current status of a survey of
faint, optically selected quasars covering 0.7 square
degrees. Candidates have been selected through multicolour
search, grism spectroscopy, and, to some extent,
variability analysis. Slit spectroscopy for almost all the
high priority candidates brighter than J=20.84 yielded 22
confirmed quasars. In the magnitude range J=21-22 our
results are in very good agreement with other existing
surveys. The extension of the survey on an adjacent field
is in progress.

1. INTRODUCTION

A number of surveys are now available at relatively bright
magnitudes (Schmidt and Green 1983; Mitchell, Warnock and
Usher 1984; Marshall et al. 1984), providing the basis for
the study of the quasars' luminosity function and its
evolution for redshift smaller than 2.2. At faint
magnitudes (B>20) the available information from complete
samples has grown up recently, with the use of 4m class
telescopes for both search and spectroscopic follow-up (Koo
et al. 1986, Zamorani et al. 1986, Shanks et al. 1986).
After the first direct indication of a significant
flattening of the counts of quasar candidates at magnitudes
fainter than B=20 (Koo and Kron 1982), these spectroscopic
surveys are providing the detailed information necessary to
a complete description of the evolution of the quasar
luminosity function at lower intrinsic luminosity. In this

Based on observations collected at the European Southern
Observatory, La Silla, Chile.

A. Hewitt et al. (eds.), Observational Cosmology, 657–659.

paper we present the current status of our work on a field
of about 0.7 square degrees, where candidates in the
magnitude range J=18-22 (J=B-0.1) have been selected
through different criteria, namely colours, appearance on
grism plates, and variability.

2. THE SELECTION OF THE CANDIDATES

From two sets of IIIa-J and IIIa-F plates (U, J, F)
obtained in 1981 and 1982 at the 3.6m ESO telescope in La
Silla and their PDS scanning, we have extracted a working
list of about 6000 objects complete to J=22.0.
     The observations and the data reductions are described
in Marano et al. 1986a and 1986b. Using three different
shape parameters we have classified all the objects of the
working list into three classes: "stellar", "intermediate"
and "extended" and we have constructed the colour-colour
diagrams for the three classes. Then we have selected our
quasar candidates following the criteria suggested by Koo
and Kron (1982). In order not to introduce a potential
bias against the selection of low redshift quasars, which
may show some fuzziness on our plates, we have added to the
list of candidates some objects selected from the classes
of intermediate and extended objects.
     In the same field we have also obtained at the 3.6m
ESO telescope three IIIa-J grism plates, with a dispersion
of about 2000 A/mm. The wavelength range (3400 - 5300 A)
is such that the $Ly\alpha$ line would not be detected at
redshifts higher than 3.3 . The two best among the three
plates were visually inspected by each of us . The merging
of the obtained lists yielded 146 objects. Candidates have
then been extracted from this list adopting the criterion
that a possible emission line was visible in the
microphometric tracing of the spectrum. We are now
planning to repeat the grism search adopting objective,
machine-defined selection criteria.
     In addition, we have searched for variability all the
stellar objects in our field and a number of variable
objects has been added to the final list of candidates.

3. RESULTS

Up to now we have obtained spectra for almost all the
candidates brighter than J=20.84 using partly the Boller
and Chivens spectrograph and partly the EFOSC at the ESO
3.6m. A summary of the results is the following:
     a) Down to J=20.84 we have 22 broad emission lines
objects, corresponding to a surface density of 32 quasars
per square degree.

b) The redshift distribution covers the range $0.6<z<2.8$, with 8 objects with redshift larger than 2.

c) Three high redshift objects ($z>2.0$) were missed by our colour selection, while three relatively low redshift quasars ($z<1.0$) were missed by our selection from the grism plates. This result shows that, within our present magnitude limit, the grism selection can be as efficient as the colour selection in finding quasars (see Crampton et al. 1985). The complementarity of the losses of the two selection techniques stresses the usefulness of applying them at the same time on the same field in order to increase the combined level of completeness.

d) Variability selection has had a much lower success rate (5 variable quasars out of the 22 confirmed); however, given the small time baseline covered by our survey, a 20% success rate should be regarded as highly encouraging.

e) In the magnitude range $J=21-22$, our counts of quasars and quasar candidates are in good agreement with the results of similar surveys (see Koo et al. 1986). At brighter magnitudes ($J=19-20$), instead, our counts are lower (by a factor 1.5-2.0) than those in the Braccesi field (see Marshall et al. 1984). This discrepancy, which at present is only marginally significant, will be checked by adding new data from an adjacent field on which we are currently working.

REFERENCES

Crampton,D., Schade,D., and Cowley,A.P. 1985, A.J., 90, 987
Koo,D.C., and Kron,R.G. 1982, Astr.Ap., 105, 107.
Koo,D.C., Kron,R.G., and Cudworth,K.M. 1986, P.A.S.P., 98, 285.
Marano,B., Zamorani,G., and Zitelli,V., 1986a, in "Structure and Evolution of Active Galactic Nuclei", G.Giuricin et al. eds., (Dordrecht:Reidel), 1985, p. 339
Marano,B., Zamorani,G., and Zitelli,V., 1986b, Proc. of the "Convegno Nazionale Astronet 1984-1985", G.Sedmak ed., p. 401.
Marshall,H.L., Avni,Y., Braccesi,A., Huchra,J.P., Tananbaum,H., Zamorani,G., and Zitelli,V. 1984, Ap.J., 283, 50.
Mitchell,K.J., Warnock III,A., and Usher,P.D. 1984, Ap.J.(Letters), 287, L3.
Schmidt,M., and Green,R.F. 1983, Ap.J., 269, 352.
Shanks,T., Fong,R., and Boyle,B.J. 1986, Proc. of IAU Symp. No. 119 on "Quasars", G.Swarup and V.K. Kapahi eds., (Dordrecht: Reidel), p. 37.
Zamorani,G., Zitelli,V., and Marano,B. 1986, Proc. of IAU Symp. No. 119 on "Quasars", G.Swarup and V.K. Kapahi eds., (Dordrecht: Reidel), p. 33.

A MULTI-COLOUR SURVEY FOR HIGH-REDSHIFT QUASARS

S.J. Warren, P.C. Hewett and M.J. Irwin
Institute of Astronomy
Madingley Road
Cambridge   CB3 OHA
United Kingdom

ABSTRACT. We describe how the application of techniques which utilise
all the information contained in multi-colour surveys of stellar-like
objects can be employed to detect quasars over an extended redshift
range. The method is particularly effective for the identification of
quasars with redshifts z>2.2 where the application of the ultra-violet
excess criterion breaks down. Spectroscopy of a small sample of
objects from the first survey field has resulted in the detection of
two new faint quasars with redshifts z=3.42 and z=4.01 (ref. 1).

1.   INTRODUCTION.

The identification of intrinsically bright, high-redshift quasars by
Hazard and coworkers (ref. 2), contrasts with the apparent lack of
large numbers of intrinsically faint high-redshift quasars established
by Koo and others (ref. 3-5). The space density and luminosity
function of quasars at high redshift remains of considerable
cosmological interest, and there is a need for a wide angle, relatively
deep survey to explore the absolute magnitude range between those
covered by Hazard and Koo.

The study of broad-band optical colours offers the potential for
identifying many types of quasar with a large range of redshift owing
to the wide wavelength coverage 3300-8500Å.   The combination of
Automated Plate Measuring Facility (APM) measures of United Kingdom
Schmidt Telescope (UKST) plates is ideally suited for such a survey.
Each plate covers a useable area of 30 square degrees, a factor 100
larger than the area of the very deep 4-metre surveys, while the
magnitude limit extends some one and a half magnitudes fainter than the
objective-prism surveys of Hazard et al. The use of direct plates
results in simpler selection effects compared to slitless spectroscopic
techniques.

A. Hewitt et al. (eds.), Observational Cosmology, 661–664.

2. THE MULTI-COLOUR SURVEY.

Object magnitudes in the U,J,V,R and I photometric bands are derived
from APM measures of UKST direct plates. Previous quasar searches have
proceeded by identifying objects lying away from the well defined
stellar sequence evident in colour-colour diagrams. Substantial
improvements in the effectiveness of the quasar search can be realised
by using the full four-dimensional (4D) colour space rather than the
conventional two-dimensional (2D) colour-colour projections.

The location of each object in the five dimensional magnitude
space can be represented by a five component vector. If the apparent
brightness of an object is not to be incorporated as a parameter in the
search for quasars - i.e. two objects with the same colours but with
different apparent magnitudes are regarded as identical -,one of the
five dimensions becomes redundant. The most familiar representation of
the resultant 4D space is in terms of the colours U-J,J-V,V-R and R-I.

The use of the U-J,J-V,V-R and R-I colours as axes describing the
4D space is not ideal because of the high degree of correlation between
the axes. The optimal representation of the 4D space is one where the
axes are normalised and mutually orthogonal. In the case of the
familiar 2D parameter space employing U,B and V data it is easy to show
that the axes $(U-V)*\sqrt{3}/2$ and $B-(U+V)/2$ satisfy these criteria. It may
be noted that Koo (ref. 6), using an astrophysical argument, arrives at
a similar conclusion concerning the representation of two-colour data:
nevertheless his axes, $(U-V)$ and $B-(U+V)/2$, whereas they are
orthogonal, have not been normalised.

The procedure is readily extended to three and four (or higher)
dimensions. Figure 1 shows a three dimensional plot of the J,V,R,I
multi-colour space, where the axes $x=(R-V)*\sqrt{2}/\sqrt{3}$, $y=(I+J-V-R)/\sqrt{3}$ and
$z=(I-J)*\sqrt{2}/\sqrt{3}$ are normalised and mutually orthogonal, together with a
2D projection of the same data. The positions of 14000 stellar objects
in a field at the South Galactic Pole field at 0h 53m, -28 03' are
plotted. In addition, the locations of some 35 known quasars in the
field are shown, including three previously identified quasars with
redshifts z>3.3 and 0046-293, the recently discovered z=4 quasar.

The vast majority of stellar objects are stars within our own
galaxy which exhibit a very restricted range of properties, and hence
are confined to a small region of the multi-colour space. Quasars, by
contrast, are widely dispersed and may be recognised by their locations
in low-density regions of the multi-colour space. By surveying all
objects in low-density portions of the parameter space all quasars
except those that mimic closely the properties of common stars can be
identified. The advantage of performing such a search in 4D rather
than in 2D lies in the observation that the fraction of the total
volume of a parameter space occupied by common stars becomes smaller
with the addition of each independent dimension. The fraction of the
4D colour space (which is more difficult to illustrate) occupied by
stars is even smaller than that in the 3D colour space shown in Figure 1.

In a project to determine much improved limits on the space
density of high-redshift z>3 quasars we have started a survey of three
UKST fields, a total of 100 square degrees to limiting magnitudes

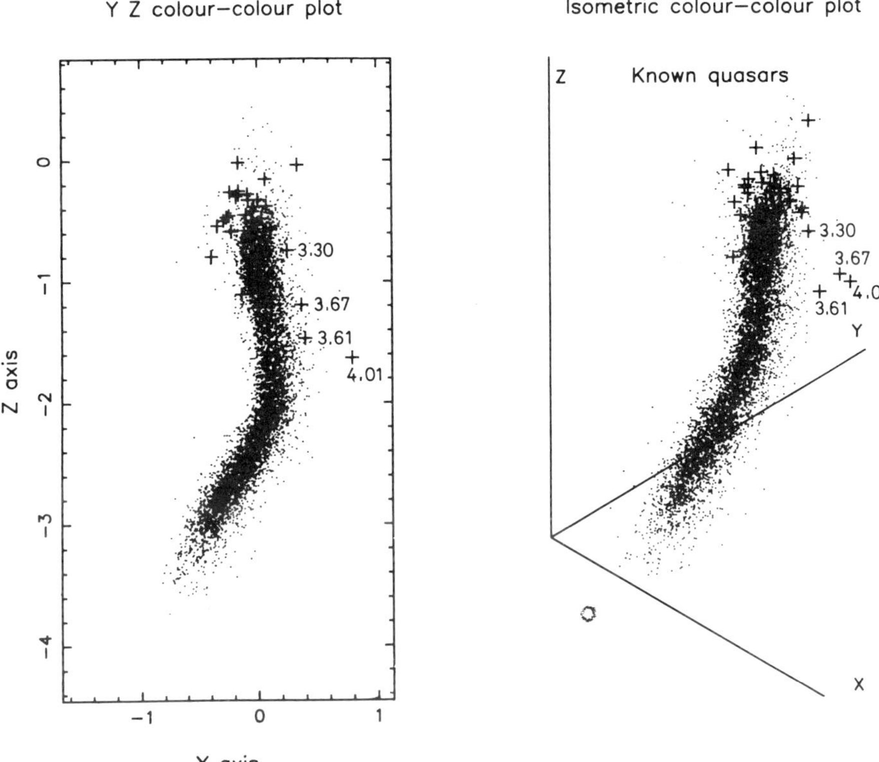

Figure 1.   2D and 3D representations of the multi-colour data, showing
the locations of known quasars in the SGP field.

m_J=21.5, m_R=20.0.   Observations of 12 candidates from our first survey
field resulted in the discovery of a redshift z=3.42 quasar, with
broad-band magnitudes J=20.7, V=19.4, R=19.7 and I>19.3, together with
a redshift z=4.01 quasar with broad-band magnitudes J=21.0, V=19.5,
R=19.2 and I=19.0.   The discovery of these two quasars demonstrates
that faint high-redshift quasars do exist, but reliable limits on the
space density of such objects will have to await the completion of the
survey.

REFERENCES

1. Warren, S.J., Hewett, P.C., Irwin, M.J., McMahon, R.G., Bridgeland,
        M.T., Bunclark, P.S. and Kibblewhite, E.J. Nature (in the press).
2. Hazard, C., McMahon, R.G. and Sargent, W.L.W. Nature 322, 38-40
        (1986).
3. Koo, D.C., Kron, R.G. and Cudworth, K.M. P.A.S.P. 98, 285-306
        (1986).
4. Osmer, P.S. Astrophys. J. 253, 28-37 (1982).
5. Schmidt, M., Scheider, D.P. and Gunn, J.E. Astrophys. J. (in the
        press).
6. Koo, D.C. Astronom. J. 90,(3),418-440 (1985).

DISCUSSION

AUDOUZE: What is the maximum redshift you can anticipate detecting with your technique?

HEWETT: The detection of quasars with redshifts $z > 4.5$ will be difficult as the objects will be red in all the optical colours. In principle, however, any object with a detectable flux in the optical, and which does not closely mimic a star, could be identified.

THE INFRARED SPECTRA OF QUASARS - A LUMINOSITY DEPENDENCE

Beverley J. Wills
McDonald Observatory and Department of Astronomy
University of Texas, RLM 15.308
Austin, Texas   78712
U.S.A.

The IR-optical-UV continua of quasars are often represented by two
components: (i) a flat spectrum component dominating in the optical-UV
(the "Big Bump") and sometimes attributed to thermal radiation from an
accretion disk with temperatures of about 20000 to 40000 K - we will
call it the "disk" component - and (ii) a near IR component
characterized by a steep rise, $\alpha \sim 1$ for $\lambda > 1$ μm, often thought to be
a synchrotron spectrum - an extrapolation of the cm or mm wavelength
radio spectrum - although some have preferred an explanation in terms
of thermal re-radiation of the ionizing continuum by hot dust (e.g.,
Hyland and Allen 1982, Neugebauer et al. 1979).

We have undertaken a survey of the near IR spectra of about 40
optically bright radio quasars ($< 17^m$) with redshifts < 1.3, using the
NASA-University of Hawaii IRTF 3.0 m telescope. We have measured most
quasars between 1.25 and 5 μm, and many at 10 and 20 μm. The rms
uncertainties range from 3 to 6 % at 1.25 to 2.2 μm, and increase with
wavelength. We have combined these with IRAS data, where available, at
12, 25, 60 and 100 μm (although these observations were often not
simultaneous with the near IR). The IR spectra of the optically violent
variable (OVV) quasars, when bright, are like the BL Lac spectra, with a
variable power-law spectrum in the near IR ($0.8 < \alpha < 2$) and are well
studied by others also. Here I report results for the non-OVV quasars.

**The near IR spectra are remarkably similar from one quasar to
another.** Typical spectra are shown in Fig. 1. F_ν vs. ν is plotted in
the upper panels. In 16 of 22 non-OVV quasars **the spectra between rest
wavelengths 1.5-3 μm have $\alpha = 1.7$** within ± 0.2 and 4 others are
consistent with this but have larger uncertainties. This is much
steeper than the canonical $\alpha \sim 1$. The lower panels show the same data
plotted as the power per unit logarithmic frequency interval (νL_ν) as a
function of ν. They show that **the near IR power output peaks
between 3 and 3.5 μm.** (Neugebauer et al. (1979) show this '3 μm bump'
for the quasar 3C273). The power in this IR component is about the
same as the sum of the power emitted in the broad emission lines.

This similarity in the shape of quasars' spectral energy
distribution has been noted before for the optical-UV continua > 1200 A
(e.g., Soifer et al. 1983) and suggests that the spectral components are

A. Hewitt et al. (eds.), Observational Cosmology, 665–668.

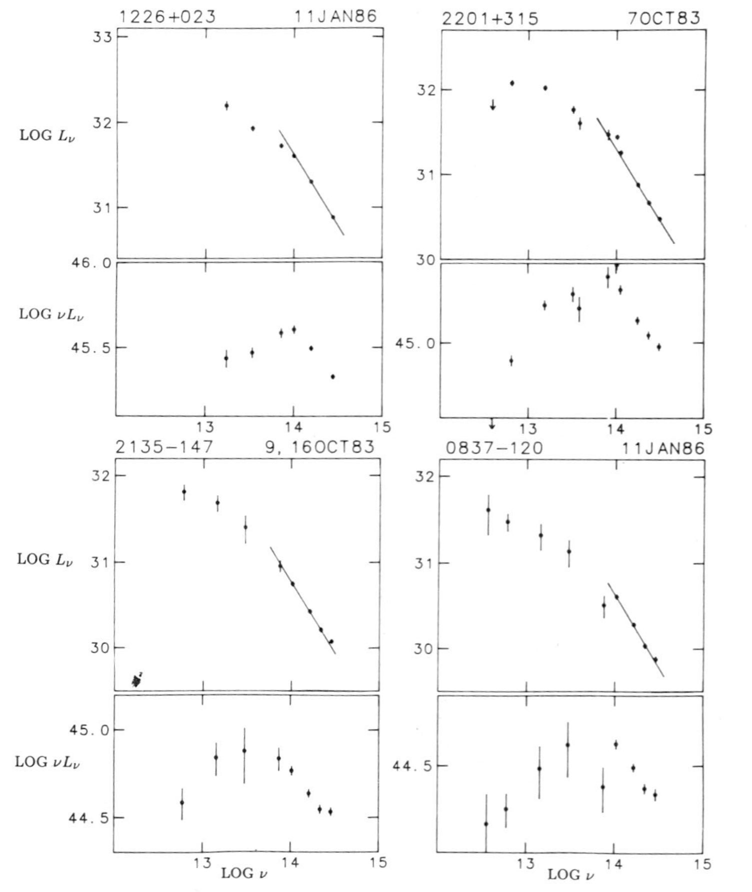

Figure 1. Typical IR spectra for non-OVV quasars. A power-law of $\alpha = 1.7$ is shown in each upper panel. Cgs units are used, frequencies are in the rest frame and error bars are $\pm 1\sigma$.

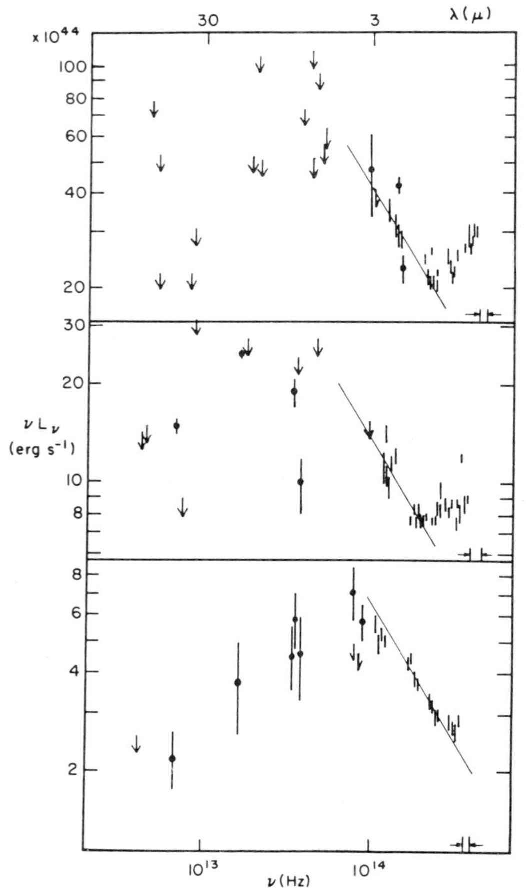

Figure 2. Composite spectra representing different average luminosities.

also very similar. Addition of a BL Lac-type power-law spectrum would destroy the observed similarity, so its contribution must be small.

The 3 μm bump may be either free-free emission from high density gas, $n_e \sim 10^{11}$ cm^{-3} (Puetter and Hubbard 1985), or thermal radiation from dust at 1200 K heated by the strong optical-UV continuum. This is close to the expected evaporation temperature of the grains, and nicely explains the similar spectra.

Figure 3 shows composite spectra for non-OVV quasars plotted on a rest wavelength scale, and for three luminosity ranges. Luminosity is defined at 1 μm and we use $H_0 = 100$ km s^{-1}Mpc^{-1} and $q_0 = 0$. The spectrum in the lower panel is representative of 12 low luminosity quasars of which 4 are shown. The middle panel combines 6 spectra, and the upper panel, 7. Compare the spectra at $\lambda_0 = 1$ μm. **The relative strength of the disk component** (with $\alpha \sim 0$, a slope of +1 on this plot) **increases with increasing 1 μm luminosity.** Extrapolating the 1.7 slope to 1 μm we can roughly separate the "3 μm component" and the disk component. If we assume a power law relation between the strengths of the two components we find

$$L_\nu (3 \ \mu\text{m component}) \propto L_\nu (\text{disk})^{+0.5}$$

Interestingly this is somewhat reminiscent of the Baldwin relation between the strength of the CIV λ1549 line and the underlying continuum-the continuum of the same disk component to which we are referring here.

Explanations for the luminosity dependence may be the same as those proposed for the Baldwin effect, i.e., differences in the disk continuum-either the result of time-variability or different inclinations of the disk to the line-of-sight (Netzer 1985, and references therein).

A more detailed account of this work may be found in Wills (1986).

I thank my collaborators: D. Wills, H. Netzer and D.F. Lester, also the support staff of the Infra Red Telescope Facility (operated by the University of Hawaii under contract to the National Aeronautics and Space Administration), and of the Infrared Processing and Analysis Center in Pasadena, for help in obtaining Infra Red Astronomy Satellite data. This research is supported by the U.S. National Science Foundation (Grant AST-8215477). My travel to Beijing was made possible by an American Astronomical Society travel grant.

REFERENCES

Hyland, A.R., and Allen, D.A. 1982, MNRAS, **199**, 943.
Netzer, H. 1985, MNRAS, **216**, 63.
Neugebauer, G., Oke, J.B., Becklin, E.E., and Matthews, K. 1979, Ap.J.,
      230, 79.
Puetter, R.C., and Hubbard, E.N. 1985, Ap.J., **295**, 394.
Soifer, B.T., Neugebauer, G., Oke, J.B., Matthews, K., and Lacy, J.H.
      1983, Ap.J., **265**, 18.
Wills, B.J. 1986, Ap.J., submitted.

DISCUSSION

ROWAN-ROBINSON: If the 3μ feature is to be interpreted as radiation
from dust then from the narrow width of the feature we can deduce that
the dust must be in the form of a shell of rather narrow linear extent
(it can also not have very high optical depth in dust). For dust
located at a fixed distance from the central source, we would also
expect a significant variation in the spectrum with the luminosity of
the quasar. For a very luminous quasar, the dust may be totally
destroyed and the 3μ feature might be expected to disappear.

SPECTROSCOPY OF QSOs FROM THE TEXAS RADIO SURVEY

D. Wills   and   Beverley J. Wills,
McDonald Observatory and Department of Astronomy,
University of Texas, Austin, Texas 78712, U.S.A.

We report briefly on our optical spectroscopy of QSOs from the radio
survey in progress at the University of Texas Radio Astronomy
Observatory (UTRAO).   The radio survey, under the direction of Dr
J. N. Douglas, covers declinations north of -36 degrees, in strips
that are 9 degrees wide in declination. Preliminary results for the
strip centered at +18 degrees have been published (Douglas et al.
1980) and other strips are currently being prepared for publication.
The complete catalogue will list 1 arcsec positions and some structure
information for about 75,000 sources stronger than about 250 mJy at
365 MHz.   Optical identifications are made from the Palomar
Observatory Sky Survey plates using an interactive laser measuring
machine (Ghigo 1977), and the accuracy of the radio source positions
allows the identifications to be made regardless of the colours of the
objects. Optical spectroscopy of a sample of about 300 QSO candidates
brighter than 19 mag, drawn from the +36, +18 and -12 degree strips,
is being carried out at McDonald Observatory using an Intensified
Dissector Scanner at the cassegrain focus of the 2.7m reflector. Two
image-tube and dissector chains are available, together covering the
range 3200 - 8500 A; a typical spectrum takes 1 hour to obtain, with
wavelength resolution of 10 A over a total range of 3000 A.

So far, we have new redshifts for 111 QSOs and probable redshifts
for 28 others, while about 20 more objects show only a single emission
line in the spectra that we have obtained so far. We are trying to be
spectroscopically complete because the goal of the study is an
improved radio-optical luminosity function for radio-selected QSOs;
this, of course, takes time because the 'difficult' redshifts need
more than one spectrum (often with both image-tube chains), while
redshifts around z = 2 take much less time to observe. The 111
redshifts have been communicated to A. Hewitt and G.R. Burbidge for
inclusion in their updated QSO catalogue. While the spectroscopy must
be completed before we can draw any firm conclusions about the radio-
optical luminosity function and its evolution with redshift, we can
make some preliminary remarks.

A. Hewitt et al. (eds.), Observational Cosmology, 669–672.
© *1987 by the IAU.*

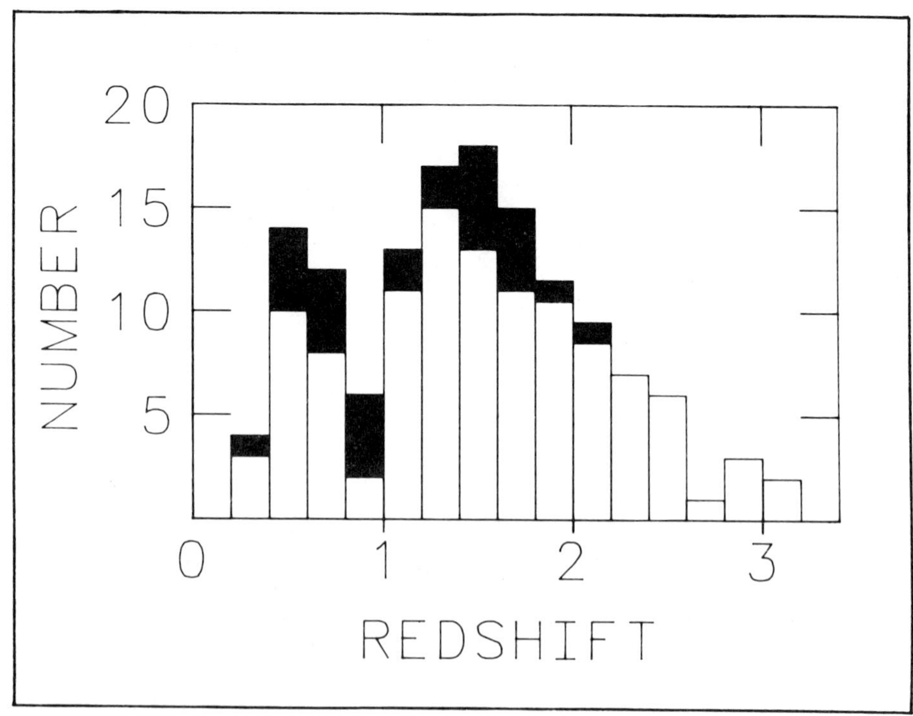

Figure 1. Histogram of new redshifts for UTRAO QSOs

     The redshift histogram to date is shown in Figure 1, with the
'probable' values shaded; from our previous experience, about 70% of
these will turn out to be correct when we obtain better spectra. It
is clear that the probable redshifts tend to be smaller than the
certain ones, at about the 1% significance level, for reasons well
known to most QSO spectroscopists: redshifts of 1.7 and above are
usually unambiguous because of the strong lines of Lyman-α and C IV
(1216, 1549 A), while at redshifts near 0.7 only Mg II is a reliable
line on survey-quality spectra (the C III] line is weaker).
     The median redshift of the objects in Figure 1 is about 1.5,
which is significantly higher than those for earlier samples of QSOs
selected at low radio frequencies, e.g. 0.9 - 1.1 for the 3C and 4C
samples discussed by Wills and Lynds (1978). The optical limits of
all these samples are similar, and part of the difference is a result
of the fainter radio limit for the UTRAO sample, but there may be some
selection effects that will be revealed only after the completion of
the spectroscopy. First, as already mentioned, the lower redshifts
are more difficult spectroscopically, and many of the one-line objects
probably fall in this category, together with some objects that have
so far revealed no lines at all in their spectra. Second, the

complete sample includes objects whose redshifts were known already and which should therefore be added to Figure 1; their inclusion actually makes little difference to the median redshift. Third, the UTRAO survey was made with an interferometer and therefore discriminates against sources with angular diameters larger than about 1 arcmin, which tend to be those of lower redshift. If we make the pessimistic assumption that the UTRAO survey misses 10% of all QSOs above 250 mJy, and that these excluded objects all have redshifts below z = 1, and we further assume that all our one-line objects are below z = 1, then the median redshift is marginally consistent with those for the 4C samples analysed by Wills and Lynds.

BL Lac objects are rather rare among our QSO candidates, as is true for most low-frequency radio surveys; we do occasionally find them by their lineless spectra and significant brightness changes, and have confirmed most of them by their high optical polarisation. There are, surprisingly, no new coincidences between UTRAO sources and objects in the General Catalogue of Variable Stars (see Wills et al. 1986), although 0716+332 brightened from 20 mag to about 15 mag earlier this year (Wills 1986).

Because at least two of the astronomers involved in the organization of this meeting (Professors Fang LiZhi and Chu YaoQuan) have investigated periodicities in QSO redshift distributions, we thought it would be appropriate to examine this new sample for such effects. We have therefore used the 111 definite redshifts, supplemented in some cases by the 28 probable redshifts and/or the 100 or so previously-known redshifts for QSOs in the whole sample, to search for significant periodicities in various functions of redshift - specifically redshift z, log (1+z), and (1-1/√(1+z)) (cf. Fang et al. 1982). In none of the samples is there any evidence for periodicities in any of these functions at the 5% significance level.

We thank our colleagues in the Texas radio astronomy group, especially Dr J.N. Douglas, for access to the radio survey and optical identifications, and we acknowledge financial support from the National Science Foundation (grant AST-8215477). B.J. Wills thanks the American Astronomical Society for a travel grant to attend this symposium.

REFERENCES

Douglas, J.N., Bash, F.N., Torrence, G.W. and Wolfe, C. 1980, Univ. Texas Publ. Astron. No. 17.
Fang, L.Z., Chu, Y.Q., Liu, Y. and Cao, Ch. 1982, Astr. Ap. **106**, 287.
Ghigo, F.D. 1977, Ap. J. Suppl. **35**, 359.
Wills, B.J. 1986, IAU Circ. No. 4195.
Wills, D., Bozyan, F.A., Douglas, J.N. and Dove, A.P. 1986, A.J. **92**, 412.
Wills, D. and Lynds, R. 1978, Ap. J. Suppl. **36**, 317.

DISCUSSION

HAWKINS: Your discovery of an OVV quasar is interesting since in the
complete sample of 400 variable quasars, none has an amplitude greater
than 1 mag, and none appears to be an OVV object.

WILLS: That's interesting. We have a few at the 2 mag level, but
0716+332 is an extreme case. We tried it perhaps five times over the
last few years, when it was about 20 mag, and would have given up on
it, except that the telescope dome was misbehaving and I decided to
clamp it in the west and observe objects that pass close to our zenith.
Incidentally, it has a flat radio spectrum, unlike most of the UTRAO
sources (it is a member of the sample discussed by us at IAU #119 - see
page 499 of those Proceedings).

A COMPLETE SAMPLE OF FLAT-SPECTRUM RADIO SOURCES FROM
THE PARKES 2.7 GHz SURVEY

A. Savage,[a] C.J. Chandler,[a] David L. Jauncey,[b] Michael J.
Batty,[b] Graeme L. White,[b] B.A. Peterson,[c] W.L. Peters,[c]
S. Gulkis,[d] and J.J. Condon[e]

ABSTRACT. We describe our complete sample of flat-spectrum radio sources
with fluxes >0.5 Jy selected from the Parkes 2700 MHz catalogue. The
sample covers all right ascensions and declinations from +10° to -45°,
but excluding the galactic plane ($b < 10°$), and contains some 400 sources.

1. INTRODUCTION

We are investigating a sample of flat-spectrum radio sources drawn
from the Parkes 2.7 GHz survey (see Bolton et al. 1979 and references
therein) complete to 0.5 Jy. The sample covers 4.5 sr of sky and comprises some 406 sources. Accurate radio positions are being measured
(McEwan et al. 1975; Condon et al. 1977, 1978; Jauncey et al. 1982) for
all sources and optical identifications are being made from the UKST
IIIa-J sky survey. Redshifts are being sought with the Anglo-Australian
telescope (see e.g. Jauncey et al. 1984 and references therein), with
results so far for 245 sources. The program is aimed at determining,
in an unbiased manner, the space distribution of quasars with redshifts
over a large area of sky.

2. THE SURVEY

All sources have been selected from the Parkes 2.7 GHz survey and
cover the declination range +10° to -45°, excluding the region within
10° of the galactic plane. The basic sample comprises both the steep-
and flat-spectrum sources, but we are concentrating initially on the
flat-spectrum sources. The compact nature of these radio sources means
that accurate radio positions can be readily measured and thus unique
optical identifications, based on positional coincidence alone, can be
made to the 22.5 mag limit of the J survey (Jauncey et al. 1982).

This is a long-term program, under way now for about 15 years, and
making use of a wide variety of telescopes, both optical and radio.

---

[a]Royal Observatory, Edinburgh; [b]Division of Radiophysics, CSIRO,
Australia; [c]Australian National University; [d]Jet Propulsion Laboratory,
California Institute of Technology; [e]National Radio Astronomy
Observatory, USA.

A. Hewitt et al. (eds.), Observational Cosmology, 673–675.
© 1987 by the IAU.

There are a number of advantages for the present sample:
1. The survey covers a large area of the sky and thus avoids small
    area effects.
2. The large sample size reduces statistical fluctuations and provides
    reliable information on the minority populations, e.g. BL Lac
    objects, etc.
3. Accurate radio positions are being measured for all objects, allowing
    reliable optical identification to the 22.5 mag limit of the UKST
    IIIa-J survey without any reliance on colour or morphology. This is
    particularly important in avoiding redshift-dependent effects of the
    type found in optically selected samples based on colour selection
    or objective prism data.
4. The identification rate is high, about 80% (Jauncey et al. 1982),
    and the objects are identified primarily with quasars.
5. Redshifts are being sought for all objects to the 22.5 mag limit
    with the AAT. This, in turn, will provide an unbiased determination
    of the space distribution of quasars to the largest redshifts yet
    found for radio selected objects (Peterson et al. 1982).

The advantage of radio selected samples over optically selected
samples cannot be over-emphasized, since the radio samples are complete
to a well-determined flux density limit and are free from the redshift-
colour selection effects that plague optical samples. For example,
Table I shows the colours that have been obtained now for four quasars
with redshifts >3.5. Three of these are found to lie very close to the
region defined by normal stars in both the J-V versus V-R and J-R versus
R-I two-colour diagrams. Thus, although quasars with $z > 3.5$ can be
found by colour selection, the high star contamination makes it very
difficult to assemble complete and statistically significant optical
samples of such objects.

TABLE I - Colours of some high-redshift quasars.

Name	z	I	R	V	B	U
PKS 2000-330[a]	3.78	16.99	17.17	17.17	19.20	>22.0
HAZ 0055-2659[b]	3.68	17.30	17.46	17.91	19.06	18.97
DHM 0054-284[c]	3.61	17.68	18.15	18.61	20.21	>22.0
PKS Anon[d]	3.78	16.47	16.77	17.47	19.05	>22.0

[a]Peterson et al. (1982); [b]Hazard and McMahon (1985); [c]Shanks et al.
(1983); [d]G.L. White et al. (unpublished data).

3.   PRELIMINARY RESULTS
     The faint flux density limit and the population of flat-spectrum
quasars allows us to penetrate a larger volume of space than the 3CR
sample of steep-spectrum low-frequency selected quasars. Recent observations on the AAT have doubled the number (from 2 to 4, or 3% of the
total number with redshifts) of quasars with redshifts >3.5 in this
sample. In all, the present flat-spectrum sample shows 43 (17.6%) quasars with redshifts >2, compared with 2% for 3CR.

No sharp cutoff is apparent in the present redshift distribution, but rather a steady turndown above a redshift of 2. It appears that the higher redshift quasars are to be found predominantly amongst the weaker radio sources and amongst the fainter identifications, since there is a weak but significant correlation (with a large scatter) of redshift with radio and optical flux density. In particular four radio quasars with redshifts above 3.5 have J magnitudes ranging from 18.6 to 22.5, a range of four magnitudes for essentially the same redshift. As redshifts are now being obtained for the predominantly fainter identifications, we expect to find more high-redshift quasars in our sample.

We also expect to see in the Hubble diagram some effect for the highest-redshift objects caused by an increasing density of absorption lines in the Lyman forest (see e.g. Hunstead et al. 1986). This may cause a dramatic decrease in the brightness of these objects in the J band.

A preliminary analysis of the present database indicates that there is no significant difference in the number-magnitude or number-redshift distributions for the flat- and steep-spectrum radio quasars, in agreement with the results for the stronger sources as found by Wall and Peacock (1985).

Acknowledgment. We would again like to thank the staff of the UK Schmidt Telescope for taking such superb plate material. This paper represents the results of one phase of research carried out at the Jet Propulsion Laboratory, California Institute of Technology, under contract No. NAS 7-100.

References

Bolton, J.G., Savage, A. and Wright, A.E., 1979. Aust. J. Phys. Astrophys.. Suppl. No. 46.

Condon, J.J., Hicks, P.D. and Jauncey, D.L., 1977. Astron. J. 82, 692.

Condon, J.J., Jauncey, D.L. and Wright, A.E., 1978, Astron. J. 83, 1036.

Hazard, C. and McMahon, R., 1985. Nature, 314, 238.

Hunstead, R.W., Murdoch, H.S., Peterson, B.A., Blades, J.C., Jauncey, D.L., Wright, A.E., Pettini, M. and Savage, A., 1986. Astrophys. J. 305, 496.

Jauncey, D.L., Batty, M.J., Gulkis, S. and Savage, A., 1982. Astron. J. 87, 763.

Jauncey, D.L., Batty, M.J., Wright, A.E., Peterson, B.A. and Savage, A., 1984. Astrophys. J. 286, 498.

McEwan, N.J., Browne, I.W.A. and Crowther, J.H., 1975. Mem. R. Astron. Soc. 80, 1.

Peterson, B.A., Savage, A., Jauncey, D.L. and Wright, A.E., 1982. Astrophys. J. 260, L27.

Shanks, T., Fong, R. and Boyle, B.J., 1983. Nature 303, 156.

Wall, J.V. and Peacock, J.A., 1985. Mon. Not. R. Astron. Soc. 216, 173.

THE EVOLUTION OF QUASARS SELECTED BY SLITLESS TECHNIQUE

Z. G. Deng
Department of Physics, Graduate School,
Academia Sinica, Beijing, China.
Y. Y. Zhou
Center for Astrophysics, University of Science and
Technology of China, Hefei, China.
Y. Z. Liu
Department of Physics, Graduate School,
Academia Sinica, Beijing, China.

ABSTRACT. After taking account of the selection effects in the identification of emission lines and choosing the sample within a narrow range of absolute magnitude, we can investigate the evolutionary function of quasars from their redshift distribution. From data given by slitless surveys with limiting apparent magnitude 19 .5, we find that the evolutionary function takes form of $\rho = \rho_0(1+z)^{6.5\pm1}$. The analysis has also showed that the observational redshift distribution of quasars is compatible with cosmological principle.

1. INTRODUCTION

Quasars have become a powerful tool to explore the space-time structure of the universe. To this end, an important task is determining their evolutionary function. Many authors have studied the evolution of quasars by different methods in various wavebands (Schmidt, 1968, 1974; Veron, 1983; Koo and Kron, 1982; Schmidt and Green, 1983). In order to obtain more reliable evolutionary function, we would consider the selection effects appearing in the identification of emission lines carefully.

In this work, we shall use the method developed before (Zhou, Deng, and Dai, 1985) and eliminate some other selection effects in the surveys to investigate the evolutionary function of quasars selected by slitless technique from their redshift distribution.

2. ANALYSES OF SELECTION EFFECTS

The observational redshift distribution of quasars can be expressed as
$$f(z) = P(z)R(z), \tag{1}$$
where, P(z) is the real redshift distribution of quasars and R(z) is a factor caused by selection effects depending on the redshift z. Under the standard model of the universe with $\Lambda = 0$, we have

A. Hewitt et al. (eds.), Observational Cosmology, 677–679.

$$P(z) = \begin{cases} \dfrac{4\pi c^3}{H_o^3} \dfrac{[zq_o+(q_o-1)(-1+\sqrt{2q_oz+1})]^2}{q_o^4(1+z)^6(1+2q_oz)^{\frac{1}{2}}} n(z), & q_o \neq 0, \\[2em] \dfrac{4\pi c^3}{H_o^3} z^2(1+z/2)^2(1+z)^{-6}n(z), & q_o = 0, \end{cases} \qquad (2)$$

where, $n(z)$ is a factor relating to the evolution of quasars. If quasars do not evolve, $n(z)$ is proportional to $(1+z)^3$. If quasars evolve, then, under the assumption of power law evolution we can write

$$n(z) \propto (1+z)^{3+\gamma}, \qquad (3)$$

where, γ is a parameter determining the evolution of quasars.

One of the selection effects is due to the existence of limiting apparent magnitude in each survey. This effect will lead to an extra inhomogeneity in the redshift distribution. If we restrict the quasars contained in our sample within a narrow range of absolute magnitude, we may avoid the influence of this selection effect.

Another important selection effect is that in the identification of emission lines. The factor $R(z)$ caused by this selection effect can be expressed into (Zhou, Deng, and Dai, 1985)

$$R(z) = \sum_i R_i(z) + \sum_{i<j} R_{ij}(z) + \sum_{i<j<k} R_{ijk}(z) + \ldots\ldots , \qquad (4)$$

where, $R_i(z)$, $R_{ij}(z),\ldots$ are the probability densities of finding a quasar with redshift z and including only the ith line, the ith and jth lines,$\ldots$ respectively. The R's depend on the optical window, the sensitivity of emulsion etc. If we use an approximate but reasonable assembly function $S(\lambda)$ as showed in fig.1, we may have

$$R_i(z) = \frac{N_i}{N} \frac{S[(1+z)\lambda_{oi}]}{\int P(z)S[(1+z)\lambda_{oi}]dz} , \qquad (5)$$

$$R_{ij}(z) = \frac{N_{ij}}{N} \frac{S[(1+z)\lambda_{oi}]S[(1+z)\lambda_{oj}]}{\int P(z)S[(1+z)\lambda_{oi}]S[(1+z)\lambda_{oj}]dz} , \qquad (6)$$

$$\ldots\ldots\ldots\ldots\ldots ,$$

where, N is the total number of quasars in the sample, and N_i, $N_{ij},\ldots$ are the numbers of quasars in which only the ith line, only the ith and ith lines,$\ldots$ are identified in the surveys, respectively.

The sample are chosen from surveys with limiting apparent magnitude about $19^m.5$ (Savage, et al, 1984; Osmer and Smith, 1980a,b; and Crampton, et al, 1985). If we take $q_o=0$ and $H_o=50$ km/s.Mpc, a quasar with apparent magnitude $19^m.5$ will have a absolute magnitude -28.5 at redshift 2.8. So, the sample will be limited to consist of quasars with absolute magnitude less than -28.5 and with redshifts less than 2.8.

Counting the N_i, N_{ij}, $\ldots$, and giving a series of values of γ, we can obtain the calculated the redshift distribution for each value of γ by using the $S(\lambda)$ given in fig.1. Comparing the calculated and observational redshift distribution by linear regression analysis, we can estimate the value of γ from the best fitting between the observational and

calculated redshift distribution. We find that both of the largeat value
of correlation coefficient and the least deviation are approximately at
about $\gamma=3.5\pm1$. In fig.2, we present the observational redshift distribution and the calculated one for $\gamma=4.5$. The regression equation between
them is

$$f_{ob}(z) = -0.82 + 1.1f_{cal}(z), \qquad (7)$$

and the correlation coefficient is 0.97.

4. CONCLUSIONS

1. When we consider any problem relating to the redshift of quasar, the
selection effects muat be taken account of carefully.
2. The observational redshift distribution of quasars can be explained
within the frame of cosmological principle provided the selection effects
have been taken account of.
3. Our analysis gives an approximate evolutionary function of quasars
selected by slitless technique $\rho=\rho_o(1+z)^{6.5\pm1}$.

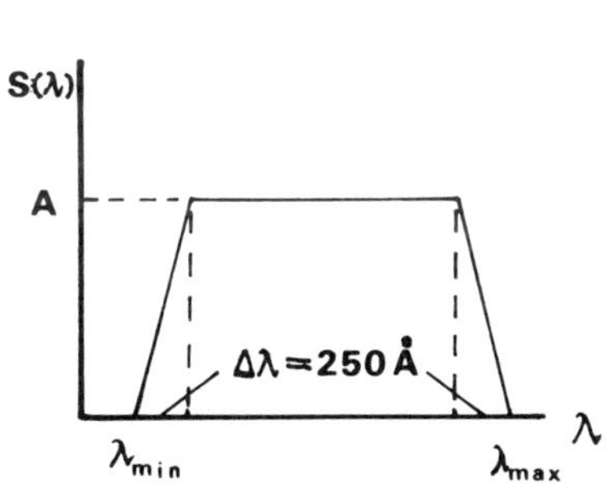

Figure 1. Sketch of the assembly
response function $S(\lambda)$. λ_{min} and
λ_{max} are the shortest and longest
wavelengths of optical windows.

Figure 2. The observational redshift
distribution (solid line) and the
calculated distribution for $\gamma=4.5$
(dashed line).

REFERENCES:

Crampton, D, et al, 1985, Astron. J., 90, 987.
Koo, D. C., and Kron, R. G. 1982, Astron. and Astrophys, 105, 107.
Osmer, P. S., and Smith, M. G. 1980a, Ap. J. Suppl., 42, 333.
Osmer, P. S., and Smith, M. G. 1980b, Ap. J. Suppl., 42, 523.
Savage, A., et al, 1984, M. N. R. A. S., 207, 39.
Schmidt, M. 1968, Ap. J., 151, 393.
Schmidt, M. 1974, ESO/SRC/CERN Conference on Research Programs for the
          New Large Telescopes, 253.
Schmidt, M., and Green, R. F. 1983, Ap. J., 269, 352.
Zhou, Y. Y., Deng, Z. G., and Dai, H. J. 1985, Ap. S. S., 112, 93.
Veron, P., 1983, Proc. of the 24th Liege IAC, 210.

AN EXPLORATION OF VOLUME TEST V/Vm UNDER VARIOUS VALUES OF DECELERATION PARAMETER q_O

S. M. Gong(Kung), H. J. Li, C. L. Xia
Purple Mountain Observatory, Academia Sinica
Nanjing
China

ABSTRACT. The problem of volume test,V/Vm,in zero pressure and matter dominated Friedmann universe model is explored under various values of deceleration parameter q_O. The following conclusions are drawn. (i) For different values of q_O, the change of V/Vm is sensitive at small values of z. Since values of V/Vm are very small at small values of z, this change exerts little effect on the average values of V/Vm. (ii) when values of z and z_m are fixed for each member of a sample, large values of q_O yield larger values of V/Vm,especially in the case of large z. (iii) For $q_O < 1$,the method of volume test is reliable. When $q_O > 2$,especially $q_O > 3$,this method has to be used with caution at large z.

1. FORMULAE AND RESULTS

Let H_O, q_O and z be the Hubble constant,deceleration parameter and the redshift of a source respectively. The value q_O has to be not smaller than zero in the matter dominated Friedmann model,then the correct formula of the luminisity distance is[1],

$$d_L = (c/H_O q_O^2)[zq_O + (q_O - 1)(\sqrt{1+2q_O z} - 1)].\tag{1}$$

Defining

$$A = d_L H_O/c = zq_O^{-1} + q_O^{-2}(q_O - 1)(\sqrt{1+2q_O z} - 1)\tag{2}$$

After some substitution and deduction,we obtain,

$$V(z) = \frac{2\pi}{QH_O^3}\begin{cases}(1-2q_O)^{-\frac{3}{2}}(P\sqrt{1+P^2}-\mathrm{sh}^{-1}P),\ K=-1(q_O<\tfrac{1}{2})\\(2q_O-1)^{-\frac{3}{2}}(\sin^{-1}P-P\sqrt{1-P^2}),K=+1(q_O>\tfrac{1}{2})\\(2/3)A^3[\tfrac{1}{2}(1+A+\sqrt{1+2A})]^{-3},K=0(q_O=\tfrac{1}{2})\end{cases}\tag{3}$$

where

$$P = A\sqrt{K(2q_O-1)}/(1+z).\tag{4}$$

A. Hewitt et al. (eds.), Observational Cosmology, 681–683.

$Q=41254°$, is the number of square degrees in the whole sky.

The ratios, $V(z)/V(z_m)$, under various values of q_0 are computed. The results of the computation are stated as (i) in the abstract.

By taking $q_0=0$, $0.2,...3.0$, the values of V/V_m and mean values, $\langle V/V_m \rangle$ are computed for 3 sampleswhich have been studied by the predecessors only for $q_0=0$ or 1. The variations of values of $\langle V/V_m \rangle$ with q_0, as the result of our computation, are shown in Fig.1 to-gether with the results of predecessors for comparison. The samples include 33 radio sources of 3CR,30 of 4C in the selected region and 60 of $\pm4°$ PKS[2],[3][4].

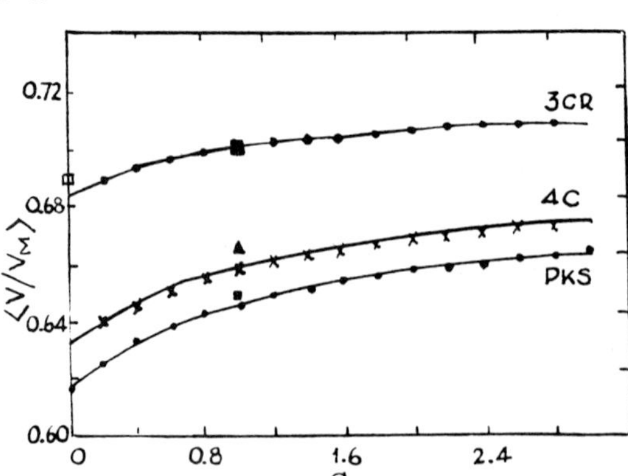

Fig. 1. Plot of $\langle V/V_m \rangle$ against q_0 for three samples.

Fig. 2. Plot of q_0 against z_{mv}.

2. ANALYTICAL EXPLORATION OF $V(q_0,z)$

Is the volume-test method equally valid for different spaces? First,we have to examine whether the maximum volume $V_{mv}(q_0,z)$ exists;if it does exist,then the value of z_{mv} corresponding to $V_{mv}(q_0,z)$ should be found. In order to get V_{mv}, as usual we perform the differentiation,$\partial V(q_0,z)/\partial z$ $=0$,for 3 cases: $q_0<\frac{1}{2}$, $q_0=0$, $q_0>\frac{1}{2}$. The result is only for the case $q_0>\frac{1}{2}$,more precisely only for $q_0>1$,there exist values of z_{mv}[5],

$$z_{mv}=(1/2q_0)\left[\frac{(2q_0-1)(\sqrt{2q_0-1}+q_0)^2}{(q_0-1)^2}-1\right].\tag{5}$$

Also,it can be shown, $\partial z_{mv}/\partial q_0 <0$, i.e.,the values of z_{mv} decrease monotonously with the increase of the values of q_0; $z_{mv}\to 1$ when $q_0\to\infty$ and $z_{mv}\to\infty$ when $q_0\to 1$. The relation between z_{mv} and q_0 is shown in Fig.2.

In order to observe clearly the variation of $V(q_0,z)$ with z under various values of q_0, we compute the values of $V(q_0,z)$ under $q_0=0,0.1,$ $0.25,0.5,...15,20$ and $z=1,2,3,...9,10$. The results are shown in Fig.3.

The purpose of volume test is to see the deviation between the mean values $\langle V/V_m \rangle$ of a sample and the expectation value $\frac{1}{2}$ in the uniform distribution. For this aim, we put

$$V_h(q_0,z)=0.5V_m(q_0,z).\tag{6}$$

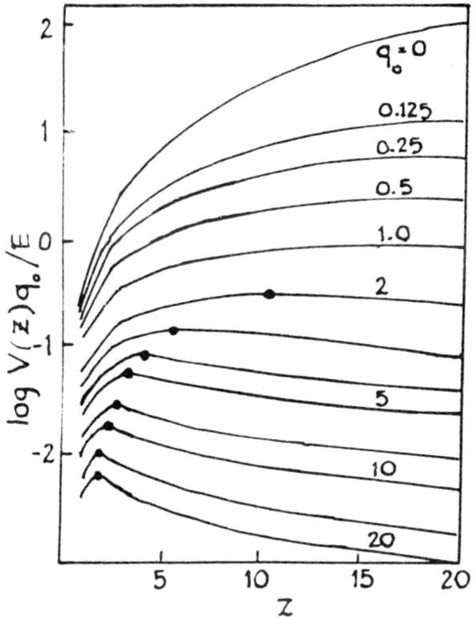

Fig. 3.   Plot log $V(z)q_o/E$
against z ($E = 2\pi/QH_o^3$).

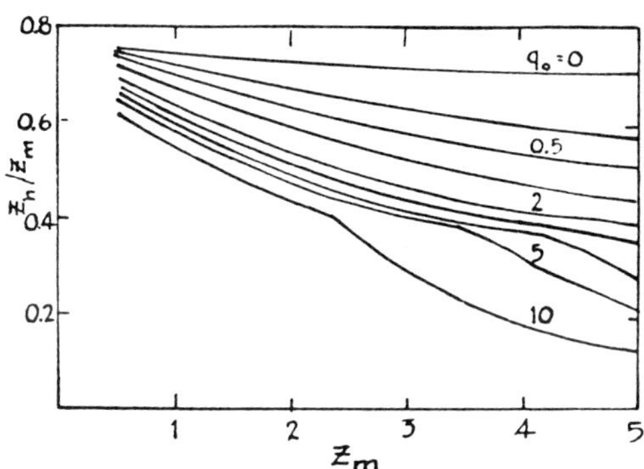

Fig. 4. Plot of z_h/z_m against z_m
for different values of q_o.

We calculate a series of (z_h/z_m)–z_m curves which are illustrated in
Fig.4.

3. DISCUSSION AND CONCLUSIONS

When z_m in Fig.3 is larger than z_{mv}, the redshift corresponding to the
maximum volume $V_{mv}(z_{mv})$, and the value of z is $\lesssim z_{mv}$, it will lead to the
absurd result, i.e., $V(z)/V(z_m)>1$.   Whenever the values of q_o and z are
large enough to cause $V(q_o,z)\simeq V_{mv}(q_o,z_{mv})$, the method is invalid.   From
Fig.4, one can easily see that using the volume-test method, with respect
to a definite value of z_m, smaller values of z_h result when larger values
of q_o are adopted , which lead to larger values of $V(z)/V_m(z_m)$ and
$\langle V/V_m\rangle$; and vice versa.
    Conculsions are stated in the abstract.

References

[1] Mattig, W., A. N.,284(1958),109; 285(1959),1.
[2]Schmidt,M.,Ap. J.,151(1968),393.
[3] Lynds, R. & Wills, D., Ap. J.,172(1972),531.
[4] Wills, D. & Lynds, R., Ap. J., Suppl.,36(1978),317.
[5] Gong,S. M., Li, H. J. & Xia, C. L., Scientia Sinica(Series A),
    29(1986), 966.

THE APPLICATION OF THE STUDY OF GALAXIES ASSOCIATED WITH QUASARS TO
OBSERVATIONAL COSMOLOGY

H.K.C. Yee
Département de Physique
Université de Montréal
Montréal, Québec   H3C 3J7
Canada

ABSTRACT. CCD direct imaging of fields around quasars is used as a
method for locating galaxy clusters and groups associated with quasars.
Average galaxy counts in the sky obtained from control fields are used
to correct for background galaxies in the quasar fields. This correction allows one to derive the luminosity function (LF) of the associated galaxies at the redshifts of the quasars. It is demonstrated that
using the derived LF and average galaxy count data, self-consistent
models of the evolution of the LF and galaxy counts can be obtained.
Current data are best fitted, with a large uncertainty, by a q_o between
0.0 and 0.5 and an evolution in M* of -0.9 ± 0.5 mag. It is found that
the average environment of radio-loud quasars at $z\sim0.6$ is about three
times richer in galaxies than that of quasars at $z\sim0.4$.

In a study of the global environments of quasars, Yee, Green and
Stockman (1986) obtained images of fields around a sample of quasars
having redshifts between 0.3 and 0.65 in Gunn $\underline{r}$ (6500 $\pm$ 500 Å) and $\underline{i}$
(8200 $\pm$ 600 Å) using a CCD camera at the Steward 2.3 m telescope.
Observations of control fields 1° N of the quasars through the $\underline{r}$ filter
were also obtained to provide self-consistent background/foreground
galaxy count corrections. In this paper, I present the analysis of the
differential luminosity function (LF) of galaxies at redshifts between
0.2 and 0.6 and the results of galaxy count models computed using the
derived LF and the control field galaxy counts.

Galaxies from fields of three subsamples of radio-loud quasars are
used to derive the LF of galaxies at redshift bins of $\langle z \rangle$ = 0.61 (9
fields), $\langle z \rangle$=0.42 (10 fields), and $\langle z \rangle$=0.24 (fields of 10 quasars with
0.15 z 0.3 from the data of Green and Yee, 1984). Throughout this
paper, a H_o of 50 km/sec/Mpc is used. For each field, galaxies are
binned according to the absolute magnitudes computed at the observed $\underline{r}$
band of all galaxies by assigning the quasar redshift to the galaxies.
The LF is determined by subtracting the background galaxy counts in the
equivalent apparent magnitude bins. LFs in the three redshift bins are
formed by summing the results. This is performed for three choices of
q_o: 0.02, representing the open universe; 0.5, the closed universe; and

A. Hewitt et al. (eds.), Observational Cosmology, 685–689.
© *1987 by the IAU.*

1.6, the value obtained by Kristian, Sandage and Westphal (1978, KSW) using first rank cluster galaxies and assuming no evolution. The results for the $\langle z \rangle = 0.42$ and $\langle z \rangle = 0.61$ subsamples are shown in Fig. 1.

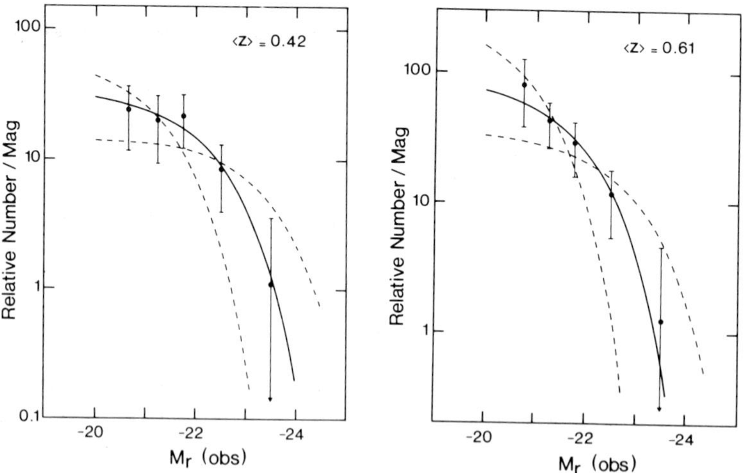

Figure 1. LFs of galaxies associated with quasars at $\langle z \rangle = 0.42$ and $\langle z \rangle = 0.61$ for $q_o = 0.02$ at the observed r band. Solid lines are best fitting Schechter functions with $\alpha = -1.0$. Dashed lines represent the fits with M^* changed by $\pm 2\sigma$.

Local galaxy LFs for different morphological types derived by King and Ellis (1985, hereafter, KE) and Sebok (1986) are used for comparison with the observed LFs. KE and Sebok fitted LFs for various morphological types to Schechter functions (Schechter, 1976) of α of -1.0 and -1.2, respectively. The observed LFs of galaxies associated with quasars are fitted to the same forms. Using the KE and Sebok LFs, model M^*'s of the total LF with no evolution are computed for the observed r band at the three redshift bins using the K-corrections for the various morphological types listed in Sebok. These model M^*'s are then compared with the observed M^*'s. The results are shown on Table 1. The differences in M^*'s can be interpreted as a combined evolution of luminosity and colours of galaxies. Note that for $q_o = 1.6$, in agreement with KSW, no evolution in M^* is observed.

If we made the assumptions that (1) the galaxies associated with the quasars are representative of the galaxies counted in our r band control fields, and (2) the shape of the LF does not evolve, then we can use the observed LFs to calculate galaxy count models; i.e.

$$N(m) = \frac{1}{4\pi} \int \sum_i \Psi_i \left[M_i^*(z) \right] dV$$

where $N(m)$ is the average galaxy counts in the sky, Ψ_i is the integral LF of the ith morphological types. For computation of the integral, linear fits to the ΔM^*'s listed on Table 1 as a function of z are used. The normalization constant of the LF is determined by matching the computed counts at $r = 20.0$ to the observed counts. The resultant fits

TABLE 1

$\langle z \rangle$	Model $M_{\underline{r}}^{*}$'s	$M_{\underline{r}}^{*}$ (observed)$-M_{\underline{r}}^{*}$ (Model)		
			q_o	
		0.02	0.5	1.6
King and Ellis (1985) $\alpha=-1.0$				
0.0	-21.88	--	--	--
0.24	-21.67	-0.19 ± 0.35	-0.15 ± 0.40	$+0.39\pm0.40$
0.42	-21.37	-0.84 ± 0.52	-0.54 ± 0.47	-0.23 ± 0.71
0.61	-20.86	-0.87 ± 0.52	-0.51 ± 0.47	-0.21 ± 0.48
Sebok (1986) $\alpha = -1.2$				
0.0	-21.79	--	--	--
0.24	-21.56	-0.48 ± 0.35	-0.44 ± 0.44	$+0.11\pm0.42$
0.42	-21.22	-1.26 ± 0.59	-0.93 ± 0.52	-0.33 ± 0.74
0.61	-20.61	-1.32 ± 0.58	-0.94 ± 0.51	-0.22 ± 0.53

are shown in Figure 2 for the KE LF and the Sebok LF. The best fitting
line for the Sebok LF is the q_o=0.5 case, and for the KE LF, the q_o
=0.02 case. Both best fitting cases give a colour+luminosity evolution of M^* at z~0.6 of -0.9 ± 0.5 mag. The large uncertainty in the
control field counts (from a total area of ~100 sq arcmin) produces a

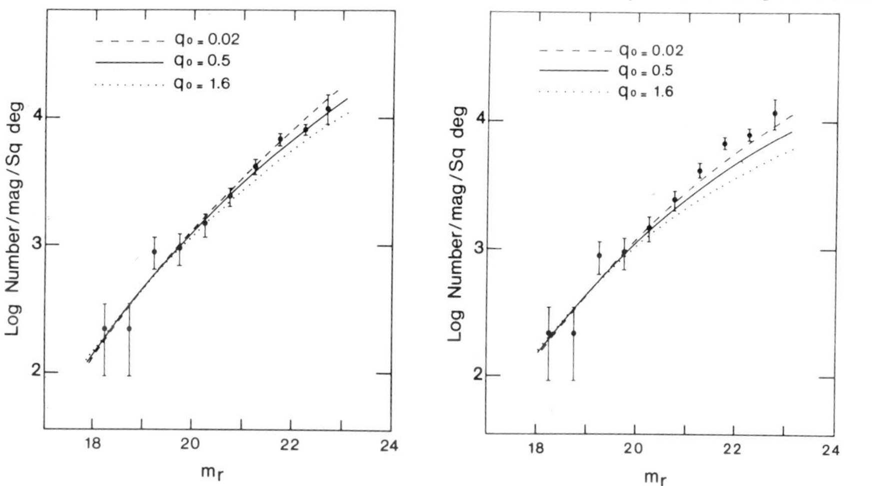

Figure 2. Background galaxy count models using the Sebok (left) and KE
(right) LF for local galaxies.

one σ uncertainty in q_o of ~0.5. The counting statistics can be easily
improved with more data. However, note that the uncertainty in α of
the LF produces a further uncertainty of $\pm\frac{1}{4}$ in q_o.

Using the method outlined in Yee and Green (1984), and the best
fitting galaxy count models, spatial covariance amplitudes (B_{gq}'s) for
the quasars are derived. It is found that the average B_{gq} for the $\langle z \rangle$

=0.61 sample is more than three times that of the low z subsamples and about 8 times that expected for field galaxies. Some radio-loud quasars at z>0.5 are found in clusters of richness similar to those of Abell class 1 clusters. Since low z radio-loud quasars are not found in clusters (e.g., Yee and Green, 1984), this suggests that physical conditions in some rich clusters have changed significantly between z~0.6 and z~0.4.

Using the average environment of quasars, we can also estimate the effect of assumption (1) above. Dressler (1980) estimated that at 10 times the density of the field, the fraction of E+S0 galaxies increases form 0.2 in the field to 0.3. Thus, given the large uncertainty in the LF determination, the effect of possibly having a different population from the field is not significant.

REFERENCE

Dressler, A. 1980, Ap.J., 236, 351.
King, C.R. and Ellis, R.S. 1985, Ap.J., 288, 456.
Kristian, J., Sandage, A., and Westphal, J.A. 1978, Ap.J., 221, 383.
Schechter, P.L. 1976, Ap.J., 203, 297.
Sebok, W.L. 1986, Ap.J.Suppl., 62, 301.
Yee, H.K.C. and Green, R.F. 1984, Ap.J., 280, 79.
Yee, H.K.C. Green, R.F., and Stockman, H.S. 1986, Ap.J.Suppl., 62, in
    press.

DISCUSSION

MILEY: Do you have enough data to draw any conclusions about possible differences in the clustering properties of lobe-dominated radio quasars (steep spectra) with those of core-dominated radio quasars (flat spectra)?

YEE: No, not at the present time. Hopefully, with the CTIO survey of the Parkes quasars, we can say something definitive concerning radio morphology and environments. There is a very tentative suggestion that the extended (i.e. resolved) objects may have a tendency to have richer environments.

SILK: Will you be able to obtain any information about any evolution in the shape of the galaxy luminosity function?

YEE: We now have over 5 times more data in the process of being reduced. I estimate that M^* with an accuracy of $\sim \pm 0.2$ mag can be obtained. However, because the background galaxy counts are always rising faster than any reasonable LF of galaxies, without redshift

information, it would be difficult to get good information on the
LF 1 or 2 magnitudes past M^*. For this reason, it could be difficult
to observe any evolution in the shape of the LF except at the very
bright end.

FILIPPENKO: In those fields which you have observed through three
filters (g,r,i), do you see any differences in the colors of the QSO
host galaxies at different redshifts? Also, do the colors of
galaxies in a given field depend on their projected distance from
the QSO?

YEE: Because our sample contains fairly bright quasars, it is
doubtful that any reliable color information of the fuzz can be
obtained at $z \gtrsim 0.3$. I have not looked at the r-i colours of the
galaxies in the quasar fields carefully yet. However, the few closest
companions all appear to have r-i colours similar to the r-i colours
expected for early type galaxies at the redshifts of the quasars.

COMPLETE SAMPLES OF VARIABLE QUASARS

M.R.S. Hawkins
Royal Observatory
Blackford Hill
Edinburgh EH9 3HJ
Scotland

ABSTRACT.   The merits of selecting samples of quasars for statistical analysis on the basis of optical variability is discussed.   Samples selected this way are essentially uncontaminated by spurious objects, and contain no serious redshift biasses.   A preliminary sample from a long term survey suggests a timescale of variation for quasars of about four years, and indicates a fall off in numbers at around B = 20.

We have heard over the last two days of a number of ways of selecting samples of quasars.   I want to describe a new approach which involves using the fact that quasars vary in luminosity to detect them.   Most methods of selecting quasars have drawbacks, and in particular, selection effects which are typically a function of redshift, and sometimes of apparent magnitude or luminosity.   For example, methods relying on objective prism or grism plates are very much more efficient at finding quasars in redshift ranges where two lines are visible, or where Lyman alpha is present, and so such surveys are of limited use in measuring quasar properties as a function of cosmological epoch.   Ultraviolet excess surveys give uniform coverage to a redshift of about 2, but give no knowledge of the quasar distribution beyond this point, and have the practical difficulty that many other objects apart from quasars have an ultraviolet excess.   Multicolour searches are now proving very useful in finding quasars, especially high redshift objects, but even here there are difficulties at certain redshifts where quasars have colours similar to stars in all optical wavebands.   For example, quasars of redshift 3 have colours very similar to A-stars.
The detection of quasars from their optical variability is a relatively new approach and offers several advantages over other methods and one or two drawbacks.   Perhaps the most important feature of the technique is that there are no preferred or inaccessible redshift ranges. There will be a time dilation effect which will tend to make high red- shift objects vary more slowly, but provided the survey is conducted over a sufficiently long time base this effect can be minimised, and in any case can be corrected for.   An additional advantage of such a survey is that the resulting sample is well suited to the study of quasar variability itself, with large samples of several hundred objects

A. Hewitt et al. (eds.), Observational Cosmology, 691–694.

observed over many years. This leads to the main drawback of the method, that many, even hundreds, of wide field plates are required over many years to detect a complete sample of quasars.

The survey which I have been undertaking was started in 1978, although archival material was available back to 1975. The observational programme consists in taking 3 or 4 sky limited plates per year in each of three passbands with the UK Schmidt Telescope. The emulsions are IIIa-J, IIIa-F and IV-N with appropriate filters, and the plates are measured with the COSMOS measuring machine at the Royal Observatory, Edinburgh. Some 200,000 objects are detected in the central 16 square degrees of the field and after calibration this data set may be used to search for variables. In order to eliminate spurious candidates an algorithm which looks for variation from year to year, but allows only small variations within each year is used, and this has been found to provide an essentially uncontaminated sample of quasars.

A subsidiary programme has been completed to look for quasars which vary on a timescale of weeks or months, and it is clear that such objects are very rare. Considerable effort has been put into verifying that candidates detected on the basis of variability are indeed quasars (e.g. Hawkins, 1986). To this end, some fifty objects have been observed spectroscopically, and essentially all have been confirmed as quasars.

The lastest sample which has been analysed was selected from a 6 year baseline and comprised about 400 objects in 16 square degrees to $B = 21$ with amplitude greater than 0.3 magnitudes. The overwhelming majority of objects showed an ultraviolet excess with $(U-B) < -0.35$. The remaining objects are red and of great interest as they would appear to be excellent candidates for very high redshift quasars. Some idea of the completeness of the survey can be gained by noting that of all UVX objects in the field, some 35% are included in the variability sample, and about 70% show significant variation, but not sufficient to fulfill the criteria for inclusion in the sample.

The magnitude range covered by the sample is $B = 18-21$, and the number magnitude relation for the sample may be measured over this interval. The slope varies from a factor of six per magnitude at the bright end to four per magnitude at the sample limit, and the turn over in quasar numbers is very evident in this critical regime. Further analysis of the evolution of the luminosity function must await the measurement of redshifts for the sample.

Since the sample contains by definition variable quasars, it is possible to examine the statistics of quasar variability. As a first step the auto correlation function for quasar magnitudes was derived for the whole sample. This shows an initial steep drop followed by a flattening at about 4 years, implying that the timescale of variation for the sample as a whole is about 4 years. This is exemplified by the fact that if a sample of quasars is selected on a one year baseline, only about 2% of the UVX objects are included. It is also consistent with the very small numbers of quasars detected on a timescale of weeks or months.

Several more years of data gathering are still necessary before a definitive sample can be expected, and the large task of obtaining several hundred redshifts remains. However, due to the unbiassed way in

which the sample will be selected, subsequent analysis should throw much
light on the question of quasar evolution.

REFERENCE

Hawkins, M.R.S., 1986. Mon. Not. R. astr. Soc. 219, 417.

DISCUSSION

FILIPPENKO: Have you checked whether there are many previously-
discovered QSOs on your plates which do not vary?

HAWKINS: There were very few previously known quasars in the field.
Examination of the percentage of UVX objects which do not vary gives
an answer to your question. The statistics are that roughly 35% were
included in the sample, and another 35% varied significantly, but not
enough to satisfy the selection criterion.

MILEY: Have you carried out an objective prism survey of the same
fields to catch the non-variable quasars? It would be interesting to
compare the properties of the non-variable to the variable QSOs.

HAWKINS: I am in the process of carrying out such a survey with Paul
Hewett. Preliminary work with the brightest quasars shows no obvious
differences.

WALSH: It would be interesting to compare these results with a sample
of bright QSOs. Allen and Keel have done CCD photometry on the PG
sample over a period of about one year to an accuracy of 0.01 mag.
They find very few that vary by more than 0.1-0.2 mag. These objects
are mostly low z, so the (1+z) correction must be remembered, but it
would be interesting to see the results of more extended observations.

HAWKINS: This appears to be consistent with my results. Only about
2% of UVX objects were detected as variables ($\delta m > 0.3$) over a year,
whereas some 30% were detected over 5 years.

WORRALL: Do you have plans to investigate the radio properties of your
objects? It would be interesting to examine whether or not the
variability criterion preferentially selects radio brighter QSOs.

HAWKINS: I have surveyed half of the sample with Lance Miller at
the VLA, and we came up with the surprising result that only three
objects out of about 200 were detected. One of these was a Parkes
source about a hundred times brighter than our threshhold, and one a
double source.

BAHCALL: Can you tell whether the strength of the variability depends
on the absolute magnitude of the QSO; e.g., do brighter objects vary
more strongly than fainter ones?

HAWKINS: I cannot answer this question for sure yet, but indications
are that there is little or no correlation between variability and
brightness.

QUASARS DISKS AND COSMOLOGY

Hagai Netzer
School of Physics and Astronomy, Tel Aviv University
Tel Aviv 69978
Israel

ABSTRACT   Much of the optical and ultraviolet radiation of bright quasars may originate in a massive accretion disk around a central black hole. Most searches for the signature of such disks gave ambiguous results but lately there are new ideas that may lead to their discovery. In particular, the apparent brightness of thin disks depend on their inclination to the observer's line of sight and this may be detected by the equivalent width of some emission lines (Netzer 1985, 1986). This idea may change our view on the inner structure of quasars and other AGN. In addition, it points to a potential selection effect that has not been taken into account so far. Magnitude limited optical quasar samples may contain, preferentially, face-on disks, thus cosmological evolution based on such samples may be biased. There are other implications, especially to the observed correlation of L_{op} with L_x in quasars.

DISCUSSION

WEBB: It has been suggested that the incidence of broad absorption in QSOs may be due to an orientation effect. If so, could there be any correlation between broad CIV absorption and the shape of the continuum in the region of the CIV emission line.

NETZER: I do not know how BAL quasars fit into this picture.

WINDHORST: One would expect that viewing a QSO accretion disk edge-on or face-on would produce different high resolution radio morphologies (at the VLA A-array or at VLBI resolution), if they are radio sources. Do you have any radio data for your QSO's that might shed further light on your suggestion?

A. Hewitt et al. (eds.), Observational Cosmology, 695–696.

NETZER: It could have been a good way if we understood the
structure of the inner radio source and how it is related to the
central disk. I can tell you what is happening if, for example,
the radio jet is perpendicular to the disk. I am not sure, however,
that this is indeed the case.

WALSH: Following on the last point, if a sample of radio QSOs is
selected by the flux from the extended emission regions, it should
be free of orientation bias, and could provide a test of your ideas.

NETZER: Certainly. In my opinion it might be the best way to select
them.

CHAPTER XII

DARK MATTER

DARK MATTER

R. Sancisi and T.S. van Albada
Kapteyn Astronomical Institute
Groningen University
Postbus 800, 9700 AV Groningen

ABSTRACT. The observational evidence for unseen matter is briefly reviewed for objects ranging from small to very large scales. The existence of large mass discrepancies is clearly recognized in individual spiral galaxies and in rich clusters of galaxies. For other systems - dwarfs, ellipticals, binaries and groups - the results are more uncertain and still rather controversial. The data on spirals indicate values of the cosmological density parameter Ω around 0.02, those on clusters $\Omega \simeq 0.2$. The spatial distribution of this dark matter is still largely unknown: while on the galaxy scale it must be located mainly in the outer parts of the stellar system, in clusters it is unclear whether it follows the distribution of the visible galaxies or not.

1. INTRODUCTION

The main questions concerning dark matter are whether it is really present in the first place and, if so, <u>how much</u> is there, <u>where</u> is it and <u>what</u> does it consist of.

<u>How much</u>. In general one wants to know the amount of dark matter relative to luminous matter. For cosmology the main issue is whether there is enough dark matter to close the universe. Is the density parameter Ω equal to 1?

<u>Where</u>. The problem of the distribution of dark matter with respect to luminous matter is fundamental for understanding its origin and composition. Is it associated with individual galaxies or is it spread out in intergalactic and intracluster space? If associated with galaxies how is it distributed with respect to the stars?

<u>What</u>. What is the nature of dark matter? Is it baryonic or non-baryonic or is it both?

We examine here the observational evidence for dark matter on all scales, from the smallest galaxies to clusters, with the purpose of getting information on these questions. We discuss separately individual galaxies - dwarfs, spirals and ellipticals - and multiple systems - binaries, groups and clusters. The present review is restricted to the more direct evidence based on the kinematics of

A. Hewitt et al. (eds.), Observational Cosmology, 699–712.
© *1987 by the IAU.*

objects in dynamical equilibrium and on X-ray data. Other evidence, e.g. that based on gravitational lensing, is not discussed.

2. OBSERVATIONAL EVIDENCE

2.1. Dwarf Galaxies

The study of the dynamics of these smallest systems offers the possibility of investigating whether dark matter is lumped on very small (~ 1 kpc) scales. Recent reviews of this subject are given by Kormendy (1987) and by Freeman (1986). Only dwarf spheroidals and irregulars are discussed here. The more regular and luminous dwarf spirals ($M_B \lesssim -16$), such as those studied by Carignan and Freeman (1985) and discussed by Kormendy (1987), give qualitatively similar results as the higher luminosity spirals and are not considered here.

Dwarf spheroidals. For these systems the evidence for the presence of dark matter comes from optical observations of the stellar radial velocity dispersion. A few nearby objects (Table 1) have been investigated recently by Aaronson (1986) and Aaronson and Olszewski (1987). The two faintest systems have large values of the M/L ratio. These large values do not seem to be explainable as being due to errors in the observations or to uncertainties in the analysis (cf. Tremaine 1986).

Table 1. Dwarf spheroidal galaxies (Aaronson 1986).

Galaxy	M_V	Velocity Dispersion (km/s)	M/L_V
Fornax	−12.6	7 ± 3	2
Sculptor	−11.1	6 ± 2	5
Carina	− 9.4	6 ± 2	8
Draco	− 8.5	11 ± 2	60
Ursa Minor	− 8.5	10 ± 2	80

Dwarf irregulars. Neutral hydrogen is generally used to trace the kinematics of these gas-rich objects. For the larger ones the analysis is based on rotation curves, as for spirals, but it is much more uncertain. For the smaller systems, where random motions dominate or are comparable to rotation, the virial theorem is applied. While for the brighter systems small values of M/L are generally found, in the fainter ones M/L reaches large values (Table 2). Perhaps some of these can be explained as due to large residual gas motions after bursts of star formation. It should be noted that for these fainter irregular systems, as for the spheroidal ones, only the "global" M/L ratios (as given in Tables 1 and 2) are known.

Table 2. Dwarf irregular galaxies (cf. Freeman 1986).

M_B	M/L
-16 to -13	1 - 10
-13 to -10	10 - 30

In conclusion, for both dwarf spheroidals and irregulars there is a clear tendency towards larger M/L values at lower luminosities. This does not necessarily prove, however, the presence of large amounts of dark matter in these objects. An alternative interpretation is that of a varying initial mass function (see Aaronson 1986 and references therein).

At present no clear answer can be given to the question of dark matter in faint dwarf galaxies. The possibility that there may be completely dark systems at the faint end of the luminosity sequence remains a fascinating speculation.

2.2. Spirals

There now exists indisputable evidence for discrepancies between the dynamically computed mass and the luminous mass of spiral galaxies (cf. review by van Albada and Sancisi 1986). The total masses and mass distributions of these systems are derived from optical and HI rotation curves, which are now available for a large number of objects (Rubin et al. 1985, Bosma 1981, Sancisi and van Albada 1987). The most secure evidence to date, however, comes from the study of the systems in Table 3. These systems have exponential luminosity profiles and HI layers extending far beyond the optical images - out to 10-12 disk scalelengths. Their velocity fields are regular and do not show signs of large deviations from axial symmetry or flows in the radial direction: in other words, there are no indications that the assumption of gas moving in circular orbits is not valid. The estimated total masses, i.e. the mass inside the last measured point on the rotation curve, lead to values of M/L_B in the range of 10 to 30. The corresponding values of Ω, 0.01 to 0.02, are, however, negligible with respect to the closure density of the universe.

The rotation curves for three of the galaxies in Table 3 are shown in Fig. 1. The distance from the centre is given in units of disk scalelength. The curves extend to 10-12 scalelengths and are flat out to the last measured point. The uncertainties in the circular velocities are approximately 2 to 5 km/s. Circular velocities can also be calculated from the luminosity profile (see Fig. 2 top) by assuming

Table 3. Spiral galaxies with large dark halos.

NGC	R_{HI} (kpc)	R_{HI}/h	$V(R_{HI})$ (km/s)	L_B ($10^{10}\ L_{B\odot}$)	M_T/L_B $R < R_{HI}$
2841	36	11.6	280	2.5	26
5055	41	10.0	180	2.8	11
2903	23	12.4	182	1.6	11
3198	30	11.1	149	0.9	16
2403	20	9.5	134	0.8	10
6503	25	12.5	115	0.7	11

Note: The values of R_{HI}, L_B and M_T/L_B are based on
      H_O = 75 km/s/Mpc.

a constant M/L ratio. If the value of M/L is chosen such as to <u>maximize</u>
the contribution by the luminous disk, one obtains an estimate of the
<u>minimum</u> amount of dark matter required to explain the observed rotation
curve (see Fig. 2 bottom). Clearly there is a large discrepancy between
the observed and predicted curves, starting around 2-3 scalelengths and
increasing in the outer parts. The local M/L ratio increases from about

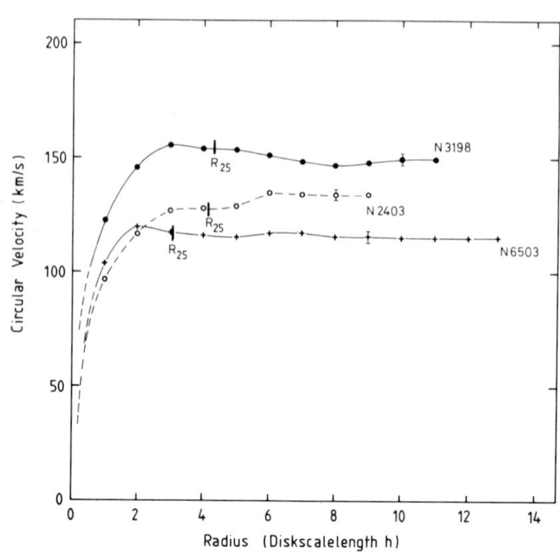

Figure 1. HI rotation
curves for three spiral
galaxies (Table 3) with
extended, symmetrical HI
disks and exponential
luminosity profiles
(from Begeman 1986).

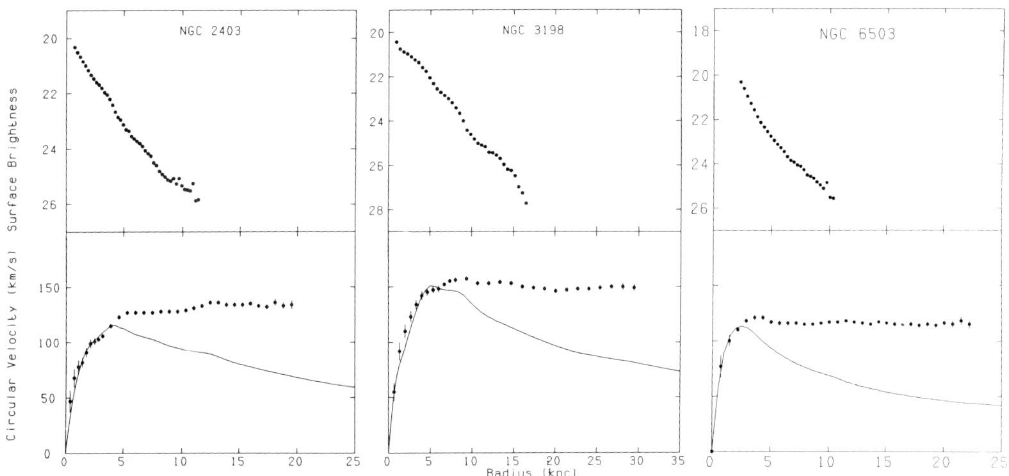

Figure 2. Light profiles and rotation curves for three spiral galaxies with extended, symmetrical HI disks. Upper panels: luminosity profiles from Wevers (1984). Lower panels: HI rotation curves (dots with error bars, Begeman 1986) and curves representing the circular velocity of stars and gas (solid line). The light contribution has been calculated from the luminosity profile by assuming that M/L is constant with radius and has the maximum value allowed by the observed rotation curve.

1-4 in the inner parts to several thousands in the outer parts (assuming that the dark matter lies in a disk; see van Albada and Sancisi 1986). The amount of dark matter inside the last measured point is about 4 times as large as the mass of the disk (assumed to have constant M/L). This should be regarded as the minimum amount of dark matter required as, in principle, the disk contribution could be much less. It is likely, however, that the estimated maximum value of the M/L ratio for the disk is close to the true value. This is based on various results and arguments (cf. Kent 1986, and van Albada and Sancisi 1986). The conclusion then is that dark matter is concentrated in the outer parts of spiral galaxies; in the inner parts, although not entirely ruled out, its presence is not required by the observations.

To sum up, dark matter is present in the outer parts of spiral galaxies. It amounts to at least 4 times the luminous material inside 2 Holmberg radii (about 10 disc scalelengths). The local M/L values are of order 5000 or larger.

Several questions remain unanswered at present:

1. What is the total mass of a galaxy? Where is the boundary of the dark halo? No convincing case has been found as yet of declining

rotation curves. The ones found so far are subject to various criticisms and are in doubt (cf. Sancisi and van Albada 1987). For our galaxy, however, Tremaine (1986) finds from a study of the kinematics of distant globular clusters and companions that a boundary is present at a radius of about 30 kpc.

2. What is the explanation of the well known "conspiracy" of disk and halo to produce a flat rotation curve through the combination of a declining rotation curve of the disk and a rising one of the halo? This seems to form at present the greatest puzzle of all and throws some serious doubts on the very hypothesis of dark halos to explain rotation curves.

3. What is the distribution of dark matter: a spherical halo or a disk which becomes darker (and perhaps thicker) in the outer parts? Although the halo hypothesis is generally favoured there is no compelling observational evidence as yet.

4. Do amount and distribution of dark matter depend on luminosity, morphological type, or environment? Some recent attempts to study these aspects are those of Carignan and Freeman (1985), Bahcall and Casertano (1985), Burstein and Rubin (1985) and Burstein et al. (1986).

5. What is the nature of the dark matter associated with galaxies? Is it all baryonic? What are its constituents?

2.3. Ellipticals

The study of total masses and mass distributions in elliptical galaxies has been vigorously pursued in recent years using optical, radio and X-ray observations.

Optical studies. The analysis of velocity dispersions from stellar absorption measurements leads to values of M/L independent of radius (Tonry 1983) and generally close to 10 out to one effective radius r_e. Such values can be attributed to a stellar population, thus no dark matter is required in the bright inner parts of ellipticals.

HI data. A few hydrogen-rich ellipticals have now been studied in detail. Maps of the HI density and velocity have been obtained and rotation curves, although less accurate than for spirals, have been derived out to several effective radii. The velocity fields indicate sufficient large-scale regularity and symmetry to make the mass determination reliable. For NGC 4278 Raimond et al. (1981) derive M/L_B values larger than 15 (H_o=75) out to $5r_e$ = 13 kpc, assuming that the inclination angle is less than 60°. From the velocity field we tentatively estimate a minimum inclination angle of 30°, which would lead to M/L_B less than 40. For NGC 1052 van Gorkom et al. (1986) estimate M/L_B = 11 (H_o=75) inside $6r_e$ = 21 kpc. This value agrees within the errors with the M/L inside one r_e derived from the stellar velocity dispersion by Davies and Illingworth (1986). This result would indicate that no dark matter is needed inside $6r_e$ = 21 kpc in this elliptical galaxy.

X-ray data. The Einstein observations of elliptical galaxies have been used to determine total masses out to several effective radii (Forman, Jones and Tucker, 1985). The analysis is based on the assumption of hydrostatic equilibrium for the hot gaseous halo surrounding the galaxies. For M 87 (Fabricant and Gorenstein 1983) and NGC 4472 (Forman et al. 1985) the density and the temperature profiles are both available, although the latter are quite uncertain. The M/L values are about 200 out to 10 r_e for M 87, and 20 out to 3 r_e for NGC 4472. Recently Fabian et al. (1986) have reanalysed the X-ray data for 14 ellipticals and have derived minimum total binding masses yielding values of M/L between 10 and 100. All these estimates, however, are dependent on the assumption of hydrostatic equilibrium, which may not be justified in the outer parts of the X-ray emitting regions as suggested by the sometimes irregular and asymmetric shape of the outer contours (see for example the X-ray map of NGC 4472 in Forman et al. 1985). A recent discussion of these general uncertainties is given by Trinchieri, Fabbiano and Canizares (1986), who find that the mass within the observed region could be as low as that derived from optical measurements. This is true also for NGC 4472. Moreover, in the case of the Sombrero galaxy (NGC 4594) the M/L values from X-ray studies seem inconsistent with that from the extrapolated, flat HI rotation curve, which indicates a much lower value. For M 87, which has perhaps the best information, the main problem arises from its central location in the Virgo cluster and from the possibility that at least part of the estimated binding mass belongs to the cluster. We conclude that the X-ray results, although certainly consistent with the presence of extended and massive halos around elliptical galaxies, do not seem to form as yet incontrovertible evidence of their existence.

Shells. The presence of ripples in the light distribution around ellipticals, revealed first by Malin and Carter (1980), has been used by Hernquist and Quinn (1986) and by Dupraz and Combes (1986a) to set constraints on the amount of dark matter around the elliptical galaxy NGC 3923. They find values of M/L of order 100. These results are based on the hypothesis of a sudden tidal disruption of a companion and the large M/L value derived is directly related to this assumption. In fact, Dupraz and Combes (1986b) argue that no dark matter is needed if a slowly decaying orbit is adopted.

In conclusion, in the inner parts of elliptical galaxies there is no need for any dark matter, as for spiral galaxies. For the outer parts there is no indisputable evidence yet for the presence of massive dark halos.

2.4. Binary Galaxies

In analogy with stars, double galaxies might be expected to give valuable information on the masses of galaxies. But expectations since early work of Holmberg and Page have not come true, despite considerable effort. The difficulties lie mainly in the interpretation (cf. White et al. 1983). The large range of M/L values found is indicative of the great uncertainties in the mass estimates.

An important requirement is a sample of well-isolated pairs, so that it is reasonable to assume that the galaxies only feel each others gravitational field. Such samples do exist (e.g. Karachentsev 1972), but the linear separation of these pairs is only a few optical diameters. In this case one learns less from the statistical analysis of a sample of pairs than from HI rotation curves of individual galaxies, which also extend over several optical diameters in a number of cases. The mass-to-light ratio of about 10 found for Karachentsev's sample (Karachentsev 1985) is consistent with M/L for galaxies with extended HI rotation curves.

Double galaxies become more interesting when their separations exceed the linear size covered by HI rotation curves (true separation > 50 kpc, or > 3 optical diameters). This restriction leads to severe problems:

1. The sample will be contaminated by non-physical pairs (e.g. sample of Peterson 1979).

2. The pairs are in general no longer isolated with respect to the galaxies surrounding them (e.g. they may belong to a group). Even though they may be physical pairs in the sense that they are isolated in space, their dynamics may be affected by other galaxies. Thus the velocity difference ΔV is perhaps not solely due to the galaxies in the pair. This is a problem for pairs in groups with crossing times less than the Hubble time.

A satisfactory solution of these problems has not yet been found. In addition, there are other problems with the mass determinations of binary galaxies:

3. An assumption must be made regarding the gravitational interaction of two galaxies in a pair. Since galaxies are surrounded by dark matter it seems unlikely that a model consisting of two point masses moving around each other in elliptic orbits is adequate.

4. Results will in general depend on the unknown ellipticity of the orbits or, alternatively, on the anisotropy of the velocity distribution. That is, the resulting mass for assumed circular orbits will differ from the resulting mass for assumed radial orbits (about a factor of 3).

White et al. (1983) try to overcome these problems by using the observations not only to solve for M/L, but also for the anisotropy of the velocity distribution and the type of gravitational interaction. To make such an analysis possible one must assume that the model (and reality) is scale-free, that is: anisotropy and type of gravitational potential are independent of radius. It seems unlikely that these assumptions could be valid; for example dynamical friction and merging must have led to a depletion of elongated orbits for small orbital dimensions.

In conclusion, binary galaxies data support the evidence for dark matter in individual spiral galaxies but they do not provide reliable results for larger linear scales.

2.5. Groups of Galaxies

In the case of groups the situation is perhaps slightly better than for binary galaxies. A large number of groups have been studied. Huchra and Geller (1982) find a wide range of M/L values, from 10 to 300. Tully

(see paper in this volume) finds a similar large spread. Williams and
Rood (1986) derive M/L values between 1 and 3 from their HI study of
the Hickson compact groups. In all these studies the main assumption
made is that the groups are uncontaminated systems in dynamical
equilibrium. This may not be always true (see e.g. Valtonen and Byrd
1986). The question remains, therefore, whether the large spread in the
M/L values is spurious or whether it reveals an intrinsic variation in
the amount of dark matter present in the various groups.

2.6. Clusters of Galaxies

The presence of dark matter in clusters of galaxies was advocated
already fifty years ago by Zwicky (1933). The conclusion is based on
the virial analysis of radial velocities of the cluster members and on
the assuption that the clusters are bound and in equilibrium. The
results for the Coma, Perseus and Virgo Clusters are summarized in
Table 4. The M/L value for the Virgo cluster core is based on the
assumption that this system is bound but not virialized. The mass-to-
light ratios in Table 4 yield values of Ω between 0.2 and 0.3.

Table 4. Mass-to-light ratios for rich clusters of galaxies
        (H_0 = 75 km/s/Mpc).

Cluster	M/L_B	Radius	Reference
COMA	350	3^0 = 5.0 Mpc	Kent and Gunn (1982)
PERSEUS	580	3^0 = 3.8 "	Kent and Sargent (1983)
VIRGO core	370	6^0 = 1.7 "	Huchra (1985)

In these analyses the distribution of dark matter is assumed to
follow that of the light. The effect on the total mass determination
when this assumption is relaxed has been investigated by The and White
(1986). A factor three smaller or larger estimate of the total mass is
obtained in the cases of respectively a concentration of dark matter in
the inner parts or a more extended distribution in the outer parts.
Despite the significantly lower mass in the case of a central
concentration, a large amount of dark matter is still needed.
    In principle, X-ray observations can also be used to trace the
binding mass of clusters. But uncertainties in the temperature
distribution of the gas have hampered the analysis of the X-ray data
and have made it difficult, as in the case of the elliptical galaxies,
to draw firm conclusions about the dark mass in clusters.

3. AMOUNT OF DARK MATTER: SUMMARY OF OBSERVATIONAL EVIDENCE

The general balance of the amount of dark matter associated with the various objects discussed above is summarized in Table 5. The cosmological density parameter Ω is based on the observed mean luminosity density of galaxies (Davis and Huchra 1982). It is clear that the strongest evidence for dark matter comes from the study of rotation curves of spiral galaxies, which require relatively small amounts of dark matter, and from the virial analysis of the velocity dispersions in clusters, which lead to cosmologically important quantities of dark matter. The known baryonic matter consists of stars ($M/L \simeq 10$) and hot intergalactic gas. The total mass of the latter as estimated from the observations of diffuse X-ray emission in clusters, is rather uncertain, but it seems to be at least as large as the stellar component (Henriksen and Mushotzky 1985), and may be represented, therefore, with M/L values ranging from 10 to 30. If we compare the M/L values of clusters with those of stars and gas together we find a factor 10 in the cluster M/L unaccounted for.

Table 5. Summary of observational evidence for dark matter
         (H_O = 75 km/s/Mpc).

Object	Radius (kpc)	M/L_B	Ω
Dwarfs	3	----	----
Spirals	30	10–30	0.01–0.02
Ellipticals	10–50	(10–100 ?)	
Binaries	20–50	10– ?	
Groups	50–250	10–300 ?	
Clusters	2000–5000	350	0.2

4. DISTRIBUTION OF DARK MATTER

From the analysis of rotation curves it appears that at least a fraction of the dark matter in clusters is associated with individual galaxies and is located in their outer parts. This is, however, only a small fraction. It is not clear where the remaining part is, whether it is all associated with galaxies or whether it is spread out in intracluster space. The former possibility seems unlikely considering the small separation between galaxies in the dense cluster regions. Dynamical arguments can be used to set a limit of about 15 % on the fraction of dark matter associated with individual galaxies (Merritt and White 1987). It is interesting to note that the dark matter in spirals, although closely coupled to the light, does not follow its distribution, as it is more extended. One may wonder whether such a segregation might also occur on the scale of clusters and whether dark matter might be generally more extended than luminous matter and perhaps not be traced at all by the light.

5. NATURE OF DARK MATTER

The observational evidence obtained so far, and briefly reviewed here, does not seem to provide clear information on the question of the nature and composition of dark matter. The amounts required would not be inconsistent with the possibility of it being exclusively baryonic material. Any hypothesis on the nature and origin of dark matter, at least for dark matter associated with spiral galaxies, should account for the large local M/L values mentioned above and for its close "coupling" with the luminous material (i.e. the disk-halo conspiracy).

6. ALTERNATIVES

Alternative explanations, less conventional than the presence of large quantities of dark matter, have been proposed for the observed mass discrepancies. Finzi (1963), and more recently Milgrom (1983), Milgrom and Bekenstein (1987 and references therein) and Sanders (1984, 1986), have made radical suggestions involving modifications of Newtonian dynamics or gravity and have compared their predictions with the observations. So far these comparisons do not seem to have led to inconsistencies which would rule out such alternatives.

We thank K. Begeman for the use of his rotation curve data prior to publication and S. Casertano and R. H. Sanders for helpful comments.

REFERENCES

Aaronson, M. 1986, Paper presented at Eighth Santa Cruz Summer Workshop Nearly Normal Galaxies.
Aaronson, M., and Olszewski, E. 1987, in IAU Symposium **117**, Dark Matter in the Universe, ed. J. Kormendy and G.R. Knapp (Dordrecht: Reidel), p. 153.
van Albada, T.S., and Sancisi, R. 1986, Phil. Trans. R. Soc. Lond., in press.
Bahcall, J.N., and Casertano, S. 1985, Astrophys. J. **293**, L7.
Begeman, K. 1986, Preprints.
Bosma, A. 1981, Astron. J., **86**, 1825.
Burstein, D., and Rubin, V.C. 1985, Astrophys. J. **297**, 423.
Burstein, D., Rubin, V.C., Ford, W.K., Jr., and Whitmore, B.C. 1986, Astrophys. J., **305**, L11.
Carignan, C., and Freeman, K.C. 1985, Astrophys. J., **294**, 494.
Davies, R.L., and Illingworth, G.D. 1986, Astrophys. J., **302**, 234.
Davis, M., and Huchra, J. 1982, Astrophys. J., **254**, 425.
Dupraz, Ch. and Combes, F. 1986a, Astron. Astrophys., **166**, 53.

Dupraz, Ch. and Combes, F. 1986b, Preprint.

Fabian, A.C., Thomas, P.A., Fall, S.M., and White III, R.E. 1986, Mon. Not. Roy. astr. Soc., **221**, 1045.

Finzi, A. 1963, Mon. Not. Roy. astr. Soc., **127**, 21.

Forman, W., Jones, C., and Tucker, W. 1985, Astrophys J., **293**, 102.

Freeman, K.C. 1986, Paper presented at Eighth Santa Cruz Summer Workshop, Nearly Normal Galaxies.

van Gorkom, J.H., Knapp, G.R., Raimond, E., Faber, S.M., and Gallagher, J.S. 1986, Astron. J., **91**, 791.

Henriksen, M.J., and Mushotzky, R.F. 1985, Astrophys. J., **292**, 441.

Hernquist, N., and Quinn, P. 1986, Preprint.

Huchra, J.P. 1985, in ESO Workshop, The Virgo Cluster of Galaxies, ed. O.-G. Richter and B. Binggeli, p. 181.

Huchra, J.P., and Geller, M.J. 1982, Astrophys. J., **257**, 423.

Karachentsev, I.D. 1972, Soob. Sp. Astr. Obs. Akad. Nauk., No. 7.

Karachentsev, I.D. 1985, Sov. Astron., **29**, 243.

Kent, S.M. 1986, Astron. J. **91**, 1301.

Kent, S.M., and Gunn, J.E. 1982, Astron. J., **87**, 945.

Kent, S.M., and Sargent, W.L.W. 1983, Astron. J., **88**, 697.

Kormendy, J., 1987, in IAU Symposium **117**, Dark Matter in the Universe, ed. J. Kormendy and G.R. Knapp (Dordrecht: Reidel), p. 139.

Malin, D.F., and Carter, D. 1980, Nature, **285**, 643.

Merritt, D. and White, S.D.M. 1987, in IAU Symposium **117**, Dark Matter in the Universe, ed. J. Kormendy and G.R. Knapp (Dordrecht: Reidel), p. 283.

Milgrom, M. 1983, Astrophys. J., **270** , 365.

Milgrom, M., and Bekenstein, J. 1987, in IAU Symposium **117**, Dark Matter in the Universe, ed. J. Kormendy and G.R. Knapp (Dordrecht: Reidel), p. 319.

Peterson, S.D. 1979, Astrophys. J. Suppl. **40**, 527.

Raimond, E., Faber, S.M., Gallagher III, J.S., and Knapp, G.R. 1981, Astrophys. J., **246**, 708.

Rubin, V.C., Burstein, D., Ford, W.K., Jr., and Thonnard, N. 1985, Astrophys. J. **289**, 81.

Sancisi, R., and van Albada, T.S. 1987, in IAU Symposium **117**, Dark Matter in the Universe, ed. J. Kormendy and G. R. Knapp (Dordrecht: Reidel), p. 67.

Sanders, R.H. 1984, Astron. Astrophys., **136**, L21.

Sanders, R.H. 1986, Mon. Not. Roy. astr. Soc., **223**, 539.

The, L.S., and White, S.D.M. 1986, Astron. J., **92**, 1248.

Tonry, J. 1983, Astrophys. J., **266**, 58.

Tremaine, S. 1986, Paper presented at Eighth Santa Cruz Summer Workshop Nearly Normal Galaxies.

Trinchieri, G., Fabbiano, G., and Canizares, C.R. 1986, Astrophys. J., **310**, 637.

Valtonen, M.J., and Byrd, G.G. 1986, Astrophys. J., **303**, 523.

Wevers, B.M.H.R. 1984, Ph. D. Thesis, Groningen University.

White, S.D.M., Huchra, J., Latham, D., and Davis, M. 1983, Mon. Not. Roy. astr. Soc., **203**, 701.

Williams, B.A., and Rood, H.J. 1986, Preprint.

Zwicky, F. 1933, Helv. Phys. Acta., **6**, 110.

DISCUSSION

RUBIN: You have discussed with great clarity the deconvolution of the
rotation curves by making use of the maximum disk. I would like to
mention recent observations which may be relevant. John Graham and I
have used the Kitt Peak 4-m Echelle spectrograph + CCD at very high
velocity resolution (0.125 Å/pix) to obtain spectra of the same
galaxies for which we had earlier obtained rotation curves. In a large
fraction of the galaxies, the rotation velocities do not go smoothly to
zero with decreasing nuclear distances, but instead go to a velocity
near 50 km/s. I have not yet made mass models, but the observations
suggest a nuclear mass which causes locally a falling density within a
few arcsec (hundreds of parsecs).

SANCISI: A very interesting result. I wonder whether it may be related
to the presence of central components in the light distribution of
several disk galaxies as shown by the photometric profiles published
recently by Kent (1986, Astr. J. 91, 1301).

N. BAHCALL: The increase of the dark matter component, or M/L, with
the increasing scale of the system – from galaxies through clusters of
galaxies– may possibly be further extended to the cores of the rich
superclusters. Our data (Bahcall, Soneira and Burgett 1986 Ap.J.)
suggest that large velocity dispersions ($>$1000 km/s) may exist in rich
superclusters. If so, the M/L ratios in superclusters, on scales of 20
h^{-1} Mpc, may be still larger than that of clusters by a factor of $\sim$
2 – 3.

SANCISI: Yes, possibly. But I am not sure about a systematic relative
increase of the dark matter component with increasing scale of the
system. As to superclusters, are they bound systems in dynamical
equilibrium?

FANG: Does the ratio of the luminous matter to dark matter depend on
the morphology of galaxies ?

SANCISI: From the small number of good cases studied so far there seems
to be no clear evidence of such a dependence.

CANIZARES: Relative to the question whether or not the dark matter in
clusters is closely tied to the galaxies, the distribution of X-ray
emitting hot gas (which is "virialized" and so traces the potential)
does not closely follow the galaxies, in general. Therefore, the dark
matter must be more widely distributed between the visible galaxies.
The distribution of X-ray gas can also set limits on the degree of
concentration of the dark matter.

BURKE: With respect to the distribution of matter in clusters of
galaxies,the data on gravitational lenses provide definite conclusions.
The original twin quasar 0957+561, an incontrovertible example of
lensing, requires the matter to be distributed in a fashion very
different from that implied by the distribution of luminous matter. It
does not seem to be directly related to the galaxies in the foreground
cluster, at least not without giving eccentric M/L values to the
indidual galaxies. It is almost certainly distributed in a diffuse
manner.

SANCISI: I am glad you mention gravitational lenses, which seem indeed
to become an important source of evidence on the presence and
distribution of dark matter in the universe. Unfortunately it has not
been possible to discuss this evidence in the present review.

AXIONS IN THE UNIVERSE

S. Tsuruta[*], and K. Nomoto[**]
(*) Department of Physics, Montana State University
(**) Department of Earth Science and Astronomy, University of Tokyo,
    and Department of Physics, Brookhaven National Laboratory

INTRODUCTION

The observational evidence for the presence of dark matter is now generally accepted, with no lack of possible candidates (see e.g. Dekel, Einasto, and Rees 1986). The proposed candidates are devided into two groups, baryonic and non-baryonic. The latter is further devided to hot and cold dark matter. For the cold dark matter, among the first to be proposed is the axion. In this paper we shall not dwell on numerous cold dark matter candidates offered by particle physicists, for there are review articles on the subject (see e.g. Turner 1986, Primack 1986). The main purpose of the present report is to suggest that neutron star cooling theory and future space satellite programs (e.g. AXAF, XAO, LXAO) have a potential for offering the best astrophysical constraint on the axion mass and hence, giving valuable insight to some cosmological problems.

Even though other candidates for cold dark matter are perfectly plausible, axions may have the following advantages. Their presence was predicted, independently of cosmology, as a natural solution to the strong CP problem (Peccei and Quinn 1977). Also, once we accept their presence, we can carry out specific calculations, to predict various properties (e.g. their emissivities).

CONSTRAINTS ON AXION MASS

The axion is a relatively light, pseudoscalar boson associated with the Peccei and Quinn symmetry, a child of the strong CP problem and grandchild of the U(1) problem (Sato and Sato 1975, Weinberg 1978, Wilczek 1978). The symmetry breaking scale F, however, is left undetermined in the theory. The axion mass m_a is related to the parameter F, roughly as $m_a \sim 10^{16}$ eV^2/F (e.g. Turner 1986).

The observational upper and lower limits to the value of F could be set by cosmological and astrophysical considerations. The upper limit, $F \sim 10^{12}$ GeV, which corresponds to $\Omega = 1$, comes from cosmological constraints, that coherent axion oscillations in the present-day universe do not produce an unacceptably large energy density. It gives the lower limit to the axion mass, as $m_a > \sim 10^{-5}$ eV. The lower limit to F (or upper limit to m_a) generally comes from stellar

A. Hewitt et al. (eds.), Observational Cosmology, 713–717.
© *1987 by the IAU.*

constraints, that various interactions in the stellar interiors involving axions would not be too fast to affect stellar evolution, to support observational evidence (Fukugita et al. 1982). The best stellar constraints to date, using the normal "invisible" axion model, are shown in Table 1 (Raffelt 1986a, b, Dearborn et al. 1986,). We note that the constraints from cooling of white dwarfs and Helium burning stars are the most severe, both giving $m_a < \sim 0.01$ eV.

It has been reported (Kim 1979,1984, Kaplan 1985) that axions should be even more "invisible" than proposed earlier. The important characteristics of this model, often referred to as the "hadronic model", is that these axions do not couple to leptons. The implication is that dominant axion emissivities in most stars (including all listed in Table 1) would be greatly reduced since their dominant axion interactions involve electrons.

	m_a(eV)	F(GeV)
Solar	$< \sim 3.2$	$> \sim 3 \times 10^6$
Red Giants	$< \sim 0.06$	$> \sim 2 \times 10^8$
White Dwarfs	$< \sim 0.01$	$> \sim 10^9$
He Burning	$< \sim 0.01$	$> \sim 10^9$

TABLE 1
Stellar constraints for
"normal" axion model

The constraints for these "hadronic axions" are shown in Table 2 (Raffelt 1986 a). A major conclusion is that if this model is correct, the constraints on the axion mass from these stars are _weakened enormously_. Note that the strongest constraint is reduced to the axion mass of _only a few eV_ (at least 10^5 times larger than the lower limit corresponding to $\Omega = 1$).

Another important implication of no axion interactions with electrons was emphasized by Kaplan (1985). That is, laboratory detecion of relic axions may become more difficult, if not impossible, than predicted earlier. Various laboratory experiments on detection of axions have been proposed (see e.g. a recent review by Smith 1986). It may be within our reach in a foreseeable future, to test by various laboratory experiments which of the invisible axion models, the "normal" or "hadronic", is correct. Therefore, it may be worthwhile to consider seriously an alternative possibility for offering a far stronger constraint on the axion mass. Here we propose that the combination of neutron star cooling theory and future X-ray satellite programs has a potential for offering such a possibility.

	m_a(eV)	F(GeV)
Solar	$< \sim 25$	$> \sim 4 \times 10^5$
Red Giants	$< \sim 1.3$	$> \sim 8 \times 10^6$
Helium Burning	$< \sim 2$	$> \sim 6 \times 10^6$

TABLE 2
Stellar constraints for
"hadronic" axion model

AXION MASS LIMIT FROM NEUTRON STAR COOLING

Since we can theoretically estimate the temperature of a neutron star of a given age through cooling theory, comparison of the theoretical temperature with observation will offer another means of stellar constraints on the axion mass, in the following sense. Any axion mass resulting in fast cooling (compared with observations) should be ruled out. At the moment the best obserational data come from the Einstein X-Ray Observtory (see e.g. Tsuruta 1986, Nomoto and Tsuruta 1986). Unfortunately, strictly speaking, several detected point sources in pulsars and supernova remnants offer only the upper limits to the stellar temperatures, mainly due to lack of spectroscopic resolution of the HRI which has the highest spacial resolution (Giacconi et al. 1979). However, there is a realistic possibility that future X-ray satellites, especially AXAF, XAO and LXAO, could offer more conclusive data. Therefore, it may be worthwhile to explore what this option could do.

Among the point sources detected by the Einstein, RCW 103 is considered to be the best candidate as the detection of direct radiation from the neutron star surface (Tuohy and Garmire 1980, Tuohy et al. 1983). Therefore, we shall examine, in the following, what could be done by comparing neutron star cooling results with this source. If we assume that the RCW 103 data is the measurement of the stellar temperature, we can choose a cooling model which is consistent with this observational data. Then, models with very fast cooling agents such as quarks and pions should be ruled out for this particular source, and Case SX (with superfluid nucleons) of the FP neutron star model should be imposed (Tsuruta 1986). The FP model is a model of an intermediate stiffness constructed by Friedman and Pandaripande (1981), which is generally considered to be the most realistic nuclear model. (See Nomoto and Tsuruta 1987 for details). In the present calculations, the star cools with emission of axions, neutrinos and photons. Axion emissivities involving nucleons in the core are taken from Iwamoto 1984. Those involving electrons in the crust are taken from Itoh et al. 1986. The latter is drastically reduced for "hadronic axions". As noted earlier, this affects axion cooling of other stars drastically. However, this does not affect neutron star cooling, because the axion interaction with nucleons in the core, the dominant cooling process in a neutron star, is unaffected. Other input physics and methods are the same as in Nomoto and Tsuruta 1987.

We obtained neutron star cooling curves for $F = 10^{10}$, 3×10^9, 10^9, and 10^8 GeV, corresponding to $m_a \sim 10^{-3}$, 3×10^{-3}, 0.01, and 0.1 eV. These curves are compared with the _Einstein_ data points. Assuming that the _Einstein_ data for RCW 103 is the detection, the comparison of our theoretical curves with this source gives the most stringent upper limit to date: $m_a < \sim 10^{-3}$ eV (Tsuruta and Nomoto 1986).

CONCLUDING REMARKS

If the Einstein data point for RCW 103 is confirmed to be the detection, neutron star cooling theory can place the strongest upper limit to the axion mass, as $m_a < \sim 10^{-3}$ eV, with the corresponding $\Omega > \sim 0.01$. Then, the axion contribution to the dark matter should be at least as large as the baryonic contribution. The future space programs such as AXSF, XAO and LXAO should be crucial, with the realistic potential for confirming this detection. They may offer even more exciting outome, by the detection of other yet undiscovered sources.

ACKNOWLEDGEMENTS
        We thank Drs. M.J. Rees, K. Sato, M.A. Ruderman, G.. Raffelt, G.
Börner, and D.E. Morris, for valuable discussions and comments. ST
acknowledges with thanks the hospitality of the directors of IOA in
Cambridge and ISAS in Tokyo, where a part of this work was carried out.
This work was supported in part by the NSF under grant AST-8602087, by
MONTS under grant 16404216, and by DOA under contract DE-AC02-
76CH00016.

REFERENCES
Dearborn, D.S.P., Schramm, D.N., and Steigman, G.: 1986, Phys. Rev.
    Lett. $\underline{56}$, 26.
Dekel, A., Einasto, J., and M. Rees: 1986, Rev. Mod. Phys., in press.
Friedman, B., and Pandharipande, V.R.: 1981, Nucl. Phys. $\underline{A361}$, 502.
Fukugita, M., Watamura, S., and Yoshimura, M.: 1982, Phys. Rev. Lett.
    $\underline{48}$, 1522.
Giacconi, R. et al.: 1979, Astrophys. J. Lett. $\underline{230}$, 540.
Iwamoto, N.: 1984, Phys. Rev. Lett. $\underline{53}$, 1198.
Itoh, N., et al. 1986: Private Communication.
Kaplan, D.N.: 1985, Nucl. Phys. $\underline{B260}$, 215.
Kim, J.E.: 1979, Phys. Rev. Lett. $\underline{43}$, 10.
___________: 1984, Seoul National Univ. Preprint SNUHE-84/02.
Nomoto, K., and Tsuruta, S: 1986, Astrophys. J. Lett. $\underline{305}$, L19.
___________: 1987, Astrophys. J., Jan. 15, Vol. 312, in press.
Peccei, R.D., and Quinn, H.R.: 1977, Phys. REv. Lett. $\underline{38}$, 1440.
Primack, J.R.: 1986, Preprint SCIPP 86/65.
Raffelt, G.G.: 1986a, Ph.D. Thesis, Univ. of München.
___________: 1986b, Phys. Rev. D, $\underline{33}$, 897, & Phys. Lett. $\underline{166B}$, 402.
Sato, K., and Sato, H.: 1975, Prog. Theor. Phys. $\underline{54}$, 1564.
Smith,P.F.: 1986: Preprint RAL 86-029.
Tsuruta, S.: 1986, Com. Astrophys. $\underline{XI}$, 151.
Tsuruta, S., and Nomoto, K.: 1986, in preperation.
Tuohy, I.R., and Garmire, G.P.: 1980, Astrophys. J. Lett. $\underline{239}$, L107.
Tuohy, I.R., Garmire, G.P., Manchester, R.N., and Dopita, M.A., 1983,
    Astrophys. J. $\underline{268}$, 778.
Turner M.S.: 1986, Preprint.
Weinberg, S.: 1978, Phys. Rev. Lett. $\underline{44}$, 223.
Wilczek, F.: 1978, Phys. Rev. Lett. $\underline{40}$, 279.

DISCUSSION

CANIZARES: If there were some heating of the neutron star, say by
accretion, wouldn't that make it impossible to determine an upper
limit to the axion mass by your method?

TSURUTA: No, what you said does not apply in our case. Any possible
heating mechanisms become important only for older neutron stars (t $\gtrsim$
~ 10^4 years) such as older radio pulsars or accreting neutron stars in

a binary system (see e.g. Tsuruta 1979, Physics Reports, or Tsuruta 1986 in the text). However, exactly because of this, these objects were purposely avoided in our comparison. What I mean is that our objects for comparison are confined only to young, isolated neutron stars in supernova remnants (all with $t < \sim 10^4$ years) where accretion is not important. For instance, the object we specifically chose, RCW 103, is only 1000 years old, and there is no possibility for its being an accreting binary star (see Tuohy et al. 1983 in the text). For all of the objects of our choice, cooling definitely determines the stellar temperature, and therefore, our method works perfectly well (see Tsuruta and Nomoto 1986 in the text, for details).

THE NEUTRINO MASS AND THE CELLULAR LARGE SCALE STRUCTURE OF THE
UNIVERSE

R. Ruffini, D.J. Song, and S. Taraglio
International Center for Relativistic Astrophysics - I.C.R.A.
Department of Physics
University of Rome
La Sapienza

ABSTRACT. We show how within the theoretical framework of a Gamow
cosmology with massive neutrinos, the observed correlation functions
between galaxies and between clusters of galaxies, naturally lead to
a "cellular" structure of the Universe. From the size of "elementary
cells" we derive constraints on the value of the masses and chemical
potentials of the cosmological "inos". We outline a procedure to
estimate the "effective" average mass density of the Universe. We
predict also the angular size of the inhomogeneities to be expected
in the cosmological black body radiation as remnants of this cellular
structure. A possible relation of our model to a fractal structure is
indicated:

That the further we look for galaxies and clusters of galaxies the
lower is the average density of the Universe, decreasing in distance
with a fixed power law, may be considered, if confirmed, one of the
major results of modern observational cosmology (de Vaucouleurs 1970).
This result is apparently in clear disagreement with the commonly
accepted view in theoretical cosmology that the further we look in
space the earlier epochs of the Universe are explored and, consequently,
the higher should be the observed average density of our Universe.
This theoretical view, first introduced by Friedmann, further developed
by Gamow, has obtained a patent observational confirmation by the
discovery of Hubble law and the cosmological black body radiation (see
e.g. F. Melchiorri and R. Ruffini 1986 and references therein).

There must exist, therefore, a cutoff distance, D_c, at which the
highly inhomogeneous component with decreasing density ends (an
"elementary cell"). On large enough scales the usual homogeneous and
isotropic distribution, to be expected in a Friedmann model, has to be
recovered.

The determination of this cutoff is the topic of this communication
(see also Ruffini et al., 1986): we show how the Gamow cosmology,
deeply modified by the existence of massive "inos", naturally leads to
the existence of a characteristic mass scale, M_c, and size, D_c, at which
this cutoff should occur. This scale is fixed by the epoch t_c (or
equivalently redshift z_{nr}) at which the "inos" become non-relativistic

A. Hewitt et al. (eds.), Observational Cosmology, 719–722.
© 1987 by the IAU.

and the mass M coincides with the mass contained within the horizon at
that epoch.

    Observations at spatial distances $D > D_c$ should reveal an overall
structure of the Universe composed of a set of such "elementary cells"
each one of mass M_c observed at different phases of cosmological
evolution, all the way up to $z = z_{nr}$ where the strict usual features of
a Friedmann universe are recovered (see Fig. 1).

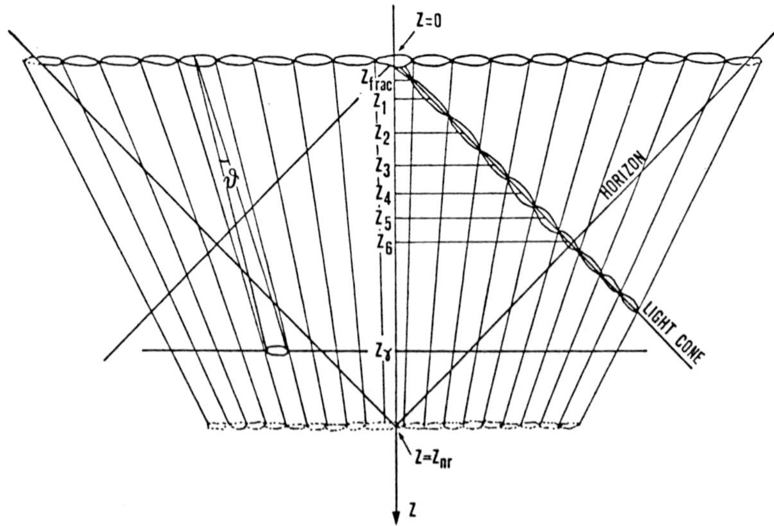

FIG. 1. A sketch of the evolution of the cellular structure of the
Universe. The vertical axis is the axis of redshifts (increasing
downwards). From the top to the bottom we can find; z = 0 which
represents the present time with the array of elementary cells as they
are nowadays; z_1, z_2, ..., z_n which are the redshifts at which we
actually see the centers of the successive cells, it is also sketched
the light cone of the observations; $z = z_\gamma$ which represents the surface
of last scattering, Θ is the angular scale of the cellular structure
at the decoupling time; finally $z = z_{nr}$ is the time at which the cells
formed via gravitational instability.

    Since z_{nr} is larger than the redshift correponding to the de-
coupling of matter and radiation, the formation of this cellular
structure may leave an imprint in the cosmological black body radiation.
The angular scales of the expected inhomogeneities are evaluated.

    In each "elementary cell" the density is a function of the volume
considered and we cannot speak of an average cosmological density. Due
care should be taken, therefore, in evaluating the "effective" average
density of the Universe, since we do our observations from within our
elementary cell, an extrapolation procedure is given (see Fig. 2)

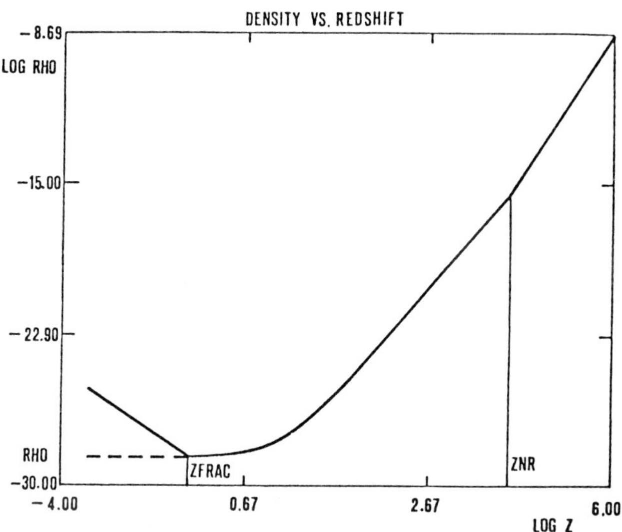

FIG 2. This is a log-log plot of the behavior of the density of the Universe as a function of redshift. The first zone, up to $z = z_{frac}$, is the self similar (fractal) part, inside a single cell. As we exit this zone we have the increasing of the density as the cube of the cosmological redshift up to $z = z_{nr}$. At this point the density will follow the fourth power of the redshift as in the usual Friedmann Universe.

The consistency of this theoretical framework with the previous cosmological considerations (Ruffini et al., 1983, Ruffini and Song 1986) as well as the requirement that the "elementary cell" be larger than a minimum size do impose constraints on the masses and chemical potentials of the cosmological "inos".

The possible relevance of these considerations to a fractal cosmological structure are discussed.

REFERENCES

Melchiorri, F., and Ruffini, R., Editors, 1986, The proceedings of the
    LXXXVI International Summer School of E. Fermi on "Gamow Cosmology".
Ruffini, R., and Song, D.J., 1986, Astron. Astrophys., in press.
Ruffini, R., Song, D.J., and Stella, L., 1983, Astron. Astrophys. 125,
    265.
Ruffini, R., Song, D.J., and Taraglio, S., 1986, Submitted for
    publication.

DISCUSSION

DEKEL: How do you compare your result of 9 ev from galaxies with the classical Tremaine-Gunn result that obtained higher neutrino masses?

RUFFINI: The approach of Tremaine-Gunn is meaningless; it is wrong to pretend that considerations on the density distribution of degenerate or semi-degenerate configurations can be inferred by merely imposing limits on the occupation number in phase space of a classical isothermal sphere, even taking into account the presence of visible matter.

DETERMINATION OF "INOS" MASSES COMPOSING GALACTIC HALOS

D. J. Song and R. Ruffini
International Center for Relativistic Astrophysics - I.C.R.A.
Department of Physics
University of Rome
La Sapienza

ABSTRACT.   One of the most important discoveries in modern cosmology
and astrophysics is that the dark matter dominates at every scale of
the Universe (Rubin et al., 1978, Forman et al., 1985, Faber and
Gallagher, 1979, Sancisi, 1987).   The problem that follows is to understand the components of the dark matter and to determine the physical
properties of these components.

In this talk we would like to establish the cosmological and
astrophysical constraints on the physical properties of "inos" which
are assumed to dominate substantially the dark matter making up the
galactic halo and the Universe. In establishing constraints two
different approaches, cosmological (Gershtein and Zel'dovich, 1966)
and astrophysical (Malagoli and Ruffini, 1986), are usually followed,
but, in this talk, we follow a new intermediate approach. We
explicitly show that, in addition to the usual cosmological arguments,
we have to impose some additional largely model independent constraints
in order for "inos" to form the galactic halos at the necessary
cosmological epoch.

The assumptions are (1) the galactic halos are virialized systems,
either quantum or classical systems and that their mean mass density
stays constant from the moment of formation until today, (2) the
details of multi-fragmentation process in superclusters and clusters
of galaxies are neglected. The galactic halos merge at $z \gtrsim z_F$ into a
non-relativistic, homogeneous Friedmann Universe, in which the
cosmological density as well as the momenta of the "inos" change with
the well-known cosmological scaling law; $\rho_{nr} \propto a(t)^{-3}$ and $p(t) \propto a(t)^{-1}$,
where a(t) is the cosmological scale factor, (3) at $z \sim z_R$, the "inos"
acquire relativistic velocities, correspondingly the density varies
with the law $\rho_r \propto a(t)^{-4}$, (4) the decoupling between "inos" and the
cosmological matter occurred at the time of the relativistic regime,
when the "inos" were in thermal equilibrium, and (5) the cosmological
density parameter is defined by $\Omega = \rho/\rho_{crit}$, where $\rho_{crit} = 1.88 \times 10^{-29}$
h^2 g cm^{-3} and $h = H_o /[100$ km/sec Mpc$]$.

With these assumptions, the cosmological energy density of "inos"
which were in thermal equilibrium before decoupling is given by
(see Ruffini et al., 1983, Ruffini and Song, 1986b).

723

A. Hewitt et al. (eds.), Observational Cosmology, 723–726.
© 1987 by the IAU.

$$\rho_r = \left[7/8 \sum_x g_x (T_x/T_e)^4 F_r(\xi_x) \right] a_B T_e^4 \tag{1}$$

for $z \geq z_R$ and

$$\rho_{nr} = \left[3/4 \sum_x g_x m_x (T_x/T_e)^3 F_{nr}(\xi_x) \right] \lambda T_e^3 \tag{2}$$

for $z < z_R$, where $\lambda = 3/4 \quad \zeta(3)/\zeta(4) a_B k$, $F_r(\xi_x) = \left[12\eta(4) \right]^{-1}$ $\left[1/4 \, \xi_x^4 + 6 \, \eta(2) \, \xi_x^2 + 12 \, \eta(4) \right]$, $F_{nr}(\xi_x) = \left[4\eta(3) \right]^{-1} \left[1/3 \, \xi_x^3 + 4\eta(2)\xi_x \right.$ $\left. + 4 \sum_{n=1}^{\alpha} (-1)^{n+1} e^{-n\xi_x}/n^3 \right]$, $\xi_x = \mu_D/kT_D$, $T_e = T_o (1+z)$, and $T_o = 2.7\emptyset^o$ K.

It is clear that the value T_x/T_e is dependent on the decoupling epochs of "inos" and clearly we must have $T_x/T_e \leq 1$ (Ruffini and Song, 1986a). Assuming that at the epoch of halo formation the mean density of the Universe was equal to that of the galactic halo, we can then determine the redshift z_F, at the epoch of formation by using the present density of galactic halos (Ruffini and Song, 1986b);

$$1 + z_F = 11.98 \, (\Omega h^2)^{-1/3} \, \tilde{M}_H^{-1/3} \, \tilde{R}_H^{-1} \tag{3}$$

The flat rotation curves of galaxies give the epoch of the relativization of "inos";

$$1 + z_R = 1.22 \times 10^4 \, (\Omega h^2)^{-1/3} \, \tilde{M}_H^{-1/6} \, \tilde{R}_H^{-1/2} \tag{4}$$

where $\tilde{M}_H$ and $\tilde{R}_H$ are mass and radius parameters of galactic halos in units of $M_H = 2 \times 10^{12} \, M_\odot$ and $R_H = 100$ kpc, respectively.

It is interesting to note that both the epoch z_R and z_F are only functions of Ωh^2 and of the macroscopic parameters, mass and radius, of the galactic halos. They are, however, not functions of the masses of the "inos" forming the galactic halo. Consequently we may express, in the different regimes, the mean mass density of the Universe in terms of the mass density of galactic halos, ρ_H, using the epochs of z_R and z_F;

$$\rho_r = \rho_H (1 + z_F)^{-3} (1 + z_R)^{-1} (1 + z)^4 \tag{5}$$

and $z > z_R$ and

$$\rho_{nr} = \rho_H \ (1 + z_F)^{-3} \ (1 + z)^3 \tag{6}$$

for $z < z_R$ and $\rho_H = 3.23 \times 10^{-26}$ g cm^{-3} $\tilde{M}_H \ \tilde{R}_H^{-3}$.

If we match expressions (5) and (6) with the cosmological "inos" energy density given in eqs. (1) and (2) by using eqs. (3) and (4), we obtain constraints on mass, T_x/T_e, g_x, and ξ_x of "inos" by the following equations:

$$\sum_x g_x m_x (T_x/T_e)^3 \ F_{nr} \ (\xi_x) = 36.24 \ ev \ \Theta^3 \ \Omega h^2 \tag{7}$$

$$\sum_x g_x (T_x/T_e)^4 \ F_r \ (\xi_x) = 3.74 \ \Theta^{-4} \ (\Omega h^2)^{-4/3} \ \tilde{M}_H^{1/6} \ \tilde{R}_H^{1/2} \ . \tag{8}$$

It is very interesting to note that, if we define the effective number of degrees of freedom of "inos" by, (see eq. (1)),

$$N_x = 7/8 \ \sum_x g_x (T_x/T_e)^4 \ F_r \ (\xi_x) \tag{9}$$

the constraints on N_x, from the cosmological and galactic halo point of view, are then

$$N_x = 3.27 \ \Theta^{-4} \ (\Omega h^2)^{-4/3} \ \tilde{M}_H^{1/6} \ \tilde{R}_H^{1/2} \ . \tag{10}$$

This value is consistent with the one obtained by Yang $\underline{et \ al.}$ (1984).

For the most canonical case with $g_x = 2$, assuming the one "inos" dominating case we made a Figure 1 for the mass and temperature ratio with varying ξ_x, the degeneracy parameter. Knowing the condition $T_x/T_e \leq 1$, we may put constraints on mass and temperature. For this case, the governing equation for mass is

$$m_x = 11.3 \ ev \ F_r^{3/4}(\xi_x) \ F_r^{-1} \ (\xi_x) \ \tilde{M}_H^{-1/8} \ \tilde{R}_H^{-3/8} . \tag{11}$$

Clearly the mass depends on the cosmological (F's in eq. (11)) and galactic halo ($\tilde{M}_H$ and $\tilde{R}_H$) components. For values of ξ_x, from zero to infinity, a very limited range of m_x is allowed and complimentary information can be obtained by the study of supercluster structure (Ruffini $\underline{et \ al.}$, 1986), and by the morphological study of galactic rotation (Malagoli and Ruffini, 1986, Arbolino and Ruffini, 1986).

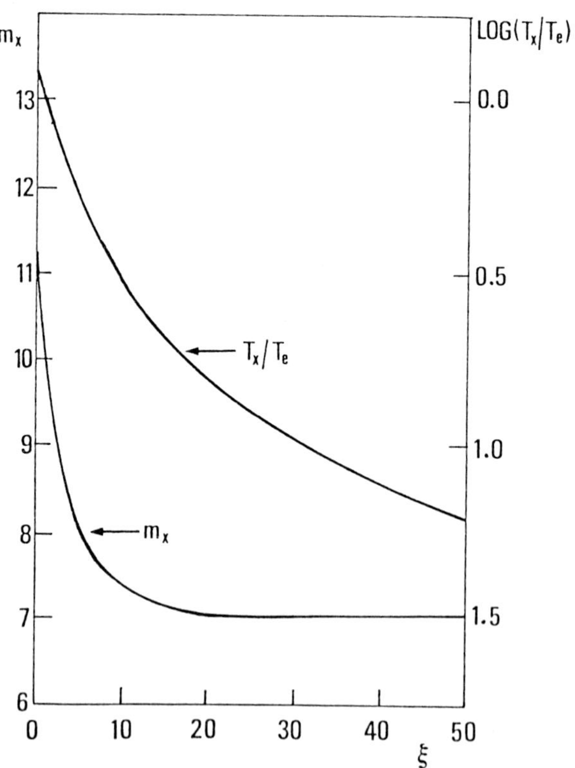

FIG. 1. The mass, m_x, and temperature ratio, T_x/T_e, are plotted by changing the degeneracy parameter ξ_x. Only the values of $T_x/T_e \leq 7.14$ x 10^{-1} is cosmologically allowed.

REFERENCES

Arbolino, M., and Ruffini, R. 1986, Astron. Astrophys., in press.
Faber, S.M., and Gallagher, J.S. 1979, Ann. Rev. Astron. Astrophys.,
      17, 135.
Forman, W., Jones, C., and Tucker, W. 1985, Ap. J. 293, 102.
Gershtein, S.S., and Zel'dovich, Ya. B. 1966, JETP Lett. 4, 120.
      See also Lee, B.W., and Weinberg, S., 1977, Phys. Rev. Lett., 38,
      1237, and Ruffini, R., Song, D.J., and Stella, L. 1983, Astron.
      Astrophys., 125, 265.
Malagoli, A., and Ruffini, R. 1986, Astron. Astrophys. 157, 283.
Rubin, V.C., Ford, W.K., and Thonnard, N. 1978, Ap. J. 228, L107.
Ruffini, R., and Song, D.J., 1986a, in the Proceedings of the LXXXVI
      Summer School of E. Fermi, Ed. F. Melchiorri and R. Ruffini.
Ruffini, R., and Song, D.J. 1986b, Astron. Astrophys., in press.
Ruffini, R., Song, D.J., and Taraglio, S. 1986, preprint.
Sancisi, R. 1987, in this volume.
Yang, J., Turner, M.S., Steigman, D., Schramm, D.N., and Olive, K.A.
      1984, Ap. J., 281, 493.

CHAPTER XIII

GRAVITATIONAL LENSES

GRAVITATIONAL LENSES AS TOOLS IN OBSERVATIONAL COSMOLOGY

Claude R. Canizares
Department of Physics and Center for Space Research
Massachusetts Institute of Technology

1. INTRODUCTION

The study of gravitational lenses is intimately tied to observational
cosmology. When we observe a gravitationally lensed quasar, we are
viewing a single object along two or more neighboring paths (null
geodesics) of cosmological dimensions (Figure 1). What we see depends
on bulk properties of the universe, such as H_0 and q_0, on the large
scale structure and inhomogeneities along the paths, and on the small
scale structure in and around the primary deflector. Furthermore, the
deflection of light depends on the gravitational field along the line
of sight, so it is sensitive to all forms of matter: luminous or dark,
baryonic or exotic. Thus the images of gravitationally lensed quasars
contain an imprint of the universe that is virtually inaccessible by
any other means. The hope of decoding this imprint has stimulated
observers and theorists to expend many thousands of hours of telescope
time, computer time and cogitation on the elucidation of gravitational
lens properties.

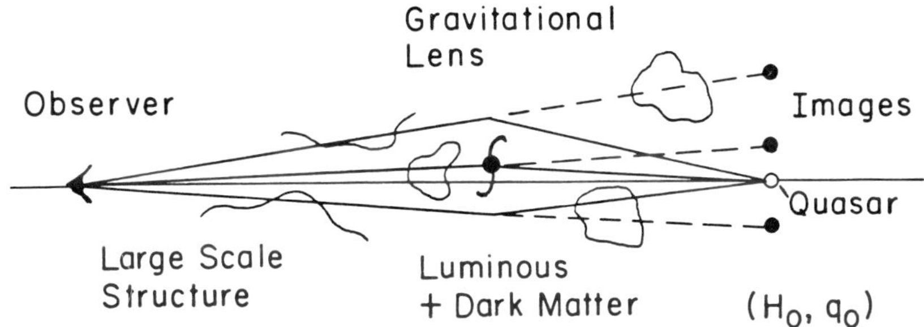

Figure 1. Schematic of a gravitational lens.

As a reviewer of the vast literature that this subject has generated,
I must be selective. I have chosen to focus on topics directly
relevant to this meeting, though this forces me to ignore much
interesting work. (See Peacock 1983, Turner 1987, Gott 1987, and
Burke 1986 for other reviews).

A. Hewitt et al. (eds.), Observational Cosmology, 729–746.

Before embarking on a discussion of the data, let me recall some of
the basics. Gravitational lensing involves the bending of light from a
distant quasar by matter along the line of sight. For the moment I
will make the usual assumption that most of the deflection takes place
in a localized region, within a galaxy or cluster, and that the
universe is otherwise reasonably homogeneous (but see below). Then a
mass concentration will bend the light sufficiently to make multiple
images of a distant object if its surface mass density Σ exceeds $\Sigma_c \approx$
0.7 g cm^{-2}, for reasonable assumptions about the distances to the lens
and quasar (Narayan, Blandford and Nityananda 1984, Turner, Ostriker
and Gott (1984), Subramanian and Cowling 1986). If the lens has a
density distribution characterized by a core radius r_{core}, this
translates to a limit on its line-of-sight velocity dispersion of $\sigma >$
1000 $(r_{core}/100$ kpc$)^{0.5}$ km s^{-1}. Individual galaxies generally satisfy
this criterion, whereas most clusters do not. The separations between
images are roughly comparable to the angular scale of the deflector,
~1-2" for a galaxy. Image separations > 1-2" can be caused by a
combination of galaxy plus cluster — in effect the galaxy makes the
images and the cluster spreads them (Turner, Ostriker and Gott 1984,
Narayan, Blandford and Nityananda 1984, Dyer 1984, Kovner 1986). Solid
objects like stars, of course, have surface densities $\gg \Sigma_c$ and so
they also can make multiple images. These so-called microlenses give
image separations ~$10^{-6}(M_L/M_\odot)^{1/2}$ arc sec, where M_L is the mass of the
lens (e.g. Press and Gunn 1973).

2. FERMAT'S PRINCIPLE

In general, the images formed by anything other than a highly
symmetric lens can be very complex. A major and elegant theoretical
advance has been the application of Fermat's principle to this
problem, which greatly facilitates the computation and visualization
of image formation (Schneider 1985, Blandford and Narayan 1986,
Narayan 1986). I remind you that Fermat's principle in classical
optics states that light will travel from a source to an observer
along paths for which the propagation time is an extremum:

$$\delta \int d\tau = 0.$$

In the relativistic extension, the propagation time for a given path
has two terms, the geometric travel time for the deflected ray and the
gravitational time delay. For a given distant source, one can compute
the light travel time for a possible ray that reaches the observer
from each position (α,δ) on the sky (Figure 2). This gives a two
dimensional "Fermat time surface" $\tau(\alpha,\delta)$. The locations of the
extrema of this surface then correspond to the locations of the images
of the source on the sky. In a transparent homogeneous universe with
no lens, there is one extremum: the path with minimum propagation time
that defines the line of sight to the source (Figure 3a). Adding mass
along the line of sight increases the propagation time for the central
rays. For example, a cylindrically symmetric transparent mass with

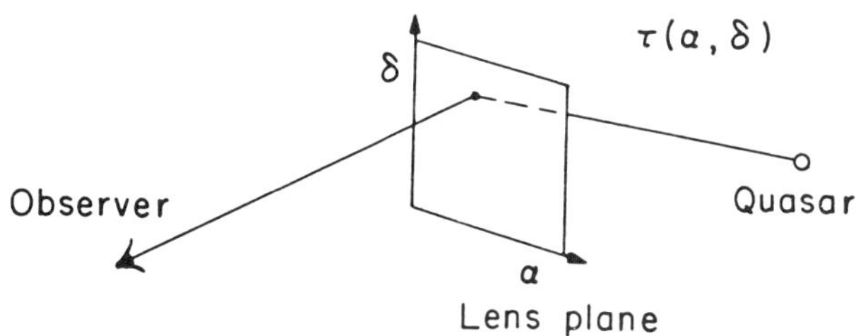

Figure 2. A possible light path from a quasar to an observer. In the "thin lens" approximation, bending can only take place in the plane containing the excess mass associated with the gravitational lens (e.g., a glaaxy of cluster). The total propagation time τ is expressed as a function of position on the sky (α · δ) as seen by the observer. Images occur where τ is an extremum (Fermat's principle).

$\Sigma > \Sigma_c$ centered on the line-of-sight introduces an inflection in the Fermat time surface. The central ray is delayed so it is no longer a minimum but rather a local maximum (Figure 3b). The surface has a trough around the line of sight, so this very special case gives a point image surrounded by a ring image. With less symmetry one would get a maximum, a minimum and a saddle point corresponding to three images on the sky (Figure 3c,d). More complex mass distributions will distort the time surface still further giving additional images.

One thing that emerges immediately is that any smooth, non-singular, transparent mass distribution must give an odd number of images (Dyer and Roeder 1980, Burke 1981, Blandford and Narayan 1986): beyond the original image, new images come in pairs corresponding to either a maximum or minimum in the time surface plus a saddle point (except for singular cases like the ring in Figure 3b). The more compact the mass distribution, the narrower its primary maximum and the fainter the corresponding image (the radius of curvature of the surface near the extremum gives the magnification of the image). For a single microlens the maximum becomes a cusp, leaving only two images (or more generally and even number). Images corresponding to maxima and minima have positive parity, whereas for saddle points they have negative parity -- the latter are mirror reflections of an extended source.

3. OBSERVATIONS

There are now more than ten candidates for gravitationally lensed quasars (see Table 1), several of which have recently emerged from a massive search program reported here by Hewitt (1986,1987), and one of which is newly reported at this meeting (Hammer 1987). The degree of public confidence that lensing is truly observed varies from object to object. Most practitioners would agree that the first four on the list are secure. The next two are less well established, 1146+111 is highly controversial, and the last three are new, unpublished candidates.

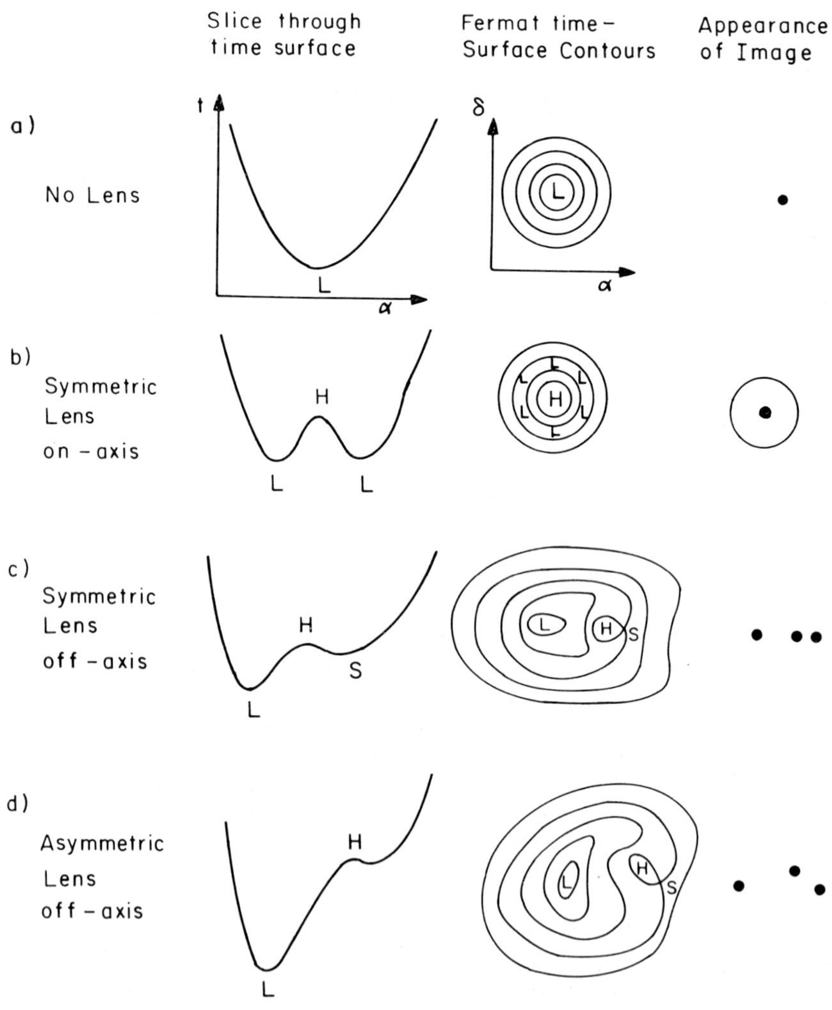

Figure 3. Illustration of the application of Fermat's principle for a distant point source seen through various possible distributions of transparent lens mass. The first column shows a slice through the Fermat time surface (τ vs. α at constant δ), the second column shows full two dimensional surface as contours of constant τ, the third columns shows the resulting images. Minima, maxima and saddlepoints are marked by L, H, and S, respectively. These figures follow Blandford and Narayan (1986) and Narayan (1986). a) No lens. b) Cylindrically symmetric lens centered on the (undeflected) line of sight to the source. c) Cylindrically symmetric lens off the line of sight. d) Asymmetric lens off the line of sight.

Table 1

GRAVITATIONAL LENS [CANDIDATES]

Name	z_Q	images	$\Delta\theta$	Lens?	Reference
0957+561	1.4	2	6"	Yes	1
1115+080	1.7	4	2	Maybe	2
2016+112	3.3	3	4	Yes?	3
2237+031	1.7	2	2	Yes	4
1635+267	2.0	2	4	No	5
2345+007	2.2	2	7	No	6
1146+111	1.0	2	157	No	7
3C324	1.2	2+	2	Yes	8
0023+171	1.0	2+	5	No	9
1042+178	0.9	4+	2	?	9

References: 1. Walsh, Carswell and Weymann (1979), Greenfield, Roberts and Burke (1985); 2. Weymann *et al.* (1980); Weymann and Folz (1983), Foy, Bonneau, and Blazit (1985), Henry and Heasley (1986); 3. Lawrence *et al.* (1984), Schneider *et al.* (1985, 1986); 4. Huchra *et al.* (1985), Tyson and Gorenstein (1985); 5. Djorgovski and Spinrad (1984); 6. Weedman *et al.* (1982), Foltz *et al.* (1984), Tyson *et al.* (1986); 7. Turner *et al.* 1986, Shaver and Cristiani 1986, Huchra 1986; 8. Hammer (1986); 9. Hewitt (1986, 1987);

I will not review each of them in detail, but will say a few words about some areas of current interest or controversy. In particular, I will focus on the degree to which the observations can be understood in the context of conventional theory. Some reasonable level of understanding is a clear prerequisite if lenses are to be used as tools of cosmology.

The first and by far the best studied and most secure lens is 0957+561 (Walsh, Carswell, and Weyman 1979, Greenfield, Roberts and Burke 1985 and references therein). The lensing mass is undoubtedly associated with a galaxy and cluster seen at $z = 0.36$. Because the imaged quasar is an extended radio source, it is possible to extract considerable information about the lens from the observations. Recently, Cohen *et al.* (1987; see also Gorenstein *et al.* 1984) have obtained exquisite VLBI maps of two radio jet images which show structure on milli arcsecond scales. The relative net magnification of the jets matches their relative brightness, as one would expect (neglecting extinction, gravitational lenses are achromatic and conserve surface brightness). Furthermore, the two jet images have identical shapes except for a parity inversion; this tells us immediately that one of the images corresponds to a saddle point in the Fermat time surface. The information from these maps should add constraints to models of the mass distribution in the lens (Greenfield, Roberts and Burke 1985, Cohen *et al.* 1987).

However, there are two major difficulties in the straightforward interpretation of the various lens candidates. These are not fundamental; they do not negate the lens hypothesis in general. But they do raise doubts about some of the candidates and complicate considerably the application of lensing as a tool of cosmology, even for 0957+561.

The Odd Image Problem As noted above, transparent lenses should generally produce an odd number of images, yet all but one of the lens candidates have an even number (Table 1). The exception is 2016+112, for which a faint ($>24^m$) third image was recently found lying practically on top of the image of a foreground galaxy (Schneider *et al.* 1986). Several explanations have been advanced for the general absence of the odd image:

(i) The odd image is unresolved or lies far from the other images (the latter is possible if clusters do most of the lensing; Narayan, Blandford and Nityananda 1984, Kovner 1986).

(ii) The lens galaxy has a compact core which, as noted above, causes one of the images to be very faint (Narasimha, Subramanian, and Chitre 1986). A core containing $\tilde{}10^{10}$ $M_\odot$ within 50 pc would reduce the image brightness by factors of 10^3-10^4.

(iii) The light of the odd image suffers extinction in the lens galaxy. This is not a viable explanation for 0957+561, because it cannot account for the absence of a radio image.

(iv) Stars in the lens galaxy act as microlenses which, under some circumstances, can preferentially dim the odd image (although they will also occasionally make it significantly brighter; Chang and Refsdal 1984, Subramanian, Chitre and Narasimha 1985, Paczynski 1986a). Again, this will not be effective for 0957+561, because the radio source is too large to be affected by a stellar microlens.

None of these explanations is entirely satisfying, although the degree of embarrassment for lens theory depends on how many of the candidates of Table 1 are actually lenses. Observations with the Hubble Space Telescope should be very useful for addressing this problem.

The Elusive Lens Problem This problem has two aspects. For several cases (e.g. 0957+561 and 2016+112) one can see foreground galaxies and clusters that are almost certainly responsible for the gravitational lensing. However, models which assign mass to these objects with constant mass-to-light ratios (M/L) do not account for the observed properties of the multiple images (Greenfield, Roberts, and Burke 1985, Schneider *et al.* 1985, 1986). Although this tells us something important about dark matter (see below), it hampers the use of straightforward lens models to constrain cosmological parameters (e.g. Falco, Gorenstein and Shapiro 1985)

For most of the candidates of Table 1, there is no evident lensing object. This is surprising. For an imaged quasar with $z_q > 1$, the most probable redshift of the lens is $z_l = 0.5 - 1$ (Canizares 1982, Turner, Ostriker and Gott 1984), so any reasonably massive galaxy that contributes (e.g. one with luminosity $\tilde{} L^*$) will have $m_R \sim 20-23$ (here I have used $M_R^* = -21.8$, $H_0 = 100$ km s^{-1}, $\Omega_0 = 1$, and $K_R = 0.7$ or 2.0 at $z = 0.5$ or 1, respectively). Therefore, one would expect the lens to be clearly visible on deep CCD pictures. Deep searches appear to have been successful for 1115+080 (Foy, Bonneau, and Blazit 1985, Henry and Heasley 1986, Shanklan and Hege 1986, Christian, Crabtree and Waddell 1986) and for 2016+112 (Schneider et al. 1985, 1986). But for 1635+267 there is no galaxy brighter than 23.5 mag (Djorgovski and Spinrad 1984), while for 2345+007 the limit is J=25.5 for any galaxy located between the quasar images (Tyson et al. 1986).

One possibility is that some of the lenses are compact poor groups of galaxies. Such groups may be observed in 2016+112 (Schneider et al. 1986), 1115+080 (Henry and Heasley 1986), and perhaps 2345+007 (Tyson et al. 1986). As pointed out by Narayan, Blandford and Nityananda (1984, also Blandford and Jarosczynski 1981, Kovner 1986) such groups could match or even exceed rich clusters in their effectiveness as lenses because of their smaller core radii. X-ray images show that some poor clusters do have well developed gravitational potentials (Kriss, Cioffi and Canizares 1983), and that potentials are not always traced by galaxy distributions (Beers, Huchra and Geller 1983).

By far the most spectacular case of an elusive lens is 1146+111, which was discovered by Arp and Hazard (1980) and advanced by Paczynski (1986b) as a potential lens candidate. Turner et al. (1986) confirmed common redshifts (to $\pm$ 300 km s^{-1}) and showed that the spectra over $\lambda 4500 - 8000$ were nearly identical. The very large separation of the two objects, 157", means that the deflector must have a mass of $\tilde{} 10^{15}$ $M_\odot$. This makes it an extraordinary object. However, there is no obvious visible candidate for the deflector: Bahcall, Bahcall and Schneider (1986) show that a conventional very rich cluster would have been seen on deep CCD images. The wide separation and the very absence of an apparent lens led Paczynski (1986b) and Turner et al. (1986) to suggest that the lens might be a "cosmic string," a relic of the GUTS phase transition in the early universe. Vilenkin (1984) and Gott (1985) had previously predicted that such strings, if they exist, would best be detected in precisely this manner. Other suggestions for the lens include a 10^{15} $M_\odot$ black hole (Paczynski 1986b), an unusual cluster (Ostriker and Vishniac 1986), clusters of unstable dark matter (Dekel and Piran 1986), and burnt out galaxies (Silk 1986).

The gravitational lens candidacy of 1146+111 has been questioned on several grounds (see Turner's contribution to this meeting for a more complete discussion of this object). Shaver and Cristiani (1986) and Huchra (1986) have obtained spectra to the red and blue (respectively) of the Turner et al. spectrum in which, these authors stress, the two images are clearly not identical. Actually, although they are correct in noting differences, **the** overall spectral shapes are remarkably similar from the near UV to the near IR. Proponents find refuge in the fact that identical spectra are not required because the difference in light travel times for two such widely spaced images would exceed a thousand years, and as Huchra has stated, nobody knows how much quasars vary over several millennia. But any lens capable of causing such large separations ought to have other detectable features (such as double images of other nearby quasars or fluctuations in the cosmic microwave background). Some of these can already be ruled out (e.g. Paczynski 1986b, Gott **1986,** Vilenkin 1986, Stark *et al.* 1986, Blandford, Narayan and Phinney 1987), and others are accessible to further ground based and space observations.

In addition to sparking excitement, the lens hypothesis for 1146+111 has provoked a kind of backlash against the reality of this and several other of the more tenuous lens candidates. Phinney and Blandford (1986) suggest that 1146+111 could well be a coincidence, an accidental pair of quasars with nearly identical redshifts. Bahcall, Bahcall and Schneider (1986) point out that if quasars are clustered (as may well be the case: see Shaver 1984), then chance coincidences could account for 1635+267 and 2345+007 as well as 1146+111. At some level all quasar spectra share common features, and unrelated quasars of comparable redshift might have very similar spectra by coincidence. Unfortunately, no one has ever quantified the probability of chance spectral similarity, so there is no way of including this in the statistical estimates.

The excitement over 1146+111 has increased the attention and the amount of telescope time being devoted to lenses, so it is likely that eventually the question of reliability will be settled. Although losing several lens candidates would be very disappointing, as a compensation it would alleviate somewhat both the odd image and elusive lens problems. And the lens searches should continue to provide new candidates.

Before leaving the observations, I want to mention the unusual case of 2237+031 (Huchra *et al.* 1985). Here the putative lens is very visible: it is a 15^m spiral galaxy at z = 0.04 with the spectrum of a z = 1.7 quasar superimposed on its nucleus. Huchra *et al.* immediately invoked lensing as the explanation, to the chagrin of Burbidge (1985), who prefers the interpretation that the quasar is local (i.e. its redshift is non-cosmological). Tyson and Gorenstein (1985) showed that the image is multiple, which confirms one test of the lens hypothesis. But the skeptics could reverse an argument often used against them by pointing out the lack of a control sample (how many nuclei of nearby galaxies have been so carefully scrutinized?), or they may suggest that all galaxies which eject quasars look like this. The Hubble Space Telescope should show if two (or even better, three) of the point sources have identical spectra, as the lens interpretation requires.

4. MEASURING COSOMOLOGICAL PARAMETERS

a) Hubble's Constant

Taking for granted the eventual improvement in observations, what are the prospects for using lenses as cosmological tools? By far the greatest attention has been given to a determination of H_0, following the work of Refsdal (1964, 1966), who pointed out that the difference in light travel time for two images of a gravitationally lensed quasar is inversely proportional to H_0. Happily, quasars are variable, so the difference in light travel time can be found simply by cross correlating the light curves of two images.

Unfortunately, as Einstein once remarked, nature is subtle, so there have been serious difficulties carrying out this important program. First, there are the observational realities: both 0957+561 (Schild and Cholfin 1986 [optical], Hewitt, Roberts and Burke 1984 and Hewitt 1986a [radio]) and 1115+080 (Vanderriest et al. 1986), the only lensed pairs that have been monitored, have displayed frustratingly bland light curves that are hard to correlate. Schild (private communication) has an additional 50-60 nights of data, which could strengthen confidence in the ~1.1 year time delay for 0957+561 reported earlier this year (Schild and Cholfin 1986). In any case, Einstein also noted that nature is not malicious, so we can hope eventually to have clean time delay measurements for several quasar image pairs.

The second class of difficulties may be less tractable, however. These have to do with our ability to extract H_0 from the timing measurements, whatever their accuracy. This has been a subject of sharp controversy in the recent literature, with opinions ranging from optimism (Borgeest and Refsdal 1984, Kayser 1986, Borgeest 1986) to varying degrees of pessimism (Falco, Gorenstein and Shapiro 1985, Alcock and Anderson 1985, 1986, Blandford and Narayan 1986).

That different images have different travel times is evident directly from the Fermat time surface (e.g. in Figure 3 the high points correspond to late times, etc.). What is also clear is that the difference depends on the details of the lens: an accurate measure of H_0 requires a reasonably accurate lens model. As noted above, direct observation of the lens galaxies and clusters does not yield straightforward models; assumptions of constant M/L, for example, (e.g. Boorgeest 1986) are not warranted.

On the other hand, it may be possible to measure enough observables (image brightness ratios, parities, or even the full magnification matrices) to provide interesting constraints on H_0 from time delay measurements. First applications of this approach to 0957+561 were made by Boorgeest and Refsdal (1984) and, in more detail, by Falco, Gorenstein and Shapiro (1985). Falco et al. find $H_0 < 100$ km s^{-1} for $\Delta t = 1$ year (Schild and Cholfin 1986). The value itself corresponds

to a "minimum mass" model that reproduces the relative position and
brightness of the A and B images. One has the freedom of adding a
uniform mass to the model, which would leave the observables unchanged
if H_O were smaller -- hence the upper limit (see Figure 4). The new
VLBI observations should provide more constraints on the minimum mass
model (Gorenstein, private communication), but it is also important to
limit the freedom in model parameters by measuring, for example, the
velocity dispersions of the primary lensing galaxy and cluster.

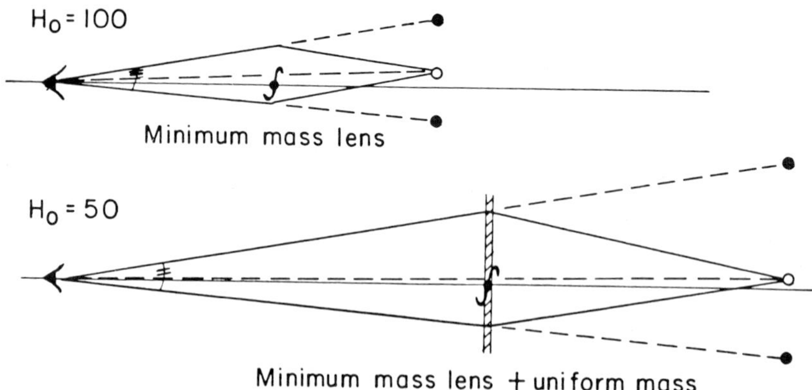

Figure 4. Illustration of the amiguity in determining H_O from the
observations of a lensed quasar. If a "minimum mass" lens model
accounts for the observed relative image separations and time delays
with H_O = 100 km s^{-1} Mpc^{-1}, then one can always construct another
model with additional uniformly distributed mass that also accounts
for the observations but with a lower value of H_O (e.g. 50 km s^{-1}
Mpc^{-1}). The *true* position of the source for the two cases is different
but is not observable. (After Falco, Gorenstein and Shapiro 1985.)

One remaining difficulty is the validity of the assumption of large
scale homogeneity of the universe, which is incorporated into every
model of a gravitational lens. Alcock and Anderson (1985, 1986) have
questioned this assumption, noting that just as a lensed quasar may be
a very rare alignment of a galaxy and/or cluster with a background
quasar, so may it signal a rare line of sight with higher than average
cosmological density. This reasoning is valid in general, but is
probably weakest for 0957+561, for which the observed galaxy+cluster
is already a plausibly sufficient lens. At any rate, the problem is
most severe for higher redshift quasars; for 0957-561 at z = 1.4, even
extreme variations by up to factors of ~2 in mean density in or near
the line sight (causing either focusing or shear of the light beams of
both images) result in < 30% uncertainties in the deduced value of H_O
(Alcock and Anderson 1985).

To summarize, gravitational lenses are no magic solution to the long
standing problem of measuring H_O. So far they have provided at best a
rather preliminary upper limit that encompasses the various values

deduced from direct distance measurements. On the other hand, the promise of an independent constraint on this fundamental cosmological parameter is still very much alive, and it has taken only seven years to approach the accuracy of the results achieved after many decades of distance measurements. What is most important to me is that each lensed quasar could eventually yield a reasonably independent value of H_0. Consistency between several such measurements will provide a check on the validity of common assumptions, such as large scale isotropy. This will be the most, and perhaps the only, convincing argument for the validity of the method.

b) Cosmological Constant

If the cosmological constant Λ is non-zero, it could affect significantly the image separations for a given gravitational lens (Paczynski and Gorski 1981, Alcock and Anderson 1986, Gott 1987). Gott (1987) used the fact that 2016+112, with a redshift of $z_q = 3.3$, has relatively "normal" images to set a limit of $q_0 > -2.3$ or $\Lambda < 6.9$ H_0^2 for $\Omega_0 = 1$.

c) Dark Matter in Galaxies and Clusters

As noted above, the models of 0957+561 and 2016+112 already tell us that the mass cannot follow the light in the lensing galaxies and clusters (Greenfield, Roberts, and Burke 1985, Schneider *et al.* 1985, 1986). As the lens models improve, they will undoubtedly contribute to our understanding of the quantity and distribution of dark matter (Gott 1987, Borgeest 1986).

d) Large Scale Structure

As mentioned above, large scale inhomogeneities in the matter distribution along the line of sight to a gravitationally lensed quasar could have a significant effect on the appearance of the images (Alcock and Anderson 1985, 1986). Large scale clumping of matter can introduce a stochastic variation in the effective distance (the angular diameter distance) for different lines of sight.

Anderson and Alcock (1986) showed that normal galaxy clustering has only a very minor effect on gravitational lens images. But there is increasing evidence that luminous matter is structured on scales > 100 Mpc (e.g. De Lapparent, Geller, and Huchra 1986, see also the contributions of Geller and Gott to this meeting). One estimate by Turner, Ostriker and Gott (1984) shows that gravitational lensing might be enhanced along some lines of sight through a universe in which most of the matter is distributed on shells. It would be interesting to see this explored in more detail. For example, I wonder under what conditions one could obtain multiple images from large scale inhomogeneities alone, in which no single galaxy would stand out as the lens.

As a long term goal, one can envision using the properties of an ensemble of lenses to set limits on the inhomogeneity of mass in the universe. Gravitational lensing may be the only way to decide whether or not the large voids are empty of mass as well as light.

e) Microlenses and the Nature of Dark Matter

Gravitational lensing may also be the only way of ascertaining if a significant fraction of the matter in the universe is in the form of compact objects.

If objects such as black holes or subluminous stars account for the dark matter in galaxy halos, then they should cause large fluctuations in the brightness of images whose light passes through the core of a galaxy (e.g. Chang and Refsdal 1979,1984, Young 1981, Gott 1981, Paczynski 1986a, Kayser, Refsdal and Stabell 1986; see also Grieger, Kayser and Refsdal 1986). The fluctuations have rather unusual signatures that should be distinguishable from other intrinsic variability. Some of the images of the objects in Table 1 should eventually show this effect, as should quasars like 0104.2+3153 (Stocke et. al 1984), which are viewed through a foreground galaxy at too large an angle to form multiple images.

Brightening of quasars by minilenses in galaxy halos will also cause enhancements of the apparent surface density of quasars near foreground galaxies (Canizares 1981, 1983, Vietri and Ostriker 1983, Zuiderwijk 1985, Schneider 1986a,b,c). Since my suggestion of this effect, several refinements have been made to the calculations, but the results are nearly the same (for a given assumed quasar luminosity function, the enhancements found by Vietri and Ostriker 1983 and Schneider 1986a,b,c differ by < 20% from those in Canizares 1981). The detectability of the effect depends on the true shape of the luminosity function. Samples of $>10^4$ galaxies, which corresponds to samples of many hundreds of quasars, are probably needed to see it (Vietri and Ostriker 1983, Canizares 1983, Schneider 1986b,c).

Brightening by microlenses may cause noticeable or even large effects on quasar luminosity functions (Turner 1980, Ostriker and Vietri 1986a, Vietri 1985, Schneider 1986d). The effect can be very strong if the intrinsic function is steep, as may well be the case (e.g. see the relevant contributions to this meeting). In fact, Ostriker and Vietri (1985) have made the intriguing suggestion that most BL Lac objects are microlensed quasars for which the small sized continuum emitting region is brightened while the larger line emitting region is not. Recently Wagoner (1986) and Schneider and Wagoner (1986) have suggested microlensing of a single evolving background supernova as a way of detecting compact dark matter.

The absence of certain lens effects is already sufficient to rule out cosmological densities in compact objects of various masses. The optical depth for microlensing is roughly equal to the density of compact objects in units of the critical density (Press and Gunn 1973). The effect of the lensing on the observed properties of quasars depends on the mass of the lenses: large masses will produce multiple images whereas smaller ones will cause variability and give differential brightening of continuum and emission lines, which will broaden any intrinsic distribution of equivalent widths (Canizares 1982). Table 2 lists limits on the contribution of compact objects to Ω_0 deduced from the apparent absence of such effects in quasar samples. The data of Hewitt (1986a,b) can also be used to limit the cosmological density of non-compact lenses (e.g. isothermal spheres), to be less than 0.1 - 1. Mpc^{-3} (depending on their central mass density). While these limits are still high, eventually this technique will tell us whether or not objects of galactic mass (from giant black holes to failed galaxies made of dark matter) are sufficiently numerous to close the universe.

Table 2

LIMITS ON COSMOLOGICAL DENSITY OF COMPACT OBJECTS

Ω_L	$M_L/M_\odot$	Reference
< 0.4	10^{11}-10^{13}	Hewitt (1986)
< 1.0	0.1-200	Canizares(1982,1983)
< 0.1	200-10^5	"

5. CONCLUSIONS

I have chosen to stress some of the problem areas of gravitational lens research because that is where the activity will be for the next several years. But I want to conclude by listing the good points about lenses. First and foremost is their existence, which is very well established in several cases. Second, there are lots of things to measure: new candidates, image location,. brightness, shape, variability, lens properties, etc. This is sure to keep observers very busy for some time to come. Third, gravitational lenses are amenable to modeling. Theoretical tools like Fermat's principle can be used to probe the considerable parameter space of each lens. This is sure to keep theorists equally busy. Fourth, gravitational lenses probe cosmology. Studies will undoubtedly contribute to our understanding of H_0, Λ, Ω_0, large scale structure, etc. Fifth, unlike some other measures of cosmological parameters, a sample of lenses could yield independent measurements which will act as a consistency check on possible systematic errors. Sixth, gravitational lens phenomena are good probes of dark matter on various scales. Lastly, gravitational lenses remain exciting and surprising. Surely there are more discoveries in store for us.

ACKNOWLEDGEMENTS

I am deeply indebted to many colleagues for useful discussions, for detailed comments on the manuscript, and for providing information or preprints prior to publication. In particular I thank Charles Alcock, Roger Blandford, Marc Gorenstein, J. Richard Gott, Jacqueline Hewitt, John Huchra, Rudy Schild, and Edwin Turner.

REFERENCES

Alcock, C. and Anderson, N. 1985, *Ap. J. (Letters)*, **291**, L29.
Alcock, C. and Anderson, N. 1986, *Ap. J.*, **302**, 43.
Anderson, N. and Alcock, C. 1986, *Ap. J.*, **300**, 56.
Arp, H. and Hazard, C. 1980, *Ap. J.*, **240**, 726.
Bahcall, J. N., Bahcall, N. A., and Schneider, D. P. 1986, preprint.
Beers, T. C., Huchra, J. P., and Geller, M. J. 1983, *Ap. J.*, **264**, 356.
Blandford. R. and Jaroszynski, M. 1981, *Ap. J.*, **246**, 1.
Blandford, R. and Narayan, R. 1986, *Ap. J.*, (in press).
Blandford, R., Narayan, R., and Phinney, S. 1987, *Ap. J.*, (in press).
Borgeest, U. 1986, *Ap. J.*, (submitted).
Borgeest, U., and Refsdal, S. 1984, *A. Ap.*, **141**, 318.
Burbidge, G. 1985, *A. J.*, **90**, 1399.
Burke, B. 1986, in Swarup, G. and Kapahi, V. K. (eds), *Proc. IAU Symp. #119: Quasars,* (Reidel), 517.
Burke, W. 1981, *Ap. J. (Letters)*, **244**, L1.
Canizares, C. R. 1981, *Nature*, **291**, 620; erratum **293**, 490.
Canizares, C. R. 1982, *Ap. J.*, **263**, 508.
Canizares, C. R. 1983, in *Proc. IAU Symp. #117: Quasars and Gravitational Lenses,* (Liege: Institut d'Astrophysique), 126.
Chang, K., and Refsdal, S. 1979, *Nature*, **282**, 561.
Chang, K., and Refsdal, S. 1984, *A. Ap.*, **132**, 168.
Christian, C., Crabtree, D., and Waddell, P., 1986, IAU Circ. No. 4182.
Cohen, N. L., Gorenstein, M. V., Shapiro, I. I., Rogers, A. E., Bonometti, R. J., Falco, E. E., Barel, N., Marcaide, J. M. 1987, in preparation.
Dekel, A. and Piran, T. 1986, *Ap. J. (Letters),* (submitted).
De Lapparent, V., Geller, M., and Huchra, J. P. 1986, *Ap. J. (Letters)*, **302**, L1.
Djorgovski, S. and Spinrad, H. 1984, *Ap. J. (Letters)*, **282**, L1.
Dyer, C. C. 1984, *Ap. J.*, **287**, 26.
Dyer, C. C. and Roeder, 1980, *Ap. J. (Letters)*, **238**, L67; erratum **242**, L53.
Falco, E., Gorenstein, M., and Shapiro, I. 1985, *Ap. J. (Letters)*, **289**, L1.
Foltz, C. B., Weyman, R. J., Roser, H.-J., and Chaffee, F. H., Jr. 1984, *Ap. J. (Letters)*, **281**, L1.
Foy, R., Bonneau, D., and Blazit, A. 1985, *A. Ap.*, **149**, L13.
Gott, J. R. 1981, *Ap. J.*, **243**, 140.
Gott, J. R. 1985, *Ap. J.*, **288**, 422.
Gott, J. R. 1986, *Nature*, **321**, 420.

Gott, J. R. 1987. in Knapp, G. (ed), *Proc. IAU Symp. #117: Dark Matter in the Universe*, (Reidel).

Gorenstein, M. V., *et al.* 1984, *Ap. J.*, **287**, 538.

Greenfield, P., Roberts, D., and Burke, B. 1985, *Ap. J.*, **293**, 370.

Grieger, B., Kayser, R., and Refsdal, S. 1986, submitted to *Nature*.

Hammer, F. **1987**, this volume.

Henry, J. P. and Heasley, J. N. 1986, *Nature*, **321**, 139.

Hewitt, J. N. **1986**, MIT PhD Thesis.

Hewitt, J. N. **1987**, this volume.

Hewitt, J. N., Roberts, D. H., and Burke, B. F. 1984, *B. A. A. S.*, **16**, 519.

Huchra, J. P. 1986, *Nature*, (submitted).

Huchra, J. P., Gorenstein, M., Kent, S., Shapiro, I., Smith, G., Horine, E., and Perley R. 1985, *A. J.*, **90**, 691.

Kayser, R. 1986, *A. Ap.*, **157**, 204.

Kayser, R., Refsdal, S., and Stabell, R. 1986, *A. Ap.*, (submitted).

Kriss, G. A., Cioffi, D. F., and Canizares, C. R. 1983, *Ap. J.*, **272**, 439.

Kovner, I. 1986, preprint.

Lawrence, C. R., Schneider, D. P., Schmidt, M., Bennett, C. L., Hewitt, J. N., Burke, B. F., Turner, E. L., and Gunn, J. E. 1984, *Science*, **223**, 46.

Narasimha, D., Subramanian, K., and Chitre S. M. 1986, *Nature*, **321**, 45.

Narayan, R., 1986, in Swarup, G. and Kapahi, V. K. (eds), *Proc. IAU Symp. #119: Quasars*, (Reidel), 529.

Narayan, R., Blandford, R., and Nityananda, R. 1984, *Nature*, **310**, 112.

Ostriker, J. P. and Vietri, M. 1985, *Nature*, **318**, 446.

Ostriker, J. P. and Vietri, M. 1986a, *Ap. J.*, **300**, 68.

Ostriker, J. P. and Vishniac, E. T. 1986, *Nature*, **322**, 804.

Paczynski, B. 1986a, *Ap. J.*, **304**, 1.

Paczynski, B. 1986b, *Nature*, **319**, 567.

Paczynski, B. and Gorski K. 1981, *Ap. J. (Letters)*, **248**, L101.

Peacock, J. 1983, in *Proc. IAU Symp. #117: Quasars and Gravitational Lenses*, (Liege: Institut d'Astrophysique), 86.

Phinney, E. S. and Blandford, R. D. 1986, *Nature*, **321**, 569.

Press, W. and Gunn, J. 1973, *Ap. J.*, **185**, 397.

Refsdal, S. 1964, *M. N. R. A. S.*, **128**, 307.

Refsdal, S. 1966, *M. N. R. A. S.*, **132**, 101.

Schild, R. and Cholfin, B. 1986, *Ap. J.*, **300**, 209.

Schneider, D. P., Gunn, J. E., Turner, E. L., Lawrence, C. R., Hewitt, J. N., Schmidt, M., and Burke, B. F. 1986, *A. J.*, **91**, 991.

Schneider, D. P., Lawrence, C. R., Schmidt, M., Gunn, J. E., Turner, E. L., Burke, B. F., and Dhawan, V. 1985, *Ap. J.*, **294**, 66.

Schneider, P. 1985, *A. Ap.*, **143**, 413.

Schneider, P. 1986a, *Ap. J. (Letters)*, **300**, L31.

Schneider, P. 1986b, *A. Ap.*, (submitted).

Schneider, P. 1986c, *A. Ap.*, (submitted).

Schneider, P. 1986d, *A. Ap.*, (submitted).

Schneider, P., and Wagoner, R. V. 1986, *Ap. J.*, (submitted).

Shaklan, S. B. and Hege, E. K. 1986, *Ap. J.*, **303**, 605.

Shaver, P. A. 1984, *A. Ap.*, **136**, L9.
Shaver, P. A. and Cristiani, S. 1986, *Nature*, **321**, 585.
Silk, J. 1986, preprint.
Stocke, J. T., Liebert, J. Schild, R. Gioia, I. M., and Maccacaro, T. 1984, *Ap. J.*, **277**, 43; erratum **295**, 685.
Stark, A. A., Dragovan, M., Wilson, R. W., and Gott, J. R. 1986, *Nature*, **322**, 805.
Subramanian, K., Chitre, S. M., and Narasimha, D. 1985, *Ap. J.*, **289**, 37.
Subramanian, K. and Cowling, S. 1986, *M.N.R.A.S.*, **219**, 333.
Tyson, J. A. and Gorenstein, M. 1985, *Sky and Telescope*, October, 319.
Tyson, J. A., Seitzer, P., Weyman, R. J., and Foltz, C. 1986, *A. J.*, **91**, 1274.
Turner, E. L. 1980, *Ap. J. (Letters)*, **242**, L135.
Turner, E. 1987 in Knapp, G. (ed), *Proc. IAU Symp. 117: Dark Matter in the Universe*, (Reidel).
Turner, E. L., *et al.*, 1986, *Nature*, **321**, 142.
Turner, E. L., Ostriker, J. P. and Gott, J. R. 1984, *Ap. J.*, **284**, 1.
Vanderriest, C., Wlerick, G., Lelievre, G., Schneider, J., Sol, H., Horville, D., Renard, L., and Servan, B. 1986, *A. Ap.*, **158**, L5.
Vietri, M. 1985, *Ap. J.*, **293**, 343.
Vietri, M. and Ostriker, J. P. 1983, *Ap. J.*, **267**, 488.
Vilenkin, A. 1984, *Ap. J. (Letters)*, **282**, L51.
Vilenkin, A. 1986, *Nature*, **322**, 613.
Walsh, D., Carswell, R. and Weyman, R. 1979, *Nature*, **279**, 381.
Wagoner, R. V. 1986, in *Proc. Vatican Obs. Conf. on Theory and Observational Limits in Cosmology* (in press).
Weedman, D. W., Weyman, R. J., Green, R. F., and Heckman, T. M. 1982, *Ap. J. (Letters)*,, **255**, L5.
Weyman, R. J. and Folz, C. B. 1983, *Ap. J. (Letters)*, **272**, L1.
Weyman, R. J., Latham, D., Angel, J. R. P., Green, R. F., Liebert, J. W., Turnshek, D. A., Turnshek, D. E., and Tyson, J. A. 1980, *Nature*, **285**, 641.
Young, P. 1981, *Ap. J.*, **244**, 756.
Zuiderwijk, E. J. 1985, *M. N. R. A. S.*, **215**, 639.

DISCUSSION

ARP: What would you expect the supposed galaxy underlying quasars to look like under magnification?

CANIZARES: That depends on the size of the lens. As a rule of thumb, if you project the lens scale back to the source, anything that falls inside will be magnified while anything outside will not. So a galaxy sized lens should be able to magnify the inner portions of a distant galaxy, whereas a cluster will magnify a whole galaxy.

ARP: But that amplified and distorted image should extend beyond the 1 or 2 arc sec limit of optical resolution. This is not observed in the optical images of the lensed quasars.

CANIZARES: Recall that gravitational lenses really just magnify the source; they conserve surface brightness. I would think that even a magnified galaxy at $z > 1$ would be very hard to observe, but I don't know if anyone has modeled this for the specific cases of the known lenses.

NORMAN: Claude, isn't there still a serious problem with respect to the significant lack of small angular separation lenses on scales of ~1".

CANIZARES: I believe the consensus on this topic is that no serious problem exists if you include both galaxies and clusters acting to-gether, as shown by Turner, Ostriker and Gott (1984) and Dyer (1984). In addition there are probably still selection effects that bias against detection of lenses with small separations. Jackie Hewitt (1986a) has been repeating the analysis for the VLA sample, where the selection effects are reasonably well understood. So far the statistics are too small to draw any conclusions. Also, it is precisely the candidates with the largest separations that are the most suspect, so some of them may go away.

LONGAIR: One important aspect of gravitational lensing for cosmo-logical studies is the impact upon the observed intensities of quasars. If the luminosity function of quasars is steeper than $\phi(L) \propto L^{-3}$, gravitational lensing could significantly enhance the luminosities of distant quasars leading to the inference of stronger cosmological evolution for these objects. The luminosity function of the most luminous quasars may well be as steep as this. Is there evidence that these large redshift, highly luminous quasars might be enhanced by gravitational lenses?

CANIZARES: There is no "evidence" that I know of, although there have been some recent suggestions that lensing does indeed have a strong effect on the luminosity function where it is steep (Ostriker and Vietri 1986a, Vietri 1985, Schneider (1986d). It is the microlensing that seems to be most important. This led Ostriker and Vietri (1985) to suggest that BL Lacs were lensed quasars.

TURNER: In response to Dr. Longair's question, I would like to report that Burke, Roberts, Gott, and I carried out a VLA survey of ~30 of the highest optical luminosity, radio quasars a few years

ago. No very good lens candidates with sizes $\gtrsim 0.3$ were found.
This would seem to rule out lensing by galaxy mass lenses as
significantly modifying the upper end of the optical luminosity
function of <u>radio selected</u> quasars, at least.

YU XIN: The classical Fermat principle is based on the "Newtonian
time". For the "lens effect" (based on general relativity) is the
Fermat principle based on the "proper time"? In this case the
(general relativistic) Fermat principle is really an assumption.
Moreover, it appears that the gravitational field has to be known a
priori before calculations can be made.

(2) For a non-Riemannian spacetime (with torsion) the quantities
$\delta\int dt = 0$ and the autoparallel curve $k^{\alpha} \wedge k^{\beta}$ are different. Would
the Fermat principle still hold in this spacetime? Or would the
Fermat principle rule out such a spacetime geometry.

CANIZARES: Yes, the General Relativistic extension of Fermat's
principle does use the proper time along the light ray. Schneider
(1985) discusses in more detail the conditions under which it appears
to be valid and shows that it reproduces the results of the previous
vector formalism. It is indeed necessary to assume a gravitational
potential in order to calculate the lens properties; the advantage
over previous methods is that the calculations are more efficient so
one can iterate many times to explore the range of possible models.
Also, certain general characteristics of lenses, such as the parity
of the various images, become much clearer (see Blandford and Narayan
1986, Narayan 1986).

(2) I am not really competent to comment on this question. My own
crude physical intuition would tell me that some form of Fermat's
principle should hold in any universe in which light propagates along
given paths. But I will leave that to the experts.

A VLA GRAVITATIONAL LENS SURVEY

J. N. Hewitt (MIT and Haystack Observatory), E. L. Turner (Princeton), B. F. Burke (MIT), C. R. Lawrence (Caltech), C. L. Bennett (NASA/GSFC), G. I. Langston (MIT), J. E. Gunn (Princeton)

Gravitational lens surveys are of cosmological interest because they provide a way to measure the gravitational field of both luminous and dark matter. Many of the other methods used to detect the presence of dark matter, such as studies of galaxy rotation curves and cluster dynamics, require that there be luminous objects in the gravitational field that act as tracers of the mass. This may introduce a selection effect. In constrast, in studies of gravitational lenses, the beacon we observe can be far (at distances of order one thousand Mpc) from the gravitational field. In this paper we describe a VLA survey designed to detect gravitational lensing on sub-arc second and arc second scales. We also present a preliminary result of the radio data: we find that the density of matter in the form of a uniform, comoving number density of 10^{11} to $10^{12} M_\odot$ compact objects, luminous or dark, must be substantially less than the critical density.

The potential of a gravitational lens survey as a cosmological tool motivated our VLA survey. The strategy of the lens survey is as follows. First we map many MG (Bennett, *et al.* 1986) sources with the VLA, and select lens candidates on the basis of radio morphology. These candidates are subjected to lens tests, for example: (1) Optical imaging - do the candidate images have similar optical and radio flux ratios? (2) Optical spectroscopy - do the candidate images have the same redshift and similar spectra? (3) VLBI - do the candidate images have milli-arc second radio morphology consistent with gravitational imaging? We are carrying out a program of optical verification of lens candidates, and some preliminary VLBI observations are planned. To our knowledge, this is the first time a large set of radio data has been collected systematically, with a good understanding of the instrumental selection biases, for the purpose of studying the properties of lensed sources.

The VLA has well defined instrumental limits that determine its ability to detect gravitational lens systems. The resolution and field of view limit the range of image separations detectable; the dynamic range limits the range of flux ratios detectable. For a given lens system, characterized by its total mass, mass distribution, and the distances to the lens and the source, the instrumental limits can be mapped into a corresponding range of source positions that produce detectable multiple images. We then assume a spatial distribution of lenses and treat the source position as a random variable within the range determined by the instrumental limits. This allows one to calculate the expected frequency of lensing and the properties of lensed images in a sample of radio sources. This method is similar

747

A Hewitt et al. (eds.), Observational Cosmology, 747–750.

to that of Turner, Ostriker, and Gott (1984); it differs in that we have included the effects of known instrumental limits.

The frequency of lensing in a sample of radio sources provides a measurement of Ω_L, where Ω_L is the density of matter in the form of lenses expressed as a fraction of the critical density. For example, in the case of point mass lenses, the cross section for the formation of multiple images scales with the mass M; the number density of lenses scales with Ω_L/M; and the frequency of lensing scales with just Ω_L (Press and Gunn 1973). If we calculate the optical depth to lensing (which for small optical depths gives the probability of lensing) for a uniform comoving distribution of point mass lenses, including the instrumental limits of the VLA in the A array, we find the results plotted in Figure 1. The VLA in the A array at an observing wavelength of 6 cm is sensitive to 10^{11} to 10^{12} $M_\odot$ lenses. We can use the radio data to constrain Ω_L. As with the microwave background, one thing that is interesting is what we do not see. We can put limits on the density of matter in the form of compact lenses by counting the number of sources that are *not* lensed on the scales detectable with the VLA. To relate this to Ω_L, we will make a few conservative assumptions.

First, we ignore the effect of a flux intensification bias on the sample. This bias causes *more* lensed images to be included in any flux limited sample. Second, we assume the redshift distribution of the sources is the same as that measured for 30 optically bright sources (to the extent that faint sources have the same redshift distribution as bright sources, as suggested at this conference by Koo, this assumption may not be conservative). Third, we make the ridiculously conservative assumption that *any* structure in a VLA map may be due to gravitational lensing. Then, for a sample of 290 sources observed at a wavelength of 6 cm with the A array of the VLA, we count the number of sources that could be lensed. The maximum value of Ω_L consistent with the data and a background metric described by $\Omega = 1$ (again a conservative assumption in the sense that it gives a large value of Ω_L), for masses in the range of 10^{11} to $10^{12} M_\odot$, is 0.7.

We can modify the third assumption by assuming, perhaps more reasonably, that resolved, collinear components represent intrinsic source structure and not lens effects. Then we find that, again for masses in the range of 10^{11} to $10^{12} M_\odot$, the maximum consistent value of Ω_L is 0.4. These maximum values of Ω_L are independent of the value of H_o.

In summary, even at this very early stage in our gravitational lens survey, the data strongly suggest that one cannot close the universe with luminous *or dark* galaxy-sized compact objects. We expect that future work will further constrain Ω_L. The observations will continue to eliminate lens candidates, and the inclusion in the calculations of the flux intensification bias will decrease the maximum density of lenses that is needed to explain a given frequency of lensing.

Finally, to demonstrate that we are not only eliminating lens candidates, but we are also finding some interesting ones, we present the radio source 1042+178 in Figure 2. This source, with four components arranged in a parallelogram, has extremely unusual radio morphology. An optical object at the position of the radio source appears to have some structure. A spectrum of the combined light from the source shows strong emission lines at a redshift of 0.921. If this source is lensed, the lensing may well be due to a single galaxy. Such a simple mass distribution, combined with the low redshift of the source that reduces cosmological distance uncertainties, may make this a good system to use to measure the Hubble constant through a measurement of the time delay.

Figure 1: Relative sensitivity of the VLA observations to compact gravitational lenses of different masses, for three values of the source redshift. A curve has been drawn through numerically calculated points to guide the eye.

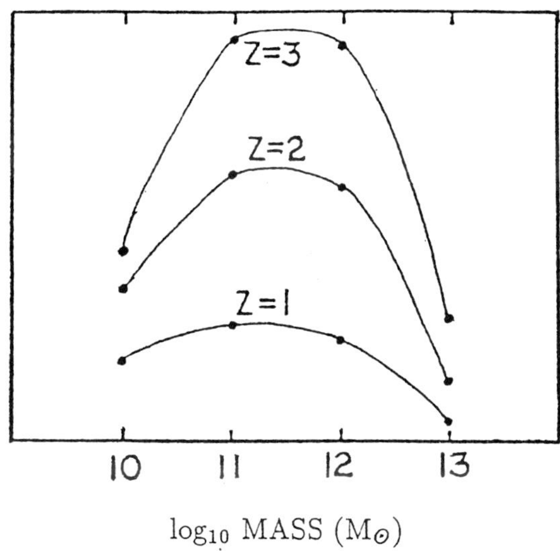

$\log_{10}$ MASS $(M_\odot)$

Figure 2: VLA map (A array; $\lambda = 6$ cm) of 1042+178.

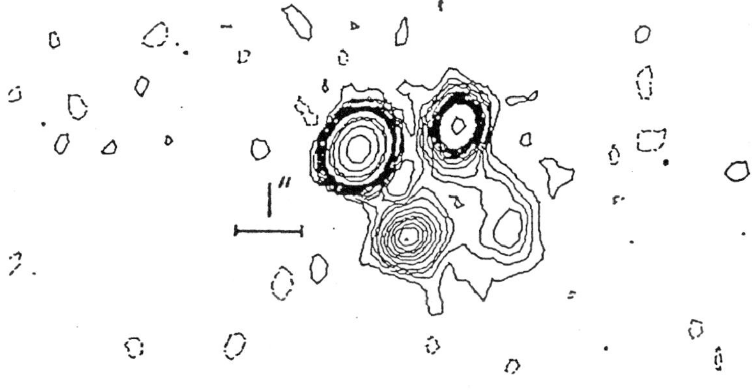

This research was supported in part by grants AST-8415677 and AST-8512598 from the National Science Foundation.

REFERENCES

Bennett, C. L., Lawrence, C. R., Burke, B. F., Hewitt, J. N., and Mahoney, J. H. 1986, *Ap. J. (Supp.)*, **61**, 1.
Press, W. H., and Gunn, J. E. 1973, *Ap. J.*, **185**, 397.
Turner, E. L., Ostriker, J. P., and Gott, J. R., III. 1984, *Ap. J.*, **284**, 1.

DISCUSSION

PECKER: How would your hypothesis that most radio structure of radio sources (or at least many of them) is due to lensing effects, affect the θ-Z relation (the classical Hubble Tolman test)?

HEWITT: A flux-limited sample would preferentially select lensed sources because of the flux intensification effect. Therefore, I would expect that the focussing due to the "extra" matter along the line of sight would tend to cause an increase in θ with increasing redshift. I would like to note that my "hypothesis" that most radio structure is due to lensing was a devil's advocate assumption I made to find a conservative limit on the density of matter in the form of compact masses.

3C324 : A PROBABLE NEW GRAVITATIONAL LENS

F. Hammer[1], O. Le fèvre[1,2], L. Nottale[1].
[1]D.A.E.C., Observatoire de Meudon, 92195 Meudon-Principal Cedex
[2]C.F.H.T. Corporation, P.O. Box 1597, Kamuela, HI-96743, USA

ABSTRACT. 3C324, a giant radiogalaxy at z = 1.206 has been suspected by us to be a gravitational lens : in a previous work, we have identified an additional line system at z = 0.845 in its spectrum. We present here new observations of 3C324 at C.F.H.T. in R broad band imaging and in adequate filters; they strongly suggest that 3C324 is the first example of a multiply imaged radiogalaxy. The 3C324 case is also a fair confirmation that gravitational lensing may well explain the overluminosity of distant radiogalaxies.

1. INTRODUCTION AND OBSERVATIONAL RESULTS

3C324 has been identified by Spinrad and Djorgovski (1984a, hereafter called SD) with a 21-22 magnitude galaxy at z = 1.206. By a careful examination of their high quality data, we have identified another distinct line system in its spectrum : i.e. H and K CaII absorption lines, $H\beta$ in absorption and [OII] emission line at z = 0.845 (Hammer et al, 1986). This new system is probably due to a foreground galaxy very close to the 3C324 line of sight. Then we have decided to get more photometric data about this source.

We have made CCD imagery with the new RCA2 CCD Camera available at C.F.H.T.. Exposure times was 4500s in R, the seeing were less than 0.7 arcsec. with a magnitude limit higher than 25.5. There are about one hundred galaxies in two arcmin² confirming the SD idea of a distant cluster lying around 3C324. The structure of 3C324 is evidently very complex with more than five components in 15 arcsec² (Le fèvre, IAU Telex 4233, 1986). Plate 1 and 2 reveals two proeminent components noted (a) and (b) separated by 1.1 arcsec., a faintest component noted (c) 0.8 arcsec. east from (a) and a probable faint component noted (d) 1.0 arcsec. NW from (a). Note also the presence of two other components : (e) 2.2 arcsec. west from (b) and (f) 3 arcsec. SW from (a). To distinguish the two distinct systems at z = 1.206 and at z = 0.845, we have pursued our observational work. Only the (a) component is detected in an image through a filter including the [OII] emission line at z = 0.845; (b) and (c) are not detected up to the level expected from the brightness ratio between (a), (b) and (c) in R broad band filters (Le fèvre et al, 1986). These two latter components are detected in the [CII emission line at z = 1.206 while (a) disappears. From these data, we deduce that the (a) component is the

A. Hewitt et al. (eds.), Observational Cosmology, 751–754.

foreground bright galaxy at z = 0.845 (m_R=22.5), the (b) and (c) components being background sources at z = 1.206.

2. NATURE OF 3C324

The angular proximity between (a) and (c) leads us to ask a question : what about the gravitational lensing of (c) due to the (a) foreground galaxy?. From a condition of Subramanian and Cowlings (1986), we show that there must be two images of (c) due to gravitational lensing by (a) only if the (a) mass is more than 1.5 10^{11} $M_{\odot}$ (for q_0=0 and H_0=50). This condition is easily verified because (a) is ten times brighter than our galaxy. Moreover (a) is just on the line joining (c) and (b) : this geometry is well explained by the simple model of Young et al (1980), where (c) and (b) are both images of a single galaxy at z = 1.206. The required (a) mass is 10^{12} $M_{\odot}$, but we have to take into account the influence of the elliptical geometry of the lens, and of a possible cluster at z = 0.845. Now we conclude on the nature of 3C324 :
*gravitational lensing of background sources at z = 1.206 by the (a) foreground galaxy at z = 0.845 occurs; a definive confirmation should wait spectroscopic observations showing the spectra of each images (Space Telescope project).
*it contributes to explain the complex configuration, but merging is not excluded to interpret the presence of the other (d), (e) and (f) components.
*this is probably the first multiple imaged galaxy ; its interpretation needs an improvement of the theory of gravitational lens to the extended source cases. The consequences are multiple, either on the determination of cosmological parameters or on the knowledge of the density profiles of distant galaxies.
*there is an overestimation of the 3C324 luminosity made by SD of more than two magnitudes; it is due to gravitational amplification of the background galaxy by the (a) deflecting galaxy <u>and</u> to the contamination of the foreground galaxy in SD 's unresolved image.

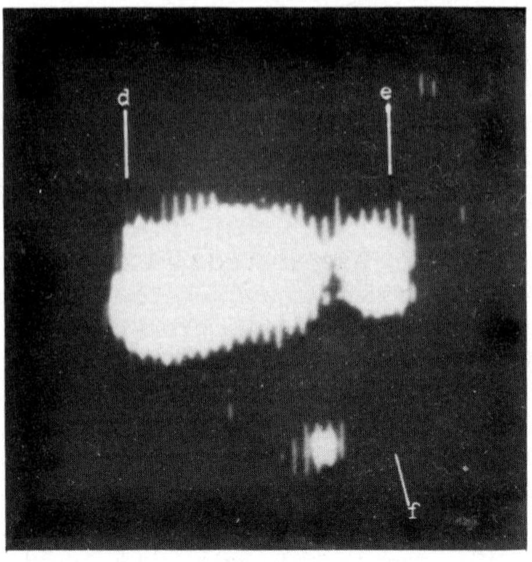

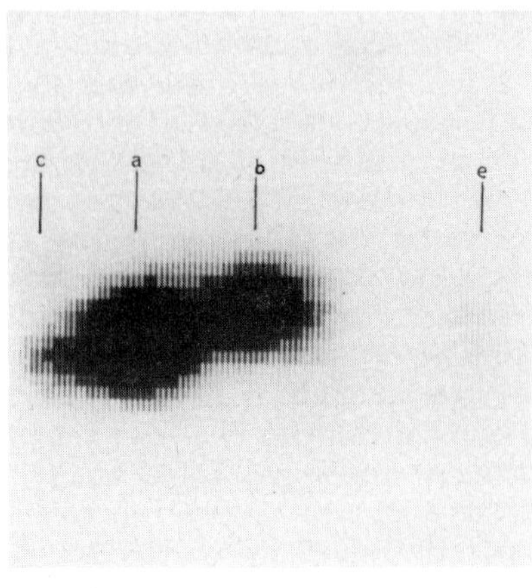

<u>Plate 1 and 2</u> : Multiple components in 3C324's structure.

3. CONCLUSION

The latter fact leads us to explain our method of work. In fact, we think that gravitational lensing is not only a phenomenon concerning very peculiar objects like gravitational multiple imaging, but concerns the whole Universe in terms of a failure of the Hubble law due to density inhomogeneities like galaxies or clusters. Then, far more extragalactic observations than currently believed are affected by gravitational amplification of light. Hence the effect of compact groups on background galaxies has been demonstrated (Hammer and Nottale, 1986a), then the statistical effect of rich clusters of galaxies on absorption-line QSOs (Nottale and Hammer, 1986, in preparation) and on brightest cluster galaxies (Hammer and Nottale, 1986b). We have shown indeed that a selection effect precisely due to gravitational amplification induces for these last sources an overluminosity of up to 0.5 magnitude. Extrapolation of this analysis to the Spinrad and Djorgovski (1984b) very distant radiogalaxies led us to propose that their large observed overluminosity was at least partly due to strong gravitational lensing (Hammer et al, 1986). These distant radiogalaxies are obviously liable to strong selection effects because all of them come from the 3C catalog (a 10 Jy flux-limited sample), and because their faint magnitudes ($V \approx 23$) are very close to the magnitude limit. Then, gravitational lensing may strongly affect the radiogalaxies luminosity diagram by the mean of selection effects. Recall that the radio properties and optical properties of a source are affected by gravitational lensing. By a case by case examination of the SD objects, we predicted an overestimation of the luminosity for at least 4 radiogalaxies with $z > 1$ (note that we have enough informations on the matter distribution near the line of sight of only two sources among 13) :
*3C13 and 3C324 lie behind foreground galaxies that evidently act as powerful deflectors; the expected magnification is more than two magnitudes.
*3C238 and 3C266 are observed behind rich Abell clusters of galaxies at $z \approx 0.2$; the predicted gravitational magnification is about one magnitude. Moreover the presence of the probable mirage 3C324 in the SD distant radiogalaxies sample is a fair confirmation of selection effects due to gravitational lensing in this 13 sources sample : without these kind of selection effects the chance of it should be less than 5 for 100,000 according to Vietri (1985). We conclude by asking a question : are overluminosities of distant radiogalaxies due to evolution or to gravitational lensing (amplification) or to both ?.

REFERENCES :
Hammer, F., and Nottale, L., 1986a, Astron. Astrophys. **155**, 420.
Hammer, F., and Nottale, L., 1986b, Astron. Astrophys. **167**, 1.
Hammer, F., Nottale, L., and Le fèvre, O., 1986,
        Astron. Astrophys. Lett., in press.
Le fèvre, O., Hammer, F., Nottale, L., and Mathez, G., 1986, submitted.
Spinrad, H., and Djorgovski, S., 1984a, Astrophys. J. Lett. **280**, L9.
Spinrad, H., and Djorgovski, S., 1984b, Astrophys. J. Lett. **285**, L49.
Subramanian, K., and Cowling, S.A., 1986, M.N.R.A.S. **219**, 333.
Vietri, M., 1985, Astrophys. J. **293**, 343.
Young, P., Gunn, J.E., Kristian, J., Oke, J.B., Westphal, J.A. 1980,
        Astrophys. J. **241**, 507.

DISCUSSION

SPINRAD: There could be some problem in your interpretation of 3C 324;
there is a velocity gradient in [OII] across the image in the SD 1984
paper.   Images are thus <u>not</u> identical in velocity space... $\Delta v \simeq 800$ km
s^{-1}.

HAMMER: In response I would like to make two points:

        First, there may be gravitational redshift effects associated
with the gravitational lensing of the possibly non-static lens galaxy.
But I think that these effects can't be very large. After seeing your
data, I believe that the estimate of the differential velocity between
C and B images requires more astrometric information. But all of your
velocity gradient data in the [OII] line are well explained by the
following argument: the angular size of your spatially resolved [OII]
line is about 6 arc sec, <u>which is</u> the angular size of our 3C 324
imagery. By placing E, B, A and C components on your diagram (v,r),
the differential velocity between B and C becomes very small, maybe
equal to zero; the velocity dispersion along the line joining B and C
is less than 300 km/s which is not surprising for a galaxy with [OII]
emission lines.

SHAVER: Is there a radio core source in 3C 324?

HAMMER: The radio map of Jenkins <u>et al.</u> reveals only two radio lobes.
But I expect to have more information from a VLA map with a smaller
noise level.

1146+111B,C: A GIANT GRAVITATIONAL LENS?

E.L. Turner
Princeton University Observatory
Peyton Hall
Princeton, NJ   08544
USA

ABSTRACT. The history and current observational status of the giant
gravitational lens candidate 1146+111B,C are reviewed. In the absence
of any new positive evidence and in the presence of reported differences in the UV and IR spectra of B and C, the lens hypothesis is
clearly much weaker than previously. Nevertheless, given the substantial similarity of the spectra over a broad range of wavelengths,
discrepancies among various spectroscopic observations of C, and the
possibility of spectral variability plus large differential time
delays, the data do not yet support any definite conclusion as to the
nature of 1146+111B,C.

1. BACKGROUND AND INTRODUCTION

Considerable attention has been attracted recently by 1146+111B,C, a
gravitational lens candidate with an unexpected and unprecedentedly
large image separation of 157". If this system is actually a gravitational lens, it has important implications for large scale structure
and dark mass in the Universe.

The two quasars in question were originally identified by C.
Hazard on prism plates among a "concentration" of quasars at a variety
of redshifts in the 1146+111 field. This "concentration" was discussed by Hazard and his collaborators H. Arp and D. Morton in a pair
of papers (1979, 1980). While noting the similarity of the redshift
and spectra of B and C, Arp and Hazard (1980) dismissed a gravitational
lens explanation on the grounds of an implausibly large redshift.
Several years later, Paczynski (1986) again called attention to the B,C
pair as a possible example of a lensing effect by a hypothetical cosmic
string.

In March 1986 we obtained very high signal-to-noise, moderate
resolution spectra of B and C in the visual and red; we also obtained
deep CCD images of the field. Examination of these data convinced us
that they strongly supported the lensing hypothesis and that it should
be taken more seriously; thus, we again called attention to the system
in the literature (Turner et al. 1986). The primary evidence which

A. Hewitt et al. (eds.), Observational Cosmology, 755–760.
© *1987 by the IAU.*

influenced us was the very striking similarity of our spectra of the
two objects. This similarity included the indistinguishable redshifts
(based on the Mg II line), the nearly identical profiles and widths of
the Mg II line, the similar strengths of the Fe II multiplet blends
producing various bumps and wiggles in the "continuum", and the absence
of [O II] emission in both objects. Although our data did show statistically significant (but quite small) differences between the two
objects, these could be accounted for in various ways (e.g., differential time delays plus variability) and seemed much less impressive than
the remarkable similarities. It was our opinion that such a degree of
similarity would be quite rare in quasars paired at random, even dis-
counting the redshift agreement which could be explained by physical
association. Certainly, no other quasar in our library of spectra
generated in other programs has line shapes, widths, and relative
strengths which would agree with those of 1146+111B,C nearly as well as
they agree with one another. A secondary bit of evidence supporting
the lens hypothesis for 1146+111B,C which influenced us was their
association in the sky with several other quasars at various redshifts:
this could have been understood rather naturally as the result of lens
magnification of all background objects in the field.

2. CURRENT OBSERVATIONAL SITUATION

Since the appearance of our spectra of 1146+111B,C in the literature,
there have been several attempts to test the lens hypothesis by extending the spectroscopic comparison into the UV and IR and by searches
for perturbations of the microwave background. Some of these have been
published, some are in preprint form, some have been communicated
privately, some we have carried out ourselves, and some may be as yet
unknown to me. None of these new observations offer strong confirmation of the lens hypothesis (as they might have), and several seem to
suggest that separate, physically associated objects are more likely.
On balance, the case for a giant gravitational lens in 1146+111B,C is
much weaker than when only our original spectra were available; nevertheless, in my opinion, the question is still an open one. All of the
observational results currently known to me are briefly reviewed
below:
    Two efforts have been made to detect microwave background distor-
tions associated with hypothetical lensing objects in 1146+111, one by
Stark et al. (1986) and the other by Lawrence et al. (1986). The
former shows no fluctuations larger than about 1 mK in a strip of sky
running through the field while the latter reports no deviations
larger than a few tenths of a mK at three selected positions in the
field. Both results suggest that there is less hot gas between B and C
than would normally have been expected for a massive galaxy cluster and
limit the transverse velocity of any cosmic string passing between B
and C to less than a few tenths of C.
    Deep VLA maps of the field show no sources to limits of somewhat
less than 0.1 mJy. This is not surprising given the fact that the
quasars in the field were optically selected and the field of view of

the VLA. Of course, if one or both of B and C had been a radio source,
this would have been quite an important constraint.
     Optical and near optical spectra have proliferated. I am aware of
Arp and Hazard's (1980) published spectra (low resolution and S/N), our
published KPNO spectra, Shaver and Cristiani's (1986) published IR
spectra, spectra obtained by Huchra (1986) on 4 separate occasions at
the MMT which extend well into the UV in three cases, spectra extending
from the near UV out to 10,000Å obtained by Neugebauer and Soifer (1986)
at the Hale 5-meter, and independent 5-meter IR spectra obtained with
the 4-Shooter spectrograph by Gunn (1986). The interpretation of these
spectra taken as a whole is far more uncertain and confusing than when
they are considered individually. The following general remarks may be
made: 1) Some pairs of spectra show a striking similarity between B
and C (e.g., our published spectra). 2) Some pairs show significant
differences, particularly in Balmer line emission (Shaver and Cristiani)
and C III] strength (Huchra's spectra). 3) Spectra obtained of
1146+111B at different epochs and by different groups appear fairly
consistent. 4) There are discrepancies between spectra of 1146+111C
obtained at different epochs and by different groups. This last point
deserves some elaboration. The two sets of Palomar spectra which
extend into the IR both show Balmer emission in C as well as B, thus
contradicting the published IR spectra; on the other hand, they do not
show Balmer lines of identical strength or width in B and C. Huchra's
spectra obtained at different epochs appear inconsistent as to the
strength and width of the C III] line. Huchra's spectrum shows a substantially broader Mg II line in C than either the published KPNO
spectrum or the Neugebauer and Soifer spectrum (which agrees very well
with the KPNO spectrum). The relative strength of Fe II blends in some
of Huchra's spectra of C appear much stronger than in other spectra
both by Huchra and others. Let me reiterate that these discrepancies
apply to various observations of the same lines in the same object!

3. DISCUSSION

Although many have reached relatively definite conclusions (of various
signs) as to whether or not 1146+111B,C is a gravitational lens or a
pair of physically associated objects, I do not believe that any certain conclusion can be supported by the available data. Clearly the
negative evidence of reported spectroscopic differences and, perhaps
equally seriously, the absence of new positive evidence makes the case
much weaker than when only the KPNO spectra were under discussion.
Nevertheless, the lens hypothesis is no more logically excluded by the
present data than the physically distinct objects hypothesis was by the
earlier data. Indeed, given the physical possibilities of neighboring
twin quasars on the one hand and variability plus long differential
time delays on the other, no definitive test is possible solely based
on spectroscopic similarities or differences. Given the (so far unexplained) difficulty in obtaining consistent spectra of C, cautious
interpretation seems even more necessary.
     The case of 1146+111 raises a number of important questions

whether or not it is eventually shown to be due to a gravitational lens:
If it is a lens, one may ask
1) What is the lensing object, or at least, how may its properties be
constrained?
2) Are there other examples of such giant lens systems, and if so, how
many and how may they be located?
If it is not a lens, one needs to ask
1) What is the explanation of the very great similarity of the spectra
near Mg II: chance or somehow the result of physical association?
2) If these are not lensed, are there other objects among the "known"
gravitational lenses (most of which have less similar spectra) which
are also physical pairs?
3) What would constitute a more reliable test than spectral similarity?
4) Quantitatively, how much variation is there in the spectra of quasars
of similar luminosity and redshift?

J.E. Gunn, J.P. Huchra, C.R. Lawrence, G. Neugebauer, D.P. Schneider,
T. Soifer, and A. Stark have all generously communicated results in
advance of publication. Those listed above and numerous other colleagues including particularly B.F. Burke, J.R. Gott, J.N. Hewitt, J.P.
Ostriker, B. Paczynski, M. Schmidt, and A. Tyson have also provided
helpful comments and insights concerning 1146+111. This work was
supported in part by NSF grant AST84-20352.

References
Arp, H., and Hazard, C. 1980, Ap. J., 240, 726.
Gunn, J.E. 1986, private communication.
Hazard, C., Arp, H., and Morton, D.C. 1979, Nature, 282, 271.
Huchra, J.P. 1986, private communication.
Lawrence, C.R., Readhead, A.C.S., Moffet, A.T., and Birkinshaw, M. 1986,
     A.J., in press.
Neugebauer, G., and Soifer, T. 1986, private communication.
Paczynski, B. 1986, Nature, 319, 567.
Shaver, P.A., and Cristiani, S. 1986, Nature, 321, 585.
Stark, A.A., Dragovan, M., Wilson, R.W., and Gott, J.R. 1986, Nature,
     in press.
Turner, E.L., Schneider, D.P., Burke, B.F., Hewitt, J.N., Langston, G.
     I., Gunn, J.E., Lawrence, C.R., and Schmidt, M. 1986, Nature, 321,
     142.

DISCUSSION

ARP: This pair is a member of a group of quasars which is very similar
to other tight groups of quasars at about $z \simeq 1$ mean redshift. It is
characteristic of these groups (see Arp, Liege Symposium 1983) that
they have pairs which are very similar in redshift. But that
similarity is generally only within 0.1 to 0.2 in redshift. This is
direct evidence that quasars exist in physically similar pairs and that
gravitational lenses are not necessarily demonstrated by quasars of
similar spectra.

TURNER: I agree that many lines of evidence, not just spectroscopy, must be mustered to "demonstrate" lensing in a particular case. I would be surprised if lensing played any role in general in the associations you mention.

WAMPLER: I want to disagree with the concept that these two spectra of 1146+111 are "remarkably similar". Quasars, like people, are similar to each other and some are very similar. Much more detail is needed before it can be claimed that a "fingerprint" has been found. In this case the structure near MgII $\lambda 2800$ is due to the multiplet structure of FeII, and not to similar intensities of independent emission lines.

TURNER: I certainly agree that much of the structure near MgII is due to FeII and that there are many fewer free physical parameters describing the gas than independent points in our spectra. Nevertheless, I do find the agreement "remarkable". I would include the similar Mg II to Fe II relative intensities, the nearly identical profile shapes in Mg II, the identical (within the measurement errors) Mg II line widths, and the absence of other lines, notably [OII] in both objects. We have examined a sample of ~20 quasars with $z \simeq 1$ and found no pair which match so precisely.

BURKE: This is a comment that in part is a summary of the positive aspects of the gravitational lens situation, which has some elements of confusion at present. The gravitational lens phenomenon is important because we know that reliable examples exist; lensing may provide the most reliable tracer for dark matter on a large scale, their frequency of occurrence sets hard lower limits on Ω, as J. Hewitt has shown, and the time delay gives information on $\delta\rho/\rho$, as Alcock and Anderson have discussed, if there is a good value for H_o. Therefore, the criteria for identification should be clear and stringent. These should be, in my opinion, an appropriate subset (preferably seven) of the following: (1) Repetition of flux variations with a well-measured time delay. This is sufficient proof, but has not yet been reliably achieved for any lensed quasars. (2) Similar spectra are needed, of course. If there are differences in the spectra, there must be a good explanation (0957+561 is a good example of differences being convincingly explained). (3) Similar milliarcsecond structure (from VLBI at present, from optical interferometry at some future time). Again differences must be explained. Relative parity is a powerful proof (the VLBI work of Gorenstein <u>et al</u>. on 0957+561 is a good example). (4) Large-scale structure gives much tighter requirements on the lens. Examples of the use of this criterion is given by the radio jet structure of 0957+561, the third image of 2016+112, and the splitting of A into A_1 and A_2 in 1115+080 (which turned the

triple quasar into a quadruple quasar, which actually should be a
quintuple quasar). Note that the radio selection method is a neutral
one, having the advantage of allowing all four tests. I suggest that
we have 2 classes of lens candidates: Class A, which are well-
established and for which the burden of proof is on the challenger
(0957+561, 1115+080, 2016+112, and 2237+035 are the four present
examples of Class A lenses); and Class B lenses, which meet the
criteria to some degree, but which still lay the burden of proof upon
the observers, with further evidence needed (1146+111 is certainly in
this category). Some have suggested that one should see visible
evidence of the lens   itself, but in our current state of knowledge
(or lack of knowledge) of dark matter, I suggest that this is not a
necessary requirement, or at least not yet.

THE "GRAVITATIONAL LENS" 3C 321: A REMARKABLE IMPOSTOR

Alexei V. Filippenko
Department of Astronomy
University of California
Berkeley, CA   94720
USA

ABSTRACT.   Recent observations of the radio galaxy 3C 321 are presented. The optical nucleus consists of two components (A, B), separated by $\sim 4''$ (~ 6 kpc), whose low-resolution spectra strongly resemble those of high-ionization type 2 Seyfert nuclei. The relative intensities of the emission lines differ in A and B by less than 1%, and their profiles are almost identical. 3C 321 appears to be a convincing example of a gravitationally lensed object.

    Careful analysis of high-quality radio and optical data, however, reveals that the system is almost certainly *not* a lens. The equivalent widths of the emission lines are roughly twice as high in B than in A, and there are significant spatial offsets between regions of bright continuum and line emission. A slight, but fundamental, difference is visible in the two Hα emission profiles. The radial velocities of A and B are discrepant by 31 ± 10 km s^{-1}. Finally, component A is nearly coincident with a flat-spectrum radio core, whereas B is next to an extended, steep-spectrum knot of radio emission.

    This object should serve as a warning to lens hunters: beware of impostors, whose true properties may be difficult to ascertain without extensive optical and radio observations.

1. INTRODUCTION

Detailed studies of extended, optical emission-line regions associated with radio galaxies can provide valuable information on the ways in which jets interact with ambient gas. "Minkowski's object," for example, consists of a dwarf galaxy in which a violent burst of star formation was recently triggered by a radio jet from an adjacent elliptical galaxy in the cluster Abell 194 (van Breugel *et al.* 1985). The interstellar medium is ionized by ultraviolet light from the hot, massive stars, as in the similar case of Cen A (Graham and Price 1981). Other objects show that jets may be deflected and disrupted when they collide with dense extranuclear material, and that synchrotron radiation from internal shocks can photoionize the surrounding gas (see van Breugel 1986 for a recent review).

    In this paper I briefly describe observations of 3C 321 ($z = 0.096$), and show that the galaxy at first sight appears to be gravitationally lensed. Careful analysis of the data, however, casts serious doubt on this interpretation. A complete discussion is given by van Breugel *et al.* (1987).

761

A. Hewitt et al. (eds.), Observational Cosmology, 761–765.
© *1987 by the IAU.*

2.　OBSERVATIONS

Visually, 3C 321 is a sixteenth magnitude galaxy with two fairly compact nuclei (A, B) separated by $\sim 4''$, or ~ 6 kpc ($H_0 = 75$ km s^{-1} Mpc^{-1}, $q_0 = 0.5$). Component A is roughly twice as bright as B, and a common envelope is present in deep photographs. Low-resolution optical spectra (FWHM ≈ 10 Å) obtained with the 3-m Shane reflector at Lick Observatory show that A and B have strong emission lines spanning a wide range of ionization levels. The spectra resemble those of high-ionization type 2 Seyfert galaxies, as indicated by the intensity ratios [O III] $\lambda5007/\mathrm{H}\beta \approx 11$ and He II $\lambda4686/\mathrm{H}\beta \approx 0.5$. Lines such as [Ne V] $\lambda3426$ are strong, but [O I] $\lambda6300$ is also prominent.

Most remarkable, however, is that the emission-line intensity ratios in components A and B are *nearly identical* — they differ by less than 1%! Although such emission-line characteristics are often found in different H II regions within a given galaxy, and in the high-excitation spectra of extragalactic H II regions, they have rarely (if ever) been seen in any two distinct Seyfert nuclei or QSOs. It seems, therefore, that 3C 321 should be added to the growing list of known gravitational lenses (see Canizares 1987 for a current review).

To explore this possibility, extensive new observations of 3C 321 were made at radio and optical wavelengths. In particular, spectra having excellent signal-to-noise ratios and moderate resolution (FWHM $\approx 1.5 - 2.5$ Å) were obtained on 29 March 1986 UT with the 5-m Hale telescope at Palomar Observatory; the long slit of width $1''$ was oriented at a position angle of $131°$ to include both components. Also, high-quality radio maps were made with the Very Large Array. These could be compared with the corresponding optical images obtained with the Shane reflector.

Figure 1 illustrates the spectrum of component A over the observed wavelength range $\sim \lambda\lambda4700 - 5600$. An effective entrance aperture of size $1'' \times 2''$, centered on A, was synthesized from the original two-dimensional (2-D) CCD data. A spectrum of component B was obtained in a similar manner, and subtracted from that of A. The difference is shown in the bottom part of the figure, with the same ordinate scale. The degree to which the emission lines are absent is astounding. Only a slight "oscillation" in the continuum level is visible at the wavelength of the strongest line, [O III] $\lambda5007$, but the *integrated* strength of this line is almost exactly zero. This confirms that the relative intensities of emission lines in the centers of A and B are nearly identical. Furthermore, the *absolute* intensities and widths must also be comparable in the central regions. The 2-D spectra show that the relative intensities are similar in the more extended regions of A and B, and in several faint, detached wisps of emission as well.

3.　3C 321: A GRAVITATIONAL LENS?

Do these observations really imply that 3C 321 is a gravitational lens? The answer, unfortunately (or fortunately, depending upon one's point of view), is no. A major obstacle to the lens interpretation is that the *equivalent widths* of the emission lines in component B are roughly twice those in A (Fig. 1). If the true nucleus of 3C 321 were lensed, the underlying continuum would be nearly the same in both images, as are the emission lines. Models involving microlensing (e.g., Kayser, Refsdal, and Stabell 1986) can produce the observed discrepancy, but this is considered unlikely here in view of the extended nature of the components, whose emission-line intensity ratios are everywhere very similar.

There are several other arguments against the lens scenario. Careful inspection of the red spectra ($\sim \lambda\lambda 6850 - 7500$), for example, reveals a very weak, broad component of Hα emission (FWZI $\gtrsim 6000$ km s^{-1}) in A, but not in B. This Hα line is reminiscent of, but much weaker than, those seen in type 1 Seyferts; it has been detected in the spectra of many bright, nearby galaxies by Filippenko and Sargent (1985). In addition, the narrow emission lines in the spectra of A and B exhibit a systematic offset of 31 ± 10 km s^{-1}. Finally, the 2-D spectra show that the region in component A having the most intense emission lines is offset by $\sim 1.1''$ from the region of brightest continuum emission, and that a displacement of $\sim 0.4''$ in the opposite spatial direction is present in component B. This has been confirmed with the narrow-band images of 3C 321.

One of the most striking discrepancies between A and B is found in the radio data. Maps at $\lambda = 6.2, 18.3$, and 21.6 cm show that optical component A is almost coincident with a flat-spectrum, compact "core" of radio emission. Component B, on the other hand, is next to a steep-spectrum, somewhat extended, radio

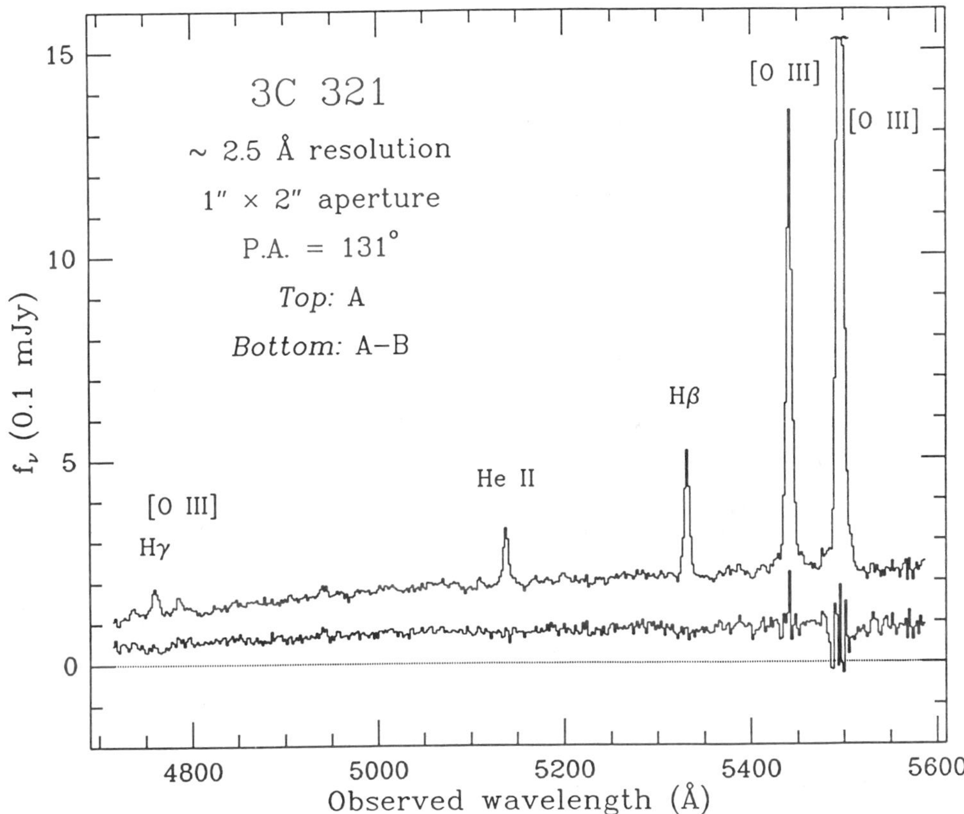

Figure 1: Blue spectrum of 3C 321 A (*top*), showing some of the high-excitation emission lines. The spectrum of 3C 321 B has been subtracted from that of A in the bottom portion of the figure, after removal of a slight velocity difference (31 km s^{-1}). A continuum devoid of emission lines remains. The equivalent widths of the lines are therefore much greater in B than in A, although their intensities are almost indistinguishable.

"knot." This knot itself lies along the path of a one-sided jet of length $\sim 20''$ which originates in the radio core. The optical images of 3C 321 show complex emission-line regions within the galaxy, aligned roughly along the jet. These characteristics are undoubtedly inconsistent with gravitational lensing. Instead, the jet and ambient medium seem to be interacting in a complicated manner, especially in the vicinity of component B.

Note that if a lens were indeed responsible for the optical appearance of 3C 321, it would probably have to be very massive ($M \approx 10^{12}$ M$_\odot$, but this is model dependent). Since 3C 321 is so nearby, the lensing object should be easily visible, unless it is a black hole or some other form of dark matter. Galactic objects such as neutron stars could produce the observed separation if the impact parameter were small, but the lens would last only a short time. Moreover, it would be difficult to account for extended emission of the type actually observed.

4. CONCLUSIONS

Although at first sight 3C 321 resembles a gravitationally lensed object, under closer scrutiny there are many arguments against this interpretation. The possibility of being fooled into thinking that two adjacent, extragalactic *H II regions* having similar spectra are lensed images of a single galaxy has already been stressed by Halpern, Marshall, and Oke (1984), in a detailed study of $1300 + 361$ A, B. Other examples of such galaxies are discussed by Chen and Shaver (1982) and by He (1987). 3C 321, however, is the first case in which separate "nuclei" with nearly identical *Seyfert 2* spectra have been identified. Despite their highly unusual spectral properties, components A and B are physically distinct entities, albeit very puzzling ones. Their origin, relationship to each other, and ionization mechanisms are not yet understood.

3C 321 should therefore serve as a warning to observers attempting to identify gravitational lenses with low-resolution optical spectra alone. Extensive optical and radio observations may be necessary to determine the true nature of lens candidates. One must beware of impostors, whose superficial characteristics may be strikingly similar to those of lenses. This, together with the intrinsic differences which can exist in the spectra of *genuinely* lensed objects (see, e.g., Kayser *et al.* 1987, and references therein), may render the search for gravitational lenses very difficult indeed.

ACKNOWLEDGMENTS

I thank Wil van Breugel, Pat McCarthy, Tim Heckman, Joe Miller, and George Miley for their happy collaboration on this project. Hy Spinrad obtained some of the first direct photographs and optical spectra of 3C 321 many years ago, and noticed the great similarity of the emission lines in the two components. Wal Sargent kindly agreed to let me procure the Palomar spectra during an observing run which was primarily devoted to other studies. The financial assistance of an AAS International Travel Grant from the NSF is greatly appreciated.

REFERENCES

Canizares, C. R. 1987, these proceedings.

Chen, J.-S., and Shaver, P. A. 1982, *ESO Messenger*, **30**, 2.
Filippenko, A. V., and Sargent, W. L. W. 1985, *Ap. J. Suppl.*, **57**, 503.
Graham, J. A., and Price, R. M. 1981, *Ap. J.*, **247**, 813.
Halpern, J. P., Marshall, H. L., and Oke, J. B. 1984, *A. J.*, **89**, 1802.
He, X.-T. 1987, these proceedings.
Kayser, R., Refsdal, S., and Stabell, R. 1986, *Astr. Ap.*, **166**, 36.
Kayser, R., Refsdal, S., and Stabell, R., and Grieger, S. 1987, these proceedings.
van Breugel, W. J. M. 1986, in *Proc. Toronto Conference on Jets from Stars and Galaxies*, ed. R. N. Henriksen and T. W. Jones (Canadian Journal of Physics, **64**), p. 392.
van Breugel, W. J. M., Filippenko, A. V., Heckman, T. M., and Miley, G. K. 1985, *Ap. J.*, **293**, 83.
van Breugel, W. J. M., Filippenko, A. V., McCarthy, P. J., Heckman, T. M., Miller, J. S., and Miley, G. K. 1987, *A. J.*, submitted.

DISCUSSION

Maccacaro: What do you think is responsible for the observed properties of 3C 321, if it is not a gravitational lens?

Filippenko: Ah, the question I was hoping for! There are many possibilities, none of which is convincing at this time. Let me explain them very briefly; full details can be found in van Breugel *et al.* (1987).

One idea is that component A is the nucleus, as suggested by the radio data, and that it emits nonstellar radiation which ionizes the surrounding gas, including that in the off-nuclear knot, B. The great similarity of the emission-line spectra of A and B, however, can only be explained if the ionization parameter is independent of location in 3C 321, and this is improbable because the gas does not exhibit large spatial density gradients. Moreover, the absorption lines in *both* A and B seem to be diluted by featureless continua, suggesting that a source of ionizing radiation is present in both.

Similar ionization parameters in A and B can be produced if the true nucleus is midway between the two components. In this case, it must be obscured by a dense cloud of dust, or by some other object for which there is no independent evidence. Furthermore, this model does not explain why the radio core is close to A, nor does it account for the nonstellar continua noted above.

Perhaps components A and B are like "cosmic mirrors" which reflect the spectrum of a hidden nucleus between them. But again, there is no evidence for thick dust or an obscuring disk in 3C 321. In addition, light scattered from A and B should be strongly polarized, whereas actually it is not.

Components A and B could both harbor sources of ionizing radiation if they are the nuclei of two merging galaxies. Although the general idea of a merger is not unlikely, the small separation between the extended radio knot and optical component B must then be a coincidence in 3C 321. Also, it is difficult to imagine why the ionization parameters should be nearly identical in A and B.

A diffuse source of ionizing photons, such as in a "thermal bath" of constant temperature, might account for the similar emission-line spectra in A, B, and surrounding regions. If so, 3C 321 is a unique object, and severe problems are encountered in attempts to explain the origin of the diffuse bath. The radio data also seem inconsistent with this hypothesis.

ASTROPHYSICAL APPLICATIONS OF GRAVITATIONAL MICRO-LENSING

R. Kayser[*], S. Refsdal[*], R. Stabell[+], B. Grieger[*]

[*] Hamburger Sternwarte
Gojenbergsweg 112
D-2050 Hamburg 80
F.R. Germany

[+] Institute of Theoretical Astrophysics
University of Oslo
P.B. 1029
Blindern
Oslo 3
Norway

ABSTRACT. Gravitational micro-lensing due to stars in the
deflecting galaxy influences the brightness and the spectra
of the macro-images. Thus differences in the spectra of
different macro-images are not automatically an argument
against gravitational lensing. Furthermore changes in the
spectra due to micro-lensing may give informations on the
quasar structure. From high amplification events the
brightness profile of the source may be obtained. The time
scale of the high amplification event is proportional to
the source radius and inverse proportional to the
transversal velocity. Due to the large brightness gradient
by a high amplification event, a "parallax-effect" occurs,
from which the transversal velocity may be obtained, and
thereby the source radius ($R = \Delta t \cdot V_T$). We roughly estimate
0.3 high amplication events per year for all gravitationally
lensed quasars. Frequent monitoring should be carried out
in order to predict high amplification events as early as
possible.

1. INTRODUCTION

Gravitational micro-lensing (GML) due to stars in an
intervening galaxy (macro-lens) may lead to a split-up of
the macro-images into several micro-images[1].
    Although the separation of this micro-images is of the
order 10^{-6} arcsec (unobservable), observable effects occur.

A. Hewitt et al. (eds.), Observational Cosmology, 767–770.

Changes in brightness and in the spectra of the macro-images may lead to important informations on quasars.

2. SPECTRAL EFFECTS

2.1. Chromatic Effect

For an extended source the mean amplification $\langle A \rangle$ must be calculated by integration over the source. If the source radius depends on the wavelength, $\langle A \rangle$ depends on the wavelength, too.[1]
     Thus, the effect of GML is stronger for compact sources (e.g. quasar continuum source) than for more extended sources (e.g. quasar emission line regions).[2]

2.2. Equivalent Width

Since the source radius is different for continuum and emission lines, $\langle A \rangle$ is also different and GML can therefore change the equivalent width of the emission lines.[2]

2.3. Emission Line Profiles

If the emission line profile varies over the source (e.g. due to rotation) and if A varies over the source, different parts of the profile are differently amplified by GML, resulting in different emission line profiles for the macro-images.[2]

3. HIGH AMPLIFICATION EVENTS

3.1. Critical Curves

The curves in the source plane where $A \to \infty$ are called critical curves. If a compact source crosses a critical curve, two micro-images appear or disappear and a typical asymmetric peak occurs in the light curve[1] (High Amplification Event (HAE)).

3.2. Brightness Profile and Source Radius

By comparison of an observed light curve with the theoretical light curve of a HAE, the brightness profile of the source (perpendicular to the critical curve) can be obtained.[3] Especially, multiple sources may show up as multiple peaks in the light curve.[4]
     Since the HAE are eclipse-like events, their time scale is proportional to the source radius.[2]

3.3. Timescale of HAE

The time scale of a HAE for a typical lensed quasar (source radius 10^{-4} pc, transversal velocity V_T=1000 km/s) is of the order 0.1 yr. For more extrem cases (e.g. 2237 +0305) time scales as short as 0.01 yr are possible.[2]

3.4. Brightness Gradient

The steepest slope in the light curve is typically of the order 1 mag/week (1 mag/day for extrem cases). This slope is due to a spatial brightness gradient at the observer : 0.25 mag/A.U. for a typical lensed quasar, 2.5 mag/A.U. for extrem cases like 2237 +0305.[2]

 This gradient is easily observable by "space telescopes" $\gtrsim$ 0.1 A.U. away from the earth. From the time lag between the light curves measured by three observers the transversal velocity perpendicular to the critical curve can be obtained[5]. Now the source radius can be obtained from the time scale of the HAE ($R = \Delta t \cdot V_T$). For very small velocities the annual motion of the earth may show up as oscillations in the light curve.[5]

3.5. Frequency of HAE and Monitoring

We roughly estimate 0.1 HAE/yr for 2237 +0305 and 0.3 HAE/yr for all known gravitationally lensed quasars.[2]

 In order to predict a HAE in time, we propose a frequent monitoring of lensed quasars. The time interval should be less than one third of the time scale of a HAE, i.e. about 10 days for a typical lensed quasar, an 2 days for 2237 +0305.

4. OBSERVATIONAL EVIDENCE FOR GML

If a lensed quasar shows intrinsic variations only on time scales much larger than the time delay between two images, the brightness ratio between this images would, without micro-lensing, be fairly constant. Thus, if changes of the brightness ratio are observed, they would be a clear evidence for micro-lensing in one (or both) macro-images. A variation of this kind has already been reported[6,7] for the components A_1 and A_2 of the multiple quasar 1115+080 (separation 0.5"[1] estimated time delay a few days). This gives evidence for GML, although further observations are needed to exclude intrinsic short-time variability of the object.

ACKNOWLEDGEMENT. This work is supported in part by the
Deutsche Forschungsgemeinschaft under Az. Re 439/3.

REFERENCES

1. K. Chang, S. Refsdal: 1979, Nature 282, 561
2. R. Kayser, S. Refsdal, R. Stabell: 1986,
    Astron.Astrophys. 166, 36
3. B. Grieger, R. Kayser, S. Refsdal: 1987, in preparation
4. K. Chang, S. Refsdal: 1984, Astron.Astrophys. 132, 168
5. B. Grieger, R. Kayser, S. Refsdal: 1986, Nature, in press
6. R. Foy, D. Bonneau, A. Blazit: 1985,
    Astron.Astrophys. 149, L13
7. C. Christian, D. Crabtree, P. Wadell: 1986,
    IAU Circular No. 4182

DISCUSSION

TURNER: In complementary work, my Princeton colleague B. Paczynski
has carried out Monte Carlo simulations of microlensing which often
show substantial step function-like changes in the amplification. If
microlensing affects continuum and emission line regions differently,
this could easily produce large spectral differences between two macro
lens images.

GRAVITATIONAL LENS EFFECTS OF DARK MATTER OF MILKY WAY

C. M. Xu and X. J. Wu
Department of Physics
Fudan University
Shanghai
P. R. China

ABSTRACT. In this paper, dark matter of Milky Way is discussed as a
transparent gravitational lens. The mass distribution of Milky Way is
composed of spheroid, disk and dark matter (corona). The trajectory of
light ray influnced by Milky Way has been calculated numerically. The
deflecting angle to a extragalactic souce is changed with different
observational direction and is about several seconds in maximum. The
brightness amplification ratio K_V of extragalactic background radiation
is also derived from the redistribution of intensity due to the gravitational lens effects of Milky Way, $\Delta K_V/K_V$ is about 10^{-5} in maximum. K_V
has been expanding to multipole with spherical harmonics by means of
K_V distribution in celestial sphere. Dipole and qudrapole are about
3×10^{-6} and 7×10^{-6} respectively, which might be added to the observational
data of microwave background radiation. A measurement method to pick
out the gravitational lens effects from real cosmological anisotropy
has been proposed. (ref. 8)

1. INTRODUCTION

Dark matter (DM) is one of the most mystery problem in astrophysics
and cosmology. Althought we do not know what is the dark matter
(Sancisi 1986), it is convinced that DM exist from rotation curve.
Any way ,every kind of DM influences the trajectory of light rays.
Since DM is dissipationless, it could be taken as a transparent
gravitational lens. The gravitational lens effects, in the case of
light source enclosed by a spherical symmetric transparent halo, have
been discussed by Xu et al. (1983) The solar system is also surrounded
by a spheroid and corona (DM) and located at the galactic disk. The
mass and potential distributions of Milky Way (spheroid, corona and
disk) have been advanced in many forms (Caldwell and Ostriker 1981;
Ostriker and Caldwell 1983) to fit the rotation curve (Knapp 1983).
A preferable model is D150 (Ostriker and Caldwell 1983). The null
geodesic equations are solved numerically for model D150.

Some new effects could be caused by the mass *dis*tribution of Milky
Way itself. The new effects predict that the observational angular

A. Hewitt et al. (eds.), Observational Cosmology, 771–773.
© *1987 by the IAU.*

positions of extragalactic sources are deflected, the observational brightness of the extragalactic background radiation is redistributed. The brightness amplification ratio K_ν and its multipolar expansion in spherical harmonics have been calculated numerically. Since the microwave background radiation is measured by Dicke Radiometer (Wilkinson 1982), in which to measure the temperature is equivalent to measure the radiation power from a standard resister at the same local oscillating frequency. Therefore, if some deviation of energy density power is observed in different point of celestial sphere, we could not distinguish weather it is caused by gravitational lens effects or by real cosmological anisotropy. Perhaps a part of anisotropy of microwave background radiation is caused by those new effect in normal measurement. But we could measure the every deviation of energy density in two different frequency at the same direction, which are located in the both sides of extremum value of Planck spectrum. Then, for the lens effect, since at the same direction the two frequency points have the same K_ν, so the energy density power of two points to compare with the standard Planck spectrum change identically, but for the real cosmological anisotropy, the variation of pure temperature may cause the Planck spectrum move along the frequency axis, the energy density power of two frequency points would change in opposite direction. The formula of the amplification ratio and the variation of temperature between two space points in terms of two different frequency can be easily obtained.

2. THE LIGHT RAY IN MILKY WAY

In Milky Way, the potential is weak (10^{-5}) and the motion of the axial symmetric disk could be omitted, then the metrics $g_{oj}=0$ (Misner et al. 1973). Since the corona, spheroid and disk are concentric, then $g_{ij}=\delta_{ij}A(\rho,\theta)$, $g_{oo}=-B(\rho,\theta)$, here ρ, θ are spherical coordinate from center. Therefore

$$dS^2 = A(\rho,\theta)\cdot(d\rho^2 + \rho^2 d\theta^2 + \rho^2 \sin^2\theta\, d\varphi^2) - B(\rho,\theta)dt^2 \tag{1}$$

In the weak field approximation,

$$A(\rho,\theta) = 1+2u, \text{ and } B = 1-2u \tag{2}$$

where u is the gravitational potential, which is composed of three parts: disk u_1, spheroid u_2 and corona u_3. u_1, u_2 and u_3 have the analogous formulas to appendix A in Caldwell and Ostriker (1981), and the parameters are taken from D150. Now the metrics can be calculated numerically. The Christoffel symbol $\{^\mu_{\nu\lambda}\}$ are obtained also. Then the four geodesic equations are

$$\frac{d^2 X^\mu}{d\rho^2} + \{^\mu_{\nu\lambda}\}\frac{dX^\nu}{d\rho}\frac{dX^\lambda}{d\rho} = 0 \tag{3}$$

The three constants of motion from the component equations of (3) are obtained. Substituties constants of motion to the above equations, there are

$$\frac{d^2\rho}{d\rho^2} + \left(\frac{1}{\rho} - \frac{2}{A}\frac{\partial u}{\partial\rho}\right)\left(\frac{\partial\rho}{\partial\rho}\right)^2 - \frac{2}{A}\frac{\partial u}{\partial\theta}\left(\frac{d\rho}{d\rho}\right)\left(\frac{d\theta}{d\rho}\right) - \frac{1}{\rho} + \frac{2}{A}\frac{\partial u}{\partial\rho} = 0$$

$$\frac{d^2\theta}{d\rho^2} + \left(\frac{2}{\rho} - \frac{2}{A}\frac{\partial u}{\partial\rho}\right)\left(\frac{d\rho}{d\rho}\right)\left(\frac{d\theta}{d\rho}\right) - \frac{2}{A}\frac{\partial u}{\partial\theta}\left(\frac{d\theta}{d\rho}\right)^2 - \frac{1}{A^2\rho^4}J^2\frac{\cos\theta}{\sin^3\theta} + \frac{2}{A^2\rho^2}\frac{\partial u}{\partial\theta} = 0 \tag{4}$$

The distance and mass are normalized with radius of corona R_{Tor} and R_{Tor}/G respectively. The position S of the solar system (8.5 KPC far from the center) is taken as the initial condition. At first we put the light source in the position of solar system. The trajectory of light ray is integrated from $f=0$ to infinity(take $f=100$ as infinity) with initial direction of spherical angular coordinate η_o, χ_o from S with the method of Merson variable step in double precision. Because of the static metrics, the optical reciprocity therem is valid. The trajectory of light ray is the same, when source exchange for observer each other.

3. THE RESULTR OF NUMERICAL CALCULATION

3.1. The Variation of Angular Position

From above calculation the total deviation of angle $d\Omega = \sqrt{d\eta^2 + \sin^2\eta\, d\chi^2}$ where $d\eta$ and dx are the angular difference between existence and non-existence of the gravitational lens. When the light ray is closer to the disk, the deflecting angle is larger ($d\Omega \approx 2.2$ Sec). But for convenient to observe, the angle to disk at least $10°$, in this case the deflecting angle is about 1.05 Sec.($\eta_o=80°$; $x_o=152°$). The distribution of deflecting angle with celestial sphere has been obtained.

3.2. Brightness amplification ratio K_V and multipole of K_V

The larger K_V-1 is about -2.49×10^{-5} at $\eta_o=74°$ and $x_o=149°$. The distribution of deflecting angle is also obtained.

K$_V$could be expanding with spherical harmonics. The coefficients of multipole are obtained as follow: The dipole $a_{l,+1}=\div 2.8 \times 10^{-6}$. The quadrupole $a_{2,+2}=3.6 \times 10^{-6}$ and $a_{2,0}=7.2 \times 10^{-6}$. Others are all equal to zero. $a_{l,+1}$ and $a_{2,+2}$ are caused by the deviation of solar system from concenter and $a_{2,0}$ by the anisotropy of the disk.

4. DISCUSSION

Some of above mentioned effects are expected to detect in future, if we could find these effects, the dark matter of Milky Way might be understood further more and the distribution of DM can be calculated.

REFERENCES

Caldwell, J.A.R. and Ostriker, J.P.: 1981, Ap. J. 251, 61
Knapp,G.R.:1983,Kinematics,Dynamics and Structure of the Milky Way,233
Misner, C. W.,Thorne, K. S. and Wheeler, J. A.:1973,Gravitation,1068.
Ostriker, J. R. and Caldwell, J. A. R. : 1983, Kinematics, Dynamics
        and Structure of the Milky Way ,249.
Sancisi, R. : 1986 ,IAU Symp. No.124.
Wilkinson, D.T. :1982, Phil. Trans. R. Soc. Lond.,A307, 55.
Xu, C. M. and Wu, X. J. : 1983, A. Ap., 120 ,15.
Xu, C. M. , Wu, X. J. and Xiang, S. :1983, Acta. Astrophy. Sinca 3,316.

CHAPTER XIV

ABSORPTION IN QUASI-STELLAR OBJECTS

QSO ABSORPTION LINES AND COSMOLOGY

Wallace L. W. Sargent
Palomar Observatory, California Institute of Technology
Pasadena, Ca. 91125 U.S.A.

ABSTRACT. The properties of the different types of QSO absorption systems are briefly summarized. An overview is given of the potential applications of absorption lines in cosmology. Recent work on the cosmological evolution of the different types of absorbers is discussed. The physical properties of the intergalactic medium are discussed in the light of recent work on the "Lyman alpha clouds". The uses of the lines for studies of the evolution of clustering in the Universe are described. Recent puzzling results on common absorption in pairs of QSOs, particularly Q1037–2704 and Q1038–2712, are summarized.

1. INTRODUCTION

It is not possible to talk both about the general properties of absorption lines in QSO spectra and their applications in Cosmology in the allotted time. Therefore, I shall only briefly review what is known about the absorption lines themselves and then concentrate most of my talk on cosmological questions. As it is, some parts of this written contribution were omitted from the talk due to time constraints.

It is currently believed that the quasar absorption lines can be divided into three types, each having a different origin:

1. *The "trough" (or BAL) systems.* A small proportion of QSOs (perhaps 5%) have very broad absorption troughs on the blue side of the corresponding emission lines. Lines of very high states of ionization–C IV, Si IV, N V, O VI–are found, in addition to Lyα absorption. The troughs have widths of up to 0.1 c and extend to equivalent velocities of up to 0.2 c from redshift of the QSO.

2. *The heavy element systems.* These contain sharp lines due to H and to heavier elements which can be shown to arise in a tenuous gas of near solar composition. Interpreted in terms of relative velocity, the observed values of $z_{em} - z_{abs}$ yield velocities ranging from infall to the quasar at a few thousand km s^{-1} to outflow speeds of 0.8 c.

3. *The Lyα systems.* All quasars in whose spectra Lyα emission can be observed show an increased density of sharp absorption lines on the blue side of the emission line as compared with that on the red side. Following the original suggestion of Lynds (1971) it is now established that most of the lines on the blue side of Lyα are single Lyα absorption lines; the stronger Lyα lines are accompanied by Lyβ and sometimes by Lyγ, but no lines of heavier elements have been observed.

Three additional types of absorption system are sometimes referred to. First there are the "Lyman limit" systems–in which the optical depth at the Lyman discontinuity is greater than or equal to unity. The H I column density is at least 10^{17} cm^{-2}. These

777

A. Hewitt et al. (eds.), Observational Cosmology, 777–792.

systems have the advantage that they are easily identified in relatively poor spectra:
they appear to be "heavy element" or, rarely, "Lyα forest" systems at the extreme of
the distribution of column densities (Tytler 1982; Snijders and Tytler 1987; Sargent
and Boksenberg 1983). Secondly, there are the "damped Lyα" systems which have
very strong Lyα absorption lines whose profiles imply H I column densities of 10^{20} to
10^{22} cm^{-2}. These systems always have associated heavy element lines of primarily low-
ionization species and are thought to arise in galactic disks (Carswell et $al.$ 1975; Wolfe
et $al.$ 1986). Lastly, there appears to be a tendency in some QSOs for clumps of heavy
element redshifts of relatively high ionization to occur close to (including, in some cases
slightly above) the emission redshift. There is a suspicion that this phenomenon may be
associated with radio QSOs (Foltz et $al.$ 1986).

It is generally believed that the "trough" systems represent material ejected
from the quasar. (for a review see Weymann and Foltz [1983].) On the other hand, the
nature of the "heavy element" and the "Lyα" systems has been more controversial.
However, it now seems generally accepted that the majority of the "heavy element"
absorption systems are produced by intervening galaxies and that the "Lyα" systems are
probably produced by primordial, intergalactic clouds. (For recent reviews see Boksenberg
and Sargent [1983] and Sargent and Boksenberg [1983].) It is possible that some other
heavy element redshift which appear to clump near the emission redshift are not due to
intervening material (Foltz et $al.$ 1986); further studies of their apparent association with
radio QSOs will settle this question.

A variety of arguments have shown that the majority of the sharp absorption lines
in QSO spectra are produced by cosmologically distributed, intervening objects. Among
these we may cite:

1. The Poissonian distribution of the number of absorption redshifts per object
in a given redshift range. This was established for the Lyα clouds by Sargent and
Boksenberg (1983) and for the C IV systems by Young, Sargent, and Boksenberg (1982).

2. The existence of common absorption systems in QSOs separated by minutes of
arc (Shaver and Robertson 1983a).

3. The upper limits on the population of excited fine structure states in ions such
as C II and Si II gives limits on the electron density of the absorbing gas which, when
combined with estimates of the degree of ionization of the gas, leads to a lower limit on its
proximity to the QSO in which the absorption is observed. Typically such distances come
out to be $\geq$ 100 kpc, thereby leading to unrealistic energy and momentum requirements
if it is supposed that the gas has been ejected from the QSO in which the absorption is
observed (Goldreich and Sargent 1976; Sargent et $al.$ 1979).

2. COSMOLOGICAL APPLICATIONS–OVERVIEW

The uses of QSO absorption lines in cosmology are only just beginning to be realized
in many cases. However, it is useful to list the potential applications for absorption line
studies:

1. $Physical$ $properties$ of the $interstellar$ gas in galaxies at large redshifts–particularly
the composition, spatial extent and the density of the gas and its division into neutral
(H I) and ionized (H II) constituents. Eventually such studies should lead to an understanding of the $evolution$ of the gas in galaxies of various morphological types. Progress in
this direction should be considerably accelerated once the Hubble Space Telescope enables
us to study absorption systems due to nearby galaxies at the same rest wavelengths as are
perforce used for the high redshift observations.

2. $Rotation$ $curves$ of the $faint$ $outer$ $parts$ of $galaxies$ by using suitably placed

background QSOs as targets. Such observations have considerable potential for work on the cosmological missing mass problem.

3. *Physical properties of the general intergalactic medium* at large redshifts. This can be done either directly through observations of the Gunn-Peterson (1965) effect, or indirectly through observations of the Lyα clouds, particularly if they are pressure confined. (The inception of the Hubble Space Telescope will permit us to obtain important limits on the properties of the intergalactic medium through observations of the Gunn-Peterson effect in He I and He II as well as H I.)

4. *Evolution of the number density* of absorbing objects–galaxies and intergalactic clouds.

5. *Primordial abundances* of D and He through studies of the compositions of the Lyα clouds.

6. *Evolution of large-scale structure* through observations of the correlation function $\xi(r)$ of galaxies and the Lyα clouds as a function of redshift.

7. *Determination of the deceleration parameter* q_0 from number density counts of particular types of absorption lines.

8. *Observations of the intergalactic radiation field* at large redshift through its effect on the physical site of the galactic halo absorbers and the Lyα clouds. In addition, it is possible to obtain direct limits on the microwave background temperature at high redshifts through studies of the populations of excited fine structure levels in suitable ions (Meyer *et al.* 1986).

3. COSMOLOGICAL EVOLUTION

3.1. Theory

Consider absorbing clouds of cross section πr_0^2 and number density Φ_0 per unit co-moving volume. It can be shown that the number density of absorption systems of a particular type per unit redshift range is given by

$$\frac{dN(z)}{dz} = N_0(1 + z)(1 + 2q_0 z)^{-1/2} \tag{1}$$

if the cosmological constant $\Lambda = 0$ (Peterson 1978; Sargent, Young, Boksenberg, and Tytler 1980). Here

$$N_0 = \pi r_0^2 \Phi_0 \frac{c}{H_0} \tag{2}$$

is the local ($z = 0$) value of $N(z)$. Expression (1) clearly holds if there is a constant size *distribution* for the absorbing objects at all redshifts. If, on the other hand, the absorbers evolve in constant time then the product $\pi R_0^2 \Phi_0$ and hence N_0 may be functions of redshift. For comparison with observations it is convenient to express equation (1) as a power law:

$$\frac{dN(z)}{dz} = \text{const.}(1 + z)^\gamma \tag{3}$$

It is easy to show that

$$\gamma = \frac{1 + q_0 z - q_0}{1 + 2q_0}. \tag{4}$$

In particular, $\gamma = 1$ for $q_0 = 0$ and $\gamma = 1/2$ if $q_0 = 1/2$. For other values of q_0, γ is a function of z.

Thus, an empirical plot of log (dN/dz) versus log (1+z) may be used to determine q_0 if one has independent evidence on the behavior of the product $\pi r_0^2 \Phi$ with redshift.

3.2. Evolution of the Lyα Clouds

The best data on redshift evolution for QSO absorption lines comes from the Lyα clouds since their density is so high. Following Peterson's (1978) initial work, all recent studies have confirmed that the Lyα clouds exhibit unquestionable cosmological evolution, although there are still small differences in the results obtained by different investigators for identical data sets which are hard to explain (Murdoch *et al.* 1986). These last authors find

$$\frac{dN}{dz} \propto (1 + z)^{2.17} \tag{5}$$

a result which is in line with several earlier studies. It will be noted from equation (4) that an exponent $\gamma = 2.17$ is too large to be accommodated with any value of q_0 and that the Lyα clouds, therefore, must evolve in cosmic time in the sense of either being more numerous or having larger effective cross sections at higher redshifts, or both. (The evolution of the density of the "Lyα forest" may easily be seen by inspecting high quality spectra of QSOs with redshifts around $z \sim 2$ and $z \sim 3$, respectively.) Carswell *et al.* (1982), Murdoch *et al.* (1986) and Tytler (1987) have recently discussed a puzzling anomaly in which the line density in most individual QSOs does not show a convincing decrease with redshift over the small range $\Delta z = 0.19$ that is accessible between the Lyα and Lyβ emission lines. (Below the Lyβ emission line the Lyα absorptions are "contaminated" by Lyβ absorptions corresponding to Lyα lines at higher redshifts and so the situation here is harder to judge.) The origin of this "inverse" effect is obscure and probably indicates that our understanding of the physics of the Lyα clouds is still very insecure. However, one possibility to explain it which has not been explored in the existing literature is that QSOs occupy regions of the Universe which are unsuitable for the formation of Lyα clouds over large volumes–for example, by having too great a pressure in the intergalactic medium.

3.2. Evolution of Heavy Element Redshift Systems

The heavy element redshift systems have a much lower density than the Lyα cloud density and so their evolution is much more difficult to study. Marginal evidence for and against evolution of the C IV redshift systems which can be observed from the ground in the redshift range $1.2 \leq z \leq 4.0$ has been discussed by Bergeron and Boissé (1983) and by Boissé and Bergeron (1985). Tytler (1986) has made an extensive study of Mg II absorption systems in the redshift range $0 < z < 2$. My interpretation of this data is that there is probably evolution if $q_0 \sim 1/2$ but that if $q_0 \sim 0$ there is no evidence for evolution. A large new homogeneous survey of C IV systems in the redshift range $1.5 < z < 3.5$ which has recently been completed by Boksenberg and the author should give valuable information on the evolution of the heavy element redshift systems. In any case, it is almost certain from existing data that the density of the Lyα clouds increases more rapidly with redshift than that of the heavy element systems.

3.4. Evolution of the Lyman Limit Systems

A particularly interesting class of QSO absorption systems is those which show a measurable discontinuity of the Lyman limit, implying that their H I column density is $N(H\ I) \geq 10^{17}$ cm^{-2}. Such systems are almost invariably associated with lines of heavier elements at the same redshift. Consequently, it is supposed that they represent the high column density "tail" of the distribution of heavy element redshift systems that are associated with galaxies. One Lyman limit system, at $z_{abs} = 2.99$ in the QSO Pks 2126–158 ($z_{em} = 3.28$), has been found with no observable heavy element lines, despite its high H I column density; this system is particularly important in attempts to set limits on the heavy element composition of the Lyα clouds (Sargent 1982; Sargent and Boksenberg 1983; Chaffee *et al.* 1985). The Lyman limit systems have the advantage that they can be detected on low-resolution spectra of relatively low S/N ratio and hence it has been possible to study their density down to redshifts as low as $z = 0.3$ using the IUE satellite. (Tytler 1982; Snijders and Tytler 1986; Bechtold *et al.* 1984). Snijders and Tytler (1986) have assembled a list of 40 Lyman limit systems covering the redshift range $0.3 < z < 3$ in 120 QSOs. The statistical study of these systems is complicated by the fact that the presence of a Lyman limit system at same redshift z_1, blots out the QSO continuum at all wavelengths below $912(1+z_1)$ Å so that only one such system can be discovered per QSO. Tytler (1982) has ingeniously modified statistical techniques called "survival statistics", developed by the Insurance industry for a different but analogous problem, to deal with this complication. The density of Lyman limit systems shows a definite increase with redshift which, however, is incompatible with $q_0 \sim 1/2$ and no evolution in the intrinsic physical properties of the absorbers. The data are compatible with $q_0 \sim 0$. It is interesting and perhaps significant that the density increases with redshift of both the Lyman limit and Mg II absorption systems are compatible with $q_0 \sim 0$ and no evolution in the product $\pi r_0^2 \Phi_0$ with redshift. If $q_0 \sim 1/2$, as the "inflationary" cosmology demands, then the product $\pi r_0^2 \Phi_0$ for the Lyman-limit absorbers must increase with z in such a manner as to mimic a $q_0 \sim 0$ Universe. On the other hand, direct determinations of q_0 from the density of luminous matter already point to a low-density Universe.

Snijders and Tytler (1986) combined measurements of the Lyman discontinuity and the equivalent widths of the associated Lyα absorption lines in order to determine H I column densities for their sample of 40 Lyman-limit redshift systems. The values are in the range $17 < \log N(H\ I) < 21.5$; the distribution log N(H I) appears to be constant over the observed redshift range $0.3 < z_{LLS} < 3.5$. Thus, there is no evidence from this independent datum that the physical properties of the absorbers change systematically with redshift.

Tytler's work shows that 50% of randomly chosen lines of sight intercept a cloud which absorbs 99% or more of the flux at the Lyman limit at a redshift of $z = 2$; the number rises to just over 90% by a redshift of $z = 4$. The corresponding numbers for systems which absorb 10% or more of the flux is 70% at a redshift of $z = 1$ and 100% at $z = 2$. Thus, such systems have a significant effect on the metagalactic flux from radiation capable of ionizing H I. Tytler estimates that the metagalactic ionizing flux from QSOs and galaxies is cut down by a factor of about 3 over the redshift range $2 < z < 4$ due to this effect.

4. ABSORPTION AND INTERGALACTIC GAS

4.1. The Gunn-Peterson Test

It was shown by Gunn and Peterson (1965) that the overlapping Lyα absorption lines from any generally distributed intergalactic neutral hydrogen would produce a depression in the level of the QSO continuum on the blue side of the Lyα emission line in a QSO as compared with the level on the red side. The optical depth of the depression is given in terms of the H I volume density by

$$\tau_\nu = \frac{8.3 \times 10^{10} n_{HI}(z)}{(1+z)(1+2q_0 z)^{1/2}} \tag{6}$$

Early observations of high redshift QSOs (e.g., 3C 9, $z_{em} = 2.016$) revealed no observable continuum depression. With a conservative estimate $\tau_\nu < 0.5$ for the optical depth, remarkably low limits could be placed on any intergalactic H I. Thus, with $q_0 = 1/2$, $n_H \leq 3 \times 10^{-11}$ cm^{-3} at $z = 2$.

Later, Davidson, Hartig, and Fastie (1977) applied the method to space-based observations of 3C 273 ($z_{em} = 0.158$). They found no evidence for a continuum depression (and no evidence for absorption *lines*) and were able to obtain a limit on the H I density at the present epoch of $n_H \leq 6 \times 10^{-12} cm^{-3}$.

The recognition of the existence of the "Lyα forest" in the early 1970s diverted attention away from the Gunn-Peterson effect because the presence of a rich line spectrum in the blue side of the Lyα emission line makes the determination of the true continuum level more difficult. Recently, however, Steidel and Sargent (1987) have accurately measured the continua of several QSOs with redshifts around $z \sim 3$ and have shown that within the errors the lines do in fact account for all the apparent continuum depression seen on the blue side of Lyα in such objects. It is possible to infer from their data that, conservatively, $\tau \leq 0.05$ at $z = 3$. This leads to limits on the density of any generally distributed H I gas of $n_H \leq 2.4 \times 10^{-12}$ cm^{-3} at $z = 3$ if $q_0 = 0$. Dividing by $(1+z)^3$ the corresponding limit at $z = 0$ would be $n_H < 3.8 \times 10^{-14}$ cm^{-3}. For $q_0 = 0$ the corresponding limits are twice these values.

4.2. The Lyα Clouds and the Intergalactic Medium

The Lyα clouds are thought to be tenuous objects of galactic dimensions which are highly ionized by the intergalactic ionizing flux due to QSOs and galaxies. Sargent, Young, Boksenberg, and Tytler (1980) suggested that they are not self–gravitating but are confined by the pressure of a general intergalactic medium. They pointed out that the clouds could only be self–gravitating if they were very large, so large in fact that they would fill a substantial fraction of the Universe. Sargent *et al.* studied the possible physical properties of the clouds and the hypothetical surrounding medium using such conditions as the requirement that the hotter and more tenuous medium should not evaporate away the clouds during the Hubble time appropriate to the epoch at which the clouds are observed. Their resulting estimates (all at $z = 2.44$) were:

For the Lyα clouds: density $n_H^c = 10^{-4}$ cm^{-3}, temperature $T_c = 3 \times 10^4$ K, ionization fraction $n_{HII}^c / n_{HI}^c = 10^5$, cloud diameter $D = 3 \times 10^{22}$ cm (10 kpc), cloud mass $M^c = 10^{41}$g $= 3 \times 10^7 M_\odot$ and space density $\Phi^c = 2 \times 10^2$ Mpc^{-3}. The contribution of the clouds to the cosmological density is small, namely $\Omega^c = 2 \times 10^{-3}$.

For the intergalactic medium: density $n_H^M = 10^{-5}$ cm^{-3}, temperature $T_M = 3 \times 10^5$ K.

It was pointed out that those parameters are very tight–the clouds would evaporate if more tenuous than the quoted value and collapse if less tenuous. More sophisticated considerations along similar lines by Ostriker and Ikeuchi (1983) and by Black (1981) led to similar results. Later, Ikeuchi and Ostriker (1986) carried out calculations of the evolution of the clouds and surrounding medium in cosmic time which accounted in a satisfactory way for the properties observed at $z \sim 2.5$.

In a beautiful and difficult observation, Foltz, Weymann, Roser, and Chaffee (1984) obtained spectra of the Lyα absorption lines in the two components of the double

QSO Q2345+007 ($z_{em} \sim 2.1$) which are separated by $7''$ and which are believed to be the result of gravitational lensing by a so far undetected galaxy or cluster. Foltz *et al.* showed that most of the stronger Lyα lines are common to the two components but that a few would appear not to be. This led them to infer a size of about 8 kpc for the Lyα clouds– the exact value depends on the unknown distance of the lensing galaxy which is assumed to have a redshift of $z \sim 1$. This diameter estimate is seen to be very close to that arrived at by purely astrophysical considerations and would at first sight appear to confirm the picture of pressure confinement for the clouds. However, Rees (1986) proposed that the Lyα clouds are baryonic condensations in "mini-halos" of cold dark matter which serves to confine them gravitationally. This idea is attractive because such mini-halos are the first objects to form in the currently popular cool dark matter dominated cosmologies.

The general intergalactic medium inferred from the pressure confinement arguments would have a temperature $T^M = 10^3$ K and a density $n^M \sim 2 \times 10^{-7}$ cm^{-3} at the present epoch ($z = 0$); it would remain highly ionized as the Universe expands despite its low temperature because the recombination time is longer than the Hubble time. Its contribution to the cosmological density parameter, $\Omega^M \sim 0.1$ for $H_0 = 50$ km s^{-1} Mpc^{-1} is somewhat higher than that inferred from visible galaxies and is consistent with the baryonic density $\Omega_B \sim 0.14$ required by current primordial nucleosynthesis data (Boesgaard and Steigman 1985). A medium with the quoted parameters is too cool and tenuous to be observed directly by presently conceivable techniques.

While the current situation is not clear, it is certain that the Lyα clouds could not co-exist with a hot, dense intergalactic gas with $T \sim 10^7$ K and $\Omega^M \sim 1$ which is required to account for the diffuse X-ray background.

5. CLUSTERING OF QSO ABSORPTION LINES

5.1. Clustering of Galaxies

The crudest and most commonly used quantitative measure of the clustering tendency of galaxies is the correlation function $\xi(r)$ introduced by Peebles. The probability dP of finding a galaxy in volume dV at distance r from any given galaxy written in the form

$$dP = \Phi[1 + \xi(r)]dV \tag{7}$$

where the first term represents the Poissonian expectation from randomly distributed galaxies and $\xi(r) > 1$ represents the effect of clustering. Empirically it is found that locally

$$\xi(r) = \left(\frac{r}{5 \text{ Mpc}}\right)^{1.77}. \tag{8}$$

Virtually nothing is known observationally about the evolution of $\xi(r)$ in cosmic time.

It has been clear for some years that in principle we can use QSO absorption lines to study the evolution of the correlation function at large redshifts beyond where we can directly observe typical galaxies. Thus if the absorbers are clustered so should their redshifts be if they are largely due to the expansion of the Universe. (The observed redshifts will, of course, be modified by the peculiar motions induced by the clustering if observations are made along the line of sight to a particular QSO). On the other hand, observations of correlated absorptions in nearby lines of sight leads to complementary information. A little consideration shows that by combining the results of the two types of observation it is possible to determine the distortions in redshift space of which the "fingers of God" are one manifestation (Sargent and Turner [1977] have discussed how such characteristic distributions may be used to determine Ω.)

Thus we are led to the idea of looking at the *correlation functions in redshift* for different types of QSO absorption lines. As a practical matter it is necessary to eliminate the epoch-dependent stretching effects on large scale clustering produced by the expansion of the Universe. Thus, consider two absorbers of redshift z_1 and z_2, respectively. We first calculate the proper distance S_0 of each as measured at the present cosmic epoch for $q_0 = 1/2$:

$$S_0(1/2, z) = \frac{2c}{H_0}[1 - (1 + z)^{-1/2}]$$ (9)

so that for any other epoch $S(z) = S_0(z)/(1 + z)$. We then calculate the spatial separation

$$s_0 = S_0(z_2) - S_0(z_1).$$ (10)

The natural unit of length is the Hubble radius $c/H_0 = 6000$ Mpc if $H_0 = 50$ km s^{-1} Mpc^{-1}.

5.2. Clustering of the Heavy Element Redshifts

The heavy element redshifts are strongly clustered on a scale of 150 km s^{-1}. It is not yet clear whether this is due to the relative motion of clouds within particular galaxies (Bahcall and Spitzer 1976) or whether it represents the tip of the galaxian correlation function (Young, Sargent, and Boksenberg 1982). The implications if the latter hypothesis is correct are discussed later in this section. Marginal evidence for clustering of C IV doublets on a scale of 2000 km s^{-1} was found by Young *et al.* (1982). This, of course, could not be due to motions of discrete clouds within galaxies; one of the most important aims of the large survey by Sargent and Boksenberg which was mentioned earlier to investigate C IV clustering on scales of 300–2000 km s^{-1}.

5.3. Clustering of Lyα Clouds

Since the Lyα clouds are much more numerous than the objects (presumably galaxies) which produce the heavy element redshifts, it is much easier to investigate their clustering properties. Sargent *et al.* (1980) showed from a correlation function analysis that there is no observable clustering in the Lyα clouds on all scales from 300 to 30,000 km s^{-1}. In a later paper, Sargent, Young, and Schneider (1982) showed that a cross correlation analysis between Lyα absorption lines in the spectra of two QSOs, Q1623+268 and Q1623+267, which are separated by 3' on the plane of the sky, also show no evidence for clustering in the same range of scales. Sargent *et al.* (1982) showed that these measurements imply that the "correlation length" for the Lyα clouds in equation (8) is $r_c \leq 0.2$ Mpc at $z \sim 2.5$. On the "hierarchial" model of the evolution of clustering which has been advocated by Peebles and his associates, and which arises naturally in the "cold dark matter dominated" cosmologies, the correlation function is expected to evolve such that the variation of the correlation length with redshift scales as

$$r_c(z) = r_{c(z=0)}(1 + z)^{-5/3}.$$ (11)

Thus, we might expect $r_c = 1$ Mpc at $z = 2.5$ on this particularly simple model. The observed upper limit is a factor of 5 less than this. As we shall see later, one possible explanation is that the Lyα clouds await galaxy clusterings. (At the present Symposium Atwood and colleagues presented evidence that the Lyα clouds cluster weakly on scales less than $\Delta v = 300$ km s^{-1}.) This result, if correct, is not in conflict with earlier work. Moreover, it still remains the case that the correlation functions of the Lyα clouds and the heavy element absorbers are remarkably different on scales $\Delta v \sim 150$ km s^{-1}; this fact alone almost certainly indicates that the two kinds of absorption lines originate in different kinds of object.

(Since this talk was given, a reprint has appeared by Rees and Carswell (1987) drawing attention to the fact that the flat correlation function of the Lyα clouds on scales of several thousand km s^{-1} imply that there is no evidence for "voids" in their distribution of the kind which are so prominent in the local distribution of galaxies.)

5.4. Implications of the Clustering Data

Salmon and Hogan (1986) have compared the data published by Sargent *et al.* (1980) regarding the clustering properties of the Lyα clouds and the heavy element absorbers with simulations of the behavior of absorption-line systems in cold dark matter cosmologies. The difference in the clustering tendencies of the two types of absorber can be understood in two possible ways:

1. The heavy element absorption systems are associated with galaxies which in turn constitute an unbiased sample of the overall mass distribution in an $\Omega = 0.2$ Universe. On this hypothesis some mechanism must be invented in order to prevent the Lyα clouds also following the mass distribution. A natural possibility, which was pointed to me by J. Ostriker in another connection, is that the intergalactic medium is clustered with the galaxies and that its pressure destroys the clouds. As Ostriker pointed out, the intergalactic pressure in the Coma cluster gas ($T \sim 10^8$ $K, n_H \sim 10^{-3} - 10^{-4}$cm^{-3}) is far in excess of the pressure inferred for the clouds. ($n_H^c T^c \sim 1$. Presumably, weaker clusterings have gas with lower values of $n_H^M T^M$.)

2. In an $\Omega = 1$ Universe the correlation at $z = 3$ is much smaller than that in an $\Omega = 0.2$ Universe. The Lyα clouds are then an unbiased sample of the overall mass distribution while the heavy element absorbers, like the galaxies, are more strongly clustered than the mass.

These considerations, inconclusive as they are, point up the enormous potential value of QSO absorption lines in giving information about the clustering of matter at redshifts inaccessible to other modes of observation.

6. PAIRS OF QSOs AND LARGE SCALE STRUCTURE

6.1 The Pair Q0307–199A,B

The existing studies of common heavy element absorption redshifts in pairs of QSOs offer tantalizing glimpses of large scale organization in the clustering of the absorbers. Shaver and Robertson (1983a) have been particularly active in this kind of investigation. A particularly interesting pair studied by Shaver and Robertson (1983b) is that comprising UM 680 (Q0307–195A; $z_{em} = 2.144$) and UM 681 (Q0308–195B; $z_{em} = 2.122$). These 19^m objects are separated by $58''$ on the plane of the sky; this corresponds to $378\ (H_0/100)^{-1}$ kpc at $z = 2.13$, assuming $q_0 = 1/2$. Shaver and Robertson (1983b) and later Sargent and Boksenberg (1987) have found heavy element absorption redshifts in UM 680 at $z_{abs} = 2.1228, 2.0919, 2.0353, 1.7042,$ and 1.5623, respectively. In UM 681 they found redshifts at $z_{abs} = 2.1220, 2.0323, 1.7057,$ and 1.7885, respectively. It will be seen that the spectra contain common redshifts at $z_{abs} = 2.12, 2.03,$ and 1.70.

Shaver and Robertson (1983b) pointed out that the common system at $z_{abs} =$

2.12 is likely to be a very large cloud associated with UM 681. The velocity difference $\Delta v = c\Delta z/1 + z$ between the emission redshift of UM 681 and the mean of the absorption redshifts observed in the two QSOs is Δv(em. - abs.) $= 29 \pm 29$ km s^{-1}. If it is supposed that the absorbing gas is in the form of a disk about 1 Mpc in diameter around UM 681 then we may estimate that the total mass of the gas is $2 \times 10^{10} M_\odot$. The velocity difference between the absorption systems at $z_{abs} = 2.1228$ in UM 680 and $z_{abs} = 2.1220$ in UM 681 turns out to be $\Delta v(A - B) = 77 \pm 14$ km s^{-1}. If this velocity difference is due to rotation of the hypothetical disk, then its total mass is

$$M_t = \frac{10^{11} - 10^{12}}{h \sin^2 i} M_\odot \qquad (12)$$

where $h = (H_0/100)$ and i is the angle of inclination of the disk to the line of sight to UM 681. This object is reminiscent of the enormous cloud of ionized gas around the radio source MR 2251–178 which was discovered by Bergeron *et al.* (1983).

The other two absorption systems, at $z_{abs} = 2.03$ and $z_{abs} = 1.70$, respectively, are clear cases of common absorption which is not associated with either QSO. They could be due to galaxies in intervening clusters or superclusters or to the extended halos of single galaxies. In either case, the velocity differences $\Delta v(A - B)$ of –297 km s^{-1} (for $z_{abs} = 2.03$) and $\Delta v = 172$ km s^{-1} (for $z_{abs} = 1.70$) point to the existence of substantial mass on large scales–as is also implied by the flat rotation curves of nearby galaxies and the large velocity differences between nearest neighbor pairs of galaxies in our immediate vicinity (Davis, Huchra, Latham and Tonry 1982).

Crotts (1985) favors the notion that these correlated absorptions are all produced by very extended absorbers. He estimates on the basis of correlation function simulations that these large objects have about the same space density as clusters in the Zwicky catalogue ($\sim 10^{-4}$ Mpc^{-3}) and a correlation scale length of about 500 kpc which does not change with redshift over the observed range.

6.2 Absorption in a QSO in the Field of Pks 0237–233

The 16.8 mag QSO Pks 0237–233 ($z_{em} = 2.220$) was one of the first to have its absorption spectrum studied in some detail, both because the QSO is abnormally bright and because its absorption spectrum is unusually rich. In particular, it has close double absorption redshifts at $z_{abs} = 1.59, 1.61, 1.63$ and 1.65 and a prominent "clump" of at least 7 redshifts at $z_{abs} = 1.67$ which extends over a range corresponding to $\Delta v \sim 5000$ km s^{-1} in velocity. In 1977, following the completion of the work on the absorption spectrum of Pks 0237–233 by Boroson, Sargent, Boksenberg, and Carswell (1978), Boksenberg and the present author realized that such redshift clumps would correspond to relatively large angles on the sky if they corresponded to large structures such as superclusters of galaxies. For example, at $z = 2$ a structure of size 50 Mpc ($H_0 = 50$ km s^{-1} Mpc^{-1}) subtends an angle of order 1 degree on the plane of the sky if $q_0 = 1/2$ and about 0.5 deg if $q_0 = 0$. Accordingly, we embarked on a systematic survey for QSOs in the field around Pks 0237–233. Two methods were used. In the first one the field around Pks 0237–233 was photographed through U,B,V filters with the Palomar 48–inch Schmidt telescope on a single plate with the telescope being moved between the three exposures which were arranged so as to give roughly equal images for typical stars. These multiple exposure plates were then searched for abnormally blue objects by C. Kowal; his resulting candidate QSOs were then observed spectroscopically at low resolution by Sargent and Kunth at the Palomar 200–inch and Las Campanas 100–inch telescopes.

In a parallel search, A. Boksenberg examined an objective prism plate of the field which had been obtained by the UK Schmidt Unit. His QSO candidates around Pks 0237–233 were also observed by Sargent and Kunth.

The result of this work (which is described in more detail in Sargent, Boksenberg, and Kunth [1987]) is that a 19 mag QSO Q0236–234 with a redshift $z_{em} = 2.180$ was discovered 30′ away from Pks 0237–233. High–resolution spectra of this object, secured by Boksenberg at the Anglo–Australian telescope, revealed a spectrum generally devoid of absorption lines longwards of the Lyα emission line. However, there is a weak C IV doublet at a redshift $z_{abs} = 1.5953$ which coincides exactly with one component $z_{abs} = 1.5950$ of one of the double absorption systems in Pks 0237–233. The separation of the lines of sight to the two QSOs on the plane of the sky at $z = 1.6$ is 19 Mpc (assuming $H_0 = 50$ km s^{-1} Mpc^{-1} and $q_0 = 1/2$) evaluated at the epoch when the light traversed the absorbers. The corresponding linear separation evaluated at the present epoch is 50 Mpc. This coincidence again points to the existence of large–scale structures at high redshifts and points up the possibility of mapping them through observations of absorption lines in suitable background QSOs.

6.3. The Wide Pair Q1037–2704, Q1038–2712

An intriguing case of several common absorption redshifts in the quasar pair Q1037–2704 and Q1038–2712 has recently been discovered by Jakobsen, Perryman, Ulrich, Macchetto, and di Serego Alighieri (1986). The pair is separated by 17′ on the plane of the sky; this corresponds to 4.3 Mpc (assuming $H_0 = 100$ km s^{-1} Mpc and $q_0 = 0.5$) at $z \sim 2$. In March 1986 the writer and C. C. Steidel obtained spectra of resolution 2 Å of the pair at the Las Campanas DuPont reflector (Sargent and Steidel 1987). The following is a brief summary of the combined results, most of the essentials of which were described on the basis of lower dispersion spectra by Jakobsen *et al.*

1. The emission redshifts are $z_{em} = 2.331$ (Q1038–2712) and $z_{em} = 2.193$ (Q1037–2704).

2. Both QSOs have exceptionally rich absorption spectra. In the recent high–resolution survey by Sargent and Boksenberg (1987) referred to earlier no QSO was found with an absorption spectrum as complex as either Q1037–2704 or Q1038–2712. The absorption redshifts in Q1037–2704 occur at $z_{abs} = 1.9125, 1.9718, 2.0279, 2.0708, 2.0825, 2.1279$, and 2.1390, respectively. Those in Q1038–2712 occur at $z_{abs} = 1.895, 1.9550, 2.0138, 2.065, 2.077$, and 2.146, respectively.

3. All the absorption systems contain sharp lines (including C IV doublets in all cases) except for the $z_{abs} = 2.065, 2.077$ complex in the spectrum of Q1038–2712 which is a broad feature with some structure, similar to the broad absorption "troughs" found the "BAL" QSOs.

4. There are six pairings: cases in which similar absorption redshifts occur in both QSOs. The calculated velocity differences $\Delta v = c\Delta z/(1 + z)$ between the various pairs of systems range from 1800 to 540 km s^{-1}; they do not all have the same sign.

5. Several other QSO candidates were discovered in the field of Q1037–2704 and Q1038–2712 by Bohuski and Weedman (1979). One of these, Q1038–2707, is only 5′ away from Q1038–2712. Sargent and Steidel (1987) obtained low–resolution spectra of this object and found that it has a redshift $z_{em} = 1.937$ and that, even at low resolution, there is a strong absorption system at $z_{abs} = 1.89$.

Jakobsen *et al.* (1986) discussed three hypotheses which might explain the facts summarized heretofore. The first is that the resemblance in the absorption spectra of the two objects is purely a chance occurrence. As they pointed out this is unlikely because both spectra are abnormally rich; however, this is hard to quantify *post hoc.*

A second possibility is that the redshifts are due to material which has been ejected from one or both QSOs. There are several arguments against this hypothesis. Based on their redshifts, Q1037–2704 is 22 Mpc closer to us than Q1038–2712. Since the absorptions seen in both spectra extend only as high as $z_{abs} = 2.146$, a little less than the emission redshift of Q1037–2704 ($z_{em} = 2.193$), it is fairly evident that Q1038–2712 is unlikely to be the source of the hypothetical ejection. Otherwise, it is somewhat curious that no absorption redshifts lying between $z_{abs} = 2.33$ and $z_{abs} = 2.19$ are observed in its spectrum. Moreover, it is easy to show that a shell of typical column density ($N(H) = 10^{19}$ cm^{-2}) with a radius greater than 22 Mpc leads to completely unreasonable energy and momentum outflows from Q1038–2712. Accordingly, let us consider the consequences of supposing that the ejection occurred instead from Q1037–2704. In this case the ejected matter must extend more than 4 Mpc from the source in order to cover both lines of sight. However, we do not observe any absorptions with redshifts larger than $z_{abs} = 2.19$ in the spectrum of the more distant QSO, Q1038–2712, which would correspond to ejection in its direction. Thus, we are left with the highly artificial picture that the ejection from Q1038–2712 is all in the direction towards us. Even if we accept this possibility, the ejected blobs corresponding to each of the six redshift systems in common must have an extension of at least 4 Mpc perpendicular to the direction of ejection. It is easy to calculate that a typical system must contain about $10^{11} M_\odot$ moving at 3200 km s^{-1} away from the source with a resulting kinetic energy of $\sim 10^{61}$ ergs (Sargent and Steidel 1987). If the ejection has been through radiation pressure then the energy requirements must be even higher, because a single photon can transmit in a single scattering at most a fraction v/c of its energy into kinetic energy. Thus, it turns out that for a typical absorption system the energy required for the corresponding blob to be radiatively driven is 10^{64}–10^{65} ergs. Thus the QSO would be required to radiate more than 10^{47} ergs s^{-1} for the age of the Universe!

Jakobsen *et al.* proposed that the common absorptions might be the result of independent ejections from the two QSOs and that some kind of line–locking might be operating between systems such that only one coincidence in redshift would be required rather than several. However, the improved spectra obtained by Sargent and Steidel (1987) show that the ratios of $(1+z)$ between adjacent systems are not the same in both QSOs. Moreover, the absence of absorption from excited fine structure states in the sharp redshift systems implies that the absorbing clouds must still be ~ 100 kpc from the two QSOs even if independent ejection has occurred. This again leads to severe energy and momentum problems, although they are not as bad as the ones encountered in the preceding arguments.

The remaining possibility is that the unusual density of absorption lines in the two spectra is due to the fact that the two lines of sight are traversing a large scale distribution of galaxies which extends roughly from $z = 2.19$ to $z = 1.89$. On this hypothesis Q1037–2704 is close to the far side of this structure, while the third QSO, Q1038–2707 is close to the near side. The object must have an extent of about 50 Mpc along the line of sight and be at least 5 Mpc across in the perpendicular direction. (The quoted dimensions are those appropriate to the epoch when the light traversed the system.) The abnormal richness of the absorption systems in the spectra of Q1037–2704 and Q1038–2712 also implies that the structure must be rare. The least implausible conclusion is that we are observing a supercluster from a particular rare direction. For example, we could be looking along a filament of galaxies in which clusters are embedded.

Although the above explanation of the absorption spectra of Q1038–2712, Q1037–2704, and Q1038–2707 is the most plausible of the possibilities, we are nevertheless left with an uncomfortable feeling that none of the proposed explanations are correct. However, if the large–scale structure explanation is essentially correct, we have the fascinating possibility of determining something about the intricate details of superclusters at large redshifts. On the other hand, the interpretation of the statistical behavior of the absorption systems observed along different lines of sight in the Universe will be much more complicated than has hitherto been realized.

7. EPILOGUE

It is clear from the considerations outlined in this talk that the full exploitation of QSO absorption lines for cosmological studies of various kinds, particularly the correlation and large–scale structure investigations, requires a large body of statistical data which are only now beginning to be obtained. Nevertheless, the absorption offer bright hopes of leading to important information on the distribution of clouds and galaxies at much earlier epochs.

Acknowledgements: I thank my co–workers A. Boksenberg, D. Kunth and C. Steidel for allowing me to quote the results of our joint work in advance of publication. I also thank D. Tytler for providing me with details of his unpublished work. The work described in this paper was supported in part by the National Science Foundation under Grant AST84-16704.

8. REFERENCES

Bechtold, J., Green ,R.F.,Weymann, R.J., Schmidt, M., Estabrook, F.B., Sherman,R.D., Wahlquist, H.D., and Heckman, T.M. 1984, *Ap.J.*, **281**, 76.

Bergeron ,J., Boksenberg, A., Dennefeld, M., and Tarenghi, M. 1983, *M.N.R.A.S.*, **202**, 125.

Bergeron, J. and Boissé, P. 1983, in *Quasars and Gravitational Lenses*, Proc. 24th Liège Symposium, p. 589.

Black, J.H. 1981, *M.N.R.A.S.*, **197**, 553.

Boesgaard, A.M., and Steigman, G. 1985, *Ann. Rev. Astr. Ap.*, **23**, 319.

Bohuski, T.J., and Weedman D.W. 1979, *Ap. J.*, **231**, 653.

Boissé, P., and Bergeron, J. 1985, *Astr. Ap.*, **145**, 59.

Boksenberg, A., and Sargent, W.L.W. 1983, in *Quasars and Gravitational Lenses*, Proc. 24th Liège Symposium, p. 500.

Boroson, T., Sargent, W.L.W., Boksenberg, A., and Carswell, R.F. 1977, *Ap. J.*, **220**, 772.

Carswell, R.F, Hilliard, R.L., Strittmatter, P.A., Taylor, D.J., and Weymann, R. 1975, *Ap. J.*, **196**, 351.

Carswell, R.F., Whelan, J.A.J., Smith, M.G., Boksenberg, A., and Tytler, D. 1982, *M.N.R.A.S.*, **198**, 91.

Chaffee, F.H, Foltz, C.B., Roser, H.-J., Weymann, R.J., and Latham, D.W. 1985, *Ap. J.*, **292**, 362.

Crotts, A.P. 1985, *Ap. J.*, **298**, 732.

Davidsen, A. , Hartig, G.F. and Fastie, W.G. 1977, *Nature*, **269**, 203.

Davis, M., Huchra, J. Latham, D.W., and Tonry J. 1982, *Ap.J.*, **253**, 423.

Foltz, C.B., Weymann, R.J., Peterson, B.M., Sun, L., Malkan, M.A., and Chaffee, F.H. 1986, *Ap. J.*, **307**, 504.

Foltz, C.B., Weymann R.J., Roser, H.-J., and Chaffee, F.H. 1984, *Ap. J. (Letters)*, **281**, L1.

Goldreich, P., and Sargent, W.L.W. 1976, *Comments on Ap.*, **6**, 133.

Gunn, J. E., and Peterson, B.A. 1965, *Ap. J.*, **142**, 1633.

Ikeuchi, S., and Ostriker, J.P. 1986, *Ap. J.*, **301**, 522.

Jakobsen, P., Perryman, M.A.C., Ulrich, M.H., Macchetto, F., and di Serego
    Alighieri, S. 1986, *Ap. J. (Letters)*, **303**, L27.

Lynds, C.R. 1971, *Ap. J. (Letters)*, **164**, L73.

Meyer, D.M., Black, J.H., Chaffee, F.H., Foltz, C.B., and York, D.G. 1986, *Ap.
    J. (Letters)*, **308**, L37.

Murdoch, H.S., Hunstead, R.W., Pettini, M., and Blades, J.C. 1986, *Ap. J.*,
    309, 19.

Ostriker, J.P., and Ikeuchi, S. 1983, *Ap. J. (Letters)*, **268**, L63.

Peterson, B.A. 1978, in *IAU Symposium No. 79, Large Scale Structure of the
    Universe*, eds. M.S. Longair and J. Einasto, (Dordrecht: Reidel), p.
    390.

Rees, M.J. 1986, *M.N.R.A.S.*, **218**, 25P.

Rees, M.J. and Carswell, R.F. 1987, *M.N.R.A.S.*, in press.

Salmon, J., and Hogan, C. 1986 *Ap. J.*, in press.

Sargent, W.L.W. 1982, *Phil. Trans Roy.Soc. London A*, **307**, 87.

Sargent, W.L.W., and Steidel, C.C. 1987, *Ap. J.*, in preparation.

Sargent, W.L.W., and Turner, E.L. 1977, *Ap. J. (Letters)*, **212**, L3.

Sargent, W. L. W., Boksenberg, A. and Kunth, D. 1987, *Ap. J.*, in preparation.

Sargent, W.L.W., and Boksenberg, A. 1983, in *Quasars and Gravitational
    Lenses*, Proc. 24th Liège Symposium, p. 518.

Sargent, W.L.W., and Boksenberg, A. 1987, *Ap. J.*, in preparation.

Sargent, W.L.W., Young, P.J., Boksenberg, A., and Tytler, D. 1980, *Ap. J.
    Suppl.*, **42**, 41.

Sargent, W.L.W., Young, P.J., and Schneider , D.P. 1982, *Ap. J.*, **256**, 374.

Sargent, W.L.W., Young, P.J., Boksenberg, A., Carswell, R.F., and Whelan,
    J.A.J. 1979, *Ap. J.*, **230**, 49.

Shaver,P.A., and Robertson, J.G. 1983*a*, in *Quasars and Gravitational Lenses*,
    Proc. 24th Liège Symposium, p. 598.

Shaver, P.A., and Robertson, J.G. 1983*b*, *Ap. J. (Letters)*, **268**, L57.

Snijders, A. and Tytler, D. 1987, in preparation.

Steidel, C.C., and Sargent, W.L.W. 1987, *Ap. J.*, in press.

Tytler, D. 1982, *Nature*, **298**, 427.

Tytler, D. 1987, in preparation.

Weymann, R.J., and Foltz, C.B. 1983 in *Quasars and Gravitational Lenses*,
    Proc. 24th Liège Symposium, p. 538.

Wolfe, A.M., Turnshek, D.A., Smith, H.E., Cohen, R.D. 1986, *Ap. J. Suppl.*,
    61, 249.

Young, P.J., Sargent, W.L.W., and Boksenberg, A. 1982, *Ap.J. Suppl.*, **48**,
    445.

DISCUSSION

BAHCALL: The large scale structure and motion suggested by the absorption lines of the close QSOs is similar to that we find in rich superclusters, with typical scale of $\sim$25 h^{-1} Mpc and velocity dispersion of $\sim$1000-2000 km/s. Since such systems have a reasonably small volume density, are they consistent with the QSO statistics of only occasionally finding such signatures in QSO spectra.

SARGENT: I think that the structure deduced from the observations of Q1037-2704 and Q1038-2712 is an exceptional and rare one.

SILK: Can you comment on the heavy element abundances and on the ionization state in the six multiple absorption-line systems common to the quasar pair with 17 arc-min separation?

SARGENT: No detailed studies have been carried out yet; however, based on my experience with similar absorption systems in other QSOs I would say that the heavy element abundances will come out solar or only slightly below.

RUBIN: What can you say concerning rotation curves at large r?

SARGENT: Some years ago Arp discovered a 16 magn. QSO, Q0958+68W1, about a degree SE of the center of M81 and on the edge of the outer contours of the 21 cm map done at Jodrell Bank. Observations by Boksenberg and I show clear evidence of CaII H and K absorption in the spectrum of the QSO at zero redshift and at about -130 km s^{-1}. The latter is due to M81 and shows that a flat rotation curve continues out to at least 4 Holmberg radii.

DEKEL: Why don't we see large lumps like the last case you've shown us in the general correlation function of Lα clouds.

SARGENT: It appears that the Lyman α clouds are distributed differently to galaxies. One possibility is that they occupy the "voids" in the galaxy distribution.

PECKER: I was (last year!) rather convinced by the paper by Leona Marshall-Libby et al., in which the Ly α forest was shown to be of an intrinsic origin. Now, it seems that the z-dependence of the forest has forced authors to accept the idea of intervening absorbers. The exponent γ of the laws dN/dz has then been used for some cosmological discussion. But is it not true that the two views are not contradictory but complementary, that there are <u>two</u> populations of absorption lines, and, as a consequence, <u>that</u> we should be very careful about using the coefficient γ in cosmological discussions.

SARGENT: Dr. Libby's work was shown to be unsound by Zuiderwijk (A.J. <u>89</u>, 1808, 1984). I believe that the evidence strongly suggests that <u>both</u> the Lyman α forest and the heavy element lines have an intervening origin.

ABSORPTION LINES IN QSOs

Jian-Sheng Chen
Beijing Astronomical Observatory
Academia Sinica
Beijing
People's Republic of China

ABSTRACT. A comparison between three types of intervening
absorption line systems: the Lyα forest, the Lyα damped systems
and the narrow metal line systems has been done for the
observational properties including the frequency of the systems,
the equivalent width distribution, the two point correlation
function, the metal abundances, and the dN/dz vs. z behavior.

From cosmological point of view, the narrow absorption line
systems, which include the Lyα forest, the narrow metal systems
and the Lyα damped systems, and which are thought by most
astronomers working in this field to be due to the absorptions of
the intervening materials distributed along the line of sight
between QSOs and the observers, are very important in probing the
Universe at high redshift. Do these systems represent different
populations of the objects at high redshift, or their distinctive
characters simply reflect the same population but the different
parts which intercept the light from QSOs or they represent the
different stages of the evolution? Detail studies have been
made for each of the individual classes of the systems in the
past few years. A comparison between them may give us some hints
about their relations and possible origin.

1) Frequency of the systems

The line number density per unit redshift for Lyα was estimated
as N(z=2.44) =60.7 and N(z=0)=17.6 respectively for q_0=0 with
rest frame equivalent width of line w_0>=0.32 Å from a uniform
sample of 6 QSOs (1-2 Å resolution, Sargent et al. 1980). The
number density increases rapidly with the decrease of the lower
limit of W_0. From the systematic Lick survey of the Lyα damped
systems Wolfe et al. (1986) found 47 candidates from 68 QSOs
with a total redshift coverage of 56.07 with z=2.21, and
confirmed 15 systems from 26 candidates . The derived number
densities per unit redshift are N(z=2.21)=0.48 and N(z=0) =0.15

A. Hewitt et al. (eds.), Observational Cosmology, 793–798.
© *1987 by the IAU.*

respectively. The data of the narrow metal systems show a strong tendency to clump into close pairs, triplets, or even more complex structures. For statistical purpose, each clump or complex of redshift systems is counted here as one system. Using the sample of Young et al. (1982) of 30 QSOs without absorption "troughs", the mean line densities are N(z=1.71)=1.49 and N(z=0)=0.55 for CIV systems, which is nearly the same as that from an extended sample used by Bergeron and Boisse (1984). There might be some uncertainties due to the size of the sample, the blending, and the contaminating effect, but these number densities could give us some hints for the relative frequencies of the occurrence of these three types of systems. If they are from different part of the galaxies in the same population conforming to the Schechter's galaxy luminosity function and the Holmberg radius-luminosity relation, their size will be proportional to the square root of the number density. Adopting the formulae given by Sargent et al. (1979) and the parameters therein, one finds radii R* as 175/h, 31/h, and 16/h kpc, for the Lyα forest, the narrow metal system, and the Lyα damped system respectively, when no cosmological evolution is assumed. The number density for Lyα damped system could be an upper limit due to the blend or contamination of other narrow Lyman lines, while the number density for the Ly forest and the narrow metal systems are much underestimated by the requirement of the lower limit of W_o>=0.32 Å in Lyα sample and W_o>=0.6 Å in CIV sample leaving a great number of the latter two systems uncounted. Therefore, the R* for the Lyα cloud and the narrow metal system will be much larger than the deduced values. This might explain the fact that some of the common absorption line systems from QSO pairs (e.g. 0307-195 A, B, Shaver and Robertson (1983)) can have a projected linear separation larger than 300/h kpc if they are due to the extended halo of the intervening objects. However, the fraction of the volume of the universe occupied by the Lyα forest will be too large (Sargent et al. 1980). The R* deduced for Lyα clouds also conflicts with the result form double QSO observation, which gives R* as 10 kpc (Foltz et al. 1984). This may indicate that Lyα clouds either constitute a new population in the high redshift universe, which are confined in pressure equilibrium by the hot intergalatic medium reheated at z< 6 (Sargent et al. 1980, Ostriker and Ikeuchi, 1983, Ikeuchi and Ostriker, 1986), or are self-confined systems. The rapid increase in number counts will give high detection probability. The latter explanation is confronted with a difficulty (Sargent et al. 1980) that the mass of the Lyman-alpha cloud is too small to be gravitationally self-confined. This difficulty, however, may be overcome with the model recently proposed by Rees(1986) which states that the Lyα cloud can be stably confined by the gravitational field of the cool dark matter, being able to neither to escape, nor to settle towards the center and become self-gravitating.

2) Equivalent width distribution

Bergeron and Boisse (1984) have studied the equivalent width
distribution of the $Ly\alpha$ lines for the $Ly\alpha$ forest as well as for
the narrow metal systems. After correction of the evolution
effect of the redshift, the overall W_0 distribution for both
types shows continuity and can be fit by an exponential form:
$dN/(dzdW) = 20.5 \exp(-1.85W)$. The index is between 0.87 for CIV
systems (Young et al. sample, 1982) and 2.76 for the $Ly\alpha$ forest
(Sargent et al. 1980). With the density per unit redshift per
rest equivalent width for $Ly\alpha$ damped system   from the sample of
Wolfe et al. (1986) being added (Chen 1986b), the plot of
$\log (dN/dzdW)$ vs. W_0 clearly shows that, within the quality of the
sample presently available, there is no single exponential form in
W_0 distribution from the $Ly\alpha$ forest through the metal system to
the $Ly\alpha$ damped system. It seems better to fit with a single power
law, although there is a systematic difference between them. This
result can be compared with that reported by Tytler (1984): the HI
column density distribution is a single power law from the weakest
$Ly\alpha$ line up to the strongest metal system.

3) Two point correlation function

The question whether the $Ly\alpha$ forest really does not show fine
velocity splitting with peak at 150 km/s as pointed out by
Sargent et al. (1980) is still being debated. Bergeron and
Boisse discussed the selection effect that the CIV line splitting
is seen preferentially in systems with large W_0 , and the rest
equivalent width W_0 $(Ly\alpha)/W_0(CIV)$ ratio is usually (much) larger
than unity, therefore the same property is expected only for $Ly\alpha$
lines with individual components of $W_0 > 1$-2 Å. The $Ly\alpha$ lines
will then be either more blended than the CIV doublets (higher
opacity) or may be saturated. On the other hand, Webb et al.
(1984) have reported that there is some evidence from high
resolution spectra that the $Ly\alpha$ absorbing clouds show velocity
clustering on scales of up to about 150km/s. If the fine line
splitting does reflect the clustering of the galaxies at the high
redshift, then the $Ly\alpha$ damped system, which are generally thought
to be due to the absorption of the disk components of the
galaxies, should exhibit the same phenomena. The broad $Ly\alpha$
absorption troughs extended as wide as few thousands km/s make
it impossible to resolve the splitting even if it does exit.
The only possible way to detect such a splitting in $Ly\alpha$ damped
systems is to observe in 21 cm absorption. The detected and well
resolved 16km/s splitting 21 cm $Ly\alpha$ damped absorption system
at z=2.03937 and 2.03953 in the QSO PKS 0458-02 (Wolfe et al.
1985) demonstrates its great capability. Therefore, it is highly
desirable to collect a large 21 cm absorption sample to answer
the question : whether the CIV fine splitting is due to the
clustering of the galaxies or it is the mere complex structure of
the individual object.

4) Metal abundances

One of the widely discussed problems concerns with the metal
abundances in the Lyα clouds. Since the primeval matter
consisted almost entirely of hydrogen and helium, and the
nucleosynthesis of heavy elements in the Lyα clouds would imply that
the abundances of the clouds have been enriched by stellar
material, and that the Lyα clouds are associated with galaxies.
Searches have been made for OVI in OQ172 and 4C05.34 by Norris et
al. (1983), in Q1623+269 by Sargent and Boksenberg (1983) and in
Q2000-330 by Norris and Peterson (Peterson , 1985), and the
results are conflicting. Chaffee et al. (1985) examined a
double Lyα absorption system in the QSO 0014+81 and marginally
detected two features which could be SiIII (1206 Å). However, in
a subsequent paper, Chaffee et al. (1986) discussed their
ionization model, which predicts that if SiIII is detectable,
CIII will be even stronger and must also be detectable. On the
basis of further observations they concluded that the 3 sigma
upper limit for CIII is log N(CIII) =12.9, five times weaker than
predicted by their model from the strength of previously found
SiIII lines. Thus, the detection of neither OVI nor SiIII has
been confirmed. But on the other hand, the available
observations are not sufficient to exclude the presence of the
lowest heavy element abundances found in the galaxies. The
problem remains open.

5) dN/dz vs. z for absorption line systems

Many authors have studied the number density of Lyα lines per
unit redshift interval, as a function of redshift (see Table II
of Chen(1986a), and Murdoch et al., 1986) and led to the result
that the Lyα lines do evolve with redshift as a power law of
(1+z), with a gamma index around 2. The statistical significance
is much improved by adding the BL Lac object 0215+015 (z=1.50)
and the QSO 2000-330 (z=3.78) extending thereby the interval of
the redshift of the absorption line sample in both directions.
In the pressure-confined model, the evolution of the number
density of Lyα clouds is explained by the spectrum of the mass of
the clouds and the evolution of the column density of the neutral
hydrogen (Atwood et al., 1985, Ikeuchi and Ostriker, 1986) and
therefore gives a constraint for the galaxy formation theory.
For metal line systems, Bergeron and Boisse (1984) indicated that
the number of CIV systems reveals a possible cosmological
evolution with gamma around 1.8 in the redshift range 1.4 - 2.0.
If this result could be confirmed and extended to redshift over
3, it would mean that the CIV systems, which is widely accepted
to be associated with haloes of galaxies, have the same evolution
law as the Lyα forest, while the QSOs themselves have a
completely different evolution behavior with number density
decreasing beyond z=3 and probably cut off around z=4. The Lyα
damped systems of Wolfe et al. (1986) sample is too small to

estimate the gamma. Bian et al. (1986) have argued that if the
dN/dz of Lyα lines for each individual QSO shows no consistent
tendency of evolution while the dN/dz for the overall QSO lyα
lines sample strongly evolves with redshift, it would mean that
dN/dz varies with Zem instead of Zabs. The statistics for the
sample presently available support this point with an even better
significance. If this can be confirmed in the future, It would
cast doubt on its intervening nature.

REFERENCES

Atwood, B., Baldwin, J.A., and Carswell, R.F., 1985, Ap.J.,
        292, 58.
Bergeron, J., and Boisse, P. 1984, Astron. and Astrophys.,
        133, 374.
Bian, Y-L., Chen, J-S., Zou, Z-L. 1986, Acta Astrophs. Sinica,
        6, 101.
Chaffee, F.H., Foltz, C.B., Roser, H.-J., Weymann, R.J., and
        Latham, D.W., 1985, Ap.J., 292, 362.
Chaffee, F.H., Foltz, C.B., Bechtold, J., and Weymann, R.J.,
        1986, Ap.J., 301, 116.
Chen, J-S. 1986a, Ap.S.S., 118, 473.
Chen, J-S. 1986b, in preparation.
Foltz, C.B., Weymann, R.J., Roser, H.-J., and Chaffee, F.H.
        1984, Ap.J. (Letter), 281, L1.
Ikeuchi, S., and Ostriker, J.P. 1986, Ap.J., 301, 522.
Murdoch, H.S., Hunstead, R.W., Pettini, M., and Blades, J.C.
        1986, preprint.
Norris, J., Hartwick, F.D.A., and Peterson, B.A. 1983, Ap.J.,
        273, 450.
Ostriker, J.P., and Ikeuchi, S. 1983, Ap.J. (Letter), 268,
        L63.
Peterson, B.A., 1985, preprint presented at IAU Symp. 119.
Rees, M., 1986, Mon. Not. R. Astr. Soc., 218, 25p.
Sargent, W.L.W., and Boksenberg, A., 1983, in the Proceedings of
        the 24th Liege International Astrophysical Colloquium,
        Quasars and Gravitational Lenses, p.518.
Sargent, W.L.W., Young, P., Boksenberg, A., Carswell, R.F., and
        Whelan, J.A.J., 1979, Ap.J., 230, 49.
Sargent, W.L.W., Young, P.J., Boksenberg, A., and Tytler, D.,
        1980, Ap.J. Suppl., 42, 41.
Shaver, P.A., and Robertson, J.G. 1983, Ap.J. (Letter), 268,
        L57.
Tytler, D. 1984, BAAS, 16, 1008.
Webb, J.K., Carswell, R.F., and Irwin, M.J. 1984, BAAS, 16,
        733.
Wolfe, A.M., Turnshek, D.A., Davis, M.M., Smith, H.E., and Cohen,
        H.D. 1985, Ap.J. (letter), 294, L67.
Wolfe, A.M., Turnshek, D.A., Smith, H.E., and Cohen, R.D. 1986,
        preprint.

Young, P., Sargent,W.L.W., and Boksenberg, A. 1982, _Ap.J.Suppl._,
     48, 455.

ABSORPTION LINE SYSTEMS AT z > 3 IN THE QSO 2000-330

R.W. Hunstead[1], M. Pettini[2], J.C. Blades[3] and H.S. Murdoch[1]

[1] School of Physics, University of Sydney, Australia
[2] Royal Greenwich Observatory, Hailsham, E. Sussex, U.K.
[3] Space Telescope Science Institute, Baltimore, U.S.A.

1. INTRODUCTION

This paper presents some preliminary results from analysis of our high-resolution (30-35 km/s FWHM) spectra of the Lyman α forest region in the z = 3.78 QSO 2000-330. These spectra were obtained at the Anglo-Australian Telescope over several observing seasons and have been analysed by fitting multiple-cloud Voigt profiles to Lyman series and heavy-element absorption lines. Two specific issues are addressed here:
  (i) The distribution of column densities, N(H I), and velocity dispersions, b, for hydrogen clouds in the interval z_{abs} = 3.43-3.78;
 (ii) Heavy-element abundances in a system at z_{abs} = 3.1723.

2. PROPERTIES OF THE LYMAN α CLOUDS

It is generally assumed that the Lyman line absorbing clouds seen in QSO spectra constitute an intergalactic population (Sargent *et al.* 1980) which evolves strongly with redshift (Murdoch *et al.* 1986). While statistical studies on samples complete to a given Ly α equivalent width can shed light on the global properties of the so-called Ly α clouds, we need to look in detail at the *intrinsic* cloud properties (N, b), as well as possible spatial correlations, to provide clues to their origin, confinement and ultimate fate.

   The most detailed studies of Lyman α clouds to date at high resolution are for the z = 2.14 QSO 1101-264 (Carswell *et al.* 1984, hereafter CMSSTW; resolution 20 km/s FWHM) and the z = 3.12 QSO 0420-388 (Atwood, Baldwin and Carswell 1985, hereafter ABC; 33 km/s FWHM). In each study the distribution of column densities was fitted to a power law of the form dn $\propto$ N$^{-\beta}$ dN, with β = 1.68±0.10 for 1101-264 (log N = 12.8-15.0) and β = 1.89±0.14 for 0420-388 (log N = 14.0-16.7). In the former, Voigt profiles were fitted to Ly α alone in the ranges z_{abs} = 1.89-1.98 and 2.04-2.13 while in the latter fitting was restricted to the interval z_{abs} = 2.72-3.12 so that at least two Lyman lines were available.

   Due to its higher redshift, 2000-330 gives us access to many Lyman lines ; indeed, for z > 3.65 the entire Lyman series is available above

A. Hewitt et al. (eds.), Observational Cosmology, 799-802.
© *1987 by the IAU.*

our short wavelength limit of 4240 Å and at z = 3.43 we still have coverage to Ly γ. The setting of continuum levels and the generation of superimposed plots of Lyman series lines followed the procedures given in Hunstead *et al.* (1986a). The plots were invaluable in revealing the presence of structure and blends in the higher-order lines.

Our spectra in the region below Ly β emission have uniformly good signal/noise ratio and this has generally imposed tight constraints on the values of z, b and N for the high-column (log N > 15) clouds. For clouds on the linear part of the curve of growth (log N < 13.6), the parameters are determined chiefly by the fit to Ly α; the signal/noise is therefore crucial in deciding, for example, whether to fit two or more narrow components or a single broad component. At intermediate column densities the reliability of b, N estimates depends on having the low-order Lyman lines free of serious blending. In a later paper we will discuss in detail our treatment of the many problems encountered in attempting to deconvolve the complex profiles in the Lyman forest.

The present line sample is preliminary; we have not attempted here to differentiate between lines with well-determined parameters and those where the parameters are less well defined. We show the column density distribution in Fig. 1(a). The best-fit exponent, β = 1.57±0.05, was determined by maximum likelihood (ML) fit to the 149 lines with log N > 13.25. The binned data in Fig. 1(a) seem to indicate possible structure in the distribution function, although a Kolmogorov test shows a good fit to the assumed power law. The influence of selection effects and incompleteness on this distribution are currently being investigated.

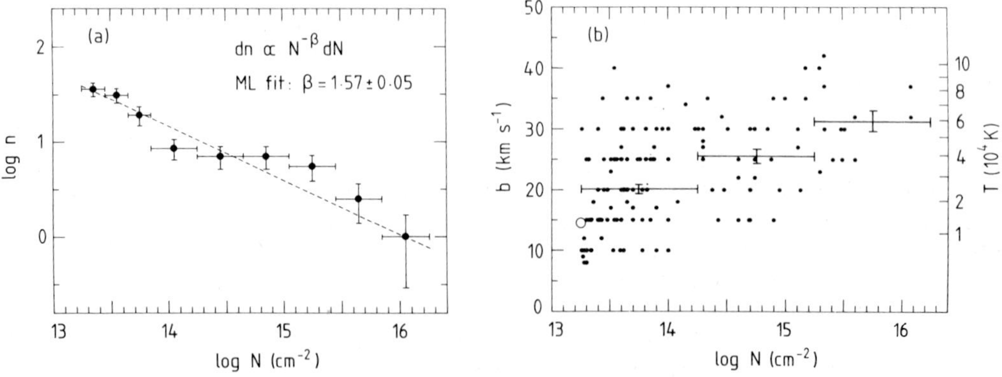

Figure 1. Lyman α clouds in 2000–330 with log N > 13.25 and ⟨z⟩ = 3.6: (a) H I column density distribution; dashed line is the ML fit to the 149 individual clouds; (b) b vs log N for individual clouds together with coarsely-grouped means and standard errors.

A plot of b versus log N is shown in Fig. 1(b). The scatter is large but the lower envelope defines a trend which is supported by that of the coarsely-grouped means. A possible correlation in the same sense was reported by CMSSTW, but ABC comment that this could be a selection effect caused by blending. Our data for 2000–330 suggest that blending is probably important at low N (refer Fig. 1b) but that the dearth of low-b, high-N clouds is real; if so, the detection of deuterium in the Lyman forest clouds may well be ruled out (Chaffee *et al.* 1983). The temperature scale shown in Fig. 1(b) assumes that the b values arise solely from thermal motions in a gas at temperature T. Chaffee *et al.* (whose measurement for a Ly α cloud towards PHL 957 is shown as an open circle in Fig. 1b), point out that temperatures as low as $1-2\times10^4$ K pose serious problems for current models for confinement of the Ly α clouds.

3. HEAVY-ELEMENT ABUNDANCES AT $z_{abs} = 3.1723$

An earlier search for heavy-element absorption systems (Hunstead *et al.* 1986b) turned up four plausible systems, all of which are confirmed by the higher-resolution data. A new system at $z_{abs} = 3.1723$, previously listed as Lyman-only, has been detected on the basis of several neutral and low-ionisation species which were too weak (or too heavily blended) to be recognised in the earlier search. This means that metals have now been found spanning the entire width ($\sim$ 1400 km/s) of the deep trough near 5080 Å (Hunstead *et al.* 1986b, Fig. 5).

Whereas the previously reported systems at $z_{abs} = 3.1881$ and 3.1914 show complex velocity structure, the system at $z_{abs} = 3.1723$ appears single, making the association of heavy-element lines with H I more straightforward. Another important factor is a good determination of N(H I): a radiation-damped Ly α profile at the redshift determined by Ly β gives an excellent fit to the blue wing of the 5080 Å trough. Changes of ±20% in N(H I) produce markedly poorer fits to the data. A final point concerns the ionisation state of the gas: the weakness or absence of high-ionisation species (e.g. Si III weaker than Si II; Si IV not detected) implies that the gas is predominantly neutral, which may indicate an origin in the disk of an intervening galaxy.

The column densities and relative abundances in this system are given in Table I. These measurements imply a more-or-less uniform underabundance in C, N, O and Si of $\sim$ 2.2 dex relative to solar values, suggesting only modest metal enrichment. If systems with comparable underabundances are found to be common at $z \sim 3$, this would provide important constraints for current models of galactic chemical evolution.

TABLE I.   Element abundances at $z_{abs} = 3.1723$

Ion	Ab$_\odot$	log N(X)	log depletion
H^o	12.00	19.78	...
C$^+$	8.57	≤13.95*	≤-2.40
N^o	8.06	≤13.65*	≤-2.18
O^o	8.83	14.40	-2.21
Al$^+$	6.40	<12.1**	<-2.1
Si$^+$	7.55	13.15	-2.18
Si^{++}	7.55	13.08	...

* Upper limit due to line blending
** Upper limit based on nearby spectral features

REFERENCES

Atwood,B., Baldwin,J.A. and Carswell,R.F., 1985. *Ap. J.*, **292**, 58 (ABC).
Carswell,R.F., Morton,D.C., Smith,M.G., Stockton,A.N., Turnshek,D.A. and
        Weymann,R.J., 1984. *Ap. J.*, **278**, 486 (CMSSTW).
Chaffee,F.H., Weymann,R.J., Latham,D.W. and Strittmatter,P.A., 1983.
        Ap. J., **267**, 12.
Hunstead,R.W., Murdoch,H.S., Blades,J.C. and Pettini,M., 1986a. In *IAU
        Symposium 119*, *'Quasars'*, ed. G. Swarup & V.K. Kapahi, p. 569.
Hunstead,R.W., Murdoch,H.S., Peterson,B.A., Blades,J.C., Jauncey,D.L.,
        Wright,A.E., Pettini,M. and Savage,A., 1986b. *Ap. J.*, **305**, 496.
Murdoch,H.S., Hunstead,R.W., Pettini, M. and Blades,J.C., 1986.
        Ap. J., **309**, 19.
Sargent,W.L.W., Young,P.J., Boksenberg,A. and Tytler,D., 1980.
        Ap. J. Suppl., **42**, 41.

DISCUSSION

SILK: How uncertain are your CNO abundance determinations for the Lyα absorption system at $z = 3.1723$. How sensitive are they to the assumed ionization state of the gas?

HUNSTEAD: Firstly, H, N and O have similar ionization potentials and little N^0 and O^0 can exist in HII regions due to efficient recombination through charge-exchange reactions with H^+. In addition, there seems to be very little HII gas anyway, judging from the relative weakness of Si III with respect to Si II.

CLUSTERING OF Lyα ABSORBING CLOUDS AT HIGH REDSHIFT

J.K. Webb
Sterrewacht
P.O. Box 9513
2300 RA LEIDEN
The Netherlands

ABSTRACT. High resolution observations of regions in the Lyα forest of
QSO spectra have been obtained using the Anglo-Australian Telescope. The
amount and quality of the data, and the analysis techniques used, have
enabled the clustering properties of these objects to be explored on
scales smaller than has previously been possible. The results suggest
that the Lyα clouds are weakly clustered with $\langle \xi \rangle = 0.32 \pm 0.08$ over the
velocity range 50-290 km/s.

1. Introduction

At present, the majority of existing spectra of the Lyα forest
absorption lines are at a resolution (~ 1Å FWHM) such that individual
absorption features are generally unresolved. Although statistical
procedures have been defined and applied to the data to obtain
homogeneous Lyα samples, (Young et al., 1979), at higher spectral
resolution most or all of the stronger features are found to break up
into more components. The absorption features can be more reliably
located and profile fitting techniques (Carswell et al., 1984) may be
used to obtain estimates of the column density of absorbing atoms, the
velocity dispersion parameter and the redshift. With such a sample, the
clustering properties of the Lyα clouds may be investigated down to
scales ~ 50 km/s compared to ~ 300 km/s for the majority of the sample
used by Sargent et al. (1980) who found their data to be consistent with
clouds randomly distributed on scales between 300 and 30,000 km/s.

2. The data

We have collected, at the Anglo-Australian Telescope, high resolution
spectra (~ ¼Å FWHM) covering regions in the Lyα forest of several QSOs.
Combining these new data with existing data on Q1101-264 (Carswell et
al., 1984) provides us with 13 Lyα regions of ~130Å each over the
absorption redshift range 1.9<z<2.8. A maximum likelihood optimization
technique was employed in profile fitting to the data to obtain

803

A. Hewitt et al. (eds.), Observational Cosmology, 803–806.
© *1987 by the IAU.*

estimates for the cloud parameters and their associated errors. It is
important to ensure that blended absorption features are not overfitted
with too many components since this could cause an apparent small scale
clustering. We have attempted to fit complex features with the minimum
number of components required to give an acceptable fit. Additionally,
we have tested the goodness of fit by comparing normalized chi=squareds
for blended and single features for the whole sample. No evidence was
found for either over or underfitting.

3. Clustering Analysis

To investigate the clustering properties of the Lyα clouds, all lines
associated with known metal-containing systems were removed. Lines
within 3000 km/s of the Lyα emission line were omitted to reduce effects
caused by the drop in number density of absorption lines approaching the
emission line (Weymann et al., 1981, Murdoch et al., 1986). Only lines
redwards of the Lyβ emission line were used to avoid confusion with
higher order Lyman transitions. The resulting sample comprised ~300
absorption lines.
     For the remaining features, the two-point correlation function was
calculated for each wavelength region (Sargent et al., 1980). The summed
normalized correlation function is shown below.

Two−Point Correlation Function

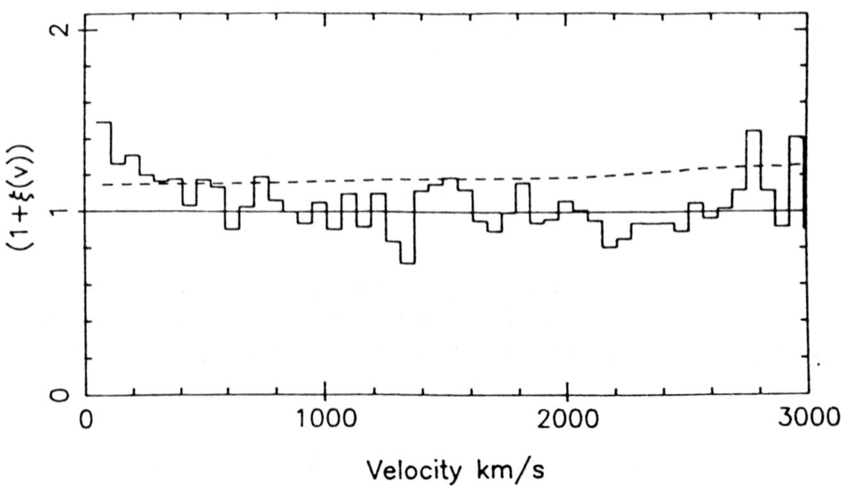

Figure 1 Two-point correlation function normalized to (1+ξ(v)). The bin
size is 60 km/s and the fist bin starts at 50 km/s, below which blending
problems are severe. The dashed line is the 1σ upper error contour
derived from Poisson statistics.

In previous analyses the correlation function was incomplete on scales below ~300 km/s. Consequently we use the range 50-290 km/s to search for small scale velocity clustering and find that $\langle \xi \rangle$ = 0.32±0.08.

A possible selection bias could occur if some of the features which have been assumed to be Lyα absorption lines are actually unidentified metal lines; these are believed to be strongly clustered (Sargent et al., 1980). However, an estimate of the number of contaminating lines that might be present suggests that there are not enough to account for the observed clustering.

Interestingly, there appears to be a weak trend for $\langle \xi \rangle$ to increase with decreasing redshift; dividing the samples into two redshift bins, 1.9<z<2.4 and 2.4<z<2.8, we find $\langle \xi \rangle$ = 0.44±0.12 and $\langle \xi \rangle$ = 0.18±0.11 respectively. Clearly, more data are required to check on this.

In conclusion, we find significant evidence for weak clustering on small velocity scales for the Lyα clouds.

References

Carswell, R.F., Morton, D.C., Smith, M.G., Stockton, A.N., Turnshek, D.A., and Weymann, R.J. 1984, Ap. J., 278, 486.
Murdoch, H.S., Hunstead, R.W., Pettini, M., and Blades, J.C. 1986, preprint submitted to Ap. J.
Sargent, W.L.W., Young, P.J., Boksenberg, A., and Tytler, D. 1980, Ap. J. Suppl., 42, 41.
Weymann, R.J., Carswell, R.F., and Smith, M.G. 1981, Ann. Rev. Astron. Astrophys., 19, 41.
Young, P.J., Sargent, W.L.W., Boksenberg, A., Carswell, R.F., and Whelan, J.A. 1979, Ap. J., 229, 891.

DISCUSSION

BAHCALL: If the clouds were moving with some Δv (~100 or 200 km s^{-1}), that would tend to weaken any intrinsic correlation function among these systems, then in principle, the correlations could be even stronger.

WEBB: I agree. Random motions could smear out the correlation function. Observations of suitable QSO pairs could eventually help us to understand how spatial and velocity correlations are related.

SILK: Could you compare the correlation length you have found for the
Lyα clouds with that found for the CIV absorption systems at a similar
redshift.

WEBB: The line of sight two-point correlation function has been
calculated for a small sample of CIV lines by Sargent et al. (1980).
They find a strong excess at ~150 km s^{-1} with an effect extending
out to ~400 km s-1. The clustering amplitude seems stronger than
the one I find for the Lyα clouds.

CHAPTER XV

CLUSTERING OF QUASI-STELLAR OBJECTS

CLUSTERING OF QUASARS FROM THE ROE/ESO LARGE-SCALE AQD SURVEY FOR QUASARS

Roger G. Clowes[1], Angela Iovino[2] and Peter Shaver[2]

[1]Royal Observatory, Edinburgh, Scotland
[2]European Southern Observatory, Garching bei München, FRG

Abstract

The new ROE/ESO large-scale AQD survey for quasars forms a connected area of ~ 200 deg^2 near the south galactic pole, and has resulted in the discovery of a total number of quasar candidates that is comparable to the number previously published from all other sources (see the poster paper by Iovino, Clowes & Shaver at this conference). In this paper we describe the first results of a three-dimensional self-clustering analysis of ~ 1100 "high-probability" candidates occupying the assigned-redshift band of 1.8 to 2.4. Although the analysis is sensitive to very weak clustering we find no evidence that quasars are distributed in any way other than randomly. The implications of this result are discussed.

Introduction

One of the main difficulties of using quasars as cosmological indicators has, for many years, been the lack of large, complete samples. However, in the ROE/ESO survey for quasars, we have now successfully completed a very important stage in the assembly of such samples.

Briefly, seven low-dispersion, IIIa-J objective-prism plates from the UK Schmidt Telescope (UKST) have been measured with the COSMOS measuring machine (MacGillivray & Stobie 1984) at ROE, and the data processed with the software for Automated Quasar Detection (AQD - Clowes, Cooke & Beard 1984, Clowes 1986) to find the quasars. The seven plates form a connected area of ~ 200 deg^2 near the south galactic pole, and the total number of newly-discovered quasar candidates is comparable to the number published from all other sources since the first quasars were discovered. Full details of the progress and aims of the ROE/ESO survey are given in the poster paper by Iovino, Clowes & Shaver at this conference.

With such a large number of new candidates the ROE/ESO survey ideally needs a long programme of follow-up spectroscopy both for confirmation and for obtaining accurate redshifts. In this paper on the clustering properties, however, we circumvent the present absence of such spectroscopy by restricting our

A. Hewitt et al. (eds.), Observational Cosmology, 809–814.
© *1987 by the IAU.*

samples to the "high-probability" candidates from the five best plates.

The ability to grade the candidates in order of probability of being quasars is a very useful feature (it has something in common with the expert-system approach used in machine intelligence). It means that we can produce high-probability subsets with surface densities of ~ 10 $\deg^{-2}$ - more than would be found in a typical visual search -, with minimal need for access to large telescopes. There are, of course, many further quasars in the lower-probability grades.

<u>The clustering analysis - 3-D power spectrum analysis</u>

We use the method of power spectrum analysis (PSA) to test for clustering. The method was developed by Webster (1976) with the particular aim of distinguishing a population that is weakly clustered from one that is randomly distributed. A brief summary of the theory of PSA is given by Clowes (1986).

To obtain the third coordinate for 3-D PSA the cosmological hypothesis of redshifts is assumed and a particular cosmological model is adopted - $q_0 = 0$, $H_0 = 100h$ $kms^{-1}Mpc^{-1}$.

First we produced the high-probability subsets in the following five fields.

ESO/SERC field	UKST plate	Plate centre (1950)		N1	N2	N3
295	UJ6536P	00 52 00	-40 00 00	399	328	284
296	UJ5406P	01 18 00	-40 00 00	219	213	171
297	UJ4514P	01 44 00	-40 00 00	407	380	308
351	UJ6528P	00 48 00	-35 00 00	275	261	208
411	UJ7307P	00 46 00	-30 00 00	216	205	168
			totals:	1516	1387	1139

N1 = number in the high-probability subset
N2 = number after rejection of obvious contaminants
N3 = number with assigned redshift between 1.8 and 2.4

Our objects cover ~ 28.3 $\deg^2$ of each field. The limiting magnitudes are unknown because of the present lack of external data for calibration, but they should be B ~ 20.5 for the best plates; consequently, no significance should be attached to the variations in surface density from field to field.

Obvious contaminants were rejected; these are usually stars or surviving overlaps which, for special reasons, have slipped through the grading process.

A high-probability candidate must have an emission line - so a redshift can always be assigned. Redshifts were assigned on the assumption that the typically single lines are Ly-α, and the objects then restricted to the range 1.8 to 2.4. The reason for the restriction is that 3-D PSA requires a comoving cuboid of points (see Webster 1982 and Clowes 1986), which can only be defined for a small range of redshifts if there is to be no serious distortion of scales. The range 1.8 to 2.4 contains a large fraction of the objects.

The way to define the comoving cuboid is given in Clowes (1986); its present-epoch volume is ~ 390 x 390 x 1000h^{-3} Mpc^3 for each of the fields, the largest dimension being in the redshift direction.

At this point it is useful to have an estimate of what fraction of the objects are definitely quasars. Visual prejudice applied to a film copy of each plate and 1-D plots of the objective-prism spectra suggests that the mean fraction corresponding to convincing quasars is ~ 80%. The true fraction could be higher still.

Thus, we expect that our final samples for 3-D PSA will contain a very low fraction of non-quasar contaminants, but we should also estimate in what fraction the assumption of Ly-α is correct. Assumed Ly-α could really be CIV λ1549, CIII] λ1909, and MgII λ2798. Hazard's (1985) histogram of redshifts is not ideal but gives the most relevant data: it suggests that the assumption will be correct in 30/43 or ~ 70% of cases. The true percentage could, perhaps, be as high as 80% simply because our candidates are of the highest grade.

The objective-prism redshifts are systematically too small by ~ 0.04 at z ~ 2.1; however, this error is small and has been ignored. The random errors on the redshifts are such that the separation in the redshift direction of two objects carries an error of ~ 30 Mpc at z ~ 2.1, which is much smaller than the mean separation, of objects.

The effect of non-quasar contaminants, misidentified lines, and redshift errors has been established from simulations.

Results of the 3-D PSA

Fig. 1 shows the results of combining the 3-D PSA statistics (the purely 1-D and 2-D terms are excluded) from the five fields. The figure plots Q$'$ against 1/λ for a bin size of 2.0 Gpc^{-1}, with error bars of $\pm\sigma$. Recall from Webster (1976) and Clowes (1986) that if the plot of Q$'$ against 1/λ suggests clustering on a scale of λ_c then the PSA statistics are combined for $\lambda > \lambda_c$ ($k < k_c$), giving Q, to establish the significance of the clustering and to estimate the mean number of objects per cluster. For a random distribution the expectation values of Q$'$ and Q are 1.

Simulations of our data show that, if we combine the PSA statistics from the five fields, we can detect clustering if only ~ 6% of all quasars are paired on small scales. This value incorporates allowances for non-quasar contaminants, misidentified lines, and redshift errors.

There is no indication of clustering in Fig. 1.

Comments

The result of this clustering analysis is that there is convincing evidence for no

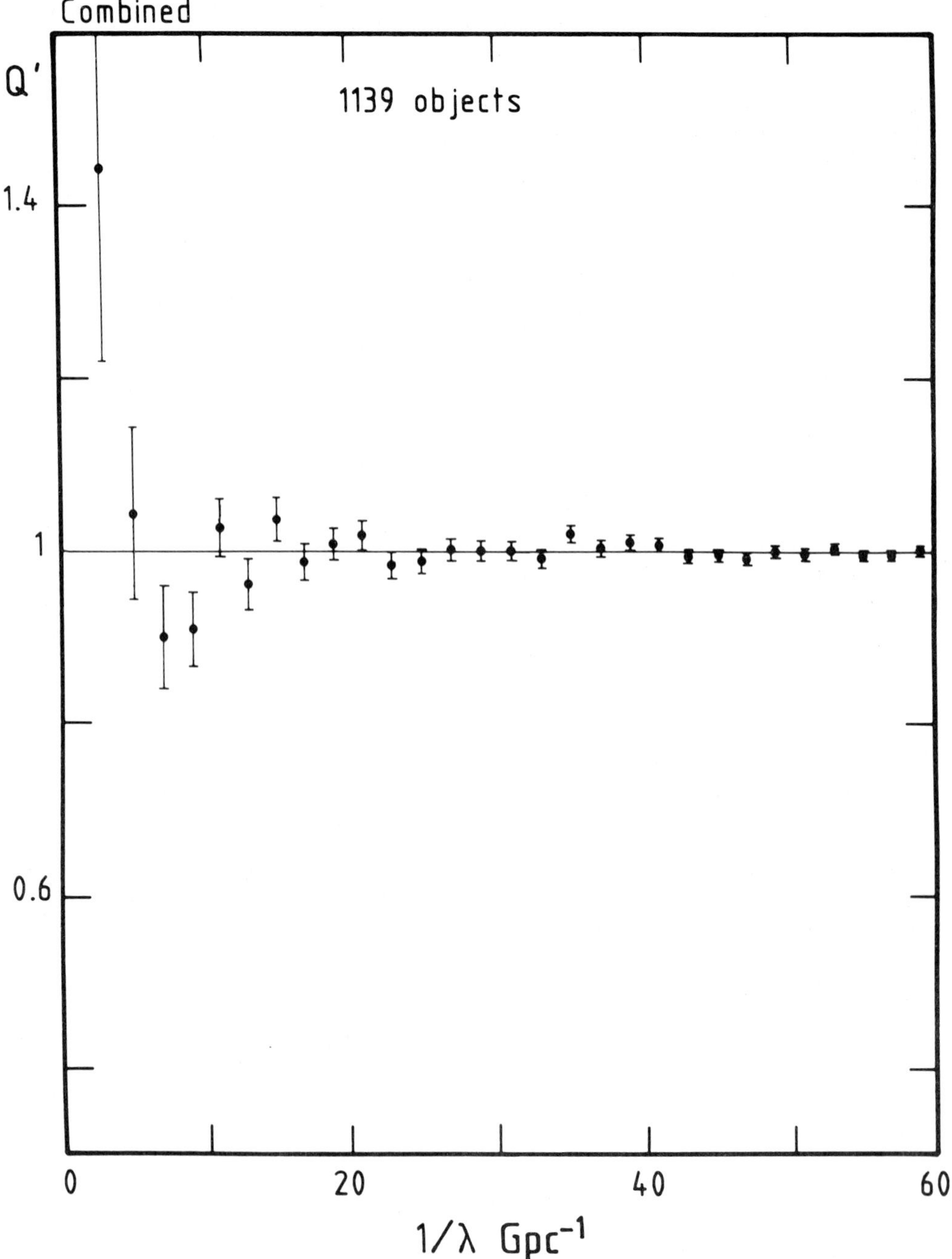

Figure 1.   The 3-D PSA plot of Q′ against $1/\lambda$, with 1-D and 2-D terms excluded, for the combined statistics. The bin size is 2.0 Gpc^{-1} and the error bars are $\pm\sigma$.

clustering. If quasars are, in fact, physically clustered as pairs on the scale of superclusters (Oort 1983) then we can deduce that the fraction of our quasars participating in pairs is less than $\sim$ 6%.

There have been other clustering analyses which cover the same small ($\sim 10h^{-1}$ Mpc) to intermediate ($\sim 100h^{-1}$ Mpc) scales that are considered here, and they have divided into results of no clustering (eg Osmer 1981, Webster 1982, Clowes 1986) and results of possible clustering on small scales but not on intermediate scales (Shaver 1984, Boyle 1986).

Shaver's (1984) approach is to use the published positions and redshifts of quasars that are collected in a large, inevitably heterogeneous, compilation (Véron-Cetty & Véron 1984). The heterogeneity makes comparison very difficult: the result could be real, and possibly attributable to low-redshift quasars only (Kruszewski 1986); it could, despite the allowances made, be caused by the selection effects of observation and publication that are inherent in a compilation.

Comparison with Boyle's (1986) result is rather easier. The quasars were discovered in an automated ultraviolet-excess survey, and the 3-D clustering analysis, using the autocorrelation method, of 169 quasars had the advantage of spectroscopic redshifts. The clustering result implies that $\sim$ 13% of quasars are paired on scales $<\sim 10h^{-1}$ Mpc (q_0 = 0.5), but it is also so noisy that a random distribution is not definitely excluded. A fraction of $\sim$ 13% would transform to $\sim$ 2-3% for our brighter quasars, and be below the detection threshold. Also, the mean redshift of Boyle's quasars is $\sim$ 1.5 compared with $\sim$ 2.1 for ours, which could be relevant.

Fewer than $\sim$ 6% of our quasars can be paired. To detect clustering still weaker than $\sim$ 6% the sensitivity of the analysis must be increased still further. In principle, this can be done quite easily by substituting spectroscopic redshifts for objective-prism redshifts. A quasar redshift survey would, in just one of our fields, improve the sensitivity to better than 3% for pairs of maximum separation $10h^{-1}$ Mpc. There would also be the opportunity to investigate variations of clustering with epoch.

<u>Acknowledgments</u>

The staff of the Image and Data Processing Unit at ROE and of the UK Schmidt Telescope Unit at ROE and Coonabarabran are thanked for their help.

<u>References</u>

Boyle, B.J., 1986. PhD thesis, University of Durham.
Clowes, R.G., Cooke, J.A. & Beard, S.M., 1984. Mon. Not. R. astr. Soc., 207, 99.
Clowes, R.G., 1986. Mon. Not. R. astr. Soc., 218, 139.
Hazard, C., 1985. In "Active Galactic Nuclei", ed. Dyson, J.E., Manchester University Press.
Kruszewski, A., 1986. Preprint.
MacGillivray, H.T. & Stobie, R.S., 1984. Vistas Astr., 27, 433.

Oort, J.H., 1983.   In "Quasars and Gravitational Lenses", 24th Liège Astrophysical Colloquium, Université de Liège.
Osmer, P.S., 1981.   Astrophys. J., 247, 762.
Shaver, P.A., 1984.   Astron. Astrophys., 136, L9.
Véron-Cetty, M.-P. & Véron, P., 1984.   ESO scientific report no. 1.
Webster, A., 1976.   Mon. Not. R. astr. Soc., 175, 61.
Webster, A., 1982.   Mon. Not. R. astr. Soc., 199, 683.

DISCUSSION

BOYLE: You said you would only expect to find 6% of your QSOs in pairs at separations less than a few Mpc. Could you quantify the word 'few'?

CLOWES: About 5.

QUASAR CLUSTERING

P.A. Shaver
European Southern Observatory
Karl-Schwarzschild-Str. 2
D-8046 Garching bei München
F.R.G.

1. THE SKY DISTRIBUTION OF QUASARS

If quasars are distributed cosmologically according to their red-shifts, then it is expected that their distribution on the sky will be very close to random, because it is in that case an average over an enormous range in distance. Most studies find no significant cluster-ing. It has recently been suggested, however, that there may be a highly significant excess of close pairs of quasars with different redshifts[1], although this has been disputed[2]. The new ROE/ESO quasar survey[3] pro-vides a very large and deep sample over a connected area of sky, ideal for a more sensitive search for clustering. The result is shown in fig. 1. The observed distributions appear to be random, on all scales from less than an arcmin to tens of degrees.

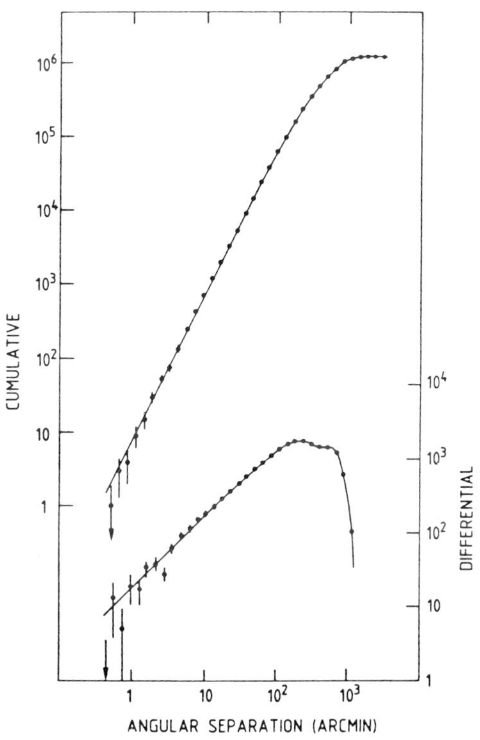

Figure 1. - Cumulative and differen-tial numbers of quasar pairs from the ROE/ESO survey, as a function of angular separation. The solid lines represent Monte Carlo simulations for a purely random distribution.

A. Hewitt et al. (eds.), Observational Cosmology, 815–819.
© 1987 by the IAU.

2. QUASAR-GALAXY CLUSTERING

Quasars at low redshifts are located preferentially in compact groups or clusters of galaxies[4-9], with relatively small ($\sim$ 60 h^{-1} kpc) and dense cores. Typical quasar-galaxy redshift differences are $\sim$ 300-400 km s^{-1}. The quasar-galaxy correlation function is stronger than the galaxy-galaxy correlation function, and strongest in the case of radio-loud quasars[6]. A large fraction of quasars have very close distorted or compact companions[5,8,9]. These facts strongly suggest that interactions are often involved in activating and fuelling quasars.

The association of quasars with galaxies appears to be even stronger at higher redshifts[10,11]. At the highest redshifts the evidence is largely indirect: narrow absorption lines near the emission redshift[12-14], highly distorted radio structures[15], and a possible quasar-galaxy pair at z = 3.2[16]. Quasars may therefore be located in groups or clusters of galaxies at all redshifts, and it has been suggested that the clustering environment may be richest at the highest redshifts[17].

3. QUASAR-QUASAR CLUSTERING

Quasar clustering has been difficult to detect and measure reliably, because of the very low space density of these objects, and complications due to selection effects and "contamination" from gravitational lensing which can in principle produce image splittings of arcminutes.[18,19] The problem of distinguishing between gravitational lenses and physical pairs has been discussed by several authors[20-23], and cannot easily be solved in individual cases. However, the contamination due to lenses is not likely to be statistically important in large-scale clustering analyses.

Two approaches have been taken in studies of quasar clustering. Homogeneous samples have been examined for evidence of physical clustering, but these have been generally too small, too shallow, or both. Larger and deeper samples are now being produced, but so far the results are contradictory: Clowes et al.[24] find no clustering at high redshifts in a large AQD sample, and Crampton et al. [25] find no clustering amongst quasars from a grens survey, but Boyle et al.[26] find tentative evidence for strong clustering in their UVX sample.

Alternatively, large quasar catalogues have been used to search for clustering - these provide large numbers of quasars, but they are heterogeneous, and a special technique has had to be used to minimize selection effects[27]. This approach gave early evidence of possible quasar clustering[27], and the same technique now gives preliminary indications that quasar clustering on large scales may increase strongly with decreasing redshift[28,29]. Efforts are underway to refine the clustering analysis of large catalogues, and to enlarge the homogeneous samples, and it is hoped that a more consistent picture of quasar clustering will soon emerge.

A knowledge of quasar clustering and its evolution may provide useful constraints on cosmological models. At high redshifts, it would provide information on whatever large-scale structure existed at early epochs, and possibly on the mass distribution of the intervening medium through considerations of gravitational lensing. And a study of the evolution of quasar clustering towards lower redshifts may yield information on the way in which structure evolved on different scales.

REFERENCES

1. Burbidge, G.R., Narlikar, J.V., Hewitt, A. 1985, Nature 317, 413.
2. Shaver, P.A. 1986, Nature 323, 185.
3. Iovino, A., Clowes, R., Shaver, P.A. 1987, this volume.
4. Stockton, A. 1978, Ap.J. 223, 747.
5. Stockton, A. 1982, Ap.J. 257, 33.
6. Yee, H.K.C., Green, R.F. 1984, Ap.J. 280, 79.
7. Heckman, T.M., et al. 1984, A.J. 89, 958.
8. Gehren, T., et al. 1984, Ap.J. 278, 11.
9. Hutchings, J.B., Crampton, D., Campbell, B. 1984, Ap.J. 280, 41.
10. Yee, H.K.C., Green, R.F. 1986, in Quasars (IAU Symp. 119, ed. G. Swarup & V.K. Kapahi; D. Reidel Publ. Co.), p. 481.
11. Tyson, A. 1986, in Quasars (IAU Symp. 119, ed. G. Swarup & V.K. Kapahi; D. Reidel Publ. Co.), p.486.
12. Shaver, P.A., Robertson, J.G. 1984, in Frontiers of Astronomy & Astrophysics (ed. R. Pallavicini), p. 201.
13. Phillipps, S. 1986, M.N.R.A.S. (in press).
14. Foltz, C.B., et al. 1986, Steward Observatory Preprint.
15. Barthel, P.D. 1986, in Quasars (IAU Symp. 119, ed. G. Swarup & V.K. Kapahi; D. Reidel Publ. Co.), p. 181.
16. Djorgovski, S., et al. 1985, Ap.J. 299, L1.
17. de Robertis, M. 1985, A.J. 90, 998.
18. Vilenkin, A. 1984, Ap.J. 282, L51.
19. Paczynski, B. 1986, Nature 319, 567.
20. Shaver, P.A., Cristiani, S. 1986, Nature 321, 585.
21. Phinney, E.S., Blandford, R.D. 1986, Nature 321, 569.
22. Turner, E.L., 1987, this volume.
23. Bahcall, J.M., Bahcall, N.A., Schneider, D.P. 1986, Nature (in press).
24. Clowes, R.G., Iovino, A., Shaver, P.A. 1987, this volume.
25. Crampton, D. 1987, this volume.
26. Boyle, B.J., 1987, this volume.
27. Shaver, P.A. 1984, Astron. Astrophys. 136, L9.
28. Kruszewski, A. 1986, Princeton Observatory Preprint.
29. present work.

DISCUSSION

ARP: It appears your maximum clustering affect comes for quasars of
redshift $z \sim 1$. I would like to remind you that at the Liege Conference
in 1983 I showed that the five densest groups of quasars known all had
redshifts near $z \sim 1$. <u>But</u>, and this point needs to be faced, these
groups of quasars contained redshifts which had a range of at least
several tenths from the mean redshift of the group.

SHAVER: A preliminary examination of the quasar pairs that comprise
the possible excess at low redshifts does not reveal any obvious bias
or preferred redshift.

BOYLE: Have you tried to model the z dependence of the QSO-QSO
correlation length with typical models for the evolution of galaxy
clustering with redshift?

SHAVER: No, we would first like to improve the data set to obtain a
consistent picture of QSO clustering and its evolution, before making
any comparisons with models.

HARTWICK: As reported by Dr. Crampton we found no evidence for widespread clustering in the CFHT survey as a whole. In addition, when
the sample was divided into two groups at the median redshift of
$z \sim 1.6$ no statistically significant evidence for clustering was
found in either the low or high redshift subsamples.

ULMER: QSO's are supposed to be in clusters. Why don't we see QSO
clustering as clusters are correlated?

SHAVER: The analyses of QSO catalogues suggest that QSO clustering
may approach that of clusters of galaxies at the lowest redshifts,
although this remains to be confirmed using large homogeneous samples.

BAHCALL: You have described 2 trends that appear to go in opposite
directions. If QSOs are more strongly associated with richer galaxy
clusters at higher redshifts, I would have expected the QSO correlation
function to be stronger at these higher z's, resembling more the
cluster-cluster correlation function, while at smaller z's the
correlation would be weaker. The results of the QSO correlation,
however, are exactly opposite to this trend. Can you comment on these
trends?

SHAVER: The QSO-galaxy correlation is on small scales (< 1 Mpc), whereas the QSO-QSO clustering may reflect structure on much larger scales. Therefore, if these trends are correct, they may indicate that structure on the largest scales was the last to form.

YEE: In regard to Neta Bahcall's question, I would like to point out that the increased quasar-galaxy co-variance amplitude at higher redshift has only been found for radio-loud quasars so far. Since most of the quasar-quasar clustering tests use optical quasar samples, the fact that no strong clustering of quasars has been found at high z is not necessarily contradictory.

TURNER: Even though the relative roles of physical clustering and gravitational lensing in producing close ($\Delta\theta \lesssim 10''$) high redshift quasar pairs will be difficult to determine on a case by case basis, there should be a straightforward statistical test. If physical clustering is important, many of the pairs should have a small but easily measurable redshift difference between the two components.

SHAVER: Yes, but the required number of such pairs may be large, in view of the uncertainty in redshifts measured from the broad emission lines.

SUMMARY

OBSERVATIONAL COSMOLOGY 1986

A summary of IAU Symposium No 124, Beijing, China

M S Longair
Royal Observatory
Blackford Hill
Edinburgh
EH9 3HJ

"This symposium marks the real beginning of observational cosmology",
Allan Sandage, 29 August 1986.

INTRODUCTION

It is a pleasure to be invited to attempt to summarise the very intense work of the last 6 days. Virtually all aspects of contemporary observational cosmology have been described and debated and it is my task to try to put this wealth of new material into context. As in all such surveys, allowance must be made for personal bias - like everyone else, I am sure I am giving an unbiased view but you must judge for yourselves!

First of all, my general impressions. I was struck by two aspects of the symposium. First, with a few important exceptions, there is now really quite remarkable agreement between observers about many of the most important observations. Let me list some of the topics on which there was substantial agreement.

- the isotropy and spectrum of the Microwave Background Radiation.
- the existence of voids on scales up to at least 60 Mpc.
- the counts of faint galaxies.
- the counts of radio sources.
- the decrease in the comoving space density of quasars at redshifts greater than about two. This is not a selection effect.
- the spatial distribution of Lyman-α absorbing clouds at large redshifts.
- quasar clustering, or rather, the lack of it.

Second, on the theoretical side, many important advances have been made but there is no complete theory for many of the most important observations and many plausible theories can account for them. I would suggest that, at this moment, there is much better agreement about the observations than there is about their interpretation. I believe this is a much healthier situation than if it were the other way around.

A. Hewitt et al. (eds.), Observational Cosmology, 823–840.
© 1987 by the IAU.

I was moved last night by the remark quoted at the beginning of this review, made to me by Dr Allan Sandage.   Coming as it does from one of the most distinguished pioneers of the subject, one can appreciate the huge leaps in observational capability and astrophysical insight which we are now gaining from the efforts of many outstanding observers and theoreticians.   I hope I will be able to do justice to his vision.

Let me make two methodological remarks which are appropriate to setting the framework for the discussion.   The first concerns *the determination of cosmological parameters*.   In principle, we would like to determine independently the five parameters, H_0 the Hubble constant, T_0 the age of the Universe, q_0 the deceleration parameter, Ω the density parameter and Λ_0 the cosmological constant. These are all independently determinable, in principle, and they provide key tests of the applicability of Einstein's General Theory of Relativity on the very largest scales accessible to us.   For example, in classical general relativity with $\Lambda_0=0$, $\Omega=2q_0$ and T_0 is a unique function to Ω and H_0.   If $\Lambda_0 \neq 0$, $\Omega=2q_0 + 2/3 \Lambda H_0^{-2}$ and T_0 again is a different unique function of H_0, Ω, q_0 and Λ.   These relations are thus falsifiable.   The determination of the parameters has, however, proved to be extraordinarily difficult and I believe it will be a very long time before these tests become possible.   I therefore believe that self-consistency is probably the best we can reasonably hope for, although we must not lose sight of the important goal of measuring all the parameters independently.

The second concerns the way in which we undertake astrophysical cosmology.   I can represent the procedure by the following flow diagram.

	$\rightarrow$	ASTROPHYSICAL	$\rightarrow$	PROVEN
OBSERVATIONS				
	$\leftarrow$	THEORY	$\leftarrow$	PHYSICS

It is important to note the second reversible interaction.   In essentially all of the best astrophysical theory, the physics used is based upon proven laboratory experiments.   This is why we can have confidence in the models of the thermal evolution of the Universe from temperatures of about 10^{13}K to 3K.   My concern is that much currently fashionable theory is based upon vast extrapolations of the physics which has been tried and tested in the laboratory.   The Grand Unified Thories of elementary particles, inflation, phase transitions in the early Universe, etc are far outside the present scope of experimental physics.   As Dekel emphasised, there are no constraints on the particle physics and I worry that we have no experimental control over the theories.   Some people rejoice that, maybe only in the very early Universe, can we test the theories of unification of the three (or four) forces of nature and regard this as a "useful" application of cosmology to the "real" world.   My view is the opposite.   The lack of constraint on the theory is for me a tragedy.   Of course, there is no lack of examples of astrophysical problems leading to new fundamental physics.   There are examples such as Kepler's laws of planetary motion leading to Newton's laws of gravity and of motion; Fred Hoyle's remarkable prediction of the ^{12}C resonance which leads to the triple-α process and so on.   I believe, however, that in none of the previous examples has the extrapolation been so enormous or so ambitious.   My worry is that theoretical internal self-consistency may be all that can be achieved but perhaps I am being unduly pessimistic.   The ideas are certainly very persuasive and have opened up new areas of astrophysical endeavour.

In the summary which follows, I will review the proceedings of the Symposium, not from the point of view of the expert but from the point of view of what I will tell my students about the impact of the new work upon the standard lecture course on Astrophysical Cosmology which I have been delivering for the last few years. It has seven chapters as follows:

1.   Fundamental Observations
2.   Classical World Models of General Relativity
3.   The Determination of Cosmological Parameters H_0, T_0, q_0, Ω and Λ
4.   Astrophysical Evolution with Cosmic Epoch
5.   Thermal History of the Universe - First Approximation to the Real Universe
6.   Second Approximation - The Universe with Small Perturbations
7.   Third Approximation - The Non-Linear Universe

I will tell my students what was *and* what was not said at the Symposium.

Chapter 1.   Fundamental Observations

1.1   *The Isotropy of the Universe.*   The Microwave Bacground Radiation still provides us with the best evidence that, on the global scale, the Universe is isotropic. **Partridge** and **Lukash** presented reviews of the large scale isotropy and the independent surveys are now in excellent agreement. Lukash's images of the whole sky showing clearly the dipole anisotropy were particularly impressive. There is agreement that on the large scale there is a dipole component with amplitude $\Delta T \approx 3mK$ with maximum intensity in the direction 11^h $-5°$. No other higher moments in the distribution of the radiation on the celestial sphere have been detected, the limits on the quadruple and octopole moments being $\Delta T/T < 5$ x 10^{-5}. The significance of this level of smoothness for possible anisotropic behaviour of the models in the early Universe was alluded to by Lukash.

There is also excellent agreement about the thermal spectrum of the radiation. **Partridge, Halpern** and **Mandolesi** showed that the observations are all now consistent with a Planckian spectrum at $T = 2.75K$. It is of astrophysical importance that independent measurements of the brightness temperature in the Rayleigh-Jeans and Wien regions of the spectrum give the same temperature. This sets important constraints on the amount of Compton heating which could have occurred since the effects of this scattering is to shift the whole spectrum to higher frequencies but conserving the numbers of photons, thus leading to a higher temperature in the Wien region as compared to the Rayleigh-Jeans region.

Restricting attention to large angular scales, there is now excellent evidence for the isotropy of the Universe as defined by discrete objects such as galaxies, quasars and radio sources. The radio sources have long been known to display an isotropic distribution on the sky, limited only by counting statistics. The faint galaxy counts now also show good agreement. **Ellis** and **Koo** showed independent galaxy counts to 24-25 magnitude in which there were small variations in the numbers of galaxies counted from area to area but both obtained the same average numbers and variations. Furthermore, the origin of these flucuations could be elegantly accounted for in terms of the clumpiness of the distribution of galaxies with redshift. The latter is associated with the "spongy" distribution of galaxies and how many galaxies are observed depends upon how many "sheets" are intersected along the line of sight.

Large unbiassed samples of quasars are now becoming available specifically to test their clustering properties. **Chu, Clowes, Shaver** and **Boyle** described the results of these studies. Thanks to the use of high speed measuring machines, it is now possible to generate large samples of quasars objectively over several Schmidt telescope fields. For example, Clowes and Shaver discussed a major programme in which over 1000 quasars were found in about 100 square degrees. Objective prism redshifts enabled their three as well as two diminsional clustering to be tested. The quasars at redshifts z>2 appear to be randomly distributed on the sky. Shaver and Chu showed that there is some evidence for small-scale clustering at redshifts z<2 but this needs further investigation because the quasars were not drawn from homogeneous catalogues, although these authors have made every attempt to make allowance for inhomogeneities and incompleteness. In their complete samples, Boyle finds evidence at the 3σ level for quasar clustering on scales less than $10h^{-1}$ Mpc but Crampton does not.

1.2   *The Homogeneity of the Universe.*   The observations described in Section 1.1 show that on large enough scales the Universe is isotropic and hence homogeneous but this is certainly not true on small scales. Holes, voids, sheets and filaments are apparent in the distribution of galaxies. The Centre for Astrophysics (CfA) redshift survey of galaxies, described by **Geller** is one of the most important surveys ever carried out. She described how the aim is to measure the velocities of over 30,000 galaxies from the Zwicky catalogue of galaxies, a task which is already more than half completed but which it is estimated will take a further 5 years to complete. The areas now completed present a quite remarkable picture of the distribution of galaxies, convincing evidence being observed for large voids. Geller made a number of important points about the interpretation of these pictures. First, it is likely that the largest of the voids observed, about 60 Mpc, is simply limited by the size of the survey - larger voids might well be present but would not be readily identifiable in the CfA survey. Second, from slices taken in neighbouring redshift intervals, it appears that the filamentary structure seen is not in linear filaments but rather in surfaces about holes or voids. This picture is fully supported by the much deeper redshift surveys described by **Ellis** and **Koo** in which pronounced clumping of galaxies in redshift along the line of sight is observed. Third, many of the sheets appear to be remarkably thin.

Many questions can be asked about the nature of the voids. For example, what is the void-probability distribution, a function introduced by **Lachièze-Rey** and his colleagues? How empty are the voids? **Dekel** quoted a galaxy density of only 10% of the surrounding distribution. **Chincarini** showed that dwarf galaxies have been found in some of the voids. These are potentially important clues which need to be substantiated by much large bodies of data.

On larger scales, **Bahcall** described the two-point correlation functions for clusters and superclusters. These indicate clearly the existence of structures on scales up to 150 Mpc in the case of supercluster-supercluster associations. The exact nature of these correlations is, at the moment, limited by the available samples of clusters and superclusters, the best data being drawn from the Abell catalogue of clusters. The clear trend exhibited by these data is that the large scale structures are more strongly correlated than the galaxies and the very largest scale structures more strongly correlated than clusters. These observations are broadly consistent with a hierarchical picture in which only a certain fraction of galaxies are associated with the cluster-cluster associations and only a certain fraction of the cluster-cluster associations with the supercluster-supercluster associations. These observations should be greatly improved over the next few years as the major galaxy and cluster surveys of the southern hemisphere become available, one of which was described by Olowin.

There remains the question of what one means by a two-point correlation function in the presence of voids. Everyone recognises that $\xi(r)$ is a crude but simple measure of the clustering tendency of objects about any given object. It remains to be seen what the best statistic for the distribution of galaxies on the large scale will turn out to be.

Another hot topic was the observation of large scale streaming velocities of galaxies. The history of these studies dates back to the much-discussed Rubin and Ford effect and is a good example of an observation, which many people wished would go away, suddenly becoming eminently respectable. Davies reported streaming velocities of ≈ 600 km s^{-1} from a study of the distances and recession velocities of elliptical galaxies. To complicate the issue, the streaming velocity is not in the direction of the maximum of the dipole component of the Microwave Background Radiation. No simple picture emerged from the debate but I found it interesting that the magnitudes of the random velocities of galaxies seemed to be creeping up, velocities of 300 - 400 km s^{-1} being typical.

It is amusing to note that these velocities have been discovered just when the voids have become an integral part of the large scale distribution of galaxies. There is a positive and negative side to this story. On the positive side, a streaming velocity of 600 km s^{-1} will move galaxies about 10 Mpc during the age of the Universe. Thus, considerable regions could be evacuated of galaxies, but not on the scale of the largest structures known. On the negative side, such velocities would destroy the rather narrow sheets of galaxies observed in the CfA survey. The situation is unclear.

1.3 *The Hubble Flow*. It is worthwhile emphasising the point made by Sandage that the Hubble flow, $v \propto r$, is very well defined in the redshift range $0.01 < z \lesssim 0.5$ by the brightest galaxies in clusters. Indeed, I believe this must be one of the best defined linear correlations in all astronomy. The same result is found if one selects strong radio galaxies. The purist would want to know *why* these objects are apparently such good standard candles. In the case of the cluster galaxies, one can make a good case on the basis of the universal form of the luminosity function of galaxies and adding on, if desired, the effects of cannibalism. In other words, the case can be made very compelling that $v \propto r$ out to $z = 0.5$. In the case of the radio galaxies, we do not understand yet why it is that it is only the brightest galaxies which become strong radio sources and so the case is not so compelling. However, empirically, it is found that the presence of strong radio emission selects galaxies with a remarkably small spread in absolute magnitude.

Another important point is that it is *only* necessary to find v $\propto$ r in the range 0 to 0.5. At larger redshifts, the Universe was much younger than it is now and hence the dynamics of the Universe and the luminosities of galaxies may well have changed as we discuss in Chapter 4.

1.4 *Conclusion.* I conclude that the basic observations that we are necessary to derive the Robertson-Walker metric are now in rather good shape. Since all participants in this Symposium can derive the classical Friedmann models of the Universe, we can proceed directly to Chapter 3. (If you really want a **Chapter 2**, see, for example, Chapters 14 and 15 of my book "Theoretical Concepts in Physics", Cambridge University Press, 1986).

Chapter 3. Determination of Cosmological Parameters.

I stress again that H_0, T_0, q_0, Ω and Λ are all in principle observable parameters.

3.1 *Hubble's Constant H_0.* **Tammann** presented the problem of determining H_0 with his usual clarity and forcefulness. He emphasised the key problems of eliminating Malmquist bias and of measuring accurate distances to nearby galaxies. His results favour values of H_0 close to 50 km s^{-1} Mpc^{-1}. **Aaronson** used the infrared Tully-Fisher relation to find a value of H_0 of about 90 km s^{-1} Mpc^{-1} in data which showed no Malmquist bias. **Giraud** found evidence from an analysis of the Tully-Fisher relation for groups at different distances for an increase in H_0 with distance, high values of H_0 being favoured at larger distances. I have simply stated what was said at the symposium. It has a familiar ring to it. As I tell my students, you should *not* take averages in this case in order to find a best estimate of H_0 because the determinations are dominated by *systematic* rather than *random* errors.

This is an unhappy situation which has lasted for many years now. I believe that part of the problem is the need to use distance indicators which are found from empirical studies of the properties of galaxies and one would dearly love to have these founded upon a firm astrophysical basis with predictive power. This will eventually come about but, at the moment, we seem to be faced with an impasse.

Tammann listed a number of reasons why H_0 = 100 km s^{-1} Mpc^{-1} is unpleasant and, indeed, I believe each of his arguments can be used to make independent estimates of H_0 which would favour values close to 50 km s^{-1} Mpc^{-1}. Among these arguments, was the determination of T_0, the age of the Universe, from the ages of globular clusters. The quoted estimates are about 17±3 billion years which is uncomfortably large if H_0 = 100 km s^{-1} Mpc^{-1} unless we consider models with $\Lambda \neq 0$. Narlikar correctly urged us not to use this argument as a way of estimating H_0 for the reasons I outlined in the introduction. I believe every astrophysicist has to make up his or her own mind about their attitude to this problem. My own approach is highly pragmatic. There is no evidence that $\Lambda \neq 0$ and I find the classical Friedmann models to have the great appeals of simplicity and elegance. Therefore, if I need H_0 in my calculations, I prefer to use H_0 = 50 km s^{-1} Mpc^{-1} as a "safe" value consistent with stellar evolution as we understand it just now. I realise this value is in conflict with the values favoured by de Vaucouleurs, Aaronson and others.

Ideally, we need new *physical* methods of estimating H_O. **Birkinshaw** described the Sunyaev-Zeldovich-Gunn test using the hot gas clouds in clusters and produced the remarkably unhelpful result $H_O > 11$ km s^{-1} Mpc^{-1}. **Canizares** and **Burke** showed that, in principle, we can find H_O if we can observe the time delays in well-modelled gravitational lenses. This still remains something for the future because, as Canizares pointed out, we have yet to observe the time delay expected in any of the gravitational lenses.

Dr Sandage suggested to me that we might propose a moritorium on estimates of H_O. I would certainly put a moritorium on Birkinshaw's estimates until there is a real prospect of getting considerably improved precision by the SZG technique. I am against a moritorium on the 50/100 dichotomy because there is important astrophysics to be unravelled and we can only hope that, as the data are better understood astrophysically, the origins of the discrepancies can be resolved.

3.2 *The decleration parameter q_O.* **Spinrad** brought us up to date on the infrared redshift-magnitude relation for clusters and radio galaxies. It will do no harm if I repeat my admiration for his perseverence in pressing on with the difficult task of measuring spectra and redshifts for the very faint radio galaxies in the 3CR and 1-Jy samples which have produced the remarkable results he showed us. I should of course declare an interest in this work, some of which I carried out in collaboration with Dr Simon Lilly. The work described by Spinrad has aroused considerably optimism about the possibilities of determining q_O and so I will use my "reviewer's privilege" to give my own opinion about some of this work.

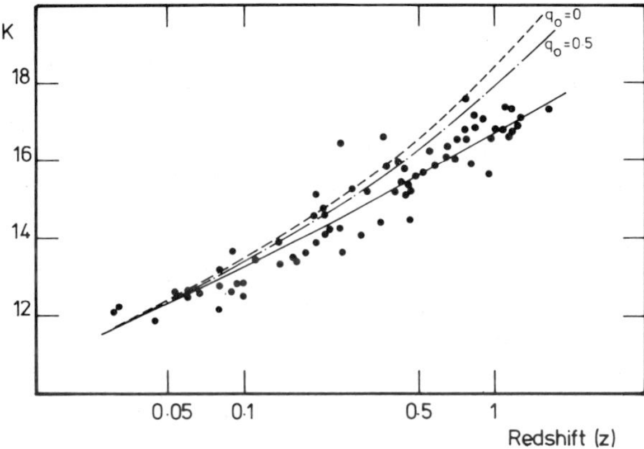

Figure 1. The infrared K-z Hubble diagram. The dashed lines indicate the predicted relation for an elliptical galaxy of constant luminosity for $q_O = 0$ and 0.5. The solid line is a model-fit to the data showing luminosity evolution of about a magnitude at z=1. (from Lilly, S.J. and Longair, M.S., 1984. Mon. Not R. astr. Soc., 211, page 849).

I have reproduced our version of the K-z relation for radio galaxies in Figure 1. I would make the following points about the diagram. First, note the small dispersion in intrinsic absolute magnitudes at small and large redshifts. At $z < 0.5$ and $z > 1$, the dispersion remains about 0.5 magnitudes showing that the galaxies display systematic behaviour over this redshift range. The small dispersion is confirmed by the new points Spinrad has added from the 1-Jy sample. Second, we have to look very carefully into the various correlations which might bias the relation. **Yates** described the tests he has made to search for correlations with radio luminosity and finds a small positive correlation which may be related to the environment of the radio galaxy. We need to be convinced about the significance of such effects before we can begin to estimate q_0. Third, the correction for stellar evolution of the parent population is relatively straightforward in the infrared waveband. Fourth, although the results are not yet of much significance, Simon Lilly found that marginally better agreement of the best-fitting line though our data was obtained with $q_0 \sim 0$-1 with the evolution correction rather than a large q_0 without evolution. This indicates that the shape of the K-z relation could be used to help remove the evolutionary corrections if we had large enough statistics in the redshift range 1-2.

Taking all of this into consideration, I believe we can say already that q_0 lies in the range 0 to 1 but I would not be as optimistic as Spinrad just yet. Notice that, with the next generation of large telescopes and infrared spectrometers with low-noise detector materials, we should be able to test the evolution directly by making spectroscopic observations in the near infrared wavebands.

Wampler described an intriguing reanalysis of the optical redshift-magnitude relation for complete samples of quasars and, after making allowance for various known correlations with the data, found a best fitting value of q_0 of about 3, almost exactly the same as we found from the K-z relation for radio galaxies uncorrected for stellar evolution. Does this suggest that the underlying luminosity evolution in quasars is the same as the optical luminosity of the stellar populations in radio galaxies? If this were true, it would be very important because there might be some direct or indirect physical relation between these very different types of activity. This will certainly repay further study and has implications for the evolution of the luminosity function of quasars.

Other methods of finding q_0 were mentioned but many are, at this stage, masked by evolutionary effects. **Kapahi** described the angular diameter-redshift test for double radio sources and **Sargent** the possibility of using absorption lines in distant quasars. **Loh** discussed number counts and claimed good results using multicolour photometric redshifts - this needs detailed scrutiny in the light of what is now known about the spectra of galaxies at redshifts greater than 0.5, a point emphasised by **Ellis**.

3.3 *The density parameter* Ω. **Tully** and **Sancisi** reviewed the various approaches to determining the local density of matter in the Universe. Much of this story was very familiar. Spiral and elliptical galaxies have mass-luminosity ratios of about 10-30, corresponding to $\Omega \sim 0.02$. The spiral galaxies must have haloes consisting of some form of dark matter to maintain the flatness of their rotation curves out to the largest radii measured. Notice that Sancisi quoted an M/L ratio of 5000 in the outermost regions of some of his galaxies. The best studied great clusters have M/L ~ 350-500 and, if this is typical of all luminous matter in the Universe, Ω ~ 0.2. Tully reported a new estimate of Ω from his studies of nearby groups of galaxies and found $\Omega \gtrsim 0.1$.

Through all these estimates runs the difficult problem of whether or not the hidden mass is distributed like the galaxies. This problem is particularly problematic for the Virgo cluster infall test described by Tully and for the cosmic virial theorem. In both cases, the velocities are driven by perturbations in the average density distribution of matter. In the case of Virgo cluster infall, the velocity is given by v $\propto$ H$_O$ r $\Omega^{0.6}$ δ where δ is the average density enhancement due to the Local Supercluster. If we assume *all* the matter is distributed like the galaxies, $\delta = 2$ and $\Omega = 0.3$. However, suppose the hidden mass is much more evenly distributed than the galaxies. Then $\delta < 2$ overall and hence Ω is greater. Exactly the same argument bedevils the cosmic virial theorem.

Rowan-Robinson reported an intriguing new result on the origin of the acceleration which might drive our peculiar velocity of 600 kms^{-1} with respect to the Microwave Background Radiation. He used the distribution of IRAS galaxies to work out the local gravitational acceleration g due to galaxies, assuming they are tracers of the mass and finds that the acceleration vector is in the same direction as the maximum of the dipole component of the microwave background. His calculations provide an estimate of $\Omega = 1$. This appears to be in contradiction with the results of the cosmic virial theorem and the Virgo cluster infall test which with the same assumptions suggest Ω ~ 0.2-0.3. The origin of the discrepancy is not understood. Many more velocities are needed for galaxies in the IRAS sample so that more secure estimates of g can be made. In all cases, the question of the distribution of the visible and dark matter is crucial. **Sancisi** and **Burke** showed from the rotation curves of galaxies and studies of gravitational lenses respectively that it is certainly not true on the scale of galaxies.

One interesting new result on the possible contribution of massive black holes to the present mass density of the Universe came from **Hewitt's** analysis of the frequency of gravitational lensing among 4000 radio sources surveyed for evidence of this phenomenon. The small number of candidates found among this sample sets limits to the number of gravitational lenses of different masses present in the Universe. The statistical results are now reaching very interesting limits, so that, for example, a value of $\Omega=1$ in lenses of mass $10^{10}-10^{12}M_\odot$ can be excluded.

Chapter 4   Astrophysical Evolution with Cosmic Epoch.

4.1   *Active galaxies and quasars.*   I will consider first the cosmological evolution of "non-thermal" sources. The radio sources, both radio galaxies and radio quasars, and the optically selected samples of quasars show similar strong cosmic evolution. **Wall** surveyed the radio data and showed the type of evolutionary behaviour required by both steep and flat spectrum radio sources. **Schmidt** and **Boyle** showed that strong *luminosity* evolution of the population of radio quiet quasars could account for the quasar counts and the redshift distributions at different magnitudes. **Deng** showed how the evolution could be inferred over small redshift ranges, carefully taking account of the selection effects which are present in objective prism surveys. There is good internal consistency concerning the secure now. The interpretation of the effect is not clear but the effect of the environment in changing the properties of the double sources at large redshifts was convincingly demonstrated by **Miley** who showed that the double structures of quasars with $z > 1.5$ are much more distorted than those at lower redshifts. According to his observations the environments of radio sources at $z > 1.5$ are less benign than those at lower redshifts.

The galactic environments of radio galaxies and quasars were addressed by **Yee**, his results showing that the quasars at redshifts $z > 0.5$ are in richer cluster environments than those at small redshifts. **Yates** speculated that the same may be true of radio galaxies. In turn the environment, both the numbers of companion galaxies and the ambient gas, can contribute to the evolutionary effects. In an elegant piece of work on Seyfert galaxies, **Byrd** showed how the close companions observed in many of them might lead to Seyfert activity.

4.2   *Lyman-α absorption clouds.*   There now seems to be agreement about the evolution of the comoving number density of Lyman-α absorbing clouds in the ultraviolet spectra of quasars. Up to the limit of observation, $z = 3$, **Sargent** and **Chen** reported that the number of clouds increases as $(1+z)^2$. It is interesting that in contrast, the quasar number density at a given luminosity is decreasing roughly as $(1+z)^{-2}$ at $z \gtrsim 2$. The cause of this behaviour is not understood, largely because the nature of Lyman-α clouds is not clear - are they expanding clouds, or are they bound? Are they being eroded by the flux of ultraviolet radiation from quasars? **Webb** reported the interesting result that the Lyman-α clouds are clustered on velocity scales less than 150 km s^{-1}.

4.3   *Galaxies in general.*   The beautiful new results of **Ellis** and **Koo** provide important constraints on galaxy evolution at redshifts less than about 0.5. The counts of galaxies show a small excess over the predictions of uniform, non-evolving world models at $m \approx 22$ but not at the faintest magnitudes. There is thus little scope for evolution in the sense that the galaxies were brighter in the past. This result is confirmed by the deep spectroscopic surveys carried out by both groups. No evidence of evolution is found in the redshift distribution at a given apparent magnitude. The redshift distributions discussed by **Ellis** and **Koo** and by **Karachentsev** show a highly non-uniform distribution and they very plausibly suggest that this reflects the fact that the galaxies are confined to sheets which are part of the "spongy" structure of the distribution of galaxies.

The consistency of the counts and redshift distributions can clearly be used to set important limits on when galaxies in general first formed and I look to forward to hearing how useful a constraint this will be.

4.4   *The Radio Galaxies Again*   The radio galaxies have already been mentioned in the context of the determination of q_0 where it was indicated that there is strong evidence that the evolution of the stellar populations of these galaxies is playing an important role. **Spinrad** has described his appraoch to these data and our own is very similar. I will indicate some of the differences - I do not believe there is any major disagreement between the actual observations. In our approach, we distinguish carefully between the radio galaxies and the brightest galaxies in clusters. Although these classes of galaxy seem to have similar luminosities at redshifts of 0.4 and greater, there may be differences at lower evolution of different classes of active system, although how the radio objects are selected from the population of quasars in general remains a thorny problem and was mentioned by **Machalski**.

There was also good agreement about the large redshift behaviour of the different populations of quasars. For the optically selected samples, the comoving number density of quasars decreases with increasing redshift beyond $z \sim 2$. This has been found in new surveys described by **Crampton** and **Schmidt** and a similar decline has been found by **Véron** who analysed data in the existing catalogues. The maximum in the comoving space density of sources occurs about a redshift of 2 and decreases to larger redshifts. A new largest-redshift quasar was reported by **Hewett**, $z = 4.01$, who had used automatic plate measuring techniques to extract candidate quasars. **Savage** reported the recent upsurge in the discovery of quasars with redshifts greater than 3.5 which has taken place over the last few months. Nonetheless, these discoveries are not sufficient to reverse the decline to large redshifts.

It is important that exactly the same behaviour has been found by Dunlop, Peacock and their colleagues in radio samples of quasars with flat radio spectra. This is a completely independent method of probing the evolution of quasars at large redshifts since the primary selection criterion is radio flux density. It should be emphasised that there is not a cut-off but rather a broad maximum in the comoving number density of quasars. I suspect a gaussian distribution of space density or of luminosity with cosmic time might be a good model which might have implications for the origin of galaxies and the strong cosmic evolution of quasars.

It will eventually be possible to undertake similar studies for discrete x-ray sources but at the moment the major surveys described by **Gioia** refer to bright source samples, similar in numbers to the 3CR survey of radio sources. Her extension of the Einstein medium flux-density survey to over 800 sources promises to be a major advance in the definition of the population of extragalactic x-ray sources. At the moment, the strongest constraint on the cosmological evolution of the properties of the x-ray population comes from the upper limit to their integrated intensity. As mentioned by **Schmidt** and **Setti**, luminosity evolution of the quasar population with a constant optical-x-ray spectral index is consistent with the limits imposed by the observed x-ray background whilst density evolution would result in too great an intensity because of the large numbers of low luminosity x-ray sources. **Boldt** surveyed the thorny problem of accounting for the intensity, spectrum and smoothness of the x-ray background. I fully endorse his point that the origin of the spectral break at 40 keV and the smoothness of the background pose major problems for most known classes of x-ray source.

Kapahi described new analyses of the redshift dependence of the sizes of double radio sources and found that all his results now require the intrinsic size of double sources to decrease with increasing redshift as $l = l_0 (1+z)^{-\beta}$ with $\beta = 1.5$-3. Although this may not look a difficult achievement, it is in fact a complicated problem because of the strong differential evolutionary effects which can strongly influence what types of source appear in different proportions at different flux densities. Up till now, I have not been wholly convinced that $\beta \neq 0$ because of the sensitivity of the test to small numbers of objects but I think the result looks redshifts which might be ascribed to dynamical evolution of the brightest galaxies in the clusters. Therefore, in the work by Lilly and myself, we are careful to consider only the radio galaxies. A powerful probe of the evolution of the galaxies is the use of the optical-infrared colour, R-K, for example. I show in Figure 2 our version of the R-K v redshift diagram. Also shown is the line "passive evolution" which corresponds to the predicted colours if galaxies formed

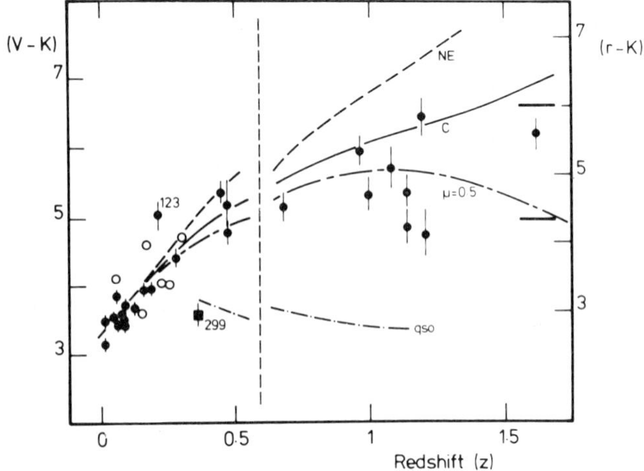

Figure 2. The optical-infrared colours (V-K) and (r-K) for 3CR radio galaxies. The C model represents a model including passive evolution of the stellar population of a standard elliptical galaxy. The model $\mu=0.5$ is from Bruzual (from Lilly and Longair, op. cit., page 845).

all their stars very early on and the subsequent evolution is simply the decay of that initial stellar population. It is important that none of the observed points lie above this line which, in my view, corresponds to a very reasonable null hypothesis for galactic evolution. Below this line, we find a significant scatter in the observed points which is interpreted as meaning that the galaxies show star formation activity at the epochs observed. Spinrad interprets this in terms of a single model of the Bruzual type with $\mu = 0.7$. We believe that the scatter is too great to be accounted for by a single model and that the evolution is associated with separate star formation events, possibly associated with the events which gave rise to the radio source. It is intriguing that some of points lie close to the "passive evolution" line suggesting that there is little on-going star formation activity in these galaxies.

Sandage discussed the observations of Spinrad and Hamilton who searched for evidence of evolution of the stellar populations of galaxies from variations in the amplitude of the 400 nm break with redshift. Spinrad finds evidence for a decrease in the break with increasing redshift, consistent with evolution of the stellar population whilst Hamilton does not. These and similar indicators of stellar populations provide crucial evidence on the evolution of the stellar populations of galaxies and, with the large telescopes of the future becoming a reality, we can hope that these studies can be extended to the largest redshift galaxies observed.

One of the most interesting results of this programme comes from the study of the 1-Jy sample of Allington-Smith which was designed to test the conclusions of the 3C observations and to extend it to larger redshifts. Figure 3 shows the R-K v K magnitude diagram, in which we have had to use K rather than redshift as a distance indicator because redshifts are not available for all the objects in the sample. It will be noted that there is a group of sources at faint magnitudes which appear to have the colours of passively evolving galaxies. We believe that these are likely to be passively evolving galaxies at redshifts in the range 1.5 to 3. If this is true they are of great cosmological interest because they must have formed at epochs significantly earlier than 3. Indeed, I believe that these data can already be used to set useful limits to the epochs at which these galaxies could

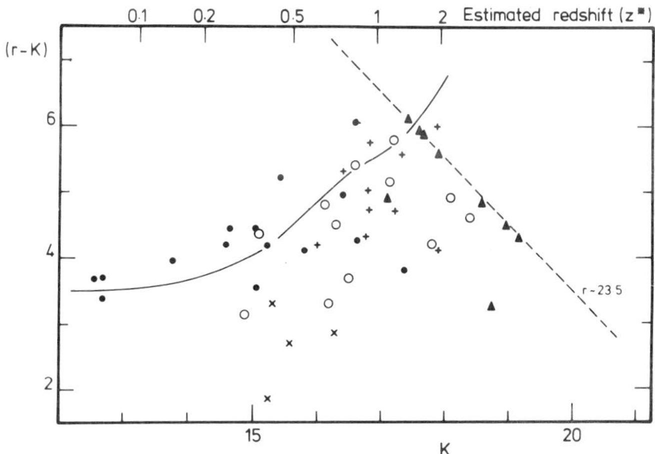

Figure 3. The (r-K)-K colour magnitude diagram for sources in the 1-Jy sample. The dashed line represents the limit r=23.5 to which identifications have been sought and the continuous curve indicates the locus of the passively evolving C model of Bruzual. An estimated redshift scale is shown along the top of the diagram (from Lilly, S.J., Longair, M.S. and Allington-Smith, J.R., 1985. Mon. Not. R. astr. Soc., 215, page 47).

have formed. For example, if the galaxies formed at redshifts of 3 or 4, then even if they undergo only passive evolution, none of them should be fainter than about K = 18 at any lower redshift. Studies of this type are of the greatest interest and much larger samples are needed to test these ideas.

Chapter 5. The Thermal History of the Universe

What has now become the standard view of the thermal history of the Universe was elegantly described by Audouze. The subject is dominated by the necessity of synthesising the light elements ^{4}He, ^{3}He, D and ^{7}Li in their observed cosmic proportions. The beautiful new observations of **Duncan** showed how the abundance of ^{7}Li is a sensitive tracer of stellar evolution and also how there seems to be a more or less constant "primordial" abundance in those stars in which it is inferred that the convection zones are narrow and hence will not have convected the ^{7}Li into temperature regimes in which it would be depleted. The result of all these studies is that too little deuterium is produced if the density parameter Ω in baryons is greater than 0.2 $(50/H_0)^2$.

This has become such a compelling result that no-one is very keen to challenge it and it has become the prime basis for assuming that if the density of the Unverse is actually greater than this, we must call upon various forms of dark matter, massive neutrinos, etc. to make up the discrepancy. Indeed, much of the galaxy formation industry thrives upon the fact that dark matter in non-conventional forms may exist. It is therefore very important to test all the ways in which the argument might be circumvented. Most of the astrophysical proposals have proved unacceptable. One possibility raised by Audouze was that it might be possible to build conventional Universes with $\Omega = 1$ in baryons if there is some way of destroying some of the ^{4}He which is overproduced in these models. To quote some figures, in a standard $\Omega = 1$ Universe, about 26% ^{4}He and about an order of magnitude too little D is produced. Therefore if only a little of the ^{4}He were converted into ^{3}He and D, we could probably produce the observed abundances of the light elements. How this could be done is not clear. One possibility mentioned by Audouze is that there might exist massive unstable particles with half lives of about 10^{5-6} seconds which would leave the nucleosynthesis intact but which, on decaying into γ-rays, can dissociated some of the ^{4}He nuclei. This is perhaps a rather inelegant solution but I wonder if it is less contrived that the introduction of other forms of matter which have yet to be discovered experimentally such as cold dark matter, axions, cosmic strings, etc.

I only mention this possibility because of the central importance which the synthesis of the light elements now plays in our present best-buy picture of the Universe. If we were to find a way of allowing $\Omega_{baryon} = 1$, this would have very important repercussions for all cosmology. For example, as **Dekel** mentioned, there are models of galaxy formation with $\Omega = 1$ which can be made consistent with all the observations, including the low level of fluctuations in the Microwave Background Radiation.

Chapter 6. The Universe with Small Perturbations

The crucial confrontation between the theory and the observations lies with the upper limits to the fluctuations in the Microwave Background Radiation described by **Partridge** and **Davies**. **Silk** and **Dekel** expounded our present understanding of the origin of structure in the Universe, Silk talking the view that most of the observations can be accomodated in a single theory involving cold dark matter, Grand Unified Theories, inflation, etc. Dekel took a somewhat less ambitious view and delineated the rather wide range of parameter space which is still available for galaxy formation.

There are important issues for observational cosmologists to consider in these debates. There is no question but that a great deal of the relevant theory is currently being driven by the predictions of the theories of elementary particles and the attempts to unify all four forces of nature. It is these theories which have led to the plethora of new particles which could exist if the strong and the electro-weak theories are unified at very high energies. The possibility of inflation in the early stages can come about if the properties of the particles are such that the phase transition from the pseudo to the real vacuum solutions for the elementary particles occurs at a late enough stage. I am not competent to discourse in a meaningful way about the speculative areas in these theories but I would emphasise that we are not talking about any single theory of fundamental interactions which particle physicists have agreed about and which produces the physics we need to understand galaxy formation. Rather, the particle physicists have produced generalised features of the theory which can be adjusted to have cosmologically interesting results. However, this is very far from a real and provable picture of the early universe. I urge astrophysicists to recall **Dekel's** remark that the theory of elementary processes at these very high energies is essentially unconstrained.

I make these remarks simply to emphsaise that whilst we must listen with close attention to developments in the theory of elementary particles, as observational cosmologists we should not be blinded by them nor constrained by them in interpreting our data. Fashions can change quite quickly in the area of elementary particle physics. There are many separate components which are necessary in the theory of the origin of structure in the Universe and many of them are astrophysical questions which fall squarely within the province of observational cosmology.

This consideration leads to a personal conviction of mine that the theory of the origin of the large scale structure in the Universe is not one but many theories. Although Silk claimed that he had great success in discussing the origin of everything in the Universe within a single theory, I believe it is in fact a number of theories put together in a particular way. Let me list some of them:

- The theory of small perturbations in the Universe
- The theory of hierarchical clustering
- The theory of dissipative processes in the cosmological context
- The theory of explosive formation of galaxies and structures
- The astrophysics of galactic evolution

I am sure that one can add to this list. The point is that I have a personal preference to try and solve them one at a time rather than all simultaneously.

The present position concerning the theory of small perturbations was beautifully summarised by **Dekel.** It is worthwhile repeating the rather extensive parameter space which should be considered by theorists and observers

- $\Omega = 0.1$ to 1; $\Lambda = ?$; $H_0 = ?$
- Dark matter: baryonic, hot, warm, cold, unstable
- Fluctuations: adiabatic, isothermal
- Their origin: inflation, strings, their spectrum and probability distribution
- Explosions - astrophysical processes
- Relative distribution of visible and dark matter

I believe that the parameter space may well expand with time.

One methodological point which I believe may be of interest concerns the way in which the theories are compared with the observations. Many of these involve trying to obtain the correct shape and amplitude of the 2-point correlation function of galaxies $\xi_{gg}(r)$. The problem is that this may not trace the correlation function of the total mass of the Universe if there is dark matter present. I believe it may become of more interest to work in terms of the formation epochs of objects of different masses, since this should be related to the time when the perturbations which gave rise to galaxies, whatever their origin, became non-linear $\delta\rho/\rho \geqslant 1$. In Chapter 4 I indicated how limits can now be set to the epochs when various types of object, galaxies and radio galaxies, formed and other arguments can be made for larger scale systems. It may be that this is a stabler indicator of the point at which perturbations on various mass scales became unstable.

It remains the case that, of the various constraints upon the epoch of galaxy formation, those imposed by the absence of fluctuations in the Microwave Background Radiation are by far the most stringent. In fact, if fluctuations are not found within a factor of 10 increase in the sensitivity of the present experiments, we may be in for a major reappraisal of the theory of galaxy formation.

We should not forget that there may well be important clues about the properties of the initial fluctuations from careful interpretation of data available now. I was particularly impressed by the remarkable analysis described (and presented in three-dimensions) by Gott. His demonstration of how the topology of the large scale distribution of galaxies can be related to the random phases of the initial fluctuations was a remarkable instance of how the observational data on the large scale structure of the Universe can be used to gain important insights. It also gives us a powerful conceptual tool for envisaging the distribution of galaxies in three dimensions. It is this analysis which has instantly converted me to using the term "sponge-like" to describe the large scale distribution of galaxies.

Chapter 7. The Non-linear Universe

The non-linear Universe is the universe we know and love. It is the non-linearities which give rise to galaxies as we know them and one could characterise the problem of the origin of structure in the Universe as being one in which we attempt to unwind the non-linearities back to their linear stages. Essentially I include in this chapter all the astrophysical tools necessary to understand the properties of galaxies, quasars, radio sources, intergalactic clouds, etc. Let me give a brief and incomplete list of some of these phenomena.

1)   *The epoch of galaxy and object formation*   I have emphasised this several times. Other observations not yet mentioned include:

a)   the observations of the near infrared background radiation by **Matsumoto** and the follow up observations by **Xie**. If this background is indeed a cosmological component, it is potentially of great importance for cosmology. Matsumoto showed that its intensity is not so different from the background intensity predicted by Peebles and Partridge long ago from young galaxies. The further investigation of the near infrared background seems a very important task.

b)   The mapping of the chemical distribution of the heavy element abundance in the metal-rich absorption systems at large redshifts promises to be an important tool for studying the formation of the general distribution of chemical elements in galaxies. **Hunstead** reported what he believed to be the first detection of low metal abundances in a large redshift absorption line system. Objects like this may help to trace the evolution of the chemical abundances of the elements in galaxies with cosmic epoch.

c)   **Swarup's** proposal to search for intergalactic hydrogen clouds associated with protoclusters is the type of observation which I believe must be pursued to give us further insight into the *terra incognita* which lies between redshifts of about 4 and 1000.

2)   *The Reheating of the Intergalactic Gas*   This remains one of the cosmological problems which attracts less attention than it deserves. Although **Setti** indicated that it would be pleasant to account for the observed intensity of the X-ray background by the thermal bremsstrahlung of intergalactic gas with $\Omega = 1$ and temperature $kT = 40$ keV, this would result in a very large intergalactic gas pressure which would compress the Lyman-α clouds. **Sargent** reported the rather stringent limits which can be placed upon the properties of the intergalactic gas if it is not to exert too great a pressure on the Lyman-α clouds. He also reported new upper limits to the number density of neutral hydrogen atoms at $z=3$ from a recent investigation of the Gunn-Peterson test for large redshift quasars. Less than a 5% decrement is observed implying $N_H \leqslant 2 \times 10^{-12}\mathrm{cm}^{-3}$ at $z=3$. **Koo** reported the problem of ionising the intergalactic gas using his best estimates of the numbers of quasars which are inferred to be present at large redshifts. Presumably, a low enough intergalactic gas density could be ionised by the Lyman continuum flux emitted by quasars and it is important to know just how low this density might be. Related to this is the question of the redshift at which it is necessary to make the chemical elements observed in the most distant quasars and in the metal-rich intergalactic gas clouds. These questions are related to the epoch of formation of the first generations of stars.

3)   *Astrophysical processes which affect $\xi(r)$ in the non-linear regime*   It seems to me that explosions, the formation of quasars, etc quite inevitably lead to the biasing of galaxy formation in one way or another. Indeed, I would quite invert the problem. It seems to me that unbiassed galaxy formation is much more difficult to achieve that biassed galaxy formation and that the bias could easily go either way. This is a purely astrophysical question but the answer

is important.   For example, although the simulations of the hot dark matter picture with neutrinos seem to do too well in producing the spongy structure of the universe, this may well be softened by astrophysical processes such as explosions and quasar formation as described by **Dekel**.

4)   *Astrophysics*   Finally, I have to make a plea for a better understanding of the tools of astrophysical cosmology.   This is not just for objects like galaxies, dwarf galaxies, clusters and quasars as would have been advocated by Baade (see **Sandage's** paper), but also the correlations which play such an important role in, for example, the determination of q_0.   This comment applies to correlations such as the Tully-Fisher relation, the velocity dispersion-diameter relation for galaxies and all the other empirical rules of observational cosmology.

CONCLUSIONS

We have been treated to five days of outstanding science and it is very gratifying that our Chinese colleagues have been able to hear so many of the key questions discussed by the eminent workers who have contributed so much to our present understanding.   There is no need to summarise the summary except to repeat the quotation of Dr Sandage which I have put at the head of this review.   We are on the brink of a new era in astrophysical cosmology and it is especially pleasing that this coincides with the great upsurge in the physical sciences in China.

We have had a wonderful week and I conclude by referring to Francesco Sansovino's delightful interpretation of the name of the city of Venice. He claimed that the name Venezia derived from the Latin phrase *veni etiam* which he interpreted as "come back again, and again, for however many times you come, you will always see new and beautiful things" (F Sansovino, "Venetia città nobilissima et singolare", 1663).   No quotation could be more appropriate to the city of Beijing and the Chinese astronomers nor to the magnificant reception we have received from our hosts.

POSTER PAPERS

TITLES AND AUTHORS

Barbieri, C., and Chu, Y.
   QSO CANDIDATES IN THE FIELD OF SA 94(III): CLUSTERING ANALYSIS

Bergeron, J., Kunth, D., and D'Odorico, S.
   THE LOW REDSHIFT MULTIPLE ABSORPTION SYSTEMS TOWARD THE QSO
   PKS 1327-206

Bi, H.G.
   FRACTAL DIMENSION OF GALAXY CLUSTERING

Bian, Y.L., Zou, Z.L., and Chen, J.S.
   MONTE CARLO SIMULATIONS OF THE MAXIMUM LIKELIHOOD METHOD
   USED TO DETERMINE dN/dZ OF Ly α CLOUDS

Chen, S., and Yu, Y.Q.
   THE PERFECT GAS IN KALUZA-KLEIN MODEL UNIVERSES

Chen, L.F., and Zhou, M.Y.
   ENTROPY PRODUCTION IN HIGH DIMENSIONAL ANISOTROPIC UNIVERSES

Cheng, F.H., Kinney, A.L., and Fang, L.Z.
   AN ARCHIVAL STUDY OF IUE OBSERVATIONS OF LOW REDSHIFT QUASARS

Cheng, F.H., You, J.H., and Grady, C.
   A REDISCUSSION OF THE BALMER DECREMENT OF BROAD LINE RADIO
   GALAXIES

Cowley, A., and Crampton, D.
   THE FREQUENCY OF BROAD ABSORPTION-LINE QUASARS

Cui, Z.X., and Chen, J.S.
   THE IDENTIFICATION OF HEAVY-ELEMENT ABSORPTION SYSTEMS IN THE
   QSO 2000-330

Djorgovski, S., Spinrad, H., McCarthy, P., and Strauss, M.
   A SEARCH FOR GALAXIES BEYOND Z = 3

Fan, X.M., and Chen, J.S.
   AUTO-CORRELATION STUDIES OF 6C RADIO SOURCES

Fang, L.Z., Chu, Y.Q., and Zhu, X.F.
   AN UPPER LIMIT TO THE DENSITY INHOMOGENEITY IN THE UNIVERSE

Fort, B., Lefevre, O., Mathez, G., Mellier, Y., Pello, R., Picat, J.P.,
and Soucail, G.
   LONG TERM PROGRAM ON DISTANT CLUSTERS

Gopal-Krishna, and Wiita, P.Z.
        COSMOLOGICAL EPOCH DEPENDENCE OF THE EXPANSION OF POWERFUL
        RADIO SOURCES

Guo, W.W., and Li, Z.W.
        THE USE OF TYPE I SUPERNOVAE TO DETERMINE THE HUBBLE CONSTANT

Gush, H., Halpern, M., and Wishnow, E.
        ROCKET MEASUREMENT OF THE MBR SPECTRUM

Hammer, F., and Nottale, L.
        GRAVITATIONAL AMPLIFICATION OF BRIGHTEST CLUSTER GALAXIES BY
        FOREGROUND CLUSTERS

Hu, F.X., Allen, R., Van der Kruit, P., and You, J.H.
        SOME PROPERTIES OF THE GALACTIC HALO IN NGC 891

Huang, W.L., and Chen, Z.X.
        NEUTRINO MASS, PANCAKE, AND FILAMENT

Huang, W.L.
        SUPPLEMENT PARTICLES AND SUPERCLUSTERS

Johnson, D.G., and Wilkinson, D.T.
        A MEASUREMENT OF THE TEMPERATURE OF THE COSMIC MICROWAVE
        RADIATION AT $\lambda = 1.2$ cm

Lachieze-Rey, M.
        FRACTALS AND GALAXY FORMATION

Li, Z.W.
        THE STATISTICAL ANALYSIS OF SUPERNOVAE

Liu, Y.L., and Liu, J.Y.
        THE LUMINOSITY FUNCTION OF QUASARS AND RADIO GALAXIES-I.
        PURE LUMINOSITY EVOLUTION

Liu, Y.L.
        THE LUMINOSITY FUNCTION OF QUASARS-II.   EVIDENCE OF EVOLUTION

Liu, R.L., Cao, L.H., You, J.H., and Cheng, F.Z.
        X-RAY AND OPTICAL RADIATION FROM ACTIVE GALACTIC NUCLEI, I

Liu, Y.Z.
        DISCONTINUOUS PHENOMENA IN THE STRUCTURE OF THE UNIVERSE
        (THE LARGE SCALE QUANTUM EFFECTS)

Liu, M.C.
        THE REDSHIFTS OF QUASARS AND THE LAW OF HUBBLE

Liu, X.D.
    ORIGIN AND EVOLUTION OF Ly α ABSORPTION CLOUDS

Lu, W.C., and Deng, Z.G.
    STATISTICAL STUDY OF THE RELATION OF MORPHOLOGICAL TYPE OF
    GALAXY WITH ITS NEAREST NEIGHBOR

Mazure, A., Mathez, G., Mellier, Y., and Capolato, H.
    A NEW INSIGHT INTO THE COMA CLUSTER

McCarthy, P., Spinrad, H., Djorgovski, S., Strauss, M., Van Breugel, W.,
and Liebert, J.
    EXTENDED Ly α EMISSION IN 3C 321.1: A 100 MPC CLOUD OF
    IONIZED GAS AT $z = 1.82$

Ozernoy, L.
    COSMIC VOIDS AT LARGE z: ARE THEY "LYMAN BUBBLES"?

Parker, Q., Beard, S., and Mac Gillivray, H.
    OBJECTIVE-PRISM SPECTRA AND THE LARGE-SCALE GALAXY
    DISTRIBUTION

Ramella, M., Giuricin, G., Mardirossian, F., and Mezzetti, M.
    MORPHOLOGICAL CONTENT AND FIRST-PANCAKE GALAXY IN LOOSE
    GROUPS

Shang, S.P.
    THE SMALL SCALE ANISOTROPY OF THE MBR COMPUTED BY MEANS OF
    THE TRANSFER FUNCTION BETWEEN $z = 2000$ and $z = 800$

Shen, Y.G.
    THE COSMICAL WAVE FUNCTION OF SYMMETRICAL HIGGS ϕ^4 FIELD

Shen, Y.G., and Yu, D.W.
    A RADIATIVE SOLUTION FOR THE COSMOLOGICAL MODEL OF THE
    5-DIMENSIONAL BIANCHI TYPE V

Sun, K.
    EVOLUTION OF THE LUMINOSITY FUNCTION OF QUASARS

Sun, X.J.
    A STATISTICAL STUDY OF CLUSTERS OF CLOUDS CONTAINING HEAVY
    ELEMENTS FROM QSO ABSORPTION SPECTRA

Tang, Z.M.
    THE DEFLECTION OF AN ELECTROMAGNETIC SIGNAL IN NHN FIELD

Wang, Y.J., and Peng, Q.H.
    THE GRAVITATIONAL PROPERTIES OF A NEUTRON STAR WITH MAGNETIC
    CHARGE AND MAGNETIC MOMENT

Xie, G.Z., Hau, P.J., Zhou, Y., Lu, R.W., Liu, X.D., and Cheng, F.Z.
        OBSERVATIONS OF RAPID LUMINOSITY VARIABILITY OF NGC 4051

Xie, G.Z., Lu, R.W., Zhou, Y., Hau, P.J., Zhang, Y., Liu, X.D., Wu, J.
X., and Li, X.Y.
        OPTICAL BEHAVIOR OF EIGHT BL LAC OBJECTS AND THE PROPERTIES
        OF THEIR CENTRAL NUCLEI

Xie, G.Z., Zhang, Y., and Wu, J.X.
        THE EFFECT OF DUST IN THE EARLY UNIVERSE ON THE COSMIC
        MICROWAVE BACKGROUND

Xie, G.Z., Zhang, Y., Hao, B.J., Liu, X.D., Lu, R.W., Li, K.H., Wu, J.
X., and Cheng, F.Z.
        ZELDOVICH EFFECT OF THE 1-5 μm INFRARED BACKGROUND RADIATION

Yang, L.T., Liu, C., and Yang, P.B.
        THE ACCRETION DISK MODEL FOR THE ULTRAVIOLET EXCESS OF
        QUASARS, HUBBLE FLOW AND THE FRIEDMANN MODEL

Yang, L.T., Liu, C., and Yang, P.B.
        THE THIN DISK MODELS FOR THE ULTRAVIOLET EXCESS OF QUASARS
        AND SEYFERT 1 GALAXIES

Yin, Q.F.
        VLA AND MI-YUN OBSERVATIONS OF M51

Yin, Q.F., and He, X.T.
        VLA AND OPTICAL OBSERVATIONS OF NGC 3434

You, J.H., Cheng, F.Z., Gong, R.S., and Liu, R.L.
        X-RAY AND OPTICAL RADIATION OF ACTIVE GALACTIC NUCLEI, II -
        MODEL CONSIDERATION OF CORRELATION OF $L_x \sim L_{op}$ AND
        "DRAG EFFECT"

Yu, D.W.
        THE CARTAN CLASSIFICATION OF COSMOLOGICAL MODELS

Zhang, S.W.
        INFLATION FROM QUANTUM CONFORMAL FLUCTUATION

Zhou, Y.Y., Gao, Y., Deng, Z.G., and Dai, H.J.
        AN EXPLANATION OF THE REDSHIFT DISTRIBUTION OF HETEROGENEOUS
        SAMPLES OF QUASARS

Zhu, S.C., Zhu, J.M., Guo, H.Y., and Guo Han-Ying
        ENTROPY PRODUCTION OF THE KALUZA-KLEIN COSMOLOGIES

AUTHOR INDEX

AARONSON, M. 187
AIZU, K. 565
AKIBA, M. 69
ALEF, W. 537
ARP, H. 479
AUDOUZE, J. 89

BAHCALL, N.A. 335
BALKOWSKI, C. 315
BATTY, M.J. 673
BATUSKI, D.J. 319,323
BENNETT, C.L. 747
BIRKINSHAW, M. 83
BLADES, J.C. 799
BOLDT, E. 611
BOYLE, B.J. 643
BURKE, B.F. 747
BURNS, J.O. 319,323
BURSTEIN, D. 223
BYRD, G.G. 503,527

CANIZARES, C.R. 729
CANNON, R.D. 531
CHAMARAUX, P. 315
CHANDLER, C.J. 673
CHEN, J.S. 535,793
CHINCARINI, G. 275,327
CHU, Y. 627
CLOWES, R.G. 809
CONDON, J.J. 673
COWLEY, A.P. 655
CRAMPTON, D. 655
CRANE, P. 59
CRUDDACE, R.G. 523

DAVIES, R.D. 55
DAVIES, R.L. 223
DE LAPPARENT, V. 301
DE SOUZA, R.E. 327
DEKEL, A. 415
DENG, Z.G. 363,677
DJORGOVSKI, S. 129
DRESSLER, A. 223,573
DUNCAN, D.K. 119

EINASTO, J. 349
ELLIS, R. 367

FABER, S.M. 223
FANG, L.Z. 461,627
FENIMORE, E. 523
FILIPPENKO, A.V. 761
FONG, R. 643
FONTANELLI, P. 315
FRITZ, G. 523

GELLER, M.J. 301
GIOIA, I.M. 593
GIOMMI, P. 601
GIRAUD, E. 199
GIURICIN, G. 515
GONG, S.M. 681
GOSSET, E. 499
GOTT,III J.R. 433
GRIEGER, B. 767
GULKIS, S. 673
GUNN, J.E. 747

HALPERN, M. 63
HAMMER, F. 751
HARTWICK, F.D.A. 655
HAWKINS, M.R.S. 691
HE, X.T. 531
HEGYI, D.J. 59
HEWETT, P.C. 661
HEWITT, J.N. 747
HOBBS, L.M. 119
HUCHRA, J.P. 301
HUNSTEAD, R.W. 799

INOUE, M. 565
IOVINO, A. 809
IRWIN, M.J. 661

JAUNCEY, D.L. 673

KAPAHI, V.K. 251
KATO, T. 565
KAYSER, R. 767
KELLERMANN, K.I. 545

WEGNER, G.   223
WHITE, G.L.   673
WILLS, B.J.   665,669
WILLS, D.   669
WINDHORST, R.A.   573
WOLTER, A.   593
WORRALL, D.M.   607
WU, X.J.   771

XIA, C.L.   681
XIA, X.Y.   363
XIAO, X.H.   531
XU, C.M.   771

YAHIL, A.   247
YATES, M.G.   143
YEE, H.K.C.   685

ZAMORANI, G.   657
ZHOU, Y.Y.   363,677
ZITELLI, V.   657
ZOU, Z.L.   535

SUBJECT INDEX

AAT faint galaxy survey 374
Abell clusters 276,319,323
Active galaxies 531
Alternative cosmologies 447
AM 0059-402 492
AM 0213-283 492
AM 0328-222 492
AM 2006-295 492
AM 2054-221 492
Angular diameter redshift test 11
Angular size-flux density relation 260
Angular size-redshift relation, optical 11
Angular size-redshift relation, radio 251
        -comparison with models 258
        -for galaxies 252
        -for quasars 252
Anisotropy
        -dipole 63
        -large-scale 77
        -small-scale 77
Axions 713

Baade 2,3
Baryon density 95
Baryons
        -non-conservation of 450
Biased galaxy formation 349,353,420
Big bang nucleosynthesis 94
Bimodal star formation model 103
BL Lac objects 593
Bright galaxies
        -distribution 275
         -morphology 294
         -structures 294
Bruzual evolution 136,141
Bubble-like structure 306

3C 120 490
3C 303 486
3C 324 751
3C radio sources 130,143
C-field 449
CFA redshift survey 301
Chromatic effect 768
Clusters of galaxies 515,519,593

```
        -correlation functions 337
        -Southern 331
Continuous creation 459
Coma cluster 276,279,281,303,308,504
Coma region 292
Cosmic microwave background
        -angular fluctuations 41,55,73
        -large-scale isotropy 37
        -motion of galaxies with respect to 226
        -polarization 48
        -spectrum 32
        -temperature 33,59
Cosmic strings 326,427
Cosmic wave function 467
Cosmological theories
        -chronometric model 454
        -empirical two-component model 450
        -G-varying model 451
        -hot big-bang model 447
        -matter-antimatter symmetric model 448
        -quantum cosmology 455,462
        -steady state model 449

Dark halos 400
Dark matter 395,415,417,421,433,740
        -in binary systems 705
        -in clusters of galaxies 707
        -in dwarf galaxies 700
        -in elliptical galaxies 702
        -in groups of galaxies 706
        -in spiral galaxies 701
        -massive neutrinos 719
De Vaucouleurs 276
Deuterium 90
Disk galaxies 515
Distance indicators
        -brightest stars 155
        -21 cm linewidths and IR magnitudes 167,187
        -classical Cepheids 153,189
        -RR Lyrae stars 154,188,197
        -supernovae 173
Dwarf galaxies 294,402

ESO-Uppsala Catalogue 279
EXOSAT survey 601

Faint galaxies
        -distribution 367
        -evolution 383
        -number counts 370
        -optical spectra 375
Faint object redshift surveys 368,374
Fermat principle 730,746
```

Galactic evolution 100
Galaxian surface density
        -Northern polar cap 316
        -Southern polar cap 316
Galaxies
        -groups of 492
        -spatial distribution 349
Galaxy correlation function 394
Galaxy count distribution 3,4
Galaxy formation 269,391,437
        -explosions 428
Galaxy redshifts
        -quantization 493
Gravitational lenses 729,751,755
        -candidates 733
        -3C 321 an impostor 761
        -criteria for 759
        -effects of dark matter 771
        -elusive lens problem 734
        -odd image problem 734
        -survey 747
Gravitational microlenses 740,767
Gravitinos 108
Gunn-Peterson test 781

Helium3 91
Helium4 91,115
Homogeneity of the universe 826
Hubble 2,3,19,21
Hubble constant 18,24,83,129,185,192,197,828
        -determination of 151,737
        -distance indicators 153,166,170,197,223
        -numerical value 177
        -selection bias 159,217
Hubble diagram 7,11,129,133,134,148,492,495,508,827
        -for quasars 147
        -infrared 135,137
Hubble flow
        -non-uniformities 223
Hydra-Centaurus region 279,283

I 2402 486
Inflationary cosmology 391
Inos 723
Interacting galaxies 492
Inverse square law 451
IRAS dipole moment 232
IRAS galaxies 229,249
IRAS sources
        -distribution of 240
Isolated galaxies 355
Isotropy of the universe 825

KOSS survey 302

Lahav study 237
Large numbers hypothesis 452
Large-scale structure 19,301,323,415
        -spatial distribution 335
        -sponge-like 307,433
Lithium7 92,119,125
Local expansion field 171
Local supercluster 285,287
Luminosity index 164
Lyman alpha clouds 399
Lynx-Ursa Major supercluster 317

M 82 488
Mach's principle 453
Malmquist bias 13,23,163,201
Markarian205 484,498
MCG 03-34-085 486
Micro-superclusters 296
Microwave dipole component 39
Missing mass 503
Molonglo deep survey 569
Multiple images 464
Multiply connected topology 470
Multiply connected universe 471

Neutrinos 95,116
NGC 53 492
NGC 470 480
NGC 520 487,499,502
NGC 622 480
NGC 1073 480
NGC 1097 487
NGC 1232 492
NGC 1275 524
NGC 3842 480
NGC 4319 484,486,498
NGC 5296 486
NGC 7603 491,492

Omega 393,418,830
        -estimate of 207,217,235,248
        -value of 309
Open universe 426

Pancake scenarios 423
Pancake-like structures 329
Parkes 2.7 GHz survey 677
Particle physics
        -and early nucleosynthesis 105
Peculiar velocities 341

Peebles 276
Percolation algorithm 279,285,320,327
Perseus cluster 523
        -iron abundance in 523
        -temperature of 525
Perseus-Pisces region 279,282,313
Phenomenological clustering model 340
Photinos 108
Photodisintegration processes 107
Pisces-Cetus region 319
PKS 1327-206 485,486
Planck time 391
Primordial abundances 90
Primordial perturbations 80
Protoclusters 441
        -radio searches for 441
Protogalaxies 396,405

Q_0 9,148
Quantum cosmology 466
Quark nuggets 110
Quasistellar object-galaxy clustering 816
Quasistellar object-quasistellar object clustering 816
Quasistellar object redshifts 470
        -distribution peaks 481
        -quantization of 481
Quasistellar objects 593
        -absorption 270,777,793,799
        -absorption-line redshifts 471
        -associations with galaxies 473,479,482,685
        -BAL systems 777
        -close pairs 472,785
        -clustering of 629,634,638,643,646,809,815
        -clustering of lines 783
        -cosmological evolution 779
        -counts 619,627,639,644,655,657
        -density evolution 620
        -disks 695
        -disk component 667
        -evolution 383,649,677
        -heavy element systems 777,780,801
        -high redshift 661
        -infrared emission 607
        -infrared spectra 665
        -luminous bridges 483
        -luminosity evolution 621,643,677
        -Lyman alpha systems 777,793,799,803
        -radio emission 650
        -selection by emission lines 621
        -spectra 669
        -structure 767
        -tidal triggering of 527
        -variable 691
        -x-ray emitting 579,582,607

Radio luminosity function 266
Radio galaxies 537
        -bending 267
        -cores of 537
        -counts 545,550,552,565,569
        -evolution 554
        -faint 556
        -luminosity-volume test 554,558
        -optical identifications 546,570,573
        -spectroscopy of 573
Ram pressure confinement 262
Redshift asymmetries 503
RELIC satellite 75

Sculptor group 489,506
Selection effects 13
Seyfert environment 535
Seyfert galaxies-tidal triggering of 527
Shane and Wirtanen 276
Shane-Wirtanen Lick counts 21
Shapley and Ames 276
Size evolution 262
Space interferometers 74
Sunyaev-Zel'dovich effect 49,83
Superclusters 277,319,349
        -correlation functions 337
        -geometrical elongation 346
        -shape 327
Surface brightness test 22

Thin sheets 302
Time-scale test 17
Tired light 22,507
Topology of the Universe 461,463
Totsuji and Kihara 276
Tully-Fisher relation 199
Two-dimensional luminosity distribution 437
Two-point correlation function 325,364

Universal correlation function 338
Ursa Major region 437

Virgo cluster 437
        -distance 175
Virgocentric flow model 247
Void probability function 359
Voids 289,302,309,319,349

X-ray background 587,611
X-ray high sensitivity survey 579
X-ray medium sensitivity survey 579,594
X-ray source counts 597

Zwicky 276